Integral Domains Inside Noetherian Power Series Rings

Mathematical Surveys and Monographs

Volume 259

Integral Domains Inside Noetherian Power Series Rings

Constructions and Examples

William Heinzer

Christel Rotthaus

Sylvia Wiegand

AMERICAN MATHEMATICAL SOCIETY

Providence, Rhode Island

2020 *Mathematics Subject Classification.* Primary 13-00, 13-01, 13-02, 13A05, 13B21, 13B22, 13B25, 13B30, 13B35, 13B40.

For additional information and updates on this book, visit
www.ams.org/bookpages/surv-259

Library of Congress Cataloging-in-Publication Data

Names: Heinzer, William J., author. | Rotthaus, Christel, 1949- author. | Wiegand, Sylvia, author.
Title: Integral domains inside Noetherian power series rings : constructions and examples / William Heinzer, Christel Rotthaus, Sylvia Wiegand.
Description: Providence, Rhode Island : American Mathematical Society, [2021] | Series: Mathematical surveys and monographs, 0076-5376 ; Volume 259 | Includes bibliographical references and index.
Identifiers: LCCN 2021021513 | ISBN 9781470466428 (paperback) | ISBN 9781470467333 (ebook)
Subjects: LCSH: Noetherian rings. | Integral domains. | AMS: Commutative algebra – General reference works (handbooks, dictionaries, bibliographies, etc.) pertaining to commutative algebra. | Commutative algebra – Instructional exposition (textbooks, tutorial papers, etc.) pertaining to commutative algebra. | Commutative algebra – Research exposition (monographs, survey articles) pertaining to commutative algebra. | Commutative algebra – General commutative ring theory – Divisibility and factorizations in commutative rings. | Commutative algebra – Commutative ring extensions and related topics – Integral dependence in commutative rings; going up, going down. | Commutative algebra – Commutative ring extensions and related topics – Integral closure of commutative rings and ideals; integrally closed rings, related rings (Japanese, etc.). | Commutative algebra – Commutative ring extensions and related topics – Polynomials over commutative rings. | Commutative algebra – Commutative ring extensions and related topics – Rings of fractions and localization for commutative rings. | Commutative algebra – Commutative ring extensions and related topics – Completion of commutative rings. | Commutative algebra – Commutative ring extensions and related topics – Étale and flat extensions; Henselization; Artin approximation.
Classification: LCC QA251.3 .H45 2021 — DDC 512/.44–dc23
LC record available at https://lccn.loc.gov/2021021513

∞ The paper used in this book is acid-free and falls within the guidelines
established to ensure permanence and durability.
Visit the AMS home page at `https://www.ams.org/`

10 9 8 7 6 5 4 3 2 1 26 25 24 23 22 21

This monograph is dedicated to Mary Ann Heinzer, to Maria Rotthaus, to Roger Wiegand, and to the past, present and future students of the authors.

Contents

Preface xi

Chapter 1. Introduction 1

Chapter 2. Tools 9
2.1. Conventions and terminology 9
2.2. Basic theorems 16
2.3. Flatness 18
Exercises 21

Chapter 3. More tools 25
3.1. Introduction to ideal-adic completions 25
3.2. Uncountable transcendence degree for a completion 29
3.3. Power series pitfalls 31
3.4. Basic results about completions 32
3.5. Chains of prime ideals, fibers of maps 34
3.6. Henselian rings 36
3.7. Regularity and excellence 38
Exercises 41

Chapter 4. First examples of the construction 45
4.1. Universality 45
4.2. Elementary examples 47
4.3. Historical examples 48
4.4. Prototypes 51
Exercises 55

Chapter 5. The Inclusion Construction 59
5.1. The Inclusion Construction and a picture 59
5.2. Approximations for the Inclusion Construction 60
5.3. Basic properties of the constructed domains 63
Exercises 71

Chapter 6. Flatness and the Noetherian property 73
6.1. The Noetherian Flatness Theorem 73
6.2. Introduction to Insider Construction 10.7 81
6.3. Nagata's example 82
6.4. Christel's Example 83
6.5. Further implications of the Noetherian Flatness Theorem 84
Exercise 85

Chapter 7. The flat locus of an extension of polynomial rings 87
7.1. Flatness criteria 87
7.2. The Jacobian ideal and the smooth and flat loci 91
7.3. Applications to polynomial extensions 98
Exercises 101

Chapter 8. Excellent rings and formal fibers 103
8.1. Background for excellent rings 103
8.2. Nagata rings and excellence 109
8.3. Henselian rings 111
Exercises 113

Chapter 9. Height-one prime ideals and weak flatness 115
9.1. The limit-intersecting condition 115
9.2. Height-one primes in extensions of Krull domains 119
9.3. Flatness for Inclusion Construction 5.3 122
Exercises 124

Chapter 10. Insider Construction details 127
10.1. Localized polynomial rings over special DVRs 128
10.2. Describing the construction 130
10.3. The non-flat locus of Insider Construction 10.7 131
10.4. Preserving excellence with the Insider Construction 136

Chapter 11. Integral closure under extension to the completion 139
11.1. Integral closure under ring extension 139
11.2. Extending ideals to the completion 142
11.3. Comments and questions 145
Exercises 147

Chapter 12. Iterative examples 149
12.1. Iterative examples and their properties 149
12.2. Residual algebraic independence 154
Exercises 162

Chapter 13. Approximating discrete valuation rings by regular local rings 165
13.1. Local quadratic transforms and local uniformization 165
13.2. Expressing a DVR as a directed union of regular local rings 167
13.3. Approximating other rank-one valuation rings 170
Exercises 171

Chapter 14. Non-Noetherian examples of dimension 3 173
14.1. A family of examples in dimension 3 173
14.2. Two cases of Examples 14.1 and their spectra 177
Exercises 183

Chapter 15. Noetherian properties of non-Noetherian rings 185
15.1. A general question and the setting for this chapter 185
15.2. More properties of Insider Constructions 187
15.3. A four-dimensional non-Noetherian domain 189
Exercises 196

Chapter 16. Non-Noetherian examples in higher dimension 197
16.1. Higher dimensional non-Noetherian examples 197
16.2. A 4-dimensional prime spectrum 200
16.3. $D + M$ constructions, coherence, finite conductor domains 204
Exercises 207

Chapter 17. The Homomorphic Image Construction 209
17.1. Two construction methods and a picture 209
17.2. Approximations for Homomorphic Image Construction 211
17.3. Basic properties of the Approximation Domains 213
17.4. Noetherian flatness for homomorphic images 215
17.5. Weak Flatness for Homomorphic Image Construction 17.2 218
17.6. Inclusion Constructions are Homomorphic Image Constructions 219
17.7. Connecting the two constructions 223
Exercises 229

Chapter 18. Catenary local rings with geometrically normal formal fibers 231
18.1. History, terminology and summary 231
18.2. Geometrically normal formal fibers and the catenary property 233
18.3. Flatness for the intersection of finitely many ideals 237
18.4. Regular maps and geometrically regular formal fibers 238
18.5. Examples that are not universally catenary 239
18.6. The depth of the constructed rings 244
Exercises 244

Chapter 19. An Ogoma-like example 245
19.1. Cohen-Macaulay formal fibers 245
19.2. Constructing the Ogoma-like example 251

Chapter 20. Multi-ideal-adic completions of Noetherian rings 257
20.1. Ideal filtrations and completions 257
20.2. Basic mechanics for the multi-adic completion 258
20.3. Preserving Noetherian under multi-adic completion 262
20.4. Preserving excellence or Henselian under multi-adic completion 265
Exercises 267

Chapter 21. Noetherian flatness and multi–adic constructions 269
21.1. Flatness lemmas 269
21.2. Multi-adic inclusion constructions 270

Chapter 22. Idealwise algebraic independence 275
22.1. Idealwise independence, weakly flat and PDE extensions 275
22.2. Primarily independent elements 280
22.3. Residually algebraically independent elements 285
Exercises 292

Chapter 23. Idealwise algebraic independence II 293
23.1. Primary independence and flatness 293
23.2. Composition, base change and polynomial extensions 295
23.3. Passing to the Henselization 300
23.4. Summary diagram for the independence concepts 304

Chapter 24. Krull domains with Noetherian x-adic completions 307
24.1. Applying Inclusion Construction 5.3 308
24.2. Limit-intersecting elements 309
24.3. Non-Noetherian intersection/approximation domains 316
24.4. A birational connection and additional examples 322
Exercises 325

Chapter 25. Inclusion Constructions over excellent normal local domains 327
25.1. Primarily limit-intersecting extensions and flatness 327
25.2. Existence of primarily limit-intersecting extensions 330
Exercise 337

Chapter 26. Weierstrass techniques for generic fiber rings 339
26.1. Terminology, background and results 339
26.2. Variations on a theme by Weierstrass 343
Exercises 353

Chapter 27. Generic fiber rings of mixed polynomial-power series rings 355
27.1. Weierstrass implications for the ring $B = k[[X]]\,[Y]_{(X,Y)}$ 355
27.2. Weierstrass implications for the ring $C = k[Y]_{(Y)}[[X]]$ 358
27.3. Subrings of the power series ring $k[[z,t]]$ 361
27.4. Weierstrass implications for the localized polynomial ring A 364
27.5. Generic fibers of power series ring extensions 368
27.6. $\mathrm{Gff}(R)$ and $\mathrm{Gff}(S)$ for S an extension domain of R 369
27.7. Formal fibers of prime ideals in polynomial rings 371
27.8. Other results on generic formal fibers 373
Exercise 374

Chapter 28. Mixed polynomial-power series rings and relations among their spectra 375
28.1. Two motivations 375
28.2. Trivial generic fiber (TGF) extensions and prime spectra 378
28.3. Spectra for two-dimensional mixed polynomial-power series rings 379
28.4. Higher dimensional mixed polynomial-power series rings 385
Exercises 390

Chapter 29. Extensions of local domains with trivial generic fiber 391
29.1. General remarks about TGF extensions 392
29.2. TGF extensions with finite residue field extension 393
29.3. The case of transcendental residue extension 397
Exercise 400

Chapter 30. Constructions and examples discussed in this book 401
30.1. Construction techniques 401
30.2. Examples considered in the book 402

Bibliography 407

Index 417

Preface

The authors have had as a long-term project the creation of examples using power series to analyze and distinguish several properties of commutative rings and their spectra. This monograph is our attempt to expose the results that have been obtained in this endeavor, to put these results in better perspective and to clarify their proofs. We hope in this way to assist current and future researchers in commutative algebra in utilizing the techniques described here.

William Heinzer, Christel Rotthaus, Sylvia Wiegand

CHAPTER 1

Introduction

When we started to collaborate on this work over twenty years ago, we were inspired by expository talks Judith Sally gave on the following question:

QUESTION 1.1. What rings lie between a Noetherian integral domain and its field of fractions?

Also Shreeram Abhyankar's research inspired us to ask the following related question:[1]

QUESTION 1.2. Let I be an ideal of a Noetherian integral domain R, and let R^* denote the I-adic completion of R. What rings lie between R and R^*? For example, if x and y are indeterminates over a field k, what rings lie between the polynomial ring $k[x,y]$ and the mixed polynomial-power series ring $k[y][[x]]$?

In this book we encounter a wide variety of integral domains fitting the descriptions of Question 1.1 and Question 1.2. In particular we have the following goals:

(1) To construct new examples of Noetherian integral domains, continuing a tradition that goes back to Akizuki and Schmidt in the 1930s and Nagata in the 1950s.
(2) To construct new non-Noetherian integral domains that illustrate recent advances in ideal theory.
(3) To study birational extensions of Noetherian integral domains as in Question 1.1.
(4) To consider, as in Question 1.2, the extension $R \hookrightarrow R^*$, and to relate the fibers of R^* over R to birational extensions of R.

These objectives are interrelated: Noetherian Flatness Theorem 6.3 gives conditions for the constructed domains to be Noetherian. The Noetherian domains constructed in (1) are used to produce non-Noetherian domains in (2) by using Insider Construction 10.7. Construction 17.2 involves birational extensions of a base ring R as in (3). The extension and the fibers mentioned in (4) are useful for the construction in (1).

Over the past eighty years, important examples of Noetherian integral domains have been constructed that arise as an intersection of a field with a homomorphic image of a power series ring. An ideal-adic completion of R with respect to a

[1] Abhyankar's work demonstrates the vastness of power series rings; a power series ring in two variables over a field k contains for each positive integer n an isomorphic copy of the power series ring in n variables over k, [**1**]. The authors have fond memories of many pleasant conversations with Ram concerning power series.

finitely generated ideal of R is a homomorphic image of a power series ring over R; see Section 3.1. [2]

The basic idea of the construction is to start with an integral domain R, usually Noetherian, such as a polynomial ring over a field. We look for more unusual Noetherian and non-Noetherian extension rings inside a ring S that is a homomorphic image of a Noetherian ideal-adic completion of R. For reference purposes we label such an intersection as *Intersection Construction* 1.3.

INTERSECTION CONSTRUCTION 1.3. Let R be an integral domain, let R^* be a Noetherian ideal-adic completion of R, and let I be an ideal of R^*. Assume that R is a subring of $S := R^*/I$ and that L is a field between the field of fractions of R and the total quotient ring of S. Define the *Intersection Domain* A:

$$(1.3.0) \qquad A \ := \ L \cap S. \text{ [3]}$$

Many of the examples of this book are produced with $I = 0$, as in Construction 5.3.

We have been captivated by these topics and have for a number of years been examining ways to create new rings from well-known ones. Several chapters of this monograph, such as Chapters 4, 5 6, 14, 17, and 24, contain a reorganized development of previous work using constructions of the form in Intersection Construction 1.3.

As presented here, Intersection Construction 1.3 is *universal* in the following sense: Assume that A is a Noetherian local domain that has a coefficient field k, and that the field of fractions L of A is finitely generated as a field extension of k. Then A is an intersection $A = L \cap S$, as in Intersection Construction 1.3, where $S = \widehat{R}/I$, for a suitable ideal I of the $\mathfrak{m}$-adic completion $\widehat{R}$ of a Noetherian local domain $(R, \mathfrak{m})$. Also k is a coefficient field for R, L is the field of fractions of R, and R is essentially finitely generated over k; see Section 4.1 of Chapter 4.

Classical examples of Noetherian integral domains with interesting properties are constructed by Akizuki, Schmidt, and Nagata. This work is continued by Brodmann-Rotthaus, Ferrand-Raynaud, Heitmann, Lequain, McAdam, Nishimura, Ogoma, Ratliff, Rotthaus, Weston and others.[4]

CLASSICAL EXAMPLES 1.4. Many of the classical examples concern integral closure. Akizuki's 1935 example is a one-dimensional Noetherian local domain R of characteristic zero such that the integral closure of R is not a finitely generated R-module [**13**]. Schmidt's 1936 example is a one-dimensional normal Noetherian local domain R of positive characteristic such that the integral closure of R in a finite purely inseparable extension field is not a finitely generated R-module [**165**, pp. 445-447]. In relation to integral closure, Nagata's classic examples include (1) a two-dimensional Noetherian local domain with a non-Noetherian birational integral extension and (2) a three-dimensional Noetherian local domain such that the integral closure is not Noetherian [**137**, Examples 4 and 5, pp. 205-207].

Example 4.15 is another example constructed by Nagata. This is the first occurence of a two-dimensional regular local domain containing a field of characteristic zero that fails to be a Nagata domain, and hence is not excellent. For the definitions

[2]Terminology used in this introduction, such as "ideal-adic completion", "coefficient field", "essentially finitely generated" and "integral closure", is defined in Chapters 2 and 3.

[3]The domain A is also referred to as the *constructed ring*.

[4]See [**13**], [**165**], and [**135**]; and [**26**], [**27**], [**49**], [**95**], [**96**], [**97**], [**109**], [**124**], [**140**], [**146**], [**147**], [**152**], [**155**], [**156**], and [**183**].

and information on Nagata rings and excellent rings, see Definitions 2.20 and 3.47 and Chapter 8. We describe in Example 4.17 a construction due to Rotthaus of a Nagata domain that is not excellent.

In the foundational work of Akizuki, Nagata and Rotthaus (and indeed in most of the papers cited above) the description of the constructed ring A as the intersection domain of Construction 1.3 is not explicitly stated. Instead A is defined as a *direct limit of subrings*; or equivalently as a *directed union* or *nested union*.[5] In Chapters 4, 5, and 17, we expand Intersection Construction 1.3 to include an additional integral domain, also associated to the ideal-adic completion of R with respect to a principal ideal. Our expanded "Intersection Construction" consists of two integral domains that fit with these examples:

Intersection Construction With Approximation 1.5. This construction consists of two integral domains described as follows:

(IC1) The "intersection" integral domain A of Intersection Construction 1.3: $A = L \cap S$, is the intersection of a field L with a homomorphic image S of a principal ideal-adic completion of R, and

(IC2) An "approximation" domain B, that is a directed union inside A that approximates A and is more easily understood; sometimes B is a nested union of localized polynomial rings over R.

The details of the construction of B as in (IC2) are given in Chapters 5 and 17. Construction Properties Theorems 5.14 and 17.11 describe essential properties of the construction and are used throughout this book.

In certain circumstances the approximation domain B of (IC2) is equal to the intersection domain A of (IC1). In this case, the intersection domain A is a directed union. This yields more information about A. The description of A as an intersection is often unfathomable! In case $A = B$, the critical elements of B that determine L are called *limit-intersecting* over R; see Chapter 5 (Definition 5.10) and Chapters 9, 24 and 25 where we discuss the limit-intersecting condition further.

To see a specific example of the construction, consider the ring $R := \mathbb{Q}[x, y]$, the polynomial ring in the variables x and y over the field $\mathbb{Q}$ of rational numbers. Let S be the formal power series ring $\mathbb{Q}[[x, y]]$ and let L be the field $\mathbb{Q}(x, y, e^x, e^y)$.[6] Then Equation 1.3 yields that

$$\alpha = \frac{e^x - e^y}{x - y} \in A = \mathbb{Q}(x, y, e^x, e^y) \cap \mathbb{Q}[[x, y]]. \tag{1.5.a}$$

It turns out that $\alpha \notin B$, the approximation domain. In this example, the intersection domain A is Noetherian, whereas the approximation domain B is not Noetherian. More details about this example are given in Example 4.11 and in Theorem 12.3 and Example 12.7.

The construction method for the approximation domain B as in (IC2) was originally introduced by Akizuki. Akizuki's method is based on the idea of adjoining so-called "endpieces" of power series to a local base ring R in order to obtain a finite transcendental extension of R contained in the completion of R. This basic

---

[5]Our use of these terms is explained in Remark 4.8.2.

[6]In this example $S = \mathbb{Q}[[x, y]]$ is not a principal ideal-adic completion of R as in (IC1) above, but S may be realized by taking first the x-adic completion R^* and then taking the y-adic completion of R^*.

principle of adjoining endpieces was then extended by Nagata and later by Rotthaus to produce more counterexamples in commutative algebra.

In 1981, Rotthaus extended the endpiece method by adjoining "multi-adic" endpieces. Then Ogoma introduced a variation of the multi-adic construction by adjoining "frontpieces" (instead of endpieces). A first hint that there might be some general principle involved came with the Brodmann-Rotthaus theorem, which showed that a wide variety of rings could be constructed via the multi-adic method. In this book we apply this method to construct a wide class of Noetherian and non-Noetherian rings.

A primary task of our study is to determine, for a given Noetherian domain R, whether the ring $A = L \cap S$ of Intersection Construction 1.3 is Noetherian. An important observation related to this task is that the Noetherian property for the associated direct limit ring B is equivalent to a flatness condition; see Noetherian Flatness Theorems 6.3 and 17.13. Whereas the original proof of the Noetherian property for Example 4.15 of Nagata took a page and a half [**137**, Example 7, pages 209-211], the original proof of the Noetherian property for Example 4.17 of Rotthaus took seven pages [**155**, pages 112-118].The results presented in Chapter 6 establish the Noetherian property rather quickly for these and other examples.

The construction of B is related to an interesting construction introduced by Ray Heitmann [**95**, page 126]. Let x be a nonzero nonunit in a Noetherian integral domain R, and let R^* denote the x-adic completion of R. Heitmann describes a procedure for associating, to an element τ in R^* that is transcendental over R, an extension ring T of $R[\tau]$ having the property that the x-adic completion of T is R^*. [7] Heitmann uses this technique to construct interesting examples of non-catenary Noetherian rings. In their 1997 article [**73**], the present authors adapt the construction of Heitmann to prove a version of Noetherian Flatness Theorem 6.3 that applies for one transcendental element τ over a semilocal Noetherian domain R: If the element τ satisfies a certain flatness condition, then τ is called *primarily limit-intersecting* and the constructed intersection domain A is equal to the approximation domain B and is Noetherian [**73**, Theorem 2.8].

This "primarily limit-intersecting" concept from [**73**] extends to more than one transcendental element τ; see Noetherian Flatness Theorem 6.3. This extends Heitmann's construction to finitely many elements of the x-adic completion R^* of R that are algebraically independent over R; see [**73**, Theorem 2.12] or Theorem 6.3.1.a.

In Chapter 5, we present Inclusion Construction 5.3, a simplified version of Intersection Construction 1.3 with Setting 5.1. In Setting 5.1, the base ring R is an integral domain that is not necessarily Noetherian, the element x is a nonzero nonunit of R, the extension ring S is the x-adic completion R^* of R and is assumed to be Noetherian, and the field L is generated by a finite set of elements of R^* that are algebraically independent over R. In Chapter 6, Noetherian Flatness Theorem 6.3 is proved for this non-Noetherian setting.

With Setting 5.1, the integral domain $A = L \cap R^*$ is sometimes Noetherian. If the approximation domain B is Noetherian, then B is equal to the intersection domain A. The converse fails however; it is possible for B to be equal to A and

[7]Heitmann remarks in [**95**] that this type of extension also occurs in [**137**, page 203]. The ring T is not finitely generated over $R[\tau]$ and no proper $R[\tau]$-subring of T has R^* as its x-adic completion. Necessary and sufficient conditions are given in order that T be Noetherian in Theorem 4.1 of [**95**].

not be Noetherian; see Example 10.15. If B is not Noetherian, we can sometimes determine the prime ideals of B that are not finitely generated; see Example 14.1. If a ring has exactly one prime ideal that is not finitely generated, that prime ideal contains all non-finitely generated ideals of the ring.

In Section 6.2 of Chapter 6 and in Chapter 12, we adjust the construction from Chapters 4 and 5. An "insider" technique is introduced in Section 6.2 and generalized in Chapter 10 for building new examples inside more straightforward examples constructed as above. Using Insider Construction 10.7, the verification of the Noetherian property for the constructed rings is streamlined. Even if one of the constructed rings is not Noetherian, the proof is simplified. We analyze classical examples of Nagata and others from this viewpoint in Section 6.3 and 6.4. Chapter 12 contains an investigation of more general rings that involve power series in two variables x and y over a field k, such as the specific example given above in Equation 1.5.a.

In Chapters 14 to 16, we use Insider Construction 10.7 to construct low-dimensional non-Noetherian integral domains that are strangely close to being Noetherian: One example is a three-dimensional local unique factorization domain B inside $k[[x,y]]$; the ring B has maximal ideal $(x,y)B$ and exactly one prime ideal that is not finitely generated; see Example 14.9.

There has been considerable interest in non-Noetherian analogues of Noetherian notions such as the concept of a "regular" ring; see the book by Glaz [**57**]. Rotthaus and Sega in [**161**] show that the approximation domains B constructed in Chapters 14 - 16 using Insider Construction 10.7, even though non-Noetherian, are *coherent regular* local rings by showing that every finitely generated submodule of a free module over B has a finite free resolution; see [**161**] and Remark 16.15.

One of our additional goals is to consider the question: "What properties of a ring extend to a completion?" Chapter 11 contains an example of a three-dimensional regular local domain $(A, \mathfrak{n})$ with a height-two prime ideal P such that the extension $P\widehat{A}$ to the $\mathfrak{n}$-adic completion of A is not integrally closed.

We consider excellence in regard to the question: "What properties of the base ring R are preserved by the construction?" Since excellence is an important property satisfied by most of our rings, we present in Chapter 8 a brief exposition of excellent rings. In some cases we determine conditions in order that the constructed ring is excellent; see Prototype Theorems 10.2 and 10.6, Remarks 10.5, and Theorem 10.17. Assume the ring R is a unique factorization domain (UFD) and R^* is the x-adic completion of R with respect to a prime element x of R. We observe that the approximation domain B is then a UFD; see Theorem 5.24 of Chapter 5.

Since the Noetherian property for the approximation domain is equivalent to the flatness of a certain homomorphism, we devote considerable time and space to exploring flat extensions. We present results involving flatness in Chapters 3, 6, 7, 9, 10, 17, 18 and 20-25.

In Chapter 17, we develop Homomorphic Image Construction 17.2 and relate it to Inclusion Construction 5.3. Homomorphic Image Construction 17.2 is used in Chapter 18 to obtain for each integer $n \geq 2$ a catenary Noetherian local integral domain having geometrically normal formal fibers that is not universally catenary. In Chapter 18, we also prove that the Henselization of a Noetherian local ring having geometrically normal formal fibers is universally catenary.

In Chapter 19 we discuss properties of a famous example of Ogoma of a 3-dimensional normal Nagata local domain whose generic formal fiber is not equidimensional. We draw connections with Cohen-Macaulay formal fibers and present in Theorem 19.15 and its proof the construction of an example with the properties of Ogoma's Example.

Let R be a Noetherian ring with Jacobson radical $\mathcal{J}$. In Chapter 20 we consider the multi-ideal-adic completion R^* of R with respect to a filtration $\mathcal{F} = \{Q_n\}_{n \geq 0}$, where $Q_n \subseteq \mathcal{J}^n$ and $Q_{nk} \subseteq Q_n^k$ for each $n, k \in \mathbb{N}$. We prove that R^* is Noetherian. If R is an excellent local ring, we prove in Theorem 20.15 that R^* is excellent.If R is a Henselian local ring, we prove that R^* is Henselian. Chapter 21 contains a multi-adic version of Inclusion Construction 5.3.

The application of Intersection Construction 1.3 in Chapters 22 and 23 yields "idealwise" examples that are of a different nature from the examples in earlier chapters. Whereas the base ring $(R, \mathfrak{m})$ is an excellent normal local domain with $\mathfrak{m}$-adic completion $(\widehat{R}, \widehat{\mathfrak{m}})$, the field L is more general than in Chapter 5. We take L to be a purely transcendental extension of the field of fractions K of R such that L is contained in the field of fractions of $\widehat{R}$; say $L = K(G)$, where G is a set of elements of $\widehat{\mathfrak{m}}$ that are algebraically independent over K. Define $D := L \cap \widehat{R}$. The set G is said to be *idealwise independent* if $K(G) \cap \widehat{R}$ equals the localized polynomial ring $R[G]_{(\mathfrak{m},G)}$. The results of Chapters 22 and 23 show that the intersection domain can sometimes be small or large, depending on whether expressions in the power series allow additional prime divisors as denominators. The consideration of idealwise independence leads us to examine other related flatness conditions. The analysis and properties related to idealwise independence are summarized in Summaries 22.6 and 23.1.

In Chapters 24 and 25, we consider properties of the constructed rings A and B in the case where R is an excellent normal local domain. We present in Chapter 24 a specific example where $A = B$ is non-Noetherian.

In Chapter 28, we study prime ideals and their relations in mixed polynomial-power series extensions of low-dimensional rings. For example, we determine the prime ideal structure of the power series ring $R[[x]]$ over a one-dimensional Noetherian domain R and the prime ideal structure of $k[[x]][y]$, where x and y are indeterminates over a field k. We analyze the generic fibers of mixed polynomial-power series ring extensions in Chapters 26 and 27. Motivated by a question of Hochster and Yao, we consider in Chapter 29 extensions of integral domains $S \hookrightarrow T$ having *trivial generic fiber*; that is, every nonzero prime ideal of T intersects S in a nonzero prime ideal.

In Chapter 30, we list the construction techniques and examples given in this book.

The topics of this book include the following:

(1) An introduction and glossary for the terms and tools used in the book: Chapters 1-3.
(2) The construction of the intersection domain A and the approximation domain B: Chapters 4- 6, 10, 17, 21, and 30.
(3) Flatness properties of maps of rings: Chapters 3, 6, 7, 9, 10, 17, 18, and 21-25.
(4) Preservation of properties of rings and ideals under passage to completion: Chapters 3, 11, 18, 20, and 30.

(5) The catenary and universally catenary property of Noetherian rings: Chapters 3, 17- 19, and 30.
(6) Excellent rings and geometrically regular and geometrically normal formal fibers: Chapters 3, 7, 8, 10, 17, 18, 25, and 30.
(7) Examples of non-Noetherian local rings having Noetherian completions: Chapters 4, 5, 10, 12, 14-22, 30.
(8) Examples of Noetherian rings: Chapters 4, 5, 10- 13, 17, 30.
(9) Prime ideal structure: Chapters 14- 16, and 26-30.
(10) Approximating a discrete rank-one valuation domain using higher-dimensional regular local rings: Chapters 13 and 30.
(11) Trivial generic fiber extensions: Chapters 26-30.
(12) Transfer of excellence: Chapters 10, 17, 20, and 30.
(13) Birational extensions of Noetherian domains: Chapters 6, 11, 14- 16, 18, 19, 24, 25, and 30.
(14) Completions and multi-ideal-adic completions: Chapters 3, 20, and 21.
(15) Exercises to engage the reader in these topics and to lead to further extensions of the material presented here.

We thank Bruce Olberding for carefully reading an earlier draft of this manuscript and for his many helpful suggestions. We also thank the reviewers and the American Mathematical Society editors for their generous assistance with the book.

The authors are grateful for the hospitality, cooperation and support of Michigan State, Nebraska, Purdue, CIRM in Luminy and MSRI in Berkeley, where we worked on this research. SW was partially supported by a 2019 UNL Emeriti & Retiree Association Wisherd Award.

CHAPTER 2

Tools

In this chapter we review conventions and terminology, state basic theorems and review the concept of flatness of modules and homomorphisms.

2.1. Conventions and terminology

We generally follow the notation of Matsumura [**122**]. Thus by a ring we mean a commutative ring with identity, and a ring homomorphism $R \to S$ maps the identity element of R to the identity element of S. For commutative rings, we write $R \subseteq S$ to mean that R is a subring of S, and that R contains the identity element of S. We use the words "map", "morphism", and "homomorphism" interchangeably.

We use $\mathbb{Z}$ to denote the ring of integers, $\mathbb{N}$ the positive integers, $\mathbb{N}_0$ the nonnegative integers, $\mathbb{Q}$ the rational numbers, $\mathbb{R}$ the real numbers and $\mathbb{C}$ the complex numbers.

The set of prime ideals of a ring R is called the *prime spectrum* of R and is denoted $\operatorname{Spec} R$. The set $\operatorname{Spec} R$ is naturally a partially ordered set with respect to inclusion. For an ideal I of a ring R, let

$$\mathcal{V}(I) \;=\; \{P \in \operatorname{Spec} R \mid I \subseteq P\ \}.$$

The *Zariski topology* on $\operatorname{Spec} R$ is obtained by defining the *closed* subsets to be the sets of the form $\mathcal{V}(I)$ as I varies over all the ideals of R. The *open* subsets are the complements $\operatorname{Spec} R \setminus \mathcal{V}(I)$. $\operatorname{Spec} R$ is said to be *Noetherian* if the closed subsets of $\operatorname{Spec} R$ in the Zariski topology satisfy the descending chain condition, or, equivalently, the open subsets satisfy the ascending chain condition. A Noetherian ring has Noetherian spectrum; see [**122**, Exercise 4.9], [**15**, Exercise 8, p.79].

Regular elements, regular sequence. An element r of a ring R is said to be a *zerodivisor* if there exists a nonzero element $a \in R$ such that $ar = 0$, and r is a *regular element* if r is not a zerodivisor.

A sequence of elements $x_1, \ldots, x_d$ in R is called a *regular sequence* if it satisfies: (i) $(x_1, \ldots, x_d)R \neq R$, and (ii) x_1 is a regular element of R, and, for i with $2 \leq i \leq d$, the image of x_i in $R/(x_1, \ldots, x_{i-1})R$ is a regular element; see [**122**, pages 123].

Localizations. Let S be a multiplicatively closed subset of a ring R such that $1 \in S$. The *localization* of R at S as defined in [**122**, pages 20-21] is a ring denoted $S^{-1}R$ along with a ring homomorphism $f : R \to S^{-1}R$ such that

(1) $f(S) \subseteq \{$ units of $S^{-1}R\}$,
(2) If $g : R \to T$ is a ring homomorphism such that $g(S) \subseteq \{$ units of $T\}$, then there is a unique ring homomorphism $h : S^{-1}R \to T$ such that $g = hf$.

If S consists of regular elements of R, then $S^{-1}R = \{\frac{r}{s} \mid r \in R, s \in S\}$, where $f(r) = \frac{r}{1}$, for every $r \in R$, and the map f is injective.

The *total ring of fractions* of R, denoted $\mathcal{Q}(R)$, is the localization of R at the set S of all the regular elements of R. There is a natural embedding $R \hookrightarrow \mathcal{S}^{-1}(R)$, where $r \mapsto \frac{r}{1}$ for every $r \in R$.

For a prime ideal P of R, the localization $(R \setminus P)^{-1}R$ is denoted R_P and is called the localization of R at P.

An *integral domain*, also called a *domain* or an *entire* ring, is a nonzero ring in which every nonzero element is a regular element. If R is a subring of an integral domain S and S is a subring of $\mathcal{Q}(R)$, then S is *birational over* R, or a *birational extension* of R.

2.1.1. Krull dimension, height. The *Krull dimension*, or briefly *dimension*, of a ring R, denoted $\dim R$, is n if there exists a chain $P_0 \subsetneq P_1 \subsetneq \cdots \subsetneq P_n$ of prime ideals of R and there is no such chain of length greater than n. We say that $\dim R = \infty$ if there exists a chain of prime ideals of R of length greater than n for each $n \in \mathbb{N}$. For a prime ideal P of a ring R, we say $\dim P = \dim(R/P)$; the *height* of P, denoted $\operatorname{ht} P$, is $\dim R_P$. The *height* of a proper ideal I, denoted $\operatorname{ht} I$, is defined to be

$$\operatorname{ht} I = \min\{\operatorname{ht} P \mid P \in \operatorname{Spec} R \text{ and } I \subseteq P\}.$$

Let I be a proper ideal of a Noetherian ring R with $\operatorname{ht} I = r$. If there exist elements $a_1, \ldots, a_r \in I$ such that $I = (a_1, \ldots, a_r)R$, then I is said to be a *complete intersection*.

Unique factorization domains. An integral domain R is a *unique factorization domain* (UFD), sometimes called a *factorial ring*, if every nonzero nonunit of R is a finite product of prime elements; a nonzero element $p \in R$ is *prime* if pR is a prime ideal.

In a UFD every height-one prime ideal is principal; this is Exercise 2.1.

Local rings. If a ring R (not necessarily Noetherian) has a unique maximal ideal $\mathfrak{m}$, we say R is *local* and write $(R, \mathfrak{m})$ to denote that R is local with maximal ideal $\mathfrak{m}$. A ring with only finitely many maximal ideals is called *semilocal.*

A *localized polynomial ring* over a local ring $(R, \mathfrak{m})$ is the localization of a polynomial ring $S := R[\{x_i \mid i \in I\}]$ at the maximal ideal $(\mathfrak{m}, \{x_i \mid i \in I\})S$ Here I is an index set, and the $\{x_i \mid i \in I\}$ are indeterminates over R.

If $(R, \mathfrak{m})$ and $(S, \mathfrak{n})$ are local rings, a ring homomorphism $f : R \to S$ is a *local homomorphism* if $f(\mathfrak{m}) \subseteq \mathfrak{n}$.

If $(R, \mathfrak{m})$ is a subring of a local ring $(S, \mathfrak{n})$, then S *dominates* R if $\mathfrak{m} = \mathfrak{n} \cap R$, or equivalently, if the inclusion map $R \hookrightarrow S$ is a local homomorphism. The local ring $(S, \mathfrak{n})$ *birationally dominates* $(R, \mathfrak{m})$ if S is an integral domain that dominates R and S is contained in the field of fractions of R.

Nilradical, reduced. For an ideal I of a ring R, the *radical* of I, denoted $\sqrt{I}$, is the ideal $\sqrt{I} = \{a \in R \mid a^n \in I \text{ for some } n \in \mathbb{N}\}$. The ideal I is a *radical ideal* if $\sqrt{I} = I$. The *nilradical* of a ring R is $\sqrt{(0)}$. The nilradical of R is the intersection of all the prime ideals of R. The ring R is *reduced* if (0) is a radical ideal. A radical ideal I of R is also called a *reduced ideal* since R/I is a reduced ring.

Jacobson radical. The *Jacobson radical* $\mathcal{J}(R)$ of a ring R is the intersection of all maximal ideals of R. An element z of R is in $\mathcal{J}(R)$ if and only if $1 + zr$ is a unit of R for all $r \in R$.

If I is a proper ideal of R, then $1 + I := \{ 1 + a \mid a \in I \}$ is a multiplicatively closed subset of R that does not contain 0. Let $(1 + I)^{-1}R$ denote the localization $R_{(1+I)}$ of R at the multiplicatively closed set $1+I$, [**122**, Section 4]. If P is a prime ideal of R and $P \cap (1+I) = \emptyset$, then $(P+I) \cap (1+I) = \emptyset$. Therefore I is contained in every maximal ideal of $(1 + I)^{-1}R$, so $I \subseteq \mathcal{J}((1 + I)^{-1}R)$. In particular for the principal ideal $I = zR$, where z is a nonunit of R, we have $z \in \mathcal{J}((1 + zR)^{-1}R)$.

Associated primes, prime divisors. A prime ideal $\mathfrak{p}$ of a ring R is *associated to* an R-module M if $\mathfrak{p}$ is the *annihilator* ideal in R of an element $x \in M$; that is, $\mathfrak{p} = \{a \in R \mid ax = 0\}$. For I an ideal of R, the associated prime ideals for I are the prime ideals $\mathfrak{p}$ of R containing I such that $\mathfrak{p}/I$ is an associated prime ideal of R/I. An associated prime ideal of I is *embedded* if it properly contains another associated prime ideal of I.

For an ideal I of a ring R, the associated primes of R/I are also called the *prime divisors* of I.

Finite, finite type, finite presentation. Let R be a ring, let M be an R-module and let S be an R-algebra.

(1) M is said to be a *finite* R-module if M is finitely generated as an R-module.
(2) S is said to be *finite* over R if S is a finitely generated R-module.
(3) S is of *finite type* over R if S is finitely generated as an R-algebra. Equivalently, S is an R-algebra homomorphic image of a polynomial ring in finitely many variables over R.
(4) S is *finitely presented* as an R-algebra if, for some polynomial ring $R[x_1, \ldots, x_n]$ in variables $x_1, \ldots, x_n$ and some surjective R-algebra homomorphism $\varphi : R[x_1, \ldots, x_n] \to S$, then the kernal of φ is a finitely generated ideal of $R[x_1, \ldots, x_n]$.
(5) S is *essentially finite* over R if S is a localization of a finite R-module.
(6) S is *essentially of finite type* over R if S is a localization of a finitely generated R-algebra. We also say that S is *essentially finitely generated* in this case.
(7) S is *essentially finitely presented* over R if S is a localization of a finitely presented R-algebra.

Symbolic powers. If P is a prime ideal of a ring R and e is a positive integer, the eth *symbolic power of* P, denoted $P^{(e)}$, is defined as

$$P^{(e)} := \{a \in R \mid ab \in P^e \text{ for some } b \in R \setminus P \}.$$

Valuation domains and valuations. An integral domain R is a *valuation domain* if, for each element $a \in \mathcal{Q}(R) \setminus R$, we have $a^{-1} \in R$. A valuation domain R is called a *discrete rank-one valuation ring* or a *discrete valuation ring* (DVR) if R is Noetherian and not a field; equivalently, R is a local principal ideal domain (PID) and not a field.

REMARKS 2.1. Let R be a valuation domain with field of fractions K.

(1) If F is a subfield of K, then $R \cap F$ is again a valuation domain and has field of fractions F [**137**, (11.5)]. If R is a DVR and the field F is not contained in R, then $R \cap F$ is again a DVR [**137**, (33.7)].

(2) The nonzero R-submodules of K are totally ordered with respect to inclusion. Let $G = \{xR \mid x \in K \text{ and } x \neq 0\}$. Define $xR \leq yR \iff xR \supseteq yR$, and $xR \circ yR = xyR$. Then G is a totally ordered abelian group; that is, $xR \geq yR, zR \geq tR \implies xzR \geq ytR$, for every $x, y, z, t \in K$. Then $G \cup \{\infty\}$ is a totally ordered set where $g < \infty$, for every $g \in G$, and G is called the *value group* of R.

(3) The valuation domain R has an associated *valuation* v, where v is a function $v : K \to G \cup \{\infty\}$ satisfying properties (i)- (iii) for every $a, b \in K$:
   (i) If $a + b \neq 0$, then $v(a + b) \geq \min\{v(a), v(b)\}$.
   (ii) $v(ab) = v(a) \circ v(b)$.
   (iii) $v(a) = \infty \iff a = 0$.

See [**122**, p. 75] or [**54**, pp. 171-182] for more information about the value group and valuation associated to a valuation domain.

DEFINITION 2.2. **Algebraic independence.** Let R be a subring of a commutative ring S.

(1) Elements $a_1, \dots, a_m \in S$ are *algebraically independent* over R if, for indeterminates $x_1, \dots, x_m$ over R, the only polynomial $f(x_1, \dots, x_m) \in R[x_1, \dots, x_m]$ with $f(a_1, \dots, a_m) = 0$ is the zero polynomial.

(2) A subset Γ of S is *algebraically independent* over R if every finite subset of Γ consists of algebraically independent elements over R.

Integral ring extensions, integral closure, normal domains. Let R be a subring of commutative ring S.

(1) An element $a \in S$ is *integral* over R if a is a root of some monic polynomial in the polynomial ring $R[x]$.

(2) The ring S is *integral* over R, or an *integral extension* of R, if every element $a \in S$ is integral over R.

(3) The *integral closure* of R in S is the set of all elements of S that are integral over R.

(4) The ring R is *integrally closed* in S if every element of S that is integral over R is in R.

(5) The ring R is *integrally closed* if R is integrally closed in its total ring of fractions $\mathcal{Q}(R)$.

(6) The *integral closure* or *derived normal ring* of an integral domain R is the integral closure of R in its field of fractions $\mathcal{Q}(R)$.

(7) As in [**122**, page 64], a ring R is a *normal ring* if for each $P \in \operatorname{Spec} R$ the localization R_P is an integrally closed domain. Since every localization of an integrally closed domain is again an integrally closed domain [**122**, Example 3, page 65], an integrally closed domain is a normal ring.

REMARK 2.3.

(1) If R is a Noetherian normal ring and $\mathfrak{p}_1, \dots, \mathfrak{p}_r$ are the minimal primes of R, then R is isomorphic to the direct product $R/\mathfrak{p}_1 \times \cdots \times R/\mathfrak{p}_r$ and each $R/\mathfrak{p}_i$ is an integrally closed domain; see [**122**, page 64]. Since a nontrivial direct product is not local, a normal Noetherian local ring is a normal domain.

(2) A Noetherian integral domain R is integrally closed if and only if R satisfies the following two conditions:

(a) R_P is a DVR for every $P \in \operatorname{Spec} R$ with $\operatorname{ht} P = 1$.

(b) All associated primes of a nonzero principal ideal of R have height 1.

See [**122**, Corollary, page 82], or [**137**, (12.9), page 41], or [**103**, Theorem 54, page 35].

We record in Theorem 2.4 an important result about the integral closure of a normal Noetherian domain in a finite separable algebraic field extension; see [**122**, Lemma 1, page 262], [**137**, (10.16)], [**192**, Corollary 1, page 265], or [**4**, page 522].

THEOREM 2.4. *Let R be a normal Noetherian integral domain with field of fractions K. If L/K is a finite separable algebraic field extension, then the integral closure of R in L is a finite R-module. Thus, if R has characteristic zero, then the integral closure of R in a finite algebraic field extension is a finite R-module.*

REMARK 2.5. Let R be a normal integral domain with field of fractions K and let L/K be a finite separable algebraic field extension. The integral closure of R in L is always contained in a finitely generated R-module. Two different proofs of this are given in [**192**, Theorem 7, page 264]; both proofs involve a vector space basis for L/K of elements integral over R. The first proof uses the discriminant of this basis, while the second proof uses the dual basis determined by the trace map of L/K.

DEFINITION 2.6. **The order function associated to an ideal.** Let I be a nonzero ideal of an integral domain R such that $\bigcap_{n=0}^{\infty} I^n = (0)$. Adopt the convention that $I^0 = R$, and for each nonzero element $r \in R$ define

$$\operatorname{ord}_{R,I}(r) := n \quad \text{if} \quad r \in I^n \setminus I^{n+1}.$$

In the case where $(R, \mathfrak{m})$ is a local ring, we abbreviate $\operatorname{ord}_{R,\mathfrak{m}}$ by ord_R.

REMARK 2.7. With R, I and $\operatorname{ord}_{R,I}$ as above, consider the following two properties for nonzero elements a, b in R:

(1) If $a + b \neq 0$, then $\operatorname{ord}_{R,I}(a + b) \geq \min\{\operatorname{ord}_{R,I}(a), \operatorname{ord}_{R,I}(b)\}$.

(2) $\operatorname{ord}_{R,I}(ab) = \operatorname{ord}_{R,I}(a) + \operatorname{ord}_{R,I}(b)$.

The function $\operatorname{ord}_{R,I}$ always satisfies property 1.

Assume $\operatorname{ord}_{R,I}$ satisfies property 2 for all nonzero a, b in R. Then the function $\operatorname{ord}_{R,I}$ extends uniquely to a function on $\mathcal{Q}(R) \setminus (0)$ by defining

$$\operatorname{ord}_{R,I}\left(\frac{a}{b}\right) := \operatorname{ord}_{R,I}(a) - \operatorname{ord}_{R,I}(b)$$

for nonzero elements $a, b \in R$, and the set

$$V := \{q \in \mathcal{Q}(R) \setminus (0) \mid \operatorname{ord}_{R,I}(q) \geq 0\} \cup \{0\}$$

is a DVR. Moreover, if $\mathfrak{m}_V$ denotes the maximal ideal of V, then $R \cap \mathfrak{m}_V = I$.

Thus, if $\operatorname{ord}_{R,I}$ satisfies property 2 for all nonzero a, b in R, then I is a prime ideal of R, the function $\operatorname{ord}_{R,I}$ is the valuation on V described in Remark 2.1.2, and the value group is the integers viewed as an additive group.

Let A be a commutative ring and let $R := A[[x]] = \{f = \sum_{i=0}^{\infty} f_i x^i \mid f_i \in A\}$, the *formal power series ring* over A in the variable x. With $I := xR$ and f a nonzero element in R, we write $\operatorname{ord} f$ for $\operatorname{ord}_{R,I}(f)$. Thus $\operatorname{ord} f$ is the least integer $i \geq 0$ such that $f_i \neq 0$. The element f_i is called the *leading form* of f.

Regular local rings. A local ring $(R, \mathfrak{m})$ is a *regular local ring*, often abbreviated *RLR*, if R is Noetherian and $\mathfrak{m}$ can be generated by $\dim R$ elements. If $\dim R = d$ and $\mathfrak{m} = (a_1, \ldots, a_d)R$, then $a_1, \ldots, a_d$ is called a *regular system of parameters*. If $(R, \mathfrak{m})$ is a regular local ring, then R is an integral domain [**122**, Theorem 14.3, p.106]; thus we may say R is a *regular local domain*.

The order function ord_R of a RLR satisfies the properties of Remark 2.7, and the associated valuation domain

$$V := \{q \in \mathcal{Q}(R) \setminus \{0\} \mid \mathrm{ord}_R(q) \geq 0\} \cup \{0\}$$

is a DVR that birationally dominates R. If $x \in \mathfrak{m} \setminus \mathfrak{m}^2$, then $V = R[\mathfrak{m}/x]_{xR[\mathfrak{m}/x]}$, where $\mathfrak{m}/x = \{y/x \mid y \in \mathfrak{m}\}$.

DEFINITION 2.8. **Serre's conditions.** Let A be a Noetherian ring, and let $i, j \in \mathbb{N}_0$. Then Serre introduced the following terminology:

(R_i) A_P is an RLR, for every $P \in \mathrm{Spec}\, A$ with $\mathrm{ht}\, P \leq i$.
(S_j) $\mathrm{depth}\, A_P \geq \min(j, \mathrm{ht}\, P)$, for every $P \in \mathrm{Spec}\, A$.[1]

The condition (S_0) always holds; (S_1) is equivalent to every associated prime ideal of A is minimal.

THEOREM 2.9. Serre's Normality Theorem. [**122**, Theorem 23.8] *A Noetherian ring A is normal if and only if A satisfies the Serre conditions (R_1) and (S_2).*

REMARKS 2.10.

(1) A regular local ring is a normal Noetherian local domain; this follows from Theorem 2.9. It also follows from item 2 of Remark 2.3.

(2) A regular local ring is a UFD. This result, first proved in 1959 by Auslander and Buchsbaum [**17**], represents a significant triumph for homological methods in commutative algebra; see [**122**, Theorem 20.3].

Krull domains. We record the definition of a Krull domain:

DEFINITION 2.11. An integral domain R is a *Krull domain* if there exists a defining family $\mathcal{F} = \{V_\lambda\}_{\lambda \in \Lambda}$ of DVRs of its field of fractions $\mathcal{Q}(R)$ such that

- $R = \bigcap_{\lambda \in \Lambda} V_\lambda$, and
- A nonzero element of $\mathcal{Q}(R)$ is a unit in all but finitely many of the V_λ.

REMARKS 2.12. We list several properties of Krull domains. See [**122**, pp. 86-88] or [**23**, pp. 475-493] for proofs of these properties.

(1) A unique factorization domain (UFD) is a Krull domain, and a Noetherian integral domain is a Krull domain if and only if it is integrally closed. An integral domain R is a Krull domain if and only if it satisfies the following three properties:
   - $R_\mathfrak{p}$ is a DVR for each prime ideal $\mathfrak{p}$ of R of height one.
   - $R = \bigcap\{ R_\mathfrak{p} \mid \mathfrak{p}$ is a height-one prime $\}$.
   - Every nonzero element of R is contained in only finitely many height-one primes of R.

(2) If R is a Krull domain, then $\mathcal{F}_0 = \{R_\mathfrak{p} \mid \mathfrak{p}$ is a height-one prime ideal $\}$ is the unique minimal family of DVRs satisfying the properties in the definition of a Krull domain [**122**, Theorem 12.3]. Every defining family

[1] For the definition of depth, see Definition 3.35

$\mathcal{F}$ contains $\mathcal{F}_0$. The family $\mathcal{F}_0$ is called the family of *essential valuation rings* of R. For each nonzero nonunit a of R the principal ideal aR has no embedded associated prime ideals and has a unique irredundant primary decomposition $aR = \mathfrak{q}_1 \cap \cdots \cap \mathfrak{q}_t$. If $\mathfrak{p}_i = \sqrt{(\mathfrak{q}_i)}$, then $R_{\mathfrak{p}_i} \in \mathcal{F}$ and $\mathfrak{q}_i$ is a symbolic power of $\mathfrak{p}_i$; that is, $\mathfrak{q}_i = \mathfrak{p}_i^{(e_i)}$, where $e_i \in \mathbb{N}$; see [**122**, Corollary, page 88].

Krull domains have an approximation property with respect to the family of DVRs and valuations[2] obtained by localizing at height-one primes.

THEOREM 2.13. Approximation Theorem. [**122**, Theorem 12.6] *For A a Krull domain with field of fractions K, let $\mathfrak{p}_1, \ldots, \mathfrak{p}_r$ be height-one primes of A, and let v_i denote the valuation with value group $\mathbb{Z}$ associated to the DVR $A_{\mathfrak{p}_i}$, for each i with $1 \leq i \leq r$. For arbitrary integers $e_1, \ldots, e_r$, there exists $x \in K$ such that*

$$v_i(x) = e_i \quad \text{for} \quad 1 \leq i \leq r \quad \text{and} \quad v(x) \geq 0,$$

for every valuation v associated to a height-one prime ideal of A that is not in the set $\{\mathfrak{p}_1, \ldots, \mathfrak{p}_r\}$.

DEFINITION 2.14. Let R be a Krull domain and let $R \hookrightarrow S$ be an inclusion map of R into a Krull domain S. The extension $R \hookrightarrow S$ satisfies the **PDE** condition ("pas d'éclatement", or in English "no blowing up") provided that for every height-one prime ideal $\mathfrak{q}$ in S, the height of $\mathfrak{q} \cap R$ is at most one; see Fossum [**48**, page 30], or Bourbaki [**22**, Chapitre 7, Proposition 14, page 18].

Divisorial ideals and the Divisor Class Group of Krull domains. Following Samuel [**162**, Ch. 1, Sections 2 and 3] and Bourbaki [**23**, Ch. VII, Sections 1.1-1.3 and 1.6], we say a submodule F of the field of fractions K of an integral domain R is a *fractional* ideal of R provided that there exists a nonzero element $d \in R$ such that $dF \subseteq R$. A fractional ideal F is *principal* if $F = Ra$, for some $a \in K$. A fractional ideal F is *divisorial* if F is nonzero and F is an intersection of principal fractional ideals. An integral domain R is *completely integrally closed* provided the following condition holds: if $a \in K$ is such that the ring $R[a]$ is a fractional ideal of R, then $a \in R$.

REMARKS 2.15. Let R be an integral domain with field of fractions K.

(1) If F is a nonzero fractional ideal of R, then there exists a unique *smallest divisorial ideal containing* F, denoted $\overline{F}$ and given by $\overline{F} = (R : (R : F))$ [**162**, Lemma 1.2].
(2) The set $\mathcal{D}(R)$ of divisorial ideals of R forms a partially ordered commutative monoid under the multiplication defined by $F \circ J = \overline{(FJ)}$ and the ordering given by inclusion [**162**, p. 3].
(3) Let $\mathcal{F}(R)$ denote the group of nonzero principal fractional ideals of R. Then $\mathcal{F}(R)$ is a subgroup of the monoid $\mathcal{D}(R)$.
(4) The monoid $\mathcal{D}(R)$ is a partially ordered group if and only if R is completely integrally closed [**162**, Theorem 2.1].
(5) If R is a Krull domain, then R is completely integrally closed [**162**, Theorem 3.1].
(6) By items 4 and 5, if R is a Krull domain, then $\mathcal{D}(R)$ is a partially ordered group.

[2]See Remarks 2.1.3)

DEFINITION 2.16. Assume that R is a completely integrally closed domain or, equivalently, that $\mathcal{D}(R)$ is a group. The *Divisor Class Group* of R is defined to be $\mathcal{C}(R) := \frac{\mathcal{D}(R)}{\mathcal{F}(R)}$, the quotient of $\mathcal{D}(R)$ by $\mathcal{F}(R)$.

PROPOSITION 2.17. [**162**, Prop. 4.3, Theorem 6.4] *If R is a Krull domain and x is an indeterminate over R, then the polynomial ring $R[x]$ is a Krull domain, and the divisor class group of $R[x]$ is isomorphic to the divisor class group of R.*

DISCUSSION 2.18. [**162**, Section 3] and [**23**, Theorems 2 & 3, pp. 480 & 485] For a Krull domain R, the partially ordered group $\mathcal{D}(R)$ is isomorphic to a direct sum of copies of the integers and this direct sum is indexed by the height-one prime ideals of R. A height-one prime $\mathfrak{p}$ of R is a divisorial ideal and $R_\mathfrak{p}$ is a DVR. The $\mathfrak{p}$-primary ideals are precisely the symbolic powers of $\mathfrak{p}$ and are precisely the divisorial ideals having radical $\mathfrak{p}$. Here $\mathcal{C}(R)$ is generated by the images in this quotient group of those divisorial ideals that are height-one prime ideals of R.

FACT 2.19. Discussion 2.18 implies that the Divisor Class Group $\mathcal{C}(R)$ of a Krull domain R is torsion if and only if every height-one prime ideal of R is the radical of a principal ideal.

Nagata rings. In the 1950s Nagata introduced and investigated a class of Noetherian rings that behave similarly to rings that arise in algebraic geometry [**130**], [**132**]. In Nagata's book, Local Rings [**137**], the rings in this class are called *pseudo-geometric*. Following Matsumura, we call these rings *Nagata rings*:

DEFINITION 2.20. A commutative ring R is called a *Nagata* ring if R is Noetherian and, for every $P \in \operatorname{Spec} R$ and every finite extension field L of $\mathcal{Q}(R/P)$, the integral closure of R/P in L is finitely generated as a module over R/P.

It is clear from the definition that a homomorphic image of a Nagata ring is again a Nagata ring. We refer to the following nontrivial theorem due to Nagata as the *Nagata Polynomial Theorem.*

THEOREM 2.21. Nagata Polynomial Theorem. [**137**, Theorem 36.5, page 132] *If A is a Nagata ring and $x_1, \ldots, x_n$ are indeterminates over A, then the polynomial ring $A[x_1, \ldots, x_n]$ is a Nagata ring. It follows that every algebra essentially of finite type over a Nagata ring is again a Nagata ring.*

By Theorem 2.21, every algebra of finite type over a field, over the ring of integers, or over a discrete valuation ring of characteristic 0, is a Nagata ring.

For more information about Nagata rings, see Chapter 8.

2.2. Basic theorems

Theorem 2.22 is a famous result proved by Krull that is now called the Krull Intersection Theorem.

THEOREM 2.22 (Krull [**122**, Theorem 8.10]). *Let I be an ideal of a Noetherian ring R.*

(1) *If I is contained in the Jacobson radical $\mathcal{J}(R)$ of R, then $\bigcap_{n=1}^{\infty} I^n = 0$, and, for each finite R-module M, we have $\bigcap_{n=1}^{\infty} I^n M = 0$.*
(2) *If I is a proper ideal of a Noetherian integral domain, then $\bigcap_{n=1}^{\infty} I^n = 0$.*

Theorem 2.23 is another famous result of Krull that is now called the Krull Altitude Theorem. It involves the concept of a minimal prime divisor of an ideal I of a ring R, where $P \in \operatorname{Spec} R$ is a *minimal prime divisor* of I if $I \subseteq P$ and if $P' \in \operatorname{Spec} R$ and $I \subseteq P' \subseteq P$, then $P' = P$.

THEOREM 2.23 (Krull [**122**, Theorem 13.5]). *Let R be a Noetherian ring and let $I = (a_1, \dots, a_r)R$ be an ideal generated by r elements. If P is a minimal prime divisor of I, then* $\operatorname{ht} P \leq r$. *Hence the height of a proper ideal of R is finite.*

Theorem 2.24 is yet another famous result that is now called the Krull-Akizuki Theorem.

THEOREM 2.24 (Krull-Akizuki [**122**, Theorem 11.7]). *Let A be a one-dimensional Noetherian integral domain with field of fractions K, let L be a finite algebraic field extension of K, and let B be a subring of L with $A \subseteq B$. Then*

(1) *The ring B is Noetherian of dimension at most one.*
(2) *If J is a nonzero ideal of B, then B/J is an A-module of finite length.*

To prove that a ring is Noetherian, it suffices by the following well-known result of Cohen to prove that every prime ideal of the ring is finitely generated.

THEOREM 2.25 (Cohen [**36**]). *If each prime ideal of the ring R is finitely generated, then R is Noetherian.*

Theorem 2.26 is another important result proved by Cohen.

THEOREM 2.26 (Cohen [**37**]). *Let R be a Noetherian integral domain and let S be an extension domain of R. For $P \in \operatorname{Spec} S$ and $\mathfrak{p} = P \cap R$, we have*

$$\operatorname{ht} P \ + \ tr.deg._{k(\mathfrak{p})} k(P) \ \leq \ \operatorname{ht} \mathfrak{p} \ + \ tr.deg._R S,$$

where $k(\mathfrak{p})$ is the field of fractions of $R/\mathfrak{p}$ and $k(P)$ is the field of fractions of S/P.

Theorem 2.27 is a useful result due to Nagata about Krull domains and UFDs.

THEOREM 2.27. [**162**, Theorem 6.3, p. 21] *Let R be a Krull domain. If S is a multiplicatively closed subset of R generated by prime elements and $S^{-1}R$ is a UFD, then R is a UFD.*

We use the following:

FACT 2.28. If D is an integral domain and c is a nonzero element of D such that cD is a prime ideal, then $D = D[1/c] \cap D_{cD}$.

PROOF. Let $\beta \in D[1/c] \cap D_{cD}$. Then $\beta = \frac{b}{c^n} = \frac{b_1}{s}$ for some $b, b_1 \in D$, $s \in D \setminus cD$ and integer $n \geq 0$. If $n > 0$, we have $sb = c^n b_1 \implies b \in cD$. Thus we may reduce to the case where $n = 0$; it follows that $D = D[1/c] \cap D_{cD}$. □

REMARKS 2.29.

(1) If R is a Noetherian integral domain and S is a multiplicatively closed subset of R generated by prime elements, then $S^{-1}R$ a UFD implies that R is a UFD [**162**, Theorem 6.3] or [**122**, Theorem 20.2].

(2) If x is a nonzero prime element in an integral domain R such that R_{xR} is a DVR and $R[1/x]$ is a Krull domain, then R is a Krull domain by Fact 2.28; and, by Theorem 2.27, R is a UFD if $R[1/x]$ is a UFD.

(3) Let R be a valuation domain with value group $\mathbb{Z} \oplus \mathbb{Z}$ ordered lexicographically; that is, for every pair $(a, b), (c, d)$ of elements of $\mathbb{Z} \oplus \mathbb{Z}$, $(a, b) > (c, d) \iff a > c$, or $a = c$ and $b > d$. Then the maximal ideal $\mathfrak{m}$ of R is principal, say $\mathfrak{m} = xR$. It follows that $R[1/x]$ is a DVR; however R is not a Krull domain.

The Eakin-Nagata Theorem is useful for proving descent of the Noetherian property.

THEOREM 2.30 (Eakin-Nagata [**122**, Theorem 3.7(i)]). *If B is a Noetherian ring and A is a subring of B such that B is a finitely generated A-module, then A is Noetherian.*

An interesting result proved by Nishimura is

THEOREM 2.31. (Nishimura [**140**, Theorem, page 397], [**122**, Theorem 12.7]). *Let R be a Krull domain. If R/P is Noetherian for every height-one prime ideal P of R, then R is Noetherian.*

REMARK 2.32. It is observed in [**71**, Lemma 1.5] that the conclusion of Theorem 2.31 still holds if it is assumed that R/P is Noetherian for all but at most finitely many of the height-one primes P of R.

Theorem 2.33 is useful for describing the maximal ideals of a power series ring $R[[x]]$. It is related to the fact that an element $f = a_0 + a_1x + a_2x^2 + \cdots \in R[[x]]$, where each $a_i \in R$, is a unit of $R[[x]]$ if and only if a_0 is a unit of R.

THEOREM 2.33 ([**137**, Theorem 15.1]). *Let $R[[x]]$ be the formal power series ring in a variable x over a commutative ring R. There is a one-to-one correspondence between the maximal ideals $\mathfrak{m}$ of R and the maximal ideals $\mathfrak{m}^*$ of $R[[x]]$, where $\mathfrak{m}^*$ corresponds to $\mathfrak{m}$ if and only if $\mathfrak{m}^*$ is generated by $\mathfrak{m}$ and x.*

As an immediate corollary of Theorem 2.33, we have

COROLLARY 2.34. *The element x is in the Jacobson radical $\mathcal{J}(R[[x]])$ of the power series ring $R[[x]]$. In the formal power series ring $S := R[[x_1, \ldots, x_n]]$, the ideal $(x_1, \ldots, x_n)S$ is contained in the Jacobson radical $\mathcal{J}(S)$ of S.*

Theorem 2.35 is an important result first proved by Chevalley.

THEOREM 2.35 (Chevalley [**34**]). *If $(R, \mathfrak{m})$ is a Noetherian local domain, then there exists a DVR that birationally dominates R.*

More generally, let P be a prime ideal of a Noetherian integral domain R. There exists a DVR V that birationally contains R and has center P on R, that is, the maximal ideal of V intersects R in P.

2.3. Flatness

The concept of flatness was introduced by Serre in the 1950s in an appendix to his paper [**167**]. Mumford writes in [**125**, page 424]: "The concept of flatness is a riddle that comes out of algebra, but which technically is the answer to many prayers."

DEFINITIONS 2.36. A module M over a ring R is *flat* over R if tensoring with M preserves exactness of every exact sequence of R-modules. The R-module M is said to be *faithfully flat* over R if, for every sequence $\mathcal{S}$ of R-modules,

$$\mathcal{S}: \quad 0 \longrightarrow M_1 \longrightarrow M_2,$$

the sequence $\mathcal{S}$ is exact if and only if its tensor product with M, $\mathcal{S} \otimes_R M$, is exact.

A ring homomorphism $\phi : R \to S$ is said to be a *flat homomorphism* if S is flat as an R-module.

Flatness is preserved by several standard ring constructions as we record in Remarks 2.37. There is an interesting elementwise criterion for flatness that is stated as item 2 of Remarks 2.37.

REMARKS 2.37. The following facts are useful for understanding flatness. We use these facts to obtain the results in Chapters 6 and 14.

(1) Since localization at prime ideals commutes with tensor products, the module M is flat as an R-module if and only if M_Q is flat as an R_Q-module, for every prime ideal Q of R.
(2) An R-module M is flat over R if and only if for every $m_1, \ldots, m_n \in M$ and $a_1, \ldots, a_n \in R$ such that $\sum a_i m_i = 0$, there exist a positive integer k, a subset $\{b_{ij}\}_{i=1,}^{n}{}_{j=1}^{k} \subseteq R$, and elements $m'_1, \ldots, m'_k \in M$ such that $m_i = \sum_{j=1}^{k} b_{ij} m'_j$ for each i and $\sum_{i=1}^{n} a_i b_{ij} = 0$ for each j; see [**122**, Theorem 7.6] or [**120**, Theorem 1]. Thus every free module is flat, and a nested union of flat modules is flat.
(3) A finitely generated module over a local ring is flat if and only if it is free [**120**, Proposition 3.G].
(4) If the ring S is a localization of R, then S is flat as an R-module [**120**, (3.D), page 19].
(5) Let S be a flat R-algebra. Then S is faithfully flat over R if and only if one has $JS \neq S$ for every proper ideal J of R; see [**120**, Theorem 3, page 28] or [**122**, Theorem 7.2].
(6) If the ring S is a flat R-algebra, then every regular element of R is regular on S [**120**, (3.F)].
(7) Let S be a faithfully flat R-algebra and let I be an ideal of R. Then $IS \cap R = I$ [**122**, Theorem 7.5].
(8) Let R be a subring of a ring S. If S is Noetherian and faithfully flat over R, then R is Noetherian; see Exercise 9 at the end of this chapter.
(9) Let R be an integral domain with field of fractions K and let S be a faithfully flat R-algebra. By item 6, every nonzero element of R is regular on S and so K naturally embeds in the total quotient ring $\mathcal{Q}(S)$ of S. By item 7, all ideals in R extend and contract to themselves with respect to S, and thus $R = K \cap S$. In particular, if $S \subseteq K$, then $R = S$ [**120**, page 31].
(10) If $\varphi : R \to S$ is a flat homomorphism of rings, then φ satisfies the *Going-down property* (or Going-down theorem) [**120**, (5.D), page 33] or [**122**, Theorem 9.5]: Let $\mathfrak{p} \subseteq \mathfrak{q}$ be prime ideals of R and let $Q \in \operatorname{Spec} S$ be such that $\varphi^{-1}(Q) = \mathfrak{q}$. Then there exists $P \in \operatorname{Spec} S$ with $P \subseteq Q$ and $\varphi^{-1}(P) = \mathfrak{p}$. It follows that the height of P in S is greater than or equal to the height of $\varphi^{-1}(P)$ in R, for each $P \in \operatorname{Spec} S$.
(11) Let $R \to S$ be a flat homomorphism of rings and let I and J be ideals of R. Then $(I \cap J)S = IS \cap JS$. If J is finitely generated, then $(I :_R J)S = IS :_S JS$; see [**122**, Theorem 7.4] or [**120**, (3.H) page 23].
(12) Consider the following short exact sequence of R-modules:

$$0 \longrightarrow A \longrightarrow B \longrightarrow C \longrightarrow 0.$$

The modules A and C are flat over R if and only if B is [**122**, Theorem 7.9].

(13) If S is a flat R-algebra and M is a flat S-module, then M is a flat R-module [**122**, page 46].
(14) If S is an R-algebra and M is faithfully flat over both R and S, then S is faithfully flat over R [**122**, page 46].
(15) Let S be an R-algebra and M an S-module. If M is faithfully flat over S and M is flat over R, then S is a flat R-algebra. If M is faithfully flat over both R and S, then S is a faithfully flat R-algebra [**122**, page 46].

The following standard result about flatness follows from what Matsumura calls "change of coefficient ring"; see [**122**, p. 46]. It is convenient to refer to both the module and homomorphism versions.

FACT 2.38. Let C be a commutative ring, let D, E and F be C-algebras.

(1) If $\psi : D \to E$ is a flat, respectively faithfully flat, C-algebra homomorphism, then $\psi \otimes_C 1_F : D \otimes_C F \to E \otimes_C F$ is a flat, respectively faithfully flat, C-algebra homomorphism.
(2) If E is a flat, respectively faithfully flat, D-module via the C-algebra homomorphism ψ, then $E \otimes_C F$ is flat, respectively faithfully flat, as a $D \otimes_C F$-module via the C-algebra homomorphism $\psi \otimes_C 1_F$.

PROOF. By the definition of flat, respectively faithfully flat, homomorphism in Definitions 2.36, the two statements are equivalent. Since E is a flat, respectively faithfully flat, D-module, $E \otimes_D (D \otimes_C F)$ is a flat, respectively faithfully flat, $(D \otimes_C F)$-module by [**122**, p. 46]. Since $E \otimes_D (D \otimes_C F) = E \otimes_C F$, Fact 2.38 holds. □

We use Remark 2.39.3 in Chapter 12.

REMARKS 2.39. Let R be an integral domain.

(1) Every flat R-module M is torsionfree, i.e., if $r \in R, x \in M$ and $rx = 0$, then $r = 0$ or $x = 0$; see [**120**, (3.F), page 21]
(2) Every finitely generated torsionfree module over a PID is free; see for example [**42**, Theorem 5, page 462].
(3) Every torsionfree module over a PID is flat. This follows from item 2 and Remark 2.37.2.
(4) Every injective homomorphism of R into a field is flat. This follows from Remarks 2.37.13 and 2.37.4.

DEFINITION 2.40. Let $\varphi : R \to S$ be a ring homomorphism. The *non-flat locus* of φ is $\mathcal{F}$, where

$$\mathcal{F} := \{Q \in \operatorname{Spec}(S) \mid \text{the map} \quad \varphi_Q : R \to S_Q \text{ is not flat } \}.$$

A subset F of S or an ideal F of S *defines* or *determines* the non-flat locus of φ, if

$$\mathcal{F} \;=\; \mathcal{V}(F) \;:=\; \{P \in \operatorname{Spec} S \mid F \subseteq P\}. \tag{2.40.0}$$

REMARKS 2.41. Let $\varphi : R \to S$ be a ring homomorphism.

(1) If $P \subseteq Q$ in $\operatorname{Spec} S$, then $P \in \mathcal{F} \implies Q \in \mathcal{F}$.
(2) If $\mathcal{F}$ is closed in the Zariski topology on $\operatorname{Spec} S$, then $F := \bigcap_{P \in \mathcal{F}} P$ defines the non-flat locus of φ.

If S is a finitely generated R-algebra defined by a homomorphism $\varphi : R \to S$, then there is an ideal of S that determines the non-flat locus of φ by Theorem 2.42.

THEOREM 2.42. [**122**, Theorem 24.3] *Let R be a Noetherian ring and let $\varphi : R \to S$ define S as a finitely generated R-algebra. For a finite S-module M, set $U := \{P \in \operatorname{Spec} S \mid M_P$ is flat over $R\}$; then U is open in* $\operatorname{Spec} S$. *In particular, the set $\mathcal{F}$ of Definition 2.40 is closed, and so $F := \bigcap_{P \in \mathcal{F}} P$ defines the non-flat locus of φ.*

PROPOSITION 2.43. *Let R be a ring and let $R \overset{\alpha}{\hookrightarrow} S \overset{\beta}{\hookrightarrow} T$ be an extension of R-algebras such that β is flat. If the non-flat locus of α is closed in* $\operatorname{Spec} S$ *and defined by a subset F of S, then the non-flat locus of $\varphi := \beta \circ \alpha$ is closed in* $\operatorname{Spec} T$ *and is also defined by F.*

PROOF. Let $\mathfrak{p} \in \operatorname{Spec} T$. Then:

$$F \subseteq \mathfrak{p} \iff F \subseteq \mathfrak{p} \cap S \iff \alpha_{\mathfrak{p} \cap S} : R \hookrightarrow S_{\mathfrak{p} \cap S} \text{ is not flat.}$$

By Remarks 2.37.13 and 2.37.5, the map $\alpha_{\mathfrak{p} \cap S}$ is not flat if and only if the composite map $\varphi_{\mathfrak{p}} : R \overset{\alpha_{\mathfrak{p} \cap S}}{\hookrightarrow} S_{\mathfrak{p} \cap S} \overset{\beta_{\mathfrak{p}}}{\hookrightarrow} T_{\mathfrak{p}}$ is not flat. □

Exercises

(1) Prove that every height-one prime ideal of a UFD is principal.

(2) Let V be a local domain with nonzero principal maximal ideal yV. Prove that V is a DVR if $\bigcap_{n=1}^{\infty} y^n V = (0)$.

COMMENT. It is not being assumed that V is Noetherian, so it needs to be established that V has dimension one.

(3) Prove as stated in Remark 2.1 that if R is a valuation domain with field of fractions K and F is a subfield of K, then $R \cap F$ is again a valuation domain and has field of fractions F; also prove that if R is a DVR and the field F is not contained in R, then $R \cap F$ is again a DVR.

(4) Prove that a unique factorization domain is a Krull domain.

(5) Let a and b be nonzero elements of an integral domain R.

An element $d \in R$ is said to be a *greatest common divisor* of a and b, denoted $\gcd(a, b) = d$, provided: (i) $d \mid a$ and $d \mid b$, and (ii) if $c \in R$ and $c \mid a$ and $c \mid b$, then $c \mid d$.[3]

An element $\ell \in R$ is said to be a *least common multiple* of a and b, denoted $\operatorname{lcm}(a, b) = \ell$, provided: (i) $a \mid \ell$ and $b \mid \ell$, and (ii) if $m \in R$ and $a \mid m$ and $b \mid m$, then $\ell \mid m$.

An integral domain R is called a *GCD-domain* if every pair of nonzero elements of R has a gcd.

(a) Prove that $\ell = \operatorname{lcm}(a, b) \iff R\ell = Ra \cap Rb$.

(b) If $\ell = \operatorname{lcm}(a, b)$ and $0 \neq t \in R$, prove that $\ell t = \operatorname{lcm}(at, bt)$.

(c) If $\ell = \operatorname{lcm}(a, b)$, prove that $\gcd(a, b)$ exists and $\frac{ab}{\ell} = \gcd(a, b)$.

(d) Give an example of nonzero elements a and b in an integral domain R for which $\gcd(a, b)$ exists but $\operatorname{lcm}(a, b)$ does not exist.

(e) Give an example of nonzero elements a, b and c in an integral domain R for which $\gcd(a, b)$ exists but $\gcd(ac, bc)$ does not exist.

(f) If $0 \neq t \in R$ and $d = \gcd(at, bt)$, prove that $\frac{d}{t} \in R$ and $\frac{d}{t} = \gcd(a, b)$.

---

[3] We use the symbol "|" to denote divides.

(g) If R is a GCD-domain, prove that every pair of elements of R has a least common multiple.

(6) Let R be a Noetherian ring. Let $P_1 \subset P_2$ be prime ideals of R. If there exists a prime ideal Q of R with Q distinct from P_1 and P_2 such that $P_1 \subset Q \subset P_2$, prove that there exist infinitely many such prime ideals Q.

SUGGESTION. Apply Krull's Altitude Theorem 2.23, and use the fact that an ideal contained in a finite union of primes is contained in one of them; see for example [**15**, Proposition 1.11, page 8].

(7) Prove as asserted in Remark 2.7 that, if $\operatorname{ord}_{R,I}(ab) = \operatorname{ord}_{R,I}(a) + \operatorname{ord}_{R,I}(b)$, for all nonzero a, b in R, and if we define $\operatorname{ord}_{R,I}(\frac{a}{b}) := \operatorname{ord}_{R,I}(a) - \operatorname{ord}_{R,I}(b)$ for nonzero elements $a, b \in R$, then:
(a) The function $\operatorname{ord}_{R,I}$ extends uniquely to a function on $\mathcal{Q}(R) \setminus (0)$ with this definition.
(b) $V := \{q \in \mathcal{Q}(R) \setminus (0) \mid \operatorname{ord}_{R,I}(q) \geq 0\} \cup \{0\}$ is a DVR, and
(c) R is an integral domain and I is a prime ideal.

(8) Let $R[[x]]$ be the formal power series ring in a variable x over a commutative ring R.
(i) Prove that $a_0 + a_1x + a_2x^2 + \cdots \in R[[x]]$, where the $a_i \in R$, is a unit of $R[[x]]$ if and only if a_0 is a unit of R.
(ii) Prove that x is contained in every maximal ideal of $R[[x]]$.
(iii) Prove Theorem 2.33 that the maximal ideals $\mathfrak{m}$ of R are in one-to-one correspondence with the maximal ideals $\mathfrak{m}^*$ of $R[[x]]$, where $\mathfrak{m}^*$ corresponds to $\mathfrak{m}$ if and only if $\mathfrak{m}^*$ is generated by $\mathfrak{m}$ and x.

(9) Prove items 4-8 of Remarks 2.37.

SUGGESTION. For the proof of item 8, use item 7.

(10) Let $f : A \hookrightarrow B$ be an injective ring homomorphism and let P be a minimal prime of A.
(i) Prove that there exists a prime ideal Q of B that contracts in A to P.
(ii) Deduce that there exists a minimal prime Q of B that contracts in A to P.

SUGGESTION. Consider the multiplicatively closed set $A \setminus P$ as a subset of B; use Zorn's Lemma.

(11) Let $f : A \to B$ be a ring homomorphism and let P be a prime ideal of A. Prove that there exists a prime ideal Q in B that contracts in A to P if and only if the extended ideal $f(P)B$ contracts to P in A, i.e., $P = f(P)B \cap A$. (Here we are using the symbol $\cap$ as in Matsumura [**122**, item (3), page xiii].)

SUGGESTION. Consider the multiplicatively closed set $A \setminus P$ of A and its image $f(A \setminus P)$ in B.

(12) Let P be a height-one prime of a Krull domain A and let v denote the valuation with value group $\mathbb{Z}$ associated to the DVR A_P. If A/P is Noetherian, prove that $A/P^{(e)}$ is Noetherian for every positive integer e.

SUGGESTION. Using Theorem 2.13, show there exists $x \in \mathcal{Q}(A)$ with $v(x) = 1$ and $1/x \in A_Q$, for every height-one prime Q of A with $Q \neq P$. Let $B = A[x]$.
(i) Show that $P = xB \cap A$ and $B = A + xB$.
(ii) Show that $A/P \cong B/xB \cong x^iB/x^{i+1}B$ for every positive integer i.

(iii) Deduce that B/x^eB is a Noetherian B-module and thus a Noetherian ring.
(iv) Prove that $x^eB \cap A \subseteq x^eA_P \cap A = P^{(e)}$ and B/x^eB is a finite $A/(x^eB \cap A)$-module generated by the images of $1, x, \ldots, x^{e-1}$.
(v) Apply Theorem 2.30 to conclude that $A/(x^eB \cap A)$ and hence $A/P^{(e)}$ is Noetherian.

(13) Let A be a Krull domain having the property that A/P is Noetherian for all but at most finitely many of the $P \in \operatorname{Spec} A$ with $\operatorname{ht} P = 1$. Prove that A is Noetherian.

SUGGESTION. By Nishimura's result Theorem 2.31, and Cohen's result Theorem 2.25, it suffices to prove each prime ideal of A of height greater than one is finitely generated. Let $P_1, \ldots, P_n$ be the height-one prime ideals of A for which A/P_i may fail to be Noetherian. For each nonunit $a \in A \setminus (P_1 \cup \cdots \cup P_n)$, observe that $aA = Q_1^{(e_1)} \cap \cdots \cap Q_s^{(e_s)}$, where $Q_1, \ldots, Q_s$ are height-one prime ideals of A not in the set $\{P_1, \ldots, P_n\}$. Consider the embedding $A/aA \hookrightarrow \prod(A/Q_i^{(e_i)})$. By Exercise 12, each $A/Q_i^{(e_i)}$ is Noetherian. Apply Theorem 2.30 to conclude that A/aA is Noetherian. Deduce that every prime ideal of A of height greater than one is finitely generated.

(14) Let R be a two-dimensional Noetherian integral domain. Prove that every Krull domain that birationally dominates R is Noetherian.

COMMENT. It is known that the integral closure of a two-dimensional Noetherian integral domain is Noetherian [**137**, (33.12)]. A proof of Exercise 14 is given in [**65**, Theorem 9]. An easier proof may be obtained using Nishimura's result Theorem 2.31.

CHAPTER 3

More tools

In this chapter we discuss ideal-adic completions. We describe results concerning completions, transcendence degree, and possible pitfalls of power series notation. We review the definitions of catenary, Henselian, and excellent rings, and record results about these rings.

3.1. Introduction to ideal-adic completions

DEFINITIONS 3.1. Let R be a commutative ring with identity. A *filtration* on R is a descending sequence $\{I_n\}_{n=0}^{\infty}$ of ideals of R. Since $I_{n+1} \subset I_n$, the natural maps $R/I_{n+1} \to R/I_n$ form an inverse system. Associated to the filtration $\{I_n\}$, there is a well-defined completion R^* that may be defined to be the inverse limit [1]

$$R^* = \varprojlim_n R/I_n. \tag{3.1.1}$$

There is a canonical homomorphism $\psi : R \to R^*$ [**144**, Chapter 9], and the map ψ induces a map $R \to R^*/I_nR^*$ such that

$$R^*/I_nR^* \cong R/I_n; \tag{3.1.2}$$

see [**144**, page 412] or [**122**, page 55] for more details.

Regarding the filtration $\{I_n\}_{n=0}^{\infty}$ as a system of neighborhoods of 0, and defining for each $x \in R$ the family $\{x + I_n\}$ to be a system of neighborhoods of x, makes R a topological group under addition. This type of topology is called a *linear topology* on R. For more details and an extension to R-modules, see [**122**, Section 8].

If $\bigcap_{n=0}^{\infty} I_n = (0)$, then this linear topology is Hausdorff [**122**, page 55] and gives rise to a metric on R: For $x \neq y \in R$, the distance from x to y is $d(x, y) = 2^{-n}$, where n is the largest n such that $x - y \in I_n$. In particular, the map ψ is injective, and R may be regarded as a subring of R^*.

In the terminology of Northcott, a filtration $\{I_n\}_{n=0}^{\infty}$ is said to be *multiplicative* if $I_0 = R$ and $I_nI_m \subseteq I_{n+m}$, for all $m \geq 0$, $n \geq 0$ [**144**, page 408].

A well-known example of a multiplicative filtration on R is the I-adic filtration $\{I^n\}_{n=0}^{\infty}$, where I is a fixed ideal of R. In this case we say $R^* := \varprojlim_n R/I^n$ is the *I-adic completion* of R. If the canonical map $R \to R^*$ is an isomorphism, we say that R is *I-adically complete*. An ideal L of R is *closed in the* I-adic topology on R if $\bigcap_{n=1}^{\infty}(L + I^n) = L$.

If R is a local ring with maximal ideal $\mathfrak{m}$, then $\widehat{R}$ denotes the $\mathfrak{m}$-adic completion of R. In this case, we say that $\widehat{R}$ is the *completion* of R. If $\mathfrak{m}$ is generated by elements $a_1, \ldots, a_n$, then $\widehat{R}$ is realizable by taking the a_1-adic completion R_1^* of

---

[1] We refer to Appendix A of [**122**] for the definition of direct and inverse limits. Also see the discussion of inverse limits in [**15**, page 103].

R, then the a_2-adic completion R_2^* of R_1^*, ..., and then the a_n-adic completion of R_{n-1}^*.

More generally, we use the notation $\widehat{R}$ for the situation where R is a semilocal ring with Jacobson radical $\mathcal{J}$. In this case, $\widehat{R}$ denotes the $\mathcal{J}$-adic completion of R.

FACT 3.2. Let $\{I_n\}_{n=0}^{\infty}$ be a filtration of ideals of a ring R, and let R^* denote the completion $R^* = \varprojlim_n R/I_n$ of R with respect to the filtration $\{I_n\}_{n=0}^{\infty}$. Then:

(1) Let L be an ideal of R with $I_n \subseteq L$ for some n. Then $R/L \cong R^*/LR^*$ and $L = LR^* \cap R$, that is, L is contracted from R^*.
(2) Let J be an ideal of R^* with $I_nR^* \subseteq J$ for some n. Then $R^*/J \cong R/(J \cap R)$, and $J = (J \cap R)R^*$, that is J is extended from R.
(3) Assume that R is semilocal, and let $\mathcal{J}$ denote the Jacobson radical of R. If $I_n \subseteq \mathcal{J}^n$, for every $n \in \mathbb{N}$, then $\widehat{R^*} = \widehat{R}$.

PROOF. For items 1 and 2, $R/I_n = R^*/I_nR^*$, by Equation 3.1.2. This implies that the ideals of R containing I_n are in one-to-one correspondence with the ideals of R^* containing I_nR^*.

For item 3, the maximal ideals of R contain each I_n, since $I_n \subseteq \mathcal{J}^n$, and so, by items 1 and 2, they correspond to the maximal ideals of R^*. Thus the Jacobson radical $J(R^*)$ satisfies $\mathcal{J} \cdot R^* = \mathcal{J}(R^*)$. The condition $I_n \subseteq \mathcal{J}^n$ also implies that $R/\mathcal{J}^n = R^*/\mathcal{J}^nR^* = R^*/\mathcal{J}(R^*)^n$. Therefore

$$\widehat{R} = \varprojlim_n R/\mathcal{J}^n = \varprojlim_n R^*/\mathcal{J}(R^*)^n = \widehat{R^*}.$$

This completes the proof. □

We record the following results about ideal-adic completions.

REMARKS 3.3. Let I be an ideal of a commutative ring R.

(1) If R is I-adically complete, then I is contained in the Jacobson radical $\mathfrak{J}(R)$; see [**122**, Theorem 8.2] or [**120**, 24.B, pages 73-74].
(2) If R is a Noetherian ring, then the I-adic completion R^* of R is flat over R [**122**, Theorem 8.8], and R^* is Noetherian by [**122**, Theorem 8.12].
(3) If R is Noetherian, then the I-adic completion R^* of R is faithfully flat over $R \iff$ for each proper ideal J of R we have $JR^* \neq R^*$.
(4) If R is a Noetherian ring and $I \subseteq \mathfrak{J}(R)$, then the I-adic completion R^* is faithfully flat over R, and $\dim R = \dim R^*$ [**120**, Theorem 56, page 172] and [**120**, pages 173-175]. Moreover, if R is an integral domain with field of fractions K, then $R = K \cap R^*$ by Remark 2.37.9.
(5) If $I = (a_1, \dots, a_n)R$ is an ideal of a Noetherian ring R, then the I-adic completion R^* of R is isomorphic to a quotient of the formal power series ring $R[[x_1, \dots, x_n]]$; namely,
$$R^* = \frac{R[[x_1, \dots, x_n]]}{(x_1 - a_1, \dots, x_n - a_n)R[[x_1, \dots, x_n]]}$$
[**122**, Theorem 8.12].
(6) If R^* is the I-adic completion of R and P is a prime ideal of R that contains I, then $(R_P)^* = (R^*_{PR^*})^*$. That is, the IR_P-adic completion of R_P is the same as the $IR^*_{PR^*}$-adic completion of $R^*_{PR^*}$. To see this,

observe that $R/I^n = R^*/I^nR^*$ for every n. Since $I \subseteq P$, the ideal PR^* is prime and $P/I^n = PR^*/I^nR^*$. Therefore

$$\frac{R_P}{I^nR_P} = \left(\frac{R}{I^n}\right)_{\left(\frac{P}{I^n}\right)} = \left(\frac{R^*}{I^nR^*}\right)_{\left(\frac{PR^*}{I^nR^*}\right)} = \frac{R^*_{PR^*}}{I^nR^*_{PR^*}},$$

for every $n \in \mathbb{N}$. Since these isomorphisms extend in a compatible way from the nth stage to the $(n+1)^{\text{st}}$ stage, we have

$$\varprojlim \frac{R_P}{I^nR_P} = \varprojlim \frac{R^*_{PR^*}}{I^nR^*_{PR^*}},$$

and the respective completions are the same.

EXAMPLE 3.4. To illustrate Remark 3.3.6, let $R = k[x, y, z]$ be a polynomial ring in the variables x, y, z over a field k. Let $I = xR$ and let $P = (x, y)R$. Then:

(1) The I-adic completion of R is $R^* = k[y, z][[x]]$, the formal power series ring in x over the polynomial ring $k[y, z]$.
(2) $R_P = k(z)[x, y]_{(x,y)}$, the localized polynomial ring in x and y over the field $k(z)$, and $R^*_{PR^*} = (R^* \setminus PR^*)^{-1}R^*$ is a two-dimensional regular local ring with maximal ideal $(x, y)R^*_{PR^*}$ that dominates R_P.
(3) The IR_P-adic completion of R_P is $(R_P)^* = k(z)[y]_{(y)}[[x]]$, the power series ring in x over the DVR $k(z)[y]_{(y)}$. By Remark 3.3.6, $(R_P)^*$ is also the completion of $R^*_{PR^*}$ in the $IR^*_{PR^*}$-adic topology.
(4) Both R^* and $(R_P)^*$ are formal power series rings in x. The power series ring R^* has coefficient ring $S = k[y, z]$, and $(R_P)^*$ has coefficient ring S_{yS}, a localization of S. By [**173**], the field of fractions of $(R_P)^*$ has infinite transcendence degree over the field of fractions of R^*.

REMARKS 3.5. Let x be a nonzero nonunit of an integral domain R, and assume that $\bigcap_{n=1}^{\infty} x^nR = (0)$. Let $R^* := \varprojlim_n (R/x^nR)$ be the *x-adic completion* of R, the completion with respect to the xR-adic topology.

(1) For each $n \in \mathbb{N}$, let $\theta_{n+1} : \frac{R}{x^{n+1}R} \to \frac{R}{x^nR}$ be the canonical R-module homomorphism. A sequence $\{\zeta_n\}_{n\in\mathbb{N}}$, with $\zeta_n \in R/x^nR$, is said to be *coherent* if $\theta_{n+1}(\zeta_{n+1}) = \zeta_n$ for each $n \in \mathbb{N}$; see [**15**, page 103]. The elements of R^* are in one-to-one correspondence with coherent sequences.

The element x is regular in R^*. To see this, observe that the coherent sequence corresponding to the element x is

$$x \mapsto \{0 + (x), x + (x^2), x + (x^3), \cdots, x + (x^{n+1}), \cdots\}$$
$$\in \frac{R}{xR} \times \frac{R}{x^2R} \times \cdots \frac{R}{x^{n+1}R} \times \cdots,$$

Suppose that $x \cdot \zeta = 0$, where ζ is the coherent sequence shown:

$$\zeta \mapsto \{a_0 + (x), a_0 + a_1 + (x^2), \cdots, a_0 + a_1 + \cdots a_n + (x^{n+1}), \cdots\}$$
$$\in \frac{R}{xR} \times \frac{R}{x^2R} \times \cdots \frac{R}{x^{n+1}R} \times \cdots,$$

where each $a_i \in R$. By coherence, $a_1 \in (x), a_2 \in (x^2), \ldots a_n \in (x^n)$, etc. Then, since R is an integral domain, $0 = x \cdot \zeta$ implies, in the second coordinate of $x \cdot \zeta$,

$$xa_0 + xa_1 \in (x^2) \implies a_0 \in (x) \implies a_0 + (x) = 0 + (x),$$

that is, $\zeta_1 = 0 + (x)$. Similarly in the third coordinate of $x \cdot \zeta$,

$$xa_0 + xa_1 + xa_2 \in (x^3) \implies a_0 + a_1 \in (x^2) \implies \zeta_2 = 0 + (x^2).$$

Repetition of this argument yields that $\zeta = 0$. Therefore x is regular in R^*.

(2) With the notation of item 1, for a coherent sequence $\{\zeta_n\}_{n\in\mathbb{N}}$, let

$$f_n = a_0 + a_1 x + \cdots + a_{n-1}x^{n-1} \in R$$

be such that the image of f_n in R/x^n is ζ_n. Then $\{f_n\}_{n\in\mathbb{N}}$ is a Cauchy sequence in R with respect to the x-adic topology as defined in [**122**, page 57]. The limit of the Cauchy sequence $\{f_n\}_{n\in\mathbb{N}}$ is an element f of R^* associated to the coherent sequence $\{\zeta_n\}_{n\in\mathbb{N}}$, and can be expressed as the power series $f = \sum_{n\in\mathbb{N}} a_{n-1}x^{n-1}$ of R^*.

Conversely every Cauchy sequence $\{f_n\}_{n\in\mathbb{N}}$ with $f_n = \sum_{i=0}^{n-1} a_i x^i \in R$ and each $a_i \in R$ determines a coherent sequence $\{\zeta_n\}_{n\in\mathbb{N}}$ with elements $\zeta_n \in R/x^nR$, and the limit of the Cauchy sequence in R^* can be identified with the power series $f = \sum_{i=0}^{\infty} a_i x^i$.

Thus each element of R^* has an expression as a power series in x with coefficients in R, but without the uniqueness of expression as power series that occurs in a formal power series ring over R.

(3) Let y be an indeterminate over R. There exists a surjective R-algebra homomorphism $\psi : R[[y]] \to R^*$ defined by

$$\psi(\sum_{i=0}^{\infty} a_i y^i) = \sum_{i=0}^{\infty} a_i x^i = f.$$

The map ψ is well defined because every Cauchy sequence in R with respect to the x-adic topology has a unique limit in R^*. Then $\psi(y-x) = 0$, and ψ induces a surjective R-algebra homomorphism $\overline{\psi} : \frac{R[[y]]}{(y-x)R[[y]]} \to R^*$.

Since $\bigcap_{n=1}^{\infty} x^nR = (0)$, the canonical map $\beta : R \to R^*$ is injective. We have the following commutative diagram:

$$\begin{array}{ccc} R & \xrightarrow[\alpha]{} & R[[y]] \\ {\scriptstyle\beta}\big\downarrow & & {\scriptstyle\varphi}\big\downarrow \\ R^* & \xleftarrow[\overline{\psi}]{} & \frac{R[[y]]}{(y-x)R[[y]]}, \end{array}$$

where $\alpha : R \hookrightarrow R[[y]]$ is the canonical inclusion map, and φ is the canonical surjection. Then $\psi = \overline{\psi} \circ \varphi$. Since $\psi(y-x) = 0$ and (0) is closed in the x-adic topology on R^*, $\ker\psi$ is the closure I of the ideal $(y-x)R[[y]]$ in the J-adic topology on $R[[y]]$, where $J := (y,x)R[[y]]$. Thus the x-adic completion R^* has the form

$$R^* = \frac{R[[y]]}{I},$$

as in [**137**, (17.5)].

(4) If R is Noetherian and y is an indeterminate, then the ideal $(y-x)R[[y]]$ is closed in the J-adic topology on $R[[y]]$. For a direct proof of this statement, let $\overline{}$ denote image in $R[[y]]/(y-x)R[[y]]$. It suffices to show that $\bigcap_{n=1}^{\infty} \overline{(y,x)^nR[[y]]} = (\overline{0})$. We have $\overline{(y,x)^nR[[y]]} = \overline{y^nR[[y]]}$, for every

$n \in \mathbb{N}$. By Corollary 2.34, the element y is in the Jacobson radical of $R[[y]]$. Hence $\overline{y}$ is in the Jacobson radical of $\overline{R[[y]]}$, a Noetherian ring. Thus

$$\bigcap_{n=1}^{\infty} \overline{(y,x)^n R[[y]]} = \bigcap_{n=1}^{\infty} \overline{y^n R[[y]]} = (\overline{0}).$$

The second equality follows from Theorem 2.22.1. Therefore $(y-x)R[[y]]$ is closed in the J-adic topology.

(5) If R^* is Noetherian and y is an indeterminate, then the ideal $(y-x)R[[y]]$ is closed in the J-adic topology on $R[[y]]$, even if R is not Noetherian. To see this, observe the following claim:

CLAIM 3.6. (a) If $w \in R^*$ and $xw \in R$, then $w \in R$.
(b) $(y-x)R[[y]] = ((y-x)R^*[[y]]) \cap R[[y]]$.
(c) $\frac{R[[y]]}{(y-x)R[[y]]}$ naturally embeds into $\frac{R^*[[y]]}{(y-x)R^*[[y]]}$.

PROOF. Fact 3.2.1 implies $xR^* \cap R = xR$. By item 1, x is regular on R. Hence part a of claim 3.6 holds.

For part b, suppose that $z \in (y-x)R^*[[y]] \cap R[[y]]$. Then $z = (y-x)w$, where $w \in R^*[[y]]$, $w = \sum_{i=0}^{\infty} w_i y^i$, $z = \sum_{i=0}^{\infty} z_i y^i$, $w_i \in R^*$, and $z_i \in R$.

The expression $z = (y-x)w$ implies equations in the coefficients of each power of y. The constant term, coefficient of y^0, in $z = (y-x)w$ implies $z_0 = -xw_0$. By part a, $w_0 \in R$.

Suppose by induction that we have shown that $w_0, \dots, w_{n-1} \in R$, for some $n \in \mathbb{N}$. The y^n term of $z = (y-x)w$ yields that $z_n = w_{n-1} - xw_n$. Thus $xw_n \in R$, and so $w_n \in R$ by part a. Thus $w \in R[[y]]$ and so part b holds.

For part c, observe that the natural embedding $R \hookrightarrow R^*$ extends to $R[[y]] \hookrightarrow R^*[[y]]$ and the composite map ψ

$$R[[y]] \hookrightarrow R^*[[y]] \to \frac{R^*[[y]]}{(y-x)R^*[[y]]}$$

has kernel $\psi = ((y-x)R^*[[y]]) \cap R[[y]] = (y-x)R[[y]]$. This proves part c and the claim. $\square$

To show that $(y-x)R[[y]]$ is closed, it suffices to show, as was done in item 4, that $\bigcap_{n=1}^{\infty} \overline{(y,x)^n R[[y]]} = (\overline{0})$. This holds by Claim 3.6.c, where R is replaced by R^*. As shown in item 4, R^* Noetherian implies $\overline{(y,x)^n R^*[[y]]} = (\overline{0})$. Now

$$\overline{R[[y]]} = \frac{R[[y]]}{(y-x)R[[y]]} \hookrightarrow \frac{R^*[[y]]}{(y-x)R^*[[y]]} = \overline{R^*[[y]]} \implies$$

$$\bigcap_{n=1}^{\infty} \overline{(y,x)^n R[[y]]} \subseteq \bigcap_{n=1}^{\infty} \overline{(y,x)^n R^*[[y]]} = (\overline{0}).$$

Thus $(y-x)R[[y]]$ is closed in the J-adic topology on $R[[y]]$.

3.2. Uncountable transcendence degree for a completion

In this section, we consider cases where the transcendence degree of completions and power series rings are uncountable over a base integral domain. These results are labeled "facts", because they appear to be well known. Brief proofs are included here to make the results more accessible.

We begin with a useful fact about uncountable Noetherian rings.

FACT 3.7. If R is an uncountable Noetherian ring, then there exists a prime ideal P of R such that R/P is uncountable. Hence there exists a minimal prime P_0 of R such that R/P_0 is uncountable.

PROOF. The ring R contains a finite chain of ideals

$$0 = I_0 \subset I_1 \subset \cdots \subset I_\ell = R$$

such that each quotient $I_{i+1}/I_i \cong R/P_i$, for some prime ideal P_i of R, [**122**, Theorem 6.4]. If each of the quotients were countable then R would be countable. Thus R/P is uncountable for some prime ideal P of R, and hence R/P_0 is uncountable, for each minimal prime P_0 contained in P. □

FACT 3.8. If R is a countable Noetherian integral domain and x is a nonzero nonunit of R, then the x-adic completion R^* of R contains an uncountable subset that is algebraically independent over R. That is, R^* has uncountable transcendence degree over R.

PROOF. We first observe that the x-adic completion

$$R^* := \varprojlim_n R/x^n R$$

of R is uncountable. For each $n \in \mathbb{N}$, let $\theta_{n+1} : \frac{R}{x^{n+1}R} \to \frac{R}{x^n R}$ be the canonical homomorphism. Elements of R^* may be identified with coherent sequences $\{\zeta_n\}_{n\in\mathbb{N}}$ in the sense that $\theta_{n+1}(\zeta_{n+1}) = \zeta_n$ for each $n \in \mathbb{N}$; see [**15**, page 103]. Since for each n and each ζ_n, there are at least two choices for the element ζ_{n+1} such that $\theta_{n+1}(\zeta_{n+1}) = \zeta_n$, the cardinality of R^* is at least $2^{\aleph_0}$ and hence is uncountable.

By Fact 3.7 there exists a minimal prime P_0 of R^* such that R^*/P_0 is uncountable. Since R is a Noetherian integral domain, R^* is flat over R by Remark 3.3.2. Thus, by Remark 2.37.9, $P_0 \cap R = 0$. Since a countably generated extension domain of R is countable and the algebraic closure of the field of fractions of a countable integral domain is countable, there exists an uncountable subset Λ of R^*/P_0 such that Λ is algebraically independent over R. Let $\Lambda^* \subset R^*$ be such that the elements of Λ^* map in a one-to-one way onto the elements in Λ under the residue class map $R^* \to R^*/P_0$. Then $R[\Lambda^*] \subset R^*$, and $R[\Lambda^*]$ is a polynomial ring over R in an uncountable set of indeterminates.[2] □

In relation to transcendence degree and filtrations, Joe Lipman brought Remark 3.9 and Fact 3.10 to our attention; he also indicated the proofs sketched below.

REMARK 3.9. Let k be a field and let R be a ring containing k. Let $\{I_a\}_{a\in A}$ be a family of ideals of R with index set A such that the family is closed under finite intersection and the intersection of all of the I_a is (0). If $m \in \mathbb{N}$ and $v_1, \dots, v_m \in R$ are linearly independent vectors over k, then for some a their images in R/I_a are linearly independent. Otherwise, if V is the vector space generated by the v_i, then

[2] It may happen that there exist nonzero elements in the polynomial ring $R[\Lambda^*]$ that are zero-divisors in R^*.

$\{V \cap I_a\}_{a \in A}$ would be an infinite family of nonzero vector subspaces of the finite-dimensional vector space V that is closed under finite intersection and such that the intersection of all of them is (0), a contradiction.

FACT 3.10. Let y be an indeterminate over a field k. Then the power series ring $k[[y]]$ has uncountable transcendence degree over k.

PROOF. We show the k-vector space dimension of $k[[y]]$ is uncountable. For this, let k_0 be the prime subfield of k. We consider the family $\{I_n := y^n k_0[[y]]\}_{n \in \mathbb{N}}$ of ideals of $k_0[[y]]$ and the corresponding family $\{I'_n := y^n k[[y]]\}_{n \in \mathbb{N}}$ of ideals of $k[[y]]$. For every $n \in \mathbb{N}$, the k-homomorphism $\varphi : k \otimes_{k_0} k_0[[y]] \to k[[y]]$ induces a map $\overline{\varphi} : k \otimes_{k_0} (k_0[[y]]/y^n k_0[[y]]) \to k[[y]]/y^n k[[y]]$ that is an isomorphism of two n-dimensional vector spaces over k.

Since $k_0[[y]]$ is uncountable and k_0 is countable, the k_0-vector space dimension of $k_0[[y]]$ is uncountable, and so there is an uncountable subset $\mathcal{B}$ of $k_0[[y]]$ that is linearly independent over k_0. Let $v_1, \ldots v_m$ be a finite subset of $\mathcal{B}$. Then by Remark 3.9 the images of $v_1, \ldots, v_m$ in $k_0[[y]]/(y^n k_0[[y]])$ are linearly independent over k_0, for some n. Since $\overline{\varphi}$ is a k-isomorphism, the images of $v_1, \ldots, v_m$ in $k[[y]]/(y^n k[[y]])$ are linearly independent over k. Thus $v_1, \ldots, v_m$ must be linearly independent over k. Therefore $\mathcal{B}$ is linearly independent over k. □

3.3. Power series pitfalls

Roger Wiegand called to our attention possible pitfalls when working with power series involving both negative and positive powers.

Here is Roger's example of a power series pitfall:

EXAMPLE 3.11. Define elements of $\mathbb{Z}[[x, x^{-1}]]$ as follows:

(1) $f(x) = \sum_{i=0}^{\infty} x^i$.
(2) $g(x) = -\sum_{i=0}^{\infty} x^{-i-1}$.
(3) $h(x) = 1 - x$.

We have

$$h(x)f(x) = 1 \quad \text{and} \quad h(x)g(x) = 1.$$

It appears that $h(x)$ has two distinct inverses. Hmm . . . .

The explanation is that the $\mathbb{Z}[x, x^{-1}]$-module $\mathbb{Z}[[x, x^{-1}]]$ is not a ring. In order to multiply some pairs in $\mathbb{Z}[[x, x^{-1}]]$ you would need to add infinitely many integers to compute some of the coefficients. Roger's comment: "That's illegal!"

The multiplications above are valid though, since $\mathbb{Z}[[x, x^{-1}]]$ is a $\mathbb{Z}[x, x^{-1}]$-module and $h(x) \in \mathbb{Z}[x, x^{-1}]$. Note that $\mathbb{Z}[[x, x^{-1}]]$ is not a torsion-free module since $h(x)$ annihilates

$$f(x) - g(x) = \cdots + x^{-2} + x^{-1} + 1 + x + x^2 + \cdots .$$

In connection with Example 3.11, Roger raised the following interesting question:

QUESTION 3.12. If $M = \mathbb{Z}[[x, x^{-1}]]$ is regarded as a module over the ring $\mathbb{Z}[x, x^{-1}]$, what is the torsion submodule of M? That is, which elements $m \in M$ are annihilated by a nonzero element of $\mathbb{Z}[x, x^{-1}]$?

COMMENTS 3.13. Roger comments about Question 3.12: If $r \in R := \mathbb{Z}[x, x^{-1}]$ and $m \in M := \mathbb{Z}[[x, x^{-1}]]$ is such that $m = \sum_{n \in \mathbb{Z}} a_n x^n$ with $a_n = 0$, for all but finitely many negative integers n, then both r and m live in $D := \mathbb{Z}[[x]][\frac{1}{x}]$, an integral domain. Similarly, if $a_n = 0$, for all but finitely many positive integers n, then both r and m live in the integral domain $E = \mathbb{Z}[[x^{-1}]][x]$. Hence the nonzero elements of $D \cup E$ are torsionfree over R.

But suppose m has $a_n \neq 0$, for infinitely many positive n AND for infinitely many negative n. Obviously, not all elements of this form can be zerodivisors, since the set of all such elements, together with 0, is not a submodule of M.

3.4. Basic results about completions

Theorem 3.14 was originally proved by Chevalley in 1943 [**33**].

THEOREM 3.14. *Let R be a semilocal ring with maximal ideals $\mathfrak{m}_1, \dots, \mathfrak{m}_r$ and Jacobson radical $\mathcal{J} = \cap_{i=1}^r \mathfrak{m}_i$. Then the $\mathcal{J}$-adic completion $\widehat{R}$ of R decomposes as a direct product*

$$\widehat{R} \;=\; \widehat{R_{\mathfrak{m}_1}} \times \cdots \times \widehat{R_{\mathfrak{m}_r}},$$

where $\widehat{R_{\mathfrak{m}_i}}$ is the completion of the local ring $R_{\mathfrak{m}_i}$.

In Proposition 3.15 we give conditions for an ideal to be closed with respect to an I-adic topology.

PROPOSITION 3.15. *Let I be an ideal in a ring R and let R^* denote the I-adic completion of R.*

(1) *Let L be an ideal of R such that LR^* is closed in the I-adic topology on R^*. Then L is closed in the I-adic topology on R if and only if $LR^* \cap R = L$.* [3]
(2) *If R is Noetherian and I is contained in the Jacobson radical of R, then every ideal L of R is closed in the I-adic topology on R.*
(3) *If R^* is Noetherian, then every ideal $\mathfrak{A}$ of R^* is closed in the I-adic topology on R^*.*

PROOF. For item 1, we have $LR^* = \bigcap_{n=1}^{\infty}(L + I^n)R^*$, since the ideal LR^* is closed in R^*. By Equation 3.1.2, $R/I^n \cong R^*/I^nR^*$, for each $n \in \mathbb{N}$. It follows that

$$R/(L + I^n) \cong R^*/(L + I^n)R^*, \quad \text{and} \quad L + I^n = (L + I^n)R^* \cap R, \tag{3.15.0}$$

for each $n \in \mathbb{N}$. By Equation 3.15.0, L is closed in R if and only if $LR^* \cap R = L$. This proves item 1. Item 2 now follows from statements 3 and 4 of Remark 3.3.

Item 3 follows from item 2, since IR^* is contained in the Jacobson radical of R^* by Remark 3.3.1. □

In Theorem 8 of Cohen's famous paper [**35**] on the structure and ideal theory of complete local rings a result similar to Nakayama's lemma is obtained without the usual finiteness condition of Nakayama's lemma [**122**, Theorem 2.2]. As formulated in [**122**, Theorem 8.4], the result is:

THEOREM 3.16. (A version of Cohen's Theorem 8) *Let I be an ideal of a ring R and let M be an R-module. Assume that R is complete in the I-adic topology and $\bigcap_{n=1}^{\infty} I^n M = (0)$. If M/IM is generated over R/I by elements $\overline{w}_1, \dots, \overline{w}_s$ and w_i is a preimage in M of $\overline{w}_i$ for $1 \le i \le s$, then M is generated over R by $w_1, \dots, w_s$.*

---

[3] Here, as in [**122**, page xiii], we interpret $LR^* \cap R$ to be the preimage $\psi^{-1}(LR^*)$, where $\psi : R \to R^*$ is the canonical map of R to its I-adic completion R^*.

Let K be a field and let $R = K[[x_1, \dots, x_n]]$ be a formal power series ring in n variables over K. It is well-known that there exists a K-algebra embedding of R into the formal power series ring $K[[y, z]]$ in two variables over K [**193**, page 219]. We observe in Corollary 3.17 restrictions on such an embedding.

COROLLARY 3.17. *Let $(R, \mathfrak{m})$ be a complete Noetherian local ring. Assume that the map $\varphi : (R, \mathfrak{m}) \to (S, \mathfrak{n})$ is a local homomorphism, and that $\bigcap_{n=1}^{\infty} \mathfrak{m}^n S = (0)$.*

(1) *If $\mathfrak{m}S$ is $\mathfrak{n}$-primary and $S/\mathfrak{n}$ is finite over $R/\mathfrak{m}$, then S is a finitely generated R-module.*
(2) *If $\mathfrak{m}S = \mathfrak{n}$ and $R/\mathfrak{m} = S/\mathfrak{n}$, then φ is surjective.*
(3) *Assume that $R = K[[x_1, \dots, x_n]]$ is a formal power series ring in $n > 2$ variables over the field K and $S = K[[y, z]]$ is a formal power series ring in two variables over K. If φ is injective, then $\varphi(\mathfrak{m})S$ is not $\mathfrak{n}$-primary.*

We record in Remarks 3.19 several consequences of Cohen's structure theorems for complete local rings. We use the following definitions.

DEFINITIONS 3.18. Let $(R, \mathfrak{m})$ be a local ring.

(1) $(R, \mathfrak{m})$ is said to be *equicharacteristic* if R has the same characteristic as its residue field $R/\mathfrak{m}$.
(2) A subfield k of R is a *coefficient field* of R if the canonical map of $R \to R/\mathfrak{m}$ restricts to an isomorphism of k onto $R/\mathfrak{m}$.

REMARKS 3.19.

(1) Every equicharacteristic complete Noetherian local ring has a coefficient field; see [**35**], [**122**, Theorem 28.3], [**137**, (31.1)].
(2) If k is a coefficient field of a complete Noetherian local ring $(R, \mathfrak{m})$ and $x_1, \dots, x_n$ are generators of $\mathfrak{m}$, then every element of R can be expanded as a power series in $x_1, \dots, x_n$ with coefficients in k; see [**137**, (31.1)]. Thus R is a homomorphic image of a formal power series ring in n variables over k.
(3) (i) Every complete Noetherian local ring is a homomorphic image of a complete regular local ring.
    (ii) Every complete regular local ring is a homomorphic image of a formal power series ring over either a field or a complete discrete valuation ring [**35**], [**137**, (31.12)].
(4) Let $(R, \mathfrak{m})$ be a complete Noetherian local domain. Then:
    (a) R is a finite integral extension of a complete regular local domain [**137**, (31.6)].
    (b) The integral closure of R in a finite algebraic field extension is a finite R-module [**137**, (32.1)].

Historically the following terminology has been used for local rings to indicate properties of the completion.

DEFINITIONS 3.20. A Noetherian local ring R is said to be

(1) *analytically unramified* if the completion $\widehat{R}$ is reduced, i.e., has no nonzero nilpotent elements;
(2) *analytically irreducible* if the completion $\widehat{R}$ is an integral domain;
(3) *analytically normal* if the completion $\widehat{R}$ is an integrally closed (i.e., normal) domain.

If a Noetherian local ring R is analytically irreducible or analytically normal, then R is analytically unramified. If R is analytically normal, then R is analytically irreducible. If $\widehat{R}$ is not reduced, we say that R is *analytically ramified*; if $\widehat{R}$ is not an integral domain, we say that R is *analytically reducible*.

A classical theorem of Rees describes necessary and sufficient conditions in order that a Noetherian local ring be analytically unramified. We refer to this result as the *Rees Finite Integral Closure Theorem*.

THEOREM 3.21. (Rees Finite Integral Closure Theorem) [**154**] *Let $(R, \mathfrak{m})$ be a reduced Noetherian local ring with total ring of fractions $\mathcal{Q}(R)$. Then the following are equivalent.*

(1) *The ring R is analytically unramified.*
(2) *For every choice of finitely many elements $\lambda_1, \dots, \lambda_n$ in $\mathcal{Q}(R)$, the integral closure of $R[\lambda_1, \dots, \lambda_n]$ in $\mathcal{Q}(R)$ is a finite $R[\lambda_1, \dots, \lambda_n]$-module.*

The following is an immediate corollary of Theorem 3.21.

COROLLARY 3.22. (Rees) [**154**] *Let $(R, \mathfrak{m})$ be an analytically unramified Noetherian local ring and let $\lambda_1, \dots, \lambda_n$ be elements of $\mathcal{Q}(R)$. For every prime ideal P of $A = R[\lambda_1, \dots, \lambda_n]$, the local ring A_P is also analytically unramified.*

REMARKS 3.23. Let R be a Noetherian local ring.

(1) If R is analytically unramified, then the integral closure of R in $\mathcal{Q}(R)$ is a finite R-module by Rees Finite Integral Closure Theorem 3.21 or [**137**, (32.2)].
(2) If $(R, \mathfrak{m})$ is one-dimensional and an integral domain, then the following two statements hold [**137**, Ex. 1 on page 122] and Katz [**104**].
    (i) The integral closure $\overline{R}$ of R is a finite R-module if and only if R is analytically unramified.
    (ii) The minimal primes of $\widehat{R}$ are in one-to-one correspondence with the maximal ideals of $\overline{R}$.

3.5. Chains of prime ideals, fibers of maps

We begin by discussing chains of prime ideals.

DEFINITIONS 3.24. Let P and Q be prime ideals of a ring A.

(1) If $P \subsetneq Q$, we say that the inclusion $P \subsetneq Q$ is *saturated* if there is no prime ideal of A strictly between P and Q.
(2) A possibly infinite chain of prime ideals $\cdots \subsetneq P_i \subsetneq P_{i+1} \subsetneq \cdots$ is called *saturated* if every inclusion $P_i \subsetneq P_{i+1}$ is saturated.
(3) A ring A is *catenary* provided for every pair of prime ideals $P \subsetneq Q$ of A, every chain of prime ideals from P to Q can be extended to a saturated chain and every two saturated chains from P to Q have the same number of inclusions.
(4) A ring A is *universally catenary* provided every finitely generated A-algebra is catenary.
(5) A ring A is said to be *equidimensional* if $\dim A = \dim A/P$ for every minimal prime P of A.
(6) A Noetherian local ring $(A, \mathfrak{n})$ is said to be *quasi-unmixed* if its completion $\widehat{A}$ is equidimensional.

Theorem 3.25 is a well-known result of Jack Ratliff that we call Ratliff's Equidimension Theorem.

THEOREM 3.25. (Ratliff's Equidimension Theorem) [**122**, Theorem 31.6] *A Noetherian local domain A is universally catenary if and only if its completion $\widehat{A}$ is equidimensional.*

Ratliff's sharper result, also called Ratliff's Equidimension Theorem, relates the universally catenary property to properties of the completion, even if the Noetherian local ring is not a domain.

THEOREM 3.26. (Ratliff's Equidimension Theorem) [**151**, Theorem 2.6] *A Noetherian local ring $(R, \mathfrak{m})$ is universally catenary if and only if the completion of $R/\mathfrak{p}$ is equidimensional for every minimal prime ideal $\mathfrak{p}$ of R.*

REMARK 3.27. Every Noetherian local ring that is a homomorphic image of a regular local ring, or even a homomorphic image of a Cohen-Macaulay local ring, is universally catenary [**122**, Theorem 17.9, page 137].

We record in Proposition 3.28 an implication of the Krull Altitude Theorem 2.23.

PROPOSITION 3.28. *Let R be a Noetherian domain and let $P \in \operatorname{Spec} R$ with $\dim R/P = n \geq 1$. Let d be an integer with $1 \leq d \leq n$, and let*

$$\mathcal{A} := \{Q \in \operatorname{Spec} R \mid P \subseteq Q \text{ and } \dim R/Q = d\}.$$

Then $P = \bigcap_{Q \in \mathcal{A}} Q$.

PROOF. If $d = n$, then $P \in \mathcal{A}$ and the statement is true. To prove the assertion for d with $1 \leq d < n$, it suffices to prove it in the case where $\dim R/P = d + 1$; for if the statement holds in the case where $n = d + 1$, then by an iterative procedure on intersections of prime ideals, the statement also holds for $n = d + 2, \cdots$.

Thus we assume $n = d + 1$, that is, $\dim R/P = d + 1 \geq 2$. Let

$$P = P_0 \subsetneq P_1 \subsetneq P_2 \subsetneq \cdots \subsetneq P_{d+1} = \mathfrak{m} \tag{3.28.0}$$

be a maximal chain of prime ideals among all chains from P to a maximal ideal $\mathfrak{m} = P_{d+1}$ in R.

Since the chain in Equation 3.28.0 is a maximal length chain, so is the chain $Q \subsetneq P_2 \subsetneq \cdots \subsetneq P_{d+1} = \mathfrak{m}$, for each prime ideal Q with $P \subsetneq Q \subsetneq P_2$. Thus $\dim R/Q = d$, and so $Q \in \mathcal{A}$. The Krull Altitude Theorem 2.23 implies there exist infinitely many such prime ideals Q. Therefore $\mathcal{A}$ is infinite.

Now $P \subseteq \bigcap_{Q \in \mathcal{A}} Q$. If there exists an element $r \in R$ such that $r \in (\bigcap_{Q \in \mathcal{A}} Q) \setminus P$, then Q is minimal over the ideal $P + rR$, for each $Q \in \mathcal{A}$. Since an ideal in a Noetherian ring has only finitely many minimal primes, we have $P = \bigcap_{Q \in \mathcal{A}} Q$. □

DISCUSSION 3.29. Let $f : A \to B$ be a ring homomorphism. The map f can always be factored as the composite of the surjective map $A \to f(A)$ followed by the inclusion map $f(A) \hookrightarrow B$. This is often helpful for understanding the relationship of A and B. If J is an ideal of B, then $f^{-1}(J)$ is an ideal of A called the *contraction* of J to A with respect to f. As in [**122**, page xiii], we often write $J \cap A$ for $f^{-1}(J)$. If Q is a prime ideal of B, then $P := f^{-1}(Q) = Q \cap A$ is a prime ideal of A. Thus associated with the ring homomorphism $f : A \to B$, there is a well-defined *spectral map* $f^* : \operatorname{Spec} B \to \operatorname{Spec} A$ of topological spaces, where for $Q \in \operatorname{Spec} B$ we define

$f^*(Q) = f^{-1}(Q) = Q \cap A = P \in \operatorname{Spec} A$. The Spec map is a contravariant functor from the catetory of commutative rings with unit to the category of topological spaces and continuous maps.[4]

Let A be a ring and let $P \in \operatorname{Spec}(A)$. The *residue field of* A at P, denoted $k(P)$, is the field of fractions $\mathcal{Q}(A/P)$ of A/P. By permutability of localization and residue class formation we have $k(P) = A_P/PA_P$.

Given a ring homomorphism $f : A \to B$ and an ideal I of A, the ideal $f(I)B$ is called the *extension* of I to B with respect to f. For $P \in \operatorname{Spec} A$, the extension ideal $f(P)B$ is, in general, not a prime ideal of B. The *fiber* over P in $\operatorname{Spec} B$ is the set of all $Q \in \operatorname{Spec} B$ such that $f^*(Q) = P$. Exercise 7 of Chapter 2 asserts that the fiber over P is nonempty if and only if P is the contraction of the extended ideal $f(P)B$. The *fiber ring* of the map f over P is the ring C defined as:

$$C := B \otimes_A k(P) = S^{-1}(B/f(P)B) = (S^{-1}B)/(S^{-1}f(P)B), \tag{3.29.0}$$

where S is the multiplicatively closed set $A \setminus P$; see [**122**, last paragraph, p. 47]. In general, the fiber over P in $\operatorname{Spec} B$ is the spectrum of the ring C. That is, the fiber of f over P in $\operatorname{Spec} B$ is the set of prime ideals of C with the Zariski topology. Notice that a prime ideal Q of B contracts to P in A if and only if $f(P) \subseteq Q$ and $Q \cap S = \emptyset$. This describes exactly the prime ideals of C as in Equation 3.29.0.

For $Q^* \in \operatorname{Spec} C$, and $Q = Q^* \cap B$, we have $P = Q \cap A$ and

$$Q^* = QC, \quad \text{and} \quad C_{Q^*} = B_Q/PB_Q = B_Q \otimes_A k(P); \tag{3.29.1}$$

see [**122**, top, p. 48].

3.6. Henselian rings

The notion of Henselian ring is due to Azumaya [**19**]; see [**137**, p. 221].

DEFINITION 3.30. [**137**, p. 103] A local ring $(R, \mathfrak{m})$ is *Henselian* provided the following holds: for every monic polynomial $f(x) \in R[x]$ satisfying

$$f(x) \equiv g_0(x)h_0(x) \pmod{\mathfrak{m}[x]},$$

where g_0 and h_0 are monic polynomials in $R[x]$ such that

$$g_0R[x] + h_0R[x] + \mathfrak{m}[x] = R[x],$$

there exist monic polynomials $g(x)$ and $h(x)$ in $R[x]$ such that $f(x) = g(x)h(x)$ and such that both

$$g(x) - g_0(x) \quad \text{and} \quad h(x) - h_0(x) \in \mathfrak{m}[x].$$

Thus Henselian rings are precisely those local rings that satisfy the property asserted for complete local rings in Hensel's Lemma 3.31.

LEMMA 3.31. Hensel's Lemma [**122**, Theorem 8.3] *Let $(R, \mathfrak{m})$ be a complete local ring, let x be an indeterminate over R, let $f(x) \in R[x]$ be a monic polynomial and let $\overline{f}$ be the polynomial obtained by reducing the coefficients mod $\mathfrak{m}$. If $\overline{f}(x)$ factors modulo $\mathfrak{m}[x]$ into two comaximal factors, then this factorization can be lifted back to $R[x]$.*

[4] In his remarkable paper [**99**], Hochster answers the difficult question: Which topological spaces actually occur as the spectrum of a commutative ring?

The concept of the Henselization of a local ring is due to Nagata [**129**], [**131**], [**136**].[5] In Remarks 3.32, we list properties of the Henselization of a local ring R and of Henselian rings in general; see [**137**] for the proofs.

REMARKS 3.32. Let $(R, \mathfrak{m})$ be a local ring.

(1) There exists an extension ring of R, denoted $(R^h, \mathfrak{m}^h)$, having the following properties:
- R^h is Henselian and local,
- R^h dominates R, R^h has the same residue field as R, and $\mathfrak{m}R^h = \mathfrak{m}^h$, [**137**, (43.3), p. 180, Theorem 43.5, p. 181].

Nagata proves that the ring R^h is unique up to an R-isomomorphism [**137**, pages 181-182]. The ring R^h is called the *Henselization* of R.

(2) The Henselization R^h of R is faithfully flat over R [**137**, Theorem 43.8].[6] It follows that $R/\mathfrak{m}^n$ is canonically isomorphic to $R^h/(\mathfrak{m}^h)^n$, for each $n \in \mathbb{N}$. Thus the $\mathfrak{m}$-adic completion $\widehat{R}$ of R is also the $\mathfrak{m}^h$-adic completion of R^h; the $\mathfrak{m}$-adic topology and completion are defined in Definitions 3.1.

(3) If $(R, \mathfrak{m})$ is Noetherian, then $(R^h, \mathfrak{m}^h)$ is a Noetherian local ring such that with respect to the topologies on R and R^h defined by $\mathfrak{m}$ and $\mathfrak{m}^h$, respectively, R is a dense subspace of R^h [**137**, Theorem 43.10]. Thus we have $R \hookrightarrow R^h \hookrightarrow \widehat{R}$. Every complete Noetherian local ring is Henselian [**137**, Theorem 30.3].

(4) If R is Henselian, then $R^h = R$ [**137**, (43.11)]. If R is normal and H is a Henselian ring that dominates R, then there exists a unique injective local R-homomorphism $\varphi : R^h \hookrightarrow H$ [**137**, Theorem 43.5].

(5) If $(R, \mathfrak{m})$ is an integral domain, then R is Henselian if and only if for every integral domain S that is an integral extension of R, S is a local domain [**137**, Theorem 43.12].

(6) If R is Henselian and R' is a local ring that is integral over R, then R' is Henselian [**137**, Corollary 43.16].

(7) If $(R', \mathfrak{m}')$ is a local ring that is integral over $(R, \mathfrak{m})$, then $R' \otimes_R R^h = (R')^h$ [**137**, Theorem 43.17].

(8) If $(R', \mathfrak{m}')$ is a local ring that dominates $(R, \mathfrak{m})$ and if R' is a localization of a finitely generated integral extension of R, then $(R')^h$ is a finitely generated module over R^h [**137**, Theorem 43.18].

(9) Assume $(R, \mathfrak{m})$ is an integral domain, and let R^h denote the Henselization of R. Then (1) a prime ideal $\mathfrak{p}^*$ of R^h is an associated prime ideal of zero if and only if $\mathfrak{p}^* \cap R = 0$, (2) the zero ideal of R^h is a radical ideal, and (3) there is a one-to-one correspondence between the maximal ideals of the integral closure of R and the associated prime ideals of zero of R^h [**137**, Theorem 43.20]. In particular, if $(R, \mathfrak{m})$ is a normal local domain, then R^h is an integral domain.

We give more information about Henselian rings and the Henselization of a local ring in Chapter 8.

---

[5]The notation in [**137**], in particular the meaning of "local ring" and "finite type", differ from the terminology of [**122**] that we are using in this book. We have adjusted these results to our terminology.

[6]The concept of faithful flatness is defined in Definitions 2.36.

3.7. Regularity and excellence

THEOREM 3.33. [**122**, Theorem 23.7 and Corollary, p. 184] *Let $(A, \mathfrak{m})$ and $(B, \mathfrak{n})$ be Noetherian local rings and $\varphi : A \to B$ a flat local homomorphism. Then*

(1) *If B is regular, normal, or reduced, then so is A.*
(2) *If A and $B/\mathfrak{m}B$ are regular, then B is regular.*
(3) *If both A and the fiber rings of φ are normal, respectively, reduced, then B is normal, respectively, reduced.*

COROLLARY 3.34. *Let $(R, \mathfrak{m})$ be a Noetherian local ring and let $\widehat{R}$ denote its $\mathfrak{m}$-adic completion. Then R is an RLR if and only if $\widehat{R}$ is an RLR.*

PROOF. By Remark 3.3.4, the extension $R \hookrightarrow \widehat{R}$ is faithfully flat. Thus Theorem 3.33 applies. □

We consider more properties of completions in Discussion 3.36. The notion of depth is relevant for that discussion[7] and is defined in Definition 3.35.

DEFINITION 3.35. Let I be an ideal in a Noetherian ring R and let M be a finitely generated R-module such that $IM \neq M$. Elements $x_1, \ldots, x_d$ in I are said to form a *regular sequence on* M, or an *M-sequence*, if x_1 is not a zerodivisor on M and for i with $2 \leq i \leq d$, the element x_i is not a zerodivisor on $M/(x_1, \ldots, x_{i-1})M$. It is known that maximal M-sequences of elements of I exist and all maximal M-sequences of elements of I have the same length n; see [**122**, Theorem 16.7] or [**103**, Theorem 121]. This integer n is called the *grade of* I on M and denoted $\operatorname{grade}(I, M)$. If R is a Noetherian local ring with maximal ideal $\mathfrak{m}$, and M is a nonzero finitely generated R-module, then the grade of $\mathfrak{m}$ on M is also called the *depth* of M. In particular the depth of R is $\operatorname{grade}(\mathfrak{m}, R)$.

DISCUSSION 3.36. Related to Corollary 3.34, we are interested in the relationship between a Noetherian local ring $(R, \mathfrak{m})$ and its $\mathfrak{m}$-adic completion $\widehat{R}$. Certain properties of the ring R may fail to hold in $\widehat{R}$. For example,

(1) The rings A/fA and D of Remarks 4.16.2 and 4.16.1 are Noetherian local domains, whereas the completion of the one-dimensional domainA/fA is not reduced and the completion of the two-dimensional normal ring D is not an integral domain.
(2) Let T be a complete Noetherian local ring of depth at least two such that no nonzero element of the prime subring of T is a zero divisor on T. Ray Heitmann has shown the remarkable result that every such ring T is the completion of a Noetherian local UFD [**96**, Theorem 8]. Let $T = k[[x, y, z]]/(z^2)$, where x, y, z are indeterminates over a field k. By Heitmann's result there exists a two-dimensional Noetherian local UFD $(R, \mathfrak{m})$ such that the completion of R is T. Thus there exists a two-dimensional normal Noetherian local domain for which the completion is not reduced.

REMARK 3.37. Shreeram Abhyankar and Ben Kravitz in [**10**, Example 3.5] use Heitmann's construction mentioned in Discussion 3.36.2 along with Rees Finite Integral Closure Theorem 3.21 to give a counterexample to an erroneous theorem on

[7]The meaning of the term "depth" in Definition 3.35 is different from the way the term is used in Zariski-Samuel [**192**] or in Nagata [**137**].

page 125 of the book *Commutative Algebra II* by Oscar Zariski and Pierre Samuel [**193**]. Abhyankar and Kravitz also note that a related lemma on the previous page of [**193**] is incorrect.

With R and $\widehat{R}$ as in Discussion 3.36, if $Q \in \operatorname{Spec} \widehat{R}$ and $P = Q \cap R$, then the natural map $\varphi : R \to \widehat{R}$ induces a flat local homomorphism $\varphi_Q : R_P \to \widehat{R}_Q$. Theorem 3.33 applies in this situation with $A = R_P$ and $B = \widehat{R}_Q$. This motivates interest in the ring $\widehat{R}_Q/P\widehat{R}_Q$.

DEFINITIONS 3.38. Let $f : A \to B$ be a ring homomorphism of Noetherian rings, let $P \in \operatorname{Spec} A$, and let $k(P) = A_P/PA_P$.

(1) The fiber over P with respect to the map f is said to be *regular* if the ring $B \otimes_A k(P)$ is a Noetherian regular ring, i.e., $B \otimes_A k(P)$ is a Noetherian ring with the property that its localization at every prime ideal is a regular local ring.
(2) The fiber over P with respect to the map f is said to be *normal* if the ring $B \otimes_A k(P)$ is a normal Noetherian ring, i.e., $B \otimes_A k(P)$ is a Noetherian ring with the property that its localization at every prime ideal is a normal Noetherian local domain.
(3) The fiber over P with respect to the map f is said to be *reduced* if the ring $B \otimes_A k(P)$ is a Noetherian reduced ring.
(4) The map f has *regular*, respectively, *normal, reduced, fibers* if the fiber over P is regular, respectively, normal, reduced, for every $P \in \operatorname{Spec} A$.

DEFINITIONS 3.39. Let $f : A \to B$ be a ring homomorphism of Noetherian rings, and let $P \in \operatorname{Spec} A$.

(1) The fiber over P with respect to the map f is said to be *geometrically regular* if for every finite extension field F of $k(P)$ the ring $B \otimes_A F$ is a Noetherian regular ring. The map $f : A \to B$ is said to have *geometrically regular* fibers if for each $P \in \operatorname{Spec} A$ the fiber over P is geometrically regular.
(2) The fiber over P with respect to the map f is said to be *geometrically normal* if for every finite extension field F of $k(P)$ the ring $B \otimes_A F$ is a Noetherian normal ring. The map $f : A \to B$ is said to have *geometrically normal* fibers if for each $P \in \operatorname{Spec} A$ the fiber over P is geometrically normal.
(3) The fiber over P with respect to the map f is said to be *geometrically reduced* if for every finite extension field F of $k(P)$ the ring $B \otimes_A F$ is a Noetherian reduced ring. The map $f : A \to B$ is said to have *geometrically reduced* fibers if for each $P \in \operatorname{Spec} A$ the fiber over P is geometrically reduced.

REMARK 3.40. Let $f : A \to B$ be a ring homomorphism with A and B Noetherian rings and let $P \in \operatorname{Spec} A$. To check that the fiber of f over P is geometrically regular as in Definition 3.39, it suffices to show that $B \otimes_A F$ is a Noetherian regular ring for every finite purely inseparable field extension F of $k(P)$, [**62**, No 20, Chap. 0, Théorème 22.5.8, p. 204]. Thus, if the characteristic of the field $k(P) = A_P/PA_P$ is zero, then, for every ring homomorphism $f : A \to B$ with B Noetherian, the fiber over P is geometrically regular if and only if it is regular. A similar statement is true with "regular" replaced by "normal" or "reduced". That is, in characteristic zero, if the homomorphism f is normal, resp. reduced, then f is geometrically normal, resp. geometrically reduced [**62**, No 24, Ch. IV, Prop. 6.7.4 and Prop. 6.7.7].

DEFINITIONS 3.41. Let $f : A \to B$ be a ring homomorphism, where A and B are Noetherian rings.

(1) The homomorphism f is said to be *regular* if it is flat with geometrically regular fibers. See Definition 2.36 for the definition of flat.
(2) The homomorphism f is said to be *normal* if it is flat with geometrically normal fibers.

REMARK 3.42. Let $f : A \to B$ be a ring homomorphism of Noetherian rings and $P \in \operatorname{Spec} A$. By Remark 2.10, every regular local ring is a normal Noetherian local domain. Thus, if the fiber over P with respect to f is geometrically regular, then the fiber over P is geometrically normal; if f has geometrically regular fibers, then f has geometrically normal fibers; and if f is a regular homomorphism, then f is a normal homomorphism.

EXAMPLE 3.43. Let x be an indeterminate over a field k of characteristic zero, and let

$$A := k[x(x-1), x^2(x-1)]_{(x(x-1),x^2(x-1))} \subset k[x]_{(x)} =: B.$$

Then $(A, \mathfrak{m}_A)$ and $(B, \mathfrak{m}_B)$ are one-dimensional local domains with the same field of fractions $k(x)$ and with $\mathfrak{m}_A B = \mathfrak{m}_B$. Hence the inclusion map $f : A \hookrightarrow B$ has geometrically regular fibers. Since $A \neq B$, the map f is not flat by Remark 2.37.8. Hence f is not a regular morphism.

We present in Chapter 7 examples of maps of Noetherian rings that are regular, and other examples of maps that are flat but fail to be regular.

The formal fibers of a Noetherian local ring as in Definition 3.44 play an important role in the concepts of excellent Noetherian rings, defined in Definition 3.47 and Nagata rings, defined in Definition 2.20.

DEFINITION 3.44. Let $(R, \mathfrak{m})$ be a Noetherian local ring and let $\widehat{R}$ be the $\mathfrak{m}$-adic completion of R. The *formal fibers* of R are the fibers of the canonical inclusion map $R \hookrightarrow \widehat{R}$.

DEFINITION 3.45. A Noetherian ring A is called a *G-ring* if, for each prime ideal P of A, the map of A_P to its PA_P-adic completion is regular, or, equivalently, the formal fibers of A_P are geometrically regular for each prime ideal P of A.

REMARK 3.46. In Definition 3.45 it suffices that, for every maximal ideal $\mathfrak{m}$ of A, the map from $A_{\mathfrak{m}}$ to its $\mathfrak{m}A_{\mathfrak{m}}$-adic completion is regular, by [**122**, Theorem 32.4]

DEFINITION 3.47. A Noetherian ring A is *excellent* if

(i) A is universally catenary,
(ii) A is a G-ring, and
(iii) for every finitely generated A-algebra B, the set $\operatorname{Reg}(B)$ of prime ideals P of B for which B_P is a regular local ring is an open subset of $\operatorname{Spec} B$.

REMARKS 3.48. The class of excellent rings includes the ring of integers as well as all fields and all complete Noetherian local rings [**122**, page 260]. All Dedekind domains of characteristic zero are excellent [**120**, (34.B)]. Every excellent ring is a Nagata ring by [**120**, Theorem 78, page 257].

The usefulness of the concept of excellent rings is enhanced by the fact that the class of excellent rings is stable under the ring-theoretic operations of localization and passage to a finitely generated algebra [**62**, Chap. IV], [**120**, (33.G) and (34.A)]. In particular, excellence is preserved under homomorphic images.

Let R be a semilocal excellent ring. Then a power series ring in finitely many variables over R is also excellent [**157**, (3.1), p.179]. If $x \in R$, then the x-adic completion of R is excellent [**157**, (3.2)]; see Remark 3.5.2.

REMARKS 3.49. As shown in Proposition 10.4, there exist DVRs in positive characteristic that are not excellent. In Corollary 18.16, we prove that the two-dimensional Noetherian local ring B of characteristic zero constructed in Example 18.15 has the property that the map $f : B \to \widehat{B}$ has geometrically regular fibers. This ring B of Example 18.15 is also an example of a catenary ring that is not universally catenary. Thus the property of having geometrically regular formal fibers does not imply that a Noetherian local ring is excellent.

REMARK 3.50. In order to discuss early examples using the techniques of this book, we have included in Chapters 2 and 3 brief definitions of deep, technically demanding concepts, such as geometric regularity and excellence. These concepts are discussed in more detail in Chapters 7 and 8.

Exercises

(1) ([**43**]) Let R be a commutative ring and let P be a prime ideal of the power series ring $R[[x]]$. Let $P(0)$ denote the ideal in R of constant terms of elements of P.
   (i) If $x \notin P$ and $P(0)$ is generated by n elements of R, prove that P is generated by n elements of $R[[x]]$.
   (ii) If $x \in P$ and $P(0)$ is generated by n elements of R, prove that P is generated by $n+1$ elements of $R[[x]]$.
   (iii) If R is a PID, prove that every prime ideal of $R[[x]]$ of height one is principal.

(2) Let R be a DVR with maximal ideal yR and let $S = R[[x]]$ be the formal power series ring over R in the variable x. Let $f \in S$. Recall that f is a unit in S if and only if the constant term of f is a unit in R by Exercise 4 of Chapter 2.
   (a) Show that S is a 2-dimensional RLR with maximal ideal $(x, y)S$.
   (b) If g is a factor of f and S/fS is a finite R-module, then S/gS is a finite R-module.
   (c) If n is a positive integer and $f := x^n + y$, then S/fS is a DVR. Moreover, S/fS is a finite R-module if and only if $R = \widehat{R}$, i.e., R is complete.
   (d) If f is irreducible and $fS \neq xS$, then S/fS is a finite R-module implies that R is complete.
   (e) If R is complete, then S/fS is a finite R-module for each nonzero f in S.

   SUGGESTION. For item (d) use that if R is not complete, then by Nakayama's lemma, the completion of R is not a finite R-module. For item (e) use Theorem 3.16.

(3) (Related to Tiberiu Dumitrescu's article [**41**]) Let R be an integral domain and let $f \in R[[x]]$ be a nonzero nonunit of the formal power series ring $R[[x]]$.

Prove that the principal ideal $fR[[x]]$ is closed in the x-adic topology, that is, $fR[[x]] = \bigcap_{m\geq 0}(f, x^m)R[[x]]$.

SUGGESTION. Reduce to the case where $c = f(0)$ is nonzero. Then f is a unit in the formal power series ring $R[\frac{1}{c}][[x]]$. If $g \in \bigcap_{m\geq 0}(f, x^m)R[[x]]$, then $g = fh$ for some $h \in R[\frac{1}{c}][[x]]$, say $h = \sum_{n\geq 0} h_n x^n$, with $h_n \in R[\frac{1}{c}]$. Let $m \geq 1$. As $g \in (f, x^m)R[[x]]$, $g = fq + x^m r$, for some $q, r \in R[[x]]$. Thus $g = fh = fq + x^m r$, hence $f(h - q) = x^m r$. As $f(0) \neq 0$, $h - q = x^m s$, for some $s \in R[\frac{1}{c}][[x]]$. Hence $h_0, h_1, \dots, h_{m-1} \in R$.

(4) Let R be a commutative ring and let $f = \sum_{n\geq 0} f_n x^n \in R[[x]]$ be a power series having the property that its leading form f_r is a regular element of R, that is, $\operatorname{ord} f = r$, so $f_0 = f_1 = \cdots = f_{r-1} = 0$, and f_r is a regular element of R. As in the previous exercise, prove that the principal ideal $fR[[x]]$ is closed in the x-adic topology.

(5) Let $f : A \hookrightarrow B$ be as in Example 3.43.
 (i) Prove as asserted in the text that f has geometrically regular fibers but is not flat.
 (ii) Prove that the inclusion map of $C := k[x(x-1)]_{(x(x-1))} \hookrightarrow k[x]_{(x)} = B$ is flat and has geometrically regular fibers. Deduce that the map $C \hookrightarrow B$ is a regular map.

(6) Let $\phi : (R, \mathfrak{m}) \hookrightarrow (S, \mathfrak{n})$ be an injective local map of the Noetherian local ring $(R, \mathfrak{m})$ into the Noetherian local ring $(S, \mathfrak{n})$. Let

$$\widehat{R} = \varprojlim_n R/\mathfrak{m}^n \quad \text{and} \quad \widehat{S} = \varprojlim_n S/\mathfrak{n}^n,$$

where $\widehat{R}$ is the $\mathfrak{m}$-adic completion of R, and $\widehat{S}$ is the $\mathfrak{n}$-adic completion of S.
 (i) Prove that there exists a map $\widehat{\phi} : \widehat{R} \to \widehat{S}$ that extends the map $\phi : R \hookrightarrow S$.
 (ii) Prove that $\widehat{\phi}$ is injective if and only if for each positive integer n there exists a positive integer s_n such that $\mathfrak{n}^{s_n} \cap R \subseteq \mathfrak{m}^n$.
 (iii) Prove that $\widehat{\phi}$ is injective if and only if for each positive integer n the ideal $\mathfrak{m}^n$ is closed in the topology on R defined by the ideals $\{\mathfrak{n}^n \cap R\}_{n\in\mathbb{N}}$, i.e., the topology on R that defines R as a *subspace* of S.

SUGGESTION. For each $n \in \mathbb{N}$, we have $\mathfrak{m}^n \subseteq \mathfrak{n}^n \cap R$. Hence there exists a map $\phi_n : R/\mathfrak{m}^n \to R/(\mathfrak{n}^n \cap R) \hookrightarrow S/\mathfrak{n}^n$, for each $n \in N$. The family of maps $\{\phi_n\}_{n\in\mathbb{N}}$ determines a map $\widehat{\phi} : \widehat{R} \to \widehat{S}$. Since $R/\mathfrak{m}^n$ is Artinian, the descending chain of ideals $\{\mathfrak{m}^n + (\mathfrak{n}^s \cap R)\}_{s\in\mathbb{N}}$ stabilizes, and $\mathfrak{m}^n$ is closed in the subspace topology if and only if there exists a positive integer s_n such that $\mathfrak{n}^{s_n} \cap R \subseteq \mathfrak{m}^n$. This holds for each $n \in \mathbb{N}$ if and only if the $\mathfrak{m}$-adic topology on R is the subspace topology from S.

(7) Let $(R, \mathfrak{m})$, $(S, \mathfrak{n})$ and $(T, \mathfrak{q})$ be Noetherian local rings. Assume there exist injective local maps $f : R \hookrightarrow S$ and $g : S \hookrightarrow T$, and let $h := gf : R \hookrightarrow T$ be the composite map. For $\widehat{f} : \widehat{R} \to \widehat{S}$ and $\widehat{g} : \widehat{S} \to \widehat{T}$ and $\widehat{h} : \widehat{R} \to \widehat{T}$ as in the previous exercise, prove that $\widehat{h} = \widehat{g}\widehat{f}$.

(8) Let $(R, \mathfrak{m})$ and $(S, \mathfrak{n})$ be Noetherian local rings such that S dominates R and the $\mathfrak{m}$-adic completion $\widehat{R}$ of R dominates S.
 (i) Prove that R is a subspace of S.

(ii) Prove that $\widehat{R}$ is an algebraic retract of $\widehat{S}$, i.e., $\widehat{R} \hookrightarrow \widehat{S}$ and there exists a surjective map $\pi : \widehat{S} \to \widehat{R}$ such that π restricts to the identity map on the subring $\widehat{R}$ of $\widehat{S}$.

(9) Let k be a field and let R be the localized polynomial ring $k[x]_{xk[x]}$, and thus $\widehat{R} = k[[x]]$. Let $n \geq 2$ be a positive integer. If $\operatorname{char} k = p > 0$, assume that n is not a multiple of p.
   (i) Prove that there exists $y \in k[[x]]$ such that $y^n = 1 + x$.
   (ii) For y as in (i), let $S := R[yx] \hookrightarrow k[[x]]$. Prove that S is a local ring integral over R with maximal ideal $(x, yx)S$. By the previous exercise, $\widehat{R} = k[[x]]$ is an algebraic retract of $\widehat{S}$.
   (iii) Prove that the integral closure $\overline{S}$ of S is not local. Indeed, if the field k contains a primitive n-th root of unity, then $\overline{S}$ has n distinct maximal ideals. Deduce that $\widehat{R} \neq \widehat{S}$, so $\widehat{R}$ is a nontrivial algebraic retract of $\widehat{S}$.

   SUGGESTION. Use Remark 3.19.3 and Remark 3.23.2ii.

(10) (Cohen) Let $(B, \mathfrak{n})$ be a local ring that is not necessarily Noetherian. If the maximal ideal $\mathfrak{n}$ is finitely generated and $\bigcap_{n=1}^{\infty} \mathfrak{n}^n = (0)$, prove that the completion $\widehat{B}$ of B is Noetherian [**35**] or [**137**, (31.7)]

   SUGGESTION. Use Theorem 3.16.

   COMMENT. In [**35**, page 56] Cohen defines $(B, \mathfrak{n})$ to be a *generalized local ring* if $\mathfrak{n}$ is finitely generated and $\bigcap_{n=1}^{\infty} \mathfrak{n}^n = (0)$. He proves that the completion of a generalized local ring is Noetherian, and that a complete generalized local ring is Noetherian [**35**, Theorems 2 and 3].Cohen mentions that he does not know whether there exists a generalized local ring that is not Noetherian. Nagata in [**128**] gives such an example of a non-Noetherian generalized local ring $(B, \mathfrak{n})$. In Nagata's example $\widehat{B} = k[[x, y]]$ is a formal power series ring in two variables over a field. Heinzer and Moshe Roitman in [**68**] survey properties of generalized local rings including this example of Nagata.

CHAPTER 4

First examples of the construction

In this chapter, we describe elementary and historical examples of Noetherian rings. In Section 4.1, we justify that Intersection Construction 1.3 is universal in the sense described in Chapter 1. In Sections 4.2, 4.3 and 4.4, several examples are described using a form of Intersection Construction 1.5.

4.1. Universality

In this section we describe in what sense Intersection Construction 1.3 can be regarded as universal for the construction of many Noetherian local domains.

Consider the following general question.

QUESTION 4.1. Let k be a field and let L/k be a finitely generated field extension. What are the Noetherian local domains $(A, \mathfrak{n})$ such that

(1) L is the field of fractions A, and
(2) k is a coefficient field for A?

Recall from Section 2.1, that k is a coefficient field of $(A, \mathfrak{n})$ if the composite map $k \hookrightarrow A \to A/\mathfrak{n}$ defines an isomorphism of k onto $A/\mathfrak{n}$.

In relation to Question 4.1, Theorem 4.2 yields the following general facts.

THEOREM 4.2. *Let $(A, \mathfrak{n})$ be a Noetherian local domain having a coefficient field k. Then there exists a Noetherian local subring $(R, \mathfrak{m})$ of A such that:*

(1) *The local ring R is essentially finitely generated over k.*
(2) *If $\mathcal{Q}(A) = L$ is finitely generated over k, then R has field of fractions L.*
(3) *The field k is a coefficient field for R.*
(4) *The local ring A dominates R and $\mathfrak{m}A = \mathfrak{n}$.*
(5) *The inclusion map $\varphi : R \hookrightarrow A$ extends to a surjective homomorphism $\widehat{\varphi} : \widehat{R} \to \widehat{A}$ of the $\mathfrak{m}$-adic completion $\widehat{R}$ of R onto the $\mathfrak{n}$-adic completion $\widehat{A}$ of A.*
(6) *For the ideal $I := \ker(\widehat{\varphi})$ of the completion $\widehat{R}$ of R from item 5, we have:*
    (a) *$\widehat{R}/I \cong \widehat{A}$, so $\widehat{R}/I$ dominates A, and*
    (b) *$P \cap A = (0)$ for every $P \in \operatorname{Ass}(\widehat{R}/I)$, and so the field of fractions $\mathcal{Q}(A)$ of A embeds in the total ring of quotients $\mathcal{Q}(\widehat{R}/I)$ of $\widehat{R}/I$, and*
    (c) *$A = \mathcal{Q}(A) \cap (\widehat{R}/I)$.*

PROOF. Since A is Noetherian, there exist elements $t_1, \dots, t_n \in \mathfrak{n}$ such that $(t_1, \dots, t_n)A = \mathfrak{n}$. For item 2, we may assume that $L = k(t_1, \dots, t_n)$, since every element of $\mathcal{Q}(A)$ has the form a/b, where $a, b \in \mathfrak{n}$. To see the existence of the integral domain $(R, \mathfrak{m})$ and to establish item 1, we set $T := k[t_1, \dots, t_n]$ and $\mathfrak{p} := \mathfrak{n} \cap T$. Define $R := T_{\mathfrak{p}}$ and $\mathfrak{m} := \mathfrak{n} \cap R$. Then $k \subseteq R \subseteq A$, $\mathfrak{m}A = \mathfrak{n}$, R is essentially finitely generated over k and k is a coefficient field for R. Thus we have established items

1- 4. Even without the assumption that $\mathcal{Q}(A)$ is finitely generated over k, there is a relationship between R and A that is realized by passing to completions. Let φ be the inclusion map $R \hookrightarrow A$. The map φ extends to a map $\widehat{\varphi} : \widehat{R} \to \widehat{A}$, and by Corollary 3.17.2, the map $\widehat{\varphi}$ is surjective; thus item 5 holds. Let $I := \ker \widehat{\varphi}$. Then $\widehat{R}/I \cong \widehat{A}$, for the first part of item 6. The remaining assertions in item 6 follow from the fact that A is a Noetherian local domain and $\widehat{A} \cong \widehat{R}/I$. Applying Remarks 3.3, we have $\widehat{R}/I$ is faithfully flat over A, and by Remark 2.37.6 the nonzero elements of A are regular on $\widehat{R}/I$.

The following commutative diagram, where the vertical maps are injections, displays the relationships among these rings:

$$
\begin{array}{ccccccc}
 & & \widehat{R} & \xrightarrow{\widehat{\varphi}} & \widehat{A} \cong \widehat{R}/I & \longrightarrow & \mathcal{Q}(\widehat{R}/I) \\
 & & \uparrow & & \uparrow & & \uparrow \\
k & \xrightarrow{\subseteq} & R & \xrightarrow{\varphi} & A := \mathcal{Q}(A) \cap (\widehat{R}/I) & \longrightarrow & \mathcal{Q}(A)
\end{array} \tag{4.2.7}
$$

This completes the proof of Theorem 4.2. □

Theorem 4.2 implies Corollary 4.3, yielding further information regarding Question 4.1.

COROLLARY 4.3. *Every Noetherian local domain* $(A, \mathfrak{n})$*having a coefficient field* k, *and having the property that the field of fractions* L *of* A *is finitely generated over* k *is realizable as an intersection* $L \cap (\widehat{R}/I)$, *where* R *is a Noetherian local domain essentially finitely generated over* k *with* $\mathcal{Q}(R) = L$, *and* I *is an ideal in the completion* $\widehat{R}$ *of* R *such that* $P \cap R = (0)$ *for each associated prime* P *of* $\widehat{R}/I$. *Moreover,* $\widehat{A} = \widehat{R}/I$, *where* $\widehat{A}$ *is the completion of* A.

Related to Corollary 4.3, it is natural to ask which of the Noetherian local domains A as in Corollary 4.3 are essentially finitely generated over k. Remark 4.4 gives a partial answer to this question.

REMARK 4.4. Let $(R, \mathfrak{m})$ be a d-dimensional Noetherian local domain that is essentially finitely generated over a field k. Heinzer, Huneke and Sally prove: Every d-dimensional Noetherian local domain S that birationally dominates R and is either normal or quasi-unmixed in the sense of Definition 3.24 is essentially finitely generated over R [**66**, Corollary 2]. Thus every such ring S is essentially finitely generated over k.

Question 4.5, is motivated by Sally's Question 1.1, and concerns the existence of a partial converse to Theorem 4.2:

QUESTION 4.5. Let R be a Noetherian integral domain. What Noetherian overrings of R exist inside the field of fractions of R?

In connection with Question 4.5, the Krull-Akizuki theorem (see Theorem 2.24) implies that every birational overring of a one-dimensional Noetherian integral domain is Noetherian and of dimension at most one. On the other hand, every Noetherian domain of dimension greater than one admits birational overrings that are not Noetherian. Indeed, if R is an integral domain with $\dim R > 1$, then by [**137**, (11.9)] there exists a valuation ring V that is birational over R with $\dim V > 1$. Since a Noetherian valuation ring has dimension at most one, if $\dim R > 1$, then there exist birational overrings of R that are not Noetherian.

REMARK 4.6. Corollary 4.3 is a first start towards a classification of the Noetherian local domains A having a given coefficient field k, and having the property that the field of fractions of A is finitely generated over k. A drawback with Corollary 4.3 is that it is not true for every triple R, L, I as in Corollary 4.3 that $L\cap(\widehat{R}/I)$ is Noetherian (see Examples 10.15 below). In order to have a more satisfying classification an important goal is to identify necessary and sufficient conditions that $L \cap (\widehat{R}/I)$ is Noetherian for R, L, I as in Corollary 4.3.

4.2. Elementary examples

We first consider examples where R is a polynomial ring over a field k. In the case of one variable the situation is well understood:

EXAMPLE 4.7. Let x be a variable over a field k, let $R := k[x]$, and let L be a subfield of the field of fractions of $k[[x]]$ such that $k(x) \subseteq L$. Then the intersection domain $A := L \cap k[[x]]$ is a rank-one discrete valuation domain (DVR) with field of fractions L (see Remark 2.1), maximal ideal xA and x-adic completion $A^* = k[[x]]$. For example, if we work with the field $\mathbb{Q}$ of rational numbers and our favorite transcendental function e^x, and we put $L = \mathbb{Q}(x, e^x)$, then A is a DVR having residue field $\mathbb{Q}$ and field of fractions L of transcendence degree 2 over $\mathbb{Q}$.

REMARKS 4.8. (1) The integral domain A of Example 4.7 with $k = \mathbb{Q}$ is perhaps the simplest example of a Noetherian local domain on an algebraic function field $L/\mathbb{Q}$ of two variables that is not essentially finitely generated over its ground field $\mathbb{Q}$, i.e., A is not the localization of a finitely generated $\mathbb{Q}$-algebra. As such, it is hard to describe all the elements of A.

(2) We show in Section 4.4 that A can be described explicitly for some choices of the field L. The technique used in Section 4.4 to compute the ring A of Example 4.7 is to express A as an infinite *directed union* of localized polynomial rings in two variables over $\mathbb{Q}$. For our purposes here the *directed union* is given by a countably infinite family of subrings $\{B_n\}_{n\in\mathbb{N}}$ of some larger ring such that $B_n \subseteq B_{n+1}$ for each n, and the directed union is defined to be $B := \bigcup_{n=1}^{\infty} B_n$. Sometimes we refer to such a countable union as a *direct limit* or a *nested union of subrings.*

The case where the base ring R involves two variables is more interesting than Example 4.7. The following theorem of Valabrega [**181**] is useful in considering this case.

THEOREM 4.9. (Valabrega) *Let C be a DVR, let y be an indeterminate over C, and let L be a subfield of $\mathcal{Q}(C[[y]])$ such that $C[y] \subset L$. Then the integral domain $D = L\cap C[[y]]$ is a two-dimensional regular local domain having completion $\widehat{D} = \widehat{C}[[y]]$, where $\widehat{C}$ is the completion of C.*

Exercise 4 of this chapter outlines a proof for Theorem 4.9. Applying Valabrega's Theorem 4.9, we see that the intersection domain is a two-dimensional regular local domain with the "right" completion in the following two examples:

EXAMPLE 4.10. Let x and y be indeterminates over $\mathbb{Q}$ and let C be the DVR $\mathbb{Q}(x, e^x) \cap \mathbb{Q}[[x]]$. Then $A_1 := \mathbb{Q}(x, e^x, y) \cap C[[y]] = C[y]_{(x,y)}$ is a two-dimensional regular local domain with maximal ideal $(x, y)A_1$ and completion $\mathbb{Q}[[x, y]]$.

EXAMPLE 4.11. This example is related to the iterative examples of Chapter 12. Let x and y be indeterminates over $\mathbb{Q}$ and let E be the DVR $\mathbb{Q}(x, e^x) \cap \mathbb{Q}[[x]]$ as in

Example 4.7. Then $A_2 := \mathbb{Q}(x, y, e^x, e^y) \cap E[[y]]$ is a two-dimensional regular local domain with maximal ideal $(x, y)A_2$ and completion $\mathbb{Q}[[x, y]]$. See Theorem 12.3.

REMARKS 4.12. (1) There is a significant difference between the integral domains A_1 of Example 4.10 and A_2 of Example 4.11. As is shown in Proposition 4.27, the two-dimensional regular local domain A_1 of Example 4.10 is, in a natural way, a nested union of three-dimensional regular local domains. It is possible therefore to describe A_1 rather explicitly. On the other hand, the two-dimensional regular local domain A_2 of Example 4.11 contains, for example, the element $\frac{e^x - e^y}{x - y}$. There is an integral domain B naturally associated with A_2 that is a nested union of four-dimensional RLRs, and the ring B is three-dimensional and is not Noetherian; see Example 12.7. Notice that the two-dimensional regular local ring A_1 is a subring of an algebraic function field in three variables over $\mathbb{Q}$, while A_2 is a subring of an algebraic function field in four variables over $\mathbb{Q}$. Since the field $\mathbb{Q}(x, e^x, y)$ is contained in the field $\mathbb{Q}(x, e^x, y, e^y)$, the local ring A_1 is dominated by the local ring A_2.

(2) It is shown in Theorem 22.20 and Corollary 22.23 of Chapter 22 that if we go outside the range of Valabrega's theorem, that is, if we take more general subfields L of the field of fractions of $\mathbb{Q}[[x, y]]$ such that $\mathbb{Q}(x, y) \subseteq L$, then the intersection domain $A = L \cap \mathbb{Q}[[x, y]]$ can be, depending on L, a localized polynomial ring in $n \geq 3$ variables over $\mathbb{Q}$ or even a localized polynomial ring in infinitely many variables over $\mathbb{Q}$. In particular, $A = L \cap \mathbb{Q}[[x, y]]$ need not be Noetherian. Theorem 12.3 describes possibilities for the intersection domain A in this setting.

4.3. Historical examples

There are classical examples, related to singularities of algebraic curves, of one-dimensional Noetherian local domains $(R, \mathfrak{m})$ such that the $\mathfrak{m}$-adic completion $\widehat{R}$ is not an integral domain, that is, R is analytically reducible. We demonstrate this in Example 4.13.

EXAMPLE 4.13. Let X and Y be variables over $\mathbb{Q}$ and consider the localized polynomial ring

$$S := \mathbb{Q}[X, Y]_{(X,Y)} \quad \text{and the quotient ring} \quad R := \frac{S}{(X^2 - Y^2 - Y^3)S}.$$

Since the polynomial $X^2 - Y^2 - Y^3$ is irreducible in the polynomial ring $\mathbb{Q}[X, Y]$, the ring R is a one-dimensional Noetherian local domain. Let x and y denote the images in R of X and Y, respectively. The principal ideal yR is primary for the maximal ideal $\mathfrak{m} = (x, y)R$, and so the $\mathfrak{m}$-adic completion $\widehat{R}$ is also the y-adic completion of R. Thus

$$\widehat{R} = \frac{\mathbb{Q}[X][[Y]]}{(X^2 - Y^2(1 + Y))}.$$

Since $1 + Y$ has a square root $(1 + Y)^{1/2} \in \mathbb{Q}[[Y]]$, we see that $X^2 - Y^2(1 + Y)$ factors in $\mathbb{Q}[X][[Y]]$ as

$$X^2 - Y^2(1 + Y) = (X - Y(1 + Y)^{1/2}) \cdot (X + Y(1 + Y)^{1/2}).$$

Thus $\widehat{R}$ is not an integral domain. Since the polynomial $Z^2 - (1 + y) \in R[Z]$ has x/y as a root and $x/y \notin R$, the integral domain R is not normal; see Section 2.1.

The birational integral extension $\overline{R} := R[\frac{x}{y}]$ has two maximal ideals,

$$\begin{aligned} \mathfrak{m}_1 &:= \left(\mathfrak{m}, \frac{x}{y} - 1\right)\overline{R} = \left(\frac{x-y}{y}\right)\overline{R} \\ \text{and} \quad \mathfrak{m}_2 &:= \left(\mathfrak{m}, \frac{x}{y} + 1\right)\overline{R} = \left(\frac{x+y}{y}\right)\overline{R}. \end{aligned}$$

To see, for example, that $\mathfrak{m}_1 = (\frac{x-y}{y})\overline{R}$, it suffices to show that $\mathfrak{m} \subset (\frac{x-y}{y})\overline{R}$. It is obvious that $x - y \in (\frac{x-y}{y})\overline{R}$. We also clearly have $\frac{x^2-y^2}{y^2} \in (\frac{x-y}{y})\overline{R}$, and $x^2 - y^2 = y^3$. Hence $\frac{y^3}{y^2} = y \in (\frac{x-y}{y})\overline{R}$, and so $\mathfrak{m}_1$ is principal and generated by $\frac{x-y}{y}$. Similarly, the maximal ideal $\mathfrak{m}_2$ is principal and is generated by $\frac{x+y}{y}$. Thus $\overline{R} = R[\frac{x}{y}]$ is a PID, and hence is integrally closed. To better understand the structure of R and $\overline{R}$, it is instructive to extend the homomorphism

$$\varphi : S \longrightarrow \frac{S}{(X^2 - Y^2 - Y^3)S} = R.$$

Let $X_1 := X/Y$ and $S' := S[X_1]$. Then S' is a regular integral domain and the map φ can be extended to a map $\psi : S' \to R[\frac{x}{y}]$ such that $\psi(X_1) = \frac{x}{y}$. The kernel of ψ is a prime ideal of S' that contains $X^2 - Y^2 - Y^3$. Since $X = YX_1$, and Y^2 is not in ker ψ, we see that ker $\psi = (X_1^2 - 1 - Y)S'$. Thus

$$\psi : S' \longrightarrow \frac{S'}{(X_1^2 - 1 - Y)S'} = R[\frac{x}{y}] = \overline{R}.$$

Notice that $X_1^2 - 1 - Y$ is contained in exactly two maximal ideals of S', namely

$$\mathfrak{n}_1 := (X_1 - 1,\ Y)S' \quad \text{and} \quad \mathfrak{n}_2 := (X_1 + 1,\ Y)S'.$$

The rings $S_1 := S'_{\mathfrak{n}_1}$ and $S_2 := S'_{\mathfrak{n}_2}$ are two-dimensional RLRs that are local quadratic transforms [1] of S, and the map ψ localizes to define maps

$$\psi_{\mathfrak{n}_1} : S_1 \to \frac{S_1}{(X_1^2 - 1 - Y)S_1} = \overline{R}_{\mathfrak{m}_1} \text{ and } \psi_{\mathfrak{n}_2} : S_2 \to \frac{S_2}{(X_1^2 - 1 - Y)S_2} = \overline{R}_{\mathfrak{m}_2}.$$

Thus the integral closure $\overline{R}$ of R is a homomorphic image of a regular domain of dimension two with precisely two maximal ideals.

REMARK 4.14. Examples given by Akizuki [**13**] and Schmidt [**165**] provide one-dimensional Noetherian local domains R such that the integral closure $\overline{R}$ is not finitely generated as an R-module; equivalently, the completion $\widehat{R}$ of R has nonzero nilpotents; see [**137**, (32.2) and Ex. 1, page 122] and the paper of Katz [**104**, Corollary 5].

If R is a normal one-dimensional Noetherian local domain, then R is a rank-one discrete valuation domain (DVR) and it is well-known that the completion of R is again a DVR. Thus R is analytically irreducible. Zariski showed that the normal Noetherian local domains that occur in algebraic geometry are analytically normal; see [**193**, pages 313-320] and Section 3.5. In particular, the normal local domains occurring in algebraic geometry are analytically irreducible.

This motivated the question of whether there exists a normal Noetherian local domain for which the completion is not a domain. Nagata produced such examples

[1] Chapter 13 contains more information about local quadratic transforms; see Definitions 13.1.

in [**135**]. He also pinpointed sufficient conditions for a normal Noetherian local domain to be analytically irreducible [**137**, (37.8)].

In Example 4.15, we present a construction of Nagata [**135**], [**137**, Example 7, pages 209-211] of a two-dimensional regular local domain A with completion $\widehat{A} = k[[x, y]]$, where k is a field with $\operatorname{char} k \neq 2$. Nagata proves that A is Noetherian, but is not excellent. Nagata also constructs a related two-dimensional normal Noetherian local domain D that is analytically reducible.[2] Although Nagata constructs A as a nested union of subrings, we give in Example 4.15 a description of A as an intersection.

Example 4.15. (Nagata) [**137**, Example 7, pages 209-211] Let x and y be algebraically independent over a field k, where $\operatorname{char} k \neq 2$, and let R be the localized polynomial ring $R = k[x, y]_{(x,y)}$. Then the completion of R is $\widehat{R} = k[[x, y]]$. Let $\tau \in xk[[x]]$ be an element that is transcendental over $k(x, y)$, e.g., if $k = \mathbb{Q}$ we may take $\tau = e^x - 1$. Let $\rho := y + \tau$ and $f := \rho^2 = (y + \tau)^2$. Now define

$$A := \ k(x, y, f) \cap k[[x, y]] \quad \text{and} \quad D := \ \frac{A[z]}{(z^2 - f)A[z]}\ ,$$

where z is an indeterminate. It is clear that the intersection ring A is a Krull domain having a unique maximal ideal. Nagata proves that f is a prime element of A and that A is a two-dimensional regular local domain with completion $\widehat{A} = k[[x, y]]$; see Proposition 6.19. Nagata also shows that D is a normal Noetherian local domain. We discuss and establish other properties of the integral domains A and D in Remarks 4.16. We show the ring A is Noetherian in Section 6.3.

Remarks 4.16. (1) The integral domain D in Example 4.15 is analytically reducible.This is because the element f factors as a square in the completion $\widehat{A}$ of A. Thus

$$\widehat{D} \ = \ \frac{k[[x, y, z]]}{(z - (y + \tau))(z + (y + \tau))},$$

which is not an integral domain. As recorded in [**66**, page 670], David Shannon observed that there exists a two-dimensional regular local domain S that birationally dominates D such that S is not essentially finitely generated over D. Moreover, S has the property that its completion is $k[[x, y]]$.

This behavior of D differs from the situation described in Remark 4.4. D is an example of a two-dimensional normal Noetherian local domain for which the version of Zariski's Main Theorem on birational transformations as stated in [**137**, Theorem 37.4, page 137] does not apply because D is analytically reducible. There exists a regular local birational extension S of D that is not essentially finitely generated over D.

(2) The two-dimensional regular local domain A of Example 4.15 is not a Nagata ring and therefore is not excellent.[3] To see that A is not a Nagata ring, notice that

[2] These concepts are defined in Sections 3.5 and 3.1.

[3] For the definition of a Nagata ring, see Definition 2.20 of Chapter 2; for the definition of excellence, see Definition 3.47 of Chapter 3. More details about these concepts are given in Sections 8.1 and 8.2 of Chapter 8.

A has a principal prime ideal generated by f that factors as a square in $\widehat{A} = k[[x,y]]$; namely f is the square of the prime element ρ of $\widehat{A}$. Therefore the one-dimensional local domain A/fA has the property that its completion $\widehat{A}/f\widehat{A}$ has a nonzero nilpotent element. This implies that the integral closure of the one-dimensional Noetherian domain A/fA is not finitely generated over A/fA by Remark 3.23.2.i. Hence A is not a Nagata ring. Moreover, the map $A \hookrightarrow \widehat{A} = k[[x,y]]$ is not a regular morphism; see Section 3.5.

The existence of examples such as the normal Noetherian local domain D of Example 4.15 naturally motivated the question: Is a Nagata domain necessarily excellent? Rotthaus shows in [**155**] that the answer is "no" as described below.

In Example 4.17, we present the construction of Rotthaus. In [**155**] the ring A is constructed as a direct limit. We show in Christel's Example 4.17 that A can also be described as an intersection. For this we use that A is Noetherian implies that its completion $\widehat{A}$ is a faithfully flat extension, and then we apply Remark 2.37.9.

EXAMPLE 4.17. (Christel) Let x, y, z be algebraically independent over a field k, where $\operatorname{char} k = 0$, and let R be the localized polynomial ring $R = k[x,y,z]_{(x,y,z)}$. Let $\sigma = \sum_{i=1}^{\infty} a_i x^i \in k[[x]]$ and $\tau = \sum_{i=1}^{\infty} b_i x^i \in k[[x]]$ be power series such that x, σ, τ are algebraically independent over k, for example, if $k = \mathbb{Q}$, we may take $\sigma = e^x - 1$ and $\tau = e^{x^2} - 1$. Let $u := y + \sigma$ and $v := z + \tau$. Define

$$A \; := \; k(x,y,z,uv) \;\cap\; (k[y,z]_{(y,z))}[[x]]).$$

We demonstrate some properties of the ring A in Remark 4.18.

REMARK 4.18. The integral domain A of Example 4.17 is a Nagata domain that is not excellent. Rotthaus shows in [**155**] that A is Noetherian and that the completion $\widehat{A}$ of A is $k[[x,y,z]]$, so A is a 3-dimensional regular local domain. Moreover she shows the formal fibers of A are reduced, but are not regular. Since u, v are part of a regular system of parameters of $\widehat{A}$, it is clear that $(u,v)\widehat{A}$ is a prime ideal of height two. It is shown in [**155**] that $(u,v)\widehat{A} \cap A = uvA$. Thus uvA is a prime ideal and $\widehat{A}_{(u,v)\widehat{A}}/uv\widehat{A}_{(u,v)\widehat{A}}$ is a non-regular formal fiber of A. Therefore A is not excellent.

Since A contains a field of characteristic zero, to see that A is a Nagata domain it suffices to show for each prime ideal P of A that the integral closure of A/P is a finite A/P-module; see Theorem 2.4. Since the formal fibers of A are reduced, the integral closure of A/P is a finite A/P-module; see Remark 3.23.1.

4.4. Prototypes

In this section we develop examples called *Prototypes*. The reason for the term "Prototype" is because these rings have a simple format, but they enable us to construct and verify properties of more intricate and sophisticated examples. As we see in Remark 4.20 and Proposition 4.27, a (Local) Prototype is just a (localized) polynomial ring over a DVR C such that $k[x] \subseteq C \subseteq k[[x]]$, for some field k and an indeterminate x.

SETTING 4.19. Let x be an indeterminate over a field k, and let s be a positive integer. By Fact 3.10, there exist elements $\tau_1, \ldots, \tau_s \in xk[[x]]$ that are algebraically independent over $k(x)$. In order to construct Prototypes associated to $\tau_1, \ldots, \tau_s$, we first construct a discrete valuation domain C_s such that

- $k[x] \subset C_s$,
- the maximal ideal of C_s is xC_s,
- the x-adic completion of C_s is $k[[x]]$,
- C_s has field of fractions $k(x, \tau_1, \ldots, \tau_s)$.

REMARK 4.20. If C_s is a DVR satisfying the properties in Setting 4.19, then by Remark 3.3.4, $C_s = k(x, \tau_1, \ldots, \tau_s) \cap k[[x]]$. Hence C_s is uniquely determined by its field of fractions.

We describe two methods to construct the integral domain C_s. They are given below as Construction 4.21 and Construction 4.22.

CONSTRUCTION 4.21. The intersection method. In this case C_s is denoted A. This method is used in Example 4.7. We show that the intersection integral domain $A = k(x, \tau_1, \ldots, \tau_s) \cap k[[x]]$ satisfies the properties in Setting 4.19.

By Exercise 3 of Chapter 3, the integral domain A is a DVR with field of fractions $k(x, \tau_1, \ldots, \tau_s)$. Furthermore, we have $x^n k[[x]] \cap A = x^n A$, for every positive integer n, and

$$\frac{k[x]}{x^n k[x]} \subseteq \frac{A}{x^n A} \subseteq \frac{k[[x]]}{x^n k[[x]]} = \frac{k[x]}{x^n k[x]}.$$

Thus the inclusions above are equalities, xA is the maximal ideal of A, and the x-adic completion of A is $\widehat{A} = k[[x]]$.

CONSTRUCTION 4.22. The approximation method: In this case, we denote the ring C_s by B.

This method is relevant for the construction of many examples later in the book. The ring B is defined as a nested union of subrings B_n of the field $k(x, \tau_1, \ldots, \tau_s)$. In order to define B we consider the last parts or the *endpieces* τ_{in} of τ_i. Suppose that for all $1 \leq i \leq s$ the power series τ_i is given by:

$$\tau_i := \sum_{j=1}^{\infty} a_{ij} x^j \in xk[[x]],$$

where $a_{ij} \in k$. The nth endpiece of τ_i is the power series:

$$\tau_{in} := \frac{1}{x^n}(\tau_i - \sum_{j=1}^{n} a_{ij} x^j) = \sum_{j=n+1}^{\infty} a_{ij} x^{j-n} \in xk[[x]]. \tag{4.22.0}$$

For each $n \in \mathbb{N}$ and each $i \in \{1, \ldots, s\}$, we have an *endpiece recursion relation:*

$$\tau_{in} = \tau_{in+1} x + a_{in+1} x \in k(x, \tau_1, \ldots, \tau_s) \cap k[[x]]. \tag{4.22.1}$$

We define

$$B_n := k[x, \tau_{1n}, \ldots, \tau_{sn}]_{(x, \tau_{1n}, \ldots, \tau_{sn})}. \tag{4.22.2}$$

Each of the rings B_n is a localized polynomial ring in $s+1$ variables over the field k. Because of the recursion relation in Equation 4.22.1, we have that $B_n \subset B_{n+1}$ for each $n \in \mathbb{N}$. We define B to be the directed union:

$$B = \bigcup_{n \in \mathbb{N}} B_n = \varinjlim_n B_n.$$

We show that B has the five properties listed in Setting 4.19. We first describe a different construction of B. For each $n \in \mathbb{N}$ define:

$$(4.22.3) \qquad U_n \; := \; k[x, \tau_{1n}, \dots, \tau_{sn}].$$

Notice that U_n is a polynomial ring in $s+1$ variables over the field k. By the recursion relation in Equation 4.22.1, we have $U_n \subset U_{n+1}$. Consider the directed union of polynomial rings:

$$U \; := \; \bigcup_{n \in \mathbb{N}} U_n \; = \; \varinjlim_{n} U_n.$$

By the recursion relation in Equation 4.22.1, each $\tau_{in} \in xU_{n+1}$; this implies that $xB \cap U_n = (x, \tau_{1n}, \dots, \tau_{sn})U_n$ is a maximal ideal of U_n, and it follows that $xB \cap U$ is a maximal ideal of U. Since each B_n is a localization of U_n, the ring B is a localization of the ring U at the maximal ideal $xB \cap U$. We show in Theorem 5.14 that B can also be expressed as $B = (1 + xU)^{-1}U$.

PROPOSITION 4.23. *With notation as in Construction 4.22, for each $\gamma \in U$ and each $t \in \mathbb{N}$, there exist elements $g_t \in k[x]$ and $\delta_t \in U$ such that:*

$$\gamma \; = \; g_t \; + \; x^t \delta_t.$$

PROOF. We have $\gamma \in U_n$, for some $n \in \mathbb{N}$. Thus we can write γ as a polynomial in $\tau_{1n}, \dots, \tau_{sn}$ with coefficients in $k[x]$:

$$\gamma \; = \; \sum a_{(j)} \tau_{1n}^{j_1} \dots \tau_{sn}^{j_s},$$

where $a_{(j)} \in k[x]$ and (j) represents the tuple $(j_1, \dots, j_s)$. Using the recursion relation in Equation 4.22.1, for all $1 \le i \le s$, we have

$$\tau_{in} \; = \; x^t \tau_{in+t} + r_i$$

where $r_i \in k[x]$. By substituting $x^t \tau_{in+t} + r_i$ for τ_{in} we can write γ as an element of U_{n+t} as follows:

$$\gamma \; = \; \sum a_{(j)} (x^t \tau_{1n+t} + r_1)^{j_1} \dots (x^t \tau_{sn+t} + r_s)^{j_t} \; = \; g_t \; + \; x^t \delta_t,$$

where $g_t \in k[x]$ and $\delta_t \in U_{n+t}$. □

PROPOSITION 4.24. *The ring B is a DVR with maximal ideal xB, the intersection $x^t k[[x]] \cap B \; = \; x^t B$, for every $t \in \mathbb{N}$, and $\widehat{B} = k[[x]]$.*

PROOF. Let $\gamma \in B$ with $\gamma \in x^t k[[x]]$. First note that $\gamma = \gamma_0 \epsilon$ where ϵ is a unit of B and $\gamma_0 \in U$. By Proposition 4.23,

$$\gamma_0 \; = \; g_{t+1} \; + \; x^{t+1} \delta_{t+1},$$

where $g_{t+1} \in k[x]$ and $\delta_{t+1} \in U$. By assumption, $\gamma \in x^t k[[x]]$; thus $g_{t+1} \in x^t k[[x]]$. Since the embedding $k[x]_{(x)} \hookrightarrow k[[x]]$ is faithfully flat, we have $g_{t+1} \in x^t k[x]_{(x)}$, and therefore $\gamma \in x^t B$. This shows that $x^t k[[x]] \cap B \; = \; x^t B$, for every $t \in \mathbb{N}$.

Since $\bigcap_{t \in \mathbb{N}} (x^t) k[[x]] \; = \; (0)$, every nonzero element $\gamma \in B$ can be written as $\gamma = x^t \epsilon$ where $\epsilon \in B$ is a unit. It follows that the ideals of B are linearly ordered and B is a DVR with maximal ideal xB.

The ring B also satisfies the five conditions of Setting 4.19. Obviously, B dominates $k[x]_{(x)}$ and is dominated by $k[[x]]$. By Proposition 4.24, B is a DVR with maximal ideal xB, and by construction $k(x, \tau_1, \dots, \tau_s)$ is the field of fractions of B. By Proposition 4.24, we have $x^t k[[x]] \cap B = x^t B$, for every $t \in \mathbb{N}$. Therefore we have:

$$\frac{k[x]}{x^t k[x]} \subseteq \frac{B}{x^t B} \subseteq \frac{k[[x]]}{x^t k[[x]]} = \frac{k[x]}{x^t k[x]}.$$

Thus the inclusions above are equalities, and so $\widehat{B} = k[[x]]$. □

NOTE 4.25. By Remark 4.20, we have $C_s = A = B$, where A is the DVR described as an intersection in Construction 4.21 and B is the DVR described as a directed union in Construction 4.22.

We extend this example to higher dimensions by adjoining additional variables.

LOCAL PROTOTYPE EXAMPLE 4.26. Assume as in Setting 4.19 that x is an indeterminate over a field k, that s is a positive integer, and that $\tau_1, \dots, \tau_s \in xk[[x]]$ are algebraically independent over $k(x)$. Let C_s be the DVR of Constructions 4.21 and 4.22 with maximal ideal xC_s. Let r be a positive integer and let $y_1, \dots, y_r$ be additional indeterminates over C_s.

We construct a regular local ring D such that

(1) $k[x, y_1, \dots y_r] \subset D$,
(2) the maximal ideal of D is $(x, y_1, \dots, y_r)D$,
(3) the x-adic completion of D is $k[y_1, \dots, y_r]_{(y_1,\dots,y_r)}[[x]]$,
(4) The completion of D with respect to its maximal ideal is

$$\widehat{D} = k[[x, y_1, \dots, y_r]].$$

(5) D has field of fractions $k(x, \tau_1, \dots, \tau_s, y_1, \dots, y_r)$, and
(6) the transcendence degree of D over k is $s + r + 1$.

PROPOSITION 4.27. *With the notation of Setting* 4.19 *and Constructions* 4.21 *and* 4.22, *we define* $D := C_s[y_1, \dots, y_r]_{(x,y_1,\dots,y_r)}$. *Then we have:*

(1) *D satisfies properties 1-6 of Local Prototype Example* 4.26.
(2) $D = k(x, \tau_1, \dots, \tau_s, y_1, \dots, y_r) \cap k[y_1, \dots, y_r]_{(y_1,\dots,y_r)}[[x]]$ *and*

$$D = k(x, \tau_1, \dots, \tau_s, y_1, \dots, y_r) \cap k[[x, y_1, \dots, y_r]].$$

(3) $D = \bigcup_{n=1}^{\infty} k[x, \tau_{1n}, \dots, \tau_{sn}, y_1, \dots, y_r]_{(x,\tau_{1n},\dots,\tau_{sn},y_1,\dots,y_r)}$ *a directed union of localized polynomial rings, where each τ_{in} is the nth endpiece of τ_i, as in Equation* 4.22.1.

PROOF. We first observe that D as defined is a regular local ring with maximal ideal $\mathfrak{m} = (x, y_1, \dots, y_r)D$, that the $\mathfrak{m}$-adic completion of D is $k[[x, y_1, \dots, y_r]]$, and that the x-adic completion of D is $k[y_1, \dots, y_r]_{(y_1,\dots,y_r)}[[x]]$. Therefore D satisfies the six properties of Local Prototype Example 4.26. Since completions of Noetherian local rings are faithfully flat, we have that D satisfies part 2 of Proposition 4.27; see Remark 3.3.4.

In order to establish that D is the directed union of localized polynomial rings of the third part of Proposition 4.27, we define for each $n \in \mathbb{N}$:

$$W_n \ := \ k[x, y_1, \dots, y_r, \tau_{1n}, \dots, \tau_{sn}] = U_n \ \otimes_k \ k[y_1, \dots, y_r]$$

and

$$D_n \ := \ B_n[y_1, \dots, y_r]_{(\mathfrak{m}_n, y_1, \dots, y_r)},$$

where $\mathfrak{m}_n = (x, \tau_{1n}, \dots, \tau_{sn})B_n$ and $B_n = k[x, \tau_{1n}, \dots, \tau_{sn}]_{(x,\tau_{1n},\dots,\tau_{sn})}$, as in Construction 4.22. Thus W_n is a polynomial ring in $s + r + 1$ variables over the field

k, and D_n is a localization of W_n at the maximal ideal of W_n generated by these $s+r+1$ variables.

We have the inclusions $W_n \subset W_{n+1} \subset k[y_1, \dots, y_r][[x]]$, and

$$D_n \subset D_{n+1} \subset k[y_1, \dots, y_r]_{(y_1,\dots,y_r)}[[x]].$$

We define

$$W := \bigcup_{n\in\mathbb{N}} W_n \quad \text{and} \quad D' := \bigcup_{n\in\mathbb{N}} D_n.$$

Since direct limits commute with tensor products, we have:

$$W = U[y_1, \dots, y_r].$$

It follows that

$$D' = W_{(x,y_1,\dots,y_r)} = C_s[y_1, \dots, y_r]_{(x,y_1,\dots,y_r)} = D,$$

as desired for the proposition. □

A regular local ring D as described in Local Prototype Example 4.26 exists for each positive integer s and each nonnegative integer r.

DEFINITION 4.28. With the notation of Local Prototype Example 4.26, the regular local ring $D = C_s[y_1, \dots, y_r]_{(x,y_1,\dots,y_r)}$ is called the *Local Prototype* or the *Local Prototype Domain associated to* $\{\tau_1, \dots, \tau_s, y_1, \dots, y_r\}$. The *Intersection Form of the Prototype* is

$$D = k(x, \tau_1, \dots, \tau_s, y_1, \dots, y_r) \cap k[y_1, \dots, y_r]_{(y_1,\dots,y_r)}[[x]]. \tag{4.28.1}$$

REMARKS 4.29. With the notation of Local Prototype Example 4.26, let R be the localized polynomial ring $R := k[x, y_1, \dots, y_r]_{(x,y_1,\dots,y_r)}$, and let R^* denote the x-adic completion of R. Thus $R^* = k[y_1, \dots, y_r]_{(y_1,\dots,y_r)}[[x]]$.

(1) Equation 4.28.1 implies that Local Prototype D of Definition 4.28 satisfies

$$D = \mathcal{Q}(R)(\tau_1, \dots, \tau_s) \cap R^*, \tag{4.29.11}$$

where $\mathcal{Q}(R)$ denotes the field of fractions of R.

(2) As mentioned at the beginning of this section, the ring D is called a "Prototype" because of its use in the construction of other examples. Later we construct more complex integral domains E that dominate R and are dominated by the local integral domain D so that we have:

$$R = k[x, y_1, \dots, y_r]_{(x,y_1,\dots,y_r)} \hookrightarrow E \hookrightarrow D \hookrightarrow k[[x, y_1, \dots, y_r]].$$

Exercises

(1) Prove that the intersection domain A of Example 4.7 is a DVR with field of fractions L and x-adic completion $A^* = \mathbb{Q}[[x]]$.

COMMENT. Exercise 2 of Chapter 2 implies that A is a DVR. With the additional hypothesis of Example 4.7, it is true that the x-adic completion of A is $\mathbb{Q}[[x]]$.

(2) Let R be an integral domain with field of fractions K.
   (i) Let F be a subfield of K and let $S := F \cap R$. For each principal ideal aS of S, prove that $aS = aR \cap S$.

(ii) Assume that S is a subring of R with the same field of fractions K. Prove that $aS = aR \cap S$ for each $a \in S \iff S = R$.

(3) Let R be a local domain with maximal ideal $\mathfrak{m}$ and field of fractions K. Let F be a subfield of K and let $S := F \cap R$. Prove that S is local with maximal ideal $\mathfrak{m} \cap S$, and thus conclude that R dominates S. Give an example where R is not Noetherian, but S is Noetherian.

REMARK. It can happen that R is Noetherian while S is not Noetherian; see Chapter 14.

(4) Assume the notation of Theorem 4.9. Thus y is an indeterminate over the DVR C and $D = C[[y]] \cap L$, where L is a subfield of the field of fractions of $C[[y]]$ with $C[y] \subset L$. Let x be a generator of the maximal ideal of C and let $R := C[y]_{(x,y)C[y]}$. Observe that R is a two-dimensional RLR with maximal ideal $(x, y)R$ and that $C[[y]]$ is a two-dimensional RLR with maximal ideal $(x, y)C[[y]]$ that dominates R. Let $\mathfrak{m} := (x, y)C[[y]] \cap D$.

(i) Using Exercise 2, prove that

$$C \cong \frac{R}{yR} \hookrightarrow \frac{D}{yD} \hookrightarrow \frac{C[[y]]}{yC[[y]]} \cong C.$$

(ii) Deduce that $C \cong \frac{D}{yD}$, and that $\mathfrak{m} = (x, y)D$.

(iii) Let $k := \frac{C}{xC}$ denote the residue field of C. Prove that $\frac{D}{xD}$ is a DVR and that

$$k[y] \hookrightarrow \frac{R}{xR} \hookrightarrow \frac{D}{xD} \hookrightarrow \frac{C[[y]]}{xC[[y]]} \cong k[[y]].$$

(iv) For each positive integer n, prove that

$$\frac{R}{(x,y)^n R} \cong \frac{D}{(x,y)^n D} \cong \frac{C[[y]]}{(x,y)^n C[[y]]}.$$

Deduce that $\widehat{R} = \widehat{D} = \widehat{C}[[y]]$, where $\widehat{C}$ is the completion of C.

(v) Let P be a prime ideal of D such that $x \notin P$. Prove that there exists $b \in P$ such that $b(D/xD) = y^r(D/xD)$ for some positive integer r, and deduce that $P \subset (b, x)D$.

(vi) For $a \in P$, observe that $a = c_1 b + a_1 x$, where c_1 and a_1 are in D. Since $x \notin P$, deduce that $a_1 \in P$ and hence $a_1 = c_2 b + a_2 x$, where c_2 and a_2 are in D. Conclude that $P \subset (b, x^2)D$. Continuing this process, deduce that

$$bD \subseteq P \subseteq \bigcap_{n=1}^{\infty} (b, x^n)D.$$

(vii) Extending the ideals to $C[[y]]$, observe that

$$bC[[y]] \subseteq PC[[y]] \subseteq \bigcap_{n=1}^{\infty} (b, x^n)C[[y]] = bC[[y]],$$

where the last equality is because the ideal $bC[[y]]$ is closed in the topology defined by the ideals generated by the powers of x on the Noetherian local ring $C[[y]]$. Deduce that $P = bD$.

(viii) Conclude by Theorem 2.25 that D is Noetherian and hence a two-dimensional regular local domain with completion $\widehat{D} = \widehat{C}[[y]]$.

(5) Let k be a field and let $f \in k[[x,y]]$ be a formal power series of order $r \geq 2$. Let $f = \sum_{n=r}^{\infty} f_n$, where $f_n \in k[x,y]$ is a homogeneous form of degree n. If the leading form f_r factors in $k[x,y]$ as $f_r = \alpha \cdot \beta$, where α and β are coprime[4] homogeneous polynomials in $k[x,y]$ of positive degree, prove that f factors in $k[[x,y]]$ as $f = g \cdot h$, where g has leading form α and h has leading form β.

SUGGESTION. Let $G = \bigoplus_{n \geq 0} G_n$ represent the polynomial ring $k[x,y]$ as a graded ring obtained by defining $\deg x = \deg y = 1$. Notice that G_n has dimension $n+1$ as a vector space over k. Let $\deg \alpha = a$ and $\deg \beta = b$. Then $a + b = r$ and for each integer $n \geq r+1$, we have $\dim(\alpha \cdot G_{n-a}) = n - a + 1$ and $\dim(\beta \cdot G_{n-b}) = n - b + 1$. Since α and β are coprime, we have

$$(\alpha \cdot G_{n-a}) \cap (\beta \cdot G_{n-b}) = f_r \cdot G_{n-r}.$$

Conclude that $\alpha \cdot G_{n-a} + \beta \cdot G_{n-b}$ is a subspace of G_n of dimension $n+1$ and hence that $G_n = \alpha \cdot G_{n-a} + \beta \cdot G_{n-b}$. Let $g_a := \alpha$ and $h_b := \beta$. Since $f_{r+1} \in G_{r+1} = \alpha \cdot G_{r+1-a} + \beta \cdot G_{r+1-b} = g_a \cdot G_{b+1} + h_b \cdot G_{a+1}$, there exist forms $h_{b+1} \in G_{b+1}$ and $g_{a+1} \in G_{a+1}$ such that $f_{r+1} = g_a \cdot h_{b+1} + h_b \cdot g_{a+1}$. Since $G_{r+2} = g_a \cdot G_{b+2} + h_b \cdot G_{a+2}$, there exist forms $h_{b+2} \in G_{b+2}$ and $g_{a+2} \in G_{a+2}$ such that $f_{r+2} - g_{a+1} \cdot h_{b+1} = g_a \cdot h_{b+2} + h_b \cdot g_{a+2}$. Proceeding by induction, assume for a positive integer s that there exist forms $g_a, g_{a+1}, \ldots, g_{a+s}$ and $h_b, h_{b+1}, \ldots, h_{b+s}$ such that the power series $f - (g_a + \cdots + g_{a+s})(h_b + \cdots + h_{b+s})$ has order greater than or equal to $r + s + 1$. Using that

$$G_{r+s+1} = g_a \cdot G_{b+s+1} + h_b \cdot G_{a+s+1},$$

deduce the existence of forms $g_{a+s+1} \in G_{a+s+1}$ and $h_{b+s+1} \in G_{b+s+1}$ such that the power series $f - (g_a + \cdots + g_{a+s+1})(h_b + \cdots + h_{b+s+1})$ has order greater than or equal to $r + s + 2$.

(6) Let k be a field of characteristic zero. Prove that both

$$xy + z^3 \quad \text{and} \quad xyz + x^4 + y^4 + z^4$$

are irreducible in the formal power series ring $k[[x,y,z]]$. Thus there does not appear to be any natural generalization to the case of three variables of the result in the previous exercise.

[4]coprime means α and β have no common factors in $k[x,y]$.

CHAPTER 5

The Inclusion Construction

This chapter introduces and describes a technique that yields the examples of Chapter 4 and also leads to more examples. This technique, Inclusion Construction 5.3, is a version of Intersection Construction 1.3. As defined in Section 5.1, Construction 5.3 gives an "Intersection Domain" $A := L \cap R^*$, where R^* is an ideal-adic completion of an integral domain R and L is a subfield of the total quotient ring of R^* that contains the field of fractions of R.

The approximation methods in Section 5.2 yield a subring B of the constructed domain A of Inclusion Construction 5.3. This subring B is helpful for describing A. The "Approximation Domain" B is a directed union of localized polynomial rings over R.

Section 5.3 includes basic properties of Inclusion Construction 5.3. With the hypotheses of Setting 5.1, Construction Properties Theorem 5.14 states that the domains A and B both have ideal-adic completion R^*. If R is a UFD and R^* is the ideal-adic completion of R with respect to a principal prime ideal of R, then B is also a UFD by Theorem 5.24.

5.1. The Inclusion Construction and a picture

We establish the following setting for Inclusion Construction 5.3:

SETTING 5.1. Let R be an integral domain with field of fractions K and let $x \in R$ be a nonzero nonunit. Assume that

- R is separated in the x-adic topology, that is, $\bigcap_{n\in\mathbb{N}} x^n R = (0)$, and
- the x-adic completion R^* of R is a Noetherian ring.

Remark 3.5.1 implies that x is a regular element of R^* in Setting 5.1.

In many of our applications, the ring R is a Noetherian integral domain. Often the ring R is a polynomial ring in one or more variables over a field.

REMARKS 5.2.

(1) If x is a nonzero nonunit of a Noetherian integral domain R, then the two conditions of Setting 5.1 hold by Krull's Theorem 2.22.2, by Remarks 3.3, parts 5 and 2, and by Remark 2.37.6.

(2) Moreover, if R is Noetherian, Remark 3.5 implies that R^* has the form

$$R^* = \frac{R[[y]]}{(y-x)R[[y]]},$$

where y is an indeterminate over R. It is natural to ask for conditions that imply R^* is an integral domain, or, equivalently, that imply $(y-x)R[[y]]$ is a prime ideal. The element $y-x$ obviously generates a prime ideal of the polynomial ring $R[y]$. Our assumption that x is a nonunit of R implies that $(y-x)R[[y]]$ is a proper ideal.

Exercise 2 of this chapter provides examples where $(y-x)R[[y]]$ is a prime ideal and examples where it is not a prime ideal.

Inclusion Construction 5.3 features an "Intersection Domain" A that is transcendental over R and is contained in a power series extension of R, in the sense of Remarks 3.5.1.

INCLUSION CONSTRUCTION 5.3. Assume Setting 5.1. Let $\tau_1, \dots, \tau_s \in xR^*$ be algebraically independent elements over R such that $K(\tau_1, \dots, \tau_s) \subseteq \mathcal{Q}(R^*)$, the total ring of fractions of R^*.[1] Thus every nonzero element of $R[\tau_1, \dots, \tau_s]$ is a regular element of R^*. Define A to be the *Intersection Domain* $A := K(\tau_1, \dots, \tau_s) \cap R^*$ inside $\mathcal{Q}(R^*)$. Thus A is a subring of R^* and A is a transcendental extension of R.

Diagram 5.3 below shows how A is situated.

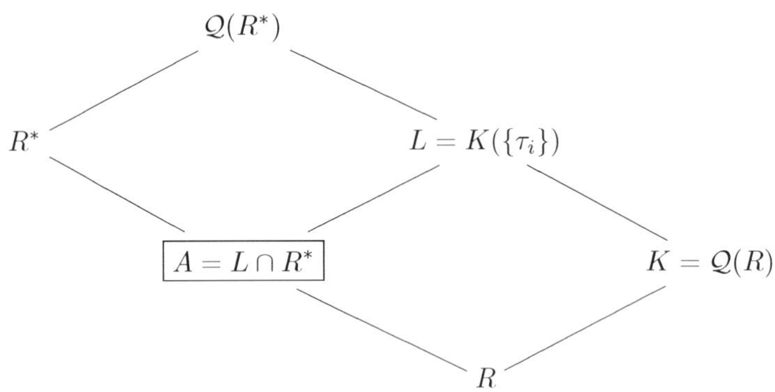

Diagram 5.3.1. $A := L \cap R^*$

The first difficulty we face with Construction 5.3 is identifying precisely what we have constructed—because, while the form of the example as an intersection as given in Construction 5.3 is wonderfully concise, sometimes it is difficult to fathom. For this reason, we construct in Section 5.2 an "Approximation Domain" B that is useful for describing A.

5.2. Approximations for the Inclusion Construction

This section contains an explicit description of the Approximation Domain B that approximates the integral domain A of Inclusion Construction 5.3. The approximation uses the last parts, the *endpieces*, of the power series $\tau_1, \dots, \tau_s$. First we describe the endpieces for a general element γ of R^*.

ENDPIECE NOTATION 5.4. Let R, x and R^* be as in Setting 5.1. By Remarks 3.5, each $\gamma \in xR^*$ has an expansion as a power series in x over R,

$$\gamma := \sum_{i=1}^{\infty} c_i x^i, \quad \text{where} \quad c_i \in R.$$

[1] Since we are interested in the polynomial ring $R[\tau_1, \dots, \tau_s]$, there is no loss of generality in the assumption that the $\tau_i \in xR^*$ rather than $\tau_i \in R^*$. Substituting $\tau_i + r$ for τ_i, where $r \in R$, does not change the ring $R[\tau_1, \dots, \tau_s]$ nor the field $K(\tau_1, \dots, \tau_s)$.

For each nonnegative integer n, define the nth *endpiece* γ_n of γ with respect to this expansion:

$$\gamma_n := \sum_{i=n+1}^{\infty} c_i x^{i-n}. \tag{5.4.1}$$

It follows that, for each nonnegative integer n, there is a basic useful relation.

$$\gamma_n = c_{n+1}x + x\gamma_{n+1}. \tag{5.4.2}$$

ENDPIECE RECURSION RELATION 5.5. With R, x and R^* as in Setting 5.1 and $\gamma = \sum_{i=1}^{\infty} c_i x^i$, where each $c_i \in R$, the following *Endpiece Recursion Relations* hold for γ:

$$\begin{aligned}
\gamma_n &= c_{n+1}x + x\gamma_{n+1}\ ; \quad \gamma_{n+1} = c_{n+2}x + x\gamma_{n+2}\ ; \\
\gamma_n &= c_{n+1}x + c_{n+2}x^2 + x^2\gamma_{n+2}\ ; \ \cdots . \\
\gamma_n &= c_{n+1}x + + \cdots + c_{n+r}x^r + x^r\gamma_{n+r} \implies \\
\gamma_n &= ax + x^r\gamma_{n+r} \quad \text{and} \quad \gamma_{n+1} = bx + x^{r-1}\gamma_{n+r}\ ,
\end{aligned} \tag{5.5.1}$$

for some $a \in (c_{n+1}, \dots, c_{n+r})R$ and $b \in (c_{n+2}, \dots, c_{n+r})R$.

Now assume that elements $\tau_1, \dots \tau_s \in xR^*$ are algebraically independent over the field of fractions K of R and have the property that every nonzero element of the polynomial ring $R[\tau_1, \dots, \tau_s]$ is a regular element of R^*. Thus $K(\tau_1, \dots, \tau_s)$ is contained in the total quotient ring $\mathcal{Q}(R^*)$. As in Inclusion Construction 5.3, define the *Intersection Domain* $A := K(\tau_1, \dots, \tau_s) \cap R^*$ inside $\mathcal{Q}(R^*)$. Set

$$U_0 := R[\tau_1, \dots, \tau_s] \subseteq A := K(\tau_1, \dots, \tau_s) \cap R^*.$$

Thus U_0 is a polynomial ring in s variables over R. Each $\tau_i \in xR^*$ has a representation $\tau_i := \sum_{j=1}^{\infty} r_{ij}x^j$, where the $r_{ij} \in R$. For each nonnegative integer n, associate with this representation of τ_i the nth endpiece,

$$\tau_{in} := \sum_{j=n+1}^{\infty} r_{ij}x^{j-n}. \tag{5.4.3}$$

Define

$$U_n := R[\tau_{1n}, \dots, \tau_{sn}] \quad \text{and} \quad B_n := (1 + xU_n)^{-1}U_n. \tag{5.4.4}$$

For each $n \in \mathbb{N}_0$, the ring U_n is a polynomial ring in s variables over R, and x is in every maximal ideal of B_n, and so $x \in \mathcal{J}(B_n)$, the Jacobson radical of B_n; see Section 2.1. By Endpiece Recursion Relation 5.5, there is a birational inclusion of polynomial rings $U_n \subset U_{n+1}$, for each $n \in \mathbb{N}_0$. Also $U_{n+1} \subset U_n[1/x]$. By Remark 3.3.1, the element x is in $\mathcal{J}(R^*)$. Hence the localization B_n of U_n is also a subring of A and $B_n \subset B_{n+1}$. Define rings U and B associated to the construction:

$$U := \bigcup_{n=0}^{\infty} U_n = \bigcup_{n=0}^{\infty} R[\tau_{1n}, \dots, \tau_{sn}] \quad \text{and} \quad B := \bigcup_{n=0}^{\infty} B_n. \tag{5.4.5}$$

REMARKS 5.6. (1) For each $n \in \mathbb{N}_0$, $U_n \subseteq U_{n+1}$. Moreover each $B_n \subseteq B_{n+1}$. The ring U is a directed union of polynomial rings over R, and the ring B, the *Approximation Domain* for the construction, is a localization of U. Then

$$B = (1 + xU)^{-1}U \quad \text{and} \quad B \subseteq A := K(\tau_1, \dots, \tau_s) \cap R^*. \tag{5.4.6}$$

Thus x is in the Jacobson radical of B.

(2) By Endpiece Recursion Relation 5.5 and Equations 5.4.4 and 5.4.5,

$$R[\tau_1,\dots,\tau_n][1/x] \;=\; U_0[1/x] \;=\; U_1[1/x] \;=\; \cdots \;=\; U[1/x]. \tag{5.4.7}$$

DEFINITION 5.7. With Setting 5.1, the ring $A = K(\tau_1,\dots,\tau_s)\cap R^*$ is called the *Intersection Domain of Inclusion Construction* 5.3 *associated to* $\tau_1,\dots,\tau_s$. The ring $B = \bigcup_{n=0}^{\infty} B_n$ is called the *Approximation Domain of Inclusion Construction* 5.3 *associated to* $\tau_1,\dots,\tau_s$. If the context is clear they are simply called the "Intersection Domain" and "Approximation Domain".

REMARK 5.8. The representation in Equation 5.4.3,

$$\tau_i \;=\; \sum_{j=1}^{\infty} r_{ij}x^j$$

of τ_i as a power series in x with coefficients in R, is not unique. Indeed, since $x\in R$, it is always possible to modify the coefficients r_{ij} in this representation. It follows that the endpiece τ_{in} is also not unique. However, as is shown in Proposition 5.9, the rings U and U_n are uniquely determined by the τ_i.

PROPOSITION 5.9. *Assume Setting* 5.1 *and the notation of Equations* 5.4.4 *and* 5.4.5. *Then the ring U and the rings U_n are independent of the representation of the τ_i as power series in x with coefficients in R. Hence also the ring B and the rings B_n are independent of the representation of the τ_i as power series in x with coefficients in R.*

PROOF. For $1\le i\le s$, assume that τ_i and $\omega_i=\tau_i$ have representations

$$\tau_i := \sum_{j=1}^{\infty} a_{ij}x^j \qquad \text{and} \qquad \omega_i := \sum_{j=1}^{\infty} b_{ij}x^j,$$

where each $a_{ij}, b_{ij}\in R$. Define the nth-endpieces τ_{in} and ω_{in} as in (5.4):

$$\tau_{in} = \sum_{j=n+1}^{\infty} a_{ij}x^{j-n} \qquad \text{and} \qquad \omega_{in} = \sum_{j=n+1}^{\infty} b_{ij}x^{j-n}.$$

Then

$$\tau_i = \sum_{j=1}^{\infty} a_{ij}x^j = \sum_{j=1}^{n} a_{ij}x^j + x^n\tau_{in} = \sum_{j=1}^{\infty} b_{ij}x^j = \sum_{j=1}^{n} b_{ij}x^j + x^n\omega_{in} = \omega_i.$$

Therefore, for $1\le i\le s$ and each positive integer n,

$$x^n\tau_{in} - x^n\omega_{in} = \sum_{j=1}^{n} b_{ij}x^j - \sum_{j=1}^{n} a_{ij}x^j, \qquad \text{and so} \qquad \tau_{in}-\omega_{in} = \frac{\sum_{j=1}^{n}(b_{ij}-a_{ij})x^j}{x^n}.$$

Thus $\sum_{j=1}^{n}(b_{ij}-a_{ij})x^j \in R$ is divisible by x^n in R^*. Then $x^nR = R\cap x^nR^*$, since x^nR is closed in the x-adic topology on R. Hence the sum $\sum_{j=1}^{n}(b_{ij}-a_{ij})x^j$ is divisible by x^n in R. Therefore $\tau_{in}-\omega_{in}\in R$. Thus the ring U_n and the ring $U=\bigcup_{n=0}^{\infty}U_n$ are independent of the representation of the τ_i. The rings B_n and the ring B are also independent of the representation of the τ_i, since $B=\bigcup_{n=0}^{\infty}B_n$ and $B_n=(1+xU_n)^{-1}U_n$. □

It is important to identify conditions in order that the Approximation Domain B equals the Intersection Domain A of Inclusion Construction 5.3. The term "limit-intersecting" of Definition 5.10 refers to this situation.

DEFINITION 5.10. Assume the setting and notation of Proposition 5.9. Then Inclusion Construction 5.3 is *limit-intersecting* over R with respect to the τ_i if $B = A$. In this case, the sequence of elements $\tau_1, \dots \tau_s \in xR^*$ are called *limit-intersecting* over R or *limit-intersecting* for A.

Observe that with the ring $R = k[x]$, the elements $\tau_1, \dots, \tau_s$ are limit-intersecting for the DVR of Constructions 4.21 and 4.22, since these constructions yield the same ring; see Note 4.25. With the ring $R = k[x, y_1, \dots, y_m]_{(x,y_1,\dots,y_m)}$, the elements $\tau_1, \dots, \tau_s$ are limit-intersecting for Local Prototype Example 4.26.

REMARK 5.11. The limit-intersecting property depends on the choice of the elements $\tau_1, \dots, \tau_s$ in the completion. For example, if R is the polynomial ring $\mathbb{Q}[x, y]$, then the x-adic completion $R^* = \mathbb{Q}[y][[x]]$. Let $s = 1$, and let

$$\tau_1 = \tau := e^x - 1 \in xR^*.$$

Then τ is algebraically independent over $\mathbb{Q}(x, y)$. Let $U_0 = R[\tau]$. Local Prototype Example 4.26 shows that τ is limit-intersecting. On the other hand, the element $y\tau$ is not limit-intersecting. If $U_0' := R[y\tau,]$, then $\mathcal{Q}(U_0) = \mathcal{Q}(U_0')$, and the Intersection Domain

$$A \;=\; \mathcal{Q}(U_0) \,\cap\, R^* \;=\; \mathcal{Q}(U_0') \,\cap\, R^*$$

is the same for τ and $y\tau$. However the Approximation Domain B' associated to U_0' does not contain τ. Indeed, $\tau \notin R[y\tau][1/x]$. Hence B' is properly contained in the Approximation Domain B associated to U_0. Thus $B' \subsetneq B = A$ and the limit-intersecting property fails for the element $y\tau$.

5.3. Basic properties of the constructed domains

The following two lemmas are useful for proving basic properties of the integral domains A and B of Construction 5.3 and Equation 5.4.5.

LEMMA 5.12. *Let S be a subring of a ring T, and let $x \in S$ be a nonunit regular element of T. The following conditions are equivalent:*

(1) *Both* (i) $xS = xT \cap S$ *and* (ii) $T = S + xT$ *hold. Equivalently,* (iii) $S/xS = T/xT$.
(2) *For each positive integer n: Both* (i) $x^nS = x^nT \cap S$, *and* (ii) $T = S + x^nT$ *hold. Equivalently,* (iii) $S/x^nS = T/x^nT$.
(3) *The rings S and T have the same x-adic completion.*
(4) *Both* (i) $S = S[1/x] \cap T$ *and* (ii) $T[1/x] = S[1/x] + T$ *hold.*

PROOF. In item 1, to see that (i) and (ii) are equivalent to (iii), let ψ denote the composite map

$$\psi : S \hookrightarrow T \twoheadrightarrow T/xT,$$

where the left map is inclusion and the right map is the natural projection. Then $\ker \psi = S \cap xT$, so that (i) is equivalent to injectivity of $\overline{\psi} : S/xS \hookrightarrow T \to T/xT$, whereas (ii) is equivalent to surjectivity of $\overline{\psi}$. Similarly, in item 2, (i) and (ii) are equivalent to (iii).

To see that item 1 implies item 2, observe that

$$x^nT \cap S = x^nT \cap xS = x(x^{n-1}T \cap S),$$

so the equality $x^{n-1}S = x^{n-1}T \cap S$ implies the equality $x^nS = x^nT \cap S$. Moreover $T = S + xT$ implies $T = S + xT = S + x(S + xT) = \cdots = S + x^nT$, so $S/x^nS = T/x^nT$ for every $n \in \mathbb{N}$. Therefore (1) implies (2).

It is clear that item 2 is equivalent to item 3.

To see that item 2 implies (4i), let $s/x^n \in S[1/x] \cap T$ with $s \in S$ and $n \geq 0$. Item 2 implies that $s \in x^nT \cap S = x^nS$ and therefore $s/x^n \in S$. To see (4ii), let $\frac{t}{x^n} \in T[1/x]$ with $t \in T$ and $n \geq 0$. Item 2 implies that $t = s + x^nt_1$ for some $s \in S$ and $t_1 \in T$. Therefore $\frac{t}{x^n} = \frac{s}{x^n} + t_1$. Thus (2) implies (4).

It remains to show that item 4 implies item 1. To see that (4) implies (1i), let $t \in T$ and $s \in S$ be such that $xt = s$. Then $t = s/x \in S[1/x] \cap T = S$, by (4i). Thus $xt \in xS$. To see that (4) implies (1ii), let $t \in T$. Then $\frac{t}{x} = \frac{s}{x^n} + t'$, for some $n \in \mathbb{N}$, $s \in S$ and $t' \in T$ by (4ii). Thus $t = \frac{s}{x^{n-1}} + t'x$. Hence by (4ii)

$$t - t'x = \frac{s}{x^{n-1}} \in S[1/x] \cap T = S.$$

□

The following lemma is a generalization of Proposition 4.23 of Chapter 4.

LEMMA 5.13. *Assume Setting* 5.1 *and the notation of Equations* 5.4.4 *and* 5.4.5. *Then:*

(1) *For every $\eta \in U$ and every $t \in \mathbb{N}$, there exist elements $g_t \in R$ and $\delta_t \in U$ such that $\eta = g_t + x^t\delta_t$.*
(2) *For each $t \in \mathbb{N}$, $x^tR^* \cap U = x^tU$.*

PROOF. Since R^* is the x-adic completion of R, we have $x^nR^* \cap R = x^nR$. For item 1, suppose that $\eta \in U_n$, for some $n \in \mathbb{N}$. Then η can be written as:

$$\eta = \sum_{(j) \in \mathbb{N}^s} r_{(j)} \tau_{1n}^{j_1} \cdots \tau_{sn}^{j_s},$$

where $r_{(j)} \in R$, each (j) represents a tuple $(j_1, \dots, j_s)$, and only finitely many of the $r_{(j)}$ are different from zero. Endpiece Recursion Relation 5.5.1 for τ_{jn} implies, for each $j \in \{1, \dots, s\}$, that:

$$\tau_{jn} = x^t\tau_{j,n+t} + h_j,$$

where $h_j \in R$. These expressions for the τ_{jn} imply:

$$\eta = \sum_{(j) \in \mathbb{N}^s} r_{(j)}(x^t\tau_{1,n+t} + h_1)^{j_1} \dots (x^t\tau_{s,n+t} + h_s)^{j_s} = g_t + x^t\delta_t,$$

where $g_t \in R$ and $\delta_t \in U_{n+t}$.

For item 2, assume that $\eta \in x^tR^* \cap U$. Then $\eta = g_t + x^t\delta_t$, where $g_t \in R$ and $\delta_t \in U$. Therefore $g_t \in x^tR^* \cap R$. Then $x^tR^* \cap R = x^tR$ implies $\eta \in x^tU$. □

Construction Properties Theorem 5.14 contains several basic properties of the integral domains associated with Inclusion Construction 5.3.

CONSTRUCTION PROPERTIES THEOREM 5.14. *Assume Setting* 5.1. *Thus R is an integral domain with field of fractions K, and $x \in R$ is a nonzero nonunit such that $\bigcap_{n \in \mathbb{N}} x^nR = (0)$ and the x-adic completion R^* of R is a Noetherian ring, Let*

$\underline{\tau} = \{\tau_1, \dots, \tau_s\}$ be a set of elements of xR^* that are algebraically independent over K and such that $K(\underline{\tau}) \subseteq \mathcal{Q}(R^*)$. The ring $R[\underline{\tau}]$ is a polynomial ring in s variables over R. As in Inclusion Construction 5.3, define $A := K(\underline{\tau}) \cap R^*$. Let U_n, B_n, B and U be defined as in Equations 5.4.4 and 5.4.5. Then:

(1) *$x^nR^* \cap R = x^nR$, $\quad x^nR^* \cap A = x^nA$, $\quad x^nR^* \cap B = x^nB$ and $x^nR^* \cap U = x^nU$, for each $n \in \mathbb{N}$.*
(2) *$R/x^nR = U/x^nU = B/x^nB = A/x^nA = R^*/x^nR^*$, for each $n \in \mathbb{N}$, and these rings are all Noetherian.*
(3) *The x-adic completions of the rings U, B and A are all equal to R^*, that is, $R^* = U^* = B^* = A^*$.*
(4) *$R[\underline{\tau}][1/x] = U[1/x]$, $\quad U = R[\underline{\tau}][1/x] \cap B = R[\underline{\tau}][1/x] \cap A$; $\quad B[1/x]$ is a localization of $R[\underline{\tau}]$ and thus B is a localization $S_n^{-1}B_n$ of B_n, for every $n \in \mathbb{N}$, where S_n is a multiplicatively closed subset of B_n. In addition:*
   (a) *The integral domains $R[\underline{\tau}], U, B$ and A all have the same field of fractions, namely $K(\underline{\tau})$.*
   (b) *$B_P = U_{P\cap U} = R[\underline{\tau}]_{P\cap R[\underline{\tau}]}$, for every $P \in \operatorname{Spec} B$ such that $x \notin P$.*
(5) *The definitions in Equation 5.4.5 of B and U are independent of the representations given in Notation 5.4 for the τ_i as power series in R^*.*
(6) *If R^* is local with maximal ideal $\mathfrak{m}_{R^*}$, then $\mathfrak{m}_R := \mathfrak{m}_{R^*} \cap R$ is a maximal ideal of R and B is local with maximal ideal $\mathfrak{m}_{R^*} \cap B$. Also,*
$$B = (1 + xU)^{-1}U = \bigcup_{n=0}^{\infty} (U_n)_{(\mathfrak{m}_R, \tau_{1n}, \dots, \tau_{sn})U_n} = U_{(\mathfrak{m}_{R^*} \cap U)},$$
where $U_n = R[\tau_{1n}, \dots, \tau_{sn}]$, as defined in Equation 5.4.4, and the τ_{in} are the nth endpieces of the τ_i, using Endpiece Notation 5.4.
(7) *The inclusions $R \hookrightarrow B_n$, for $n \in \mathbb{N}$, and $R \hookrightarrow B$ are flat. If $x \in \mathcal{J}(R)$, then all these inclusions are faithfully flat. Thus, if R is a local integral domain, then all these inclusions are faithfully flat.*

PROOF. For item 1, $x^nR^* \cap R = x^nR$, by Fact 3.2, $x^nR^* \cap U = x^nU$, by Lemma 5.13, and $x^nR^* \cap A = x^nA$, by Exercise 3 at the end of this chapter. If $\eta \in x^nR^* \cap B$, then $\eta = \eta_0\epsilon$, where $\eta_0 \in U$ and ϵ is a unit in B. Since x is in the Jacobson radical of R^*, ϵ is also a unit in R^*, and therefore $\eta_0 \in x^nR^* \cap U = x^nU$. Thus $\eta \in x^nB$.

To prove item 2, observe that from item 1, there are embeddings:
$$R/x^nR \hookrightarrow U/x^nU \hookrightarrow B/x^nB \hookrightarrow A/x^nA \hookrightarrow R^*/x^nR^*.$$
Since $R/x^nR \hookrightarrow R^*/x^nR^*$ is an isomorphism, for every $n \in \mathbb{N}$, all the equalities follow. Since R^* is Noetherian, so are R^*/x^nR^* and all of the rings isomorphic to R^*/x^nR^*.

Item 3 follows from item 2.

For item 4, Remark 5.6.2 implies $U[1/x] = R[\underline{\tau}][1/x]$. Then
$$U = U[1/x] \cap B = R[\underline{\tau}][1/x] \cap B = R[\underline{\tau}][1/x] \cap A,$$
by applying item 3 and Lemma 5.12.4 with U for the ring S and B or A for the ring T in Lemma 5.12.4. By Remark 5.6.1, B is a localization of U. Since $U[1/x] = R[\underline{\tau}][1/x]$, it follows that $B[1/x]$ is a localization of $R[\underline{\tau}]$. This implies the fields of fractions of U, B, and A are all contained in the field of fractions $K(\underline{\tau})$ of $R[\underline{\tau}]$, and so statement a of item 4 holds. For statement b of item 4, since $B[1/x]$

is a localization of $R[\underline{\tau}]$ and of U and B_P is a localization of $B[1/x]$, it follows that B_P is a localization of $R[\underline{\tau}]$ and of U. By Exercise 1, $B_P = R[\underline{\tau}]_{P \cap R[\underline{\tau}]} = U_{P \cap U}$.

Item 5 is Proposition 5.9.

For item 6, notice that $x \in \mathfrak{m}_R$. By Remark 5.6.1, $x \in \mathcal{J}(B)$, that is, x is in every maximal ideal of B. By item 2, $B/xB = R^*/xR^*$. Since R^* is local with maximal ideal $\mathfrak{m}_{R^*}$, it follows that B is local with maximal ideal $\mathfrak{m}_{R^*} \cap B$. The first equality of the displayed equation of item 6 is by Remark 5.6.1.

We show that B is also the directed union of the localized polynomial rings $C_n := (U_n)_{P_n}$, where $P_n := (\mathfrak{m}_R, \tau_{1n}, \dots, \tau_{sn})U_n$ and $U_n = R[\tau_{1n}, \dots, \tau_{sn}]$. Note that P_n is a maximal ideal of U_n with $\mathfrak{m}_{R^*} \cap U_n = P_n$, Then $C_n \subseteq C_{n+1}$. Also $P_n \cap (1 + xU_n) = \emptyset$ implies that $B_n \subseteq C_n$. We show that $C_n \subseteq B$: Let $\frac{a}{d} \in C_n$, where $a \in U_n$ and $d \in U_n \setminus P_n$. Then $a \in B$ and $d \in B \setminus (\mathfrak{m}_{R^*} \cap B)$. Since B is local with maximal ideal $\mathfrak{m}_{R^*} \cap B$, d is a unit in B. Hence $a/d \in B$.

For item 7, $R \hookrightarrow U_n = R[\underline{\tau}_n]$, where $\underline{\tau}_n = \{\tau_{1n}, \dots, \tau_{sn}\}$ are the nth endpieces of the τ_i, is faithfullly flat because $R[\underline{\tau}_n]$ is a polynomial ring in n indeterminates over R. Since B_n is a localization of U_n, $R \hookrightarrow B_n$ is also flat, for every $n \in \mathbb{N}$. Therefore $R \hookrightarrow B$ is flat; see [**22**, Chap.1, Sec.2.3, Prop.2, p.14]. If $x \in \mathcal{J}(R)$, then $1 + xu$ is a unit for every $u \in U$. Thus, for every maximal ideal $\mathfrak{m}$ of R, the intersection $\mathfrak{m} \cap (1 + xU) = \emptyset$. Therefore $\mathfrak{m}(1 + xU)^{-1}U \neq (1 + xU)^{-1}U = B$, and so $R \hookrightarrow B$ is faithfully flat, and so is $R \hookrightarrow B_n$, for every $n \in \mathbb{N}$. If R is local, then $x \in \mathcal{J}(R)$, and so the result holds.

This completes the proof of Theorem 5.14. □

COROLLARY 5.15. *Assume the setting and notation of Theorem* 5.14, *and assume R is Noetherian. Then:*

(1) *B/xB and $B[1/x]$ are Noetherian.*
(2) *The extension $R \hookrightarrow B[1/x]$ has regular fibers.*

PROOF. For item 1, since R is Noetherian, the polynomial ring $U_0 := R[\underline{\tau}]$ is Noetherian. By Theorem 5.14.4, $B[1/x]$ is a localization of $R[\underline{\tau}]$, and so $B[1/x]$ is Noetherian. By Theorem 5.14.2, B/xB is Noetherian.

For item 2, $B[1/x]$ is a localization of U_0 by Theorem 5.14.4. Since U_0 is a polynomial ring over R, the composite map $R \hookrightarrow U_0 \hookrightarrow B[1/x]$ has regular fibers. □

REMARK 5.16. In items 1 and 3 below we apply part 6 of Construction Properties Theorem 5.14 to the case that the base ring R is a localized polynomial ring over a field in variables $x, y_1, \dots, y_r$, where $r \in \mathbb{N}_0$, and the completion is taken with respect to the variable x. For one variable, the idea is quite simple, as shown in items 1 and 2 below. Let $\tau_1, \dots, \tau_s \in xR^*$ be algebraically independent elements over R as in Construction 5.3.

(1) In the special case $R = k[x]_{(x)}$, the x-adic completion of R is $R^* = k[[x]]$ and $U_n = k[x]_{(x)}[\tau_{1n}, \dots, \tau_{sn}]$. Then, for $P_n := (x, \tau_{1n}, \dots, \tau_{sn})k[x]_{(x)}[\tau_{1n}, \dots, \tau_{sn}]$,

$$B = \bigcup_{n=0}^{\infty} k[x]_{(x)}[\tau_{1n}, \dots, \tau_{sn}]_{P_n} = \bigcup_{n=0}^{\infty} k[x, \tau_{1n}, \dots, \tau_{sn}]_{(x, \tau_{1n}, \dots, \tau_{sn})},$$

by Theorem 5.14.6.

(2) If $R = k[x]$, then $U_n = k[x, \tau_{1n}, \dots, \tau_{sn}]$ and $R^* = k[[x]]$ is the x-adic completion of R. Then $B = \bigcup_{n=0}^{\infty} k[x, \tau_{1n}, \dots, \tau_{sn}]_{(x, \tau_{1n}, \dots, \tau_{sn})}$, by the proof of

part 6 of Theorem 5.14. That is, the ring B is the same for $R = k[x]$ as for $R = k[x]_{(x)}$.

(3) Let R be the localized polynomial ring $k[x, y_1, \dots, y_r]_{(x,y_1,\dots,y_r)}$ over a field k with variables $x, y_1, \dots, y_r$, and let $\mathfrak{m} := (x, y_1, \dots, y_r)R$. Then the x-adic completion of R is $R^* = k[y_1, \dots, y_r]_{(y_1,\dots,y_r)}[[x]]$. Let $\tau_1, \dots, \tau_s$ be elements of xR^* that are algebraically independent over R. By part 6 of Theorem 5.14,

$$B = \bigcup_{n=0}^{\infty} R[\tau_{1n}, \dots, \tau_{sn}]_{(\mathfrak{m},\tau_{1n},\dots,\tau_{sn})}.$$

Since R is the localization of $k[x, y_1, \dots, y_r]$ at the maximal ideal generated by $x, y_1, \dots, y_r$,

$$R[\tau_{1n}, \dots, \tau_{sn}]_{(\mathfrak{m},\tau_{1n},\dots,\tau_{sn})} = k[x, y_1, \dots, y_r, \tau_{1n}, \dots, \tau_{sn}]_{(x,y_1,\dots,y_r,\tau_{1n},\dots,\tau_{sn})}.$$

(4) With the base ring $R = k[x, y_1, \dots, y_r]_{(x,y_1,\dots,y_r)}$ as in item 3, if $\tau_1, \dots, \tau_s$ are elements of $xk[[x]]$ that are algebraically independent over $k(x)$, then Proposition 4.27 implies the ring B is the ring D of Local Prototype Example 4.26.

Proposition 5.17 concerns the extension to R^* of a prime ideal of either A or B that does not contain x, and provides information about the maps from $\operatorname{Spec} R^*$ to $\operatorname{Spec} A$ and to $\operatorname{Spec} B$. We use Proposition 5.17 in Chapters 12 and 14–16.

PROPOSITION 5.17. *With the notation of Construction Properties Theorem* 5.14*:*

(1) *x is in the Jacobson radical of each of the rings B, A and R^*. If P is a prime ideal in either $\operatorname{Spec} B$ or $\operatorname{Spec} A$, then $PR^* \neq R^*$.*
(2) *Let $\mathfrak{q}$ be a prime ideal of R. Then*
   (a) *$\mathfrak{q}U$ is a prime ideal in U.*
   (b) *Either $\mathfrak{q}B = B$ or $\mathfrak{q}B$ is a prime ideal of B.*
   (c) *If $\mathfrak{q}B \neq B$, then $\mathfrak{q}B \cap U = \mathfrak{q}U$ and $U_{\mathfrak{q}U} = B_{\mathfrak{q}B}$.*
   (d) *If $x \notin \mathfrak{q}$, then $\mathfrak{q}U \cap U_n = \mathfrak{q}U_n$ and $U_{\mathfrak{q}U} = (U_n)_{\mathfrak{q}U_n}$.*
   (e) *If $x \notin \mathfrak{q}$ and $\mathfrak{q}B \neq B$, then $\mathfrak{q}B \cap B_n = \mathfrak{q}B_n$ and*

$$(U_n)_{\mathfrak{q}U_n} = U_{\mathfrak{q}U} = B_{\mathfrak{q}B} = (B_n)_{\mathfrak{q}B_n}.$$

(3) *Let I be an ideal of B or of A and let $t \in \mathbb{N}$. Then $x^t \in IR^* \iff x^t \in I$.*
(4) *Let $P \in \operatorname{Spec} B$ or $P \in \operatorname{Spec} A$ with $x \notin P$. Then x is a nonzerodivisor on R^*/PR^*. Thus $x \notin Q$ for each associated prime of R^*/PR^*. Since x is in the Jacobson radical of R^*, it follows that PR^* is contained in a non-maximal prime ideal of R^*.*
(5) *If R is local, then R^*, A and B are local. Let $\mathfrak{m}_R$, $\mathfrak{m}_{R^*}, \mathfrak{m}_A$ and $\mathfrak{m}_B$ denote the maximal ideals of R, R^*, A and B, respectively. In this case*
   (a) *$\mathfrak{m}_B = \mathfrak{m}_R B$, and $\mathfrak{m}_A = \mathfrak{m}_R A$. Each prime ideal P of B such that $\operatorname{ht}(\mathfrak{m}_B/P) = 1$ is contracted from R^*, and each prime ideal P of A such that $\operatorname{ht}(\mathfrak{m}_A/P) = 1$ is contracted from R^*.*
   (b) *$\widehat{R} = \widehat{B} = \widehat{A}$, where $\widehat{R}, \widehat{B}$, and $\widehat{A}$ are the $\mathfrak{m}_R, \mathfrak{m}_B$, and $\mathfrak{m}_A$-adic completions of R, B, and A, respecitvely.*
   (c) *Let I be an ideal of B. Then IR^* is primary for $\mathfrak{m}_{R^*} \iff I$ is primary for $\mathfrak{m}_B$. In this case, $IR^* \cap B = I$ and $B/I \cong R^*/IR^*$. Let I be an ideal of A. Then IR^* is primary for $\mathfrak{m}_{R^*} \iff I$ is primary for $\mathfrak{m}_A$. In this case, $IR^* \cap A = I$ and $A/I \cong R^*/IR^*$.*

PROOF. For item 1, since $B_n = (1 + xU_n)^{-1}U_n$, it follows that $1 + xb$ is a unit of B_n for each $b \in B_n$. Therefore x is in the Jacobson radical of B_n for each n and thus x is in the Jacobson radical of B. By Remark 3.3.1, x is in the Jacobson radical of R^*. Hence $1 + ax$ is a unit of R^* for every $a \in R^*$. Since $A = \mathcal{Q}(A) \cap R^*$, an element of A is a unit of A if and only if it is a unit of R^*. Thus x is in the Jacobson radical of A.

By Theorem 5.14.2, $B/xB = R^*/xR^*$. If $P \in \operatorname{Spec} B$, then P is contained in a maximal ideal $\mathfrak{m}$ of B and $x \in \mathfrak{m}$. Therefore $\mathfrak{m}/xB = \mathfrak{m}R^*/xR^*$, and $PR^* \neq R^*$. Similarly, if $P \in \operatorname{Spec} A$, then $PR^* \neq R^*$.

For item 2, since each U_n is a polynomial ring over R, the ideal $\mathfrak{q}U_n$ is a prime ideal of U_n and thus $\mathfrak{q}U = \bigcup_{n=0}^{\infty} \mathfrak{q}U_n$ is a prime ideal of U. Since B is a localization of U, either $\mathfrak{q}B = B$, or $\mathfrak{q}B$ is a prime ideal of B such that $\mathfrak{q}B \cap U = \mathfrak{q}U$ and $U_{\mathfrak{q}U} = B_{\mathfrak{q}B}$.

For part d of item 2, since $U_n[1/x] = U[1/x]$ and the ideals $\mathfrak{q}U_n$ and $\mathfrak{q}U$ are prime ideals in U_n and U that do not contain x, the localizations $(U_n)_{\mathfrak{q}U_n}$ and $U_{\mathfrak{q}U}$ are both further localizations of $U[1/x]$. Moreover, they both equal $U[1/x]_{\mathfrak{q}U[1/x]}$. Thus $U_{\mathfrak{q}U} = (U_n)_{\mathfrak{q}U_n}$. Since $U_n \subset U$, also $\mathfrak{q}U \cap U_n = \mathfrak{q}U_n$. Since B is a localization of U, the assertions in part e follow as in the proof of part d.

To see item 3, let I be an ideal of B. The proof for A is identical. Observe that there exist elements $b_1, \ldots, b_s \in I$ such that $IR^* = (b_1, \ldots, b_s)R^*$. If $x^t \in IR^*$, there exist $\alpha_i \in R^*$ such that

$$x^t = \alpha_1 b_1 + \cdots + \alpha_s b_s.$$

Then $\alpha_i = a_i + x^{t+1}\lambda_i$ for each i, where $a_i \in B$ and $\lambda_i \in R^*$. Thus

$$x^t[1 - x(b_1\lambda_1 + \cdots + b_s\lambda_s)] = a_1 b_1 + \cdots + a_s b_s \in B \cap x^t R^* = x^t B.$$

Hence $\gamma := 1 - x(b_1\lambda_1 + \cdots + b_s\lambda_s) \in B$. Thus $x(b_1\lambda_1 + \cdots + b_s\lambda_s) \in B \cap xR^* = xB$, and so $b_1\lambda_1 + \cdots + b_s\lambda_s \in B$. By item 1, the element x is in the Jacobson radical of B. Therefore γ is invertible in B. Since $\gamma x^t \in (b_1, \cdots, b_s)B$, it follows that $x^t \in I$. If $x^t \in I$, then $x^t \in IR^*$. This proves item 3.

For item 4, assume that $P \in \operatorname{Spec} B$. The proof for $P \in \operatorname{Spec} A$ is identical. Observe that

$$P \cap xB = xP \quad \text{and so} \quad \frac{P}{xP} = \frac{P}{P \cap xB} \cong \frac{P + xB}{xB}$$

By Construction Properties Theorem 5.14.3, B/xB is Noetherian. Hence the B-module P/xP is finitely generated. Let $P = (g_1, \ldots, g_t)B + xP$, with $g_1, \ldots, g_t \in P$. Then also $PR^* = (g_1, \ldots, g_t)R^* + xPR^* = (g_1, \ldots, g_t)R^*$, the last equality by Nakayama's Lemma.

Let $\widehat{f} \in R^*$ be such that $x\widehat{f} \in PR^*$. We show that $\widehat{f} \in PR^*$.

Since $\widehat{f} \in R^*$, we have $\widehat{f} := \sum_{i=0}^{\infty} c_i x^i$, where each $c_i \in R$. For each $m > 1$, let $f_m := \sum_{i=0}^{m} c_i x^i$, the first $m + 1$ terms of this expansion of $\widehat{f}$. Then $f_m \in R \subseteq B$ and there exists an element $\widehat{h}_1 \in R^*$ so that.

$$\widehat{f} = f_m + x^{m+1}\widehat{h_1}.$$

Since $x\widehat{f} \in PR^*$,

$$x\widehat{f} = \widehat{a_1}g_1 + \cdots + \widehat{a_t}g_t,$$

where $\widehat{a_i} \in R^*$. The $\widehat{a_i}$ have power series expansions in x over R, and thus there exist elements $a_{im} \in R$ such that $\widehat{a_i} - a_{im} \in x^{m+1}R^*$. Thus

$$x\widehat{f} = a_{1m}g_1 + \cdots + a_{tm}g_t + x^{m+1}\widehat{h_2},$$

where $\widehat{h_2} \in R^*$, and

$$xf_m = a_{1m}g_1 + \cdots + a_{tm}g_t + x^{m+1}\widehat{h_3},$$

where $\widehat{h_3} = \widehat{h_2} - x\widehat{h_1} \in R^*$. The g_i are in B, and so $x^{m+1}\widehat{h_3} \in x^{m+1}R^* \cap B = x^{m+1}B$, the last equality by Construction Properties Theorem 5.14.1.Therefore $\widehat{h_3} \in B$. By rearranging the last displayed equation above,

$$x(f_m - x^m\widehat{h_3}) = a_{1m}g_1 + \cdots + a_{tm}g_t \in P.$$

Since $x \notin P$, we have $f_m - x^m\widehat{h_3} \in P$. It follows that $\widehat{f} \in P + x^mR^* \subseteq PR^* + x^mR^*$, for each $m > 1$. Hence $\widehat{f} \in PR^*$, as desired.

For item 5, if R is local, then B is local, A is local, R^* is local, and $\mathfrak{m}_B = \mathfrak{m}_R B$, $\mathfrak{m}_A = \mathfrak{m}_R A$, and $\mathfrak{m}_{R^*} = \mathfrak{m}_R R^*$, since $R/xR = B/xB = A/xA = R^*/xR^*$ and x is in the Jacobson radical of B and of A.

We complete the proof of item 5.a for the ring B; the same proof works for A. If $x \notin P \in \operatorname{Spec} B$, then item 4 implies that no power of x is in PR^*. Hence PR^* is contained in a prime ideal Q of R^* that does not meet the multiplicatively closed set $\{x^n\}_{n=1}^{\infty}$. Thus $P \subseteq Q \cap B \subsetneq \mathfrak{m}_B$. Since $\operatorname{ht}(\mathfrak{m}_B/P) = 1$, we have $P = Q \cap B$, so P is contracted from R^*. If $x \in P$, then $B/xB = R^*/xR^*$ implies that PR^* is a prime ideal of R^* and $P = PR^* \cap B$.

For part b of item 5, Fact 3.2 implies that $\widehat{R^*} = \widehat{R}$. Also, by Theorem 5.14.3, $A^* = B^* = R^*$. Thus the $\mathfrak{m}_A$-adic completion of A and $\mathfrak{m}_B$-adic completion of B are equal to the completion of R^*, which is $\widehat{R}$.

For part c of item 5, let I be an ideal of B. By item 3, for each $t \in \mathbb{N}$, we have $x^t \in IR^* \iff x^t \in I$. If either IR^* is $\mathfrak{m}_{R^*}$-primary or I is $\mathfrak{m}_B$-primary, then $x^t \in I$ for some $t \in \mathbb{N}$. By Theorem 5.14.3, $B/x^tB = R^*/x^tR^*$. Hence the $\mathfrak{m}_B$-primary ideals containing x^t are in one-to-one inclusion preserving correspondence with the $\mathfrak{m}_{R^*}$-primary ideals that contain x^t. This completes the proof of item 5. □

THEOREM 5.18. *Let the notation be as in Inclusion Construction 5.3, and assume that $(R, \mathfrak{m}_R)$ is a Noetherian local integral domain of dimension d. Let U and B be the Approximation Domains as in Equation 5.4.5 corresponding to elements $\underline{\tau} = \tau_1, \ldots, \tau_s$ of xR^* that are algebraically independent over R. Then, for each $P \in \operatorname{Spec} B$ such that P is maximal with respect to $x \notin P$,*

(1) $\dim(B/P) = 1 = \dim(R/(P \cap R))$.
(2) *If R is catenary, then* $\operatorname{ht}(P \cap R) = d - 1$, *and* $d - 1 \leq \operatorname{ht} P \leq d + s - 1$.

PROOF. By Proposition 5.17.5.a and Theorem 5.14.6, $B = U_{(\mathfrak{m}_B \cap U)}$, where $\mathfrak{m}_B$ is the maximal ideal of B. Since B is local and $P \in \operatorname{Spec} B$ is maximal with respect to $x \notin P$, the ideal $x(B/P)$ is in every nonzero prime of B/P. Therefore $(B/P)[1/x] = B[1/x]/PB[1/x]$ is a field and $PB[1/x]$ is a maximal ideal of $B[1/x]$. Let $Q \in \operatorname{Spec} B$ with $P \subsetneq Q \subseteq \mathfrak{m}_B$, then $x \in Q$. By Theorem 5.14.2, $Q = (Q \cap R)B$. Therefore the prime ideals of B that properly contain P are in one-to-one inclusion preserving correspondence with the prime ideals of R that properly contain $P \cap R$. Since $R/(P \cap R)$ is a Noetherian local domain, it follows that $\dim R/(P \cap R) = 1$.

For each prime Q of B that properly contains P, we have $Q \cap R = \mathfrak{m}_R$, and so $Q = \mathfrak{m}_B$, by Proposition 5.17.5. Thus $\dim(B/P) = 1$. This proves item 1.

For item 2, if R is catenary of dimension d, then $\operatorname{ht}(P \cap R) = d - 1$. Since $x \notin P$,

$$\begin{aligned} \operatorname{ht}_B P = \operatorname{ht}_{B[1/x]} PB[1/x] &= \operatorname{ht}_{R[\underline{\tau}][1/x]}(P \cap R[\underline{\tau}])R[\underline{\tau}][1/x] \\ &= \operatorname{ht}_{R[\underline{\tau}]}(P \cap R[\underline{\tau}])R[\underline{\tau}] \geq \operatorname{ht}_R(P \cap R) = d - 1. \end{aligned}$$

Since $R[\underline{\tau}]$ is a polynomial ring in $s \geq 1$ variables over R, $\operatorname{ht}_B P \leq d + s - 1$. □

Proposition 5.19 is used to compute the dimensions of the Approximation Domains B resulting from Inclusion Construction 5.3.

PROPOSITION 5.19. *Let $C = \bigcup_{n \in \mathbb{N}} C_n$ be a nested union of integral domains such that $\dim C_n \leq d$ for each $n \in \mathbb{N}$. Then*

(1) *For $P \in \operatorname{Spec} C$ and $h \in \mathbb{N}$, if $\operatorname{ht} P \geq h$, then $\operatorname{ht}(P \cap C_n) \geq h$ for all sufficiently large $n \in \mathbb{N}$.*
(2) *$\dim C \leq d$.*

PROOF. Assume that $0 = P_0 \subsetneq P_1 \subsetneq \cdots \subsetneq P_h = P$ is a strictly ascending chain of prime ideals in C. For each $j \in \{1, \ldots, h\}$, there exists an element $a_j \in P_j \setminus P_{j-1}$. Since $C = \bigcup_{n \in \mathbb{N}} C_n$, we have $\{a_j\}_{j=1}^h \subseteq C_i$ for some i. Then for all $n \geq i$

$$P_0 \cap C_n \subsetneq P_1 \cap C_n \subsetneq \cdots \subsetneq P_h \cap C_n = P \cap C_n$$

is a strictly ascending chain of prime ideals of C_n of length h. This proves item 1. Item 2 follows from item 1. □

Proposition 5.20 is useful for establishing that the prime spectra of rings constructed with Construction 5.3 are often Noetherian.

PROPOSITION 5.20. *Let x be an element in a ring C. If the rings C/xC and $C[1/x]$ have Noetherian spectrum, then $\operatorname{Spec} C$ is Noetherian. Thus, if the rings C/xC and $C[1/x]$ are Noetherian, then $\operatorname{Spec} C$ is Noetherian.*

PROOF. $\operatorname{Spec} C$ is the union of $\operatorname{Spec}(C/xC)$ and $\operatorname{Spec}(C[1/x])$. □

COROLLARY 5.21. *With the setting and notation of Inclusion Construction 5.3, assume in addition that R is a Noetherian domain. Then the Approximation Domain B of Construction 5.3 has Noetherian spectrum.*

PROOF. Let x be the nonzero nonunit of R from Construction 5.3. By Corollary 5.15, B/xB and $B[1/x]$ are Noetherian. Proposition 5.20 implies that $\operatorname{Spec} B$ is Noetherian. □

REMARK 5.22. The more general Intersection Construction 1.5 sometimes produces an intersection domain that fails to have Noetherian spectrum, even if the base ring R is Noetherian. In Corollary 22.20 with $R = k[x,y]_{(x,y)}$, we prove the existence of elements $\tau_1, \ldots, \tau_n, \ldots \in \widehat{R} = k[[x,y]]$ such that, for the field $L = k(\tau_1, \ldots, \tau_n, \ldots)$, the Intersection Domain $A = L \cap k[[x,y]]$ is a localized polynomial ring over k in infinitely many variables. Thus $\operatorname{Spec} A$ is not Noetherian.

In Proposition 5.23, we establish that the non-flat loci of the inclusion maps into $R^*[1/x]$ are the same for the two rings $S := R[\underline{\tau}]$ and B of Inclusion Construction 5.3.

PROPOSITION 5.23. *Assume the notation of Inclusion Construction 5.3. Then the non-flat locus of the extension* $\alpha : S := R[\underline{\tau}] \hookrightarrow R^*[1/x]$ *equals the non-flat locus of the extension* $\beta : B \hookrightarrow R^*[1/x]$. *It follows that the non-flat locus of the map* α *is defined by an ideal of* $R^*[1/x]$ *if and only if the non-flat locus of the map* β *is defined by the same ideal of* $R^*[1/x]$.

PROOF. By Definition 2.40, it suffices to show for each $Q^* \in \operatorname{Spec}(R^*[1/x])$ that:

$$\alpha_{Q^*} : S \hookrightarrow R^*[1/x]_{Q^*} \text{ is flat} \iff \beta_{Q^*} : B \hookrightarrow R^*[1/x]_{Q^*} \text{ is flat.}$$

By Remarks 2.37.1,

$$\alpha_{Q^*} \text{ is flat} \iff S_{Q^* \cap S} \hookrightarrow R^*[1/x]_{Q^*} \text{ is flat.}$$

Similarly,

$$\beta_{Q^*} \text{ is flat} \iff B_{Q^* \cap B} \hookrightarrow R^*[1/x]_{Q^*} \text{ is flat.}$$

By Construction Properties Theorem 5.14.4.b, $S_{Q^* \cap S} = B_{Q^* \cap B}$. This completes the proof. □

In many of the examples constructed in this book, the ring R is a polynomial ring (or a localized polynomial ring) in finitely many variables over a field; such rings are UFDs. We observe in Theorem 5.24 that the constructed ring B is a UFD if R is a UFD and x is a prime element of R.

THEOREM 5.24. *With the notation of Construction Properties Theorem 5.14:*

(1) *If* R *is a UFD and* x *is a prime element of* R, *then* xU *and* xB *are principal prime ideals of* U *and* B *respectively, and* U *and* B *are UFDs.*
(2) *If* R *is a regular Noetherian UFD, then* $B[1/x]$ *is also a regular Noetherian UFD.*

PROOF. By Proposition 5.17.2, parts a and b, xU and xB are prime ideals. Since R is a UFD and $R[\underline{\tau}]$ is a polynomial ring over R, it follows that $R[\underline{\tau}]$ is a UFD. By Theorem 5.14.2, the rings $U[1/x]$ and $B[1/x]$ are localizations of $R[\underline{\tau}]$ and thus are UFDs; moreover $B[1/x]$ is regular if R is regular. It suffices to prove U is a UFD for the remaining assertion, since B is a localization of U. By Theorem 5.14.4, the x-adic completion of U is R^*. By Proposition 5.17.1, x is in the Jacobson radical of R^*. Since R^* is Noetherian, $\bigcap_{n=1}^{\infty} x^n R^* = (0)$. Thus $\bigcap_{n=1}^{\infty} x^n U = (0)$. It follows that U_{xU} is a DVR [**137**, (31.5)].

By Fact 2.28, the ring $U = U[1/x] \cap U_{xU}$. Therefore U is a Krull domain. Since $U[1/x]$ is a UFD and U is a Krull domain, Theorem 2.27 implies that U is a UFD. Then also B is a UFD and the proof is complete. □

Exercises

(1) If R is a subring of an integral domain B, $P \in \operatorname{Spec} B$, S is a multiplicatively closed subset of R, and $B_P = S^{-1}R$, prove that $R_{P \cap R} = B_P$.

    SUGGESTION. Observe that every element of S is a unit of B_P.

(2) Let x be a nonzero nonunit of a Noetherian integral domain R, let y be an indeterminate, and let $R^* = \frac{R[[y]]}{(x-y)R[[y]]}$ be the x-adic completion of R.

(i) If $x = ab$, where $a, b \in R$ are nonunits such that $aR + bR = R$, prove that there exists a factorization

$$x - y = (a + a_1 y + \cdots) \cdot (b + b_1 y + \cdots) = (\sum_{i=0}^{\infty} a_i y^i) \cdot (\sum_{i=0}^{\infty} b_i y^i),$$

where the $a_i, b_i \in R$, $a_0 = a$ and $b_0 = b$.

(ii) If R is a principal ideal domain (PID), prove that R^* is an integral domain if and only if xR has prime radical.

(3) Let A be an integral domain with field of fractions F. Let C be an extension ring of A such that every nonzero element of A is a regular element of C. If $A = C \cap F$, prove that $xA = xC \cap F$, for every $x \in A$.

(4) Let x be a nonzero nonunit of a Noetherian local domain R and let R^* denote the x-adic completion of R. Assume that L is a subfield of the total quotient ring of R^* with $R \subset L$, and let $A := L \cap R^*$. Prove:

(a) For each $n \in \mathbb{N}$, $x^n A = x^n R^* \cap A$ and $R/x^n R \cong A/x^n A \cong R^*/x^n R^*$.

(b) R^* is the completion of A in the xA-adic topology on A.

(c) If R is one-dimensional, then A is Noetherian local and one-dimensional.

SUGGESTION. For the second part of item a, consider the composition map $A \hookrightarrow R^* \to R^*/x^n R^*$.

(5) Prove item 2 of Remark 5.16, that is, with R a polynomial ring $k[x]$ over a field k and $R^* = k[[x]]$ the x-adic completion of R, show that, with the notation of Construction 5.3, the ring $B = \bigcup_{n=0}^{\infty} C_n$, where $C_n = (U_n)_{P_n}$ and $P_n := (x, \tau_{1n}, \dots, \tau_{sn})U_n$. Conclude that B is a DVR that is the directed union of a birational family of localized polynomial rings in $n + 1$ indeterminates.

CHAPTER 6

Flatness and the Noetherian property

This chapter includes Noetherian Flatness Theorem 6.3. Theorem 6.3 is fundamental for determining if a constructed ring B is Noetherian. The ring B of Inclusion Construction 5.3 is Noetherian if and only if a certain map is flat. To describe this precisely, we formulate as Theorem 6.1 the following implication of Theorem 6.3. Theorem 6.3 is proved in Section 6.1.

THEOREM 6.1. *Let R be a Noetherian integral domain with field of fractions K. Let x be a nonzero nonunit of R and let R^* denote the x-adic completion of R. Let $\tau_1, \ldots, \tau_s \in xR^*$ be algebraically independent elements over K such that the field $K(\tau_1, \ldots, \tau_s)$ is a subring of the total quotient ring of R^*. As in Equations 5.4.4, 5.4.5 and 5.4.6, define*

$$U_n := R[\tau_{1n}, \ldots, \tau_{sn}], \quad U := \bigcup_{n=1}^{\infty} U_n, \quad \text{and} \quad B := (1 + xU)^{-1}U.$$

Then B is Noetherian if and only the extension $R[\tau_1, \ldots \tau_s] \hookrightarrow R^*[1/x]$ is flat.

The crucial Lemma 6.2 relates flatness and the Noetherian property. Local Flatness Theorem 6.13 gives conditions on prime ideals P of the ring B in order that B_P is Noetherian.

In Section 6.2, motivated by Noetherian Flatness Theorem 6.3, we examine the embedding $\psi : U_0 \hookrightarrow R^*[1/x]$ and seek necessary and sufficient conditions for ψ to be flat. Theorem 6.17 gives a sufficient condition for ψ to be flat. The "Insider Construction" combines Local Prototype 4.28 with Inclusion Construction 5.3.

In Sections 6.3 and 6.4, we apply Theorem 6.3 and Theorem 6.17 to show that Nagata's Example 4.15 and Christel's Example 4.17 are Noetherian.

6.1. The Noetherian Flatness Theorem

Lemma 6.2 is used in the proof of Noetherian Flatness Theorem 6.3. We thank Roger Wiegand for observing Lemma 6.2 and its proof.

LEMMA 6.2. *Let S be a subring of a ring T and let $x \in S$ be a regular element of T. Assume that $xS = xT \cap S$ and $S/xS = T/xT$. Then*

(1) *$T[1/x]$ is flat over S $\iff$ T is flat over S.*
(2) *If T is flat over S, then $D := (1 + xS)^{-1}T$ is faithfully flat over $C := (1 + xS)^{-1}S$.*
(3) *If T is Noetherian and T is flat over S, then $C = (1 + xS)^{-1}S$ is Noetherian.*
(4) *If T and $S[1/x]$ are both Noetherian and T is flat over S, then S is Noetherian.*

PROOF. For item 1, if T is flat over S, then, by transitivity of flatness, Remark 2.37.13, the ring $T[1/x]$ is flat over S. For the converse, Lemma 5.12 implies that $S = S[1/x] \cap T$ and $T[1/x] = S[1/x] + T$. Thus the following sequence is exact.

$$0 \to S = S[1/x] \cap T \xrightarrow{\quad\alpha\quad} S[1/x] \oplus T \xrightarrow{\quad\beta\quad} T[1/x] = S[1/x] + T \to 0,$$

where $\alpha(b) = (b, -b)$, for all $b \in S$, and $\beta(c, d) = c + d$, for all $c \in S[1/x]$, $d \in T$. Since the two end terms are flat S-modules, the middle term $S[1/x] \oplus T$ is also S-flat by Remark 2.37.12. By Definition 2.36, a direct summand of a flat S-module is S-flat. Hence T is S-flat .

For item 2, since the map $S \to T$ is flat, the embedding

$$C \;=\; (1 + xS)^{-1}S \;\hookrightarrow\; (1 + xS)^{-1}T \;=\; D$$

is flat. Since $C/xC = S/xS = T/xT = D/xD$ and xC is in the Jacobson radical of C, each maximal ideal of C is contained in a maximal ideal of D, and so D is faithfully flat over C. This establishes item 2.

If T is Noetherian, then D is Noetherian. Since D is faithfully flat over C, it follows that C is Noetherian by Remark 2.37.8, and thus item 3 holds.

For item 4, let J be an ideal of S. By item 3, C is Noetherian, and by hypothesis $S[1/x]$ is Noetherian. Thus there exists a finitely generated ideal $J_0 \subseteq J$ such that $J_0S[1/x] = JS[1/x]$ and $J_0C = JC$. To show $J_0 = J$, it suffices to show for each maximal ideal $\mathfrak{m}$ of S that $J_0S_{\mathfrak{m}} = JS_{\mathfrak{m}}$. If $x \notin \mathfrak{m}$, then $S_{\mathfrak{m}}$ is a localization of $S[1/x]$, and so $J_0S_{\mathfrak{m}} = JS_{\mathfrak{m}}$, while if $x \in \mathfrak{m}$, then $S_{\mathfrak{m}}$ is a localization of C, and so $JS_{\mathfrak{m}} = J_0S_{\mathfrak{m}}$. Therefore $J = J_0$ is finitely generated. It follows that S is Noetherian. □

NOETHERIAN FLATNESS THEOREM 6.3. (Inclusion Version) *As in Setting* 5.1, *assume that R is an integral domain with field of fractions K, $x \in R$ is a nonzero nonunit, $\bigcap_{n \in \mathbb{N}} x^nR = (0)$, and the x-adic completion R^* of R is a Noetherian ring. Let $\tau_1, \ldots, \tau_s \in xR^*$ be algebraically independent elements over K such that the field $K(\tau_1, \ldots, \tau_s)$ is a subring of the total quotient ring of R^*. As in Equations* 5.4.4, 5.4.5 *and* 5.4.6 *of Notation* 5.4, *define*

$$U_n := R[\tau_{1n}, \ldots, \tau_{sn}], \qquad U := \bigcup_{n=1}^{\infty} U_n, \qquad B_n := (1 + xU_n)^{-1}U_n,$$

$$A := K(\tau_1, \ldots, \tau_s) \cap R^*, \qquad \text{and} \qquad B := \bigcup_{n=1}^{\infty} B_n = (1 + xU)^{-1}U.$$

Then:

(1) *The following statements are equivalent:*
   (a) *The extension $\psi : U_0 := R[\tau_1, \ldots \tau_s] \hookrightarrow R^*[1/x]$ is flat.*
   (a′) *The extension $\psi' : B \hookrightarrow R^*[1/x]$ is flat.*
   (b) *The ring B is Noetherian.*
   (c) *The extension $B \hookrightarrow R^*$ is faithfully flat.*
   (d) *The ring A is Noetherian and $A = B$.*
   (e) *The ring A is Noetherian, and A is a localization of a subring of $U_0[1/x] = U[1/x]$.*

(2) *The equivalent conditions of item 1 imply the map $R \hookrightarrow R^*$ is flat.*

(3) *If x is an element of the Jacobson radical $\mathcal{J}(R)$ of R, e.g. if R is a local domain, the equivalent conditions of item 1 imply that R is Noetherian.*

(4) *If R is assumed to be Noetherian, then items a-e are equivalent to the ring U being Noetherian.*

PROOF. For item 1, (a) $\implies$ (a$'$) $\implies$ (b), if $\psi : U_0 = R[\tau_1, \ldots, \tau_s] \hookrightarrow R^*[1/x]$ is flat, then $U[1/x] = U_0[1/x] = R[\tau_1, \ldots, \tau_s][1/x] \hookrightarrow R^*[1/x]$ is flat, and so $U \hookrightarrow R^*[1/x]$ is flat. Since B is a localization of U formed by inverting elements of $(1+xU)$, it follows that $B \hookrightarrow R^*[1/x]$ is flat, that is, (a$'$) holds. By Lemma 6.2.3 with $S = U$ and $T = R^*$, the ring B is Noetherian.

For (b) $\implies$ (c), since B is Noetherian, the extension $B^* = R^*$ is flat over B by Remark 3.3.2. By Proposition 5.17.1, the element $x \in \mathcal{J}(B)$. Thus $B^* = R^*$ is faithfully flat over B by Remark 3.3.4.

For (c) $\implies$ (d), assume $B^* = R^*$ is faithfully flat over B. Then

$$B = \mathcal{Q}(B) \cap R^* = \mathcal{Q}(A) \cap R^* = K(\tau_1, \ldots, \tau_s) \cap R^* = A,$$

by Remark 2.37.9, and so $A = B$ is Noetherian.

For (d) $\implies$ (e), since $B = A$, the ring A is a localization of U, and U is a subring of $R[\tau_1, \ldots, \tau_s][1/x] = U_0[1/x]$.

For (e) $\implies$ (a), since A is a localization of a subring D of $R[\tau_1, \ldots, \tau_s][1/x]$, it follows that $A := \Gamma^{-1}D$, where Γ is a multiplicatively closed subset of D. Now

$$R[\tau_1, \ldots, \tau_s] \subseteq A = \Gamma^{-1}D \subseteq \Gamma^{-1}R[\tau_1, \ldots, \tau_s][1/x] \subseteq \Gamma^{-1}A[1/x] = A[1/x],$$

and so $A[1/x]$ is a localization of $R[\tau_1, \ldots, \tau_s]$. That is, to obtain $A[1/x]$ we localize $R[\tau_1, \ldots, \tau_s]$ by the elements of Γ and then localize by the powers of x. Since A is Noetherian, $A \hookrightarrow A^* = R^*$ is flat by Remark 3.3.2. Thus $A[1/x] \hookrightarrow R^*[1/x]$ is flat. Since $A[1/x]$ is a localization of $R[\tau_1, \ldots, \tau_s]$, it follows that $R[\tau_1, \ldots, \tau_s] \hookrightarrow R^*[1/x]$ is flat. This completes the proof of item 1.

For item 2, since U_0 is flat over R, condition a of item 1 implies that $R^*[1/x]$ is flat over R. By Lemma 6.2.1 with $S = R$ and $T = R^*$, if $R \hookrightarrow R^*[1/x]$ is flat, then $R \hookrightarrow R^*$ is flat.

For item 3, assume the equivalent conditions of item 1 hold and $x \in \mathcal{J}(R)$. The extension $R \hookrightarrow R^*$ is flat by item 2. If P is a maximal ideal of R, then $x \in P$ and $R/xR = R^*/xR^*$. Hence $PR^* \neq R^*$. Therefore $R \hookrightarrow R^*$ is faithfully flat. By Remark 2.37.8, R is Noetherian.

For item 4, assume the equivalent conditions of item 1 hold and R is Noetherian; then $U_0[1/x] = U[1/x]$ is Noetherian. The composite embedding

$$U \hookrightarrow B = A \hookrightarrow B^* = A^* = R^*$$

is flat because B is a localization of U and $B^* = R^*$ is faithfully flat over B. By Lemma 5.12, parts 1 and 4, with $S = U$ and $T = R^*$, it follows that U is Noetherian. If U is Noetherian, then the localization B of U is Noetherian, and so condition b holds. This completes the proof of Theorem 6.3. $\square$

In later chapters, it is useful to have a name for the constructed domain of Inclusion Construction 5.3 for the situation where the Intersection Domain and the Approximation Domain are equal and Noetherian, that is, the conditions of Theorem 6.3.1 hold. The terminology "Noetherian Limit Intersection Domain" of Definition 6.4 is an extension of the concept of "limit-intersecting" elements in Definition 5.10, for the situation where the Intersection Domain and the Approximation Domain are equal, but not necessarily Noetherian.

DEFINITION 6.4. Assume the notation of Noetherian Flatness Theorem 6.3. If condition 1.d holds, that is, the Approximation Domain B equals the Intersection Domain A and is Noetherian, then $B = A$ is called a *Noetherian Limit Intersection Domain.*

REMARK 6.5. By Theorem 6.3, B is a Noetherian Limit Intersection Domain if and only if the extension $R[\tau_1, \dots, \tau_s] \hookrightarrow R^*[1/x]$ is flat. Thus it follows that a Local Prototype D as in Definition 4.28 is a Noetherian Limit Intersection Domain.

COROLLARY 6.6. *Assume notation as in Noetherian Flatness Theorem* 6.3. *If* $\dim R^* = 1$, *then* $B = A$ *is a Noetherian Limit Intersection Domain.*

PROOF. We show the map $\psi : R[\tau_1, \dots, \tau_s] \hookrightarrow R^*[1/x]$ is flat. Since $\dim R^* = 1$ and x is a regular element in R^* with $x \in \mathcal{J}(R^*)$, it follows that $\dim(R^*[1/x]) = 0$. Hence $R^*[1/x]$ is the total quotient ring of R^*, and the map ψ factors as the composition of the inclusion maps $R[\tau_1, \dots, \tau_s] \hookrightarrow K(\tau_1, \dots, \tau_s) \hookrightarrow R^*[1/x]$. Modules over a field are free and hence flat, and compositions of flat maps are flat by Remarks 2.37, parts 2 and 13. Hence the map ψ is flat. By Theorem 6.3.1, the result holds. □

In the setting of Theorem 6.3, if $\dim R^* = 2$, Example 6.7 shows the Approximation Domain B may not equal the Intersection Domain A; that is, B is not a Noetherian Limit Intersection Domain for the construction in Example 6.7. Thus, by Noetherian Flatness Theorem 6.3.1, $\psi' : B \hookrightarrow R^*[1/x]$ is not flat and B is non-Noetherian.[1] We are motivated to give Example 6.7 by a question of Guillaume Rond, who has been studying constructions similar to Inclusion Construction 5.3.

EXAMPLE 6.7. Let $R := \mathbb{C}[[t]][x]$, where $\mathbb{C}$ is the complex numbers and t and x are variables. Then the x-adic completion of R is $R^* = \mathbb{C}[[t, x]]$. Let

$$\tau := \sum_{k=1}^{\infty} f_k(t) x^k \in \mathbb{C}[[t, x]] \quad \text{and} \quad c := \sum_{k=1}^{\infty} f_k(t) t^k \in R,$$

where each $f_k(t)$ is a power series in $\mathbb{C}[[t]]$ and τ is a power series in $\mathbb{C}[[t, x]]$ that is not algebraic over R. Then $\tau - c = \sum_{k=1}^{\infty} f_k(t)(x^k - t^k) \in R[\tau]$, and

$$\theta := \frac{\tau - c}{x - t} \in A := \mathcal{Q}(\mathbb{C}[[t]][x, \tau]) \cap \mathbb{C}[[t, x]].$$

By Theorem 4.9, A is a two-dimensional regular local ring. By Theorem 5.14.4, $B[1/x]$ is a localization of $U_0 = \mathbb{C}[[t]][x, \tau]$. We show the element θ is not in B, and so $B \subsetneq A$. Suppose that θ is an element of B. Then

$$\tau - c \in (x - t)B \cap U = (x - t)U,$$

since $(x - t)B \subseteq (x, t)B = \mathfrak{m}_B$ and $U_{\mathfrak{m}_B \cap U} = B$. Also

$$U \subset U[1/x] = R[\tau][1/x] \subset R[\tau]_{(x-t)R[\tau]}.$$

The last inclusion follows because $x \notin (x - t)R[\tau]$. Therefore

$$\tau - c \in (x - t)R[\tau]_{(x-t)R[\tau]} \cap R[\tau] = (x - t)R[\tau].$$

Since $U_0 = R[\tau]$ is a polynomial ring over R in τ, this contradicts the fact that $\tau - c$ is algebraically independent over R. Thus $\theta \notin B[1/x]$, and so $\theta \notin B$. Therefore $A \neq B$ and by Theorem 6.3.1, B is not Noetherian.

[1] Example 12.7 describes another example where B is not Noetherian and $B \subsetneq A$.

Corollary 6.8.3 gives a simplified flatness property for Local Prototypes.

COROLLARY 6.8. *Let* $R = k[x, y_1, \dots, y_r]_{(x,y_1,\dots,y_r)}$, *where* $x, y_1, \dots, y_r$ *are variables over a field* k, *let* $R^* = k[y_1, \dots, y_r]_{(y_1,\dots,y_r)}[[x]]$ *denote the* x*-adic completion of* R *and let* $\tau_1, \dots, \tau_s$ *be elements of* xR^* *that are algebraically independent over* R. *Then:*

(1) *The ring* B *of Noetherian Flatness Theorem 6.3 may or may not be Noetherian, and is equal to the following directed union:*

$$B = \bigcup_{n=0}^{\infty} k[x, y_1, \dots, y_r, \tau_{1n}, \dots, \tau_{sn}]_{(x,y_1,\dots,y_r,\tau_{1n},\dots,\tau_{sn})},$$

where τ_{in} *is the* n^{th}*-endpiece of* τ_i *for each* i *with* $1 \le i \le s$.

(2) *The ring* B *is a Noetherian Limit Intersection Domain if and only if the map*

$$\psi' : U_0' := k[x, y_1, \dots, y_r, \tau_1, \dots, \tau_s] \hookrightarrow R^*[1/x]$$

is flat.

(3) *If* $\tau_1, \dots, \tau_s$ *are elements of* $xk[[x]]$ *and* $V = k(x, \tau_1, \dots, \tau_s) \cap k[[x]]$, *then*
   (a) *The ring* $D = V[y_1, \dots, y_r]_{(x,y_1,\dots,y_r)}$ *is the Local Prototype of Definition 4.28, and*
   (b) *The maps* $\psi : R[\tau_1, \dots, \tau_s] \hookrightarrow R^*[1/x]$ *and* $\psi' : U_0' \hookrightarrow R^*[1/x]$ *are flat.*

PROOF. Item 1 is Remark 5.16.3. For item 2, the map ψ' is the composition

$$U_0' \hookrightarrow U_0 \overset{\psi}{\hookrightarrow} R^*[1/x],$$

and $U_0' \hookrightarrow U_0$ is flat by 2.37.4. Thus flatness of $U_0 \hookrightarrow R^*[1/x]$ implies ψ' is flat, by Remark 2.37.13. If ψ' is flat, then $U_0 \hookrightarrow R^*[1/x]$ is flat by Remark 2.37.1, and so item 2 holds.

For item 3, by Remark 4.29.1, the ring $D = \mathcal{Q}(R)(\tau_1, \dots, \tau_s) \cap R^*$. That is, D is the Intersection Domain of Inclusion Construction 5.3 for R with respect to the τ_i. Thus Noetherian Flatness Theorem 6.3 applies. By Proposition 4.27, D equals its Approximation Domain, given in item 1. Since D is an RLR, the equivalent conditions of item 1 of Theorem 6.3 hold, and so the map ψ is flat. Equivalently, by item 2, the map ψ' is flat. □

REMARK 6.9. The original proof given for Noetherian Flatness Theorem 6.3 in [**77**] is an adaptation of a proof given by Heitmann in [**95**, page 126]. Heitmann considers the case where there is one transcendental element τ and defines the corresponding extension U to be a *simple PS-extension of* R *for* x. Heitmann proves in this case that a certain monomorphism condition on a sequence of maps is equivalent to U being Noetherian [**95**, Theorem 1.4].

REMARK 6.10. Examples where $A = B$ and A is not Noetherian show that it is possible for A to be a localization of U and yet for A, and therefore also U, to fail to be Noetherian, even if R is Noetherian; see Example 16.4 and Theorem 16.6. Thus the equivalent conditions of Noetherian Flatness Theorem 6.3 are not implied by the property that A is a localization of U.

The following diagram displays the situation concerning possible implications among certain statements for Inclusion Construction 5.3 and the approximations in Section 5.2:

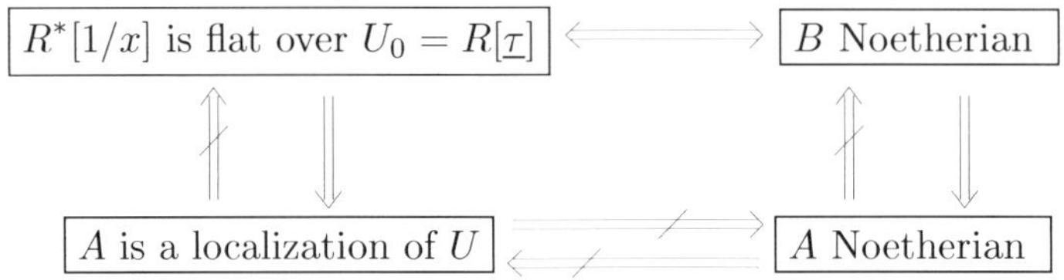

REMARKS 6.11. Let R, $x \in R$, $\underline{\tau} = \{\tau_1, \dots, \tau_s\}$ and $\psi : R[\underline{\tau}] \hookrightarrow R^*[1/x]$ be as in Inclusion Construction 5.3 and Noetherian Flatness Theorem 6.3.

(1) It is sometimes difficult to determine whether the map ψ is flat. One helpful fact is given in Remark 2.37.10: If there exists a prime ideal P of $R^*[1/x]$ such that $\operatorname{ht} P < \operatorname{ht}(P \cap R[\underline{\tau}])$, then $R[\underline{\tau}] \hookrightarrow R^*[1/x]$ is not flat, and hence U is not Noetherian.

(2) Assume that the equivalent conditions of item 1 of Theorem 6.3 hold. Let Δ be a multiplicatively closed subset of R and let $R' := \Delta^{-1}R$. Assume that $xR = R \cap xR'$, and consider the following diagram:

$$\begin{array}{ccccc} R^* & \longrightarrow & \Delta^{-1}R^* & \longrightarrow & (\Delta^{-1}R^*)^* \\ \uparrow & & \uparrow & & \uparrow \\ R & \longrightarrow & \Delta^{-1}R = R' & \longrightarrow & (R')^*, \end{array}$$

where $(\Delta^{-1}R^*)^*$ is the $(x\Delta^{-1}R^*)$-adic completion of $\Delta^{-1}R^*$, and $(R')^*$ is the (xR')-adic completion of R'. It follows from the diagram that the $\tau_i = \sum a_{ij}x^j$ can be considered as elements of $(R')^*$ and, for each i and n, the nth-endpiece τ_{in} of τ_i is the same in $(R')^*$.

Let B be the Approximation Domain associated to R and the τ_i and and B' be the Approximation Domain associated to R' and the τ_i. Then $B' = \Delta^{-1}B$. To see this, as in Theorem 6.3, $U_n = R[\tau_{1n}, \dots, \tau_{sn}]$ and $B_n = (1 + xU_n)^{-1}U_n$. Similarly

$$B'_n := (1 + xU'_n)^{-1}U'_n,$$

where $U'_n := R'[\tau_{1n}, \dots, \tau_{sn}] = (\Delta^{-1}R)[\tau_{1n}, \dots, \tau_{sn}] = \Delta^{-1}U_n$. Then $xU_n = U_n \cap xU'_n$, and so

$$B'_n = (1 + x\Delta^{-1}U_n)^{-1}\Delta^{-1}U_n = \Delta^{-1}(1 + xU_n)^{-1}U_n = \Delta^{-1}B_n,$$

since doing these two localizations yields the same result in the reverse order. Therefore $B' = \bigcup_{n=1}^{\infty} B'_n = \Delta^{-1}B$.

Since $B = A$ is Noetherian, B' is Noetherian. Hence the equivalent conditions of item 1 of Theorem 6.3 hold for the construction over R'.

(3) In certain cases the non-flat locus of $\psi : R[\underline{\tau}] \hookrightarrow R^*[1/x]$ is closed. This is true for the cases considered in Corollary 6.8. But it is unclear that this holds in general.

QUESTION 6.12. If the Approximation Domain B of Noetherian Flatness Theorem 6.3 is not Noetherian, for which prime ideals P of B is B_P Noetherian?

The first part of Local Flatness Theorem 6.13 answers Question 6.12 for the prime ideals P of B with $x \in P$. If the non-flat locus of the map ψ of Theorem 6.3 is defined by an ideal of $R^*[1/x]$, then a more complete answer is given in Local Flatness Theorem 6.13.2. Remark 6.14 applies if R is Noetherian and $x \notin P$.

LOCAL FLATNESS THEOREM 6.13. *Assume the notation of Noetherian Flatness Theorem 6.3, and let P be a prime ideal of B such that $x \in P$. Then*

(1) *The following statements are equivalent:*
   (a) $\psi_{PR^*} : R[\underline{\tau}] \hookrightarrow R^*_{PR^*}[1/x]$ *is flat.*
   (b) $B_P \hookrightarrow R^*_{PR^*}$ *is faithfully flat.*
   (c) B_P *is Noetherian.*
(2) *Assume that there exists an ideal F of R^* such that $FR^*[1/x]$ defines the non-flat locus of the map $\psi : R[\underline{\tau}] \hookrightarrow R^*[1/x]$. If ψ is flat, set $F = R^*$. Then the equivalent conditions in item d are each equivalent to statements a-c of part 1:*
   (d) $(F,x)R^* \cap R \not\subseteq P$; *equivalently,* $(F,x)R^* \cap B \not\subseteq P$; *equivalently,* $(F,x)R^* \not\subseteq PR^*$; *equivalently,* $(F,x)R^* \cap R \not\subseteq P \cap R$.

PROOF. Let $\mathfrak{p} = P \cap R$. By Construction Properties Theorem 5.14.2,

$$R/xR = B/xB = R^*/xR^*, \tag{6.13.1}$$

and so $P = \mathfrak{p}B$ and $PR^* = \mathfrak{p}R^*$ is a prime ideal of R^*.

For part 1 of Theorem 6.13, Noetherian Flatness Theorem 6.3.1 implies that the statement $B_{\mathfrak{p}B} = B_P$ is Noetherian is equivalent to each of the following conditions:
(i) the map $B_{\mathfrak{p}B} \hookrightarrow (R_{\mathfrak{p}})^*$ is faithfully flat, and
(ii) the map $R_{\mathfrak{p}}[\underline{\tau}] \hookrightarrow (R_{\mathfrak{p}})^*[1/x]$ is flat.
By Remark 3.3.6, the completion $(R_{\mathfrak{p}})^*$ of $R_{\mathfrak{p}}$ in the $xR_{\mathfrak{p}}$-adic topology is also the completion of $R^*_{\mathfrak{p}R^*}$ in the $xR^*_{\mathfrak{p}R^*}$-adic topology. Hence the canonical maps

$$R_{\mathfrak{p}} \hookrightarrow R^*_{\mathfrak{p}R^*} \overset{\theta}{\hookrightarrow} (R_{\mathfrak{p}})^*$$

are faithfully flat. Since the map θ is faithfully flat, condition i is equivalent to faithful flatness of the map $B_P = B_{\mathfrak{p}B} \hookrightarrow R^*_{\mathfrak{p}R^*} = R^*_{PR^*}$, that is, statement b. Also condition ii is equivalent to flatness of the map $R_{\mathfrak{p}}[\underline{\tau}] \hookrightarrow R^*_{\mathfrak{p}R^*}[1/x]$ or, equivalently, to flatness of the map $\psi_{PR^*} : R[\underline{\tau}] \hookrightarrow R^*_{PR^*}[1/x]$. Thus statements a,b, and c are equivalent.

For part 2, Equation 6.13.1 implies $((F,x)R^* \cap R)B = (F,x)R^* \cap B$ and also

$$(F,x)R^* \subseteq PR^* \iff ((F,x)R^* \cap R)B \subseteq P \iff (F,x)R^* \cap R \subseteq \mathfrak{p},$$

since $x \in \mathfrak{p}$. Thus the four conditions of item d are equivalent. In addition

$$F \subseteq PR^* \iff (F,x)R^* \subseteq PR^*. \tag{6.13.2}$$

If $x^n \in F$, for some $n \in \mathbb{N}$, then $FR^*[1/x] = R^*[1/x]$, and so ψ is flat. Then $F = R^*$, by assumption. Also statement a holds and $(F,x)R^* \not\subseteq PR^*$. Thus part 2 holds in the case $x^n \in F$.

Assume $x^n \notin F$ for all $n \in \mathbb{N}$. To show, as in part 2, that the conditions of item d are equivalent to the statements of part 1 of Theorem 6.13, first assume $(F,x)R^* \subseteq PR^*$, and show statement a fails. Since $x \in PR^* \setminus F$, F is properly contained in PR^*. Let $\mathfrak{q}$ be an ideal of R^* such that $F \subseteq \mathfrak{q} \subseteq PR^*$ and $\mathfrak{q}$ is maximal with $\mathfrak{q} \cap \{x^n \mid n \in \mathbb{N}\} = \emptyset$. Then $\mathfrak{q}$ is a prime ideal properly contained in PR^* and the map

$$\psi_{\mathfrak{q}} : R[\underline{\tau}] \hookrightarrow R^*[1/x]_{\mathfrak{q}R^*[1/x]} = R^*_{\mathfrak{q}}$$

is not flat. Since $R^*_{\mathfrak{q}}$ is a localization of $R^*_{PR^*}[1/x]$, the map ψ_{PR^*} is not flat. This shows statement a $\implies$ the conditions of item d.

For the other direction, assume the condition $(F, x)R^* \not\subseteq PR^*$ from item d. To show statement a, that $\psi_{PR^*} : R[\underline{\tau}] \hookrightarrow R^*_{PR^*}[1/x]$ is flat, let $Q_1 \in \operatorname{Spec}(R^*_{PR^*}[1/x])$. Then $Q_1 = QR^*_{PR^*}[1/x]$, where $Q \in \operatorname{Spec} R^*, Q \subseteq PR^*$, and $x \notin Q$. Let ψ_Q denote the map $R[\underline{\tau}] \hookrightarrow R^*_Q$, and let ψ_{Q_1} denote the map $R[\tau] \hookrightarrow (R^*_{PR^*}[1/x])_{Q_1}$. Since $(R^*_{PR^*}[1/x])_{Q_1} = R^*_Q$, the map $\psi_{Q_1} = \psi_Q$. By Equation 6.13.2, there exists an element $h \in F \setminus PR^*$, and $h \notin PR^*$ implies $h \notin Q$. Therefore the non-flat locus $FR^*[1/x] \not\subseteq QR^*[1/x]$, and so $\psi_Q = \psi_{Q_1}$ is flat. This holds for every prime ideal Q_1 of $R^*_{PR^*}[1/x]$. It follows that $\psi_{PR^*} : R[\underline{\tau}] \hookrightarrow R^*_{PR^*}[1/x]$ is flat. This completes the proof of Theorem 6.13. □

REMARK 6.14. Assume the notation of Noetherian Flatness Theorem 6.3, and assume that R is Noetherian. Then B_P is Noetherian for every prime ideal P such that $x \notin P$. This follows because B_P is a localization of $B[1/x]$, and $B[1/x]$ is Noetherian by Corollary 5.15.

By Proposition 5.23, the non-flat locus of the map $\psi : R[\underline{\tau}] \hookrightarrow R^*[1/x]$ is the same as the non-flat locus of the map $\varphi : B \hookrightarrow R^*[1/x]$. If the non-flat locus of ψ, or, equivalently, of φ is closed, then Theorem 6.15 implies that certain homomorphic images of B are Noetherian.

THEOREM 6.15. *Let R be a Noetherian integral domain, let $x \in R$ be a nonzero nonunit, and let R^* denote the x-adic completion of R. Let $\underline{\tau} = \{\tau_1, \ldots, \tau_s\}$ be a set of elements of R^* that are algebraically independent over R, and are such that the field $K(\tau_1, \ldots, \tau_s)$ is a subring of the total quotient ring of R^*. Let B be the Approximation Domain associated to $\underline{\tau}$ as in Definition 5.7. Assume there exists an ideal F of $R^*[1/x]$ that defines the non-flat locus of the map $\varphi : B \to R^*[1/x]$. Let I be an ideal of B such that $IR^* \cap B = I$ and x is regular on R^*/IR^*.*

(1) *If $IR^*[1/x] + F = R^*[1/x]$, then the tensor product map below is flat:*

$$\varphi \otimes_B (B/I) : B/I \cong B \otimes_B (B/I) \to R^*[1/x] \otimes_B (B/I) \cong R^*[1/x]/IR^*[1/x].$$

(2) *If $R^*[1/x]/IR^*[1/x]$ is flat over B/I, then R^*/IR^* is flat over B/I.*

(3) *If $IR^*[1/x] + F = R^*[1/x]$, then B/I is Noetherian.*

PROOF. For item 1, φ_P is flat for each $P \in \operatorname{Spec} R^*[1/x]$ with $I \subseteq P$ by hypothesis. Hence for each such P we have $\varphi_P \otimes_B (B/I)$ is flat. Since flatness is a local property, it follows that $\varphi \otimes_B (B/I)$ is flat.

For items 2 and 3, apply Lemma 6.2 with $S = B/I$ and $T = R^*/IR^*$; the element x of Lemma 6.2 is the image in B/I of the element x. Since $IR^* \cap B = I$, the ring B/I embeds into R^*/IR^*. Since $B/xB = R^*/xR^*$, we have

$$R^*/(I,x)R^* = B/(I,x)B \quad \text{and} \quad x(R^*/IR^*) \cap (B/I) = x(B/I).$$

Item 2 of Theorem 6.15 follows from item 1 of Lemma 6.2. Item 3 of Theorem 6.15 follows from item 2 of Theorem 6.15 and item 4 of Lemma 6.2. □

6.2. Introduction to Insider Construction 10.7

In this section we demonstrate that Inclusion Construction 5.3 can be used to construct a variety of examples that are contained inside a Local Prototype domain. The technique we use is called "Insider Inclusion Construction", or more briefly, the "Insider Construction". Basically we use Inclusion Construction 5.3 twice: First we use it to build a more standard simple "Prototype" example using some variables τ_i, then we iterate the procedure using variables f_j which are polynomials in the τ_i. This technique is defined formally in Insider Construction 10.7.

We present in this chapter several examples using the Insider Construction, including two classical examples of Nagata and Rotthaus. The Insider Construction simplifies the verification of flatness for the maps associated to examples constructed using Inclusion Construction 5.3.

For the examples considered in this chapter, we use Setting 6.16:

SETTING 6.16. Let k be a field, let $s \in \mathbb{N}$ and $r \in \mathbb{N}_0$, let $x, y_1, \dots y_r$ be variables over k, and let $R = k[x, y_1, \dots, y_r]_{(x, y_1, \dots, y_r)}$ be the localized polynomial ring in these variables. Let the elements $\tau_1, \dots, \tau_s \in xk[[x]]$ be algebraically independent over $k(x)$. Define $V = k(x, y_1, \dots, y_r) \cap k[[x]]$. As in Corollary 6.8.3, the ring R is the base ring of a Local Prototype domain $D = V[y_1, \dots, y_r]_{(x, y_1, \dots, y_r)}$. The map

$$\psi : R[\tau_1, \dots, \tau_s] \hookrightarrow R^*[1/x]$$

is flat. We construct two "insider" integral domains A and B inside the Local Prototype D, where A is an Intersection Domain as in Construction 5.3, and B is an integral domain that "approximates" A as in Section 5.2.

Let $f \in R[\tau_1, \dots, \tau_s] \subseteq R^*$ be transcendental over $\mathcal{Q}(R) = k(x, y_1, \dots, y_r)$. As in Inclusion Construction 5.3, let $A = \mathcal{Q}(R)(f) \cap R^*$. Define endpieces f_n as in Equation 5.4.1 of Notation 5.4, and define the Approximation Domain B associated with f, as in Equation 5.4.5 and Noetherian Flatness Theorem 6.3. By Corollary 6.8.1, B is a directed union

$$B = \bigcup_{n=1}^{\infty} R[f_n]_{(\mathfrak{m}, f_n)},$$

where $\mathfrak{m}$ is the maximal ideal of R. Let $S = R[f]$ and let $T = R[\tau_1, \dots, \tau_s]$.

As we describe in Theorem 6.17, the condition that the insider Approximation Domain B is Noetherian is related to flatness of the extension $S \hookrightarrow T$, an extension of polynomial subrings of R^*. We apply Theorem 6.17 to conclude that Nagata's Example 4.15 and Christel's Example 4.17 are Noetherian.

THEOREM 6.17. *In the notation of Setting 6.16, if the extension*

$$S := R[f] \overset{\varphi}{\hookrightarrow} T := R[\tau_1, \dots, \tau_s]$$

is flat, then B is Noetherian and A equals B. Hence A is Noetherian.

PROOF. By Corollary 6.8 the map $\psi : T \hookrightarrow R^*[1/x]$ is flat. By hypothesis, $\varphi : S := R[f] \hookrightarrow R[\tau_1, \dots, \tau_s]$ is flat.

(6.18.1)

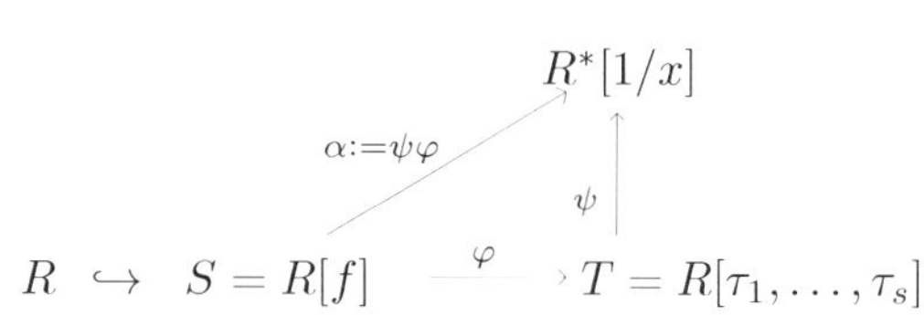

Since the composition of flat maps is again flat (Remark 2.37.13), we conclude that $\alpha : S \hookrightarrow R^*[1/x]$ is flat. By Noetherian Flatness Theorem 6.3, we have that $A = B$, as desired. $\square$

This idea is the basis for Insider Construction 10.7. The same argument goes through for several elements $f_1, \dots, f_t \in T$ that are algebraically independent over $\mathcal{Q}(R)$. Moreover, non-flatness of the extension φ sometimes implies non-flatness of the extension $U_0 \hookrightarrow R^*[1/x]$; see Theorem 10.9. In Corollary 7.7 we show $\varphi : S \hookrightarrow T$ is flat if and only if $\operatorname{ht} Q \geq \operatorname{ht}(Q \cap S)$ for every $Q \in \operatorname{Spec} T$.

6.3. Nagata's example

In Proposition 6.19 we use Theorem 6.17 to prove that Nagata's Example 4.15 is Noetherian.

SETTING 6.18. Let k be a field, let x and y be indeterminates over k, and set

$$R := k[x,y]_{(x,y)} \qquad \text{and} \qquad R^* := k[y]_{(y)}[[x]].$$

The power series ring R^* is the xR-adic completion of R. Let $\tau \in xk[[x]]$ be a transcendental element over $k(x)$. Since R^* is an integral domain, every nonzero element of the polynomial ring $R[\tau]$ is a regular element of R^*. Thus the field $k(x,y,\tau)$ is a subfield of $\mathcal{Q}(R^*)$. The Local Prototype domain D corresponding to τ is $D := k(x,y,\tau) \cap R^*$, as in Definition 4.28. By Proposition 4.27, D is a two-dimensional regular local domain and is a directed union of localized polynomial rings in three variables over the field k.

Let f be a polynomial in $R[\tau]$ that is algebraically independent over $\mathcal{Q}(R)$, for example, $f = (y+\tau)^2$, as in Nagata's example. Let $A := \mathcal{Q}(R[f]) \cap R^*$ be the *Intersection Domain* corresponding to f. Since $R[f] \subseteq R[\tau]$, we have $k(x,y,f) = \mathcal{Q}(R[f]) \subseteq \mathcal{Q}(R^*)$. The Intersection Domain A is a subring of the Local Prototype domain D.

By Corollary 6.8.1, the natural Approximation Domain B associated to A is

$$B = \bigcup_{n \in \mathbb{N}} k[x,y,f_n]_{(x,y,f_n)}, \tag{6.18.0}$$

where the f_n are the nth endpieces of f.

By Corollary 6.8.3b, the extension $T := R[\tau] \overset{\psi}{\hookrightarrow} R^*[1/x]$ is flat, where ψ is the inclusion map. Let $S := R[f] \subseteq R[\tau]$ and let φ be the embedding

$$\varphi : S := R[f] \overset{\varphi}{\hookrightarrow} T = R[\tau]. \tag{6.18.e}$$

Put $\alpha := \psi \circ \varphi : S \to R^*[1/x]$. Then we have the following commutative diagram:

$$\begin{array}{ccccc} & & & & R^*[1/x] \\ & & & \overset{\alpha:=\psi\varphi}{\nearrow} & \uparrow \psi \\ R & \hookrightarrow & S = R[f] & \xrightarrow{\varphi} & T = R[\tau] \end{array} \tag{6.18.1}$$

The proof in Proposition 6.19 of the Noetherian property for Nagata's Example 4.15 is different from the proof given in [**137**, Example 7, pp.209-211].

PROPOSITION 6.19. *With the notation of Setting* 6.18, *let* $f := (y+\tau)^2$. *In Nagata Example* 4.15, *the ring* $B = A$ *and* B *is Noetherian with completion* $k[[x,y]]$. *By Theorem* 3.33, B *is a two-dimensional regular local domain.*

PROOF. The ring $T = R[\tau]$ is a free S-module with free basis $\langle 1, y+\tau\rangle$. By Remark 2.37.2, the map φ is flat. By Theorem 6.17, B is Noetherian and $B = A$. $\square$

REMARKS 6.20. (1) In Nagata's original example [**137**, Example 7,pp. 209-211], the field k has characteristic different from 2. This assumption is not necessary for showing that the domain B of Proposition 6.19 is a two-dimensional regular local domain.

(2) Whether or not the ring B is Noetherian depends upon the polynomial f. In Example 6.24.2, the ring B is constructed in a similar way to the ring B of Proposition 6.19, but the ring B of Example 6.24.2 is not Noetherian.

6.4. Christel's Example

In this section and Section 6.5, we present additional examples using the techniques of Section 6.3. Consider elements σ and τ in $xk[[x]]$ that are algebraically independent over the field $k(x)$. To describe these examples, modify Setting 6.18 as follows.

SETTING 6.21. Let k be a field, let x, y, z be indeterminates over k, and set

$$R := k[x,y,z]_{(x,y,z)} \qquad \text{and} \qquad R^* := k[y,z]_{(y,z)}[[x]].$$

The power series ring R^* is the xR-adic completion of R. Let σ and τ in $xk[[x]]$ be algebraically independent over $k(x)$. We use the Local Prototype Domain D corresponding to σ, τ as in Definition 4.28, that is,

$$D := k(x,y,z,\sigma,\tau) \cap k[y,z]_{(y,z)}[[x]].$$

In the examples of this section, f is an element of $R[\sigma,\tau]$ such that f is transcendental over $K = \mathcal{Q}(R)$. The *Intersection Domain* of Inclusion Construction 5.3 corresponding to f is

$$A = K(f) \cap R^* = k(x,y,z,f) \cap k[y,z]_{(y,z)}[[x]].$$

Thus A is an "insider" Intersection Domain contained in the Local Prototype Domain D. As in Setting 6.18 for the Nagata Example, the Approximation Domain B associated to f is a directed union of localized polynomial rings over k in four variables.

REMARK 6.22. With Setting 6.21, let $T := R[\sigma,\tau]$ and let $S := R[f]$, where f is a polynomial in $R[\sigma,\tau]$ that is algebraically independent over $\mathcal{Q}(R)$. Let $\varphi : S \hookrightarrow T$ denote the inclusion map from S to T. Since σ and τ are algebraically independent over R, the element f in $R[\sigma,\tau]$ has a unique expression

$$f = c_{00} + c_{10}\sigma + c_{01}\tau + \cdots + c_{ij}\sigma^i\tau^j + \cdots + c_{mn}\sigma^m\tau^n,$$

where the $c_{ij} \in R$. The c_{ij} with at least one of i or j nonzero are the *nonconstant coefficients* of f. The ideal $L := (c_{10}, c_{01}, \ldots, c_{mn})R$ is the *ideal generated by the nonconstant coefficients* of f. We show in Theorem 7.28 of Chapter 7 that

$$\varphi \text{ is flat} \iff LR = R. \qquad (6.22.\text{b})$$

Theorem 6.17 implies the Noetherian property for the following example of Rotthaus [**155**], Example 4.17 of Chapters 4.

EXAMPLE 6.23. (Christel) This is the first example of a Nagata ring that is not excellent. With Setting 6.21, let $f := (y+\sigma)(z+\tau)$ and consider the intesection domain $A = k(x,y,z,f) \cap R^*$ contained in Local Prototype $D = k(x,y,z,\sigma,\tau) \cap R^*$. The nonconstant coefficients of $f = yz + \sigma z + \tau y + \sigma\tau$ as a polynomial in $R[\sigma,\tau]$ are $\{1, z, y\}$. They do generate the unit ideal of R, and so, since we assume Remark 6.22.4 for now, it follows that φ is flat. Thus, by Theorem 6.17, the associated nested union domain B is Noetherian and is equal to A.

6.5. Further implications of the Noetherian Flatness Theorem

Noetherian Flatness Theorem 6.3 also yields examples that are not Noetherian even if the Approximation Domain B is equal to the Intersection Domain A.

EXAMPLES 6.24. (1) With Setting 6.21, let $f := y\sigma + z\tau$. We show in Examples 10.15 that the map $R[f] \hookrightarrow R[\sigma,\tau]$ is not flat and that $A = B$, i.e., A is "limit-intersecting" as in Definition 5.10, but is not Noetherian. Thus we have a situation where the Intersection Domain equals the Approximation Domain, but is not Noetherian.

(2) The following is a related simpler example: Again with the notation of Setting 6.21, let $f := y\tau + z\tau^2 \in R[\tau] \subseteq D = k(x,y,z,\tau) \cap R^*$, the Prototype. Then the constructed Approximation Domain B (using f) is not Noetherian by Theorem 10.12. Moreover, B is equal to the intersection domain $A := R^* \cap k(x,y,z,f)$ by Corollary 9.19.

In dimension two (the two variable case), an immediate consequence of Valabrega's Theorem 4.9 is the following.

THEOREM 6.25. (Valabrega) *Let x and y be indeterminates over a field k and let $R = k[x,y]_{(x,y)}$. Then $\widehat{R} = k[[x,y]]$ is the completion of R. If L is a field between the field of fractions of R and the field of fractions of $k[y]_{(y)}[[x]]$, then $A = L \cap \widehat{R}$ is a two-dimensional regular local domain with completion $\widehat{R}$.*

Example 6.24 shows that the three-dimensional analog to Valabrega's result fails. With $R = k[x,y,z]_{(x,y,z)}$ the field $L = k(x,y,z,f)$ is between $k(x,y,z)$ and the fraction field of $k[y,z]\,[[x]]$, but $L \cap \widehat{R} = L \cap R^*$ is not Noetherian.

EXAMPLE 6.26. The following example is continued in Example 19.6. With the notation of Setting 6.21, let $f = (y+\sigma)^2$ and $g = (y+\sigma)(z+\tau)$. It is shown in Chapter 25 that the Intersection Domain $A := R^* \cap k(x,y,z,f,g)$ properly contains its associated Approximation Domain B and that both A and B are non-Noetherian.

We use Ratliff's Equidimension Theorem 3.25 to show that the universally catenary property is preserved by Inclusion Construction 5.3, if the constructed domain is Noetherian.

THEOREM 6.27. *Assume the notation of Noetherian Flatness Theorem 6.3, and assume that* $(R, \mathfrak{m})$ *is a universally catenary Noetherian local domain. Then:*

(1) *If* A *is Noetherian, then* A *is a universally catenary Noetherian local domain.*

(2) *If* B *is Noetherian, then* $B = A$ *and* B *is a universally catenary local domains.*

PROOF. By Construction Properties Theorem 5.14.4, $R^* = B^* = A^*$. By Proposition 5.17.5, A and B are local and their maximal ideals are $\mathfrak{m}A$ and $\mathfrak{m}B$, respectively. The $\mathfrak{m}$-, $\mathfrak{m}A$- and $\mathfrak{m}B$-adic completions of R, A and B, respectively, all equal the $\mathfrak{m}R^*$-adic completion of R^*, and so $\widehat{R} = \widehat{A} = \widehat{B}$. Ratliff's Equidimension Theorem 3.25 states that a Noetherian local domain is universally catenary if and only if its completion is equidimensional. By assumption R is universally catenary, and so $\widehat{R}$ is equidimensional by Ratliff's Theorem 3.25. Thus, if A is Noetherian, then A is also universally catenary. If B is Noetherian, then $B = A$, by Noetherian Flatness Theorem 6.3, and so B is universally catenary. □

Exercise

(1) Let x be a regular element of a commutative ring S. Prove that S is Noetherian if $S[1/x]$ and $(1 + xS)^{-1}S$ are both Noetherian.

SUGGESTION. Apply the proof of item 4 of Lemma 6.2.

CHAPTER 7

The flat locus of an extension of polynomial rings

Let R be a Noetherian ring, let n be a positive integer and let $z_1, \dots, z_n$ be indeterminates over R. This chapter concerns the flat locus of an extension φ of polynomial rings of the form

$$(7.01) \qquad S := R[f_1, \dots, f_m] \overset{\varphi}{\hookrightarrow} R[z_1, \dots, z_n] =: T,$$

where the f_j are polynomials in $R[z_1, \dots, z_n]$ that are algebraically independent over R, as in Definition 2.2. Our aim is to provide a self-contained discussion of the Jacobian ideal (Definition 7.17) and related topics, and their relation to the flat locus of φ in Equation 7.01. We are motivated to examine the flat locus of the extension φ by the flatness condition of Theorem 6.17 in the Insider Construction of Section 6.2.

Section 7.1 contains a general result on flatness. Section 7.2 concerns the Jacobian ideal of the map $\varphi : S \hookrightarrow T$ of (7.01) and the non-smooth and non-flat loci of this map. In Section 7.3 we discuss applications to polynomial extensions. Related results are given in the papers of Picavet [**149**] and Wang [**182**].

7.1. Flatness criteria

Recall that a Noetherian local ring $(R, \mathfrak{m})$ of dimension d is *Cohen-Macaulay* if there exist elements $x_1, \dots, x_d$ in $\mathfrak{m}$ that form a regular sequence as defined in Chapter 2.

The following definition is useful in connection with what is called the "local flatness criterion" [**122**, page 173].

DEFINITION 7.1. Let I be an ideal of a ring A.

(1) An A-module N is *separated* for the I-adic topology if $\bigcap_{n=1}^{\infty} I^n N = (0)$.
(2) An A-module M is said to be *I-adically ideal-separated* if $\mathfrak{a} \otimes M$ is separated for the I-adic topology for every finitely generated ideal $\mathfrak{a}$ of A.

REMARK 7.2. In Theorem 7.3, we use the following result on flatness. Let I be an ideal of a Noetherian ring A and let M be an I-adically ideal-separated A-module. By [**122**, part (1) $\iff$ (3) of Theorem 22.3], M is A-flat $\iff$ the following two conditions hold: (a) $I \otimes_A M \cong IM$, and (b) M/IM is (A/I)-flat.

Theorem 7.3 is a general result on flatness involving the Cohen-Macaulay property and a trio of Noetherian local rings.

THEOREM 7.3. *Let $(R, \mathfrak{m}), (S, \mathfrak{n})$ and $(T, \mathfrak{c})$ be Noetherian local rings, and assume there exist local maps:*

$$R \longrightarrow S \longrightarrow T,$$

such that

(i) $R \to T$ *is flat and* $T/\mathfrak{m}T$ *is a Cohen-Macaulay ring, and*
(ii) $R \to S$ *is flat and* $S/\mathfrak{m}S$ *is a regular local ring.*

Then the following statements are equivalent:

(1) $S \to T$ *is flat.*
(2) *For each prime ideal* $\mathfrak{q}$ *of* T, *we have* $\operatorname{ht}\mathfrak{q} \geq \operatorname{ht}(\mathfrak{q} \cap S)$.
(3) *For each prime ideal* $\mathfrak{q}$ *of* T *such that* $\mathfrak{q}$ *is minimal over* $\mathfrak{n}T$, *we have* $\operatorname{ht}(\mathfrak{q}) \geq \operatorname{ht}(\mathfrak{n})$.

PROOF. The implication (2) $\Longrightarrow$ (3) is obvious and the implication (1) $\Longrightarrow$ (2) is clear by Remark 2.37.10. To prove (3) $\Longrightarrow$ (1), we observe that T is an $\mathfrak{m}S$-adically ideal-separated S-module, since T is a Noetherian local ring; see Definition 7.1 and Krull's Intersection Theorem 2.22. Hence, by Remark 7.2 with $A = S$, $I = \mathfrak{m}S$ and $M = T$, it suffices to show:

(a) $\mathfrak{m}S \otimes_S T \cong \mathfrak{m}T$.
(b) The map $S/\mathfrak{m}S \longrightarrow T/\mathfrak{m}T$ is faithfully flat.

PROOF OF (a). Since $R \hookrightarrow S$ is flat, we have $\mathfrak{m}S \cong \mathfrak{m}R \otimes_R S$. Therefore

$$\mathfrak{m}S \otimes_S T \cong (\mathfrak{m} \otimes_R S) \otimes_S T \cong \mathfrak{m} \otimes_R T \cong \mathfrak{m}T,$$

where the last isomorphism follows because the map $R \to T$ is flat.

PROOF OF (b). By assumption, $T/\mathfrak{m}T$ is Cohen-Macaulay and $S/\mathfrak{m}S$ is a regular local ring. We also have $T/\mathfrak{n}T = (T/\mathfrak{m}T) \otimes_{S/\mathfrak{m}S} (S/\mathfrak{n})$. By [**122**, Theorem 23.1], if

$$\dim(T/\mathfrak{m}T) = \dim(S/\mathfrak{m}S) + \dim(T/\mathfrak{n}T), \tag{7.3.c}$$

then $S/\mathfrak{m}S \to T/\mathfrak{m}T$ is flat. Thus to prove (3) $\Longrightarrow$ (1), it suffices to establish Equation 7.3.c.

In order to prove Equation 7.3.c, we may reduce to the case where $\mathfrak{m} = 0$. Thus we may assume that R is a field, S is an RLR and T is a Cohen-Macaulay local ring. Let $\mathfrak{q} \in \operatorname{Spec} T$ be such that $\mathfrak{n}T \subseteq \mathfrak{q}$. Since the map $S \to T$ is a local homomorphism, we have $\mathfrak{q} \cap S = \mathfrak{n}$. By [**122**, Theorem 15.1i] we have

$$\operatorname{ht}(\mathfrak{q}) \;\leq\; \operatorname{ht}(\mathfrak{n}) \;+\; \dim(T_{\mathfrak{q}}/\mathfrak{n}T_{\mathfrak{q}}).$$

If $\mathfrak{q}$ is minimal over $\mathfrak{n}T$, then $\dim(T_{\mathfrak{q}}/\mathfrak{n}T_{\mathfrak{q}}) = 0$, and hence $\operatorname{ht}\mathfrak{q} \leq \operatorname{ht}\mathfrak{n}$. By statement 3, $\operatorname{ht}(\mathfrak{q}) \geq \operatorname{ht}(\mathfrak{n})$, and therefore $\operatorname{ht}(\mathfrak{q}) = \operatorname{ht}(\mathfrak{n})$, for every minimal prime divisor $\mathfrak{q}$ of $\mathfrak{n}T$. Thus $\operatorname{ht}(\mathfrak{n}) = \operatorname{ht}(\mathfrak{n}T)$.

Since T is Cohen-Macaulay and hence is catenary, we have

$$\dim(T/\mathfrak{n}T) = \dim(T) - \operatorname{ht}(\mathfrak{n}T) = \dim(T) - \operatorname{ht}(\mathfrak{n})$$

Thus $\dim T = \dim S + \dim(T/\mathfrak{n}T)$, as desired. $\square$

In Theorem 7.4 we present a result closely related to Theorem 7.3 with a Cohen-Macaulay hypothesis on all the fibers of $R \to T$ and a regularity hypothesis on all the fibers of $R \to S$. A ring homomorphism $f : A \to B$ of Noetherian rings has *Cohen-Macaulay fibers* with respect to f if, for every $P \in \operatorname{Spec} A$, the ring $B \otimes_A k(P)$ is Cohen-Macaulay, where $k(P)$ is the field of fractions of A/P. For more information about the fibers of a map, see Discussion 3.29 and Definition 3.38.

THEOREM 7.4. *Let $(R, \mathfrak{m}), (S, \mathfrak{n})$ and $(T, \mathfrak{e})$ be Noetherian local rings, and assume there exist local maps:*

$$R \longrightarrow S \longrightarrow T,$$

such that

(i) *$R \to T$ is flat with Cohen-Macaulay fibers, and*
(ii) *$R \to S$ is flat with regular fibers.*

Then the following statements are equivalent:

(1) *$S \to T$ is flat with Cohen-Macaulay fibers.*
(2) *$S \to T$ is flat.*
(3) *For each prime ideal $\mathfrak{q}$ of T, we have $\operatorname{ht}\mathfrak{q} \geq \operatorname{ht}(\mathfrak{q} \cap S)$.*
(4) *For each prime ideal $\mathfrak{q}$ of T such that $\mathfrak{q}$ is minimal over $\mathfrak{n}T$, we have $\operatorname{ht}\mathfrak{q} \geq \operatorname{ht}\mathfrak{n}$.*

PROOF. The implications (1) $\Longrightarrow$ (2) and (3) $\Longrightarrow$ (4) are obvious and the implication (2) $\Longrightarrow$ (3) is clear by Remark 2.37.10. By Theorem 7.3, item 4 implies that $S \to T$ is flat.

To show Cohen-Macaulay fibers for $S \to T$, it suffices to show, for each prime ideal $\mathfrak{q}$ of T, if $\mathfrak{p} := \mathfrak{q} \cap S$ then $T_\mathfrak{q}/\mathfrak{p}T_\mathfrak{q}$ is Cohen-Macaulay. Let $\mathfrak{q} \cap R = \mathfrak{a}$. By passing to $R/\mathfrak{a} \subseteq S/\mathfrak{a}S \subseteq T/\mathfrak{a}T$, we may assume $\mathfrak{q} \cap R = (0)$. Let $\operatorname{ht}\mathfrak{p} = n$. Since $R \to S_\mathfrak{p}$ has regular fibers and $\mathfrak{p} \cap R = (0)$, $S_\mathfrak{p}$ is an RLR, and the ideal $\mathfrak{p}S_\mathfrak{p}$ is generated by n elements. Moreover, faithful flatness of the map $S_\mathfrak{p} \to T_\mathfrak{q}$ implies that the ideal $\mathfrak{p}T_\mathfrak{q}$ has height n by Remark 2.37.10. Since $T_\mathfrak{q}$ is Cohen-Macaulay, a set of n generators of $\mathfrak{p}S_\mathfrak{p}$ forms a regular sequence in $T_\mathfrak{q}$. Hence $T_\mathfrak{q}/\mathfrak{p}T_\mathfrak{q}$ is Cohen-Macaulay [**122**, Theorems 17.4 and 17.3]. □

REMARK 7.5. Let $\varphi : C \hookrightarrow E$ be a faithfully flat local homomorphism of Cohen-Macaulay local rings. Then φ has Cohen-Macaulay fibers. To see this, let $\mathfrak{q} \in \operatorname{Spec} E$ and let $\mathfrak{p} = \mathfrak{q} \cap C \in \operatorname{Spec} C$; then the fiber $E_\mathfrak{q}/\mathfrak{p}E_\mathfrak{q}$ is Cohen-Macaulay, by [**122**, Corollary, page 181].

Since flatness is a local property by Remark 2.37.4, the following two corollaries are immediate from Theorem 7.4; see also [**149**, Théorème 3.15].

COROLLARY 7.6. *Let T be a Noetherian ring and let $R \subseteq S$ be Noetherian subrings of T. Assume that $R \to T$ is flat with Cohen-Macaulay fibers and that $R \to S$ is flat with regular fibers. Then $S \to T$ is flat if and only if, for each prime ideal $\mathfrak{q}$ of T, we have $\operatorname{ht}\mathfrak{q} \geq \operatorname{ht}(\mathfrak{q} \cap S)$.*

As a special case of Corollary 7.6, we have:

COROLLARY 7.7. *Let R be a Noetherian ring and let $z_1, \dots, z_n$ be indeterminates over R. Assume that $f_1, \dots, f_m \in R[z_1, \dots, z_n]$ are algebraically independent over R, and let $\varphi : S := R[f_1, \dots, f_m] \hookrightarrow T := R[z_1, \dots, z_n]$. Then:*

(1) *φ is flat if and only if $\operatorname{ht}(\mathfrak{q}) \geq \operatorname{ht}(\mathfrak{q} \cap S)$, for every prime ideal $\mathfrak{q}$ of T.*
(2) *For $\mathfrak{q} \in \operatorname{Spec} T$, $\varphi_\mathfrak{q} : S \to T_\mathfrak{q}$ is flat if and only if $\operatorname{ht}\mathfrak{p} \geq \operatorname{ht}(\mathfrak{p} \cap S)$, for every prime ideal $\mathfrak{p} \subseteq \mathfrak{q}$ of T.*

PROOF. Since S and T are polynomial rings over R, and $T_\mathfrak{q}$ is a localization of T, the maps $R \to S$, $R \to T$ and $R \to T_\mathfrak{q}$ are flat with regular fibers. Hence both assertions follow from Corollary 7.6. □

REMARK 7.8. Assume the notation of Inclusion Construction 5.3 and Definition 5.7; thus R is an integral domain and B is the constructed Approximation Domain.

(1) If R is a Noetherian local domain and B is Noetherian, then $B^* = R^*$ is faithfully flat over B, and so $\dim B = \dim R^* = \dim R$.
(2) If R is a Cohen-Macaulay local domain and $\dim B = \dim R$, then B is Noetherian by Theorem 7.9 below.

Theorem 7.9 is proved by applying Theorem 7.4 to Inclusion Construction 5.3, and using Noetherian Flatness Theorem 6.3.1.

THEOREM 7.9. *Let x be a nonzero nonunit of a Cohen-Macaulay local domain R. Let $\tau_1, \dots, \tau_s \in xR^*$ be algebraically independent over R as in Inclusion Construction* 5.3, *and let B be the Approximation Domain associated to $\tau_1, \dots, \tau_s$, as in Definition* 5.7. *Then* $\dim B = \dim R \iff B$ *is Noetherian.*

PROOF. The $\Longleftarrow$ direction is shown in Remark 7.8.1. For the $\Longrightarrow$ direction, it suffices to show $B \hookrightarrow R^*[1/x]$ is flat, by Noetherian Flatness Theorem 6.3.1. Consider the inclusion maps α and β:

$$R \overset{\alpha}{\hookrightarrow} B[1/x] \overset{\beta}{\hookrightarrow} R^*[1/x].$$

By Corollary 5.14, $B[1/x]$ is Noetherian and $\alpha : R \hookrightarrow B[1/x]$ is flat with regular fibers.

Let $\mathfrak{p} \in \operatorname{Spec} B$ be maximal with respect to $x \notin \mathfrak{p}$, and set $\mathfrak{p}_0 = \mathfrak{p} \cap R$. Then $\dim(B/\mathfrak{p}) = 1 = \dim(R/\mathfrak{p}_0)$ by Theorem 5.18.1. Since $x \notin \mathfrak{p}_0$ and $\dim(R/\mathfrak{p}_0) = 1$, $\mathfrak{p}_0$ is maximal in R with respect to $x \notin \mathfrak{p}_0$. Let $\mathfrak{q} \in \operatorname{Spec} R^*$ be such that $\mathfrak{q} \cap B = \mathfrak{p}$. Then $x \notin \mathfrak{q}$. Since $\mathfrak{q} \cap R = \mathfrak{p}_0$ and $R^*/\mathfrak{p}_0 R^*$ is the x-adic completion of $R/\mathfrak{p}_0$, it follows that $\dim(R/\mathfrak{p}_0) = 1 = \dim(R^*/\mathfrak{p}_0 R^*)$. Therefore $\mathfrak{q}$ is minimal over $\mathfrak{p}_0 R^*$, for every $\mathfrak{q} \in \operatorname{Spec} R^*$ such that $\mathfrak{q} \cap B = \mathfrak{p}$.

The maps $\alpha_{\mathfrak{p}}$ and $\beta_{\mathfrak{q}} \circ \alpha_{\mathfrak{p}}$ shown below are faithfully flat local homomorphisms:

$$R_{\mathfrak{p}_0} \overset{\alpha_{\mathfrak{p}}}{\hookrightarrow} B_{\mathfrak{p}} \overset{\beta_{\mathfrak{q}}}{\hookrightarrow} R^*_{\mathfrak{q}}.$$

Since $R_{\mathfrak{p}_0}$ and $R^*_{\mathfrak{q}}$ are Cohen-Macaulay, Remark 7.5 implies that the map $\beta_{\mathfrak{q}} \circ \alpha_{\mathfrak{p}}$ has Cohen-Macaulay fibers. Since $\alpha : R \hookrightarrow B[1/x]$ is flat with regular fibers, $\alpha_{\mathfrak{p}}$ is faithfully flat with regular fibers.

Now Theorem 7.4, (2) $\iff$ (4), implies that $\beta_{\mathfrak{q}} : B_{\mathfrak{p}} \hookrightarrow R^*_{\mathfrak{q}}$ is flat if and only if $\operatorname{ht} \mathfrak{q}_1 \geq \operatorname{ht} \mathfrak{p}$, for each prime ideal $\mathfrak{q}_1 \in \operatorname{Spec}(R^*_{\mathfrak{q}})$ that is minimal over $\mathfrak{p}R^*_{\mathfrak{q}}$. From the above paragraph, $\mathfrak{q}$ itself is minimal over $\mathfrak{p}R^*_{\mathfrak{q}}$. Therefore it is enough to check that $\operatorname{ht} \mathfrak{q} \geq \operatorname{ht} \mathfrak{p}$.

Let $d = \dim R = \dim B = \dim R^*$. Since R is catenary, Theorem 5.18.2 implies that $\operatorname{ht} \mathfrak{p}_0 = d - 1 = \operatorname{ht} \mathfrak{q}$ and $d - 1 \leq \operatorname{ht} \mathfrak{p}$. Since $\dim B = d$ and $x \notin \mathfrak{p}$, it follows that $\operatorname{ht} \mathfrak{p} = d - 1$. That is, $\operatorname{ht} \mathfrak{q} = \operatorname{ht} \mathfrak{p}$.

This holds for every $\mathfrak{q} \in \operatorname{Spec} R^*$ minimal over $\mathfrak{p}R^*$. Thus $B \hookrightarrow R^*[1/x]$ is flat, as desired. □

QUESTION 7.10. Does the conclusion of Theorem 7.9 hold for every Noetherian local domain (not necessarily Cohen-Macaulay)?

7.2. The Jacobian ideal and the smooth and flat loci

We use the following definitions as in Swan [**174**].

DEFINITION 7.11. Let R be a ring. An R-algebra A is said to be *quasi-smooth* over R if for every R-algebra B and ideal N of B with $N^2 = 0$, every R-algebra homomorphism $g : A \to B/N$ lifts to an R-algebra homomorphism $f : A \to B$. In the commutative diagram below, let the maps $\theta : R \to A$ and $\psi : R \to B$ be the canonical ring homomorphisms that define A and B as R-algebras and let the map $\pi : B \to B/N$ be the canonical quotient ring map

(7.11.1)

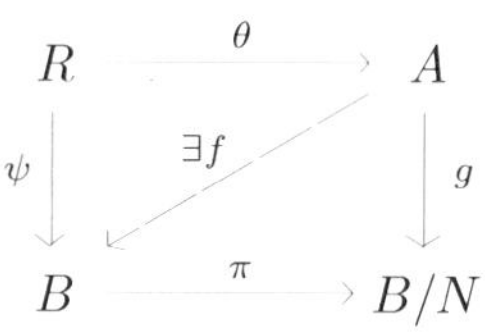

If A is quasi-smooth over R, then there exists an R-algebra homomorphism f from A to B such that $\pi \circ f = g$. If A is finitely presented and quasi-smooth over R, then A is said to be *smooth* over R. If A is essentially finitely presented and quasi-smooth over R, then A is said to be *essentially smooth* over R; see Chapter 2 for the definitions of finitely presented and essentially finitely presented.

The terminology for smoothness varies. Matsumura [**122**, p. 193] uses the term 0-*smooth* for what Swan calls "quasi-smooth". Others such as Tanimoto [**176**], [**177**] use *smooth* for "quasi-smooth".

Recall from Definition 3.41 that a homomorphism $\sigma : R \to A$ of Noetherian rings is regular if σ is flat and has geometrically regular fibers. To avoid any possible confusion in the case where R is a field, Swan in [**174**] calls such a homomorphism σ-geometrically regular.

Regularity and smoothness are the same for morphisms of Noetherian rings of finite type, as is stated in Theorem 7.12.

THEOREM 7.12. [**174**, Corollary 1.2] *Let $\sigma : R \to A$ be a homomorphism of Noetherian rings with A a finitely generated R-algebra. Then the following are equivalent:*

(1) *σ is regular.*
(2) *σ is smooth, that is, A is a smooth R-algebra.*

PROOF. This follows from [**174**, Corollary 1.2]. □

Swan's article [**174**] gives a detailed presentation of D. Popescu's proof that a regular morphism of Noetherian rings is a filtered colimit of smooth morphisms. However, even if R is a field and the R-algebra A is a Noetherian ring, the map $\sigma : R \hookrightarrow A$ may be a regular morphism but not be quasi-smooth. Tanimoto shows in [**176**, Lemma 2.1] that, for a field k and an indeterminate x over k, the regular morphism $k \longrightarrow k[[x]]$ is quasi-smooth as in Definition 7.11 if and only if k has characteristic $p > 0$ and $[k : k^p] < \infty$.

DEFINITIONS 7.13. Let R be a ring and let A be an R-algebra; say $A = R[Z]/I$, where $Z = \{z_\gamma\}_{\gamma \in \Gamma}$ is a set of indeterminates over R indexed by a possibly infinite index set Γ and I is an ideal of the polynomial ring $R[Z]$.

(1) We define $F := \bigoplus_{\gamma\in\Gamma} Adz_\gamma$ to be the free A-module on a basis $\{dz_\gamma\}_{\gamma\in\Gamma}$; this basis is to be in 1−1 correspondence with the set $\{z_\gamma\}_{\gamma\in\Gamma}$. Define $D : R[Z] \to F$ by $D(f) = \sum_{\gamma\in\Gamma} \frac{\partial f}{\partial z_\gamma} dz_\gamma$, for every $f \in I$, where $\frac{\partial}{\partial z_\gamma}$ is the usual partial derivative function on $R[Z]$, with elements of $R[Z \setminus \{z_\gamma\}]$ considered to be "constants". The map D is a *derivation* in the sense that D is an R-module homomorphism and

$$D(fg) \;=\; gD(f) \;+ fD(g), \qquad \text{for every} \qquad f, g \in I.$$

We have $D(I^2) = (0)$, since $D(I^2) \subseteq IF = (0)$. Hence D induces a map d, called the *differential morphism* on I/I^2, such that

$$d : I/I^2 \;\to\; F = \bigoplus_{\gamma\in\Gamma} Adz_\gamma \qquad \text{and} \qquad d(f + I^2) \;=\; \sum_{\gamma\in\Gamma} \frac{\partial f}{\partial z_\gamma} dz_\gamma.$$

The differential morphism d is an A-linear map, since, for each $a \in A$, each $f \in I$, and each z_γ, we have $\frac{\partial(af)}{\partial z_\gamma} = a\frac{\partial f}{\partial z_\gamma} + f\frac{\partial a}{\partial z_\gamma}$, and $f\frac{\partial a}{\partial z_\gamma}$ is in $IF = (0)$. See [**122**, p.190-2] for more discussion about derivations and differentials.

(2) Let $Z = \{z_1, \dots, z_n\}$ be a finite set of inderminates and let $g_1, \dots, g_s$ be elements of I, where $n, s \in \mathbb{N}$. Define the *Jacobian matrix* of the g_i with respect to the z_j to be the $s \times n$ matrix

$$\mathfrak{J}(g_1, \dots, g_s; z_1, \dots, z_n) \;:=\; \left(\frac{\partial g_i}{\partial z_j}\right)_{1\le i\le s,\ 1\le j\le n}.$$

For $s \le n$, define the *Delta ideal* of the g_i, $\Delta(g_1, \dots, g_s)$, to be the ideal of A generated by the images in A of the $s \times s$ minors of the Jacobian matrix $\mathfrak{J}(g_1, \dots, g_s; z_1, \dots, z_n)$. If $s = 0$, we set $\Delta(\ ) = A$.

(3) Assume that A is a finitely presented R-algebra. Then we may assume that $Z = \{z_1, \dots, z_n\}$ is a finite set, and there exist $f_1, \dots, f_m$ in $R[Z]$ such that $I = (f_1, \dots, f_m)R[Z]$.

Define the *Elkik ideal* $\widetilde{H}$ of the ring A as an R-algebra to be

$$\text{(7.13.a)} \quad \widetilde{H} \;:=\; \widetilde{H}_{A/R} := \sqrt{\Big(\sum_{g_1,\dots,g_s} (\Delta(g_1, \dots, g_s) \cdot \big[(g_1, \dots, g_s) :_{R[Z]} I\big] + I)A\Big)},$$

where $\sqrt{\ }$ denotes the radical of the enclosed ideal, and the sum is taken over all choices of s polynomials $g_, \dots, g_s$ from the ideal I, for all $s \le n$; see Elkik[**46**, p. 555] and Swan [**174**, Section 4]. Swan mentions that the Elkik ideal provides a "very explicit definition" for the non-smooth locus of A as an R-algebra; see Theorem 7.15.

Define a simpler ideal H that is similar to $\widetilde{H}$ as follows:

$$\text{(7.13.b)} \quad H \;:=\; H_{A/R} \;:=\; \sqrt{\Big(\sum_{g_1,\dots,g_s} (\Delta(g_1, \dots, g_s) \cdot \big[(g_1, \dots, g_s) :_{R[Z]} I\big] + I)A\Big)},$$

where the sum is taken over all subsets $\{g_1, \dots, g_s\}$, for all $1 \le s \le \min(m, n)$, of the given finite set $\{f_1, \dots, f_m\}$ of generators of I. It is clear that $H \subseteq \widetilde{H}$. We show in Theorem 7.15 that $H = \widetilde{H}$.

The following theorem from Swan's article [**174**] connects quasi-smoothness of an R-algebra A to the differential morphism d of Definition 7.13.1 being a split monomorphism.

THEOREM 7.14. [**174**, Parts of Theorem 3.4] *Let R be a ring and let A be an R-algebra $A := R[Z]/I$, where $Z = \{z_\gamma\}_{\gamma\in\Gamma}$ is a possibly infinite set of indeterminates over R, and I is an ideal of the polynomial ring $R[Z]$. Then the following two statements are equivalent:*

(1) *$R \to A$ is quasi-smooth.*
(2) *The differential morphism $d : I/I^2 \to \bigoplus_{\gamma\in\Gamma} Adz_\gamma$ is a split monomorphism.*

Theorem 7.15 is a modification of [**174**, Theorem 4.1], with the Elkik ideal $\widetilde{H}$ replaced by the simpler ideal H of Equation 7.13.b. The proof of Theorem 7.15 shows that H defines the non-smooth locus of A and that $\widetilde{H} = H$. For the proof we adapt Swan's elegant argument. We call this theorem the *Elkik-Swan* Theorem.

THEOREM 7.15. The Elkik-Swan Theorem. [**174**, Theorem 4.1] *Let A be a finitely presented algebra over a ring R. Write $A = R[Z]/I$, where $Z = \{z_1, \dots, z_n\}$ and $I = (f_1, \dots, f_m)R[Z]$ are as in Definition 7.13. Let H be the ideal of A defined in Definition 7.13.b, and let P be a prime ideal of A. Then:*

(1) *A_P is essentially smooth over R if and only if H is not contained in P.*
(2) *H is the intersection of all $P \in \operatorname{Spec} A$ such that A_P is not essentially smooth over R.*
(3) *H is independent of the choice of presentation.*
(4) *The Elkik ideal $\widetilde{H}$ describes the non-smooth locus of A as an R-algebra and $\widetilde{H} = H$.*

PROOF. Let $P \in \operatorname{Spec} A$ and assume that H is not contained in P. Then some summand in the expression for H is not contained in P. By reordering the set $\{f_1, \dots, f_m\}$, we let $\{f_1, \dots, f_r\}$ denote the subset associated with the summand not contained in P, where $r \le m$. Thus $(\Delta(f_1, \dots, f_r)\cdot[(f_1, \dots, f_r) :_{R[Z]} I])A$ is not contained in P. Let Q be the pre-image of P in $R[Z]$. Then $[(f_1, \dots, f_r) :_{R[Z]} I]$ is not contained in Q. Therefore $(f_1, \dots, f_r)R[Z]_Q = IR[Z]_Q = I_Q$, and so the images of $f_1, \dots, f_r$ generate $(I/I^2)_P = I_Q/I_Q^2$. Also $\Delta(f_1, \dots, f_r)A$ is not contained in P. Hence the image of some $r\times r$ minor of $\mathfrak{J}(f_1, \dots, f_r; z_1, \dots, z_n)$ is not contained in P. By reordering $\{z_1, \dots, z_n\}$, we may assume that the image in A of $\det(\frac{\partial f_i}{\partial z_j})_{1\le i,j\le r}$ is not contained in P. Thus $\det(\frac{\partial f_i}{\partial z_j})_{1\le i,j\le r}$ is a unit of A_P and so the matrix $(\frac{\partial f_i}{\partial z_j})_{1\le i,j\le r}$ is invertible as a matrix with coefficients in A_P.

Consider the following diagram:

(7.15.1)
$$\begin{array}{ccc} & & \bigoplus_{j=1}^n A_P dz_j \\ & \overset{d_P}{\nearrow} & \downarrow p \\ (I/I^2)_P & \xrightarrow{\widetilde{d}=p\circ d_P} & \bigoplus_{j=1}^r A_P dz_j \end{array}$$

Here d is the map of Definition 7.13.1 so that $d_P(f + I^2) = \sum_{j=1}^n \frac{\partial f}{\partial z_j} dz_j$, for every $f \in I$, and p is the projection $p(\sum_{i=1}^n a_i dz_i) = \sum_{i=1}^r a_i dz_i$ on the first r summands. We let $\widetilde{d} = p \circ d_P$.

CLAIM 7.16. The map d_P has a left inverse, and so d_P is a split monomorphism.

PROOF. *Of claim* 7.16. The definition $\widetilde{d} = p \circ d_P$ implies $\widetilde{d}(f_i) = \sum_{i=1}^{r} \frac{\partial f_i}{\partial z_j} dz_j$. Since $(\frac{\partial f_i}{\partial z_j})_{1 \le i,j \le r}$ is an invertible matrix over A_P, the set $\{\widetilde{d}(f_1), \dots, \widetilde{d}(f_r)\}$ forms a basis of $\bigoplus_{i=1}^{r} A_P dz_i$. Let id_r denote the identity map on $\bigoplus_{i=1}^{r} A_P dz_i$.

There exists an A_P-linear map $g : \bigoplus_{i=1}^{r} A_P dz_i \to (I/I^2)_P$ defined on this basis; set $g(\widetilde{d}(f_i)) = f_i$, for every i with $1 \le i \le r$. Thus $\mathrm{id}_r = g \circ \widetilde{d} = g \circ p \circ d_P$, and so $g \circ p$ is a left inverse of d_P. Therefore d_P is a split monomorphism. □

Return to the proof of Theorem 7.15. By Theorem 7.14, A_P is an essentially smooth R-algebra.

Conversely, if A_P is an essentially smooth R-algebra, d_P is a split monomorphism by Theorem 7.14. Thus $(I/I^2)_P$ is free, say of rank r. By relabeling, we assume that $f_1, \dots, f_r$ map to a basis of $(I/I^2)_P = I_Q/I_Q^2$. By Nakayama's lemma, these elements generate I_Q, and so $[(f_1, \dots, f_r) :_{R[Z]} I]$ is not contained in Q.

We identify $(A_P)^r$ and $(I/I^2)_P$ by the isomorphism $\alpha(a_1, \dots, a_r) = \sum a_i f_i$. Then the map $d_p : (I/I^2)_P \longrightarrow \oplus A_P dz_i$ can be identified with the linear map

$$d_P : (A_P)^r \longrightarrow \oplus A_P dz_i$$

given by the Jacobian matrix $(\frac{\partial f_i}{\partial z_j})_{1 \le i \le r; 1 \le j \le n}$. Since d_p is split, the induced map

$$\overline{d_P} : \ (A_P/PA_P)^r \longrightarrow \oplus (A_P/PA_P) dz_i$$

remains injective. Thus some $r \times r$-minor of $(\frac{\partial f_i}{\partial z_j})_{1 \le i \le r; 1 \le j \le n}$ is invertible in A_P.

Since H is a radical ideal, and every prime ideal $P \in \operatorname{Spec} A$ containing H is such that A_P is not essentially smooth, we see that H equals the intersection given in the second statement of Theorem 7.15. Since every presentation ideal I and generating set $f_1, \dots, f_m$ of I yield that H equals the same intersection of prime ideals, the ideal H is independent of presentation.

For item 4, Swan's Theorem in [**174**, Theorem 4.1] shows that $\widetilde{H}$ is the same intersection as H. Thus H is equal to the Elkik ideal $\widetilde{H}$. □

We return to the extension φ of polynomial rings from Equation 7.01

$$S \ := \ R[f_1, \dots, f_m] \ \overset{\varphi}{\hookrightarrow} \ R[z_1, \dots, z_n] \ =: \ T,$$

where the f_j are polynomials in $R[z_1, \dots, z_n]$ that are algebraically independent over R.

DEFINITIONS AND REMARKS 7.17. (1) The *Jacobian ideal* J of the extension $S \hookrightarrow T$ is the ideal of T generated by the $m \times m$ minors of the $m \times n$ matrix $\mathfrak{J}$ defined as follows:[1]

$$\mathfrak{J} := \left(\frac{\partial f_i}{\partial z_j} \right)_{1 \le i \le m, \ 1 \le j \le n}.$$

(2) As in Definition 2.40, the non-flat locus of $\varphi : S \hookrightarrow T$ is the set

$$\mathcal{F} := \{ \mathfrak{q} \in \operatorname{Spec}(T) \mid \text{the map} \quad \varphi_{\mathfrak{q}} : S \to T_{\mathfrak{q}} \text{ is not flat } \}.$$

We define the set $\mathcal{F}_{\min}$ and the ideal F of T as follows:

$$\mathcal{F}_{\min} := \{ \text{ minimal elements of } \mathcal{F} \} \ \text{ and } \ F := \bigcap \{ \mathfrak{q} \mid \mathfrak{q} \in \mathcal{F} \}.$$

[1]For related information on the Jacobian ideal of an algebra over a ring, see [**175**, Section 4.4, p. 65].

By Theorem 2.42, the set $\mathcal{F}$ is closed in the Zariski topology on $\operatorname{Spec} T$. Hence

$$\mathcal{F} \;=\; \mathcal{V}(F) \;:=\; \{\mathfrak{q} \in \operatorname{Spec} T \mid F \subseteq \mathfrak{q}\ \}.$$

Thus the set $\mathcal{F}_{\min}$ is a finite set and is equal to the set $\operatorname{Min} F$ of minimal primes of the ideal F of T. The ideal F defines the non-flat locus of φ.

By Remark 2.37.10, a flat homomorphism has the Going-Down property. Corollary 7.7 implies that

(i) $\operatorname{Min} F \subseteq \{\mathfrak{q} \in \operatorname{Spec} T \mid \operatorname{ht}\mathfrak{q} < \operatorname{ht}(\mathfrak{q} \cap S)\}$, and

(ii) If $\mathfrak{q} \in \operatorname{Min} F$, then every prime ideal $\mathfrak{q}' \subsetneq \mathfrak{q}$ satisfies $\operatorname{ht}\mathfrak{q}' \geq \operatorname{ht}(\mathfrak{q}' \cap S)$.

EXAMPLE AND REMARKS 7.18. (1) Let k be a field, let x and y be indeterminates over k and set $f = x$, $g = (x-1)y$. Then $k[f,g] \overset{\varphi}{\hookrightarrow} k[x,y]$ is not flat.

PROOF. For the prime ideal $P := (x-1) \in \operatorname{Spec}(k[x,y])$, we see that $\operatorname{ht}(P) = 1$, but $\operatorname{ht}(P \cap k[f,g]) = 2$; thus the extension is not flat by Corollary 7.7. □

(2) The Jacobian ideal J of f and g in (1) is given by:

$$J \;=\; (\det \begin{pmatrix} \frac{\partial f}{\partial x} & \frac{\partial f}{\partial y} \\ \frac{\delta g}{\partial x} & \frac{\partial g}{\partial y} \end{pmatrix})k[x,y] \;=\; (\det \begin{pmatrix} 1 & 0 \\ y & x-1 \end{pmatrix})k[x,y] \;=\; (x-1)k[x,y].$$

(3) In the example of item 1, the non-flat locus is equal to the set of prime ideals Q of $k[x,y]$ that contain the Jacobian ideal $(x-1)k[x,y]$, thus $J = F$.

(4) One can also describe the example of item 1 by taking the base ring R to be the polynomial ring $k[x]$ rather than the field k. Then both $T = R[y]$ and $S = R[g]$ are polynomial rings in one variable over R with $g = (x-1)y$. The Jacobian ideal J is the ideal of T generated by $\frac{\partial g}{\partial y} = x - 1$, so is the same as in item 1.

We record in Theorem 7.19 connections between the Jacobian ideal of the morphism $\varphi : S \hookrightarrow T$ of Equation 7.01 and the smoothness or flatness of localizations of φ.

THEOREM 7.19. *Let R be a Noetherian ring, let $z_1, \ldots, z_n$ be indeterminates over R, and let $f_1, \ldots, f_m \in R[z_1, \ldots, z_n]$ be algebraically independent over R. Consider the embedding $\varphi : S := R[f_1, \ldots, f_m] \hookrightarrow T := R[z_1, \ldots, z_n]$. Let J denote the Jacobian ideal of φ, and let F and* $\operatorname{Min} F$ *be as in Definitions and Remarks* 7.17. *Then:*

(1) $\mathfrak{q} \in \operatorname{Spec} T$ *does not contain J $\iff$ $\varphi_{\mathfrak{q}} : S \to T_{\mathfrak{q}}$ is essentially smooth. Thus J defines the non-smooth locus of φ.*

(2) *If $\mathfrak{q} \in \operatorname{Spec} T$ does not contain J, then $\varphi_{\mathfrak{q}} : S \to T_{\mathfrak{q}}$ is flat. Thus $J \subseteq F$.*

(3) $\operatorname{Min} F \subseteq \{\mathfrak{q} \in \operatorname{Spec} T \mid \operatorname{ht}(\mathfrak{q} \cap S) > \operatorname{ht}\mathfrak{q}\}$.

(4) $\operatorname{Min} F \subseteq \{\mathfrak{q} \in \operatorname{Spec} T \mid \operatorname{ht}\mathfrak{q} < \dim S$ *and* $\operatorname{ht}(\mathfrak{q} \cap S) > \operatorname{ht}\mathfrak{q}\}$.

(5) *φ is flat $\iff$ for every $\mathfrak{q} \in \operatorname{Spec}(T)$ such that* $\operatorname{ht}(\mathfrak{q}) < \dim S$, *we have* $\operatorname{ht}(\mathfrak{q} \cap S) \leq \operatorname{ht}(\mathfrak{q})$.

(6) *If* $\operatorname{ht} J \geq \dim S$, *then φ is flat.*

PROOF. For item 1, we show that, for our definition of the Jacobian ideal J given in Definition 7.17, the radical of J is the Elkik ideal of an extension given in Theorem 7.15. Using Theorem 7.15, we work with the simpler description H of the Elkik ideal given in Equation 7.13.b.

Let $u_1, \ldots, u_m$ be indeterminates over $R[z_1, \ldots, z_n]$ and identify

$$R[z_1, \ldots, z_n] \quad \text{with} \quad \frac{R[u_1, \ldots, u_m][z_1, \ldots, z_n]}{(\{u_i - f_i\}_{i=1,\ldots,m})}.$$

Since $u_1, \ldots, u_m$ are algebraically independent, the Jacobian ideal of the extension φ is the ideal J generated by the $m \times m$ minors of $\mathfrak{J} = \left(\frac{\partial f_i}{\partial z_j}\right)_{m,n}$, using this identification. We make this more explicit as follows.

Let $B := R[u_1, \ldots, u_m, z_1, \ldots, z_n]$ and $I = (\{f_i - u_i\}_{i=1,\ldots,m})B$. Consider the following commutative diagram

$$\begin{array}{ccc} S := R[f_1, \ldots, f_m] & \longrightarrow & T := R[z_1, \ldots, z_n] \\ \cong\big\downarrow & & \cong\big\downarrow \\ S_1 := R[u_1, \ldots, u_m] & \longrightarrow & T_1 := B/I \end{array}$$

To show that $H \subseteq \sqrt{(J)}$, let $g_1, \ldots, g_s \in \{f_1 - u_1, \ldots, f_m - u_m\}$. Notice that $f_1 - u_1, \ldots, f_m - u_m$ is a regular sequence in B. Thus, if $s < m$, we have $[(g_1, \ldots, g_s) :_B I] = (g_1, \ldots, g_s)B$. Thus the $m \times m$-minors of $\left(\frac{\partial f_i}{\partial z_j}\right)$ generate H up to radical, and so $H = \sqrt{J}$. Hence, by Theorem 7.15, for every prime ideal $\mathfrak{q}$ of T, $T_\mathfrak{q}$ is essentially smooth over S if and only if $\mathfrak{q}$ does not contain J.

For item 2, suppose $\mathfrak{q} \in \operatorname{Spec} T$ and $J \not\subseteq \mathfrak{q}$. Choose $h \in J \setminus \mathfrak{q}$ and consider the extension $\varphi_h : S \hookrightarrow T[1/h]$. By item 1, φ_h is smooth. Since a smooth map is flat [**174**, page 2], φ_h is flat. Thus $\varphi_\mathfrak{q} : S \hookrightarrow T_\mathfrak{q}$ is flat.

In view of Corollary 7.7 and Remarks 7.17.2, item 3 holds.

If $\operatorname{ht} \mathfrak{q} \geq \dim S$, then $\operatorname{ht}(\mathfrak{q} \cap S) \leq \dim S \leq \operatorname{ht} \mathfrak{q}$. Hence the set

$$\begin{aligned} &\{\mathfrak{q} \in \operatorname{Spec} T \mid \operatorname{ht}(\mathfrak{q} \cap S) > \operatorname{ht} \mathfrak{q}\} \\ = \ &\{\mathfrak{q} \in \operatorname{Spec} T \mid \operatorname{ht} \mathfrak{q} < \dim S \text{ and } \operatorname{ht}(\mathfrak{q} \cap S) > \operatorname{ht} \mathfrak{q}\}. \end{aligned}$$

Thus item 3 is equivalent to item 4.

The ($\implies$) direction of item 5 is clear [**122**, Theorem 9.5]. For the ($\impliedby$) direction of item 5 and for item 6, it suffices to show $\operatorname{Min} F$ is empty, and this holds by item 4. □

QUESTIONS AND REMARKS 7.20. Let notation be as in Theorem 7.19.

(1) Questions: What is the set $\mathcal{F}_{\min} = \operatorname{Min} F$? In particular, when is $J = F$ and when is $J \subsetneq F$?
(2) Example 7.18 is an example where $J = F$, whereas Examples 7.32 contains several examples where $J \subsetneq F$.
(3) If R is an integral domain, then the zero ideal is not in $\operatorname{Min} F$ and $F \neq \{0\}$.

COROLLARY 7.21. *Assume the setting and notation of Theorem 7.19. Then:*

(1) $\operatorname{Min} F = \{\mathfrak{q} \in \operatorname{Spec} T | \operatorname{ht}(\mathfrak{q} \cap S) > \operatorname{ht} \mathfrak{q}$ *and* $\operatorname{ht}(\mathfrak{p} \cap S) \leq \operatorname{ht}(\mathfrak{p})$, *for every* $\mathfrak{p} \in \operatorname{Spec} T$ *with* $\mathfrak{p} \subsetneq \mathfrak{q}\}$.
(2) *For each prime ideal* $\mathfrak{q}$ *of* $\operatorname{Min} F$, *there exist prime ideals* $\mathfrak{p}'$ *and* $\mathfrak{p}$ *of* S *with* $\mathfrak{p}' \subsetneq \mathfrak{p}$ *such that* $\mathfrak{q}$ *is a minimal prime of both* $\mathfrak{p}'T$ *and* $\mathfrak{p}T$.

PROOF. Item 1 follows from item 3 of Theorem 7.19, and the definition of the non-flat locus.

For item 2, let $\mathfrak{q} \in \operatorname{Min} F$, let $h := \operatorname{ht} \mathfrak{q}$, and let $\mathfrak{p} := \mathfrak{q} \cap S$. Then $\operatorname{ht} \mathfrak{p} > h$. There exists a prime ideal $\mathfrak{p}'$ of S with $\mathfrak{p}' \subsetneq \mathfrak{p}$ and $\operatorname{ht} \mathfrak{p}' = \operatorname{ht} \mathfrak{p} - 1 \geq h$. Then

$\mathfrak{p}'T \subseteq \mathfrak{q}' \subseteq \mathfrak{q}$, for some prime ideal $\mathfrak{q}'$ of T minimal over $\mathfrak{p}'T$. Suppose $\mathfrak{q}' \neq \mathfrak{q}$. Since $\mathfrak{q} \in \operatorname{Min} F$, $\operatorname{ht}(\mathfrak{q}' \cap S) \leq \operatorname{ht} \mathfrak{q}'$. Since $\mathfrak{p}' \subseteq \mathfrak{q}' \cap S$, it follows that

$$h \leq \operatorname{ht} \mathfrak{p}' \leq \operatorname{ht}(\mathfrak{q}' \cap S) \leq \operatorname{ht} \mathfrak{q}' < \operatorname{ht} \mathfrak{q} = h,$$

a contradiction. Thus $\mathfrak{q}$ is a minimal prime of $\mathfrak{p}'T$ and $\mathfrak{p}T$. □

Corollary 7.22 is immediate from Theorem 7.19.

COROLLARY 7.22. *Let k be a field, let $z_1, \ldots, z_n$ be indeterminates over k and let $f, g \in k[z_1, \ldots, z_n]$ be algebraically independent over k. Consider the embedding $\varphi : S := k[f, g] \hookrightarrow T := k[z_1, \ldots, z_n]$. Assume that the associated Jacobian ideal J is nonzero.*[2] *Then:*

(1) *For every* $\mathfrak{q} \in \operatorname{Min} F, \operatorname{ht} \mathfrak{q} = 1$.
(2) $\operatorname{Min} F \subseteq \{$*minimal primes* $\mathfrak{q}$ *of* J *with* $\operatorname{ht}(\mathfrak{q} \cap S) > \operatorname{ht} \mathfrak{q}\}$.
(3) φ *is flat* $\iff$ *for every height-one prime ideal* $\mathfrak{q} \in \operatorname{Spec} T$ *such that* $J \subseteq \mathfrak{q}$ *we have* $\operatorname{ht}(\mathfrak{q} \cap S) \leq 1$.
(4) *If* $\operatorname{ht} J \geq 2$, *then* φ *is flat.*

PROOF. Item 1 holds since $\dim S = 2$ and $\operatorname{ht} \mathfrak{q} < \dim S$ for every $\mathfrak{q} \in \operatorname{Min} F$. □

COROLLARY 7.23. *Assume the notation of Theorem 7.19. Then:*

(1) *If* $\mathfrak{q} \in \operatorname{Min} F$, *then* $\mathfrak{q}$ *is a non-maximal prime of* T.
(2) $\operatorname{Min} F \subseteq \{\mathfrak{q} \in \operatorname{Spec} T \mid J \subseteq \mathfrak{q}, \dim(T/\mathfrak{q}) \geq 1$ *and* $\operatorname{ht}(\mathfrak{q} \cap S) > \operatorname{ht} \mathfrak{q}\}$.
(3) φ *is flat* $\iff$ $\operatorname{ht}(\mathfrak{q} \cap S) \leq \operatorname{ht}(\mathfrak{q})$ *for every non-maximal* $\mathfrak{q} \in \operatorname{Spec}(T)$ *with* $J \subseteq \mathfrak{q}$.
(4) *If* $\dim R = d$ *and* $\operatorname{ht} J \geq d + m$, *then* φ *is flat.*

PROOF. For item 1, suppose $\mathfrak{q} \in \operatorname{Min} F$ is a maximal ideal of T. Then $\operatorname{ht} \mathfrak{q} < \operatorname{ht}(\mathfrak{q} \cap S)$ by Theorem 7.19.3. By localizing at $R \setminus (R \cap \mathfrak{q})$, we may assume that R is local with maximal ideal $\mathfrak{q} \cap R := \mathfrak{m}$. Since $\mathfrak{q}$ is maximal, $T/\mathfrak{q}$ is a field finitely generated over $R/\mathfrak{m}$. By the Hilbert Nullstellensatz [**122**, Theorem 5.3], $T/\mathfrak{q}$ is algebraic over $R/\mathfrak{m}$ and $\operatorname{ht} \mathfrak{q} = \operatorname{ht}(\mathfrak{m}) + n$. It follows that $\mathfrak{q} \cap S = \mathfrak{p}$ is maximal in S and $\operatorname{ht} \mathfrak{p} = \operatorname{ht} \mathfrak{m} + m$. The algebraic independence hypothesis for the f_i implies that $m \leq n$, and therefore that $\operatorname{ht} \mathfrak{p} \leq \operatorname{ht} \mathfrak{q}$. This contradiction proves item 1. Item 2 follows from Theorem 7.19.3 and item 1.

Item 3 follows from Theorem 7.19.5 and item 1, and item 4 follows from Theorem 7.19.6. □

As an immediate corollary to Theorem 7.19 and Corollary 7.23, we have:

COROLLARY 7.24. *Let R be a Noetherian ring, let $z_1, \ldots, z_n$ be indeterminates over R and let $f_1, \ldots, f_m \in R[z_1, \ldots, z_n]$ be algebraically independent over R. Consider the embedding $\varphi : S := R[f_1, \ldots, f_m] \hookrightarrow T := R[z_1, \ldots, z_n]$, let J be the Jacobian ideal of φ and let F be the radical ideal that describes the non-flat locus of φ as in Definition 7.17.2. Then:*

(1) $J \subseteq F$.
(2) *Either* $F = T$, *that is,* φ *is flat, or* $\dim(T/\mathfrak{q}) \geq 1$, *for every* $\mathfrak{q} \in \operatorname{Min} F$.
(3) $\operatorname{ht} \mathfrak{q} < \dim S$, *for every* $\mathfrak{q} \in \operatorname{Min} F$.

[2] This is automatic if the field k has characteristic zero.

7.3. Applications to polynomial extensions

The analysis of polynomial extensions of this section uses Setting 7.25.

SETTING 7.25. Let R be a commutative ring, let $z_1, \dots, z_n$ be indeterminates over R, and let $f_1, \dots, f_m \in R[z_1, \dots, z_n]$ be algebraically independent over R. Consider the embedding $\varphi : S := R[f_1, \dots, f_m] \hookrightarrow T := R[z_1, \dots, z_n]$.

Proposition 7.26 concerns the behavior of the extension $\varphi : S \hookrightarrow T$ with respect to prime ideals of R in Setting 7.25.

PROPOSITION 7.26. *Assume Setting* 7.25.

(1) *If* $\mathfrak{p} \in \operatorname{Spec} R$ *and* $\varphi_{\mathfrak{p}T} : S \to T_{\mathfrak{p}T}$ *is flat, then* $\mathfrak{p}S = \mathfrak{p}T \cap S$ *and the images* $\overline{f_i}$ *of the* f_i *in* $T/\mathfrak{p}T \cong (R/\mathfrak{p})[z_1, \dots, z_n]$ *are algebraically independent over* $R/\mathfrak{p}$.
(2) *If* φ *is flat, then for each* $\mathfrak{p} \in \operatorname{Spec}(R)$ *we have* $\mathfrak{p}S = \mathfrak{p}T \cap S$ *and the images* $\overline{f_i}$ *of the* f_i *in* $T/\mathfrak{p}T \cong (R/\mathfrak{p})[z_1, \dots, z_n]$ *are algebraically independent over* $R/\mathfrak{p}$.

PROOF. Item 2 follows from item 1, so it suffices to prove item 1. Assume that $T_{\mathfrak{p}T}$ is flat over S. Since $\mathfrak{p}T \neq T$, Remark 2.37.10 implies that $\mathfrak{p}T \cap S = \mathfrak{p}S$. Suppose the $\overline{f_i}$ are algebraically dependent over $R/\mathfrak{p}$, then there exist indeterminates $t_1, \dots, t_m$ and a polynomial $G \in R[t_1, \dots, t_m] \setminus \mathfrak{p}R[t_1, \dots, t_m]$ such that $G(f_1, \dots, f_m) \in \mathfrak{p}T$. This implies $G(f_1, \dots, f_m) \in \mathfrak{p}T \cap S = \mathfrak{p}S$. But $f_1, \dots, f_m$ are algebraically independent over R and $G(t_1, \dots, t_m) \notin \mathfrak{p}R[t_1, \dots, t_m]$ implies $G(f_1, \dots, f_m) \notin \mathfrak{p}S$, a contradiction. □

PROPOSITION 7.27. *Assume Setting* 7.25 *and in addition assume that* R *is a Noetherian integral domain containing a field of characteristic zero. Let* J *be the associated Jacobian ideal and let* F *be the reduced ideal of* T *defining the non-flat locus of* $\varphi : S \hookrightarrow T$. *Then:*

(1) *If* $\mathfrak{p} \in \operatorname{Spec} R$ *and* $J \subseteq \mathfrak{p}T$, *then* $\varphi_{\mathfrak{p}T} : S \to T_{\mathfrak{p}T}$ *is not flat. Thus we also have* $F \subseteq \mathfrak{p}T$.
(2) *If the embedding* $\varphi : S \hookrightarrow T$ *is flat, then for every* $\mathfrak{p} \in \operatorname{Spec} R$ *we have* $J \nsubseteq \mathfrak{p}T$.

PROOF. Item 2 follows from item 1, so it suffices to prove item 1. Let $\mathfrak{p} \in \operatorname{Spec} R$ with $J \subseteq \mathfrak{p}T$, and suppose $\varphi_{\mathfrak{p}T}$ is flat. Let $\overline{f_i}$ denote the image of f_i in $T/\mathfrak{p}T$. Consider

$$\overline{\varphi} : \overline{S} := (R/\mathfrak{p})[\overline{f_1}, \dots, \overline{f_m}] \to \overline{T} := (R/\mathfrak{p})[z_1, \dots, z_n].$$

By Proposition 7.26, $\overline{f_1}, \dots, \overline{f_m}$ are algebraically independent over $\overline{R} := R/\mathfrak{p}$. Since the Jacobian ideal commutes with homomorphic images, the Jacobian ideal of $\overline{\varphi}$ is zero. Thus for each $Q \in \operatorname{Spec} \overline{T}$ the map $\overline{\varphi}_Q : \overline{S} \to \overline{T}_Q$ is not smooth. But taking $Q = (0)$ gives $\overline{T}_Q$ is a field separable over the field of fractions of $\overline{S}$ and hence $\overline{\varphi}_Q$ is a smooth map. This contradiction completes the proof. □

Theorem 7.28 follows from [**149**, Proposition 2.1] in the case of one indeterminate z, so in the case where $T = R[z]$.

THEOREM 7.28. *Assume Setting* 7.25. *In addition assume that* R *is a Noetherian integral domain and that* $m = 1$, *so that there is exactly one polynomial* f.

Thus $z_1, \ldots, z_n$ *are indeterminates over* R, $T = R[z_1, \ldots z_n]$, *and* $f \in T \setminus R$. *Then the following are equivalent:*

(1) $R[f] \to T$ *is flat.*
(2) *For each prime ideal* $\mathfrak{q}$ *of* R, *we have* $\mathfrak{q}T \cap R[f] = \mathfrak{q}R[f]$.
(3) *For each maximal ideal* $\mathfrak{q}$ *of* R, *we have* $\mathfrak{q}T \cap R[f] = \mathfrak{q}R[f]$.
(4) *The nonconstant coefficients of* f *generate the unit ideal of* R.
(5) $R[f] \to T$ *is faithfully flat.*

PROOF. Since $f \in T \setminus R$ and R is an integral domain, $R[f]$ is a polynomial ring in the indeterminate f over R. By Proposition 7.26.2, (1) $\Longrightarrow$ (2). To show (2) $\Longrightarrow$ (1): by Corollary 7.7, $R[f] \hookrightarrow T$ is flat $\Longleftrightarrow$ for each $Q \in \operatorname{Spec} T$ we have $\operatorname{ht} Q \geq \operatorname{ht}(Q \cap R[f])$. Let $\mathfrak{q} := Q \cap R$. We have $\operatorname{ht} \mathfrak{q} = \operatorname{ht}(\mathfrak{q}R[f]) = \operatorname{ht}(\mathfrak{q}T)$. If $P := Q \cap R[f]$ properly contains $\mathfrak{q}R[f]$, then $\operatorname{ht} P = 1 + \operatorname{ht} \mathfrak{q}$, while if Q properly contains $\mathfrak{q}T$, then $\operatorname{ht} Q \geq 1 + \operatorname{ht} \mathfrak{q}$. In every case, $\operatorname{ht} Q \geq \operatorname{ht}(Q \cap R[f])$, and so (2) $\Longrightarrow$ (1). It is obvious that (2) $\Longrightarrow$ (3).

(3) $\Longrightarrow$ (4): Let $a \in R$ be the constant term of f. If the nonconstant coefficients of f are contained in a maximal ideal $\mathfrak{q}$ of R, then $f - a \in \mathfrak{q}T \cap R[f]$. Since R is an integral domain, the element $f - a$ is transcendental over R and $f - a \notin \mathfrak{q}R[f]$ since $R[f]/\mathfrak{q}R[f]$ is isomorphic to the polynomial ring $(R/\mathfrak{q})[x]$. Therefore $\mathfrak{q}T \cap R[f] \neq \mathfrak{q}R[f]$ if the nonconstant coefficients of f are in $\mathfrak{q}$.

(4) $\Longrightarrow$ (2): Let $\mathfrak{q} \in \operatorname{Spec} R$ and consider the map

$$R[f] \otimes_R \tfrac{R}{\mathfrak{q}} = \tfrac{R[f]}{\mathfrak{q}R[f]} \xrightarrow{\ \varphi\ } T \otimes_R \tfrac{R}{\mathfrak{q}} = \tfrac{T}{\mathfrak{q}T} \cong (\tfrac{R}{\mathfrak{q}})[z_1, \ldots, z_n]. \tag{7.28.a}$$

Since the nonconstant coefficients of f generate the unit ideal of R, the image of f in $(R/\mathfrak{q})[z_1, \ldots, z_n]$ has positive degree. This implies that φ is injective and $\mathfrak{q}T \cap R[f] = \mathfrak{q}R[f]$.

This completes a proof that items (1), (2), (3) and (4) are equivalent. It is obvious that (5) $\Longrightarrow$ (1).

To show (4) $\Longrightarrow$ (5), it suffices to show for $P \in \operatorname{Spec}(R[f])$ that $PT \neq T$. Let $\mathfrak{q} = P \cap R$, and let $\kappa(\mathfrak{q})$ denote the field of fractions of $R/\mathfrak{q}$. Let $\overline{f}$ denote the image of f in $R[f]/\mathfrak{q}R[f]$. Then $R[f]/\mathfrak{q}R[f] \cong (R/\mathfrak{q})[\overline{f}]$, a polynomial ring in one variable over $R/\mathfrak{q}$, since the nonconstant coefficients of f are not in $\mathfrak{q}$. Tensoring the map φ of equation 7.28.a with $\kappa(\mathfrak{q})$ gives an embedding of the polynomial ring $\kappa(\mathfrak{q})[\overline{f}]$ into $\kappa(\mathfrak{q})[z_1, \ldots, z_n]$. The image of P in $\kappa(\mathfrak{q})[\overline{f}]$ is either zero or a maximal ideal of $\kappa(\mathfrak{q})[\overline{f}]$. In either case, its extension to $\kappa(\mathfrak{q})[z_1, \ldots, z_n]$ is a proper ideal. Therefore $PT \neq T$. This completes the proof of Theorem 7.28. □

COROLLARY 7.29. *Assume the setting of Theorem* 7.28 *and let* L *denote the ideal of* R *generated by the nonconstant coefficients of* f. *Then* LT *defines the non-flat locus of the map* $R[f] \hookrightarrow T$.

PROOF. Let $Q \in \operatorname{Spec} T$ and let $\mathfrak{q} = Q \cap R$. Tensoring the map $R[f] \to T$ with $R_\mathfrak{q}$, we see that $R[f] \hookrightarrow T_Q$ is flat if and only if $R_\mathfrak{q}[f] \hookrightarrow T_Q$ is flat. Consider the extensions:

$$R_\mathfrak{q}[f] \overset{\theta}{\hookrightarrow} R_\mathfrak{q}[z_1, \ldots, z_n] \ := T_\mathfrak{q} \overset{\psi}{\hookrightarrow} T_Q.$$

Since ψ is a localization the composite $\psi \circ \theta$ is flat if θ is flat.

Assume $L \nsubseteq Q$. Then $L \nsubseteq \mathfrak{q}$, and so $LR_\mathfrak{q} = R_\mathfrak{q}$. By (4) $\Longrightarrow$ (1) of Theorem 7.28, we have $R[f] \hookrightarrow T_Q$ is flat.

Assume $L \subseteq Q$. Then $L \subseteq \mathfrak{q}$, and we have $f - a \in \mathfrak{q}T \cap R[f]$, where a is the constant term of f. This implies $\mathfrak{q}T_{\mathfrak{q}} \cap R[f] \neq \mathfrak{q}R[f]$. Since $\mathfrak{q}$ is a prime ideal of R, $\operatorname{ht}(\mathfrak{q}R[f]) = \operatorname{ht}(\mathfrak{q}R_{\mathfrak{q}}[f])) = \operatorname{ht}\mathfrak{q}$. But $\operatorname{ht}(\mathfrak{q}T_{\mathfrak{q}} \cap R[f]) > \operatorname{ht}\mathfrak{q}R[f]$. Therefore the extension $\theta : R_{\mathfrak{q}} \to T_{\mathfrak{q}}$ is not flat. We conclude that L defines the non-flat locus of the map $R[f] \hookrightarrow T$. $\square$

REMARK 7.30. A different proof that (4) $\Longrightarrow$ (1) in Theorem 7.28 is as follows: Let v be another indeterminate and consider the commutative diagram

$$\begin{array}{ccc} R[v] & \longrightarrow & T[v] = R[z_1, \dots, z_n, v] \\ \pi \downarrow & & \pi' \downarrow \\ R[f] & \xrightarrow{\varphi} & \frac{R[z_1,\dots,z_n,v]}{(v - f(z_1,\dots,z_n))}. \end{array}$$

where π maps $v \to f$ and π' is the canonical quotient homomorphism. By [**120**, Corollary 2, p. 152] or [**122**, Theorem 22.6 and its Corollary, p. 177], φ is flat if the coefficients of $f - v$ generate the unit ideal of $R[v]$. Moreover, the coefficients of $f - v$ as a polynomial in $z_1, \dots, z_n$ with coefficients in $R[v]$ generate the unit ideal of $R[v]$ if and only if the nonconstant coefficients of f generate the unit ideal of R. For if $a \in R$ is the constant term of f and $a_1, \dots, a_r$ are the nonconstant coefficients of f, then $(a_1, \dots, a_r)R = R$ clearly implies that $(a - v, a_1, \dots, a_r)R[v] = R[v]$. On the other hand, if $(a - v, a_1, \dots, a_r)R[v] = R[v]$, then setting $v = a$ implies that $(a_1, \dots, a_r)R = R$.

We observe in Proposition 7.31 that item 1 implies item 4 of Theorem 7.28 also holds for more than one polynomial f; see also [**149**, Theorem 3.8] for a related result concerning flatness.

PROPOSITION 7.31. *Assume Setting 7.25 and assume that R is an integral domain. If the inclusion map $\varphi : S := R[f_1, \dots f_m] \to T$ is flat, then the nonconstant coefficients of each of the f_i generate the unit ideal of R.*

PROOF. The algebraic independence of the f_i implies that the inclusion map $R[f_i] \hookrightarrow R[f_1, \dots, f_m]$ is flat, for each i with $1 \leq i \leq m$. If $S \longrightarrow T$ is flat, then so is the composition $R[f_i] \longrightarrow S \longrightarrow T$, and the statement follows from Theorem 7.28. $\square$

EXAMPLES 7.32. Let k be a field of characteristic different from 2 and let x, y, z be indeterminates over k.

(1) With $f = x$ and $g = xy^2 - y$, consider $S := k[f, g] \overset{\varphi}{\hookrightarrow} T := k[x, y]$. Then $J = (2xy - 1)T$. This example may also be described by taking $R = k[x]$. We then have $\varphi : S := R[xy^2 - y] \hookrightarrow T := R[y]$. The Jacobian $J = (2xy - 1)T$ is the same but is computed now as just a derivative. Then the coefficients of $xy^2 - y$ in the variable y generate the unit ideal of R. By the implication (4) $\Longrightarrow$ (1) of Theorem 7.28, φ is flat. But φ is not smooth, since J defines the non-smooth locus and $J \neq T$; see Theorem 7.19.1. Here we have $J \subsetneq F = T$.

(2) With $f = x$ and $g = yz$, consider $S := k[f, g] \overset{\varphi}{\hookrightarrow} T := k[x, y, z]$. Then $J = (y, z)T$. Since $\operatorname{ht} J \geq 2$, φ is flat by Corollary 7.22.3. Again φ is not smooth since $J \neq T$.

(3) The example of item 2 may also be described by taking $R = k[x]$. Then $S := R[yz] \hookrightarrow T := R[y, z]$. The Jacobian $J = (y, z)T$ is now computed by taking the partial derivatives $\frac{\partial(yz)}{\partial y}$ and $\frac{\partial(yz)}{\partial z}$.

(4) Let $R = k[x]$ and $S = R[xyz] \hookrightarrow R[y,z] =: T$. Then $J = (xz, xy)T$. Thus J has two minimal primes xT and $(y,z)T$. Notice that $xT \cap S = (x, xyz)S$ is a prime ideal of S of height two, while $(y,z)T \cap S$ has height one. Therefore $J \subsetneq F = xT$.

(5) Let $R = k[x]$ and $S = R[xy + xz] \hookrightarrow R[y,z] =: T$. Then $J = xT$. The map φ is not flat, since $xT \cap S = (x, xy + xz)S$.

(6) Let $R = k[x]$ and $S = R[xy + z^2] \hookrightarrow R[y,z] =: T$. Then $J = (y,z)T$. Hence $S \hookrightarrow T$ is flat but not smooth.

(7) Let $R = k[x]$ and $S = R[xy + z] \hookrightarrow R[y,z] =: T$. Then $J = T$. Hence $S \hookrightarrow T$ is a smooth map.

Exercises

(1) Let k be a field and let T denote the polynomial ring $k[x]$. Let $f \in T$ be a polynomial of degree $d \geq 1$ and let $S := k[f]$.
  (i) Prove that the map $S \hookrightarrow T$ is free and hence flat.
  (ii) Prove that the prime ideals $Q \in \operatorname{Spec} T$ for which $S \to T_Q$ is not a regular map are precisely the prime ideals Q such that the derivative $\frac{df}{dx} \in Q$.

  Assume that the field k has characteristic $p > 0$.
  (iii) If $f = x + x^p$, prove that $S \hookrightarrow T$ is smooth.
  (iv) If $f = x^p$, prove that $S \hookrightarrow T_Q$ is not smooth, for each $Q \in \operatorname{Spec} T$.

  Assume that the field k has characteristic 0.
  (v) If $\deg f = d \geq 2$, prove that there exists a finite nonempty set of maximal ideals Q of T such that $S \hookrightarrow T_Q$ is not smooth.

(2) Let k be a field and let $T = k[[u,v,w,z]]$ be the formal power series ring over k in the variables u, v, w, z. Define a k-algebra homomorphism φ of T into the formal power series ring $k[[x,y]]$ by defining

$$\varphi(u) = x^4, \quad \varphi(v) = x^3y, \quad \varphi(w) = xy^3, \quad \varphi(z) = y^4.$$

  Let $P = \ker(\varphi)$ and let $I = (v^3 - u^2w,\ w^3 - z^2v)T$. Notice that $I \subset P$. Let $A = T/I$, and let $R = k[[u,z]] \subset T$.
  (a) Prove that $\varphi|_R$ is injective, i.e., $P \cap R = (0)$.
  (b) Prove that the ring $B := \varphi(T) = k[[x^4, x^3y, xy^3, y^4]]$ is not Cohen-Macaulay.
  (c) Prove that $A = T/I$ is Cohen-Macaulay and is a finite free R-module.
  (d) Prove that PA is the unique minimal prime of A, and A/PA is not flat over R.

  SUGGESTION. To see that A is module finite over R, observe that

$$\frac{A}{(u,z)A} = \frac{T}{(u, z, v^3 - u^2w, w^3 - z^2v)T},$$

  and the ideal $(u, z, v^3 - u^2w, w^3 - z^2v)T$ is primary for the maximal ideal of T. By Theorem 3.16, A is a finite R-module.

  COMMENT. The ring A of Exercise 2 above is a complete Cohen-Macaulay local ring with $\dim A = 2$ such that $A/\mathfrak{n}$ is not Cohen-Macaulay, where $\mathfrak{n}$ is the nilradical of A.

(3) Let k be a field and let $A = k[x, xy] \subset k[x,y] = B$, where x and y are indeterminates. Let $R = k[x] + (1 - xy)B$.
  (a) Prove that R is a proper subring of B that contains A.
  (b) Prove that B is a flat R-module.

(c) Prove that B is contained in a finitely generated R-module.
(d) Prove that R is not a Noetherian ring.
(e) Prove that $P = (1-xy)B$ is a prime ideal of both R and B with $R/P \cong k[x]$ and $B/P \cong R[x, 1/x]$.
(f) Prove that the map $\operatorname{Spec} B \to \operatorname{Spec} R$ is one-to-one but not onto.

QUESTION. What prime ideals of R are not finitely generated?

(4) With $S = k[x, xy^2 - y] \hookrightarrow T = k[x, y]$ and $J = (2xy-1)T$ as in Examples 7.32.1, prove that $\operatorname{ht}(J \cap S) = 1$.

SUGGESTION. Show that $J \cap S \cap k[x] = (0)$ and use that, for A an integral domain, prime ideals of the polynomial ring $A[y]$ that intersect A in (0) are in one-to-one correspondence with prime ideals of $K[y]$, where $K = \mathcal{Q}(A)$ is the field of fractions of A.

(5) Let $z_1, \dots, z_n$ be indeterminates over a ring R, and let $T = R[z_1, \dots z_n]$. Fix an element $f \in T \setminus R$. Modify the proof of (3) implies (4) of Theorem 7.28 to prove that $\mathfrak{q}T \cap R[f] = \mathfrak{q}R[f]$ for each maximal ideal $\mathfrak{q}$ of R implies that the nonconstant coefficients of f generate the unit ideal of R without the assumption that the ring R is an integral domain.

SUGGESTION. Assume that the nonconstant coefficients of f are contained in a maximal ideal $\mathfrak{q}$ of R. Observe that one may assume that f as a polynomial in $R[z_1, \dots, z_n]$ has zero as its constant term and that the ring R is local with maximal ideal $\mathfrak{q}$. Let M be a monomial in the support of f of minimal total degree and let $b \in R$ denote the coefficient of M for f. Then b is nonzero, but $f \in \mathfrak{q}R[f]$ implies that $b \in \mathfrak{q}$ and this implies, by Nakayama's lemma, that $b = 0$.

CHAPTER 8

Excellent rings and formal fibers

The concept of "excellence" is deep and significant. It has ramifications and relationships to other chapters of this book, and more generally to much of commutative algebra. Section 8.1 contains an initial discussion of attributes that might be considered desirable for a Noetherian ring labeled "excellent". This motivates discussions of the singular locus and the Jacobian criterion in Section 8.1 and of Henselian rings and the Henselization of a Noetherian local ring in Section 8.3. Nagata rings are considered in Section 8.2. For a Noetherian local ring $(R, \mathfrak{m})$ with $\mathfrak{m}$-adic completion $\widehat{R}$, the fibers of the inclusion map $R \hookrightarrow \widehat{R}$ play an important role in determining whether R is excellent or a Nagata ring. For more information about excellent rings see [**62**], [**122**], [**160**].

8.1. Background for excellent rings

In the 1950s, Nagata constructed an example in characteristic $p > 0$ of a normal Noetherian local domain $(R, \mathfrak{m})$ such that the $\mathfrak{m}$-adic completion $\widehat{R}$ is not reduced [**137**, Example 6, p.208], [**133**]. He constructed another example of a normal Noetherian local domain $(R, \mathfrak{m})$ that contains a field of characteristic zero and has the property that $\widehat{R}$ is not an integral domain [**137**, Example 7, p.209]; see Example 4.15, Remarks 4.16 and Section 6.3 for information about this example. The existence of these examples motivated the search for conditions on a Noetherian local ring R that imply good behavior with respect to completion.

We consider the following questions:

QUESTIONS 8.1.

(1) What properties should a "nice" Noetherian ring have?
(2) What properties of a Noetherian local ring ensure good behavior with respect to completion?
(3) What properties of a Noetherian ring ensure "nice" properties of finitely generated algebras over the given ring?

In the 1960s, Grothendieck systematically investigated Noetherian rings that are exceptionally well behaved. He called these rings "excellent". The intent of his definition of excellent rings is that these rings should have the same nice properties as the rings in classical algebraic geometry. [1]

[1]Christel's 1972 Master's Thesis on excellent rings under the supervision of H. J. Nastold in Münster was based on lecture notes by Nastold. In these notes Nastold used R. Kiehl's ideas of what the definition of an "excellent" ring should be, and Kiehl's proofs of the main properties. Kiehl's definition is equivalent to, but somewhat different from, Grothendieck's definition. The notes were never published. Some of the regularity criteria used in Kiehl's proofs are in [**21**].

Among the rings studied in classical algebraic geometry are the *affine rings*

$$A = k[x_1, \ldots, x_n]/I,$$

where k is a field and I is a ideal of the polynomial ring $S := k[x_1, \ldots, x_n]$.

There are four fundamental properties of affine rings that are relevant for the definition of excellent rings. The third property involves the concept of the singular locus as in Definition 8.2.

DEFINITION 8.2. Let A be a Noetherian ring. The *singular locus* of A is:

$$\operatorname{Sing} A = \{P \in \operatorname{Spec} A \mid A_P \text{ is not a regular local ring}\}.$$

Let $\mathcal{A}$ be a class of Noetherian rings that satisfy the following:

PROPERTY $\mathcal{A}$.1. If $A \in \mathcal{A}$ and B is an algebra of finite type over A, then $B \in \mathcal{A}$.

PROPERTY $\mathcal{A}$.2. If $A \in \mathcal{A}$, then A is universally catenary.

PROPERTY $\mathcal{A}$.3. If $A \in \mathcal{A}$, then the singular locus $\operatorname{Sing} A$ is *closed* in the Zariski topology of $\operatorname{Spec} A$, that is, there is an ideal $J \subseteq A$ such that $\operatorname{Sing} A = \mathcal{V}(J)$.[2]

PROPERTY $\mathcal{A}$.4. If $A \in \mathcal{A}$, then, for every maximal ideal $\mathfrak{m} \in \operatorname{Spec} A$ and for every prime ideal $Q \in \operatorname{Spec}(\widehat{A_{\mathfrak{m}}})$, we have:

$$(\widehat{A_{\mathfrak{m}}})_Q \text{ is regular} \iff A_{Q \cap A} \text{ is regular.}$$

We discuss these properties in the remainder of this section. Properties $\mathcal{A}$.1-$\mathcal{A}$.4 hold for the class of affine rings. It is straightforward that affine rings satisfy the first two properties, since an algebra of finite type over an affine ring is again an affine ring, and every affine ring is universally catenary; see Remark 3.27. The third and fourth properties are not as obvious for affine rings; see Remark 8.17. They are, however, important properties for excellence.

David Mumford and John Tate discuss how Grothendieck's work revolutionized classical algebraic geometry in [**126**]. In particular, they write: Algebraic geometry "is the field where one studies the locus of solutions of sets of polynomial equations ...". One combines "the algebraic properties of the rings of polynomials with the geometric properties of this locus, known as a *variety*."

To apply this to the discussion of Property $\mathcal{A}$.3, let k be an algebraically closed field. For n a positive integer, let k^n denote affine n-space. An *affine algebraic variety* is a subset $\mathcal{Z}(I)$ of k^n, where $\mathcal{Z}(I)$ is the zero set of an ideal I of the polynomial ring $S = k[x_1, \ldots, x_n]$:

$$\mathcal{Z}(I) = \{a \in k^n \mid f(a) = 0, \text{ for all } f \in I\}.$$

It is clear that $\mathcal{Z}(I) = \mathcal{Z}(\sqrt{I})$. Let $A = S/\sqrt{I}$. The *singular locus* of $\mathcal{Z}(I)$ is defined to be $\operatorname{Sing} A$.

The singular locus of a reduced affine ring A over an algebraically closed field is a proper closed subset of $\operatorname{Spec} A$; see for example Hartshorne's book [**64**, Theorem 5.3, page 33]. Thus Property $\mathcal{A}$.3 is satisfied for such a ring A.

Again quoting Mumford and Tate in [**126**]: Grothendieck "invented a class of geometric structures generalizing varieties that he called *schemes*". This applies to any commutative ring, and thus includes fields that are not algebraically closed and ideals that are not reduced.

[2]Notation from Section 2.1.

Property $\mathcal{A}.3$ is related to the Jacobian criterion for smoothness.

JACOBIAN CRITERION 8.3. Let $A = S/I$ be an affine ring, where I is an ideal of the polynomial ring $S = k[x_1, \ldots, x_n]$ over the field k. Let P be a prime ideal of S with $I \subseteq P$, let $\mathfrak{p} = P/I$, and let r be the height of I_P in S_P. Assume that $I = (f_1, \ldots, f_s)S$. The *Jacobian criterion for smoothness* asserts the equivalence of the following statements:

(1) The map $\psi : k \hookrightarrow A_{\mathfrak{p}}$ is smooth, or equivalently a regular homomorphism.
(2) rank $(\partial f_i/\partial x_j) = r \pmod{P}$.
(3) The ideal generated by the $r \times r$ minors of $(\partial f_i/\partial x_j)$ is not contained in P.

These equivalent conditions imply that $A_{\mathfrak{p}}$ is an RLR.

The rank of $(\partial f_i/\partial x_j) \pmod{P}$ is at most r; see Eisenbud's book [**45**, 16.19.a]. The Jacobian criterion for smoothness is proved in [**122**, Theorem 30.3].

REMARKS 8.4. Let $A = S/I$, where $S = k[x_1, \ldots, x_n]$ is a polynomial ring over a field k, and I is an ideal of S. Let the notation be as in Criterion 8.3.

(1) By Theorem 7.12, the morphism $\psi : k \hookrightarrow A_{\mathfrak{p}}$ is a regular morphism if and only if ψ is smooth, or equivalently $A_{\mathfrak{p}}$ is a smooth k-algebra.[3] Since A is an affine k-algebra, $A_{\mathfrak{p}}$ is essentially of finite type over k. Regularity of ψ is equivalent to ψ being flat with geometrically regular fibers. Equivalently, ψ is flat and, for each prime ideal Q of A and each finite algebraic field extension L of k, the ring $A_Q \otimes_k L$ is a regular local ring. Since k is a field, A_Q is a free k-module and so the extension ψ is flat by Remark 2.37.2.

(2) If every prime ideal of S minimal over I has the same height r, then the Elkik ideal of A as a k-algebra is equal to the radical of the ideal generated by the $r \times r$ minors of the Jacobian matrix of I. To see this, by Theorem 7.15.3, the Elkik ideal of A as a k-algebra defines the non-smooth locus of $k \hookrightarrow A$. By Jacobian Criterion 8.3, the $r \times r$ minors of the Jacobian matrix of I are not contained in P if and only if the map $\psi : k \hookrightarrow A_P$ is smooth, if and only if the map ψ is regular. It follows that the Elkik ideal is equal to the radical of the ideal generated by the $r \times r$ minors of the Jacobian matrix of I.

More generally, if T is a localization of S at a multiplicatively closed set and every prime ideal of T minimal over IT has the same height r, then the Elkik ideal of T/IT as a k-algebra is equal to the radical of the ideal generated by the $r \times r$ minors of the Jacobian matrix of I.

(3) If k is a perfect field, then $A_{\mathfrak{p}}$ is a regular local ring if and only if the equivalent conditions of Criterion 8.3 hold. This follows because every algebraic extension of k is separable algebraic. Thus A_Q is a regular local ring if and only if $A_Q \otimes_k L$ is an RLR, for every $Q \in \operatorname{Spec} A_{\mathfrak{p}}$ and every finite algebraic field extension L of k. Hence the map $k \hookrightarrow A_{\mathfrak{p}}$ is regular if and only if $A_{\mathfrak{p}}$ is a regular local ring.

(4) If k is a perfect field, A is equidimensional and $\operatorname{ht} I = r$, then the Jacobian criterion defines the singular locus of A. In this case the singular locus of A is $\mathcal{V}(J)$ where J is the ideal of S generated by I and the $r \times r$ minors of the Jacobian matrix $(\partial f_i/\partial x_j)$.

(5) If k is not a perfect field, then the equivalent conditions of Criterion 8.3 are stronger than the statement that $A_{\mathfrak{p}}$ is a regular local ring [**122**, Theorem 30.3].

[3] Regularity is defined in Definition 3.41. For smoothness see Definition 7.11.

Example 8.5 is an example of a Noetherian local ring over a non-perfect field k that is a regular local ring, but is not smooth over k.

EXAMPLE 8.5. Let k be a field of characteristic $p > 0$ such that k is not perfect, that is, k^p is properly contained in k. Let $a \in k \setminus k^p$ and let $f = x^p - a$. Then $L = k[x]/(f)$ is a proper purely inseparable extension field of k. Since $\partial f/\partial x = 0$, the Jacobian criterion for smoothness implies L is not smooth over k. However, L is a field and thus a regular local ring.

REMARK 8.6. Zariski's Jacobian criterion for regularity in polynomial rings applies in the case where the ground field is not perfect; see [**122**, Theorem 30.5].[4] Assume the notation of Criterion 8.3. The singular locus of A is closed in $\operatorname{Spec} A$ and is defined by an ideal J of A; that is, $\operatorname{Sing}(A) = \mathcal{V}(J)$. In Criterion 8.3, the ideal J is generated by the $r \times r$ minors of the Jacobian matrix, whereas in Zariski's Jacobian criterion for regularity in polynomial rings if k has characteristic p and is not perfect, then the Jacobian matrix is extended by certain k^p-derivations of S, and J is generated by appropriate minors of the extended matrix. For Example 8.5, there exists a k^p-derivation $D : k[x] \to k[x]$ with $D(f) \neq 0$; see for example [**122**, page 202].

We return to properties for excellence. A first approach towards obtaining a class $\mathcal{A}$ of Noetherian rings that satisfy Properties $\mathcal{A}.1$, $\mathcal{A}.2$, $\mathcal{A}.3$ and $\mathcal{A}.4$ might be to consider the rings satisfying "Jacobian criteria", similar to the conditions of Criterion 8.3. Unfortunately this class is rather small. Example 8.7 is an excellent Noetherian local domain that fails to satisfy Jacobian criteria. This example is related to Theorem 12.18.

EXAMPLE 8.7. [**160**, p. 319] Let $\sigma = e^{(e^x - 1)} \in \mathbb{Q}[[x]]$. By a result of Ax [**18**, Corollary 1, p. 253], σ and $\partial\sigma/\partial x$ are algebraically independent over $\mathbb{Q}(x)$; see the proof of item 1 of Theorem 12.18. As in Example 4.7, consider the intersection domain

$$A := \mathbb{Q}(x, \sigma) \cap \mathbb{Q}[[x]].$$

By Remark 2.1, A is a DVR with maximal ideal xA and field of fractions $\mathbb{Q}(x, \sigma)$. Then $\mathbb{Q}[x]_{(x)} \subset A \subset \mathbb{Q}[[x]]$. For every derivation $d : A \hookrightarrow \mathbb{Q}[[x]]$, it follows that $d(\sigma) = dx(\partial\sigma/\partial x)$. Since $\partial\sigma/\partial x \notin \mathbb{Q}(x, \sigma)$, we have $d(\sigma) \notin A$ whenever $d(x) \neq 0$. Hence there is only the trivial derivation $d = 0$ from A into itself. Since every DVR containing a field of characteristic 0 is excellent, the ring A is excellent; see Remarks 3.48.

There is an important class of Noetherian local rings that admit Jacobian and regularity criteria, namely, the class of complete Noetherian local rings. These criteria were established by Nagata and Grothendieck and are similar to the above mentioned criterion. A principal objective of the theory of excellent rings is to exploit the Jacobian criteria for the completion $\widehat{A}$ of an excellent local ring A in order to describe certain properties of A, even if the ring A itself may fail to satisfy Jacobian criteria. This theory requires considerable theoretical background. The goal of Grothendieck's theory of formal smoothness and regularity is to determine the connection between a local ring A and its completion $\widehat{A}$; see [**62**, No 24, (6.8), pp. 150-153].

[4] Historically this goes back to Zariski's influential paper [**190**].

REMARK 8.8. Let $(A, \mathfrak{m})$ be a Noetherian local ring. By Cohen's structure theorems, the $\mathfrak{m}$-adic completion $\widehat{A}$ of A is the homomorphic image of a formal power series ring over a ring K, where K is either a field or a complete discrete valuation ring, that is, $\widehat{A} \cong K[[x_1, \dots, x_n]]/I$; see Remarks 3.19.3. The singular locus $\operatorname{Sing} \widehat{A}$ of $\widehat{A}$ is closed by the Jacobian criterion on complete Noetherian local rings [**122**, Corollary to Theorem 30.10].

The following discussion relates to Properties $\mathcal{A}.3$ and $\mathcal{A}.4$ and the definition of excellence.

DISCUSSION 8.9. Let $\varphi : A \hookrightarrow C$ be a faithfully flat homomorphism of Noetherian rings. For example, let $(A, \mathfrak{m})$ be a Noetherian local ring and let C be the $\mathfrak{m}$-adic completion of A. We observe connections between the singular loci $\operatorname{Sing} A$ and $\operatorname{Sing} C$. Consider the following two conditions regarding $\operatorname{Sing} A$ and $\operatorname{Sing} C$ and regularity of localizations of A and C

$$\text{(8.9.a)} \qquad \operatorname{Sing} A = \mathcal{V}(J) \quad \text{and} \quad \operatorname{Sing} C = \mathcal{V}(JC), \qquad \text{for some ideal } J \text{ of } A.$$

$$\text{(8.9.b)} \qquad \text{For every } Q \in \operatorname{Spec} C, \quad A_{Q \cap A} \text{ is regular} \iff C_Q \text{ is regular.}$$

Condition 8.9.a implies that $\operatorname{Sing} A$ and $\operatorname{Sing} C$ are closed. We show in Theorem 8.10 below if $\operatorname{Sing} C$ is closed, then Condition 8.9.a is equivalent to Condition 8.9.b. We first make some observations about Condition 8.9.b.

(8.9.1) " $\Leftarrow$ " of Condition 8.9.b is always satisfied.

PROOF. The induced morphism $A_{Q \cap A} \longrightarrow C_Q$ is faithfully flat. Since flatness descends regularity by Theorem 3.33.1, C_Q is regular implies $A_{Q \cap A}$ is regular. $\square$

(8.9.2) If $\varphi : A \to C$ has regular fibers as in Definition 3.38, then Condition 8.9.b holds.

PROOF. Let $P = Q \cap A$. Since the fiber over P is regular, the ring C_Q/PC_Q is regular. By Theorem 3.33.2, if A_P and C_Q/PC_Q are both regular, then the ring C_Q is regular. Thus " $\Rightarrow$ " of Condition 8.9.b holds. By statement 8.9.1, " $\Leftarrow$ " of Condition 8.9.b always holds. $\square$

THEOREM 8.10. *Let $\varphi : A \hookrightarrow C$ be a faithfully flat homomorphism of Noetherian rings. Assume* $\operatorname{Sing} C$ *is closed. Then:*

1. *Condition 8.9.a is equivalent to Condition 8.9.b.*
2. *If in addition the fibers of φ are regular, then* $\operatorname{Sing} A$ *is closed.*

PROOF. For item 1, it is clear that Condition 8.9.a implies Condition 8.9.b. Assume Condition 8.9.b and let I be the radical ideal of C such that $\operatorname{Sing} C = \mathcal{V}(I)$. Then $I = \bigcap_{i=1}^n Q_i$, where the Q_i are prime ideals of C. Let $P_i = Q_i \cap A$ for each i and let $I \cap A = J$. Then $J = \bigcap_{i=1}^n P_i$. We observe that $\operatorname{Sing} A = \mathcal{V}(J)$. Since C_{Q_i} is not regular, Condition 8.9.b implies that A_{P_i} is not regular. Let $P \in \operatorname{Spec} A$. If $J \subseteq P$, then $P_i \subseteq P$ for some i, and $P_i \subseteq P$ implies that A_{P_i} is a localization of A_P. Therefore A_P is not regular.

Assume that $J \not\subseteq P$. There exists $Q \in \operatorname{Spec} C$ such that $Q \cap A = P$, and it is clear that $I \not\subseteq Q$. Hence C_Q is regular, and thus by Condition 8.9.b, the ring A_P is regular. Therefore $\operatorname{Sing} A = \mathcal{V}(J)$.

It remains to observe that $\sqrt{JC} = I$. Clearly $\sqrt{JC} \subseteq I$. Let $Q \in \operatorname{Spec} C$ with $JC \subseteq Q$. Then $J \subseteq Q \cap A := P$ and A_P is not regular. By Condition 8.9.b, the ring C_Q is not regular, so $I \subseteq Q$.

For item 2, Condition 8.9.b holds for A, by (8.9.2). By item 1, Condition 8.9.b implies Condition 8.9.a. Hence the singular locus of A is closed in $\operatorname{Spec} A$. □

COROLLARY 8.11. *Let $\varphi : A \hookrightarrow C$ be a faithfully flat homomorphism of Noetherian rings. Let B be an essentially finite A-algebra such that* $\operatorname{Sing}(B \otimes_A C)$ *is closed. If the fibers of φ are geometrically regular, then* $\operatorname{Sing} B$ *is closed.*

PROOF. By Fact 2.38, the map $1_B \otimes_A \varphi : B \hookrightarrow B \otimes_A C$ is faithfully flat. Let $Q \in \operatorname{Spec}(B \otimes_A C)$, and let P' and P denote the contractions of Q to B and A, respectively. Since B is essentially finite over A, the field $k(P') = (B \setminus P')^{-1} B_{P'}$ is a finite algebraic extension of the field $k(P) = (A \setminus P)^{-1} A_P$. The fiber over P of the map φ is $\operatorname{Spec}(k(P) \otimes_A C)$; see Discussion 3.29. Since φ has geometrically regular fibers, $\operatorname{Spec}(k(P') \otimes_A C)$ is regular, that is, $(k(P') \otimes_A C)_{Q'}$ is a regular local ring for every prime ideal Q' of $k(P') \otimes_A C$.

Also the fiber over P' of the map $1_B \otimes_A \varphi$ is $\operatorname{Spec}(k(P') \otimes_B (B \otimes_A C))$. Since $k(P') \otimes_B (B \otimes_A C) = k(P') \otimes_A C$, the map $1_B \otimes_A \varphi$ has regular fibers.

By Theorem 8.10.2, $\operatorname{Sing} B$ is closed. □

COROLLARY 8.12. *Let $(A, \mathfrak{m})$ be a Noetherian local ring and let $\varphi : A \hookrightarrow \widehat{A}$ be the canonical map from A to its $\mathfrak{m}$-adic completion $\widehat{A}$.*

(1) *Condition 8.9.a is equivalent to Condition 8.9.b.*
(2) *If the formal fibers of A are regular, then* $\operatorname{Sing} A$ *is closed.*

PROOF. By Remark 8.8, Theorem 8.10 applies. □

REMARK 8.13. Let A be a Noetherian local ring with regular formal fibers. By Corollary 8.12, $\operatorname{Sing} A$ is closed. In order to obtain that every algebra essentially of finite type over A also has the property that its singular locus is closed, the stronger condition that the formal fibers of A are geometrically regular as in Definition 3.39 is needed. This is demonstrated by an example of Rotthaus of a regular local ring A that is a Nagata ring and has the property that its formal fibers are not geometrically regular; the example is described in Remark 20.8. In the example of Rotthaus, the ring A contains a prime element ω such that the singular locus of the quotient ring $A/(\omega)$ is not closed.

The following two theorems are due to Nagata.

THEOREM 8.14. [**120**, Theorem 73]. *Let A be a Noetherian ring. Then the following two statements are equivalent:*

(1) *For every A algebra B that is essentially finite over A, the singular locus* $\operatorname{Sing} B$ *is closed in* $\operatorname{Spec} B$.
(2) *For every A algebra B that is essentially of finite type over A, the singular locus* $\operatorname{Sing} B$ *is closed in* $\operatorname{Spec} B$.

THEOREM 8.15. [**120**, Theorem 74]. *If A is a complete Noetherian local ring, then A satisfies the equivalent conditions of Theorem* 8.14

From Theorems 8.14 and 8.15, we have:

COROLLARY 8.16. *Let A be a Noetherian local ring. If the formal fibers of A are geometrically regular, then for every A-algebra B essentially of finite type over A, the singular locus* $\operatorname{Sing} B$ *is closed in* $\operatorname{Spec} B$.

PROOF. Let B be an A-algebra that is essentially finite over A. Then $B \otimes_A \widehat{A}$ is an essentially finite $\widehat{A}$-algebra. By Theorem 8.15, $\operatorname{Sing}(B \otimes_A \widehat{A})$ is closed. By Corollary 8.11 with C replaced by $\widehat{A}$, we have $\operatorname{Sing} B$ is closed. This holds for every A-algebra B that is essentially finite over A. Thus by Theorem 8.14, $\operatorname{Sing} B$ is closed for every A-algebra B that is essentially of finite type over A. □

REMARK 8.17. Let A be an affine k-algebra, $\mathfrak{m}$ a maximal ideal of A, and $\widehat{A_{\mathfrak{m}}}$ the $\mathfrak{m}$-adic completion of A. By Jacobian Criterion 8.3 and Remarks 8.4.4 and 8.6, the singular locus of A is closed and $\operatorname{Sing} A = \mathcal{V}(J)$, for an ideal J defined by partial derivatives and derivations. For every maximal ideal $\mathfrak{m}$ of A, $\operatorname{Sing} \widehat{A_{\mathfrak{m}}} = \mathcal{V}(J\widehat{A_{\mathfrak{m}}})$, since the partial derivates $\partial f_i / \partial x_j$ and the k^p derivations on A and $A_{\mathfrak{m}}$ extend to derivations of $\widehat{A_{\mathfrak{m}}}$. That is, every $A_{\mathfrak{m}}$ satisfies Condition 8.9.a, and, by Theorem 8.10.1, Condition 8.9.b holds. Therefore every affine algebra A satisfies Property $\mathcal{A}.3$ and Property $\mathcal{A}.4$.

8.2. Nagata rings and excellence

Classical developments leading to the concept of excellent rings were made by Zariski, Cohen, Chevalley, Abhyankar, Nagata, Rees, Tate, Hironaka, Grothendieck and Kiehl among others over the two decades from the early 1940s to the 1960s. These authors were investigating ideal-theoretic properties of rings, the behavior of these properties under certain kinds of extension, and the relations among these properties.

For the class of Nagata rings[5], Nagata Polynomial Theorem 2.21 implies that algebras essentially of finite type over Nagata rings are again Nagata. Another classical result is Rees Finite Integral Closure Theorem 3.21; this gives a connection between the integral closure of a reduced Noetherian local ring $(R, \mathfrak{m})$ and the completion of R.

Another classical result of Nagata is:

THEOREM 8.18. [**120**, Theorem 70] [**137**, 36.4, p. 132, p. 219] *Let R be a Noetherian local Nagata domain. Then R is analytically unramified.*

Theorem 8.18 implies every Noetherian local Nagata ring R satisfies:

($*$) For every $P \in \operatorname{Spec} R$, the ring R/P is analytically unramified.

There exist Noetherian local domains $(R, \mathfrak{m})$ that are not Nagata, but satisfy condition ($*$). Proposition 10.4 and Remark 10.5 describe examples of DVRs and other Noetherian regular rings that are not Nagata rings.

Condition ($*$) requires only that the formal fibers of R are reduced. It does not require for a finite field extension L of $k(P)$ that the fibers of the map $R \otimes_R L \hookrightarrow \widehat{R} \otimes_R L$ are reduced.

A necessary and sufficient condition for a Noetherian local ring $(R, \mathfrak{m})$ to be Nagata, is that the formal fibers of R are *geometrically* reduced.

[5]For the definition of a Nagata ring see Definition 2.20.

THEOREM 8.19. [**62**, No 24, (7.6.4)] *A Noetherian local ring R is a Nagata ring if and only if the formal fibers of R are geometrically reduced.*

If R is a Nagata ring, the normal locus:

$$\operatorname{Nor} R = \{P \in \operatorname{Spec} R \mid R_P \text{ is a normal ring}\}$$

is open in $\operatorname{Spec} R$. Theorem 8.19 implies that every Noetherian local ring with geometrically reduced formal fibers has an open normal locus. For Noetherian non-local rings this is no longer true. Nishimura constructed an example of a Noetherian ring R with geometrically regular formal fibers so that $\operatorname{Nor} R$ is not open in $\operatorname{Spec} R$. Theorem 8.20 characterizes Nagata rings in general:

THEOREM 8.20. [**62**, No 24, (7.6.4), (7.7.2)] *A Noetherian ring R is a Nagata ring if and only if the following two conditions are satisfied:*
(*a*) *The formal fibers of R are geometrically reduced.*
(*b*) *For every finite R-algebra S that is a domain* $\operatorname{Nor} S$ *is open in* $\operatorname{Spec} S$.

Theorem 8.21 stated below is another way to deduce that the examples described in Proposition 10.4 and Remark 10.5 are non-Nagata rings.

THEOREM 8.21. [**120**, Theorem 71] *Let $(R, \mathfrak{m})$ be a Nagata local domain, let $\widehat{R}$ be the $\mathfrak{m}$-adic completion of R, and let $\widehat{P}$ be a minimal prime ideal of $\widehat{R}$. Then $k(\widehat{P}) = \mathcal{Q}(\widehat{R}/\widehat{P})$ is separable over the field of fractions $\mathcal{Q}(R)$ of R.*

Grothendieck defined excellence for a Noetherian local ring as follows:

DEFINITION 8.22. Let A be a Noetherian local ring. Then A is *excellent* if
(a) The formal fibers of A are geometrically regular, that is, for every prime ideal P of A and, for every finite purely inseparable field extension L of the field of fractions $k(P)$ of A/P, the ring $\widehat{A} \otimes_A L$ is regular.
(b) A is universally catenary.

For a non-local Noetherian ring A an additional condition is needed in the definition of excellence: the singular locus of every finitely generated algebra over A is closed. This condition is not included in Definition 8.22; by Corollary 8.12.2, the singular locus is closed for a Noetherian local ring that has geometrically regular formal fibers.

If A is an excellent local ring, then its completion $\widehat{A}$ inherits many properties from A. In particular, Theorem 8.23 is proved in [**62**, No 24,(7.8.3.1), p. 215].

THEOREM 8.23. *Let $(A, \mathfrak{m})$ be an excellent local ring with $\mathfrak{m}$-adic completion $\widehat{A}$. Let $Q \in \operatorname{Spec} \widehat{A}$, and let $P = Q \cap A$. Then the ring A_P is regular (normal, reduced, Cohen-Macaulay, respectively) if and only if the ring $\widehat{A}_Q$ is regular (normal, reduced, Cohen-Macaulay, respectively).*

If A is not a local ring, the formal fibers of A are the formal fibers of the local rings $A_{\mathfrak{m}}$, where $\mathfrak{m}$ is a maximal ideal of A. We say that A has geometrically regular formal fibers if the local rings $A_{\mathfrak{m}}$ for all maximal ideals $\mathfrak{m}$ of A have geometrically regular formal fibers. If A is a semilocal ring with geometrically regular formal fibers, then $\operatorname{Sing} A$ is again closed in $\operatorname{Spec} A$. If A is a non-semilocal ring with geometrically regular formal fibers then it is possible that $\operatorname{Sing} A$ is no longer closed in $\operatorname{Spec} A$; see the example of Nishimura, [**141**]. Therefore an additional condition is needed for the singular locus of A and of all algebras of finite type over A to be closed. See Definition 3.47.

8.3. Henselian rings

Let $(R, \mathfrak{m})$ be a local ring. Recall from Definition 3.30 that R is *Henselian* if Hensel's Lemma holds for R.

The Henselian property was first observed in algebraic number theory around 1910 for the ring of p-adic integers. Many popular Noetherian local rings fail to be Henselian; see for example Exercise 8.4.

In this section we describe an approach to the construction of the Henselization of the local ring R developed by Raynaud in [**153**] and discussed in [**160**]. This approach is different from that used in Nagata's book [**137**] and discussed in Remarks 3.32. Raynaud defines a local ring R to be Henselian if every finite R-algebra B is a finite product of local rings [**153**, Definition 1, p.1]. Raynaud's approach uses the concept of an étale morphism as in Definitions 8.24.[6]

DEFINITIONS 8.24. Let $(R, \mathfrak{m})$ be a local ring.

(1) Let $\varphi : (R, \mathfrak{m}) \to (A, \mathfrak{n})$ be a local homomorphism with A essentially finite over R; that is A is a localization of an R-algebra that is a finitely generated R-module. Then A is *étale* over R if the following condition holds: for every R-algebra B and ideal N of B with $N^2 = 0$, every R-algebra homomorphism $\beta : A \to B/N$ has a unique lifting to an R-algebra homomorphims $\alpha : A \to B$. Thus A is *étale* over R if for every commutative diagram of the form below, where the maps from $R \to A$ and $R \to B$ are the canonical ring homomorphisms that define A and B as R-algebras and the map $\pi : B \to B/N$ is the canonical quotient ring map

(8.24.1)

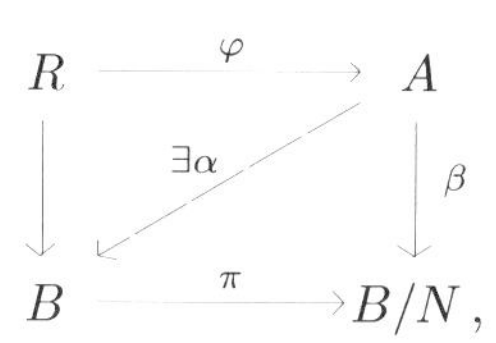

there exists a unique R-algebra homomorphism $\alpha : A \to B$ that preserves commutativity of the diagram.

(2) A local ring $(A, \mathfrak{n})$ is an *étale neighborhood* of R if A is étale over R and $R/\mathfrak{m} \cong A/\mathfrak{n}$; that is, there is no residue field extension.

Raynaud proves that Henselian local rings are closed under étale neighborhoods.

THEOREM 8.25. [**153**, Corollary 2, p. 84] *Let R be a local Henselian ring. Then R is closed under étale neighborhoods, that is, for every étale neighborhood $\phi : R \longrightarrow A$, we have that $R \cong A$ considered as R-algebras.*

Structure Theorem 8.26 is essential for Raynaud's approach to the construction of the Henselization.

THEOREM 8.26. *(Structure Theorem for étale neighborhoods)* [**153**, Theorem 1, p. 51] *Let $\varphi : (R, \mathfrak{m}) \longrightarrow (A, \mathfrak{n})$ be a local morphism with A essentially finite over*

[6]David Mumford mentions that the word *étale* "refers to the appearance of the sea at high tide under a full moon in certain types of weather" [**125**, p. 344]. Another meaning for *étale*, given on dictionary.revers.net, is "slack" and *étaler* is translated as "spread or display". A sentence given there "Il s'est étale de tout son long"—is translated as "He fell flat on his face."

R. Then A is étale over R if and only if

$$A \cong (R[x]/(f))_Q$$

where $R[x]$ is the polynomial ring over R in one variable and

(a) *$f \in R[x]$ is a monic polynomial.*

(b) *$Q \in \operatorname{Spec}(R[x])$ is a prime ideal with $Q \cap R = \mathfrak{m}$.*

(c) *$f' \notin Q$, that is, the derivative of f is not in Q.*

The proof of the structure theorem involves a form of Zariski's Main Theorem [**148**], [**47**].

Using the structure theorem Raynaud defines a representative set of étale neighborhoods of R:

$$\Lambda = \{(f, Q) \mid f \in R[x] \text{ monic}, \ Q \in \operatorname{Spec}(R[x]), \ f \in Q, \\ f' \notin Q, \ Q \cap R = \mathfrak{m}, \ (R[x]/Q)_Q = R/\mathfrak{m}\}.$$

The set Λ is a subset of the product set $R[x] \times \operatorname{Spec}(R[x])$.

Raynaud defines the Henselization of R via a direct limit over the set Λ. Let $\lambda_1 = (f_1, Q_1)$ and $\lambda_2 = (f_2, Q_2)$ be elements of Λ, and let $S_1 = (R[x]/(f_1))_{Q_1}$, respectively, $S_2 = (R[x]/(f_2))_{Q_2}$, denote the corresponding étale neighborhoods. We define a partial order on Λ by $\lambda_1 \leq \lambda_2$ if and only if there is a local R-algebra morphism $\tau : S_1 \longrightarrow S_2$. In order to define a direct limit over the set Λ two conditions must be satisfied. First, the set of R-algebra morphisms between S_1 and S_2 has to be rather small in order to restrict each choice of R-algebra morphisms to one for which "it all fits together". Second, the partially ordered set Λ must be directed, that is, for every λ_1 and $\lambda_2 \in \Lambda$, there must be a third element $\lambda_3 \in \Lambda$ with $\lambda_1 \leq \lambda_3$ and $\lambda_2 \leq \lambda_3$. The following result is what is needed:

THEOREM 8.27. [**153**, Proposition 2, p. 84] *Let $\lambda_1, \lambda_2 \in \Lambda$ with corresponding étale neighborhoods $S_i = (R[x]/(f_i))_{Q_i}$. Then:*

(a) *There is at most one R-algebra morphism $\tau : S_1 \longrightarrow S_2$.*

(b) *There is an element $\lambda_3 \in \Lambda$ with corresponding étale neighborhood S_3 that contains S_1 and S_2, i.e. $\lambda_1 \leq \lambda_3$ and $\lambda_2 \leq \lambda_3$.*

Theorem 8.27 implies that the set

$$\{(R[x]/(f))_Q \mid (f, Q) \in \Lambda\}$$

is directed in a natural way. Raynaud defines the direct limit of this system to be the Henselization of R:

$$R^h = \varinjlim_{\lambda=(f,Q)\in\Lambda} (R[x]/(f))_Q.$$

We list several properties of the Henselization:

REMARKS 8.28.

(1) A local ring R is Noetherian if and only if its Hensel-ization R^h is Noetherian [**153**, Chapitre VIII].

(2) If R is a Noetherian local ring, then the natural injection $R \hookrightarrow R^h$ is a regular map with zero-dimensional fibers [**62**, No 32, (18.6.9), p. 139].

(3) The formal fibers of a Noetherian local ring R are geometrically regular, respectively, geometrically normal, geometrically reduced, if and only if the formal fibers of R^h are geometrically regular, respectively, geometrically normal, geometrically reduced. Moreover, R is a Nagata ring if and

only if R^h is a Nagata ring. In addition, if R is excellent so is R^h. These results are in [**62**, No. 32, (18.7.4), (18.7.2), (18.7.3), and (18.7.6), pp. 143-144]; see also Remark 18.5.

(4) The Henselization R^h of a Noetherian local ring R is in general much smaller than its completion $\widehat{R}$. The Henselization R^h of R is an algebraic extension of R whereas the completion $\widehat{R}$ is usually of infinite (uncountable) transcendence degree over R, if R is a domain; see Fact 3.8

(5) If R is a Nagata, analytically normal local domain, in particular, if R is an excellent normal local domain, then its Henselization R^h is the algebraic closure of R in $\widehat{R}$, that is, every element of R^h is algebraic over R and every element of $\widehat{R} \setminus R^h$ is transcendental over R. This result is given in Nagata's book [**137**, Corollary 44.3]. Let $\overline{R}$ be the integral closure of R in $\widehat{R}$ and let $\widehat{\mathfrak{m}}$ denote the maximal ideal of $\widehat{R}$. Then $R^h = \overline{R}_{\widehat{\mathfrak{m}} \cap \overline{R}}$.

Theorem 8.29 is an extension of Remark 3.23.2 to integral domains of dimension bigger than one that have geometrically normal formal fibers.

THEOREM 8.29. [**153**, Corollaire, p. 99] *Let R be a Noetherian local domain with geometrically normal formal fibers. Then there is a one-to-one correspondence between the maximal ideals of the integral closure of R in its field of fractions $\mathcal{Q}(R)$ and the minimal prime ideals of its completion $\widehat{R}$.*

Exercises

(1) Let $\varphi : A \hookrightarrow C$ be a faithfully flat homomorphism of Noetherian rings.
   (a) If the fibers of φ are regular and for each $Q \in \operatorname{Spec} C$ the formal fiber over Q is regular, prove that for each $P \in \operatorname{Spec} A$, the formal fiber over P is regular.
   (b) If the fibers of φ are geometrically regular and for each $Q \in \operatorname{Spec} C$ the formal fiber over Q is geometrically regular, prove that for each $P \in \operatorname{Spec} A$, the formal fiber over P is geometrically regular.

(2) Let A be a Nagata ring and let $S \subset A$ be a multiplicatively closed subset of A. Show that $S^{-1}A$ is a Nagata ring.

(3) Let $A \hookrightarrow B$ be Noetherian rings with B a finite A-module. If B is a Nagata ring prove that A is also a Nagata ring.

(4) Let x be an indeterminate over a field k and let R denote the localized polynomial ring $k[x]_{(xk[x])}$. Show that R is *not* Henselian.

   SUGGESTION. Consider the polynomial $f(y) = y^2 + y + x \in R[y]$.

(5) Let $(R, \mathfrak{m})$ be a Henselian local integral domain with field of fractions K.
   (a) If V is a DVR on K, prove that $R \subseteq V$.
   (b) If A is a Noetherian domain with field of fractions K, prove that R is contained in the integral closure of A.

   COMMENT. Berger, Kiehl, Kunz and Nastold in [**21**, Satz 2.3.11, p. 60] attribute this result to F. K. Schmidt [**164**]. The result has many interesting applications. It is used by Gilmer and Heinzer in [**56**, Example 3.13]. It is developed by Abhyankar in [**5**, (3.10), pp. 121-123]. Abhyankar points out connections with the concept of root-closed fields and Newton's binomial theorem for fractional exponents. With k a field, the field of rational functions

$k(x_1, \ldots, x_n)$ does not determine the polynomial ring $k[x_1, \ldots, x_n]$, however, as Abhyankar notes in [**5**, Section 9.6], the field of fractions $k((x_1, \ldots, x_n))$ of the power series ring $k[[x_1, \ldots, x_n]]$ does uniquely determine the power series ring. We thank Tom Marley for asking us a question that motivated us to include this exercise.

CHAPTER 9

Height-one prime ideals and weak flatness

Let x be a nonzero nonunit of an integral domain R and let R^* denote the Noetherian x-adic completion of R. In this chapter, we consider the structure of a subring A of R^* of the form $A := \mathcal{Q}(R)(\tau_1, \tau_2, \dots \tau_s) \cap R^*$. We assume Setting 5.1 and the conditions of Inclusion Construction 5.3. Thus $\tau_1, \tau_2, \dots, \tau_s$ are elements of xR^* that are algebraically independent element over R and every nonzero element of $R[\tau_1, \tau_2, \dots, \tau_s]$ is regular on R^*. In this chapter, R is usually a Krull domain.

If the intersection ring A can be expressed as a directed union B of localized polynomial extension rings of R as in Definition 5.7, then the computation of A is easier. Recall that $\tau_1, \tau_2, \dots, \tau_s$ are called *limit-intersecting* for A if the ring A is such a directed union; see Definition 5.10.

The main result of Section 9.1 is Weak Flatness Theorem 9.9. In this theorem we give criteria for $\tau_1, \tau_2, \dots, \tau_s$ to be limit-intersecting for A. In Section 9.2 with the setting of extensions of Krull domains, we continue to analyze the properties of height-one primes considered in Section 9.1. We examine flatness in the setting of Inclusion Construction 5.3 in Section 9.3.

Weak Flatness Theorem 9.9 is used in Examples 10.15 to obtain a family of examples where the Approximation Domain B of Definition 5.7 is equal to the Intersection Domain A of Inclusion Construction 5.3 and is not Noetherian.

9.1. The limit-intersecting condition

In this section we prove Weak Flatness Theorem 9.9. This theorem gives conditions for the intersection domain A to be equal to the approximation domain B; that is, the construction is limit-intersecting. For this purpose, we consider the following properties of an extension of commutative rings:

DEFINITIONS 9.1. Let $\varphi : S \hookrightarrow T$ be an extension of commutative rings.

(1) The extension $\varphi : S \hookrightarrow T$ is *weakly flat*, or T is *weakly flat* over S, if every height-one prime ideal P of S with $PT \neq T$ satisfies $PT \cap S = P$. Equivalently, there exists a prime ideal Q of T such that $Q \cap S = P$; see Exercise 11 of Chapter 2.

(2) The extension $\varphi : S \hookrightarrow T$ is *height-one preserving*, or T is a *height-one preserving* extension of S, if, for every height-one prime ideal P of S with $PT \neq T$, $\operatorname{ht}(PT) = 1$; that is, there exists a height-one prime ideal Q of T with $PT \subseteq Q$.

(3) For $d \in \mathbb{N}$, the extension $\varphi : S \hookrightarrow T$ satisfies LF_d (*locally flat in height* d) if, for each $Q \in \operatorname{Spec} T$ with $\operatorname{ht} Q \leq d$, the composite map $S \to T \to T_Q$ is flat.

REMARK 9.2. Let $\varphi : S \hookrightarrow T$ be an extension of commutative rings, and let $Q \in \operatorname{Spec} T$. With $P := Q \cap S$, the composite map $S \to T \to T_Q$ factors through S_P, and the map $S \to T_Q$ is flat if and only if the map $S_P \to T_Q$ is faithfully flat.

PROPOSITION 9.3. *Let $\varphi : S \hookrightarrow T$ be an extension of commutative rings where S is a Krull domain.*

(1) *If every nonzero element of S is regular on T and every height-one prime ideal of S is the contraction of an ideal of T, then $S = \mathcal{Q}(S) \cap T$.*
(2) *If $S \hookrightarrow T$ is a birational extension and each height-one prime of S is contracted from T, then $S = T$.*
(3) *If T is a Krull domain and $T \cap \mathcal{Q}(S) = S$, then each height-one prime of S is the contraction of a height-one prime of T, and the extension $S \hookrightarrow T$ is height-one preserving and weakly flat.*

PROOF. Item 1 follows from item 2. For item 2, recall from Remark 2.12.2 that $S = \cap\{S_\mathfrak{p} \mid \mathfrak{p}$ is a height-one prime ideal of $S\}$. We show that $T \subseteq S_\mathfrak{p}$, for each height-one prime ideal $\mathfrak{p}$ of S. Since $\mathfrak{p}$ is contracted from T, there exists a prime ideal $\mathfrak{q}$ of T such that $\mathfrak{q} \cap S = \mathfrak{p}$; see Exercise 11 of Chapter 2. Then $S_\mathfrak{p} \subseteq T_\mathfrak{q}$ and $T_\mathfrak{q}$ birationally dominates $S_\mathfrak{p}$. Since $S_\mathfrak{p}$ is a DVR, we have $S_\mathfrak{p} = T_\mathfrak{q}$. Therefore $T \subseteq S_\mathfrak{p}$, for each $\mathfrak{p}$. It follows that $T = S$.

For item 3, since T is a Krull domain, Remark 2.12.1 implies that

$$T = \bigcap\{T_\mathfrak{q} \mid \mathfrak{q} \text{ is a height-one prime ideal of } T\}.$$

Hence

$$S = \mathcal{Q}(S) \cap T = \bigcap\{T_\mathfrak{q} \cap \mathcal{Q}(S) \mid \mathfrak{q} \text{ is a height-one prime ideal of } T\}.$$

Since each $T_\mathfrak{q}$ is a DVR, Remark 2.1 implies that $T_\mathfrak{q} \cap \mathcal{Q}(S)$ is either the field $\mathcal{Q}(S)$ or a DVR birational over S. By Remark 2.12.2, for each height-one prime $\mathfrak{p}$ of S, the localization $S_\mathfrak{p}$ is a DVR of the form $T_\mathfrak{q} \cap \mathcal{Q}(S)$. It follows that each height-one prime ideal $\mathfrak{p}$ of S is contracted from a height-one prime ideal $\mathfrak{q}$ of T, and that T is height-one preserving and weakly flat over S. □

Corollary 9.4 demonstrates the relevance of the weak flatness property for an extension of a Krull domain.

COROLLARY 9.4. *Let $\varphi : S \hookrightarrow T$ be an extension of commutative rings where S is a Krull domain such that every nonzero element of S is regular on T and $PT \neq T$ for every height-one prime ideal P of S.*

(i) *If $\varphi : S \hookrightarrow T$ is weakly flat, then $S = \mathcal{Q}(S) \cap T$.*
(ii) *If T is Krull, then T is weakly flat over $S \iff S = \mathcal{Q}(S) \cap T$.*
(iii) *If T is Krull, the equivalent conditions of item ii imply that $\varphi : S \hookrightarrow T$ is height-one preserving.*

PROOF. For item i, the weak flatness condition implies that each height-one prime ideal of S is contracted from T. Thus $S = \mathcal{Q}(S) \cap T$, by Proposition 9.3.1.

For items ii and iii, apply Proposition 9.3.3. □

REMARKS 9.5. Let $\varphi : S \hookrightarrow T$ be an extension of commutative rings.

(a) If $S \hookrightarrow T$ is flat, then $S \hookrightarrow T$ is weakly flat; see [**122**, Theorem 9.5].

(b) Let G be a multiplicative system in S consisting of units of T. Then $S \hookrightarrow G^{-1}S$ is flat and every height-one prime ideal of $G^{-1}S$ is the extension of a height-one prime ideal of S. Thus $S \hookrightarrow T$ is weakly flat $\iff G^{-1}S \hookrightarrow T$ is weakly flat.

REMARKS 9.6. Let $S \hookrightarrow T$ be an extension of Krull domains.

(a) If $S \hookrightarrow T$ is flat, then $S \hookrightarrow T$ is height-one preserving and satisfies PDE; that is, for every height-one prime ideal Q of T, $\operatorname{ht}(Q \cap S) \leq 1$; see Definition 2.14 and Bourbaki, [**22**, Chapitre 7, Proposition 15, page 19].

(b) Let G be a multiplicative system in S consisting of units of T. It follows as in Remarks 9.5.b that:

  (i) $S \hookrightarrow T$ is height-one preserving $\iff G^{-1}S \hookrightarrow T$ is height-one preserving.

  (ii) $S \hookrightarrow T$ satisfies PDE $\iff G^{-1}S \hookrightarrow T$ satisfies PDE.

(c) If each height-one prime ideal of S is the radical of a principal ideal, in particular, if S is a UFD, then the extension $S \hookrightarrow T$ is height-one preserving. To see this, let P be a height-one prime of S and suppose that P is the radical of the principal ideal xS. Then $PT \neq T$ if and only if xT is a proper principal ideal of T. Every proper principal ideal of a Krull domain is contained in a height-one prime. Hence if $PT \neq T$, then PT is contained in a height-one prime of T.

With these results in hand, we return to the investigation of the structure of the Intersection Domain A of Inclusion Construction 5.3. Theorem 9.7 gives conditions that imply the intermediate rings B and A are Krull domains.

THEOREM 9.7. *Assume the setting and notation of Inclusion Construction* 5.3. *In addition assume that R is a Krull domain and R^* is a normal Noetherian domain. Then the intermediate rings A and B have the following properties:*

(1) *A is a Krull domain.*

(2) *For $\mathfrak{p} \in \operatorname{Spec} A$ minimal over xA, let $\mathfrak{q} = \mathfrak{p} \cap B$ and $\mathfrak{a} = \mathfrak{p} \cap R$. Then*
  - $\mathfrak{p} = \mathfrak{q}A = \mathfrak{a}A$ *and* $\mathfrak{q} = \mathfrak{a}B$.
  - $R_{\mathfrak{a}} \subseteq B_{\mathfrak{q}} = A_{\mathfrak{p}}$, *and all three localizations are DVRs.*

(3) $B = B[1/x] \cap B_{\mathfrak{q}_1} \cap \cdots \cap B_{\mathfrak{q}_r}$, *where $\mathfrak{q}_1, \ldots, \mathfrak{q}_r$ are the prime ideals of B minimal over xB.*

(4) *B is a Krull domain.*

(5) $xB \cap B_n = (x, \tau_{1n}, \ldots, \tau_{sn})B_n$ *is an ideal of B_n of height $s+1$, for every $n \in \mathbb{N}$.*

If R is a UFD and x is a prime element of R, then B is a UFD.

PROOF. For property 1, Remark 2.12.1 implies that A is the intersection of the Krull domain R^* with a subfield of $\mathcal{Q}(R^*)$, and so A is a Krull domain.

The first part of property 2 follows since $R/xR = B/xB = A/xA$, from Construction Properties Theorem 5.14.2. Since A is Krull, $\mathfrak{p}$ has height one and $A_{\mathfrak{p}}$ is a DVR. Since R is a Krull domain and $\mathfrak{a}$ is a minimal prime of xR, $R_{\mathfrak{a}}$ is a DVR and the maximal ideal of $R_{\mathfrak{a}}$ is generated by $u \in R$. It also follows from Theorem 5.14.2 that the maximal ideals of $B_{\mathfrak{q}}$ and $A_{\mathfrak{p}}$ are principal generated by u. Since $A_{\mathfrak{p}}$ is a DVR, $\bigcap_{n=1}^{\infty} u^n A_{\mathfrak{p}} = (0)$. It follows that $\bigcap_{n=1}^{\infty} u^n B_{\mathfrak{q}} = (0)$, and so $B_{\mathfrak{q}}$ is a DVR; see Exercise 2 of Chapter 2. Since $A_{\mathfrak{p}}$ has the same field of fractions as $B_{\mathfrak{q}}$, $B_{\mathfrak{q}}$ is birationally dominated by $A_{\mathfrak{p}}$. Thus $B_{\mathfrak{q}}$ is the same DVR as $A_{\mathfrak{p}}$.

For property 3, suppose $\beta \in B[1/x] \cap (B_{\mathfrak{q}_1} \cap \cdots \cap B_{\mathfrak{q}_r}) = B[1/x] \cap (B \setminus (\cup\, \mathfrak{q}_i))^{-1} B$. There exist $t \in \mathbb{N}, a, b, c \in B$ with $c \notin \mathfrak{q}_1 \cup \cdots \cup \mathfrak{q}_r$ such that $\beta = a/x^t = b/c$. If $t = 0$, property 3 holds.

Therefore suppose that $t > 0$ and $a \notin xB$. Then $\mathfrak{p}_1 = \mathfrak{q}_1 A \ldots, \mathfrak{p}_r = \mathfrak{q}_r A$ are the minimal prime ideals of xA in A, by Theorem 5.14.2. Since A is a Krull domain, $A = A[1/x] \cap A_{\mathfrak{p}_1} \cap \cdots \cap A_{\mathfrak{p}_r}$. Hence $\beta \in A$, and $a = x^t \beta \in xA \cap B = xB$, a contradiction. Thus $t = 0$ and $\beta = a \in B$.

For property 4, $B[1/x]$ is a localization of B_0. Since B_0 is a Krull domain, it follows that $B[1/x]$ is a Krull domain. By property 3, B is the intersection of $B[1/x]$ and the DVRs $B_{\mathfrak{q}_1}, \ldots, B_{\mathfrak{q}_r}$, and so B is a Krull domain.

For property 5, let $f \in xB \cap B_n$. After multiplication by a unit of B_n, we may assume that $f \in U_n = R[\tau_{1n}, \ldots, \tau_{sn}]$, and hence f has the form

$$f = \sum_{(i) \in \mathbf{N}^s} a_{(i)} \tau_{1n}^{i_1} \cdots \tau_{sn}^{i_s}$$

with $a_{(i)} \in R$. Since $f \in xB$ and every $\tau_{jn} \in xB$, it follows that the constant term $a_{(0)}$ of f satisfies $a_{(0)} \in xB \cap R \subseteq xR^* \cap R = xR$, by Theorem 5.14.1. Therefore $a_{(0)} \in xR$, so that $f \in (x, \tau_{1n}, \ldots, \tau_{sn})B_n$. Furthermore if $g \in (x, \tau_{1n}, \ldots, \tau_{sn})B_n$, then $\tau_{in} \subseteq xB \cap B_n$, and so $g \in xB \cap B_n$. Thus $xB \cap B_n = (x, \tau_{1n}, \ldots, \tau_{sn})B_n$.

The last statement of Theorem 9.7 holds by Theorem 5.24.1. □

Weak Flatness Theorem 9.9 asserts that if the base ring R and the ring B are both Krull domains, and the extension

$$R[\tau_1, \ldots, \tau_s] \hookrightarrow R^*[1/x]$$

is weakly flat, then A is equal to B; that is, $\tau_1, \ldots, \tau_s$ are limit-intersecting in the sense of Definition 5.10. If R^* is a normal Noetherian domain, the converse also holds.

Observe that, even for the general setting of Inclusion Construction 5.3, weak flatness of the extension $R[\tau_1, \ldots, \tau_s] \hookrightarrow R^*[1/x]$ holds simultaneously with weak flatness of several other related extensions, as Proposition 9.8 shows.

PROPOSITION 9.8. *Assume the setting and notation of Inclusion Construction 5.3. Let B be the Approximation Domain of Definition 5.7. Then the following statements are equivalent:*

(1) *The extension $U_0 := R[\tau_1, \ldots, \tau_s] \hookrightarrow R^*[1/x]$ is weakly flat.*
(2) *The extension $B \hookrightarrow R^*[1/x]$ is weakly flat.*
(3) *The extension $B \hookrightarrow R^*$ is weakly flat.*

If R is a lcoal ring with maximal ideal $\mathfrak{m}$, then statements 1-3 are equivalent to statement 4:

(4) *The extension $B_0 := R[\tau_1, \ldots, \tau_s]_{(\mathfrak{m}, \tau_1, \ldots, \tau_s)} \hookrightarrow R^*[1/x]$ is weakly flat.*

PROOF. For the first part, show statement 3 $\Longrightarrow$ statement 2 $\Longrightarrow$ statement 1 $\Longrightarrow$ statement 3. To see that statement 3 $\Longrightarrow$ statement 2, consider

$$B \overset{\text{w.f.}}{\hookrightarrow} R^* \overset{\text{flat}}{\hookrightarrow} R^*[1/x].$$

Let $\mathfrak{p}$ be a height-one prime ideal of B with $\mathfrak{p}R^*[1/x] \neq R^*[1/x]$. Then $\mathfrak{p}R^* \neq R^*$ and $x \notin \mathfrak{p}$, and so $\mathfrak{p}R^*[1/x] \cap B = \mathfrak{p}R^* \cap B = \mathfrak{p}$, where the last equality uses $B \overset{\text{w.f.}}{\hookrightarrow} R^*$. Therefore statement 2 holds.

Statement 2 $\implies$ statement 1: By Remark 9.5.b, the inclusion $B \overset{\text{w.f.}}{\hookrightarrow} R^*[1/x]$ implies $B[1/x] \overset{\text{w.f.}}{\hookrightarrow} R^*[1/x]$. By Construction Properties Theorem 5.14.2, $B[1/x]$ is a localization of $R[\tau_1, \dots, \tau_s][1/x] = U[1/x]$ at elements of $1+xU$. By Remark 9.5.b, $R[\tau_1, \dots, \tau_s] \overset{\text{w.f.}}{\hookrightarrow} R^*[1/x]$.

To see that statement 1 $\implies$ statement 3, let $\mathfrak{p} \in \operatorname{Spec} B$ have height one and suppose $\mathfrak{p}R^* \neq R^*$. If $x \in \mathfrak{p}$, then Theorem 5.14.3 implies $\mathfrak{p}/xB = \mathfrak{p}R^*/xR^*$, and so $\mathfrak{p}R^* \cap B = \mathfrak{p}$ in this case.

For the case $x \notin \mathfrak{p}$: First observe that $B[1/x] \hookrightarrow R^*[1/x]$ is weakly flat. To see this, since $U_0 = R[\tau_1, \dots, \tau_s] \hookrightarrow R^*[1/x]$ is weakly flat by hypothesis, Remark 9.5.b implies $U[1/x] = U_0[1/x] \hookrightarrow R^*[1/x]$ is weakly flat. Since $B[1/x]$ is a localization of U at elements of $1 + xU$, Remark 9.5.b implies that $B[1/x] \hookrightarrow R^*[1/x]$ is weakly flat, as observed.

Now $\mathfrak{p}R^*[1/x] \neq R^*[1/x]$ and $\mathfrak{p}B[1/x] \cap B = \mathfrak{p}$. Therefore

$$\mathfrak{p}R^* \cap B \subseteq \mathfrak{p}R^*[1/x] \cap B[1/x] \cap B = \mathfrak{p}B[1/x] \cap B = \mathfrak{p}.$$

This completes the proof of Proposition 9.8.

By Remark 9.5.b, statement 1 $\iff$ statement 4. □

WEAK FLATNESS THEOREM 9.9. *Assume Setting* 5.1 *and also assume that R is a Krull domain. Thus $x \in R$ is a nonzero nonunit such that the x-adic completion R^* of R is Noetherian. Also $\tau_1, \dots, \tau_s \in xR^*$ are algebraically independent over R, and every nonzero element of the polynomial ring $R[\tau_1, \dots, \tau_s]$ is regular on R^*. Let $A = \mathcal{Q}(R)(\tau_1, \dots, \tau_s) \cap R^*$, as in Construction* 5.3, *and let B be the Approximation Domain of Definition* 5.7.

1. *If B is a Krull domain and $R[\tau_1, \dots, \tau_s] \hookrightarrow R^*[1/x]$ is weakly flat, then $A = B$; that is, $\tau_1, \dots, \tau_s$ are limit-intersecting as in Definition* 5.10.
2. *If R^* is a normal Noetherian domain, then weak flatness of $R[\tau_1, \dots, \tau_s] \hookrightarrow R^*[1/x]$ is equivalent to $A = B$.*

PROOF. For item 1: Since B is a Krull domain and the extension $B \hookrightarrow A$ is birational, by Proposition 9.3.2, it suffices to show that every height-one prime ideal $\mathfrak{p}$ of B is contracted from A. If $x \in \mathfrak{p}$, then by Construction Properties Theorem 5.14.3, $\mathfrak{p}/xB = \mathfrak{p}A/xA$, and so $\mathfrak{p}$ is contracted from A.

Let $\mathfrak{p}$ be a height-one prime ideal of B that does not contain xB. By Proposition 9.8 the extension $B \hookrightarrow R^*$ is weakly flat, and by Proposition 5.17.1, $\mathfrak{p}R^* \neq R^*$. Hence there exists a prime ideal $\mathfrak{a}$ of R^* with $\mathfrak{a} \cap B = \mathfrak{p}$. It follows that $(\mathfrak{a} \cap A) \cap B = \mathfrak{p}$. Therefore every height-one prime ideal of B is is contracted from A. Thus $A = B$ as desired.

To prove item 2, since R^* is a normal Noetherian domain, B is a Krull domain, and so one direction holds.

For the other direction, R^* is a Krull domain; see Remark 2.12.1. Since $B = A = \mathcal{Q}(B) \cap R^*$, Proposition 9.3 implies the extension $B \hookrightarrow R^*$ is weakly flat. □

REMARK 9.10. In Theorem 9.9, the map $\psi : B \hookrightarrow R^*$ being weakly flat does *not* imply that $R[\tau_1, \dots, \tau_s] \hookrightarrow R^*$ is weakly flat; see Exercise 1.

9.2. Height-one primes in extensions of Krull domains

We observe in Proposition 9.11 that a weakly flat extension of Krull domains is height-one preserving.

PROPOSITION 9.11. *If $\varphi : S \hookrightarrow T$ be a weakly flat extension of Krull domains. For every height-one prime ideal P of S with $PT \neq T$, there exists a height-one prime ideal Q of T with $Q \cap S = P$. Thus φ is height-one preserving.*

PROOF. Let $P \in \operatorname{Spec} S$ with $\operatorname{ht} P = 1$ be such that $PT \neq T$. By Exercise 11 of Chapter 2, there exists a prime ideal Q' of T such that $Q' \cap S = P$. Let a be a nonzero element of P and let $Q \subseteq Q'$ be a minimal prime divisor of aT. Since T is a Krull domain, Q has height one. We have $a \in Q \cap S$. Hence $(0) \neq Q \cap S \subseteq P$. Since $\operatorname{ht} P = 1$, we have $Q \cap S = P$. □

EXAMPLES 9.12. The height-one preserving condition does *not* imply weak flatness. We present two examples: Let x and y be variables over a field k.

(1) Let $\varphi : S := k[x, y] \hookrightarrow T := k[x, \frac{y}{x}]$. Since S is a UFD, the map φ is height-one preserving by Remark 9.6. To see that φ is not weakly flat, let $P = xS$.

(2) The second example arises from Inclusion Construction 5.3. Let $R := k[[x]][y]_{(x,y)}$ and let $C = k[[x, y]]$. There exists an element $\tau \in \mathfrak{n} = (x, y)C$ that is algebraically independent over $\mathcal{Q}(R)$. Fix such an element τ, and let $S := R[\tau]_{(\mathfrak{m},\tau)}$. Since R is a UFD, the ring S is also a UFD and the local inclusion map $\varphi : S \hookrightarrow C$ is height-one preserving. There exists a height-one prime ideal P of S such that $P \cap R = 0$. Since the map $S \hookrightarrow C$ is a local map, we have $PC \neq C$. Because φ is height-one preserving, there exists a height-one prime ideal Q of C such that $PC \subseteq Q$. Fix such a prime ideal Q of C. Since C is the $\mathfrak{m}$-adic completion $\widehat{R}$ of R and the generic formal fiber of R is zero-dimensional, $\dim(C \otimes_R \mathcal{Q}(R)) = 0$; see Discussion 3.29 and Exercise 1 of Chapter 12. Hence $Q \cap R \neq 0$. We have $P \subseteq Q \cap S$ and $P \cap R = (0)$. It follows that P is strictly smaller than $Q \cap S$, so $Q \cap S$ has height greater than one. Therefore the extension $\varphi : S \hookrightarrow C$ is not weakly flat.

Proposition 9.13 describes weakly flat and PDE (pas d'éclatment[1]) extensions.

PROPOSITION 9.13. *Let $\varphi : S \hookrightarrow T$ be an extension of Krull domains.*

(1) *φ is weakly flat $\iff$ for every height-one prime ideal $P \in \operatorname{Spec} S$ such that $PT \neq T$ there is a height-one prime ideal $Q \in \operatorname{Spec} T$ with $P \subseteq Q \cap S$ such that the induced map on the localizations*

$$\varphi_Q : S_{Q \cap S} \longrightarrow T_Q$$

is faithfully flat.

(2) *φ satisfies PDE $\iff$ for every height-one prime ideal $Q \in \operatorname{Spec} T$, the induced map on the localizations*

$$\varphi_Q : S_{Q \cap S} \longrightarrow T_Q$$

is faithfully flat.

PROOF. For the proof of item 1, to see ($\Longleftarrow$), we use that φ_Q a ot flat map implies φ_Q satisfies the Going-down property; see Remark 2.37.10. Hence $\operatorname{ht}(Q \cap S) = 1$, and so $P = Q \cap S$; thus $PT \cap S = P$. For ($\Longrightarrow$), assume $P \in \operatorname{Spec} S$ has height one, $PT \neq T$, and φ is weakly flat. Then Proposition 9.11 implies the existence of $Q \in \operatorname{Spec} T$ of height one such that $Q \cap S = P$. By Remarks 2.12.1 and 2.37.2, φ_Q is flat. Since the rings are local the extension is faithfully flat.

For the proof of item 2, ($\Longrightarrow$), let $Q \in \operatorname{Spec}(T)$ have height one. Then PDE implies that $P = Q \cap S$ has height at most one, and so S_P is a DVR or a field. The

[1] See Definition 2.14

extension is flat by Remarks 2.39, and so the extension is faithfully flat. The ($\Longleftarrow$) direction follows from the fact that a faithfully flat map satisfies the Going-down property. □

Corollary 9.14 is an immediate consequence of Proposition 9.13:

COROLLARY 9.14. *Let $\varphi : S \hookrightarrow T$ be an extension of Krull domains. Then φ satisfies PDE if and only if φ satisfies LF_1.*

EXAMPLE 9.15. A weakly flat extension of Krull domains $\varphi : S \hookrightarrow T$ need not satisfy LF_1. Let x, y and z be variables over a field k, let $S = k[x, y]$, and let $T = S[z, x/z, y/z]$. Then

(1) $T[1/z]$ is a free S-module.
(2) The extension $S \hookrightarrow T[1/z] = S[z, 1/z]$ is faithfully flat.
(3) $S = T[1/z] \cap \mathcal{Q}(S) = T \cap \mathcal{Q}(S)$; by Proposition 9.3.3, φ is weakly flat.
(4) $Q = zT \in \operatorname{Spec} T$, $\operatorname{ht} Q = 1$, and $Q \cap S = (x, y)S \implies \operatorname{ht}(Q \cap S) = 2$.
(5) $S \hookrightarrow T_Q$ is not flat.

We show in Proposition 9.16 that an extension of Krull domains satisfying both the LF_1 condition and the height-one preserving condition is weakly flat. Example 9.17 shows that LF_1 alone does not imply weak flatness.

PROPOSITION 9.16. *Let $\varphi : S \hookrightarrow T$ be an extension of Krull domains that is height-one preserving and satisfies PDE (equivalently, LF_1). Then φ is weakly flat.*

PROOF. Let $P \in \operatorname{Spec} S$ be such that $\operatorname{ht}(P) = 1$ and $PT \neq T$. Since $S \hookrightarrow T$ is height-one preserving, PT is contained in a prime ideal Q of T of height one. The PDE hypothesis on $S \hookrightarrow T$ implies that $Q \cap S$ has height one. It follows that $Q \cap S = P$, and so $PT \cap T = P$. Thus φ is weakly flat. □

Without the assumption that $\varphi : S \hookrightarrow T$ is height-one preserving, it can happen that φ satisfies PDE and yet is not weakly flat.

EXAMPLE 9.17. For extensions of Krull domains, PDE does not imply weakly flat. Since PDE and height-one preserving imply weak flatness, this example also shows that PDE does not imply height-one preserving. Let X, Y, Z, W be indeterminates over a field k and define

$$S := k[x, y, z, w] = \frac{k[X, Y, Z, W]}{(XY - ZW)} \quad \text{and} \quad T := S[\frac{x}{z}].$$

Since $w = \frac{yx}{z}$, the ring $T = k[y, z, \frac{x}{z}]$. Since $\mathcal{Q}(T)$ has transcendence degree 3 over k, the elements $y, z, \frac{x}{z}$ are algebraically independent over k and $T = k[y, z, \frac{x}{z}]$ is a polynomial ring in three variables over k. Let $A = k[X, Y, Z, W]$ and let $F = XY - ZW$. Then $S = A/FA$ and and the partials of F generate a maximal ideal of A. It follows that $S_{\mathfrak{p}}$ is regular for each non-maximal prime ideal $\mathfrak{p}$ of S; see for example [**122**, Theorem 30.3]. Since S is Cohen-Macaulay, it follows from Serre's Normality Theorem 2.9 that S is a normal Noetherian domain. By Remark 2.12.1, S is a Krull domain. The ideal $P := (y, z)S$ is a height-one prime ideal of S because $F \in (Y, Z)A$ and P is isomorphic to the height-one prime ideal $(Y, Z)A/FA$. Since $PT = (y, z)T$ and $\frac{w}{y} = \frac{x}{z}$, we have $(y, z)T \cap S = (y, z, x, w)S$, a maximal ideal of S. Thus the extension $S \hookrightarrow T$ is not weakly flat.

Another way to realize this example is to let r, s, t be indeterminates over the field k, and let $S = k[r, s, rt, st] \hookrightarrow k[r, s, t] = T$. Here we set $r = y, s = z, rt = w$ and $st = x$. Then $P = (r, s)S$. We observe that

$$
\begin{aligned}
T &\subseteq \bigcap \{S_Q \mid Q \in \operatorname{Spec} S\ ,\ \operatorname{ht} Q = 1 \ \text{ and } \ Q \neq P\ \} \\
&\subseteq \bigcup_{n=1}^{\infty} (S :_{\mathcal{Q}(S)} P^n) \ \subseteq \ S\,[\frac{1}{r}] \,\cap\, S\,[\frac{1}{s}] \\
&\subseteq T\,[\frac{1}{r}] \,\cap\, T\,[\frac{1}{s}] \ \subseteq \ T.
\end{aligned}
\tag{9.17.0}
$$

To see the first inclusion in Equation 9.17.0, let Q be a height-one prime ideal of S with $Q \neq P$. Then $r \notin Q$ or $s \notin Q$ and so $T \subseteq S_Q$. For the inclusions in the second line, see Exercise 4 at the end of this chapter and [**24**]. The first inclusion in the third line is obvious, and the last inclusion follows because r and s are non-associate prime elements of the UFD T. It follows that, if $\mathcal{F}(T)$ is the family of essential valuations for T and $\mathcal{F}(S)$ is the family of essential valuation rings for S, then $\mathcal{F}(T) = \mathcal{F}(S) \setminus \{S_P\}$; see Remarks 2.12.2. Therefore every height-one prime ideal of T lies over a height-one prime ideal of S, and so the extension $\varphi : S \hookrightarrow T$ satisfies PDE.

9.3. Flatness for Inclusion Construction 5.3

In this section we examine flatness in the setting of Inclusion Construction 5.3. We use the non-flat locus as in Definition 2.40.

REMARKS 9.18. Assume the setting of Weak Flatness Theorem 9.9. Thus R is a Krull domain. Let $\underline{\tau} = \{\tau_1, \ldots, \tau_s\}$.

(1) Assume that R^* is a normal Noetherian domain. Then the extension $R[\underline{\tau}] \hookrightarrow R^*[1/x]$ is weakly flat $\iff A = B$. By Proposition 9.11 and Proposition 9.8, if $B \hookrightarrow R^*$ is weakly flat, then $B \hookrightarrow R^*$ is height-one preserving. By Theorem 9.9.2, $A = B$ implies $B \hookrightarrow R^*$ is height-one preserving.

(2) If the ring B is Noetherian, then, by Noetherian Flatness Theorem 6.3, $A = B$ and $R[\underline{\tau}] \hookrightarrow R^*[1/x]$ is flat. Since flatness implies weak flatness, we have $R[\underline{\tau}] \hookrightarrow R^*[1/x]$ is weakly flat. By Theorem 9.9, we also have that $B \hookrightarrow R^*[1/x]$ and $B \hookrightarrow R^*$ are weakly flat.

(3) Examples 10.15 describes examples where the constructed rings A and B are equal, but are not Noetherian. The limit-intersecting property holds for these examples.

(4) By Remark 22.39, Examples 22.36 and 22.38 yield extensions of Krull domains that are weakly flat but do not satisfy PDE.

PROPOSITION 9.19. *Assume the setting of Weak Flatness Theorem 9.9, so that R is a Krull domain. Let $\alpha : S := R[\underline{\tau}] \hookrightarrow R^*[1/x]$. Then:*

(1) *Assume that there exists an ideal L of R^* such that the non-flat locus of the extension α is determined by $LR^*[1/x]$; that is, for every Q^* of $\operatorname{Spec}(R^*[1/x])$, $LR^*[1/x] \subseteq Q^* \iff \alpha_{Q^*} : S := R[\underline{\tau}] \hookrightarrow (R^*[1/x])_{Q^*}$ is not flat. Then $\operatorname{ht}(LR^*[1/x]) > 1 \iff \alpha : S \to R^*[1/x]$ satisfies LF_1.*

(2) *Assume that R^* is a normal Noetherian domain, that each height-one prime of R is the radical of a principal ideal, and that α satisfies LF_1. Then $B = A$.*

PROOF. Item 1 follows from the definition of LF_1; see Definition 9.1.3.

For item 2, assume R^* is a normal domain and each height-one prime of R is the radical of a principal ideal. Proposition 2.17 and Fact 2.19 imply that every height-one prime ideal of S is the radical of a principal ideal. Then the extension $R \hookrightarrow R^*$ is height-one preserving by Remark 9.6.c. By Proposition 9.16 , the extension is weakly flat. Theorem 9.9.2 implies that $B = A$. $\square$

Proposition 9.19 leads to the following questions:

QUESTION 9.20. Assume the setting of Weak Flatness Theorem 9.9, and that both R and R^* are normal Noetherian domains. If $\alpha : S = R[\underline{\tau}] \to R^*[1/x]$ satisfies LF_1, then is $A = B$, or, equivalently, is α necessarily weakly flat?

QUESTION 9.21. Assume the setting of Weak Flatness Theorem 9.9, and that both R and R^* are normal Noetherian domains. Then $\beta : R \hookrightarrow R^*[1/x]$ is height-one preserving since it is flat (a composition of a completion map with a localization). Is the local inclusion map $\alpha : S := R[\underline{\tau}]_{(\mathfrak{m},\underline{\tau})} \hookrightarrow R^*[1/x]$ height-one preserving?

REMARK 9.22. Assume the setting of Question 9.20.

(1) If we assume in addition that every height-one prime ideal of R is the radical of a principal ideal, then the answer to both Question 9.20 and 9.21 is "Yes". For Question 9.20, this follows from Proposition 9.19.2. For Question 9.21, this follows since the height-one prime ideals of $R[\underline{\tau}]$ are also radicals of principal ideals by Proposition 2.17. By Remark 9.6.c, we have that $\alpha : S := R[\underline{\tau}]_{(\mathfrak{m},\underline{\tau})} \hookrightarrow R^*[1/x]$ is height-one preserving.

(2) If the answer to Question 9.21 is "Yes", then the answer to Question 9.20 is "Yes", since if $\alpha : S = R[\underline{\tau}] \to R^*[1/x]$ satisfies LF_1 and also is height-one preserving, then Proposition 9.16 implies that α is weakly flat.

The notation for Diagram 9.23.0 is given in Notation 9.23.

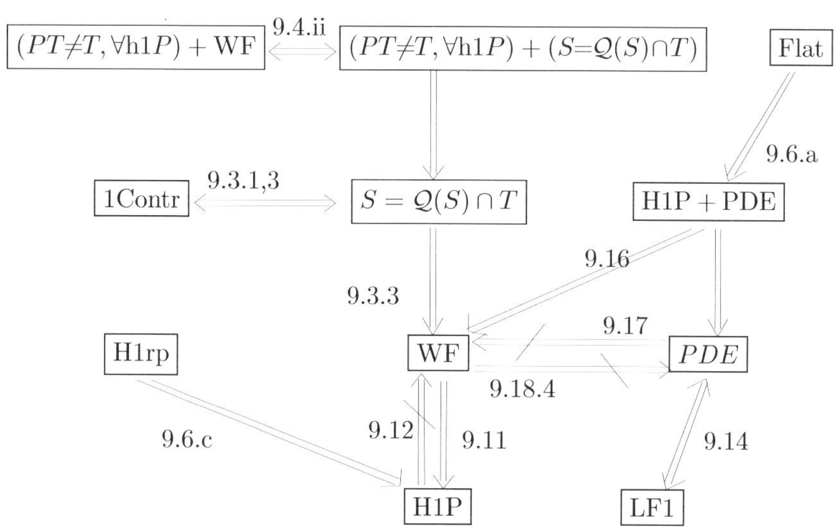

Diagram 9.23.0

NOTATION 9.23. Let $S \hookrightarrow T$ be an extension of Krull domains. Diagram 9.23.0 shows the relationships among the various properties of the extension discussed in this chapter. We use that S and T Krull domains implies the regularity condition of Proposition 9.3.1 and Corollary 9.4; that is, $0 \neq s \in S \implies s$ is a regular element of T. For convenience, we abbreviate other properties as follows: "$PT \neq T, \forall \text{h}1P$" means, "for every height-one prime ideal P of S, $PT \neq T$; "WF":="$S \to T$ is weakly flat" [Definition 9.1.1]; "1Contr":=Every height-one prime ideal P of S is the contraction of a height-one prime ideal Q of T." "Flat":="$S \to T$ is flat"; "H1rp":=Every height-one prime ideal of S is the radical of a principal ideal"; ' 'H1P":="height-one preserving" [Definition 9.1.2]; "LF_1":="Locally flat in height one" [Definition 9.1.3]; "PDE":="No blowing-up" [Definition 2.14]".

Exercises

(1) The following construction gives an example for Remark 9.10. Let $R = k[x, y]_{(x,y)}$ and let $\tau = \sum_{i=1} c_i x^i \in xk[[x]]$ be algebraically independent over $k(x)$, with each $c_i \in k$. Let the Approximation Domain B be a Local Prototype as in Local Prototype Example 4.26. Justify the following assertions:
  (a) B is Noetherian.
  (b) $\psi : B \hookrightarrow R^*$ is weakly flat.
  (c) The extension $R[\tau] \hookrightarrow R^*$ is not weakly flat.
  (d) The extension $R[\tau] \hookrightarrow R^*$ is *never* weakly flat.

  SUGGESTION. For the last statement, consider the prime ideal $P = xR[\tau]$.

(2) Let $T = k[x, y, z]$ be a polynomial ring in the 3 variables x, y, z over a field k, and consider the subring $S = k[xy, xz, yz]$ of T.
  (a) Prove that the field extension $\mathcal{Q}(T)/\mathcal{Q}(S)$ is algebraic with $[\mathcal{Q}(T) : \mathcal{Q}(S)] = 2$.
  (b) Deduce that xy, xz, yz are algebraically independent over k, so S is a polynomial ring in 3 variables over k.
  (c) Prove that the extension $S \hookrightarrow T$ is height-one preserving, but is not weakly flat.
  (d) Prove that $T \cap \mathcal{Q}(S) = S[x^2, y^2, z^2]$ is a Krull domain that properly contains S.
  (e) Prove that the map $S \hookrightarrow T[\frac{1}{xyz}]$ is flat.
  (f) Prove that $S[\frac{1}{xyz}] = T[\frac{1}{xyz}]$. (Notice that $S[\frac{1}{xyz}]$ is not a localization of S since xyz is not in $\mathcal{Q}(S)$.)

(3) In the case where T is also a Krull domain, give a direct proof using primary decomposition of the assertion in Corollary 9.4 that $S = \mathcal{Q}(S) \cap T$ implies T is weakly flat over S.

  SUGGESTION. Let $\mathfrak{p}$ be a height-one prime ideal of S and let $0 \neq a \in \mathfrak{p}$. Since T is a Krull domain, the principal ideal aT has an irredundant primary decomposition

$$aT = Q_1 \cap \cdots \cap Q_s,$$

  where each Q_i is primary for a height-one prime ideal P_i of T.
  (b) Show that $aS = \mathcal{Q}(S) \cap aT$.

(c) Show that after relabeling there exists an integer $t \in \{1, \ldots, s\}$ such that the ideal $Q_1 \cap \ldots \cap Q_t \cap S$ is the $\mathfrak{p}$-primary component of aS. Conclude that $P_i \cap S = \mathfrak{p}$, for some i.

(4) Let A be an integral domain and let I be an ideal generated by the nonzero elements $a_1, \ldots, a_r$ of A. Let $\mathcal{F} = \{P \in \operatorname{Spec} A \mid I \not\subseteq P\}$. For each $n \in \mathbb{N}$ define $I^{-n} := (A :_{\mathcal{Q}(A)} I^n)$. Prove that

$$T := \bigcup_{n=1}^{\infty} I^{-n} = \bigcap_{i=1}^{r} A\left[\frac{1}{a_i}\right] = \bigcap_{P \in \mathcal{F}} A_P.$$

COMMENT. Exercise 4 is a result proved by Jim Brewer [**24**, Prop. 1.4 and Theorem 1.5]. The ring T is called the *I-transform of A*. Ideal transforms were introduced by Nagata [**138**, pp. 35-50] to study the Zariski problem related to the Fourteenth Problem of Hilbert. For a nonzero ideal I of a Krull domain A, the I-transform of A is again a Krull domain.

(5) Let A be an integral domain and let I be a nonzero proper finitely generated ideal of A. Let T be the I-transform of A as in Exercise 4, and let

$$S = \{1 + a \mid a \in I\}.$$

Prove that S is a multiplicatively closed subset of A and $A = S^{-1}A \cap T$.

(6) Let $(R, \mathfrak{m})$ be a 3-dimensional regular local domain with $\mathfrak{m} = (x, y, z)R$, let $\mathfrak{p} = xR$ and let $V = R_{\mathfrak{p}}$. Then V is an essential valuation ring for the Krull domain R, and $R/\mathfrak{p}$ is a 2-dimensional regular local domain. Let $w = \frac{x-y^2}{z}$ and let $T = R[w]_{(y,z,w)R[w]}$.

(a) Prove that T is a 3-dimensional regular local domain that birationally dominates R and is such that $T \subset V$.

(b) Prove that V is an essential valuation ring for the Krull domain T.

(c) Let $\mathfrak{q}$ be the height-one prime ideal of T such that $T_{\mathfrak{q}} = V$. Find an element in T that generates $\mathfrak{q}$.

(d) Prove that $T/\mathfrak{q}$ is a 2-dimensional local domain that birationally dominates the 2-dimensional regular local domain $R/\mathfrak{p}$ and that $T/\mathfrak{q}$ is not regular.

CHAPTER 10

Insider Construction details

In this chapter we continue the development of Insider Construction begun in Section 6.2. Insider Construction 10.7 is a more general construction than the version given in Section 6.2. The base ring R is a Noetherian domain that is not necessarily a polynomial ring over a field.

An Intersection Domain D constructed using Inclusion Construction 5.3 is called a *Noetherian Limit Intersection Domain* if the equivalent conditions of Noetherian Flatness Theorem 6.3.1 hold. Inside a Noetherian Limit Intersection Domain D, it is possible to "iterate" the construction procedure so that inside D there are two integral domains: an intersection A of a field with an ideal-adic completion of R and a domain B that is a nested union of localized polynomial rings over R that "approximates" A. We show that B is Noetherian and equal to A if a certain map of polynomial rings over R is flat.

In Section 10.1 we present *Prototypes*; they are Intersection Domains E obtained using Inclusion Construction 5.3 in the standard Setting 10.1, where R is a polynomial ring over a field k, R^* is the completion of R with respect to the variable x, and $\tau_1, \ldots, \tau_s \in xk[[x]]$ are algebraically independent over $k(x)$. The Noetherian Flatness Theorem 6.3 holds for Prototypes, and so E equals the associated Approximation Domain. We show in Prototype Theorem 10.2 that Prototypes are localized polynomial rings over DVRs. As such, they are always excellent if the underlying DVR is excellent.

Prototype Theorem 10.2 is used in many of our examples. It is vital to the Insider Construction and the rest of this chapter. Proposition 10.4 domonstrates the importance of requiring characteristic zero in order to obtain excellence.

In Section 10.2, we describe background and notation for Insider Construction 10.7. Theorem 10.9 of Section 10.3 gives necessary and sufficient conditions for the integral domains constructed with Insider Construction 10.7 to be Noetherian and equal. We apply the analysis of flatness for polynomial extensions from Chapter 7 to obtain a general flatness criterion for Insider Construction 10.7. This yields examples where the constructed domains A and B are equal and are not Noetherian.

In Section 10.4 we discuss the preservation of excellence for Insider Construction 10.7. Assume the Intersection Domain A and the Approximation Domain B that result from the Insider Construction are equal and Noetherian and the base ring R is excellent. Theorem 10.17 gives necessary and sufficient conditions for A to be excellent.

Insider Construction 10.7 is a maneuver for constructing examples. We use it in Chapters 11, 14, 15, and 16.

10.1. Localized polynomial rings over special DVRs

As stated above, "Prototypes" are intersection domains A obtained using Inclusion Construction 5.3 with the "standard" Setting 10.1. The Prototypes that arise with this setting are polynomial rings over special DVRs that are equal to their approximation domains.

For convenience, the definitions of the intersection and approximation domains corresponding to the construction from Section 5.2 are given again.

SETTING 10.1. Let x be an indeterminate over a field k. Let r be a nonnegative integer and s a positive integer. Assume that $\tau_1, \dots, \tau_s \in xk[[x]]$ are algebraically independent over $k(x)$ and let $y_1, \dots, y_r$ be additional indeterminates. We define the following rings:

(10.1.a) $R := k[x, y_1, \dots, y_r]$, $R^* = k[y_1, \dots, y_r][[x]]$, $V = k(x, \tau_1, \dots, \tau_s) \cap k[[x]]$.

Notice that R^* is the x-adic completion of R and V is a DVR.

The "Prototype" is described using the Intersection Domain of Inclusion Construction 5.3 and the Approximation Domain of Section 5.2. Its development is similar to that of the Local Prototype of Definition 4.28:

$$D := k(x, y_1, \dots, y_r, \tau_1, \dots \tau_s) \cap R^*, \quad E := (1 + xU)^{-1}U, \tag{10.1.b}$$

where $U := \bigcup U_{j \in \mathbb{N}_0}$, each $U_j = R[\tau_{1j}, \dots \tau_{sj}]$, each τ_{ij} is the jth endpiece of τ_i and each $\tau_i \in R^*$, for $1 \le i \le s$. By Construction Properties Theorem 5.14.3, the ring R^* is the x-adic completion of each of the rings D, E and U.

PROTOTYPE THEOREM 10.2. *Assume Setting* 10.1. *Thus* $R := k[x, y_1, \dots, y_r]$, *and* $R^* = k[y_1, \dots, y_r][[x]]$. *Let* V, D *and* E *be as defined in Equations* 10.1.*a and* 10.1.*b*. *Then:*

(1) *The canonical map* $\alpha : R[\tau_1, \dots \tau_s] \hookrightarrow R^*[1/x]$ *is flat.*
(2) $D = E$ *is Noetherian of dimension* $r + 1$ *and is the localization* $(1 + xV[y_1, \dots, y_r])^{-1}V[y_1, \dots, y_r]$ *of the polynomial ring* $V[y_1, \dots, y_r]$ *over the DVR* V. *Thus* D *is a regular integral domain.*
(3) E *is a directed union of localizations of polynomial rings in* $r + s + 1$ *variables over* k.
(4) *If* k *has characteristic zero, then the ring* $E = D$ *is excellent.*

PROOF. The map $k[x, \tau_1, \dots \tau_s] \hookrightarrow k[[x]][1/x]$ is flat by Remark 2.37.4 since $k[[x]][1/x]$ is a field. By Fact 2.38

$$k[x, \tau_1, \dots \tau_s] \otimes_k k[y_1, \dots, y_r] \hookrightarrow k[[x]][1/x] \otimes_k k[y_1, \dots, y_r]$$

is flat. We also have $k[[x]][1/x] \otimes_k k[y_1, \dots, y_r] \cong k[[x]][y_1, \dots, y_r][1/x]$ and

$$k[x, \tau_1, \dots \tau_s] \otimes_k k[y_1, \dots, y_r] \cong k[x, y_1, \dots, y_r][\tau_1, \dots \tau_s].$$

Hence the natural inclusion map

$$k[x, y_1, \dots, y_r][\tau_1, \dots \tau_s] \overset{\beta}{\hookrightarrow} k[[x]][y_1, \dots, y_r][1/x]$$

is flat. Also $k[[x]][y_1, \dots, y_r] \hookrightarrow k[y_1, \dots, y_r][[x]]$ is flat since it is the map taking a Noetherian ring to an ideal-adic completion; see Remark 3.3.2. Therefore

$$k[[x]][y_1, \dots, y_r][1/x] \overset{\delta}{\hookrightarrow} k[y_1, \dots, y_r][[x]][1/x]$$

is flat. It follows that the map

$$k[x, y_1, \dots, y_r][\tau_1, \dots \tau_s] \overset{\delta \circ \beta}{\hookrightarrow} R^*[1/x] = k[y_1, \dots, y_r][[x]][1/x]$$

is flat. Thus the Noetherian Flatness Theorem 6.3 implies $E = D$ and E is Noetherian, and so we have proved items 1, 3 and part of item 2.

To see E is the localization described in item 2, we use that $V[y_1, \dots, y_r] \subseteq D$ and that x is in the Jacobson radical of D by Construction Properties Theorem 5.14. Thus every element of $1 + xV[y_1, \dots, y_r]$ is invertible in D. Hence

$$(1 + xV[y_1, \dots, y_r])^{-1}V[y_1, \dots, y_r] \subseteq D.$$

Since each U_n is contained in $V[y_1, \dots, y_r]$, we have $U \subseteq V[y_1, \dots, y_r]$. We also have $E = (1 + xU)^{-1}U$, and so $E \subseteq (1 + xV[y_1, \dots, y_r])^{-1}V[y_1, \dots, y_r]$. This completes item 2.

For item 4, if k has characteristic zero, then V is excellent by Remark 3.48; hence item 4 follows from item 2 since excellence is preserved under localization of a finitely generated algebra by Remark 3.48. For more details see [**120**, (34.B),(33.G) and (34.A)], [**62**, Chap. IV]. □

DEFINITION 10.3. For integers r and s, indeterminates $x, y_1, \dots, y_r$ over a field k, and elements $\tau_1, \dots \tau_s \in k[[x]]$ that are algebraically independent over $k(x)$, we refer to the ring

$$\begin{aligned} D &:= k(x, y_1, \dots, y_r, \tau_1, \dots \tau_s) \cap k[y_1, \dots, y_r][[x]] \\ &= (1 + xV[y_1, \dots, y_r])^{-1}V[y_1, \dots, y_r], \end{aligned}$$

where $k[x, y_1, \dots, y_r]$ and V is the DVR $k(x, \tau_1, \dots, \tau_s) \cap k[[x]]$, as a *Prototype*. The ring D depends upon the field k, the integers r and s, and the choice of $\tau_1, \dots \tau_s$, and D is also called an *Inclusion Construction Prototype*.

We observe in Proposition 10.4 that over a perfect field k of characteristic $p > 0$ (so that $k = k^{1/p}$) a one-dimensional form of the construction in Prototype Theorem 10.2 yields a DVR that is not a Nagata ring, defined in Definition 2.20, and thus is not excellent; see Remark 3.48, [**122**, p. 264], [**120**, Theorem 78, Definition 34.A].

PROPOSITION 10.4. *Let k be a perfect field of characteristic $p > 0$, let the element τ of $xk[[x]]$ be such that x and τ are algebraically independent over k and set $V := k(x, \tau) \cap k[[x]]$. Then V is a DVR for which the integral closure $\overline{V}$ of V in the purely inseparable field extension $k(x^{1/p}, \tau^{1/p})$ is not a finitely generated V-module. Hence V is not a Nagata ring and so is not excellent.*

PROOF. It is clear that V is a DVR with maximal ideal xV. Since x and τ are algebraically independent over k, $[k(x^{1/p}, \tau^{1/p}) : k(x, \tau)] = p^2$. Let W denote the integral closure of V in the field extension $k(x^{1/p}, \tau)$ of degree p over $k(x, \tau)$. Notice that

$$W = k(x^{1/p}, \tau) \cap k[[x^{1/p}]] \quad \text{and} \quad \overline{V} = k(x^{1/p}, \tau^{1/p}) \cap k[[x^{1/p}]]$$

are both DVRs having residue field k and maximal ideal generated by $x^{1/p}$. Thus $\overline{V} = W + x^{1/p}\overline{V}$. If $\overline{V}$ were a finitely generated W-module, then by Nakayama's Lemma it would follow that $W = \overline{V}$. This is impossible because $\overline{V}$ is not birational over W. It follows that $\overline{V}$ is not a finitely generated V-module, and hence V is not a Nagata ring. □

REMARK 10.5. Let $V = k(x,\tau) \cap k[[x]]$, let D be as in Setting 10.1 with $s = r = 1$, and suppose that k is a perfect field with characteristic $p > 0$. By Proposition 10.4, the ring V is not excellent. By Prototype Theorem 10.2.2, $D = (1+xV[y])^{-1}V[y]$, and so the ring V is a homomorphic image of D. Since excellence is preserved by taking homomorphic images, the two-dimensional regular ring D is not excellent in this situation; see Remark 3.48. The same argument applies if we put more variables in place of y, that is, $y_1, \dots, y_r$, as in Theorem 10.2. In general, over a perfect field of characteristic $p > 0$, the Noetherian regular ring $D = E$ obtained in Prototype Theorem 10.2 fails to be excellent.

We give below a localized form of Prototype Theorem 10.2, with the rings R, D, and E local. This works out to be the same as our first version of the Local Prototype given in Local Prototype Example 4.26, Proposition 4.27 and Definition 4.28, because of Remark 5.16.4.

More explicitly, consider two versions of the construction:

$$\bigcup_{j=1}^{\infty}(U_j)_{\mathfrak{m}_j} = U_{\mathfrak{m}_U} = \bigcup_{j=1}^{\infty}(U'_j)_{\mathfrak{m}'_j} = U'_{\mathfrak{m}_{U'}}, \text{ where } U = \bigcup_{j=1}^{\infty} U_j,\ U' = \bigcup_{j=1}^{\infty} U'_j,$$

$$\begin{aligned} U_j &= k[x, y_1, \dots, y_r]_{(x,y_1,\dots,y_r)}[\tau_{1j}, \dots, \tau_{sj}], \quad \mathfrak{m}_j = (x, y_1, \dots, y_r, \tau_{1j}, \dots, \tau_{sj})U_j, \\ U'_j &= k[x, y_1, \dots, y_r, \tau_{1j}, \dots, \tau_{sj}], \quad \mathfrak{m}'_j = (x, y_1, \dots, y_r, \tau_{1j}, \dots, \tau_{sj})U'_j, \\ \mathfrak{m}_U &= (x, y_1, \dots, y_r)U \quad \text{and} \quad \mathfrak{m}'_U = (x, y_1, \dots, y_r)U'. \end{aligned}$$

Then, with Setting 10.1, the ring E is a localization of $U = \bigcup_{j=1}^{\infty} U_j$, where each $U_j = R[\{\tau_{ij}\}_{i=1}^{s}]$, and E is also a localization of $U' = \bigcup_{j=1}^{\infty} U'_j$, where each $U'_j = k[x, \{y_s\}_{s=1}^{r}, \{\tau_{ij}\}_{i=1}^{s}]$. This simpler second form U' of U is used in Chapters 14 and 16.

LOCAL PROTOTYPE THEOREM 10.6. *If we adjust Setting* 10.1 *so that the base ring is the regular local ring* $R := k[x, y_1, \dots, y_r]_{(x,y_1,\dots,y_r)}$, *then, for the Localized Prototype (Intersection Domain)*

$$D := k(x, y_1, \dots, y_r, \tau_1, \dots, \tau_s) \cap k[y_1, \dots, y_r]_{(y_1,\dots,y_r)}[[x]]$$

of Inclusion Construction 5.3, *the conclusions of Prototype Theorem* 10.2 *are still valid; i.e.:*

(1) *With* $E = (1 + xU)^{-1}U$ *and* $V = k(x, \tau_1, \dots, \tau_s) \cap k[[x]]$, *we have*

$$D = E = V[y_1, \dots, y_r]_{(x,y_1,\dots,y_r)}$$

is a Noetherian regular local ring, and the extension $R[\tau_1, \dots, \tau_s] \to R^*[1/x]$ *is flat. In addition,* $E = \bigcup_{j=1}^{\infty}(U_j)_{\mathfrak{m}_j} = U_{\mathfrak{m}_U} = \bigcup_{j=1}^{\infty}(U'_j)_{\mathfrak{m}'_j} = U'_{\mathfrak{m}_{U'}}$, *where* U *and* U' *are as defined above.*

(2) *If* k *has characteristic zero, then Local Prototype* D *is excellent.*

PROOF. The proof of Theorem 10.2 applies to the localized polynomial rings. The statements about the rings U and U' follow from Remark 5.16. □

10.2. Describing the construction

The setting for Insider Construction 10.7 includes Noetherian domains that are not necessarily local. Thus it generalizes Settings 6.18 and 6.21.

INSIDER CONSTRUCTION 10.7. Let R be a Noetherian integral domain with field of fractions $\mathcal{Q}(R) = K$. Let x be a nonzero nonunit of R and let R^* be the x-adic completion of R. Let $\tau_1, \dots, \tau_s \in xR^*$ be algebraically independent over K, and let $\underline{\tau}$ abbreviate the list $\tau_1, \dots, \tau_s$. Assume that $K(\underline{\tau}) \subseteq \mathcal{Q}(R^*)$, the total ring of fractions of R^*, and that

$$T \ := \ R[\underline{\tau}] \ = \ R[\tau_1, \dots, \tau_s] \ \overset{\psi}{\hookrightarrow} \ R^*[1/x]$$

is flat. Define $D := K(\underline{\tau}) \cap R^*$. Since the map ψ is flat, Noetherian Flatness Theorem 6.3 implies the integral domain D is a Prototype.

We construct "insider" examples inside D as follows: Choose polynomials $f_1, \dots, f_m$ in $T := R[\underline{\tau}]$ that are algebraically independent over K. Assume that each $f_i \in (\underline{\tau})T \subseteq xR^*$, and abbreviate $f_1, \dots, f_m$ by $\underline{f}$. We have $m \le n$.

Define the Intersection Domain $A := K(\underline{f}) \cap R^*$, as in Inclusion Construction 5.3, and let B be the approximation domain corresponding to$\underline{f}$, as in Section 5.2. We have $B \subseteq D$. By Construction Properties Theorem 5.14.4, the integral domains $D[1/x]$ and $B[1/x]$ are localizations of $R[\underline{\tau}]$ and $R[\underline{f}]$, respectively,

Set $S := R[\underline{f}] = R[f_1, \dots, f_m]$, let φ be the embedding

(10.7.1) $$\varphi : S : = \ R[\underline{f}] \overset{\varphi}{\hookrightarrow} T \ := \ R[\underline{\tau}],$$

and let ψ be the inclusion map: $R[\underline{\tau}] \hookrightarrow R^*[1/x]$. Put $\alpha := \psi \circ \varphi : S \to R^*[1/x]$. Then we have

(10.7.2)
$$\begin{array}{ccccc} & & & & R^*[1/x] \\ & & & \overset{\alpha:=\psi\varphi}{\nearrow} & \uparrow \psi \\ R & \subseteq & S := R[\underline{f}] & \xrightarrow{\varphi} & T := R[\underline{\tau}]. \end{array}$$

Theorem 6.17 of Chapter 6 shows in a special case if $\varphi_x : S \hookrightarrow T[1/x]$ is flat, then A is Noetherian and is equal to the approximation domain B. In Section 10.3 we show this more generally.

REMARK 10.8. If R is a Noetherian local domain, then R^* is local and the intersection domains D and A are also local with A possibly non-Noetherian. By Construction Properties Theorem 5.14.6, the approximation domain B is also local, and D dominates B.

10.3. The non-flat locus of Insider Construction 10.7

Assume the notation of Insider Construction 10.7. [1] Let

(10.8.1) $$F := \cap\{P \in \operatorname{Spec} T \mid \varphi_P : S \to T_P \text{ is not flat }\},$$

where $\varphi : S := R[\underline{f}] \overset{\varphi}{\hookrightarrow} T := R[\underline{\tau}]$ is as in Equation 10.7.1. By Remark 7.17.2, the ideal F defines the non-flat locus of the map φ. Two advantages of going inside the Prototype are:

(1) The non-flat locus of $\alpha : T := R[\underline{\tau}] \hookrightarrow R^*[1/x]$ is known to be closed.

(2) An investigation of $\varphi : S := R[\underline{f}] \hookrightarrow T := R[\underline{\tau}]$ might give information concerning flatness for $\psi : S \hookrightarrow R^*[1/x]$.

[1] For details concerning the approximation domain B, see Section 5.2.

Theorem 10.9 relates flatness of the maps $\alpha : S \hookrightarrow R^*[1/x]$ and $\varphi : S \to T$ in Diagram 10.7.2. For $\mathfrak{q}^* \in \operatorname{Spec}(R^*[1/x])$, consider flatness of the localization $\varphi_{\mathfrak{q}^* \cap T}$ of φ :

$$\varphi_{\mathfrak{q}^* \cap T} : S \longrightarrow T_{\mathfrak{q}^* \cap T} \tag{10.8.2}$$

THEOREM 10.9. *Let R be a Noetherian domain, let x be a nonzero nonunit of R and let R^* be the x-adic completion of R. Assume the notation of Insider Construction 10.7. Let F be an ideal of T such that F defines the non-flat locus of $\varphi : S \hookrightarrow T$ as in Equation 10.8.1 above. Then:*

(1) *For $\mathfrak{q}^* \in \operatorname{Spec}(R^*[1/x])$, the map $\alpha_{\mathfrak{q}^*} : S \to (R^*[1/x])_{\mathfrak{q}^*}$ is flat if and only if the map $\varphi_{\mathfrak{q}^* \cap T}$ in Equation 10.8.2 is flat.*
(2) *$FR^*[1/x]$ defines the non-flat locus of the map $\alpha : S \to R^*[1/x]$.*
(3) *The following are equivalent:*
   (i) *The ring A is Noetherian and $A = B$.*
   (ii) *The ring B is Noetherian.*
   (iii) *For every maximal $\mathfrak{q}^* \in \operatorname{Spec}(R^*[1/x])$, the map $\varphi_{\mathfrak{q}^* \cap T}$ in Equation 10.8.2 is flat.*
   (iv) *$FR^*[1/x] = R^*[1/x]$, where F is the ideal of T given in Equation 10.8.1.*
(4) *The map $\varphi_x : S \hookrightarrow T[1/x]$ is flat if and only if $FT[1/x] = T[1/x]$. Moreover, either of these equivalent conditions implies B is Noetherian and $B = A$. It then follows that $A[1/x]$ is a localization of S.*
(5) *If x is in the Jacobson radical of R and the conditions of item 3 or item 4 hold, then $\dim R = \dim A = \dim R^*$.*

PROOF. For item 1, we have $\alpha_{\mathfrak{q}^*} = \psi_{\mathfrak{q}^*} \circ \varphi_{\mathfrak{q}^* \cap T} : S \to T_{\mathfrak{q}^* \cap T} \to (R^*[1/x])_{\mathfrak{q}^*}$, where $\psi : T \hookrightarrow R^*[1/x]$ is as in Diagram 10.7.2. Since the map $\psi_{\mathfrak{q}^*}$ is faithfully flat, the composition $\alpha_{\mathfrak{q}^*}$ is flat if and only if $\varphi_{\mathfrak{q}^* \cap T}$ is flat; see Remarks 2.37.13 and 2.37.15.

For item 2, since $T \hookrightarrow R^*[1/x])$ is flat, and the non-flat locus of $\varphi : S \hookrightarrow T$ is defined by the ideal F, Proposition 2.43 implies that the non-flat locus of α is closed and defined by the subset F or, equivalently, defined by the ideal $FR^*[1/x]$ of $R^*[1/x]$.

For item 3, the equivalence of (i) and (ii) is part of Theorem 6.3. The equivalence of (ii) and (iii) follows from item 1 and Theorem 6.3. For the equivalence of (iii) and (iv), we use $FR^* \neq R^* \iff F \subseteq \mathfrak{q}^* \cap T$, for some $\mathfrak{q}^*$ maximal in $\operatorname{Spec}(R^*[1/x]) \iff$ the map in Equation 10.8.2 fails to be flat.

The first statement of item 4 follows from the definition of F and the fact that the non-flat locus of $\varphi : S \to T$ is closed. Noetherian Flatness Theorem 6.3 implies the final statement of item 4.

Item 5 follows by Remark 3.3.4. □

Corollary 10.10 gives the standard format in which Theorem 10.9 is usually applied.

COROLLARY 10.10. *Let k be a field and let $x, \underline{y} = y_1, \dots, y_r$ be indeterminates over k. Let $R = k[x, \underline{y}]_{(x,\underline{y})}$ or $R = k[x, \underline{y}]$, and let R^* be the x-adic completion of R. Let $\underline{\tau} = \tau_1, \dots, \tau_s$ be elements of $xk[[x]]$ that are algebraically independent over $k[x]$. Let $D = k(x, \underline{y}, \underline{\tau}) \cap R^*$ be the Protoype. Let $\underline{f} = f_1, \dots f_m \in (\underline{\tau})R[\underline{\tau}]$ be algebraically independent over R. Let $A = k(x, \underline{y}, \underline{f}) \cap R^*$ be the Intersection Domain* *corresponding to the* $\underline{f}$ *and let* B *be the Approximation Domain corresponding to the set* $\underline{f}$. *Let* F *be an ideal of* T *such that* F *defines the non-flat locus of* $\varphi : S \hookrightarrow T$ *as in Equation 10.8.1 above. Then:*

(1) *For* $\mathfrak{q}^* \in \operatorname{Spec}(R^*[1/x])$, *the map* $\alpha_{\mathfrak{q}^*} : S \to (R^*[1/x])_{\mathfrak{q}^*}$ *is flat if and only if the map* $\varphi_{\mathfrak{q}^* \cap T}$ *in Equation 10.8.2 is flat.*

(2) $FR^*[1/x]$ *defines the non-flat locus of the map* $\alpha : S \to R^*[1/x]$.

(3) *The following are equivalent:*
   (i) *The ring* A *is Noetherian and* $A = B$.
   (ii) *The ring* B *is Noetherian.*
   (iii) *For every maximal* $\mathfrak{q}^* \in \operatorname{Spec}(R^*[1/x])$, *the map* $\varphi_{\mathfrak{q}^* \cap T}$ *in Equation 10.8.2 is flat.*
   (iv) $FR^*[1/x] = R^*[1/x]$, *where* F *is the ideal of* T *given in Equation 10.8.1.*

(4) *The map* $\varphi_x : S \hookrightarrow T[1/x]$ *is flat if and only if* $FT[1/x] = T[1/x]$. *Moreover, either of these equivalent conditions implies* B *is Noetherian and* $B = A$. *It then follows that* $A[1/x]$ *is a localization of* S.

(5) *If* x *is in the Jacobson radical of* R *and the conditions of item 3 or item 4 hold, then* $\dim R = \dim A = \dim R^*$.

Corollary 10.11 demonstrates the power of Local Flatness Theorem 6.13. It leads to item 8 of Theorem 10.12, and is useful for the analysis of Examples 10.15.

COROLLARY 10.11. *Let* R *be a Noetherian domain, let* x *be a nonzero nonunit of* R *and let* R^* *be the* x*-adic completion of* R. *Assume the notation of Theorem 10.9. Then we have:*

(1) *If* $\varphi : S \to T$ *is flat, then the ring* B *is Noetherian and* $B = A$.

(2) *Let* $G = (F, x)R^* \cap R$ *and let* $P \in \operatorname{Spec} B$. *Then* B_P *is Noetherian* $\iff G \not\subseteq P$.

PROOF. Item 1 follows from item 4 of Theorem 10.9. For item 2, if $x \notin P$, then B_P is Noetherian by Remark 6.14. If $x \in P$, apply item 2 of Theorem 10.9 and Local Flatness Theorem 6.13. □

Theorem 10.12 concerns a case where the non-flat locus can be described more precisely.

THEOREM 10.12. *Let* R *be a Noetherian integral domain, let* x *be a nonzero nonunit of* R *and let* R^* *be the* x*-adic completion of* R. *With the notation of Insider Construction 10.7, assume* $m = 1$, *set* $f_1 = f \in (\underline{\tau})T := (\underline{\tau})R[\tau_1, \ldots, \tau_s]$, *where* f *is a nonconstant polynomial, and let* $S := R[f]$. *Let* B *and* A *be the approximation domain and intersection domain associated to* f *over* R, *and let* L *be the ideal in* R *generated by the nonconstant coefficients of* f *as a polynomial in* T. *Then:*

(1) *The ideal* LT *defines the non-flat locus of* $\varphi : S \hookrightarrow T$.

(2) *The ideal* $LR^*[1/x]$ *defines the non-flat locus of* $\alpha : S \hookrightarrow R^*[1/x]$.

(3) *The ideal* $LR^*[1/x]$ *defines the non-flat locus of* $\beta : B \hookrightarrow R^*[1/x]$.

(4) *The following are equivalent:*
   (a) B *is Noetherian.*
   (b) B *is Noetherian and* $B = A$.
   (c) *The extension* $\alpha : S \hookrightarrow R^*[1/x]$ *is flat.*
   (d) *For each* $Q^* \in \operatorname{Spec}(R^*[1/x])$, *we have* $LR^*[1/x]_{Q^*} = R^*[1/x]_{Q^*}$.
   (e) *For each* $Q^* \in \operatorname{Spec}(R^*[1/x])$, *we have* $LR_q = R_q$, *where* $q = Q^* \cap R$.

(5) *If* ht $LR^*[1/x] = d$, *then the map* $\alpha : S \hookrightarrow R^*[1/x]$ *satisfies* LF_{d-1}, *but not* LF_d, *as defined in Definition* 9.1.3.
(6) $\varphi_x : S \hookrightarrow T[1/x]$ *is flat* $\iff LT[1/x] = T[1/x]$
$\iff LR[1/x] = R[1/x] \iff LR^*[1/x] = R^*[1/x]$.
(7) *The equivalent conditions in item 6 imply the insider approximation domain B is Noetherian and is equal to the insider intersection domain A.*
(8) *Let* $G = (L, x)R^* \cap R$ *and let* $P \in \operatorname{Spec} B$. *Then* B_P *is Noetherian* $\iff G \not\subseteq P$.

PROOF. Item 1 is Corollary 7.29, and item 2 is Theorem 10.9.2.

By Proposition 5.23, the non-flat loci for the two maps α and β are the same. Thus item 2 implies item 3.

For item 4, (a), (b) and (c) are equivalent by Noetherian Flatness Theorem 6.3. By item 2, (c) and (d) are equivalent. Since L is an ideal of R, (d) is equivalent to (e); that is $L \not\subseteq Q^* \iff L \not\subseteq Q^* \cap R = q$.

For item 5, assume that $\operatorname{ht}(LR^*[1/x]) = d$. Let $Q^* \in \operatorname{Spec}(R^*[1/x])$. The map $\alpha_{Q^*} : S \hookrightarrow (R^*[1/x])_{Q^*}$ is flat $\iff L \not\subseteq Q^*$ by item 1. Thus α_{Q^*} is flat for every Q^* with $\operatorname{ht} Q^* < d$, and so α satisfies LF_{d-1}. On the other hand, there exists $Q^* \in \operatorname{Spec}(R^*[1/x])$ such that $L \subseteq Q^*$ and $\operatorname{ht} Q^* = d$. By item 1, the map α_{Q^*} is not flat. Thus α does not satisfy LF_d.

For item 6, item 1 states that LT defines the non-flat locus of the map $\varphi : S \hookrightarrow T$. Thus $S \hookrightarrow T[1/x]$ is flat $\iff LT[1/x] = T[1/x]$. Since L is an ideal of R, and $T[1/x]$ is a polynomial ring over $R[1/x]$, we have $LT[1/x] = T[1/x] \iff LR[1/x] = R[1/x]$. This also holds if and only if $LR^*[1/x] = R^*[1/x]$, by items 1 and 2 and Theorem 10.9.1.

If $S \hookrightarrow T[1/x]$ is flat, then Theorem 10.9.4 implies that B is Noetherian and $B = A$. Thus item 7 holds.

Item 8 follows from Corollary 10.11.2. □

Corollary 10.13 is a restatement of Theorem 10.12 using the standard setting for Prototype Examples.

COROLLARY 10.13. *Let* k *be a field and let* $x, \underline{y} = y_1, \ldots, y_r$ *be indeterminates over* k. *Let* $R = k[x, \underline{y}]_{(x,\underline{y})}$ *or* $R = k[x, \underline{y}]$, *and let* R^* *be the* x*-adic completion of* R. *Let* $\underline{\tau} = \tau_1, \ldots, \tau_s$ *be elements of* $xk[[x]]$ *that are algebraically independent over* $k[x]$. *Let* $D = k(x, \underline{y}, \underline{\tau}) \cap R^*$ *be the Protoype. Let* $f \in (\underline{\tau})R[\underline{\tau}]$ *be algebraically independent over* R. *Let* $A = k(x, \underline{y}, \underline{f}) \cap R^*$ *be the Intersection Domain corresponding to* f *and let* B *be the Approximation Domain corresponding to* f. *Let* F *be an ideal of* T *such that* F *defines the non-flat locus of* $\varphi : S \hookrightarrow T$ *as in Equation* 10.8.1 *above. Let* B *and* A *be the Approximation Domain and Intersection Domain associated to* f *over* R, *and let* L *be the ideal in* R *generated by the nonconstant coefficients of* f *as a polynomial in* T. *Then:*

(1) *The ideal* LT *defines the non-flat locus of* $\varphi : S \hookrightarrow T$.
(2) *The ideal* $LR^*[1/x]$ *defines the non-flat locus of* $\alpha : S \hookrightarrow R^*[1/x]$.
(3) *The ideal* $LR^*[1/x]$ *defines the non-flat locus of* $\beta : B \hookrightarrow R^*[1/x]$.
(4) *The following are equivalent:*
 (a) *B is Noetherian.*
 (b) *B is Noetherian and* $B = A$.
 (c) *The extension* $\alpha : S \hookrightarrow R^*[1/x]$ *is flat.*
 (d) *For each* $Q^* \in \operatorname{Spec}(R^*[1/x])$, *we have* $LR^*[1/x]_{Q^*} = R^*[1/x]_{Q^*}$.

(e) *For each* $Q^* \in \operatorname{Spec}(R^*[1/x])$, *we have* $LR_q = R_q$, *where* $q = Q^* \cap R$.

(5) *If* $\operatorname{ht} LR^*[1/x] = d$, *then the map* $\alpha : S \hookrightarrow R^*[1/x]$ *satisfies* LF_{d-1}, *but not* LF_d, *as defined in Definition* 9.1.3.

(6) $\varphi_x : S \hookrightarrow T[1/x]$ *is flat* $\iff LT[1/x] = T[1/x]$
$\iff LR[1/x] = R[1/x] \iff LR^*[1/x] = R^*[1/x]$.

(7) *The equivalent conditions in item 6 imply the insider approximation domain* B *is Noetherian and is equal to the insider intersection domain* A.

(8) *Let* $G = (L,x)R^* \cap R$ *and let* $P \in \operatorname{Spec} B$. *Then* B_P *is Noetherian* $\iff G \not\subseteq P$.

Example 10.14 illustrates that in Theorem 10.12 the map $\varphi_x : S \hookrightarrow T[1/x]$ may fail to be flat even though the map $\alpha : S \hookrightarrow R^*[1/x]$ is flat.

EXAMPLE 10.14. Let $R = k[x]$, where x is an indeterminate over a field k. Let $\tau \in xk[[x]]$ be such that x and τ are algebraically independent over k. Let $T = R[\tau]$, let $f = (1-x)\tau$, and let $S = R[f]$. The ideal L of R generated by the nonconstant coefficients of f is $L = (1-x)R$. The map $\varphi_x : S \hookrightarrow T[1/x]$ is not flat, but the map $\alpha : S \hookrightarrow R^*[1/x]$ is flat since $R^*[1/x]$ is a field.

Examples 10.15 generalizes Example 6.24.

EXAMPLES 10.15. Let $d \in \mathbb{N}$ be greater than or equal to 2, and let $x, y_1, \dots, y_d$ be indeterminates over a field k. Let R be either

(1) The polynomial ring $R := k[x, y_1, \dots, y_d]$ with x-adic completion $R^* = k[y_1, \dots, y_d][[x]]$, or

(2) The localized polynomial ring $R := k[x, y_1, \dots, y_d]_{(x,y_1,\dots,y_d)}$ with x-adic completion $R^* = k[y_1, \dots, y_d]_{(y_1,\dots,y_d)k[y_1,\dots,y_d]}[[x]]$.

Let $f := y_1\tau_1 + \cdots + y_d\tau_d$, where $\tau_1, \dots, \tau_d \in xk[[x]]$ are algebraically independent over $k(x)$. Let $S := R[f]$ and let $T := R[\tau_1, \dots, \tau_d]$. Regard f as a polynomial in $\tau_1, \dots, \tau_d$ over R. By Theorem 10.12.5, the map $\varphi_x : S \hookrightarrow T[1/x]$ satisfies LF_{d-1}, but fails to satisfy LF_d because the ideal $L = (y_1, \dots, y_d)R[1/x]$ of nonconstant coefficients of f has height d. Since $d \geq 2$, the map $\varphi_x : S \hookrightarrow T[1/x]$ satisfies LF_1. Since S is a UFD, Proposition 9.19 implies $A = B$. Since φ_x does not satisfy LF_d, the map φ_x is not flat and thus B is not Noetherian by Theorem 10.9.3.

Let $P \in \operatorname{Spec} B$. By Theorem 10.12, we have $(y_1, \dots, y_d, x) \not\subseteq P$ if and only if B_P is Noetherian. Thus B_P is Noetherian for every non-maximal prime ideal of B.

Theorem 10.16 demonstrates that it is often the case that every nonzero ideal in $R[1/x]$ defines the non-flat locus of an insider construction.

THEOREM 10.16. *Let* R *be a Noetherian domain with field of fractions* K, *and let* $D = K(\tau_1, \dots, \tau_s) \cap R^*$ *be a Prototype, as in Insider Construction* 10.7. *Let* I *be a nonzero ideal of* $R[1/x]$ *and let* $L := I \cap R$. *If the radical of* L *is the radical of an ideal* L' *generated by* $d \leq n$ *elements, then there exists an approximation domain* B *such that:*

(1) $LR^*[1/x]$ *defines the non-flat locus of the inclusion map* $B \hookrightarrow R^*[1/x]$.

(2) *For* $P \in \operatorname{Spec} B$, *the ring* B_P *is Noetherian* $\iff ((L,x)R^* \cap R) \not\subseteq P$.

PROOF. Let $L' = (a_1, \dots, a_d)R$, and define $f := a_1\tau_1 + \cdots + a_d\tau_d$. Let B denote the approximation domain associated to f, as in Construction 10.7. By Theorem 10.12.2, $L'R^*[1/x]$ defines the non-flat locus of $\psi : S = R[f] \hookrightarrow R^*[1/x]$, and so item 1 holds. The second item of Theorem 10.16 follows from Theorem 10.12.8. $\square$

10.4. Preserving excellence with the Insider Construction

Theorem 10.17 describes conditions for Insider Construction 10.7 to preserve excellence.

THEOREM 10.17. *Let $(R, \mathfrak{m})$ be an excellent normal local domain with field of fractions K. Let x be a nonzero element of $\mathfrak{m}$ and let R^* denote the x-adic completion of R. Assume the s elements $\tau_1, \dots, \tau_s \in xR^*$ are algebraically independent over K, that $T := R[\tau_1, \dots, \tau_s] \longrightarrow R^*[1/x]$ is flat, and $D := K(\tau_1, \dots, \tau_s) \cap R^*$. Let $f_1, \dots, f_m \in T := R[\underline{\tau}]$, considered as polynomials in the τ_i with coefficients in R. Assume $f_1, \dots, f_m$ are algebraically independent over K; thus $m \le s$. Let $S := R[f_1, \dots, f_m]$ and $\varphi : S \hookrightarrow T$, and let J be the Jacobian ideal of φ as in Definition 7.17.1. Define $A := K(f_1, \dots, f_m) \cap R^*$, and define B to be the Approximation Domain corresponding to $\underline{f}$ as in Construction 10.7. If D is excellent, then the following are equivalent:*

(a) *The ring B is excellent.*
(b) *$JR^*[1/x] = R^*[1/x]$.*
(c) *$\alpha : S \to R^*[1/x]$ is a regular morphism.*

Moreover, if either of the following equivalent conditions holds, then B is excellent:

(b′) *$JT[1/x] = T[1/x]$.*
(c′) *$\varphi_x : S \to T[1/x]$ is a regular morphism.*

PROOF. That conditions (b′) and (c′) are equivalent follows from Theorem 7.19.1. Since T is a subring of R^* and J is an ideal of T, condition (b′) implies condition (b). For the other implications, consider the embeddings:

$$B \xrightarrow{\ \Phi\ } D \xrightarrow{\ \Psi\ } R^* \xrightarrow{\ \Gamma\ } \widehat{R}.$$

By Theorem 5.14.4, we have $B[1/x]$ is a localization of S, and $D[1/x]$ is a localization of T. Thus, for $Q \in \operatorname{Spec} R^*$ with $x \notin Q$, we have
(10.17.1)

$$\alpha_Q : S \longrightarrow S_{Q\cap S} = B_{Q\cap B} \xrightarrow{\ \Phi'\ } D_{Q\cap D} = T_{Q\cap T} \xrightarrow{\ \Psi'\ } R^*_Q.$$

(a)$\Longrightarrow$(b): Since B, D and R^* are all excellent with the same completion $\widehat{R}$, [**122**, Theorem 32.1] implies Φ is regular. Let $Q \in \operatorname{Spec}(R^*)$ with $x \notin Q$. The map $\Phi' : B_{Q\cap B} \hookrightarrow D_{Q\cap D}$ is also regular. It follows from Equation 10.17.1 that $\varphi_{Q\cap T} : S \hookrightarrow T_{Q\cap T}$ is regular. Thus $J \not\subseteq Q \cap T$. Since J is an ideal of T, we have $J \subseteq Q \cap T \iff J \subseteq Q$. We conclude that $JR^*[1/x] = R^*[1/x]$.

(b) $\iff$ (c): We show for every $Q \in \operatorname{Spec} R^*$ with $x \notin Q$ that

$$(*) \qquad J \not\subseteq Q \cap T \iff \alpha_Q \quad \text{is regular.}$$

If $J \not\subseteq Q\cap T$, then α_Q is a composition of regular maps as shown in Equation 10.17.1. If α_Q is regular, then Ψ' faithfully flat implies $S \hookrightarrow T_{Q\cap T}$ is regular [**122**, Theorem 32.1 (ii)]. Thus $J \not\subseteq Q \cap T$.

(b)$\Longrightarrow$(a) By Theorems 10.9.3 and 7.19.2, the ring B is Noetherian with x-adic completion R^*. Therefore the completion of B with respect to the powers of its maximal ideal is $\widehat{R}$. Therefore B is formally equidimensional. Hence by Ratliff's Equidimension Theorem 3.25, B is universally catenary.

To show B is excellent, it remains to show that B is a G-ring. Consider the morphisms

$$B \xrightarrow{\ \Phi\ } D \quad \text{and} \quad B \xrightarrow{\ \Gamma\circ\Psi\circ\Phi\ } \widehat{R}.$$

Since B and D are Noetherian, $\widehat{R}$ is faithfully flat over both B and D. Hence the map Φ is faithfully flat by Remark 2.37.14. A straightforward argument using Definition 3.45 of G-ring shows that B is a G-ring if the map Φ is regular in the sense of Definition 3.41; see [**122**, Theorem 32.2].

To see that Φ is regular, let $P \in \operatorname{Spec}(B)$. If $x \in P$, then we use that

$$B/xB = R^*/xR^* = R/xR = D/xD$$

from Construction Properties Theorem 5.14.2. The ring $\widehat{R} \otimes_B k(P)$ is geometrically regular over $k(P) = B_P/PB_P$, since R is excellent.

If $x \notin P$, we show that the ring $D \otimes_B L$ is regular, for every finite field extension L of $k(P)$. Let $\overline{W}$ be a prime ideal in $D \otimes_B L$ and let W be the preimage in D of $\overline{W}$. Then $W \cap B = P$. By the faithful flatness of R^* over D, there exists $Q \in \operatorname{Spec}(R^*)$ such that $Q \cap D = W$. Then $P = W \cap B = Q \cap B$. Thus $x \notin Q$. Since $JR^*_Q = R^*_Q$, we have $J \not\subseteq Q$. Hence the morphism Φ' in Equation 10.17.1, is regular. Therefore Φ is a regular morphism and B is excellent. □

COROLLARY 10.18. *Let k be a field of characteristic zero, let $x, \underline{y} = \{y_1, \dots, y_r\}$ be indeterminates over k and let D be the Local Protoype $D := k(x, \underline{y}, \underline{\tau}) \cap R^*$ of Theorem 10.6, where the base ring $R = k[x, \underline{y}]_{(x,\underline{y})}$, R^* is the x-adic completion of R, and $\underline{\tau} = \tau_1, \dots, \tau_s$ are elements of $xk[[x]]$ that are algebraically independent over $k(x)$. Assume that $\underline{f} = \{f_1, \dots, f_m\} \subseteq (\underline{\tau})T := (\underline{\tau})R[\underline{\tau}]$ are algebraically independent over $k(x, \underline{y})$. The f_i are considered as polynomials in the τ_i with coefficients in R; thus $m \leq s$. Let $S := R[\underline{f}]$, let $\varphi : S \hookrightarrow T$, and let J be the Jacobian ideal of φ as in Definition 7.17.1. Define B to be the Approximation Domain associated to $\underline{f}$ and $A = k(x, \underline{y}, \underline{f}) \cap R^*$. The following are equivalent:*

(a) *The ring B is excellent.*
(b) $JR^*[1/x] = R^*[1/x]$.
(c) *$\varphi_x : S \to T[1/x]$ is a regular morphism.*

PROOF. Apply Theorem 10.17. □

In relation to Corollary 10.18, we consider the historical examples of Nagata and Christel discussed in Chapters 4 and 6.

REMARK 10.19. (1) For the example of Nagata described in Example 4.15 and in Proposition 6.19, we have $R = k[x, y]_{(x,y)}$, $R^* = k[y][[x]]_{(x,y)}$, $r = s = m = 1$, and k is a field of characteristic zero. Let $\tau = \tau_1$, let $T := R[\tau]$, and let D be the Local Prototype $D = k(x, y, \tau) \cap R^*$. Let $f = f_1 = (y + \tau)^2$ and let $A = k(x, yf) \cap R^*$. The Jacobian ideal J of the inclusion map $\varphi : S := R[f] \to T = R[\tau]$ is the ideal of T generated by $\partial(y^2 + 2y\tau + \tau^2)/\partial\tau) = 2y + 2\tau$. We see that $JR^*[1/x] \neq R^*[1/x]$. Thus, by Proposition 6.19 and Corollary 10.18, the approximation domain $B = A$ is not excellent, and the map φ_x is not regular.

(2) For Christel's example described in Examples 4.17 and 6.23, we have k a field of characteristic zero, $R := k[x, y, z]_{(x,y,z)}$, $r = s = 2$, and $m = 1$. The elements σ and $\tau \in xk[[x]]$ are algebraically independent over $k(x)$, and we have $f := (y+\sigma)(z+\tau)$. The Jacobian ideal J of the inclusion map $\varphi : S := R[f] \to T = R[\sigma, \tau]$ is the ideal of T generated by $x + \tau$ and $y + \sigma$. Again $JR^*[1/x] \neq R^*[1/x]$. Thus, by Example 6.23 and Corollary 10.18, the approximation domain $B = A$ is not excellent, and the map φ_x is not regular.

Examples 10.20 illustrates other applications of Corollary 10.18.

EXAMPLES 10.20. As in Corollary 10.18, let k be a field of characteristic zero, and let D be the Local Protoype $D := k(x,y,z,\sigma,\tau) \cap R^*$ of Local Prototype Theorem 10.6, where the base ring $R = k[x,y,z]_{(x,y,z)}$, the ring R^* is the x-adic completion of R, and σ, τ are elements of $xk[[x]]$ that are algebraically independent over $k(x)$. The ring D is a three-dimensional regular local domain that is a directed union of five-dimensional regular local domains by Theorem 10.6.

With this setting we consider two intersection domains $A := k(x,y,z,f,g) \cap R^*$ formed from pairs of elements f and $g \in (\sigma,\tau)T := (\sigma,\tau)R[\sigma,\tau]$ that are algebraically independent over $k(x,y,z)$. By Construction Properties Theorem 5.14.4, the rings D and A have x-adic completion R^*. Let $S := R[f,g]$, let $\varphi : S \hookrightarrow T$, and let J be the Jacobian ideal of φ as in Definition 7.17.1.

(1) Let $f = (y-\sigma)^2$, $g = (z-\tau)^2$, and $A = k(x,y,z,f,g) \cap R^*$. Since $T = R[\sigma,\tau]$ is a free module over S with $\{1,\sigma,\tau,\sigma\tau\}$ as a free basis, the map $\varphi : S \to T$ is flat. By Corollary 10.11.1, A is Noetherian and is equal to its approximation domain. It follows that A is a 3-dimensional regular local domain.

Since the field k has characteristic zero, the Jacobian ideal of the map φ is $J = (\sigma - y)(\tau - z)T$, and $JR^*[1/x] \neq R^*[1/x]$. Hence by Corollary 10.18, the ring A is not excellent.

(2) Let $f = \sigma^2 + x\tau$, $g = \tau^2 + x\sigma$, and $A = k(x,y,z,f,g) \cap R^*$. The Jacobian ideal of the map $\varphi : S \hookrightarrow T$ is $J = (4\sigma\tau - x^2)T$, and $JR^*[1/x] = R^*[1/x]$. Hence by Corollary 10.18, the ring $B = A$ is excellent. However, $JT[1/x] \neq T[1/x]$. Thus it may happen that B is excellent, but conditions (b$'$) and (c$'$) of Theorem 10.17 do not hold.

CHAPTER 11

Integral closure under extension to the completion

This chapter relates to the general question:

QUESTION 11.1. What properties of ideals of a Noetherian local ring $(A, \mathfrak{n})$ are preserved under extension to the $\mathfrak{n}$-adic completion $\widehat{A}$?

Our focus here is the integral closure property; see Definition 11.2.4.

Using Insider Construction 10.7 of Chapter 10, we present in Example 11.9 a height-two prime ideal P of a 3-dimensional regular local domain such that the extension $P\widehat{A}$ of P to the $\mathfrak{n}$-adic completion $\widehat{A}$ of A is not integrally closed. The ring A in Example 11.9 is a nested union of 5-dimensional regular local domains.

More generally, we use this same technique to establish, for each integer $d \geq 3$ and each integer h with $2 \leq h \leq d-1$, the existence of a d-dimensional regular local domain $(A, \mathfrak{n})$ having a prime ideal P of height h with the property that the extension $P\widehat{A}$ is not integrally closed, where $\widehat{A}$ is the $\mathfrak{n}$-adic completion of A. A regular local domain having a prime ideal with this property is necessarily not a Nagata ring and is not excellent; see item 7 of Remark 11.8.

Section 11.1 contains conditions in order that integrally closed ideals of a ring R extend to integrally closed ideals of R', where R' is an R-algebra. In particular, we consider conditions for integrally closed ideals of a Noetherian local ring A to extend to integrally closed ideals of the completion $\widehat{A}$ of A.

11.1. Integral closure under ring extension

The concept of "integrality over an ideal" is related to "integrality over a ring", defined in Section 2.1. For properties of integral closure of ideals, rings and modules we refer to the book of Swanson and Huneke [**175**].

DEFINITIONS AND REMARKS 11.2. Let I be an ideal of a ring R.

(1) An element $r \in R$ is *integral over* I if there exists a monic polynomial $f(x) = x^n + \sum_{i=1}^{n} a_i x^{n-i}$ such that $f(r) = 0$ and such that $a_i \in I^i$ for each i with $1 \leq i \leq n$.
(2) The *integral closure* $\overline{I}$ of I is the set of elements of R integral over I; the set $\overline{I}$ is an ideal.
(3) The integral closure of $\overline{I}$ is equal to $\overline{I}$ [**175**, Corollary 1.3.1].
(4) If $I = \overline{I}$, then I is said to be *integrally closed.*
(5) The ideal I is said to be *normal* if I^n is integrally closed for every $n \geq 1$.
(6) If J is an ideal contained in I and $JI^{n-1} = I^n$ for some integer $n \geq 1$, then J is said to be a *reduction* of I.

REMARKS 11.3. Let I be an ideal of a ring R.

(1) An element $r \in R$ is integral over I if and only if I is a reduction of the ideal $L = (I, r)R$. To see this equivalence, observe that for a monic polynomial $f(x)$ as in Definition 11.2.1, we have

$$f(r) = 0 \iff r^n = -\sum_{i=1}^{n} a_i r^{n-i} \in IL^{n-1} \iff L^n = IL^{n-1}.$$

(2) An ideal is integrally closed if and only if it is not a reduction of a properly bigger ideal.

(3) A prime ideal is always integrally closed. More generally, a radical ideal is always integrally closed. This is Exercise 1.

(4) Let a, b be elements in a Noetherian ring R and let $I := (a^2, b^2)R$. The element ab is integral over I. If a, b form a regular sequence, then $ab \notin I$ and thus I is not integrally closed; see Exercise 2. More generally, if $h \geq 2$ and $a_1, \dots, a_h$ form a regular sequence in R and $I := (a_1^h, \dots, a_h^h)R$, then I is not integrally closed.

The Rees Algebra is relevant to the discussion of integral closure.

DEFINITION AND REMARKS 11.4. Let I be an ideal of a ring R, and let t be a variable over R.

(1) The *Rees algebra of* I is the subring of $R[t]$ defined as

$$R[It] := \{ \sum_{i=0}^{n} a_i t^i \mid n \in \mathbb{N} \ ; \ a_i \in I^i \} = \bigoplus_{n \geq 0} I^n t^n,$$

where $I^0 = R$.

(2) An element $a \in R$ is integral over I if and only if $at \in R[t]$ is integral over the subring $R[It]$.

(3) If R is a normal domain, then I is a normal ideal of R if and only if the Rees algebra $R[It]$ is a normal domain; see Swanson and Huneke [**175**, Prop. 5.2.1, p.95].

Our work in this chapter is motivated by the following questions:

QUESTIONS 11.5.

(1) Craig Huneke: "Does there exist an analytically unramified Noetherian local ring $(A, \mathfrak{n})$ that has an integrally closed ideal I for which the extension $I\widehat{A}$ to the $\mathfrak{n}$-adic completion $\widehat{A}$ is not integrally closed?"

(2) Sam Huckaba: "If there is such an example, can the ideal of the example be chosen to be a normal ideal?" See Definition 11.2.6.

Related to Question 11.5.1, we present in Theorem 11.11 a 3-dimensional regular local domain A having a height-two prime ideal $I = P = (f, g)A$ such that $I\widehat{A}$ is not integrally closed. Thus the answer to Question 11.5.1 is "yes". This example also shows that the answer to Question 11.5.2 is again "yes". Since f, g form a regular sequence and A is Cohen-Macaulay, the powers P^n of P have no embedded associated primes and therefore are P-primary [**120**, (16.F), p. 112], [**122**, Ex. 17.4, p. 139]. Since the powers of the maximal ideal of a regular local domain are integrally closed, the powers of P are integrally closed, that is, P is a normal ideal.

Thus, by Remarks 11.3.6, the Rees algebra $A[Pt] = A[ft, gt]$ is a normal domain while the Rees algebra $\widehat{A}[ft, gt]$ is not normal. We observe in Remarks 11.8 that this implies that A fails to have geometrically normal formal fibers.

REMARKS 11.6. Without the assumption that A is analytically unramified, there exist examples even in dimension one where an integrally closed ideal of a Noetherian local domain A fails to extend to an integrally closed ideal in $\widehat{A}$. If A is reduced but analytically ramified, then the zero ideal of A is integrally closed, but its extension to $\widehat{A}$ is not integrally closed.

Examples of reduced analytically ramified Noetherian local rings have been known for a long time. By Remark 3.19.5, the examples of Akizuki and Schmidt mentioned in Classical Examples 1.4 of Chapter 1 are analytically ramified Noetherian local domains. Another example due to Nagata is given in [**137**, Example 3, pp. 205-207]. (See also [**137**, (32.2), p. 114], and Remarks 4.16.2.)

Let R be a commutative ring and let R' be an R-algebra. In Remark 11.8 we list cases where extensions to R' of integrally closed ideals of R are again integrally closed. In this connection we use the following definition as in Lipman [**111**, page 799].

DEFINITION 11.7. An R-algebra. R' is said to be *quasi-normal* over R if R' is flat over R and the condition $N_{R,R'}$ holds:

($N_{R,R'}$): If C is any R-algebra, and D is a C-algebra in which C is integrally closed[1], then also $C \otimes_R R'$ is integrally closed in $D \otimes_R R'$.

If condition $N_{R,R'}$ holds, we also say the map $R \to R'$ is *quasi-normal.*

REMARKS 11.8. Let R be a commutative ring and let R' be an R-algebra.

(1) By a result of Lipman [**111**, Lemma 2.4], if R' satisfies ($N_{R,R'}$) and I is an integrally closed ideal of R, then IR' is integrally closed in R'.
(2) A regular homomorphism of Noetherian rings is normal by Remark 3.42, and a normal homomorphism of Noetherian rings is quasi-normal [**62**, IV,(6.14.5)]. Hence a regular homomorphism of Noetherian rings is quasi-normal.
(3) Assume that R and R' are Noetherian rings and that R' is a flat R-algebra. Let I be an integrally closed ideal of R. The flatness of R' over R implies every $P' \in \text{Ass}(R'/IR')$ contracts in R to some $P \in \text{Ass}(R/I)$ [**122**, Theorem 23.2]. Thus by the previous item, if the map $R \to R'_{P'}$ is normal or regular for each $P' \in \text{Ass}(R'/IR')$, then IR' is integrally closed.
(4) Principal ideals of an integrally closed domain are integrally closed. This is Exercise 3.i of this chapter.
(5) If I is an ideal of the Noetherian local domain A and $I\widehat{A}$ is integrally closed, then faithful flatness of the extension $A \hookrightarrow \widehat{A}$ implies that I is integrally closed.
(6) In general, integral closedness of ideals is a local condition. If R' is an R-algebra that is a normal ring in the sense that for every prime ideal P' of R', the local ring $R'_{P'}$ is an integrally closed domain, then the extension to R' of every principal ideal of R is integrally closed by item 4. In

[1]Let $h : C \to D$ be the structural map defining D as a C-algebra. Then "C is integrally closed in D" means the subring $h(C)$ of D is integrally closed in D.

particular, if $(A, \mathfrak{n})$ is an analytically normal Noetherian local domain, then every principal ideal of A extends to an integrally closed ideal of $\widehat{A}$.

(7) Let $(A, \mathfrak{n})$ be a Noetherian local ring and let $\widehat{A}$ be the $\mathfrak{n}$-adic completion of A. Since $A/\mathfrak{q} \cong \widehat{A}/\mathfrak{q}\widehat{A}$ for every $\mathfrak{n}$-primary ideal $\mathfrak{q}$ of A, the $\mathfrak{n}$-primary ideals of A are in one-to-one inclusion preserving correspondence with the $\widehat{\mathfrak{n}}$-primary ideals of $\widehat{A}$. It follows that an $\mathfrak{n}$-primary ideal I of A is a reduction of a properly larger ideal of A if and only if $I\widehat{A}$ is a reduction of a properly larger ideal of $\widehat{A}$. Therefore an $\mathfrak{n}$-primary ideal I of A is integrally closed if and only if $I\widehat{A}$ is integrally closed.

(8) If R is an integrally closed domain, then $\overline{xI} = x\overline{I}$, for every ideal I and element x of R; see Exercise 3.ii of this chapter. If $(A, \mathfrak{n})$ is analytically normal and also a UFD, then every height-one prime ideal of A extends to an integrally closed ideal of $\widehat{A}$ by item 4. In particular if A is a regular local domain, then A is a UFD by Remark 2.10.2, and so $P\widehat{A}$ is integrally closed for every height-one prime ideal P of A.

(9) If $(A, \mathfrak{n})$ is a 2-dimensional local UFD, then every non-principal integrally closed ideal of A has the form xI, where I is an $\mathfrak{n}$-primary integrally closed ideal and $x \in A$; see Exercise 4. In particular, this is the case if $(A, \mathfrak{n})$ is a 2-dimensional regular local domain. It follows from items 7 and 8 that every integrally closed ideal of A extends to an integrally closed ideal of $\widehat{A}$ in the case where A is a 2-dimensional regular local domain.

(10) If $(A, \mathfrak{n})$ is an excellent local ring, then the map $A \hookrightarrow \widehat{A}$ is quasi-normal by [**62**, (7.4.6) and (6.14.5)], and in this case every integrally closed ideal of A extends to an integrally closed ideal of $\widehat{A}$. If $(A, \mathfrak{n})$ is a Nagata local ring, then for each prime ideal P of A, the ideal $P\widehat{A}$ is reduced, and hence integrally closed [**137**, Theorem 36.4].

(11) Let $(A, \mathfrak{n})$ be a Noetherian local domain and let A^h denote the Henselization of A. Every integrally closed ideal of A extends to an integrally closed ideal of A^h. This follows because A^h is a filtered direct limit of étale A-algebras; see Lipman [**111**, (i), (iii), (vii) and (ix), pp. 800-801]. Since the map from A to its completion $\widehat{A}$ factors through A^h, every integrally closed ideal of A extends to an integrally closed ideal of $\widehat{A}$ if every integrally closed ideal of A^h extends to an integrally closed ideal of $\widehat{A}$.

11.2. Extending ideals to the completion

We present an example of a height-two prime ideal $I = (f, g)A$ of the 3-dimensional RLR $(A, \mathfrak{n})$ such that the extension $I\widehat{A}$ to the $\mathfrak{n}$-adic completion is not integrally closed. We use Example 10.20 and results from Chapters 5, 6, and 10 to justify that $I\widehat{A}$ is not integrally closed.

In Example 11.9 we review the setting and basic description of the ring A of Example 10.20.1.

EXAMPLE 11.9. In the notation of Example 10.20.1, k is a field of characteristic zero, x, y and z are indeterminates over k, and the base ring $R := k[x, y, z]_{(x,y,z)}$. The x-adic completion of R is $R^* = k[y, z]_{(y,z)}[[x]]$, and σ, τ are elements of $xk[[x]]$ that are algebraically independent over $k(x)$. As in Local Prototype Theorem 10.6, the Local Protoype $D := k(x, y, z, \sigma, \tau) \cap R^*$ is a three-dimensional regular local domain and a directed union of five-dimensional regular local domains.

With $f = (y-\sigma)^2$ and $g = (z-\tau)^2$, we define $A = k(x,y,z,f,g) \cap R^*$. By Example 10.20.1, the ring A is Noetherian, A is equal to its approximation domain, A is a 3-dimensional regular local domain, and A is not a Nagata ring.

The following commutative diagram, where all the labeled maps are the natural inclusions and B is as in Equation 11.10.1, displays the situation of Example 11.9:

$$\begin{array}{ccccc}
B = A = R^* \cap \mathcal{Q}(S) & \xrightarrow{\gamma_1} & D = R^* \cap \mathcal{Q}(T) & \xrightarrow{\gamma_2} & R^* = A^* \\
\delta_1 \uparrow & & \delta_2 \uparrow & & \psi \uparrow \\
S = R[f,g] & \xrightarrow{\varphi} & T = R[\sigma,\tau] & = & T
\end{array}$$

Commutative Diagram 11.9.1 for Example 11.9.

In order to better understand the structure of A, we recall some of the details of the approximation domain B associated to f and g.

APPROXIMATION TECHNIQUE 11.10. With k, x, y, z, f, g, R and R^* as in Example 11.9,

$$f = y^2 + \sum_{j=1}^{\infty} b_j x^j, \qquad g = z^2 + \sum_{j=1}^{\infty} c_j x^j,$$

where $b_j \in k[y]$ and $c_j \in k[z]$. The rth *endpieces* for f and g are the sequences $\{f_r\}_{r=1}^{\infty}$, $\{g_r\}_{r=1}^{\infty}$ of elements in R^* defined for each $r \geq 1$ by:

$$f_r := \sum_{j=r+1}^{\infty} \frac{b_j x^j}{x^r} \qquad \text{and} \qquad g_r := \sum_{j=r+1}^{\infty} \frac{c_j x^j}{x^r}.$$

Then $f = y^2 + xb_1 + xf_1 = y^2 + xb_1 + x^2 b_2 + x^2 f_2 = \ldots$ and similar equations hold for g. Thus we have:

$$(11.10.0) \quad f = y^2 + xb_1 + x^2b_2 + \ldots x^t b_t + x^t f_t; \quad g = y^2 + xc_1 + x^2 c_2 + \ldots x^t c_t + x^t g_t,$$

for each $t \geq 1$.

For each integer $r \geq 1$, we define:

$$(11.10.1) \qquad \begin{aligned} U_r &:= k[x,y,z,f_r,g_r], \quad \mathfrak{m}_r := (x,y,z,f_r,g_r)U_r, \\ B_r &:= (U_r)_{\mathfrak{m}_r} \quad \text{and} \quad B := \bigcup_{r=1}^{\infty} B_r. \end{aligned}$$

The ring B is the Approximation Domain associated to f and g, as in Definition 5.7.

THEOREM 11.11. *With the notation of Example* 11.9 *and Approximation Technique* 11.10, *let* $P = (f,g)A$. *Then*

(1) *The ring* $A = B$ *is a 3-dimensional regular local domain that has x-adic completion* $A^* = R^* = k[y,z]_{(y,z)}[[x]]$. *Moreover A is a nested union of five-dimensional regular local domains.*
(2) *The ideal P is a height-two prime ideal of A.*
(3) *The ideal PA^* is not integrally closed in A^*.*
(4) *The completion $\widehat{A}$ of A is $\widehat{R} = k[[x,y,z]]$ and $P\widehat{A}$ is not integrally closed.*

PROOF. Item 1 follows from Example 11.9 and Theorem 10.9, parts 3 and 4.

For item 2, it suffices to observe that P has height two and that, for each positive integer r, $P_r := (f,g)U_r$ is a prime ideal of U_r. We have $f = y^2 + xb_1 + xf_1$ and $g = z^2 + xc_1 + xg_1$. It is clear that $(f,g)k[x,y,z,f,g]$ is a height-two prime ideal.

Since $U_1 = k[x, y, z, f_1, g_1]$ is a polynomial ring over k in the variables x, y, z, f_1, g_1, we see that

$$P_1U_1[1/x] = (xb_1 + xf_1 + y^2, xc_1 + xg_1 + z^2)U_1[1/x]$$

is a height-two prime ideal of $U_1[1/x]$. Indeed, setting $f = g = 0$ is equivalent to setting $f_1 = -b_1 - y^2/x$ and $g_1 = -c_1 - z^2/x$. Therefore the residue class ring $(U_1/P_1)[1/x]$ is isomorphic to the integral domain $k[x, y, z][1/x]$. Since U_1 is Cohen-Macaulay and f, g form a regular sequence, and since $(x, f, g)U_1 = (x, y^2, z^2)U_1$ is an ideal of height three, we see that x is in no associated prime ideal of $(f, g)U_1$ (see, for example [**122**, Theorem 17.6]). Therefore $P_1 = (f, g)U_1$ is a height-two prime ideal, and so the same holds for P_1B_1.

For $r > 1$, by Equation 11.10.0, there exist elements $u_r \in k[x, y]$ and $v_r \in k[x, z]$ such that $f = x^rf_r + u_rx + y^2$ and $g = x^rg_r + v_rx + z^2$. An argument similar to that given above shows that $P_r = (f, g)U_r$ is a height-two prime ideal of U_r. Since U is the nested union of the U_r we have that $(f, g)U$ is a height-two prime ideal of U. Since B is a localization of U we see that $(f, g)B$ is a height-two prime ideal of $B = A$.

For items 3 and 4, $R^* = B^* = A^*$ by Example 10.20 and it follows that $\widehat{A} = k[[x, y, z]]$. To see that $PA^* = (f, g)A^*$ and $P\widehat{A} = (f, g)\widehat{A}$ are not integrally closed, observe that $\xi := (y - \sigma)(z - \tau)$ is integral over PA^* and $P\widehat{A}$ since $\xi^2 = fg \in P^2$. On the other hand, $y - \sigma = u$ and $z - \tau = v$ form a regular sequence in A^* and $\widehat{A}$. Since $P = (u^2, v^2)A$, an easy computation shows that $uv \notin PA^* = (u^2, v^2)A^*$; see Exercise 2. Since $PA^* = P\widehat{A} \cap A^*$, this completes the proof. □

In Example 11.13, we generalize the technique of Example 11.9 to obtain RLRs A with $\dim A = n+1$ such that A has a prime ideal P with $\operatorname{ht} P = r$, where r may be chosen to be any integer in the set $\{2, \dots, n\}$, and $P\widehat{A}$ is not integrally closed. The examples are complete intersection prime ideals P, that is, $\operatorname{ht} P = r$ and P is generated by r elements. We use the following interesting result of Shiro Goto that characterizes integrally closed complete intersection ideals of a Noetherian ring.

THEOREM 11.12 (Goto [**59**, Theorem 1.1]). *Let I be a proper ideal in a Noetherian ring A such that $\operatorname{ht} I = r$ and I is generated by r elements. Then the following conditions are equivalent:*

(1) *I is integrally closed.*
(2) *I^n is integrally closed for all $n \in \mathbb{N}$.*
(3) *For every $\mathfrak{p} \in \operatorname{Ass}(A/I)$, the local ring $A_\mathfrak{p}$ is regular and the length of the module $(IA_\mathfrak{p} + \mathfrak{p}^2A_\mathfrak{p})/\mathfrak{p}^2A\mathfrak{p}$ is at least $r - 1$.*

When this is the case, $\operatorname{Ass}(A/I) = \operatorname{Min}(A/I)$ *and I is generated by a regular sequence.*

EXAMPLE 11.13. For each integer $n \geq 2$, and every integer r with $2 \leq r \leq n$, there exists a regular local domain A with $\dim A = n + 1$ having a prime ideal P with $\operatorname{ht} P = r$ such that the extension of P to the completion of A is not integrally closed.

To see this, let k be a field of characteristic zero and, for an integer $n \geq 2$, let $x, y_1, \dots, y_n$ be indeterminates over k. Let r be an integer with $2 \leq r \leq n$, and let $\tau_1, \dots, \tau_r \in xk[[x]]$ be algebraically independent over $k(x)$. Let $R := k[x, y_1, \dots, y_n]_{(x,y_1,\dots,y_n)}$. Then R is an RLR with $\dim R = n + 1 =: d$. For each i

with $1 \le i \le$r, define $f_i = (y_i - \tau_i)^r$, and set $u_i = y_i - \tau_i$. The rings

$$S := R[f_1, \ldots, f_r] \quad \text{and} \quad T := R[\tau_1, \ldots, \tau_r] = R[u_1, \ldots, u_r].$$

are polynomial rings in r variables over R, and T is a finite free integral extension of S. The set

$$\{u_1^{e_1} \cdot u_2^{e_2} \cdot \ldots \cdot u_r^{e_r} \mid 0 \le e_i \le r-1\}$$

is a free module basis for T as an S-module. Therefore the map $S \hookrightarrow T[1/x]$ is flat. Let R^* denote the x-adic completion of R, and define D to be the Local Prototype $D := k(x, y_1, \ldots, y_n, \tau_1, \ldots, \tau_h) \cap R^*$ of Theorem 10.6. Let $A := k(x, y_1, \ldots, y_n, f_1, \ldots, f_h) \cap R^*$. By Construction Properties Theorem 5.14.4, the rings D and A have x-adic completion R^*. Since the map $S \hookrightarrow T$ is flat, Theorem 10.9 implies that the ring A is a d-dimensional regular local ring and is equal to its approximation domain B; thus A is a directed union of $(d+h)$-dimensional regular local domains.

The situation for Example 11.13 can be displayed in a commutative diagram similar to Diagram 11.9.1, where R, B, A, R^*, S, T are adjusted to fit Example 11.13.

$$(11.13.1)\qquad \begin{array}{ccccc} B = A = R^* \cap \mathcal{Q}(S) & \xrightarrow{\gamma_1} & D = R^* \cap \mathcal{Q}(T) & \xrightarrow{\gamma_2} & R^* = A^* \\ {\scriptstyle\delta_1}\uparrow & & {\scriptstyle\delta_2}\uparrow & & {\scriptstyle\psi}\uparrow \\ S = R[f_1, \ldots, f_r] & \xrightarrow{\varphi} & T = R[\tau_1, \ldots, \tau_r] & = & T \end{array}$$

Since the field k has characteristic zero and $r \geq 2$, the Jacobian ideal of the map $\varphi : S \hookrightarrow T$ has radical $\sqrt{(J)} = \Pi_{i=1}^r (y_i - \tau_i)T$, and $JR^*[1/x] \neq R^*[1/x]$. By Corollary 10.18, the ring A is not excellent. Let $P := (f_1, \ldots, f_r)A$; an argument similar to that given in Theorem 11.11 shows that P is a prime ideal of A of height r. Therefore P is an integrally closed complete intersection ideal of A. Since each f_i is an r-th power in A^* and $\widehat{A}$, Theorem 11.12 implies that PA^* and $P\widehat{A}$ are not integrally closed.

It can also be seen directly that PA^* and $P\widehat{A}$ are not integrally closed. We have $y_i - \tau_i = u_i \in A^*$. Let $\xi = \prod_{i=1}^r u_i$. Then $\xi^r = f_1 \cdots f_r \in P^r$ implies ξ is integral over PA^* and $P\widehat{A}$. Since $u_1, \ldots, u_h$ are a regular sequence in A^* and $\widehat{A}$, it follows that $\xi \notin P\widehat{A}$; see for example the thesis of Taylor [**178**, Theorem 1]. Therefore the extended ideals PA^* and $P\widehat{A}$ are not integrally closed.

11.3. Comments and questions

In connection with Theorem 11.11 it is natural to ask the following question.

QUESTION 11.14. For P and A as in Theorem 11.11, is P the only prime ideal of A that does not extend to an integrally closed ideal of $\widehat{A}$?

COMMENTS 11.15. In relation to Example 11.9 and to Question 11.14, observe the following, refering to Diagram 11.9.1 for Example 11.9:

(1) Theorem 10.9 implies that $A[1/x]$ is a localization of S and $D[1/x]$ is a localization of T. By Prototype Theorem 10.6 of Chapter 10, D is excellent. Notice, however, that A is not excellent since there exists a prime ideal P of A such that $P\widehat{A}$ is not integrally closed by Remark 11.8.10. The excellence of D implies the map $\gamma_2 : D \to A^*$ is regular [**62**, (7.8.3 v)]. Thus, for each $Q^* \in \operatorname{Spec} A^*$ with $x \notin Q^*$, the map $\psi_{Q^*} : T \to A^*_{Q^*}$ is regular. It follows that $\psi_x : T \to A^*[1/x]$ is regular.

(2) Let $Q^* \in \operatorname{Spec} A^*$ be such that $x \notin Q^*$ and let $\mathfrak{q}' = Q^* \cap T$. Assume that $\varphi_{\mathfrak{q}'} : S \to T_{\mathfrak{q}'}$ is regular. By item 1 and [**122**, Theorem 32.1], the map $S \to A^*_{Q^*}$ is regular. Thus $(\gamma_2 \circ \gamma_1)_{Q^*} : A \to A^*_{Q^*}$ is regular.

(3) Let I be an ideal of A. Since D and A^* are excellent and both have completion $\widehat{A}$, Remark 11.8.10 shows that the ideals ID, IA^* and $I\widehat{A}$ are either all integrally closed or all fail to be integrally closed.

(4) In this setting, the Jacobian ideal of $\varphi : S \hookrightarrow T$ gives information about the smoothness and regularity of φ by Theorems 7.12 and 7.19.1. The Jacobian ideal of $\varphi : S := k[x, y, z, f, g] \hookrightarrow T := k[x, y, z, \sigma, \tau]$ is the ideal of T generated by the determinant of the matrix

$$\mathcal{J} := \begin{pmatrix} \frac{\partial f}{\partial \sigma} & \frac{\partial g}{\partial \sigma} \\ \frac{\partial f}{\partial \tau} & \frac{\partial g}{\partial \tau} \end{pmatrix}.$$

Since the characteristic of the field k is zero, this ideal is $(y - \sigma)(z - \tau)T$.

In Proposition 11.16, we relate the behavior of integrally closed ideals in the extension $\varphi : S \to T$ to the behavior of integrally closed ideals in the extension $\gamma_2 \circ \gamma_1 : A \to A^*$.

PROPOSITION 11.16. *With the setting of Theorem* 11.11, *let I be an integrally closed ideal of A such that $x \notin Q$ for each $Q \in \operatorname{Ass}(A/I)$. Let $L = I \cap S$. If LT is integrally closed, respectively a radical ideal, then IA^* is integrally closed, respectively a radical ideal.*

PROOF. Since the map $A \to A^*$ is flat, Remark 11.8.3 implies that x is not in any associated prime ideal of IA^*. Therefore IA^* is contracted from $A^*[1/x]$ and it suffices to show $IA^*[1/x]$ is integrally closed (resp. a radical ideal). Our hypothesis implies $I = IA[1/x] \cap A$. By Comment 11.15.1, the ring $A[1/x]$ is a localization of S. Thus every ideal of $A[1/x]$ is the extension of its contraction to S. It follows that $IA[1/x] = LA[1/x]$. Thus $IA^*[1/x] = LA^*[1/x]$.

By Comment 11.15.1, the map $T \to A^*[1/x]$ is regular. If LT is integrally closed, then Remark 11.8.3 implies that $LA^*[1/x]$ is integrally closed. If LT is a radical ideal, then the zero ideal of $\frac{T}{JT}$ is integrally closed. The regularity of the map $\frac{T}{JT} \to \frac{A^*[1/x]}{JA^*[1/x]}$ implies that the zero ideal of $\frac{A^*[1/x]}{LA^*[1/x]}$ is integrally closed. Since the integral closure of the zero ideal is the nilradical, it follows that $LA^*[1/x]$ is a radical ideal. □

PROPOSITION 11.17. *With the setting of Theorem* 11.11 *and Comment* 11.15, *let $Q \in \operatorname{Spec} A$ be such that QA^* (or equivalently $Q\widehat{A}$) is not integrally closed. Then*

(1) *Q has height two and $x \notin Q$.*
(2) *There exists a minimal prime ideal Q^* of QA^* such that with $\mathfrak{q}' = Q^* \cap T$, the map $\varphi_{\mathfrak{q}'} : S \to T_{\mathfrak{q}'}$ is not regular.*
(3) *Q contains $f = (y - \sigma)^2$ or $g = (z - \tau)^2$.*
(4) *Q is contained in $\mathfrak{n}^2$, where $\mathfrak{n}$ is the maximal ideal of A.*

PROOF. We have $\dim A = 3$, the maximal ideal of A extends to the maximal ideal of A^*, and principal ideals of A^* are integrally closed by Remark 11.8.8. Thus the height of Q is two. By Construction Properties Theorem 5.14, we have $A^*/xA^* = A/xA = R/xR$. Hence $x \notin Q$. This proves item 1.

By Remark 11.8.3, there exists a minimal prime ideal Q^* of QA^* such that $(\gamma_2 \circ \gamma_1)_{Q^*} : A \to A^*_{Q^*}$ is not regular. Thus item 2 follows from Comment 11.15.2.

For item 3, let Q^* and $\mathfrak{q}'$ be as in item 2. Since $(\gamma_2 \circ \gamma_1)_{Q^*}$ is not regular it is not essentially smooth [**62**, 6.8.1]. By Theorem 7.19.1, $(y - \sigma)(z - \tau) \in \mathfrak{q}'$. Hence $f = (y - \sigma)^2$ or $g = (z - \tau)^2$ is in $\mathfrak{q}'$ and thus in Q. This proves item 3.

For item 4, suppose $w \in Q$ is a regular parameter for A; that is $w \in \mathfrak{n} \setminus \mathfrak{n}^2$. Then A/wA and A^*/wA^* are two-dimensional regular local domains, A/wA is a UFD, and $\operatorname{ht}(Q/wA) = 1$. By Remark 11.8.8, QA^*/wA^* is integrally closed, but this implies that QA^* is integrally closed, a contradiction to the hypothesis that QA^* is not integrally closed. This proves item 4. □

QUESTION 11.18. In the setting of Theorem 11.11 and Comment 11.15, let $Q \in \operatorname{Spec} A$ with $x \notin Q$ and let $\mathfrak{q} = Q \cap S$. If QA^* is integrally closed, does it follow that $\mathfrak{q}T$ is integrally closed?

QUESTION 11.19. In the setting of Theorem 11.11 and Comment 11.15, if a prime ideal Q of A contains f or g, but not both, does it follow that QA^* is integrally closed?

If the prime ideal Q in Question 11.19 contains a regular parameter of A, then QA^* is integrally closed by Proposition 11.17.4.

In Example 11.9, the three-dimensional regular local domain A contains height-one prime ideals P such that $\widehat{A}/P\widehat{A}$ is not reduced. This motivates us to ask:

QUESTION 11.20. Let $(A, \mathfrak{n})$ be a three-dimensional regular local domain and let $\widehat{A}$ denote the $\mathfrak{n}$-adic completion of A. If for each height-one prime ideal P of A, the extension $P\widehat{A}$ is a radical ideal, i.e., the ring $\widehat{A}/P\widehat{A}$ is reduced, does it follow that $Q\widehat{A}$ is integrally closed for each $Q \in \operatorname{Spec} A$?

REMARK 11.21. Loepp and Rotthaus in [115] give other results related to Question 11.1 about the behavior of ideals under extension to completion. They construct Noetherian local domains that demonstrate that tight closure need not commute with completion.

Exercises

(1) Let I be a radical ideal of a ring R. Prove that I is an integrally closed ideal of R.

(2) Let u, v be a regular sequence in a commutative ring R. Prove that $uv \notin (u^2, v^2)R$.

SUGGESTION. Use that if a, b are in R and $au = bv$, then $b \in uR$.

(3) Let R be an integrally closed domain.
  (i) Prove that every principal ideal in R is integrally closed.
  (ii) Let $0 \neq x \in R$ and let I be an ideal of R. Prove that $x\overline{I} = \overline{xI}$.

SUGGESTION. Show that if $a \in \overline{xI}$, then a/x is in R.

(4) (i) Let A be a UFD. Prove that every ideal of A has the form xI, where $x \in A$ and I is an ideal of A that is not contained in any proper principal ideal of A.

(ii) Let $(A, \mathfrak{n})$ be a two-dimensional local UFD. Prove that every non-principal integrally closed ideal of A has the form xI, where $x \in A$ and I is an $\mathfrak{n}$-primary integrally closed ideal of A.

CHAPTER 12

Iterative examples

This chapter contains a family of examples of subrings of the power series ring $k[[x,y]]$, where k is a field and x and y are indeterminates. Certain values of the parameters that occur in the construction yield an example of a 3-dimensional local Krull domain $(B,\mathfrak{n})$ such that B is not Noetherian, $\mathfrak{n}=(x,y)B$ is 2-generated and the $\mathfrak{n}$-adic completion $\widehat{B}$ of B is a two-dimensional regular local domain; see Example 12.7. The examples constructed are *iterative* in the sense that they arise from applying the inclusion construction twice, first using an x-adic completion and then using a y-adic completion; see Remarks 12.9.

Let R be the localized polynomial ring $R:=k[x,y]_{(x,y)}$. If $\sigma,\tau\in\widehat{R}=k[[x,y]]$ are algebraically independent over R, then the polynomial ring $R[t_1,t_2]$ in two variables t_1,t_2 over R, can be identified with a subring of $\widehat{R}$ by means of an R-algebra isomorphism mapping $t_1\to\sigma$ and $t_2\to\tau$. The structure of the local domain $A=k(x,y,\sigma,\tau)\cap\widehat{R}$ depends on the residual behavior of σ and τ with respect to prime ideals of $\widehat{R}$. Theorem 12.3 illustrates this in the special case where $\sigma\in k[[x]]$ and $\tau\in k[[y]]$; Example 4.11 is the specific case: $k=\mathbb{Q},\sigma=e^x$, $\tau=e^y$.

REMARK 12.1. In examining properties of subrings of the formal power series ring $k[[x,y]]$ over the field k, we use that the subfields $k((x))$ and $k((y))$ of the field $\mathcal{Q}(k[[x,y]])$ are linearly disjoint over k as defined for example in [**192**, page 109]. It follows that if $\alpha_1,\dots,\alpha_n\in k[[x]]$ are algebraically independent over $k(x)$ and $\beta_1,\dots,\beta_m\in k[[y]]$ are algebraically independent over $k(y)$, then the elements $\alpha_1,\dots,\alpha_n,\beta_1,\dots,\beta_m$ are algebraically independent over $k(x,y)$.

12.1. Iterative examples and their properties

The following notation is used for the examples. The remarks are used in the proof of Theorem 12.3.

NOTATION AND REMARKS 12.2. Let k be a field, let x and y be indeterminates over k, and let

$$\sigma:=\sum_{i=1}^{\infty}a_ix^i\in xk[[x]]\qquad\text{and}\qquad\tau:=\sum_{i=1}^{\infty}b_iy^i\in yk[[y]]$$

be formal power series that are algebraically independent over the fields $k(x)$ and $k(y)$, respectively. Let $R:=k[x,y]_{(x,y)}$, and let σ_n, τ_n be the nth endpieces of σ,τ

defined as in Equation 5.41, for $n \in \mathbb{N}_0$. Define the following rings:

$$(12.2.0)\qquad \begin{array}{rlrl} C_n := & k[x,\sigma_n]_{(x,\sigma_n)}, & C := & k(x,\sigma) \cap k[[x]] = \cup_{n=1}^{\infty} C_n; \\ D_n := & k[y,\tau_n]_{(y,\tau_n)}, & D := & k(y,\tau) \cap k[[y]] = \cup_{n=1}^{\infty} D_n; \\ U_n := & k[x,y,\sigma_n,\tau_n], & U := & \cup_{n=1}^{\infty} U_n; \\ B_n := & k[x,y,\sigma_n,\tau_n]_{(x,y,\sigma_n,\tau_n)} & B := & \cup_{n=1}^{\infty} B_n; \\ A := & k(x,y,\sigma,\tau) \cap k[[x,y]]. & & \end{array}$$

To explain the equalities, observe:

(i) Since $k[[x,y]]$ is the (x,y)-adic completion of the Noetherian ring R, Remark 3.3.4 implies that $k[[x,y]]$ is faithfully flat over R. By Remark 2.37.7,

$$(x,y)^n k[[x,y]] \cap R = (x,y)^n R,$$

for each $n \in \mathbb{N}_0$. Endpiece Recursion Relation 5.5 implies the inclusions

$$C_n \subset C_{n+1}, \quad D_n \subset D_{n+1}, \quad \text{and} \quad B_n \subset B_{n+1}.$$

For each of these inclusions, the larger local ring birationally dominates the smaller; the local rings C_n, D_n, B_n are all dominated by $k[[x,y]] = \widehat{R}$.

(ii) $(x,y)B \cap B_n = (x,y,\sigma_n,\tau_n)B_n$, for each $n \in \mathbb{N}_0$: To see this, let $h \in (x,y,\sigma_n,\tau_n)U_n$. By Equation 5.5,

$$\sigma_n = -xa_{n+1} + x\sigma_{n+1} \quad \text{and} \quad \tau_n = -yb_{n+1} + y\tau_{n+1}.$$

Hence $h \in (x,y)U_{n+1} \cap U_n \subseteq (x,y)U \cap U_n$. Since $(x,y,\sigma_n,\tau_n)U_n$ is a maximal ideal of U_n and is contained in $(x,y)U$, a proper ideal of U, it follows that

$$(x,y)U \cap U_n = (x,y,\sigma_n,\tau_n)U_n.$$

Thus $(x,y)B \cap B_n = (x,y,\sigma_n,\tau_n)B$, for each $n \in \mathbb{N}_0$.

(iii) $U_n[\frac{1}{xy}] = U[\frac{1}{xy}]$, for each $n \in \mathbb{N}_0$: By Equation 5.4.2,

$$\sigma_{n+1} \in U_n[\frac{1}{x}] \subseteq U_n[\frac{1}{xy}] \text{ and } \tau_{n+1} \in U_n[\frac{1}{y}] \subseteq U_n[\frac{1}{xy}].$$

Hence $U_{n+1} \subseteq U_n[\frac{1}{xy}]$, $U \subseteq U_n[\frac{1}{xy}]$, and $U_n[\frac{1}{xy}] = U[\frac{1}{xy}]$, for each $n \in \mathbb{N}$.

(iv) By Remark 2.1, the rings C and D are rank-one discrete valuation domains; as in Remark 4.20 they are the asserted directed unions. The rings B_n are four-dimensional regular local domains that are localized polynomial rings over the field k. Thus the approximation domain B is the directed union of a chain of four-dimensional regular local rings, with each ring birational over the previous ring.

Theorem 12.3 gives other properties of the rings A and B.

THEOREM 12.3. *Assume Notation 12.2. Then the ring A is a two-dimensional regular local domain that birationally dominates the ring B; A has maximal ideal $(x,y)A$ and completion $\widehat{A} = k[[x,y]]$. Moreover:*

1. *The rings U and B are UFDs, and $B = U_{(x,y)}$.*
2. *B is a local Krull domain with maximal ideal $\mathfrak{n} = (x,y)B$.*
3. *B is Hausdorff in the topology defined by the powers of $\mathfrak{n}$.*
4. *The $\mathfrak{n}$-adic completion $\widehat{B}$ of B is canonically isomorphic to $k[[x,y]]$.*
5. *The dimension of B is either 2 or 3.*
6. *The following statements are equivalent:*
   (a) *$B = A$.*

(b) *B is a two-dimensional regular local domain.*
(c) $\dim B = 2$.
(d) *B is Noetherian.*
(e) *In the $\mathfrak{n}$-adic topology every finitely generated ideal of B is closed.*
(f) *In the $\mathfrak{n}$-adic topology every principal ideal of B is closed.*

The proof of the asserted properties of the ring A of Theorem 12.3 uses the following consequence of the useful result of Valabrega given in Theorem 4.9.

PROPOSITION 12.4. *With notation as in Theorem 12.3, let $C = k(x,\sigma) \cap k[[x]]$ and let L be the field of fractions of $C[y,\tau]$. Then the ring $A = L \cap C[[y]]$ is a two-dimensional regular local domain with maximal ideal $(x,y)A$ and completion $\widehat{A} = k[[x,y]]$.*

PROOF. The ring C is a rank-one discrete valuation domain with completion $k[[x]]$, and the field $k(x,y,\sigma,\tau) = L$ is an intermediate field between the fields of fractions of the rings $C[y]$ and $C[[y]]$. By Theorem 4.9, $A = L \cap C[[y]]$ is a regular local domain with completion $k[[x,y]]$. □

PROOF. Now prove Theorem 12.3. The assertions about A follow from Proposition 12.4. Also $U_0 = k[x,y,\sigma,\tau] \subseteq U_n \subseteq U \subseteq B \subseteq A$, for each $n \in \mathbb{N}_0$. Let $\mathcal{Q}(A)$ denote the field of fractions of A and $\mathcal{Q}(U_0)$, the field of fractions of U_0. Then $\mathcal{Q}(A) \subseteq k(x,y,\sigma,\tau) = \mathcal{Q}(U_0)$, and the extensions $U_0 \subseteq U_n \subseteq U \subseteq B \subseteq A$ are birational. By Remark 12.2.ii above, $(x,y)U \cap U_n = (x,y,\sigma_n,\tau_n)U_n$. Thus $(x,y)U$ is a maximal ideal of U, and $B = U_{(x,y)}$ is local with maximal ideal $\mathfrak{n} = (x,y)B$. Since B and A are both dominated by $k[[x,y]]$, the ring A dominates B.

For item 1, to see that U and B are UFDs, Equation 12.2.0 and Remark 12.2.iii imply $U_0[\frac{1}{xy}] = U[\frac{1}{xy}]$. Thus the ring $U[\frac{1}{xy}]$ is a UFD and a Krull domain. For each $n \in \mathbb{N}$, the principal ideals xU_n and yU_n are prime ideals in the polynomial ring U_n. Therefore xU and yU are principal prime ideals of U. Moreover $U_{xU} = B_{xB}$ and $U_{yU} = B_{yB}$ are DVRs, since

$$\cap_{n\in\mathbb{N}} x^n U = \cap_{n\in\mathbb{N}}(x^n k[x,y,\sigma,\tau]) \subseteq \cap_{n\in\mathbb{N}} x^n(k[[x,y]]) = 0,$$

and similarly for yU; see Exercise 2 of Chapter 2.

Since x is a unit of U_{yU}, it follows that $U_{yU} = U[1/x]_{(yU[1/x])}$; see Exercise 1 of this chapter. By Fact 2.28 with $D = U[1/x]$ and $c = y$,

$$U[1/x] = U[\frac{1}{xy}] \cap U[1/x]_{(yU[1/x])} = U[\frac{1}{xy}] \cap U_{yU}, \tag{12.4.0}$$

By the previous paragraph, U_{yU} is a DVR and $U[\frac{1}{xy}]$ is a Krull domain. It follows from Equation 12.4.0, Definition 2.11 and Remarks 2.12 that $U[1/x]$ is a Krull domain, By Nagata's Theorem 2.27, $U[1/x]$ is a UFD. Fact 2.28, with $D = U$ and $c = x$, implies $U = U[\frac{1}{x}] \cap U_{xU}$. As above, U is a Krull domain. By Nagata's Theorem 2.27 again, U is a UFD. Since B is a localization of U, the ring B is a UFD. This completes the proof of items 1 and 2.

For item 3, since B is dominated by $k[[x,y]]$, the intersection $\bigcap_{n=1}^{\infty} \mathfrak{n}^n = (0)$. Thus B is Hausdorff in the topology defined by the powers of $\mathfrak{n}$ [**144**, Proposition 4, page 381], as in Definitions 3.1. For item 4, consider the local injective maps $R \hookrightarrow B \hookrightarrow \widehat{R}$; then $\mathfrak{m}^n B = \mathfrak{n}^n$, $\mathfrak{m}^n\widehat{R} = \widehat{\mathfrak{m}}^n$ and $\widehat{\mathfrak{m}}^n \cap R = \mathfrak{m}^n$, for each positive integer n. Since the natural map $R/\mathfrak{m}^n \to \widehat{R}/\mathfrak{m}^n\widehat{R} = \widehat{R}/\widehat{\mathfrak{m}}^n$ is an isomorphism, the map $R/\mathfrak{m}^n \to B/\mathfrak{m}^n B = B/\mathfrak{n}^n$ is injective and the map $B/\mathfrak{n}^n \to \widehat{R}/\mathfrak{n}^n\widehat{R} = \widehat{R}/\widehat{\mathfrak{m}}^n$

is surjective. Since $B/\mathfrak{n}^n$ is a finite length R-module, it follows that, for each $n \in \mathbb{N}$, the length of $R/\mathfrak{m}^n$ is the same as the length of $R/\mathfrak{n}^n$, and

$$R/\mathfrak{m}^n \cong B/\mathfrak{n}^n \cong \widehat{R}/\widehat{\mathfrak{m}}^n.$$

Hence $\widehat{B} = \widehat{R} = k[[x, y]]$.

For item 5, notice that B is a birational extension of the three-dimensional Noetherian domain $C[y, \tau]$. The dimension of B is at most 3 by Cohen's Theorem 2.26; also see [**122**, Theorem 15.5]. The elements x and y are in the maximal ideal $\mathfrak{n}$ of B_n for each $n \in \mathbb{N}$. If $\dim B = 1$, then the local UFD $B = B_{xB}$ is a DVR, and Remark 3.3.4 implies

$$1 = \dim B = \dim \widehat{B} = \dim(k[[x, y]]) = 2,$$

a contradiction. Hence $\dim B \geq 2$.

For item 6, Proposition 12.4 implies A is a two-dimensional RLR. Thus $(a) \implies (b)$. Clearly $(b) \implies (c)$. By items 1 and 2, B is a local UFD with maximal ideal $\mathfrak{n} = (x, y)B$. Hence every prime ideal of B is finitely generated. Cohen's Theorem 2.25 yields that $(c) \implies (d)$. By Remark 3.3.4, the completion of the Noetherian local ring B is a faithfully flat extension. Remark 2.37.7 implies $(d) \implies (e)$. It is clear that $(e) \implies (f)$. To complete the proof of Theorem 12.3, it suffices to show that $(f) \implies (a)$. Since A birationally dominates B, the ring $B = A$ if and only if $bA \cap B = bB$ for every element $b \in \mathfrak{n}$; see Exercise 2.ii of Chapter 4. The principal ideal bB is closed in the $\mathfrak{n}$-adic topology on B if and only if $bB = b\widehat{B} \cap B$. Also $\widehat{B} = \widehat{A}$ and $bA = b\widehat{A} \cap A$, for every $b \in B$. Thus (f) implies, for every $b \in B$,

$$bB = b\widehat{B} \cap B = b\widehat{A} \cap B = b\widehat{A} \cap A \cap B = bA \cap B,$$

and so $B = A$. This completes the proof of Theorem 12.3. □

REMARK 12.5. With σ, τ and B as in Notation 12.2, items 5 and 6 of Theorem 12.3 establish that either the approximation domain B has dimension two and is Noetherian or B has dimension three and is not Noetherian. In the remainder of this chapter we establish the existence of both types for B, and illustrate the effect of the choice of σ and τ on the resulting approximation domain B;

THEOREM 12.6. *With σ, τ and B as in Notation* 12.2, *either the ring B is non-Noetherian and strictly smaller than $A := k(x, y, \sigma, \tau) \cap k[[x, y]]$, or $B = A$. Both cases are possible.*

PROOF. By construction, $B \subseteq A$. Thus one of these two cases occurs by $(a) \iff (d)$ of Theorem 12.3(6). The first case is established in Example 12.7 and the second case in Example 12.21. □

EXAMPLE 12.7. With Notation 12.2, let $\tau \in k[[y]]$ be defined to be $\sigma(y)$, that is, set $b_i := a_i$, for every $i \in \mathbb{N}$. We then have that $\theta := \frac{\sigma-\tau}{x-y} \in A$. To see this, write

$$\sigma - \tau = a_1(x - y) + a_2(x^2 - y^2) + \cdots + a_n(x^n - y^n) + \cdots,$$

and so $\theta = \frac{\sigma-\tau}{x-y} \in k[[x, y]] \cap k(x, y, \sigma, \tau) = A$. As a specific example, one may take $k := \mathbb{Q}$ and set $\sigma := e^x - 1$ and $\tau := e^y - 1$.

Claim 12.8 below and Theorem 12.3 above together imply that, if $\tau = \sigma(y)$, then the approximation domain B is non-Noetherian and strictly contained in the corresponding intersection domain A.

CLAIM 12.8. The element θ is not in B, and so $B \subsetneq A$.

PROOF. If θ is an element of B, then

$$\sigma - \tau \subset (x - y)B \cap U = (x - y)U.$$

As above, $U_0 = k[x, y, \sigma, \tau]$ and

$$U = \bigcup_{n \in \mathbb{N}} U_n \subseteq U_0[\frac{1}{xy}] \subset (U_0)_{(x-y)U_0},$$

where the last inclusion is because $xy \notin (x - y)U_0$. Thus $\theta \in B$ implies that

$$\sigma - \tau \in ((x - y)U_0)_{(x-y)U_0} \cap U_0 = (x - y)U_0,$$

but this contradicts the fact that x, y, σ, τ are algebraically independent over k, and U_0 is a polynomial ring over k in x, y, σ, τ. □

In contrast to Example 12.7, a Krull domain that birationally dominates a two-dimensional Noetherian local domain is Noetherian; see Exercise 14 in Chapter 2.

In Remarks 12.9 we justify using the words "Iterative Example" in the title of this chapter to describe the construction of the rings B and A of Notation 12.2.

REMARKS 12.9. Assume Notation 12.2. Thus $R = k[x, y]_{(x,y)}$; $C_n = k[x, \sigma_n]_{(x,\sigma_n)}$; $C = k(x, \sigma) \cap k[[x]] = \cup_{n=1}^{\infty} C_n$; $B_n = k[x, y, \sigma_n, \tau_n]_{(x,y,\sigma_n,\tau_n)}$; $B = \cup_{n=1}^{\infty} B_n$; and $A = k(x, y, \sigma, \tau) \cap k[[x, y]]$.

(1) The rings B and A may be obtained by "iterating" Inclusion Construction 5.3 and the approximation in Section 5.2. To see this, define a ring T associated with A and B:

$$T_n := k[x, y, \sigma_n]_{(x,y,\sigma_n)} = C_n[y]_{(x,y,\sigma_n)}, \quad T := \bigcup_{n=1}^{\infty} T_n.$$

The ring T is a Local Prototype, and so $T = k[y]_{(y)}[[x]] \cap k(x, y, \sigma) = C[y]_{(x,y)}$, a two-dimensional regular local domain, as in Local Prototype Theorem 10.6. If $\operatorname{char} k = 0$, then T is excellent. The ring T is the result of one iteration of the construction, using the power series $\sigma \in xk[[x]]$ in the x-adic completion of R.

For each positive integer n, $B_n \subset T[\tau_n]_{(x,y,\tau_n)} \subset B$. Hence by definition $B = \bigcup_{n=1}^{\infty} T[\tau_n]_{(x,y,\tau_n)}$. Thus, as in Construction Properties Theorem 5.14.6, B is the approximation domain obtained using the power series τ and applying the construction with T as the base ring.

(2) Remark 12.9.1 yields alternate proofs of parts of Theorem 12.3. By Theorem 5.24 and its proof, T, U and B are UFDs and items 1 and 2 hold. By Construction Properties Theorem 5.14, item 4 holds. Moreover part d of item 6 implies part a, by Noetherian Flatness Theorem 6.3.

(3) In addition, item 1 justifies our use of the results of Chapters 5, 6 and 10 in the remainder of this chapter to show there exist σ and τ such that $A = B$.

DISCUSSION 12.10. As stated in Remark 12.5, the ring B may be Noetherian for certain choices of σ and τ. To obtain an example of a triple σ, τ and B fitting Notation 12.2 where B is Noetherian, we first establish in Example 12.21 below with $k = \mathbb{Q}$ that the elements $\sigma := e^x - 1$ and $\tau := e^{(e^y - 1)} - 1$ give an example where $B = A$. By Proposition 12.14, the critical property of τ used to prove B is Noetherian and $A = B$ is that, for $T = C[y]_{(x,y)}$, the image of τ in $\widehat{R}/Q$ is algebraically independent over $T/(Q \cap T)$, for each height-one prime ideal Q of

$\widehat{R} = \mathbb{Q}[[x,y]]$ such that $Q \cap T \neq (0)$ and $xy \notin Q$. We use Noetherian Flatness Theorem 6.3 to prove Proposition 12.14. In order to show that the property of Proposition 12.14 holds for $\tau = e^{(e^y-1)} - 1$ in the proof of Theorem 12.18, we use results of Ax that yield generalizations of Schanuel's conjectures regarding algebraic relations satisfied by exponential functions [**18**, Corollary 1, p. 253].

REMARK 12.11. In Notation 12.2, It seems natural to consider the ring compositum $\widehat{C}[\widehat{D}]$ of $\widehat{C} = k[[x]]$ and $\widehat{D} = k[[y]]$. We outline in Exercise 5 of this chapter a proof due to Kunz that the subring $\widehat{C}[\widehat{D}]$ of $k[[x,y]]$ is not Noetherian.

12.2. Residual algebraic independence

The iterative examples of Section 12.1 lead us to consider in this section "residual algebraic independence", the critical property of an extension of Krull domains $R \hookrightarrow S$ from Discussion 12.10. [1]

DEFINITION 12.12. Let $R \hookrightarrow S$ denote an extension of Krull domains. An element $\nu \in S$ is *residually algebraically independent with respect to S* over R if ν is algebraically independent over R and, for every height-one prime ideal Q of S such that $Q \cap R \neq 0$, the image of ν in S/Q is algebraically independent over $R/(Q \cap R)$.

REMARK 12.13. If $(R, \mathfrak{m})$ is a regular local domain, or more generally an analytically normal Noetherian local domain, it is natural to consider the extension of Krull domains $R \hookrightarrow \widehat{R}$, where $\widehat{R}$ is the $\mathfrak{m}$-adic completion of R, and to ask about the existence of an element $\nu \in \widehat{R}$ that is residually algebraically independent with respect to $\widehat{R}$ over R. If the dimension of R is at least two and R has countable cardinality, for example, if $R = \mathbb{Q}[x,y]_{(x,y)}$, then a cardinality argument implies the existence of an element $\nu \in \widehat{R}$ that is residually algebraically independent with respect to $\widehat{R}$ over R.[2]

If $\nu \in \widehat{\mathfrak{m}}$ is residually algebraically independent with respect to $\widehat{R}$ over R, then the intersection domain $A = \widehat{R} \cap \mathcal{Q}(R[\nu])$ is the localized polynomial ring $R[\nu]_{(\mathfrak{m},\nu)}$.[3] Therefore A is Noetherian and the completion $\widehat{A}$ of A is a formal power series ring in one variable over $\widehat{R}$. As in Exercise 6 of Chapter 3, the local inclusion maps $R \hookrightarrow A \hookrightarrow \widehat{R}$ determine a surjective map of $\widehat{A}$ onto $\widehat{R}$. Since $\dim \widehat{A} > \dim \widehat{R}$, this surjective map has a nonzero kernel. Hence A is not a a *subspace* of $\widehat{R}$; that is, the topology on A determined by the powers of the maximal ideal of A is not the same as the *subspace topology* on A defined by intersecting the powers of the maximal ideal of $\widehat{R}$ with A.

The existence of an element ν that is almost residually algebraically independent is important in completing the proof of the iterative examples of Section 12.1, as is shown in Proposition 12.14 and Theorem 12.18. The proofs of these results use Noetherian Flatness Theorem 6.3 of Chapter 6. The proof of Proposition 12.14 shows that with the notation of Remarks 12.9 the existence of ν fits Inclusion Construction 5.3 and the approximation procedure of Section 5.2. Thus Theorem 6.3 implies that the intersection domain equals the approximation domain and is Noetherian provided a certain extension is flat.

---

[1] The residually algebraically independent property is analyzed in more depth in Chapter 22. We use results from Chapter 22 to verify Example 12.7.

[2] This existence is proved in Theorems 22.20 and 22.30.

[3] This statement is proved in Propositions 22.15 and Theorem 22.30.

PROPOSITION 12.14. *With Notation 12.2, let* $T = C[y]_{(x,y)C[y]}$. *Thus* T *is a two-dimensional regular local domain with completion* $\widehat{T} = k[[x,y]] = \widehat{R}$. *If the image of* τ *in* $C[[y]]/Q$ *is algebraically independent over* $T/(Q \cap T)$ *for each height-one prime ideal* Q *of* $C[[y]]$ *such that* $Q \cap T \neq (0)$ *and* $xy \notin Q$, *then* B *is Noetherian and* $B = A$.

PROOF. As observed in Remark 12.9, B is obtained from T by Inclusion Construction 5.3, and so Noetherian Flatness Theorem 6.3 applies. Thus, in order to show that B is Noetherian and $B = A$, it suffices to show that the map

$$\phi_y : T[\tau] \longrightarrow C[[y]][1/y]$$

is flat; see Definition 2.36. By Remark 2.37.1, flatness is a local property. Hence it suffices to show for each prime ideal Q of $C[[y]]$ with $y \notin Q$ that the induced map $\phi_Q : T[\tau]_{Q \cap T[\tau]} \longrightarrow C[[y]]_Q$ is flat. If $\text{ht}(Q \cap T[\tau]) \leq 1$, then $T[\tau]_{Q\cap T[\tau]}$ is either a field or a DVR. The integral domain $C[[y]]_Q$ is torsionfree over $T[\tau]_{Q\cap T[\tau]}$, since $r \in T[\tau]_{Q\cap T[\tau]}$, $w \in C[[y]]_Q$ with $rw = 0 \implies r = 0$ or $w = 0$. If $T[\tau]_{Q\cap T[\tau]}$ is a field, then $C[[y]]_Q$ is free and hence flat. If $T[\tau]_{Q\cap T[\tau]}$ is a DVR, Remark 2.39.3 implies ϕ_Q is flat.

Therefore it suffices to show that $\text{ht}(Q \cap T[\tau]) \leq 1$. This is clear for $Q = xC[[y]]$. On the other hand, if $xy \notin Q$, then by hypothesis, the image $\bar{\tau}$ of τ in $C[[y]]/Q$ is algebraically independent over $T/(Q \cap T)$. Let α and β be the maps shown

$$\frac{T}{Q \cap T}[\bar{\tau}] = \frac{T[\tau]}{(Q \cap T)T[\tau]} \overset{\alpha}{\twoheadrightarrow} \frac{T[\tau]}{Q \cap T[\tau]} \overset{\beta}{\hookrightarrow} \frac{C[[y]]}{Q},$$

where α is surjective and the composition $\beta \circ \alpha$ is injective. Then $C[[y]]$ is the y-adic completion of T, and so is faithfully flat over T. Thus $\text{ht}(Q \cap T) \leq 1$. If $\text{ht}(Q \cap T[\tau]) = 2$, then the image of τ in $T[\tau]/(Q \cap T[\tau])$ is algebraic over $T/(Q \cap T)$, a contradiction. Therefore $\text{ht}(Q \cap T[\tau]) \leq 1$. Thus B is Noetherian and $B = A$. □

REMARKS 12.15. (1) To establish the existence of examples applicable to Proposition 12.14, take k to be the field $\mathbb{Q}$ of rational numbers. Thus $R := \mathbb{Q}[x,y]_{(x,y)}$ is the localized polynomial ring, and the completion of R with respect to its maximal ideal $\mathfrak{m} := (x,y)R$ is $\widehat{R} := \mathbb{Q}[[x,y]]$, the formal power series ring in the variables x and y. Let $\sigma := e^x - 1 \in \mathbb{Q}[[x]]$, and $C := \mathbb{Q}[[x]] \cap \mathbb{Q}(x, \sigma)$. Thus C is an excellent DVR [4] with maximal ideal xC, and $T := C[y]_{(x,y)C[y]}$ is an excellent countable two-dimensional regular local ring with maximal ideal $(x,y)T$ and with $(y)-$adic completion $C[[y]]$. The UFD $C[[y]]$ has maximal ideal $\mathfrak{n} = (x,y)$.

(2) Theorem 12.16 below shows that there exists $\tau \in C[[y]]$ such that the image $\bar{\tau}$ of τ in $C[[y]]/Q$ is transcendental over $T/(Q \cap T)$, for each height-one prime Q of $C[[y]]$ with $Q \cap T \neq (0)$ and $y \notin Q$. The proof is elementary and uses that T is countable. In contrast, for Example 12.7, with $\sigma = e^x - 1$, $\tau = e^y - 1$ and $Q = (x - y)C[[y]]$, the element $\bar{\tau}$ is not transcendental over $T/(Q \cap T)$.

(3) If there exists an element $\tau \in \mathbb{Q}[[y]]$ such that the image of τ in $C[[y]]/Q$ is transcendental over $T/(Q \cap T)$, for each height-one prime Q of $C[[y]]$ with $Q \cap T \neq (0)$ and $y \notin Q$, then Proposition 12.14 implies B is Noetherian and $B = A$, for this choice of $\sigma \in \mathbb{Q}[[x]]$ and for $\tau \in \mathbb{Q}[[y]]$, as in Theorem 12.3. More generally, if there exists an element $\tau \in C[[y]]$ such that (a) the image of τ in $C[[y]]/xC[[y]] = (C/xC)[[y]]$ is transcendental over $T/(xT) = (C/xC)[[y]]$, and (b) the image of τ

[4] Every Dedekind domain of characteristic zero is excellent [**120**, (34.B)]. See also Remark 3.48.

in $C[[y]]/Q$ is transcendental over $T/(Q \cap T)$, for each height-one prime Q of $C[[y]]$ with $Q \cap T \neq (0)$ and $y \notin Q$, then Proposition 12.14 implies B is Noetherian and $B = A$, for this choice of $\sigma \in \mathbb{Q}[[x]]$.

THEOREM 12.16. *Let C be an excellent countable rank-one DVR with maximal ideal xC and let y be an indeterminate over C. Let $T = C[y]_{(x,y)C[y]}$. Then there exists an element $\tau \in C[[y]]$ for which the image of τ in $C[[y]]/Q$ is transcendental over $T/(Q \cap T)$, for every height-one prime ideal Q of $C[[y]]$ such that $Q \cap T \neq (0)$ and $y \notin Q$. Moreover τ is transcendental over T.*

PROOF. Since C is a DVR, C is a UFD, and so are $T = C[y]_{(x,y)C[y]}$ and $C[[y]]$. Hence every height-one prime ideal Q_i of T is principal and is generated by an irreducible polynomial of $C[y]$, say $f_i(y)$. There are countably many of these prime ideals.

Let $\mathcal{U}$ be the countable set of all height-one prime ideals of $C[[y]]$ that are generated by some irreducible factor in $C[[y]]$ of some irreducible polynomial $f(y)$ of $C[y]$ *other* than y; that is, $yC[[y]]$ is *not* included in $\mathcal{U}$. Let $\{P_i\}_{i=1}^{\infty}$ be an enumeration of the prime ideals of $\mathcal{U}$. Let $\mathfrak{n} := (x, y)C[[y]]$ denote the maximal ideal of $C[[y]]$.

CLAIM 12.17. For each $i \in \mathbb{N}$, there are uncountably many distinct cosets in $((P_1 \cap \cdots \cap P_{i-1} \cap y^{i+1}C[[y]]) + P_i)/P_i$.

PROOF. Since $y \notin P_i$, the image of y in the one-dimensional local domain $C[[y]]/P_i$ generates an ideal primary for the maximal ideal. Also $C[[y]]/P_i$ is a finite $C[[y]]$-module. Since $C[[y]]$ is (y)-adically complete it follows that $C[[y]]/P_i$ is a (y)-adically complete local domain [**122**, Theorem 8.7]. Hence, if we let $\mathcal{H}$ denote a subset of $C[[y]]$ that is a complete set of distinct coset representatives of $C[[y]]/P_i$, then $\mathcal{H}$ is uncountable.

Let a_i be an element of $P_1 \cap \cdots \cap P_{i-1} \cap y^{i+1}C[[y]]$ that is not in P_i. Then the set $a_i\mathcal{H} := \{a_i\beta \mid \beta \in \mathcal{H}\}$ represents an uncountable set of distinct coset representatives of $C[[y]]/P_i$, since, if $a_i\beta$ and $a_i\gamma$ are in the same coset of P_i and $\beta, \gamma \in \mathcal{H}$, then

$$a_i\beta - a_i\gamma \in P_i \implies \beta - \gamma \in P_i \implies \beta = \gamma,$$

Thus there are uncountably many distinct cosets of $C[[y]]/P_i$ of the form $a_i\beta + P_i$, where β ranges over $\mathcal{H}$, as desired for Claim 12.17. □

To return to the proof of Theorem 12.16, use that

$$((P_1 \cap \cdots \cap P_{i-1} \cap y^{i+1}C[[y]]) + P_i)/P_i$$

is uncountable for each i: Choose $f_1 \in y^2C[[y]]$ so that the image of $y - f_1$ in $C[[y]]/P_1$ is not algebraic over $T/(P_1 \cap T)$; this is possible since $C[[y]]/P_i$ is uncountable, and so some cosets are transcendental over the countable set $T/(P_1 \cap T)$. Then the element $y - f_1 \notin P_1$, and $f_1 \notin P_1$, since $y \in T$.

Choose $f_2 \in P_1 \cap y^3C[[y]]$ so that the image of $y - f_1 - f_2$ in $C[[y]]/P_2$ is not algebraic over $T/(P_2 \cap T)$. Note that $f_2 \in P_1$ implies the image of $y - f_1 - f_2$ is the same as the image of $y - f_1$ in $C[[y]]/P_1$ and so it is not algebraic over P_1.

Successively by induction, for each positive integer n, we choose f_n so that

$$f_n \in P_1 \cap P_2 \cap \cdots \cap P_{n-1} \cap y^{n+1}C[[y]]$$

and so that the image of $y-f_1-\ldots-f_n$ in $C[[y]]/P_n$ is transcendental over $T/(T\cap P_i)$ for each i with $1 \leq i \leq n$. Then we have a Cauchy sequence $\{f_1 + \cdots + f_n\}_{n=1}^{\infty}$ in $C[[y]]$ with respect to the $(yC[[y]])$-adic topology, and so it converges to an element $a \in y^2C[[y]]$. Now

$$y - a = (y - f_1 - \ \ldots \ - f_n) - (f_{n+1} + \ldots),$$

where the image of $(y-f_1-\ \ldots\ -f_n)$ in $C[[y]]/P_n$ is transcendental over $T/(P_n\cap T)$ and $f_i \in yC[[y]]$ for all $1 \leq i \leq n$ and $(f_{n+1} + \ldots) \in \cap_{i=1}^n P_i \cap yC[[y]]$. Therefore the image of $y - a$ in $C[[y]]/P_n$ is transcendental over $T/(P_n \cap T)$, for every $n \in \mathbb{N}$, and we have $y - a \in yC[[y]]$, as desired.

For the "Moreover" statement, suppose that $\tau - a$ is a root of a polynomial $f(z)$ with coefficients in T. For each prime ideal Q such that the image of τ is transcendental over $T/(T \cap Q)$, the coefficients of $f(z)$ must all be in $T \cap Q$. Since this is true for infinitely many height-one primes $T \cap Q$, and the intersection of infinitely many height-one primes in a Noetherian domain is zero, $f(z)$ is the 0 polynomial, and so τ is transcendental over T. □

Theorem 12.18 yields explicit examples for which B is Noetherian and $B = A$ in Theorem 12.3.

THEOREM 12.18. *Let x and y be indeterminates over $\mathbb{Q}$, the field of rational numbers. Then:*

(1) *There exist elements $\sigma \in x\mathbb{Q}[[x]]$ and $\tau \in y\mathbb{Q}[[y]]$ such that the following two conditions are satisfied:*
   (i) *σ is algebraically independent over $\mathbb{Q}(x)$ and τ is algebraically independent over $\mathbb{Q}(y)$.*
   (ii) $\operatorname{trdeg}_{\mathbb{Q}} \mathbb{Q}(y, \tau, \{\frac{\partial^n \tau}{\partial y^n}\}_{n\in\mathbb{N}}) > r := \operatorname{trdeg}_{\mathbb{Q}} \mathbb{Q}(x, \sigma, \{\frac{\partial^n \sigma}{\partial x^n}\}_{n\in\mathbb{N}})$, *where $\{\frac{\partial^n \tau}{\partial y^n}\}_{n\in\mathbb{N}}$ is the set of partial derivatives of τ with respect to y and $\{\frac{\partial^n \sigma}{\partial x^n}\}_{n\in\mathbb{N}}$ is the set of partial derivatives of σ with respect to x.*
(2) *If $\sigma \in x\mathbb{Q}[[x]]$ and $\tau \in y\mathbb{Q}[[y]]$ satisfy conditions i and ii and $T = C[y]_{(x,y)}$, where $C = \mathbb{Q}(x, \sigma) \cap \mathbb{Q}[[x]]$, as in Notation 12.2 and Remark 12.9, then the image of τ in $C[[y]]/Q$ is algebraically independent over $T/(Q \cap T)$, for every height-one prime ideal Q of $C[[y]]$ such that $Q \cap T \neq (0)$ and $xy \notin Q$.*
(3) *If $\sigma \in x\mathbb{Q}[[x]]$ and $\tau \in y\mathbb{Q}[[y]]$ satisfy conditions i and ii, then the ring B of Theorem 12.3 defined for this choice of σ and τ is Noetherian and $B = A$.*

PROOF. For item 1, to establish the existence of elements σ and τ satisfying properties (i) and (ii) of Theorem 12.18, let $\sigma = e^x - 1 \in \mathbb{Q}[[x]]$ and choose for τ a hypertranscendental element in $\mathbb{Q}[[y]]$. A power series $\tau = \sum_{i=0}^{\infty} b_i y^i \in \mathbb{Q}[[y]]$ is called *hypertranscendental* over $\mathbb{Q}[y]$ if the set of partial derivatives $\{\frac{\partial^n \tau}{\partial y^n}\}_{n\in\mathbb{N}}$ is infinite and algebraically independent over $\mathbb{Q}(y)$. Two examples of hypertranscendental elements are the Gamma function and the Riemann Zeta function. (The exponential function is, of course, far from being hypertranscendental.) Thus there exist elements σ, τ that satisfy conditions (i) and (ii) of Theorem 12.18.

Another way to obtain such elements is to set $\sigma = e^x - 1$ and $\tau = e^{(e^y-1)} - 1$. In this case, conditions i and ii of Theorem 12.18 follow from [**18**, Conjecture Σ, p. 252], a generalization of Schanuel's conjectures, which is established in Ax's paper [**18**, Corollary 1, p. 253]. To see that conditions i and ii hold, it is convenient to restate Conjecture Σ of [**18**] with different letters for the power series; let y be a variable, and use only one or two power series $s, t \in \mathbb{C}[[y]]$. Thus Conjecture Σ states that, if s and t are elements of $\mathbb{C}[[y]]$ that are $\mathbb{Q}$-linearly independent, then

$$\text{(12.18.0)} \qquad \begin{aligned} \operatorname{trdeg}_{\mathbb{Q}}(\mathbb{Q}(s, e^s)) &\geq 1 + \operatorname{rank}\begin{bmatrix} \frac{\partial s}{\partial y} \end{bmatrix}. \\ \operatorname{trdeg}_{\mathbb{Q}}(\mathbb{Q}(s, t, e^s, e^t)) &\geq 2 + \operatorname{rank}\begin{bmatrix} \frac{\partial s}{\partial y} & \frac{\partial t}{\partial y} \end{bmatrix}. \end{aligned}$$

Since the rank of the matrix $\begin{bmatrix} \frac{\partial y}{\partial y} \end{bmatrix}$ is 1, $\operatorname{trdeg}_{\mathbb{Q}}(\mathbb{Q}(y, e^y)) \geq 2$, by Equation 12.18.0. By switching the variable to x, $\operatorname{trdeg}_{\mathbb{Q}}(\mathbb{Q}(x, e^x)) \geq 2$. Thus $\sigma = e^x - 1$ satisfies condition i.

Since just two transcendental elements generate the field $\mathbb{Q}(x, e^x)$ over $\mathbb{Q}$, we have $\operatorname{trdeg}_{\mathbb{Q}}(\mathbb{Q}(x, e^x)) = 2$. Furthermore $\partial^n \sigma / \partial x^n = e^x$ for every $n \in \mathbb{N}$, and so

$$\operatorname{trdeg}_{\mathbb{Q}}(\mathbb{Q}(x, \sigma, \{\frac{\partial^n \sigma)}{\partial x^n}\}_{n \in \mathbb{N}})) = 2;$$

that is, for r as in condition ii with this σ, it follows that $r = 2$.

The rank of the matrix $\begin{bmatrix} \frac{\partial y}{\partial y} & \frac{\partial (e^y - 1)}{\partial y} \end{bmatrix}$ is 1, and so

$$\operatorname{trdeg}_{\mathbb{Q}}(\mathbb{Q}(y, e^y, e^{(e^y-1)})) = \operatorname{trdeg}_{\mathbb{Q}}(\mathbb{Q}(y, e^y - 1, e^y, e^{(e^y-1)})) \geq 3,$$

by Equation 12.18.0 with $s = y$ and $t = e^y - 1$.

For τ, taking the partial with respect to y yields

$$\begin{aligned} &\partial \tau / \partial y = \partial (e^{(e^y-1)} - 1)/\partial y = e^{(e^y-1)} \cdot e^y \implies \\ &\operatorname{trdeg}_{\mathbb{Q}}(\mathbb{Q}(y, \tau, \{\frac{\partial^n \tau}{\partial y^n}\}_{n \in \mathbb{N}})) \geq \operatorname{trdeg}_{\mathbb{Q}}(\mathbb{Q}(y, e^y, e^{(e^y-1)})) > 2, \end{aligned}$$

by the computation above, and so conditions i and ii both hold for τ. Thus item 1 is proved.

Item 3 follows from item 2 by Proposition 12.14.

For item 2, observe that the ring $T = C[y]_{(x,y)}$ is an overring of $R = \mathbb{Q}[x, y]_{(x,y)}$ and a subring of $\widehat{R}$, and T has completion $\widehat{T} = \widehat{R}$:

$$R = \mathbb{Q}[x, y]_{(x,y)} \longrightarrow T = C[y]_{(x,y)} \longrightarrow \widehat{R} = \widehat{T} = \mathbb{Q}[[x, y]].$$

The relationships among these rings are shown in Diagram 12.18.a.

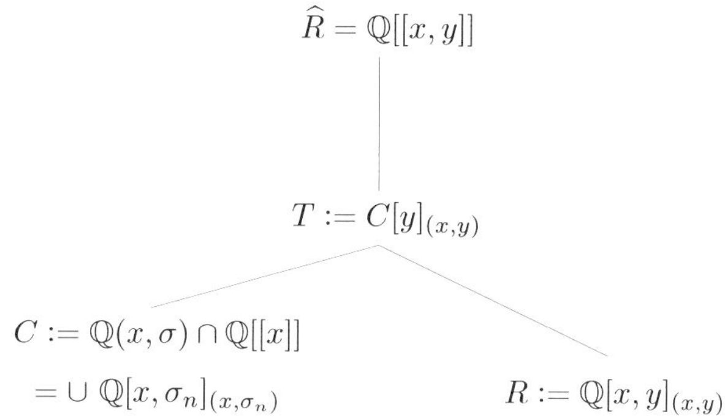

Diagram 12.18.a.The rings of the example

Let $\widehat{P}$ be a height-one prime ideal of $\widehat{R}$, let bars (for example, $\bar{x}$), denote images in $\overline{\widehat{R}} = \widehat{R}/\widehat{P}$ and set $P := \widehat{P} \cap R$ and $P_1 := \widehat{P} \cap T$. Assume that $P_1 \neq 0$ and that $xy \notin \widehat{P}$.

In the following commutative diagram, identify $\mathbb{Q}[[x]]$ with $\mathbb{Q}[[\bar{x}]]$, identify $\mathbb{Q}[[y]]$ with $\mathbb{Q}[[\bar{y}]]$, etc.

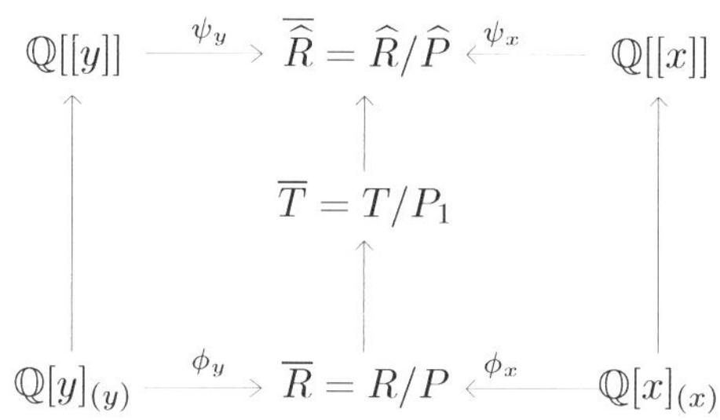

All maps in the diagram are injective and $\overline{\widehat{R}}$ is finite over both of the rings $\mathbb{Q}[[x]]$ and $\mathbb{Q}[[y]]$. There are two cases: (i) $P \neq (0)$, and (ii) $P = (0)$; in each case we show that $\overline{T} \subseteq \mathbb{Q}(\bar{x}, \bar{\sigma})^a$, the algebraic closure of $\mathbb{Q}(\bar{x}, \bar{\sigma})$.
Case i: $P = R \cap \widehat{P} \neq (0)$. Since $\operatorname{trdeg}_{\mathbb{Q}} \mathcal{Q}(\overline{R}) = 1$, the ring $\overline{R}$ is algebraic over $\mathbb{Q}[x]_{(x)}$, Thus $\bar{y}$ is algebraic over $\mathbb{Q}[x]_{(x)}$, and so $\overline{T} \subseteq \mathbb{Q}(\bar{x}, \bar{\sigma})^a$.
Case ii: $P = R \cap \widehat{P} = P_1 \cap R = (0)$. Then $P_1 \cap C = (0)$; otherwise $P_1 \cap C = xC$, since xC is the unique maximal ideal of the DVR C, and this would contradict $P_1 \cap R = (0)$. The integral domain T is a UFD since C is. Therefore the height-one prime ideal P_1 of T is generated by an element $f(y)$, which may be chosen in $C[y]$. Since $P_1 \cap C = (0)$, we have $\deg f(y) \geq 1$, where deg refers to the degree in y. Therefore $\bar{f}(\bar{y}) = 0$ in $\overline{T}$. Since the field of fractions of C is $\mathbb{Q}(x, \sigma)$, $\bar{y}$ is algebraic over the field $\mathbb{Q}(\bar{x}, \bar{\sigma})$. Hence $\overline{T}$ is contained in $\mathbb{Q}(\bar{x}, \bar{\sigma})^a$.

Let L denote the field of fractions of $\overline{\widehat{R}}$. Consider $\mathbb{Q}(y, \tau, \{\frac{\partial^n \tau}{\partial y^n}\}_{n \in \mathbb{N}})$ and $\mathbb{Q}(x, \sigma, \{\frac{\partial^n \sigma}{\partial x^n}\}_{n \in \mathbb{N}})$ as subfields of L, where

$$\operatorname{trdeg}_{\mathbb{Q}} \mathbb{Q}(y, \tau, \left\{\frac{\partial^n \tau}{\partial y^n}\right\}_{n \in \mathbb{N}})^a > \operatorname{trdeg}_{\mathbb{Q}} \mathbb{Q}(x, \sigma, \left\{\frac{\partial^n \sigma}{\partial x^n}\right\}_{n \in \mathbb{N}})^a.$$

Let d denote the partial derivative map $\frac{\partial}{\partial x}$ on $\mathbb{Q}((x))$. Since the extension L of $\mathbb{Q}((x))$ is finite and separable, d extends uniquely to a derivation $\widehat{d}: L \to L$, [**192**, Corollary 2, p. 124]. Let H denote the algebraic closure (shown in Picture 12.19.1 by a small upper a) in L of the field $\mathbb{Q}(x, \sigma, \{\frac{\partial^n \sigma}{\partial x^n}\}_{n\in\mathbb{N}})$. Let $\widehat{p}(x,y) \in \mathbb{Q}[[x,y]]$ be a prime element generating $\widehat{P}$. Claim 12.19 asserts that the images of H and $\overline{\overline{R}}$ under $\widehat{d}$ are inside H and $(1/p')\overline{\overline{R}}$, respectively, as shown in Picture 12.19.1.

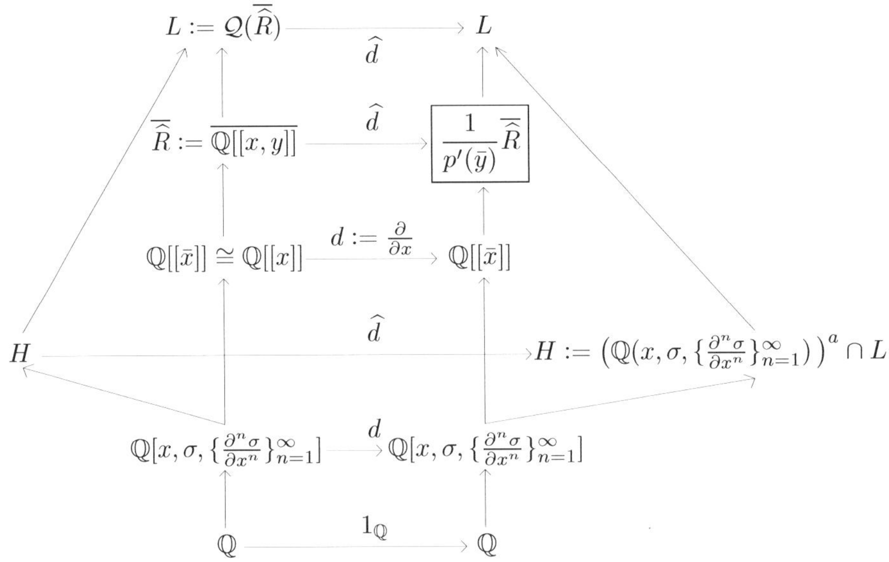

Picture 12.19.1. The image of subrings of L via the extension $\widehat{d}$ of $d := \frac{\partial}{\partial x}$.

Claim 12.19. With the notation above:
(i) $\widehat{d}(H) \subseteq H$.
(ii) There exists a polynomial $p(x,y) \in \mathbb{Q}[[x]][y]$ with $p\mathbb{Q}[[x,y]] = \widehat{P}$ and $p(\bar{y}) = 0$.
(iii) $\widehat{d}(\bar{y}) \neq 0$ and $p'(\bar{y})\widehat{d}(\bar{y}) \in \overline{\overline{R}}$, where $p'(y) := \frac{\partial p(x,y)}{\partial y}$.
(iv) For every element $\lambda \in \overline{\overline{R}}$, we have $p'(\bar{y})\widehat{d}(\lambda) \in \overline{\overline{R}}$, and so $\widehat{d}(\overline{\overline{R}}) \subseteq (1/p'(\bar{y}))\overline{\overline{R}}$.

Proof. For item i, since d maps $\mathbb{Q}(x, \sigma, \{\frac{\partial^n \sigma}{\partial x^n}\}_{n\in\mathbb{N}})$ into itself, $\widehat{d}(H) \subseteq H$.

For item ii, the elements x and y are not contained in $\widehat{P}$, and the element $\widehat{p}(x,y) \in \mathbb{Q}[[x,y]]$ generates $\widehat{P}$ and is *regular in* y as a power series in $\mathbb{Q}[[x,y]]$ (in the sense of Zariski-Samuel [**193**, p.145]); that is, $\widehat{p}(0,y) \neq 0$. Thus by [**193**, Corollary 1, p.145] the element $\widehat{p}(x,y)$ can be written as:

$$\widehat{p}(x,y) = \epsilon(x,y)(y^n + c_{n-1}(x)y^{n-1} + \ldots + c_0(x)),$$

where $\epsilon(x,y)$ is a unit of $\mathbb{Q}[[x,y]]$ and each $c_i(x) \in \mathbb{Q}[[x]]$. Hence $\widehat{P}$ is also generated by

$$p(x,y) = p(y) := \epsilon^{-1}\widehat{p} \; = \; y^n + c_{n-1}y^{n-1} + \cdots + c_0,$$

and the $c_i \in \mathbb{Q}[[x]]$. Since $p(y)$ is the minimal polynomial of $\bar{y}$ over the field $\mathbb{Q}((x))$, it follows that $0 = p(\bar{y}) := \bar{y}^n + c_{n-1}\bar{y}^{n-1} + \cdots + c_1\bar{y} + c_0$.

For item iii, observe that

$$p'(y) = ny^{n-1} + c_{n-1}(n-1)y^{n-2} + \cdots + c_1,$$

and $p'(\bar{y}) \neq 0$ by minimality. Now

$$
\begin{aligned}
0 = \widehat{d}(p(\bar{y})) &= \widehat{d}(\bar{y}^n + c_{n-1}\bar{y}^{n-1} + \cdots + c_1\bar{y} + c_0) \\
&= n\bar{y}^{n-1}\widehat{d}(\bar{y}) + c_{n-1}(n-1)\bar{y}^{n-2}\widehat{d}(\bar{y}) + d(c_{n-1})\bar{y}^{n-1} + \cdots \\
&\qquad\qquad + c_1\widehat{d}(\bar{y}) + d(c_1)\bar{y} + d(c_0) \\
&= \widehat{d}(\bar{y})(n\bar{y}^{n-1} + c_{n-1}(n-1)\bar{y}^{n-2} + \cdots + c_1) + \\
&\qquad\qquad + d(c_{n-1})\bar{y}^{n-1} + \cdots + d(c_1)\bar{y} + d(c_0) \\
&= \widehat{d}(\bar{y})(p'(\bar{y})) + \sum_{i=0}^{n-1} d(c_i)\bar{y}^i
\end{aligned}
$$

$$
\Longrightarrow \widehat{d}(\bar{y})(p'(\bar{y})) = -\Big(\sum_{i=0}^{n-1} d(c_i)\bar{y}^i\Big) \quad \text{and} \quad \widehat{d}(\bar{y}) = \frac{-1}{p'(\bar{y})}\sum_{i=0}^{n-1} d(c_i)\bar{y}^i.
$$

In particular, $p'(\bar{y})\widehat{d}(\bar{y}) \in \overline{\widehat{R}}$. If $d(c_i) = 0$ for every i, then $c_i \in \mathbb{Q}$ for every i; this would imply that $p(x, y) \in \mathbb{Q}[[y]]$ and either $c_0 = 0$ or c_0 is a unit of $\mathbb{Q}$. If $c_0 = 0$, $p(x, y)$ could not be a minimal polynomial for $\bar{y}$, a contradiction. If c_0 is a unit, then $p(y)$ is a unit of $\mathbb{Q}[[y]]$, and so $\widehat{P}$ contains a unit, another contradiction. Thus $\widehat{d}(\bar{y}) \neq 0$, as desired for item iii.

For item iv, observe that every element $\lambda \in \overline{\widehat{R}}$ has the form:

$$
\lambda = e_{n-1}(x)\bar{y}^{n-1} + \cdots + e_1(x)\bar{y} + e_0(x), \text{ where } e_i(x) \in \mathbb{Q}[[x]].
$$

Therefore:

$$
\widehat{d}(\lambda) = \widehat{d}(\bar{y})\big[(n-1)e_{n-1}(x)\bar{y}^{n-2} + \cdots + e_1(x)\big] + \sum_{i=0}^{n-1} d(e_i(x))\bar{y}^i.
$$

The sum expression on the right is in $\overline{\widehat{R}}$ and, as established above, $p'(\bar{y})\widehat{d}(\bar{y}) \in \overline{\widehat{R}}$, and so $p'(\bar{y})\widehat{d}(\lambda) \in \overline{\widehat{R}}$. □

The next claim shows that every power series $\gamma \in \mathbb{Q}[[y]]$ has an expression for $\widehat{d}(\bar{\gamma})$ in terms of the image in $\overline{\widehat{R}}$ of the partial derivative $\frac{\partial\gamma}{\partial y}$ of γ with respect to y.

CLAIM 12.20. *If $\gamma \in \mathbb{Q}[[y]]$, then $\widehat{d}(\bar{\gamma}) = \widehat{d}(\bar{y})\overline{\big(\frac{\partial\gamma}{\partial y}\big)}$.*

PROOF. For every $m \in \mathbb{N}$, the series $\gamma = \sum_{i=0}^m b_i y^i + y^{m+1}\gamma_m$, where each $b_i \in \mathbb{Q}$ and each $\gamma_m \in \mathbb{Q}[[y]]$ is an mth endpiece of γ, as in Equation 5.4.1. Therefore

$$
\widehat{d}(\bar{\gamma}) = \widehat{d}(\bar{y}) \cdot \Big(\sum_{i=1}^m i b_i \bar{y}^{i-1}\Big) + \widehat{d}(\bar{y})(m+1)\bar{y}^m\bar{\gamma}_m + \bar{y}^{m+1}\widehat{d}(\bar{\gamma}_m).
$$

Thus, for the polynomial $p(x, y) \in \mathbb{Q}[[x]][y]$ of Claim 12.19,

$$
(12.20.0)\quad p'(\bar{y})\widehat{d}(\bar{\gamma}) = p'(\bar{y})\widehat{d}(\bar{y}) \cdot \sum_{i=1}^m i b_i \bar{y}^{i-1} + \bar{y}^m[p'(\bar{y})\widehat{d}(\bar{y})(m+1)\bar{\gamma}_m + \bar{y}p'(\bar{y})\widehat{d}(\bar{\gamma}_m)].
$$

Since $\gamma = \sum_{i=0}^{\infty} b_i y^i$ with $b_i \in \mathbb{Q}$,

$$
(12.20.1)\qquad \overline{\left(\frac{\partial\gamma}{\partial y}\right)} = \sum_{i=1}^m i b_i \bar{y}^{i-1} + \bar{y}^m \sum_{i=m+1}^{\infty} i b_i \bar{y}^{i-m-1}.
$$

Multiply Equation 12.20.1 by $p'(\bar{y})\widehat{d}(\bar{y})$ to obtain

$$(12.20.2)\quad p'(\bar{y})\widehat{d}(\bar{y})\overline{\left(\frac{\partial\gamma}{\partial y}\right)} = p'(\bar{y})\widehat{d}(\bar{y})\sum_{i=1}^{m} ib_i\bar{y}^{i-1} + p'(\bar{y})\widehat{d}(\bar{y})\bar{y}^m \sum_{i=m+1}^{\infty} ib_i\bar{y}^{i-m-1}.$$

Then subtracting Equation 12.20.2 from Equation 12.20.0 for $p'(\bar{y})\widehat{d}(\bar{\gamma})$ yields

$$p'(\bar{y})\widehat{d}(\bar{\gamma}) - p'(\bar{y})\widehat{d}(\bar{y})\overline{\left(\frac{\partial\gamma}{\partial y}\right)} \in \bar{y}^m(\overline{\widehat{R}}),$$

for every $m \in \mathbb{N}$. Therefore $p'(\bar{y})\widehat{d}(\bar{\gamma}) - p'(\bar{y})\widehat{d}(\bar{y})\overline{\left(\frac{\partial\gamma}{\partial y}\right)} \in \cap y^m(\overline{\widehat{R}}) = 0$, by Krull's Intersection Theorem 2.22. Thus $\widehat{d}(\gamma) = \widehat{d}(\bar{y})\overline{\left(\frac{\partial\gamma}{\partial y}\right)}$, since $p'(\bar{y}) \neq 0$ and $\overline{\widehat{R}}$ is an integral domain. That is, Claim 12.20 holds. □

Completion of proof of Theorem 12.18. From above, in either case i or case ii, $\overline{T} \subseteq H$, where H is the algebraic closure of the field $\mathcal{Q}(\bar{x}T, \{\overline{\frac{\partial^n\sigma}{\partial x^n}}\}_{n\in\mathbb{N}})$ in L. Then $\bar{\tau} \notin H$ if and only if $\bar{\tau}$ is transcendental over H. By hypothesis, the transcendence degree of $H/\mathbb{Q}$ is r. If $\bar{\tau}$ were in H, then Claim 12.20 implies $\overline{\frac{\partial^n\tau}{\partial y^n}} \in H$, for all $n \in \mathbb{N}$, since $\widehat{d}(H) \subseteq H$, and then the field $\mathbb{Q}(y, \tau, \{\frac{\partial^n\tau}{\partial y^n}\}_{n\in\mathbb{N}})$ is contained in H, a contradiction to our hypothesis that $\operatorname{trdeg}_{\mathbb{Q}} \mathbb{Q}(y, \tau, \{\frac{\partial^n\tau}{\partial y^n}\}_{n\in\mathbb{N}}) > r$. Therefore the image of τ in $\widehat{R}/Q$ is algebraically independent over $T/(Q\cap T)$, for each height-one prime ideal Q of $\widehat{R}$ such that $Q \cap T \neq (0)$ and $xy \notin Q$. This completes the proof of Theorem 12.18. □

As mentioned in Discussion 12.10 and the second paragraph of the proof of Theorem 12.18, Ax's results in [**18**] together with the arguments of the proof imply that the elements $\sigma = e^x - 1 \in \mathbb{Q}[[x]]$ and $\tau = e^{(e^y-1)} - 1 \in \mathbb{Q}[[y]]$ satisfy the conditions of Theorem 12.18. Thus we have the following example:

EXAMPLE 12.21. For $\sigma = e^x - 1 \in \mathbb{Q}[[x]]$ and $\tau = e^{(e^y-1)} - 1 \in \mathbb{Q}[[y]]$ in Theorem 12.3, the ring B is Noetherian and $B = A$.

This completes the proof of Theorem 12.6.

Exercises

(1) Let P be a prime ideal in a commutative ring U, and let S be a nonempty multiplicatively closed subset of $U \setminus P$. Prove:
   (a) For every $s \in S$ and $u \in U$, the element $u \in P \iff u/s \in PS^{-1}U$.
   (b) $U_P = (S^{-1}U)_{(PS^{-1}U)}$; that is, $(U \setminus P)^{-1}U = (S^{-1}U)_{(PS^{-1}U)}$.

(2) Let x and y be indeterminates over a field k and let R be the two-dimensional RLR obtained by localizing the mixed power series-polynomial ring $k[[x]][y]$ at the maximal ideal $(x, y)k[[x]][y]$.
   (i) For each height-one prime ideal P of R different from xR, prove that R/P is a one-dimensional complete local domain.
   (ii) For each nonzero prime ideal Q of $\widehat{R} = k[[x, y]]$ prove that $Q \cap R \neq (0)$. Conclude that the generic formal fiber of R is zero-dimensional.

   SUGGESTION. For part (ii), use Theorem 3.16. For more information about the dimension of the formal fibers, see the articles of Matsumura and Rotthaus [**121**] and [**158**].

(3) Let x and y be indeterminates over a field k and let $R = k[x,y]_{(x,y)}$. As in Remark 12.13, assume that $\nu \in \widehat{\mathfrak{m}}$ is residually algebraically independent with respect to $\widehat{R} = k[[x,y]]$ over R. Thus $A = \widehat{R} \cap \mathcal{Q}(R[\nu])$ is the localized polynomial ring $R[\nu]_{(\mathfrak{m},\nu)}$. Let $\mathfrak{n} = (\mathfrak{m},\nu)A$ denote the maximal ideal of A. Give a direct proof that A is not a subspace of $\widehat{R}$.

SUGGESTION. Since $\nu \in \widehat{\mathfrak{m}}$ is a power series in $\widehat{R} = k[[x,y]]$, for each positive integer n, there exists a polynomial $f_n \in k[x,y]$ such that $\nu - f_n \in \widehat{\mathfrak{m}}^n$. Since A is a 3-dimensional regular local ring with $\mathfrak{n} = (x,y,\nu)A$, the element $\nu - f_n \notin \mathfrak{n}^2$. Hence for each positive integer n, the ideal $\widehat{\mathfrak{m}}^n \cap A$ is not contained in $\mathfrak{n}^2$.

(4) Let x and y be indeterminates over a field k and let $f \in k[x,y]$ be a nonconstant polynomial.
   (i) If the subfield $k(f)$ of $k(x,y)$ is relatively algebraically closed in the extension field $k(x,y)$, prove that there are only finitely many constants $c \in k$ such that the polynomial $f - c$ is reducible in $k[x,y]$
   (ii) If A is a k-subalgebra of $k[x,y]$ with $\dim A = 1$, prove that A is finitely generated as a k-algebra and the integral closure of A is a polynomial ring in one variable over k.
   (iii) If the field $k(f)$ is not relatively algebraically closed in $k(x,y)$, prove that the polynomial $f - c$ is reducible in $k[x,y]$ for every constant $c \in k$.

COMMENT. For item i of Exercise 4, see the paper of Abhyankar, Heinzer and Sathaye [**8**]. Items ii and iii of Exercise 4 relate to a ring-theoretic version of a famous result of Lüroth about subfields of transcendence degree one of a purely transcendental field extension; see the paper of Abhyankar, Eakin and Heinzer [**6**], and the paper of Igusa [**100**].

(5) (Kunz) Let L/k be a field extension with L having infinite transcendence degree over k. Prove that the ring $L \otimes_k L$ is not Noetherian. Deduce that the ring $k[[x]] \otimes_k k[[x]]$, which has $k((x)) \otimes_k k((x))$ as a localization, is not Noetherian.

SUGGESTION. Let $\{x_\lambda\}_{\lambda \in \Lambda}$ be a transcendence basis for L/k and consider the subfield $F = k(\{x_\lambda\})$ of L. The ring $L \otimes_k L$ is faithfully flat over its subring $F \otimes_k F$, and if $F \otimes_k F$ is not Noetherian, then $L \otimes_k L$ is not Noetherian. Hence it suffices to show that $F \otimes_k F$ is not Noetherian. The module of differentials $\Omega^1_{F/k}$ is known to be infinite dimensional as a vector space over F [**108**, 5.4], and $\Omega^1_{F/k} \cong I/I^2$, where I is the kernel of the map $F \otimes_k F \to F$, defined by sending $a \otimes b \mapsto ab$. Thus the ideal I of $F \otimes_k F$ is not finitely generated.

CHAPTER 13

Approximating discrete valuation rings by regular local rings

Let k be a field of characteristic zero and let $(V, \mathfrak{n})$ be a rank-one discrete valuation domain (DVR) containing k and having residue field $V/\mathfrak{n} \cong k$. If the field of fractions L of V has finite transcendence degree s over k, we prove that, for every positive integer $d \leq s$, the ring V can be realized as a directed union of regular local rings that are d-dimensional k-subalgebras of V. This construction uses an adaptation of Inclusion Construction 5.3.

13.1. Local quadratic transforms and local uniformization

The concepts of local quadratic transforms and local uniformization are relevant to this chapter.

DEFINITIONS 13.1. Let $(R, \mathfrak{m})$ be a Noetherian local domain and let $(V, \mathfrak{n})$ be a valuation domain that birationally dominates R.

(1) The *first local quadratic transform* of $(R, \mathfrak{m})$ along $(V, \mathfrak{n})$ is the ring

$$R_1 = R[\mathfrak{m}/a]_{\mathfrak{m}_1},$$

where $a \in \mathfrak{m}$ is such that $\mathfrak{m}V = aV$ and $\mathfrak{m}_1 := \mathfrak{n} \cap R[\mathfrak{m}/a]$. If $R = V$, then $R = R_1$. The ring R_1 is also called the *dilatation of* R *by the ideal* $\mathfrak{m}$ *along* V [**137**, page 141].

(2) More generally, if $I \subseteq \mathfrak{m}$ is a nonzero ideal of R, the *dilatation of* R *by* I *along* V is the ring $R[I/a]_{\mathfrak{m}_1} = R_1$, where $a \in I$ is such that $IV = aV$ and $\mathfrak{m}_1 = \mathfrak{n} \cap R[I/a]$; moreover, R_1 is uniquely determined by R, V and the ideal I [**137**, page 141].

(3) For each positive integer i, the $(i+1)^{\text{st}}$ *local quadratic transform* R_{i+1} *of* R *along* V is defined inductively: R_{i+1} is the first local quadratic transform of R_i along V.

REMARKS 13.2. Let $(R, \mathfrak{m})$ be a regular local ring and let $(V, \mathfrak{n})$ be a valuation domain that birationally dominates R. Let $\{R_i\}_{i \in \mathbb{N}}$ be the sequence of local quadratic transforms of R along V, defined in Definition 13.1.3.

(1) It is well known that the local quadratic transform R_1 of R along V is again a regular local ring [**137**, 38.1].

(2) With the notation of Definition 13.1.3, we have the following relationship among iterated local quadratic transforms:

$$R_{i+j} = (R_i)_j \quad \text{for all} \quad i, j \geq 0.$$

Associated with the set $\{R_i\}_{i \in \mathbb{N}}$, it is natural to consider the subring $R_\infty := \bigcup_{i=1}^{\infty} R_i$ of V.

(3) If $\dim R = 2$, then $R_\infty = \bigcup_{n=1}^\infty R_n = V$, by a classical result of Zariski and Abhyankar [**2**, Lemma 12].
(4) For $\dim R \geq 3$ and certain V, the union $\bigcup_{n=1}^\infty R_n$ is strictly smaller than V; see Shannon [**170**, 4.13]. In [**60**, Theorem 13], Granja gives necessary and sufficient conditions in order that $R_\infty = V$. Shannon proves in many cases that V is a directed union of iterated monoidal transforms of R, where a *monoidal transform of* R is a dilatation of R by a prime ideal P for which the residue class ring R/P is regular [**170**, (4.5), page 308].
(5) Assume that $R \subseteq S \subseteq V$, where S is a regular local ring birationally dominating R and V is a valuation domain birationally dominating S. Using monoidal transforms, Cutkosky shows in [**38**] and [**39**] that there exists an iterated local monoidal transform T of S along V such that T is also an iterated local monoidal transform of R.

If V is a DVR that birationally dominates a regular local ring, the following useful result is proved by Zariski [**191**, pages 27-28] and Abhyankar [**2**, page 336]. For a related result, see Remark 4.20.

PROPOSITION 13.3. *Let $(V, \mathfrak{n})$ be a DVR that birationally dominates a regular local ring $(R, \mathfrak{m})$, and let R_n be the nth local quadratic transform of R along V. Then:*
(1) $R_\infty = \bigcup_{n=1}^\infty R_n = V$.
(2) *If V is essentially finitely generated over R, then $R_n = V$ for some positive integer n, and thus $R_{n+i} = R_n$ for all $i \geq 0$.*

PROOF. A nonzero element η of V has the form $\eta = b/c$, where $b, c \in R$. If $(b, c)V = V$, then $b/c \in V$ implies $cV = V$. Since V dominates R, it follows that $cR = R$, so $b/c \in R$ in this case. If $\eta = b/c$, with $b, c \in R$ and $(b, c)V = \mathfrak{n}^n$, we prove by induction on n that $\eta \in R_n$. The case where $n = 0$ is done.

Assume $\beta/\gamma \in S_j$, for every j with $0 \leq j < n$, every regular local domain $(S, \mathfrak{p})$ birationally dominated by V and every nonzero element $\beta/\gamma \in V$ with $\beta, \gamma \in S$ and $(\beta, \gamma)V = \mathfrak{n}^j$, where S_j is the j-th iterated local quadratic transform of S along V. Suppose $\beta, \gamma \in S$, $\beta/\gamma \in V$ and $(\beta, \gamma)V = \mathfrak{n}^n$. Let $S_1 = S[\mathfrak{p}/a]_{\mathfrak{p}_1}$, where $a \in \mathfrak{p}$ is such that $\mathfrak{p}V = aV$ and $\mathfrak{p}_1 := \mathfrak{n} \cap S[\mathfrak{p}/a]$; that is, S_1 is the first local quadratic transform of S along V. Then $\beta_1 := \beta/a$ and $\gamma_1 := \gamma/a$ are in S_1. Thus $a \in \mathfrak{n}$ implies $(\beta_1, \gamma_1)V = \mathfrak{n}^j$, where $0 \leq j < n$, and so by induction

$$\beta/\gamma = \beta_1/\gamma_1 \in (S_1)_j = S_{j+1} \subseteq S_n.$$

This completes the proof of statement 1 of Proposition 13.3.

Statement 2 follows from [**1**, Proposition 3, p. 336]. There exists an integer n such that the finite number of "essential" generators for V are contained in R_n. $\square$

DEFINITION 13.4. Let $(R, \mathfrak{m})$ be a Noetherian local domain that is essentially finitely generated over a field k and let $(V, \mathfrak{n})$ be a valuation domain that birationally dominates R. In algebraic terms, *local uniformization of* R *along* V asserts the existence of a regular local domain extension S of R such that S is essentially finitely generated over R and S is dominated by V.

If R is a regular local ring and P is a prime ideal of R, *embedded local uniformization of R along V* asserts the existence of a regular local domain extension S of R such that S is essentially finitely generated over R and is dominated by V, and has the property that there exists a prime ideal Q of S with $Q \cap R = P$ such that the residue class ring S/Q is a regular local ring.

DISCUSSION 13.5. The classical approach for obtaining embedded local uniformization, introduced by Zariski in the 1940s [**189**], uses local quadratic transforms of R along V. Let $(R, \mathfrak{m})$ be a s-dimensional regular local ring. If $(V, \mathfrak{n})$ is a DVR that birationally dominates R and $V/\mathfrak{n}$ is algebraic over $R/\mathfrak{m}$, then by Proposition 13.3 the classical method of taking local quadratic transforms of R along V gives a representation for V as a directed union $R_\infty = \bigcup_{n \in \mathbb{N}} R_n = V$, where each R_n is an iterated local quadratic transform of R. For each n, R_n is essentially finitely generated over R, and the dimension formula [**122**, page 119] implies that $\dim R_n = \dim R = s$. If $s > 1$, the DVR V is an infinite directed union of s-dimensional RLRs. We prove in Theorem 13.6 that certain DVRs can be represented as a directed union of regular local domains of dimension d for every positive integer d less than or equal to $s = \dim R$.

13.2. Expressing a DVR as a directed union of regular local rings

Theorem 13.6 is the main result of this chapter:

THEOREM 13.6. *Let k be a field of characteristic zero and let $(V, \mathfrak{n})$ be a DVR containing k with $V/\mathfrak{n} = k$. Assume that the field of fractions L of V has finite transcendence degree s over k. Let d be an integer with $1 \leq d \leq s$. Then:*
1. *If L is finitely generated over k, then V is a countable union $\bigcup_{n=1}^{\infty} C_n$, where, for each $n \in \mathbb{N}$,*
   (a) *C_n is a d-dimensional regular local k-subalgebra of V,*
   (b) *C_{n+1} dominates C_n, and*
   (c) *V dominates C_n.*
2. *If L is not finitely generated over k, then there exists an index set Γ and a nested family $\{C_n^{(\alpha)} : n \in \mathbb{N}, \alpha \in \Gamma\}$ such that*
   (a) *V is the directed union of the $C_n^{(\alpha)}$,*
   (b) *Every $C_n^{(\alpha)}$ is a d-dimensional regular local k-subalgebras of V,*
   (c) *Every $C_n^{(\alpha)}$ has field of fractions L,*
   (d) *V dominates each $C_n^{(\alpha)}$.*

The proof of Theorem 13.6 is given after the proof of Theorem 13.11. Corollary 13.7 follows from Theorem 13.6.

COROLLARY 13.7. *Let k be a field of characteristic zero and let $(R, \mathfrak{m})$ be a local domain essentially of finite type over k with coefficient field $k \cong R/\mathfrak{m}$ and field of fractions L. Let $s = trdeg_k(L)$, and let $(V, \mathfrak{n})$ be a DVR birationally dominating R with $V/\mathfrak{n} = k$. Then, for every integer d with $1 \leq d \leq s$, there exists a sequence of d-dimensional regular local k-subalgebras C_n of V such that*
1. *$V = \bigcup_{n=1}^{\infty} C_n$,*
2. *For each n, C_{n+1} dominates C_n, and V dominates C_n.*
3. *C_n dominates R, for all sufficiently large n.*

DISCUSSION 13.8. (1) If L/k is finitely generated of transcendence degree s, then the fact that V is a directed union of s-dimensional regular local rings follows from classical theorems of Zariski. The local uniformization theorem of Zariski

[**189**] implies the existence of a regular local domain $(R, \mathfrak{m})$ containing the field k such that V birationally dominates R. Since k is a coefficient field for V, we have

- $k \hookrightarrow V \twoheadrightarrow V/\mathfrak{n} \cong k$, and k is relatively algebraically closed in L.
- $R/\mathfrak{m} = k$, since V dominates R.
- Every iterated local quadratic transform of R along V has dimension s.

By Proposition 13.3, V is a directed union of s-dimensional RLRs.

(2) If $d = 1$, Theorem 13.6 is trivially true by taking each $C_n = V$. If L/k is finitely generated of transcendence degree $s = 2$, then item 1 implies Theorem 13.6 is saying nothing new.

(3) If $s > 2$, then the classical local uniformization theorem says nothing about expressing V as a directed union of d-dimensional RLRs, where $2 \le d \le s-1$. If $(S, \mathfrak{p})$ is a Noetherian local domain containing k and birationally dominated by V with $\dim S = d < s$, then $S/\mathfrak{p} = k$, and S does not satisfy the dimension formula. It follows that S is not essentially finitely generated over k [**122**, page 119].

Remark 13.9 and Notation 13.10 are used in the proof of Theorem 13.6.

REMARK 13.9. With the notation of Theorem 13.6, let $y \in \mathfrak{n}$ be such that $yV = \mathfrak{n}$. Then the $\mathfrak{n}$-adic completion $\widehat{V}$ of V is $k[[y]]$, and we have

$$k[y]_{(y)} \subseteq V \subseteq k[[y]].$$

Then $V = L \cap k[[y]]$, since $V \hookrightarrow k[[y]]$ is flat. Since the transcendence degree of L over $k(y)$ is $s-1$, there are $s-1$ elements $\sigma_1, \ldots, \sigma_{s-d}, \tau_1, \ldots, \tau_{d-1} \in yV$ such that L is algebraic over $F := k(y, \sigma_1, \ldots, \sigma_{s-d}, \tau_1, \ldots, \tau_{d-1})$.

NOTATION 13.10. To continue the terminology of Remark 13.9, let

$$K := k(y, \sigma_1, \ldots, \sigma_{s-d}) \quad \text{and} \quad R := K \cap V = K \cap k[[y]].$$
$$F : := k(y, \sigma_1, \ldots, \sigma_{s-d}, \tau_1, \ldots, \tau_{d-1}).$$

Thus R is a DVR and the (y)-adic completion of R is $R^* = k[[y]]$. Then the ring $B_0 := R[\tau_1, \ldots, \tau_{d-1}]_{(y, \tau_1, \ldots, \tau_{d-1})}$ is a d-dimensional regular local ring and $V_0 := F \cap V$ is a DVR that birationally dominates B_0 and has y-adic completion $\widehat{V_0} = k[[y]]$. The following diagram displays these integral domains:

$$\begin{array}{ccccccccc}
k & \xrightarrow{\subseteq} & K & \xrightarrow{\subseteq} & F & = & F & \xrightarrow{\subseteq} & L = \mathcal{Q}(V) \\
\Vert\uparrow & & \cup\uparrow & & \cup\uparrow & & \cup\uparrow & & \cup\uparrow \\
k & \xrightarrow{\subseteq} & R := K \cap V & \xrightarrow{\subseteq} & B_0 & \xrightarrow{\subseteq} & V_0 := F \cap V & \xrightarrow{\subseteq} & V,
\end{array}$$

where $\mathcal{Q}(V)$ denotes the field of fractions of V. The elements $\tau_1, \ldots, \tau_{d-1} \in yR^*$ are regular elements of R^* that are algebraically independent over K. As in Notation 5.4, represent each of the τ_i by a power series expansion in y; these representations yield for each positive integer n the nth-endpieces τ_{in} and corresponding n^{th}-localized polynomial ring B_n. For $1 \le i \le d-1$, for each $n \in \mathbb{N}$, and for $\tau_i := \sum_{j=1}^{\infty} r_{ij} y^j$, where the $r_{ij} \in R$, set:

$$\tau_{in} := \sum_{j=n+1}^{\infty} r_{ij} y^{j-n}, \qquad B_n := R[\tau_{1n}, \ldots, \tau_{d-1,n}]_{(\mathfrak{m}, \tau_{1n}, \ldots, \tau_{d-1,n})}$$
$$B := \bigcup_{n=0}^{\infty} B_n = \varinjlim B_n, \qquad A := K(\tau_1, \ldots, \tau_{d-1}) \cap R^* = F \cap R^* = V_0. \tag{13.10.1}$$

Recall that A birationally dominates B. By Proposition 5.9, the definition of B_n is independent of the representations of the τ_i. Also $V_0 = F \cap k[[y]]$ implies $\tau_{in} \in V_0$. It follows that, for each $n \in \mathbb{N}$, $B_n = R[\tau_{1n}, \dots, \tau_{d-1,n}]_{(y,\tau_{1n},\dots,\tau_{d-1,n})}$ is the first quadratic transform of B_{n-1} along V_0.

Here $V_0 = B = A$, since $B_0 \hookrightarrow k[[y]][1/y]$ is flat.

The proof of Theorem 13.6 uses Noetherian Flatness Theorem 6.3 of Chapter 6 and also Theorem 13.11:

THEOREM 13.11. *Assume Notation 13.10, and for each positive integer n, let B_n^h denote the Henselization of B_n. Then $\bigcup_{n=1}^{\infty} B_n^h = V_0^h = V^h$.*

PROOF. Since $R^*[1/y]$ is a field, it is flat as an $R[\tau_1, \dots, \tau_{d-1}]$-module. By Theorem 6.3, $V_0 = \bigcup_{n=1}^{\infty} B_n$. An alternate way to justify this description of V_0 is to use Proposition 13.3, where the ring R is B_0, V is V_0, and each $R_n = B_n$. Then

$$V_0 \underset{\text{alg}}{\longrightarrow} V \longrightarrow k[[y]],$$

where V_0 and V are DVRs of characteristic zero having completion $k[[y]]$. Since V_0 and V are excellent, their Henselizations V_0^h and V^h are the set of elements of $k[[y]]$ algebraic over V_0 or V [**137**, (44.3)]. Thus $V_0^h = V^h$ and V is a directed union of étale extensions of V_0; see Definition 8.24.

The ring $C := \bigcup B_n^h$ is Henselian and contains V_0, so $V_0^h = V^h \subseteq C$. Moreover, the inclusion map $V_0 \to C = \bigcup B_n^h$ extends to a map $V^h \overset{\sigma}{\to} C = \bigcup B_n^h$. On the other hand, the maps $B_n \to V$ extend to maps: $B_n^h \to V^h$ yielding a map $\rho : C \to V^h$ with $\sigma\rho = 1_C$, and $\rho\sigma = 1_{V^h}$. Thus $\bigcup_{n=1}^{\infty} B_n^h = V^h$. □

The proofs of the two statements of Theorem 13.6 are presented separately.

Proof of statement 1 of Theorem 13.6.

PROOF. Assume that L/k is a finitely generated field extension. Since L is algebraic over F, it follows that L is finite algebraic over F. Since $\bigcup_{n=1}^{\infty} B_n^h = V^h$, we have $\bigcup_{n=1}^{\infty} \mathcal{Q}(B_n^h) = \mathcal{Q}(V^h)$ and $L \subseteq \mathcal{Q}(V^h)$. Since L/F is finite algebraic, $L \subseteq \mathcal{Q}(B_n^h)$ for all sufficiently large n. By relabeling, we may assume $L \subseteq \mathcal{Q}(B_n^h)$ for all n. Let $C_n := B_n^h \cap L$. Since B_n is a regular local ring, C_n is a regular local ring with $C_n^h = B_n^h$ [**159**, (1.3)].

Observe that for every n, C_{n+1} dominates C_n and V dominates C_n. Also $\bigcup_{n=1}^{\infty} C_n = V$. Indeed, since B_{n+1} dominates B_n, we have B_{n+1}^h dominates B_n^h and hence $C_{n+1} = B_{n+1}^h \cap L$ dominates $C_n = B_n^h \cap L$. Since $C_n = B_n^h \cap L \subseteq V^h \cap L = V$, it follows that V dominates C_n and $V_0 \subseteq \bigcup_{n=1}^{\infty} C_n \subseteq V$. Since V birationally dominates $\bigcup_{n=1}^{\infty} C_n$, it suffices to show that $\bigcup_{n=1}^{\infty} C_n$ is a DVR.

But by the same argument as before, $\bigcup_{n=1}^{\infty} C_n^h = (\bigcup_{n=1}^{\infty} C_n)^h = V^h$. This shows that $\bigcup_{n=1}^{\infty} C_n$ is a DVR, and therefore $\bigcup_{n=1}^{\infty} C_n = V$. This completes the proof of statement 1 of Theorem 13.6. □

REMARK 13.12. An alternate approach to the definition of C_n is as follows. Since V is a directed union of étale extensions of V_0 and $\mathcal{Q}(V) = L$ is finite algebraic over $\mathcal{Q}(V_0) = F$, V is étale over V_0 and therefore $V = V_0[\theta] = V_0[X]/(f(X))$, where $f(X)$ is a monic polynomial such that $f(\theta) = 0$ and $f'(\theta)$ is a unit of V. Let B_n' denote the integral closure of B_n in L and let $C_n = (B_n')_{(\mathfrak{n} \cap B_n')}$. Since $\cup_{n=1}^{\infty} B_n = V_0$, it follows that $\cup_{n=1}^{\infty} C_n = V$. Moreover, for all sufficiently large n, $f(X) \in B_n[X]$

and $f'(\theta)$ is a unit of C_n. Therefore C_n is a regular local ring for all sufficiently large n [**137**, (38.6)]. As we note in Remark 13.13 below, this allows us to deduce a version of Theorem 13.6 also in the case where k has characteristic $p > 0$ provided the field F can be chosen so that L/F is separable.

Proof of statement 2 of Theorem 13.6.

PROOF. If L is not finitely generated over k, we choose a nested family of fields L_α, with $\alpha \in \Gamma$, such that

(1) $F \subseteq L_\alpha$, for all α.
(2) L_α is finite algebraic over F.
(3) $\cup_{\alpha\in\Gamma} L_\alpha = L$.

The rings $V_\alpha = L_\alpha \cap V$ are DVRs with $\bigcup_{\alpha\in\Gamma} V_\alpha = V$ and $V_\alpha^h = V^h$, since $V_0 \subseteq V_\alpha$, for each $\alpha \in \Gamma$.

As above,

$$\bigcup_{n=1}^{\infty} B_n^h \;=\; V^h, \quad \bigcup_{n=1}^{\infty} \mathcal{Q}(B_n^h) \;=\; \mathcal{Q}(V^h), \quad L \subseteq \mathcal{Q}(V^h).$$

Thus we see that for each $\alpha \in \Gamma$, there is an $n_\alpha \in \mathbb{N}$ such that $L_\alpha \subseteq \mathcal{Q}(B_n^h)$ for all $n \geq n_\alpha$.

Set $C_n^{(\alpha)} := L_\alpha \cap B_n^h$, for each $n \geq n_\alpha$. Then $V_\alpha = \cup_{n=n_\alpha}^{\infty} C_n^{(\alpha)}$ and V_α birationally dominates $C_n^{(\alpha)}$. Hence

$$V = \bigcup_{\alpha\in\Gamma, n\geq n_\alpha} C_n^{(\alpha)}.$$

This completes the proof of Theorem 13.6. □

REMARK 13.13. If the characteristic of k is $p > 0$ then the Henselization V_0^h of $V_0 = F \cap k[[y]]$ may not equal the Henselization V^h of $V = L \cap k[[y]]$, because the algebraic field extension L/F may not be separable. But in the case where L/F is separable algebraic, the fact that the DVRs V and V_0 have the same completion implies that V is a directed union of étale extensions of V_0 (see, for example, [**7**, Theorem 2.7]). Therefore in the case where L/F is separable algebraic, V is a directed union of regular local rings of dimension d.

Thus, for a local domain $(R, \mathfrak{m})$ essentially of finite type over a field k of characteristic $p > 0$, a result analogous to Corollary 13.7 is true provided there exists a subfield F of L such that F is purely transcendental over k, L/F is separable algebraic, and F contains a generator for the maximal ideal of V.

In characteristic $p > 0$, with V excellent and the extension separable, the ring V_0 need not be excellent (see for example Proposition 10.4 or [**71**, (3.3) an d (3.4)]).

13.3. Approximating other rank-one valuation rings

A useful method for constructing rank-one valuation rings is to use generalized power series rings as in [**193**, page 101].

DEFINITION 13.14. Let k be a field and let $e_0 < e_1 < \cdots$ be real numbers such that $\lim_{n\to\infty} e_n = \infty$. For a variable t and elements $a_i \in k$, consider the *generalized power series expansion*

$$z(t) := a_0 t^{e_0} + a_1 t^{e_1} + \cdots + a_n t^{e_n} + \cdots$$

The *generalized power series ring* $k\{t\}$ is the set of all generalized power series expansions $z(t)$ with the usual addition and multiplication extended to exponents in $\mathbb{R}$.[1]

REMARKS 13.15. Assume the notation of Definition 13.14. Then:

(1) The generalized power series ring $k\{t\}$ is a field.
(2) The field $k\{t\}$ admits a valuation v of rank one defined by setting $v(z(t))$ to be the order of the generalized power series $z(t)$. Thus $v(z(t)) = e_0$ if a_0 is a nonzero element of k.
(3) The valuation ring V of v is the set of generalized power series of non-negative order together with zero. The value group of v is the additive group of real numbers.
(4) If $x_1, \ldots, x_r$ are variables over k, then every k-algebra isomorphism of the polynomial ring $k[x_1, \ldots, x_r]$ into $k\{t\}$ determines a valuation ring of rank one on the field $k(x_1, \ldots, x_r)$. Moreover, every such valuation ring has residue field k.
(5) Thus if $z_1(t), \ldots, z_r(t) \in k\{t\}$ are algebraically independent over k, then the k-algebra isomorphism defined by mapping $x_i \mapsto z_i(t)$ determines a valuation on the field $k(x_1, \ldots, x_r)$ of rank one. MacLane and Schilling prove in [**116**] a result that implies for a field k of characteristic zero the existence of a vaulation on $k(x_1, \ldots, x_r)$ of rank one with any preassigned value group of rational rank less than r. In particular, if $r \geq 2$, then every additive subgroup of the group of rational numbers is the value group of a suitable valuation on the field of rational functions in r variables over k.
(6) As a specific example, let k be a field of characteristic zero and consider the k-algebra isomorphism of the polynomial ring $k[x, y]$ into $k\{t\}$ defined by mappping $x \mapsto t$ and $y \mapsto \sum_{n=1}^{\infty} t^{e_1 + \cdots + e_n}$, where $e_i = 1/i$ for each positive integer i. The result of MacLane and Schilling [**116**] mentioned above implies that the value group of the valuation ring V defined by this embedding is the group of all rational numbers.

Exercises

(1) Let $(R, \mathfrak{m})$ be a two-dimensional regular local ring with $\mathfrak{m} = (x, y)R$ and let $a \in R \setminus \mathfrak{m}$. Define:

$$S := R\left[\frac{y}{x}\right] = R\left[\frac{\mathfrak{m}}{x}\right], \quad \mathfrak{n} := \left(x, \frac{y}{x} - a\right)S \quad \text{and} \quad R_1 := S_{\mathfrak{n}} = \left(R\left[\frac{y}{x}\right]\right)_{\mathfrak{n}}.$$

Thus R_1 is a first local quadratic transform of R. Prove that there exists a maximal ideal $\mathfrak{n}'$ of the ring $S' := R\left[\frac{x}{y}\right]$ such that $R_1 = S'_{\mathfrak{n}'}$, and describe generators for $\mathfrak{n}'$.

SUGGESTION. Notice that $\frac{y}{x}$ is a unit of R_1.

(2) Let $(R, \mathfrak{m})$ be a two-dimensional regular local ring with $\mathfrak{m} = (x, y)R$. Define:

$$S := R\left[\frac{x}{y}\right] = R\left[\frac{\mathfrak{m}}{y}\right], \quad \mathfrak{n} := \left(y, \frac{x}{y}\right)S \quad \text{and} \quad R_1 := S_{\mathfrak{n}} = \left(R\left[\frac{x}{y}\right]\right)_{\mathfrak{n}},$$

and define $P = (x^2 - y^3)R$.

[1] The power series $z(t)$ with finite support are also generalized power series.

(a) Prove that R/P is a one-dimensional local domain that is not regular.
(b) Prove that there exists a prime ideal Q of R_1 such that $Q \cap R = P$ and R_1/Q is a DVR and hence is regular.

COMMENT. This is an example of embedded local uniformization.

CHAPTER 14

Non-Noetherian examples of dimension 3

In this chapter we use Insider Construction 10.7 of Section 10.2 to construct examples where the Approximation Domain B is local and non-Noetherian, but is very close to being Noetherian. The localizations of B at all non-maximal prime ideals are Noetherian, and most prime ideals of B are finitely generated. Sometimes just one prime ideal is not finitely generated.

In Section 14.1 we describe, for each positive integer m, a three-dimensional local unique factorization domain B such that the maximal ideal of B is two-generated, B has precisely m prime ideals of height two, each prime ideal of B of height two is not finitely generated and all the other prime ideals of B are finitely generated. In Section 14.2 we give more details about a specific case where there is precisely one non-finitely generated prime ideal. We describe the prime spectrum obtained in the case where there are exactly two non-finitely generated prime ideal. John David develops a similar construction in [**40**].

In Chapters 15 and 16 we present generalizations of these examples to higher dimensions.

14.1. A family of examples in dimension 3

EXAMPLES 14.1. For each positive integer m, we construct an example of a non-Noetherian local integral domain $(B, \mathfrak{m}_B)$ such that:

(1) $\dim B = 3$.
(2) The ring B is a UFD that is not catenary, as defined in Definition 3.24.3.
(3) The maximal ideal $\mathfrak{m}_B$ of B is generated by two elements.
(4) The $\mathfrak{m}_B$-adic completion of B is a two-dimensional regular local domain.
(5) For every non-maximal prime ideal P of B, the ring B_P is Noetherian.
(6) The ring B has precisely m prime ideals of height two.
(7) Every prime ideal of B of height two is not finitely generated; all other prime ideals of B are finitely generated.

To establish the existence of the examples in Examples 14.1, we use the following notation:

NOTATION 14.2. Let k be a field, let x and y be indeterminates over k, and set

$$R := k[x,y]_{(x,y)}, \quad \mathfrak{m}_R = (x,y)R, \quad K := k(x,y) \quad \text{and} \quad R^* := k[y]_{(y)}[[x]].$$

The power series ring R^* is the xR-adic completion of R. Let $\tau \in xk[[x]]$ be transcendental over $k(x)$. Let

$$D := k(x,y,\tau) \cap k[y]_{(y)}[[x]] = K(\tau) \cap R^*$$

be the Local Prototype of Example 4.26 and Definition 4.28.

For each integer i with $1 \le i \le m$, let $p_i \in \mathfrak{m}_R \setminus xR$ be such that $p_1R^*, \dots, p_mR^*$ are m distinct prime ideals. For example, if each $p_i \in \mathfrak{m}_R \setminus (\mathfrak{m}_R^2 \cup xR)$, then each p_iR^* is a prime ideal in R^*. In particular one could take $p_i = y - x^i$. Let $e_1, \dots, e_m$ be positive integers and set $p := p_1^{e_1} \cdot p_2^{e_2} \cdots p_m^{e_m}$. We set $f := p\tau$ and consider the injective R-algebra homomorphism

$$\varphi : \ S \ := \ R[f] \hookrightarrow R[\tau] \ =: \ T.$$

In this construction the polynomial rings S and T have the same field of fractions $K(f) = K(\tau)$. Hence the two Intersection Domains of Inclusion Construction 5.3, the one associated to f and the one associated to τ, are equal:

$$(14.2.0) \qquad D \ = \ K(\tau) \cap R^* \ := \ K(f) \cap R^* = \ A.$$

By Valabrega's Theorem 4.9 or by Local Prototype Theorem 10.6.1, D is a two-dimensional regular local domain with maximal ideal $\mathfrak{m}_D = (x, y)D$ and the $\mathfrak{m}_D$-adic completion of D is $k[[x, y]]$.

Let $\tau := c_1x + c_2x^2 + \cdots + c_ix^i + \cdots \in xk[[x]]$, where the $c_i \in k$, and define for each $n \in \mathbb{N}_0$ the "n^{th} endpiece" τ_n of τ by

$$(14.2.a) \qquad \tau_n \ := \sum_{i=n+1}^{\infty} c_i x^{i-n} \ = \ \frac{\tau - \sum_{i=1}^{n} c_i x^i}{x^n}.$$

As in Endpiece Recursion Relation 5.5.1, we have the following equation relating the nth and $(n+1)^{\text{st}}$ endpieces τ_n and τ_{n+1}:

$$(14.2.b) \qquad \tau_n \ = \ c_{n+1}x \ + \ x\tau_{n+1}.$$

For each $n \in \mathbb{N}$, set $f_n \ := \ p\tau_n$, and define U_n, B_n, V_n and D_n as in Equation 14.2.c:

$$(14.2.c) \qquad \begin{aligned} U_n &:= k[x, y, f_n], \quad & B_n &:= \ (U_n)_{(x,y,f_n)} \ &= \ k[x, y, f_n]_{(x,y,f_n)} \\ V_n &:= k[x, y, \tau_n], \quad & D_n &:= \ (V_n)_{(x,y,\tau_n)} \ &= \ k[x, y, \tau_n]_{(x,y,\tau_n)}. \end{aligned}$$

Then each $U_n \subseteq V_n$, the rings U_n and V_n are three-dimensional polynomial rings over R, and the rings B_n and D_n are three-dimensional localized polynomial rings. Let U, B and V be the nested union approximation domains below, by Remarks 5.16.3. Then:

$$(14.2.d) \quad U \ := \bigcup_{n=0}^{\infty} U_n \ \subseteq \ V \ := \bigcup_{n=0}^{\infty} V_n; \qquad B \ := \bigcup_{n=0}^{\infty} B_n \ \subseteq \ \bigcup_{n=0}^{\infty} D_n \ = \ D.$$

The last equality follows from Proposition 4.27, since D is a Local Prototype. By Local Prototype Theorem 10.6.1, we have $D = C[y]_{(x,y)}$, where $C = k(x, \tau) \cap k[[x]]$ is a DVR.

In Theorem 14.3 we establish the properties asserted in Examples 14.1 and other properties of the ring B.

THEOREM 14.3. *As in Notation 14.2, let* $R \ := \ k[x, y]_{(x,y)}$*, where* k *is a field, and* x *and* y *are indeterminates. Set* $R^* = k[y]_{(y)}[[x]]$*, let* $\tau \in xk[[x]]$ *be transcendental over* $k(x)$*, and, for each integer* i *with* $1 \le i \le m$*, let* $p_i \in R \setminus xR$ *be such that* $p_1R^*, \dots, p_mR^*$ *are* m *distinct prime ideals. Set* $p := p_1^{e_1} \cdots p_m^{e_m}$ *and* $f := p\tau$*, and let the approximation domain* B *and the Local Prototype domain* D *be* *defined as in Notation 14.2. Set* $Q_i := p_iR^* \cap B$, *for each* i *with* $1 \leq i \leq m$. *Then:*

(1) *The ring* B *is a three-dimensional non-Noetherian local UFD with maximal ideal* $\mathfrak{m}_B = (x,y)B$, *and the* $\mathfrak{m}_B$*-adic completion of* B *is the two-dimensional regular local ring* $k[[x,y]]$.

(2) *The rings* $B[1/x]$ *and* B_P, *for each non-maximal prime ideal* P *of* B, *are regular Noetherian UFDs, and the ring* B/xB *is a DVR.*

(3) *The ring* D *is a two-dimensional regular local domain with maximal ideal* $\mathfrak{m}_D := (x,y)D$, *and* D *is a nested union* $\bigcup_{i=1}^{\infty} R[\tau_n]_{(x,y,\tau_n)}$ *of localized polynomial rings. The ring* D *is excellent if the field* k *has characteristic zero. If* k *is a perfect field of positive characteristic, then* D *is not excellent.*

(4) *The ideal* $\mathfrak{m}_D$ *is the only prime ideal of* D *lying over* $\mathfrak{m}_B$.

(5) $L = pR^*[1/x]$ *defines the non-flat locus of the map*

$$\beta : B \hookrightarrow R^*[1/x].$$

(6) *The ring* B *has exactly* m *height-two prime ideals, namely* $Q_1, \dots, Q_m$.

(7) *For each* $i \in \{1,\dots,m\}$, *the ideal* $Q_i = (p_i, \{f_n\}_{n=1}^{\infty})B$, *and* Q_i *is not finitely generated. The* Q_i *are the only non-finitely generated prime ideals of* B.

(8) *The ring* B *has saturated chains of prime ideals from* (0) *to* $\mathfrak{m}_B$ *of length two and of length three, and hence is not catenary.*

PROOF. For item 1, Proposition 5.19 implies $\dim B \leq 3$. By Proposition 5.17.5, B is local with maximal ideal $\mathfrak{m}_B = (x,y)B$; xB and p_iB are prime ideals; and, by Construction Properties Theorem 5.14.3, the (x)-adic completion of B is equal to R^*, the (x)-adic completion of R. Thus the $\mathfrak{m}_B$-adic completion of B is $k[[x,y]]$. Since each $Q_i = \bigcup_{i=1}^{\infty} Q_{in}$, where $Q_{in} = p_iR^* \cap B_n$, each Q_i is a prime ideal of B with $p_i, f \in Q_i$ and $x \notin Q_i$. Since $p_iB = \bigcup p_iB_n$, we have $f \notin p_iB$. Thus

$$(0) \subsetneq p_iB \subsetneq Q_i \subsetneq (x,y)B.$$

This chain of prime ideals of length at least three yields that $\dim B = 3$ and that the height of each Q_i is 2.

The prime ideal $p_iR^*[1/x]$ has height one, whereas $p_iR^*[1/x] \cap S = (p_i, f)S$ has height two. Since flat extensions satisfy the going-down property, by Remark 2.37.10, the map $S = R[f] \to R^*[1/x]$ is not flat. Therefore Noetherian Flatness Theorem 6.3 implies that the ring B is not Noetherian. By Theorem 5.24, B is a UFD, and so item 1 holds.

For item 2, by Construction Properties Theorem 5.14.2, $B/xB = R/xR$, and so B/xB is a DVR. By Theorem 5.24, $B[1/x]$ is a regular Noetherian UFD. If $x \in P$ and P is non-maximal, then, again by Theorem 5.14.2, $P = xB$ and so B_P is a DVR and a regular Noetherian UFD. If $x \notin P$, the ring B_P is a localization of $B[1/x]$ and so is a regular Noetherian UFD. Thus item 2 holds.

The first statements in item 3 concerning D are justified above using Local Prototype Theorem 10.6.1. If the field k has characteristic zero, then D is also excellent by Theorem 10.6.2. If the field k is perfect with positive characteristic, then the ring D is not excellent by Remark‘10.5. This completes the proof of item 3.

By Theorem 5.14.2, $D/xD = B/xB = R/xR$, and so $\mathfrak{m}_D = (x,y)D$, a height-two prime ideal of D, is the unique prime ideal of D lying over the height-three prime ideal $\mathfrak{m}_B = (x,y)B$ of B. Thus item 4 holds.

For item 5, Theorem 10.12.2 implies $L = pR^*[1/x]$, the ideal generated by the nonconstant coefficients of f, defines the non-flat locus of the map $\beta : B \hookrightarrow R^*[1/x]$.

For item 6, observe that x is not in any non-maximal height-two prime ideal of B. There are m distinct prime ideals Q_i with $1 \le i \le m$, since there are m distinct prime ideals of form $p_i R^*$, and $p_i \notin p_j R^*$, if $1 \le i < j \le m$.

To complete the proof of item 6, let P be a nonzero prime ideal of B with $x \notin P$ and $\operatorname{ht} P > 1$. We show that $P = Q_i$, for some i with $1 \le i \le m$. By Proposition 5.17.3, we have $x^n \notin PR^*$ for each $n \in \mathbb{N}$. Thus $\operatorname{ht}(PR^*) \le 1$. By Proposition 5.17.4, choose P' a height one prime ideal of R^* with $x \notin P'$ such that $PR^* \subseteq P'$. Then $P \subseteq P' \cap B \subsetneq \mathfrak{m}_B$ implies that $P = P' \cap B = P'R^*[1/x] \cap B$.

If $p \notin P$, then $pR^*[1/x] \nsubseteq P'R^*[1/x]$. By item 5,

$$\beta_{P'R^*[1/x]} \ : \ B \hookrightarrow R^*[1/x]_{P'R^*[1/x]}$$

is flat. We have that $P = P'R^*[1/x] \cap B$, and so $B_P \hookrightarrow R^*[1/x]_{P'R^*[1/x]}$ is faithfully flat. Since $\operatorname{ht}(P'R^*[1/x]) = 1$, flatness implies that P also has height one, a contradiction to $\operatorname{ht} P > 1$. Thus the proof of item 6 is complete for $p \notin P$.

If $p \in P$, then $p_i \in P$ for some i, and this implies $p_i R^*$ is a height-one prime ideal contained in PR^*. By the above paragraph, $\operatorname{ht}(PR^*) \le 1$, and so $p_i R^* = PR^*$. Hence $p_i B \subseteq P \subseteq Q_i = p_i R^* \cap B \neq (x, y)B$. Since $\dim B = 3$, either P has height one or $P = Q_i$ for some i. This completes the proof of item 6.

For item 7, observe $p_i B_n \subsetneq (p_i, f_n)B_n \subseteq Q_i \cap B_n \subsetneq \mathfrak{m}_{B_n}$. Since $\dim B_n = 3$ and $(p_i, f_n)B_n$ is a prime ideal, we have $(p_i, f_n)B_n = Q_i \cap B_n$. To show each Q_i is not finitely generated, we show that $f_{n+1} \notin (p_i, f_n)B$ for each $n \ge 0$. For this, observe $f = p\tau$ and thus $f_n = p\tau_n$. It follows that $f_n = xf_{n+1} + pxc_{n+1}$, by Endpiece Recursion Relation 5.5.1. Assume that $f_{n+1} \in (p_i, f_n)B$. Then $f_{n+1} = ap_i + bf_n$, for some $a, b \in B$, and so

$$\begin{aligned} xap_i + xbf_n = xf_{n+1} = f_n - pxc_{n+1} \quad &\Longrightarrow \\ f_n(1 - xb) = xap_i + pxc_{n+1} \in xp_i B. & \end{aligned}$$

By Proposition 5.17.1, $x \in \mathcal{J}(B)$, and so $1 - xb$ is a unit of B and $f_n \in p_i B$. Then $f_n \in p_i B \cap B_n = p_i B_n$, by Proposition 5.17.2, a contradiction, since p_i is an element of $(x, y)k[x, y]_{(x,y)}$ and B_n is the localization of the polynomial ring $U_n = k[x, y, f_n]$ at the maximal ideal $(x, y, f_n)U_n$. Thus Q_i is not finitely generated.

Since B is a UFD, the height-one prime ideals of B are principal and, since the maximal ideal of B is two-generated, every non-finitely generated prime ideal of B has height two and thus is in the set $\{Q_1, \ldots, Q_m\}$. This completes the proof of item 7.

For item 8, the chain $(0) \subset xB \subset (x, y)B = \mathfrak{m}_B$ is saturated and has length two, while the chain $(0) \subset p_1 B \subset Q_1 \subset \mathfrak{m}_B$ is saturated and has length three. □

REMARK 14.4. Theorem 14.3, parts 2 and 8, show that the ring B is an example with $B[1/x]$ and B/xB both universally catenary, whereas B itself is not catenary. By Remark 3.27, RLRs are universally catenary.

In Proposition 14.5, we list additional properties of the prime ideals of Examples 14.1

PROPOSITION 14.5. *Assume the notation of Theorem 14.3. Then:*

(1) *Let P be a nonzero non-maximal prime ideal of B. Then*
  (a) $\operatorname{ht}(PR^*) = \operatorname{ht}(PD) = 1$.

   (b) *P is contained in a non-maximal prime ideal of D.*
   (c) *If $\dim(B/P) = 1$, then P is the contraction of a prime ideal of D.*
(2) *Let w be a prime element of B such that $w \notin \bigcup_{i=1}^{m} Q_i$. Then wD is the unique height-one prime ideal of D lying over wB.*
(3) *If $P \in \operatorname{Spec} B$ is such that $P \cap R = (0)$, then $\operatorname{ht}(P) \leq 1$ and P is principal.*
(4) *If $P \in \operatorname{Spec} B$, $\operatorname{ht} P = 1$ and $P \cap R \neq 0$, then $P = (P \cap R)B$.*
(5) *Let p_i be one of the prime factors of p. Then p_iB is a prime ideal of B. Moreover the ideals p_iB and $Q_i := p_iD \cap B = (p_i, f_1, f_2, \ldots)B$ are the only non-maximal prime ideals of B that contain p_i. Thus they are the only prime ideals of B that lie over p_iR in R.*
(6) *The constructed ring B has Noetherian spectrum.*

PROOF. For item 1.a, if $P = Q_i$ for some i, then $PR^* \subseteq p_iR^*$ and $\operatorname{ht} PR^* = 1$. Assume P is not one of the Q_i. By Theorem 14.3 parts 1 and 6, B is a UFD and $\operatorname{ht} P = 1$. Hence P is a principal height-one prime ideal. Since D and R^* are Noetherian, $\operatorname{ht}(PD) = \operatorname{ht}(PR^*) = 1$ by Krull Altitude Theorem 2.23. Item 1.b now follows.

For item 1.c, observe that $PD \subseteq P'$, where P' is a non-maximal prime ideal of D, by item 1.b. Since $\mathfrak{m}_BD = \mathfrak{m}_D$, we have $P \subseteq P' \cap B \subsetneq \mathfrak{m}_B$. Now $\dim(B/P) = 1$ implies $P = P' \cap B$.

For w as in item 2, we have $\dim(B/wB) = 1$ by Theorem 14.3.6. By item 1.c, $wB = \mathfrak{p} \cap B$, where $\mathfrak{p} \in \operatorname{Spec} D$. The DVR B_{wB} is birationally dominated by $D_{\mathfrak{p}}$, and thus $B_{wB} = D_{\mathfrak{p}}$. This implies that $\mathfrak{p}$ is the unique prime ideal of D lying over wB. Also $wB_{wB} = \mathfrak{p}D_{\mathfrak{p}}$. Since D is a UFD and $\mathfrak{p}$ is the unique minimal prime ideal of wD, it follows that $wD = \mathfrak{p}$.

For item 3, $\operatorname{ht} P \leq 1$ because the field of fractions $K(f)$ of B has transcendence degree one over the field of fractions K of R; see Cohen's Theorem 2.26. Since B is a UFD, P is principal.

For item 4, if $x \in P$, then $P = xB$ and the statement is clear. Assume $x \notin P$. By Construction Properties Theorem 5.14. 4, $B[1/x]$ is a localization of B_n, and so $\operatorname{ht}(P \cap B_n) = 1$ for all integers $n \geq 0$. Thus $(P \cap R)B_n = P \cap B_n$, for each n, and so $P = (P \cap R)B$.

For item 5, p_iB is a prime ideal by Proposition 5.17.2. By Theorem 14.3, $\dim B = 3$ and the Q_i are the only height-two prime ideals of B. Since the ideal $p_iR + p_jR$ is $\mathfrak{m}_R$-primary for $i \neq j$, it follows that $p_iB + p_jB$ is $\mathfrak{m}_B$-primary, and hence p_iB and Q_i are the only non-maximal prime ideals of B that contain p_i.

Item 6 holds by Corollary 5.21. □

14.2. Two cases of Examples 14.1 and their spectra

We use the following lemma.

LEMMA 14.6. *Assume Notation 14.2 and the notation of Theorem 14.3.*

(1) *For every element $c \in \mathfrak{m}_R \setminus xR$ and every $t \in \mathbb{N}$, the element $c + x^tf$ is a prime element of the UFD B.*
(2) *For every fixed element $c \in \mathfrak{m}_R \setminus xR$, the set $\{c + x^tf\}_{t \in \mathbb{N}}$ consists of infinitely many non-associate prime elements of B, and so there exist infinitely many distinct height-one prime ideals of B of the form $(c + x^tf)B$.*

PROOF. For the first item, since $f = p\tau$, Equation 14.2.b implies that

$$f_r = pc_{r+1}x + xf_{r+1}$$

for each $r \geq 0$. In $B_0 = k[x, y, f]_{(x,y,f)}$, the polynomial $c + x^t f$ is linear in the variable $f = f_0$ and the coefficient x^t of f is relatively prime to the constant term c of $c + x^t f$. Thus $c + x^t f$ is a prime element of B_0. Since $f = f_0 = pc_1x + xf_1$ in $B_1 = k[x, y, f_1]_{(x,y,f_1)}$, the polynomial $c + x^t f = c + x^t pc_1x + x^{t+1} f_1$ is linear in the variable f_1 and the coefficient x^{t+1} of f_1 is relatively prime to the constant term c. Thus $c + x^t f$ is a prime element of B_1. To see that this pattern continues, observe that in B_2,

$$\begin{aligned} f &= pc_1x + xf_1 = pc_1x + pc_2x^2 + x^2 f_2 \implies \\ c + x^t f &= c + pc_1x^{t+1} + pc_2x^{t+2} + x^{t+2} f_2, \end{aligned}$$

a linear polynomial in the variable f_2. Thus $c + x^t f$ is a prime element of B_2 and a similar argument shows that $c + x^t f$ is prime in B_r for each positive integer r. Therefore the element $c + x^t f$ is prime in B, for each $t \in \mathbb{N}$.

For item 2, we prove that $(c+x^t f)B \neq (c+x^m f)B$, for positive integers $t > m$. Assume that $\mathfrak{q} := (c + x^t f)B = (c + x^m f)B$ is a height-one prime ideal of B. Then

$$(x^t - x^m)f = x^m(x^{t-m} - 1)f \in \mathfrak{q}.$$

Observe that $\mathfrak{q} \neq xB$, since $c \notin xB$. Thus $x^m \notin \mathfrak{q}$. Since B is local, $x^{t-m} - 1$ is a unit of B. It follows that $f \in \mathfrak{q}$ and thus $(c, f)B \subseteq \mathfrak{q}$. By Construction Properties Theorem 5.14.3, $B[1/x]$ is a localization of $R[f] = S$, and $x \notin \mathfrak{q}$ implies that $B_{\mathfrak{q}} = S_{\mathfrak{q} \cap S}$. This is a contradiction since the ideal $(c, f)S$ has height two.

Thus there exist infinitely many distinct height-one primes of the form $(c + x^t f)B$. □

Recall the following terminology [**193**, page 325].

DEFINITION 14.7. If a ring C is a subring of a ring E, a prime ideal P of C is *lost* in E if $PE \cap C \neq P$.

Lemma 14.8 is useful for giving a more precise description of Spec B for B as in Examples 14.1. For each nonempty finite subset $\mathcal{H}$ of $\{Q_1, \ldots, Q_m\}$, there exist infinitely many height-one prime ideals contained in each $Q_i \in \mathcal{H}$, but not contained in Q_j if $Q_j \notin \mathcal{H}$.

LEMMA 14.8. *Assume Notation 14.2 and the notation of Theorem 14.3, and also assume* $p = p_1 \cdots p_m$. *Let* G *be a nonempty subset of* $\{1, \ldots, m\}$, *let* $\mathcal{H} = \{Q_i \mid i \in G\}$, *and let* $p_G = \prod\{p_i \mid i \in G\}$. *Then for each* $t \in \mathbb{N}$:

(1) $(p_G + x^t f)B$ *is a prime ideal of* B *that is lost in* D.
(2) $(p_G^2 + x^t f)B$ *is a prime ideal of* B *that is not lost in* D.

The sets $\{(p_G + x^t f)B\}_{t\in\mathbb{N}}$ *and* $\{(p_G^2 + x^t f)B\}_{t\in\mathbb{N}}$ *are both infinite. Moreover, the prime ideals in both item 1 and item 2 are contained in each* Q_i *such that* $Q_i \in \mathcal{H}$, *but are not contained in* Q_j *if* $Q_j \notin \mathcal{H}$.

PROOF. For item 1,

$$(14.8.1) \qquad (p_G + x^t f)D \cap B = p_G(1 + x^t\tau \prod_{j\notin G} p_j)D \cap B = p_G D \cap B = \bigcap_{i\in G} Q_i.$$

Thus each prime ideal of B of the form $(p_G + x^t f)B$ is lost in D and R^*. By the second item of Lemma 14.6, infinitely many height-one prime ideals $(p_G + x^t f)B$ of B are lost in D and R^*.

For item 2,

$$\begin{aligned}(p_G^2 + x^t f)D \cap B &= (p_G^2 + x^t p_G(\prod_{j\notin G} p_j)\tau)D \cap B \\ &= p_G(p_G + x^t(\prod_{j\notin G} p_j)\tau)D \cap B \subsetneq p_G D \cap B = \bigcap_{i\in G} Q_i.\end{aligned} \tag{14.8.2}$$

The last inclusion is strict since $p_G + x^t(\prod_{j\notin G} p_j)\tau \in \mathfrak{m}_D$. This implies that prime ideals of B of form $(p_G^2 + x^t f)B$ are not lost. By Lemma 14.6 there are infinitely many distinct prime ideals of that form.

The "moreover" statement for the prime ideals in item 1 follows from Equation 14.8.1. Equation 14.8.2 implies that the prime ideals in item 2 are contained in each $Q_i \in \mathcal{H}$. For $j \notin G$, if $p_G^2 + x^t f \in Q_j$, then $p_j + x^t f \in Q_j$ implies that $p_G^2 - p_j \in Q_j$ by subtraction. Since $p_j \in Q_j$, this would imply that $p_G^2 \in Q_j$, a contradiction. This completes the proof of Lemma 14.8. □

We use Theorem 14.3, Proposition 14.5 and Lemmas 14.6 and 14.8 to describe a special case of the ring B of Examples 14.1

EXAMPLE 14.9. Assume Notation 14.2 and take $m = 1$ and $p = p_1 = y$. Then:

$$R = k[x,y]_{(x,y)}, \quad f = y\tau, \quad f_n = y\tau_n, \quad B_n = R[y\tau_n]_{(x,y,y\tau_n)}, \quad B = \bigcup_{n=0}^{\infty} B_n.$$

By Theorem 14.3, the ideal $Q := yD \cap B = (y, \{y\tau_n\}_{n=0}^{\infty})B$ is the unique prime ideal of B of height 2. Moreover, Q is not finitely generated and is the only prime ideal of B that is not finitely generated. Also $Q \cap B_n = (y, y\tau_n)B_n$ for each $n \geq 0$.

To identify the ring B up to isomorphism, observe that $\tau_n = c_{n+1}x + x\tau_{n+1}$, by Equation 14.2.b. Thus

$$f_n \;=\; xf_{n+1} \;+\; yxc_{n+1}. \tag{14.9.1}$$

The family of equations (14.9.1) uniquely determines B as a nested union of the three-dimensional RLRs $B_n = k[x, y, f_n]_{(x,y,f_n)}$.

Discussion 14.10 gives more details for Example 14.9. The prime spectrum for Example 14.9 is displayed in Diagram 14.10.d.

DISCUSSION 14.10. By Theorem 14.3, if $\mathfrak{q}$ is a height-one prime ideal of B, then $B/\mathfrak{q}$ is Noetherian if and only if $\mathfrak{q}$ is not contained in Q. This is true because: (1) B is a UFD implies that $\mathfrak{q}$ is principal, (2) Q is the unique prime ideal of B that is not finitely generated, and (3) a ring is Noetherian if every prime ideal of the ring is finitely generated, by Cohen's Theorem 2.25.

The height-one prime ideals $\mathfrak{q}$ of B may be separated into several types as follows:

TYPE I. The height-one prime ideals $\mathfrak{q} \not\subseteq Q$. These prime ideals have the property that $B/\mathfrak{q}$ is a one-dimensional Noetherian local domain. By Proposition 14.5.1.c, these prime ideals are contracted from D; that is, they are not lost in D.

Every element of $\mathfrak{m}_B \setminus Q$ is contained in a prime ideal $\mathfrak{q}$ of type I. Thus $\mathfrak{m}_B \subseteq Q \cup \bigcup\{\mathfrak{q} \text{ of Type I}\}$. Since $\mathfrak{m}_B$ is not the union of finitely many strictly smaller prime ideals, there are infinitely many prime ideals $\mathfrak{q}$ of Type I.

TYPE I* 1. The prime ideal xB. Among the prime ideals of Type I, the prime ideal xB is special since it is the unique height-one prime ideal $\mathfrak{q}$ of B such that $R^*/\mathfrak{q}R^*$ is not complete. If $\mathfrak{q}$ is a height-one prime ideal of B such that $x \notin \mathfrak{q}R^*$, then $x \notin \mathfrak{q}$. Thus $R^*/\mathfrak{q}R^*$ is complete with respect to the powers of the nonzero principal ideal generated by the image of x mod $\mathfrak{q}R^*$. Notice that $R^*/xR^* \cong k[y]_{yk[y]}$.

If $\mathfrak{q}$ is a height-one prime ideal of B not of Type I, then $\overline{B} = B/\mathfrak{q}$ has precisely three prime ideals. These prime ideals form a chain: $(\overline{0}) \subset \overline{Q} \subset \overline{(x,y)B} = \overline{\mathfrak{m}_B}$.

TYPE II. The height-one prime ideals $\mathfrak{q} \subset Q$, contracted from D. That is, $\mathfrak{q} = \mathfrak{p} \cap B$, for some $\mathfrak{p} \in \operatorname{Spec} D$, where $D = k(x,y,f) \cap R^*$. The Type II prime ideals are not lost in D. For example, the prime ideal $y(y+\tau)B$ is Type II by Lemma 14.8. For $\mathfrak{q}$ of Type II, the domain $B/\mathfrak{q}$ is dominated by the one-dimensional Noetherian local domain $D/\mathfrak{p}$. Thus $B/\mathfrak{q}$ is a non-Noetherian *generalized* local ring in the sense of Cohen; that is, the unique maximal ideal $\overline{\mathfrak{n}}$ of $B/\mathfrak{q}$ is finitely generated and $\cap_{i=1}^{\infty} \overline{\mathfrak{n}}^{\,i} = (0)$, [**35**].

For $\mathfrak{q}$ of Type II, the maximal ideal of $B/\mathfrak{q}$ is not principal. This follows because a generalized local domain having a principal maximal ideal is a DVR [**137**, (31.5)].

There are infinitely many height-one prime ideals of Type II, for example, $y(y + x^t\tau)B$ for each $t \in \mathbb{N}$ by Lemma 14.6. [1] For $\mathfrak{q}$ of Type II, the DVR $B_{\mathfrak{q}}$ is birationally dominated by $D_{\mathfrak{p}}$. Hence $B_{\mathfrak{q}} = D_{\mathfrak{p}}$, and

$$\sqrt{\mathfrak{q}D} \;=\; \mathfrak{p} \,\cap\, yD. \tag{14.10.0}$$

TYPE III. The height-one prime ideals $\mathfrak{q} \subset Q$, not contracted from D. That is, $\mathfrak{q}$ is lost in D. For example, the prime ideal yB and the prime ideal $(y+x^t y\tau)B$ for $t \in \mathbb{N}$ are Type III by Lemma 14.8. The elements y and $y + x^t y\tau$ are prime because they are in $\mathfrak{m}_B$ and are not in $\mathfrak{m}_B^2$ and B is a UFD. There are infinitely many prime ideals of Type III by Lemma 14.6. If $\mathfrak{q}$ has Type III, then $\sqrt{qD} = yD$.

If $\mathfrak{q} = yB$ or $\mathfrak{q} = (y + x^t y\tau)B$, then the image $\overline{\mathfrak{m}_B}$ of $\mathfrak{m}_B$ in $\overline{B} := B/\mathfrak{q}$ is $\overline{xB}$, a principal ideal. It follows that the intersection of the powers of $\overline{\mathfrak{m}_B}$ is $Q/\mathfrak{q}$ and $B/\mathfrak{q}$ is not a generalized local ring. To see that $\bigcap_{i=1}^{\infty} \overline{\mathfrak{m}_B}^{\,i} \neq (0)$, argue as follows: If P is a principal prime ideal of a ring and P' is a prime ideal properly contained in P, then P' is contained in the intersection of the powers of P; see [**103**, page 7, ex. 5] and Exercise 14.4.

The picture of Spec(B) is shown below.

[1]Bruce Olberding pointed out that the existence of prime ideals $\mathfrak{q}$ of Type II answers a question asked by Anderson-Matijevic-Nichols in [**16**, page 17]. Their question asks whether in an integral domain every nonzero finitely generated prime ideal P that satisfies $\bigcap_{n=1}^{\infty} P^n = (0)$ and that is minimal over a principal ideal has $\operatorname{ht} P = 1$. For $\mathfrak{q}$ of Type II, the ring $\overline{B} = B/\mathfrak{q}$ is a generalized local domain with precisely three prime ideals. An element in the maximal ideal $\overline{\mathfrak{m}_B}$ not in the other nonzero prime ideal generates an ideal primary for $\overline{\mathfrak{m}_B}$. Since $\operatorname{ht}\overline{\mathfrak{m}_B} = 2$, this yields a negative answer to the question.

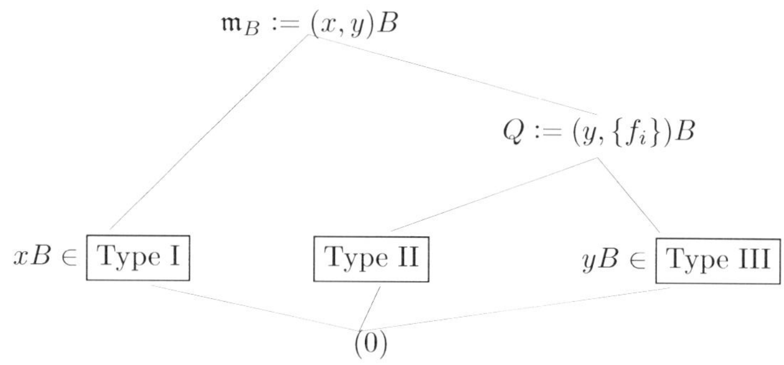

Diagram 14.10.d

In Remarks 14.11 we examine the height-one prime ideals $\mathfrak{q}$ of the ring B of Example 14.9 from a different perspective.

REMARKS 14.11. Assume the notation of Example 14.9. Then $R = k[x,y]_{(x,y)}$, $D = K(f) \cap R^*$, $f = y\tau$, $f_n = y\tau_n$, , $B_n = R[y\tau_n]_{(x,y,y\tau_n)}$, $B = \bigcup_{n=0}^{\infty} B_n$.

(1) If w is a nonzero prime element of B such that $w \notin Q = (y, \{y\tau_n\}) = yD \cap B$, then wD is a prime ideal in D and is the unique prime ideal of D lying over wB, by Proposition 14.5.2. In particular, $\mathfrak{q} = wB$ is not lost in D.

If $\mathfrak{q}$ is a height-one prime ideal of B that is contained in Q, then yD is a minimal prime of $\mathfrak{q}D$, and $\mathfrak{q}$ is of Type II or III depending on whether or not $\mathfrak{q}D$ has other minimal prime divisors.

To see this, observe that, if yD is the only prime divisor of $\mathfrak{q}D$, then $\mathfrak{q}D$ has radical yD and $yD \cap B = Q$ implies that Q is the radical of $\mathfrak{q}D \cap B$. Thus $\mathfrak{q}$ is lost in D and $\mathfrak{q}$ is of Type III.

On the other hand, if there is a minimal prime ideal $\mathfrak{p} \in \operatorname{Spec} D$ of $\mathfrak{q}D$ that is different from yD, then y is not in $\mathfrak{p} \cap B$ and hence $\mathfrak{p} \cap B \neq Q$. Since Q is the only prime ideal of B of height two, it follows that $\mathfrak{p} \cap B$ is a height-one prime ideal and thus $\mathfrak{p} \cap B = \mathfrak{q}$. Thus $\mathfrak{q}$ is not lost in D and $\mathfrak{q}$ is of Type II.

By Equation 14.10.0, for every Type II prime ideal $\mathfrak{q}$ of B, there are exactly two minimal prime ideals of $\mathfrak{q}D$; one of these is yD and the other is a height-one prime ideal $\mathfrak{p}$ of D such that $\mathfrak{p} \cap B = \mathfrak{q}$. If $\mathfrak{p}$ is a height-one prime ideal of D such that $\mathfrak{p} \cap B = \mathfrak{q}$, then $B_{\mathfrak{q}}$ is a DVR that is birationally dominated by $D_{\mathfrak{p}}$, and hence $B_{\mathfrak{q}} = D_{\mathfrak{p}}$. The uniqueness of $B_{\mathfrak{q}} = D_{\mathfrak{p}}$ as a DVR overring of D implies that there is precisely one such prime ideal $\mathfrak{p}$ of D.

An example of a height-one prime ideal $\mathfrak{q}$ of B of Type II is $\mathfrak{q} := (y^2 + y\tau)B$. The ideal $\mathfrak{q}D = (y^2 + y\tau)D$ has the two minimal prime ideals yD and $(y + \tau)D$.

(2) The ring B/yB is a rank 2 valuation domain. This can be seen directly or else one may apply a result of Heinzer and Sally [**93**, Prop. 3.5(iv)]; see Exercise 14.4. For other prime elements g of B with $g \in Q$, it need not be true that B/gB is a valuation domain. If g is a prime element contained in $\mathfrak{m}_B^2$, then the maximal ideal of B/gB is 2-generated but is not principal, and B/gB is not a valuation domain. For a specific example over the field $\mathbb{Q}$, let $g = y^2 + xy\tau$.

The following example where $m = 1$ and $e_1 = 2$ in Example 14.1 gives a proper subring of the constructed ring of Example 14.9.

EXAMPLE 14.12. Assume Notation 14.2 with $m = 1$ and $e_1 = 2$; that is, $p = y^2$. Then:

$$R = k[x,y]_{(x,y)}, \quad f = y^2\tau, \quad f_n = y^2\tau_n, \quad B_n = R[y^2\tau_n]_{(x,y,y^2\tau_n)}, \quad B = \bigcup_{n=0}^{\infty} B_n.$$

By Theorem 14.3, $\dim B \leq 3$, the maximal ideal is $\mathfrak{m}_B := (x,y)B$, and the proper chain $(0) \subsetneq yB \subsetneq yD \cap B \subsetneq \mathfrak{m}_B$ shows that $\dim B = 3$ and $\operatorname{ht}(yD \cap B) = 2$. The ideal $Q := yD \cap B = (y, \{y^2\tau_n\}_{n=0}^{\infty})B$ is the unique prime ideal of B of height 2. Moreover, Q is not finitely generated and is the only prime ideal of B that is not finitely generated. We also have $Q \cap B_n = (y, y^2\tau_n)B_n$ for each $n \geq 0$.

We return to the general case of Example 14.1.

REMARK 14.13. With Notation 14.2, consider the birational inclusion $B \hookrightarrow D$ and the faithfully flat map $D \hookrightarrow R^*$. The following statements hold concerning the inclusion maps $R \hookrightarrow B \hookrightarrow D \hookrightarrow R^*$, and the associated maps in the opposite direction of their spectra: [2]

(1) The map $\operatorname{Spec} R^* \to \operatorname{Spec} D$ is surjective, since every prime ideal of D is contracted from a prime ideal of R^*, while the maps $\operatorname{Spec} R^* \to \operatorname{Spec} B$ and $\operatorname{Spec} D \to \operatorname{Spec} B$ are not surjective. All the induced maps to $\operatorname{Spec} R$ are surjective since the map $\operatorname{Spec} R^* \to \operatorname{Spec} R$ is surjecive.

(2) By Lemma 14.8, each of the prime ideals Q_i of B contains infinitely many height-one prime ideals of B that are the contraction of prime ideals of D and infinitely many that are not.

An ideal contained in a finite union of prime ideals is contained in one of the prime ideals; see [**15**, Prop. 1.11, page 8] or [**122**, Ex. 1.6, page 6]. Thus there are infinitely many non-associate prime elements of the UFD B that are not contained in the union $\bigcup_{i=1}^{m} Q_i$. We observe that for each prime element q of B with $q \notin \bigcup_{i=1}^{m} Q_i$ the ideal qD is contained in a height-one prime ideal $\mathfrak{q}$ of D and $\mathfrak{q} \cap B$ is properly contained in $\mathfrak{m}_B$ since $\mathfrak{m}_D$ is the unique prime ideal of D lying over $\mathfrak{m}_B$. Hence $\mathfrak{q} \cap B = qB$. Thus each qB is contracted from D and R^*.

In the four-dimensional example B of Theorem 16.6, each height-one prime ideal of B is contracted from R^*, but there are infinitely many height-two prime ideals of B that are lost in R^*, in the sense of Definition 14.7; see Section 14.2.

(3) Among the prime ideals of the domain B of Examples 14.1 that are not contracted from D are the p_iB. Since $p_iD \cap B = Q_i$ properly contains p_iB, the prime ideal p_iB is lost in D.

(4) Since x and y generate the maximal ideals $\mathfrak{m}_B$ and $\mathfrak{m}_D$ of B and D, it follows that $\mathfrak{m}_D$ is *isolated over* its intersection $\mathfrak{m}_B = \mathfrak{m}_D \cap B$; that is, $\mathfrak{m}_D$ is both maximal and minimal with respect to the prime ideals of D lying over $\mathfrak{m}_B$. Since B is integrally closed in D and $B \neq D$, the version of Zariski's Main Theorem given by Evans in [**47**] implies that D is not integral over an essentially finitely generated B-subalgebra of D; see also Peskine's article [**148**]. [3]

Using the information above, we display below a picture of $\operatorname{Spec}(B)$ in the case $m = 2$ and $p = p_1p_2$.

[2] See Discussion 3.29 for information concerning the spectral maps.

[3] "Essentially finitely generated" is defined in Section 2.1.

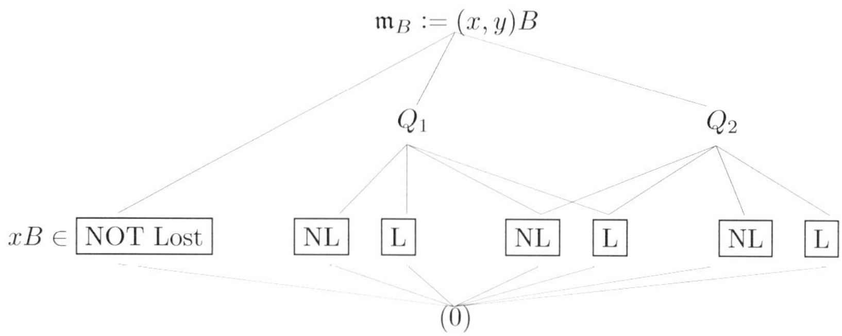

Diagram 14.13.0

Comments on Diagram 14.13.0. Here $Q_1 = p_1R^* \cap B$ and $Q_2 = p_2R^* \cap B$, and each box represents an infinite set of height-one prime ideals. We label a box "NL" for "not lost" and "L" for "lost". An argument similar to that given for the Type I prime ideals in Example 14.9 shows that the height-one prime ideals $\mathfrak{q}$ of B such that $\mathfrak{q} \not\subseteq Q_1 \cup Q_2$ are not lost. That the other boxes are infinite follows from Lemma 14.8.

Exercises

(1) Let $R = k[x, y]_{(x,y)}$ be the localized polynomial ring in the variables x, y over a field k, and let $S = R[\frac{y}{x}]$.
  (a) Prove that S is a UFD, and that the localization of S at every maximal ideal is a RLR.
  (b) Prove that the prime ideal xR is lost in S in the terminology of Definition 14.7.
  (c) Prove that xR is the only prime ideal of R that is lost in S.

  SUGGESTION. If $\mathfrak{q}$ is a prime ideal of R that does not contain x, then $S \subset R[1/x] \subset R_{\mathfrak{q}}$.

  (d) (Suggested by Bruce Olberding.) Show that there exists a height-one prime ideal of R that is the contraction of a height-one maximal ideal of S. Hence "Going-up" fails for the extension $R \hookrightarrow S$.

  SUGGESTION. Let $\mathfrak{p} = (y^2 - x)R$. Show that $\mathfrak{p}$ is a height-one prime ideal of R such that $R/\mathfrak{p}$ is a DVR with maximal ideal $x(R/\mathfrak{p})$. Let $y_1 := \frac{y}{x}$. Then $\mathfrak{p}S = x(x(y_1)^2 - 1)S$. Show that $\mathfrak{q} = (x(y_1)^2 - 1)S$ is a height-one prime ideal of S such that $\mathfrak{q} \cap R = \mathfrak{p}$. Deduce that $S/\mathfrak{q}$ is a field.

(2) As in Notation 14.2, with $p = y$, let $R = k[x, y]_{(x,y)}$, let $R^* = k[y]_{(y)}[[x]]$, let $\tau \in xk[[x]]$ be algebraically independent over $k(x)$ and let $f = y\tau$. Then $D = k(x, y, \tau) \cap k[y]_{(y)}[[x]]$ is a Local Prototype. Let B be the Approximation Domain $B = \bigcup_{n=1}^{\infty} k[x, y, f_n]_{(x,y,f_n)}$ to $A = k(x, y, f) \cap k[y]_{(y)}[[x]]$.
  (a) Prove that $D = A$ is a two-dimensional RLR that birationally dominates the local domain B.
  (b) Prove that $D[1/x]$ is a localization of $T := R[\tau]$ and $B[1/x]$ is a localization of $S := R[f]$.

(c) For $P \in \operatorname{Spec} D$ with $x \notin P$, prove that the following are equivalent:

(i) $D_P = B_{P\cap B}$ **(ii)** $\tau \in B_{P\cap B}$ **(iii)** $y \notin P$.

SUGGESTION. For part b, use Remark 5.16.3 and Construction Properties Theorem 5.14.

For part c, (ii) $\implies$ (iii), one way to show that $y \in P \implies \tau \notin B_{P\cap B}$ would be to follow this outline: Show $B[1/x]$ a localization of $R[f] \implies B_{P\cap B}$ is a localization of $R[f] \implies B_{P\cap B} = (R[f])_{P\cap R[f])}$. Then $y \in P \implies (R[f])_{P\cap R[f])} \subseteq V := R[f]_{yR[f]}$ and f is a unit in the DVR V. Conclude that $\tau = f/y \notin V$.

For (iii) $\implies$ (i) (outline) show that $R[f][1/xy] = R[\tau][1/xy]$, that $R[\tau][1/xy]$ is a localization of $R[f]$, and that $R[\tau]_{(P\cap R[\tau])}$ is a localization of $R[f]$. Show this implies $R[\tau]_{(P\cap R[\tau])} = R[f]_{(P\cap R[f])}$. Also show that $D[1/xy]$ is a localization of $R[\tau][1/xy] = R[f][1/xy]$, and so $D[1/xy]$ is a localization of $R[f]$. Since $B[1/x]$ is a localization of $R[f]$, show that D_P and $B_{P\cap B}$ are both localizations of $R[f]$. Deduce that $D_P = B_{P\cap B}$.

(3) Let $R = k[x,y]_{(x,y)}$ be the localized polynomial ring in the variables x, y over a field k. Consider the local quadratic transform $S := R[\frac{y}{x}]_{(x,\frac{y}{x})R[\frac{y}{x}]}$ of the 2-dimensional RLR R.
   (a) Prove that there are infinitely many height-one prime ideals of R that are lost in S.
   (b) Prove that there are infinitely many height-one prime ideals of R that are not lost in S.
   (c) Describe precisely the height-one prime ideals of R that are lost in S, and the prime ideals of R that are not lost in S.

(4) In connection with Remarks 14.11.2, let $(R, \mathfrak{m})$ be a local domain with principal maximal ideal $\mathfrak{m} = tR$.
   (a) If Q is a prime ideal properly contained in $\mathfrak{m}$, prove that $t^nQ = Q$, for every $n \in \mathbb{N}$.
   (b) Prove that $\bigcap_{n=1}^{\infty} \mathfrak{m}^n$ is a prime ideal that is properly contained in $\mathfrak{m}$.
   (c) Let $P := \bigcap_{n=1}^{\infty} \mathfrak{m}^n$. Prove that every prime ideal of R properly contained in $\mathfrak{m}$ is contained in P.
   (d) Prove that R/P is a DVR.
   (e) Prove that $P = PR_P$.
   (f) Prove that R is a valuation domain if and only if R_P is a valuation domain [**93**, Prop. 3.5(iv)].
   (g) Construct an example of a local domain $(R, \mathfrak{m})$ with principal maximal ideal $\mathfrak{m}$ such that R is not a valuation domain.

COMMENT. Since $P = PR_P$ by part e, there is an embedding $R/P \hookrightarrow R_P/P$. It follows that R may be regarded as a pullback as in the paper of Gabelli and Houston [**55**] or the book of Leuschke and R. Wiegand [**110**, p. 42].

SUGGESTION. To construct an example for part g, let x, y be indeterminates over a field k, let $U = k(x)[y]$, let W be the DVR U_{yU}, and let $P := yW$ denote the maximal ideal of W. Then $W = k(x) + P$. Let $R = k[x^2]_{(x^2k[x^2])} + P$. This is an example of a "$D + M$" construction, as outlined in Remark 16.12. Another example is Example 16.13.

CHAPTER 15

Noetherian properties of non-Noetherian rings

In this chapter we establish Noetherian properties that hold for integral domains constructed using Inclusion Construction 5.3 and Insider Construction 10.7. These results are helpful for determining Noetherian properties of non-Noetherian rings.

15.1. A general question and the setting for this chapter

Consider the question:

QUESTION 15.1. Let x be a non-nilpotent element of a ring C. If the rings C/xC and $C[1/x]$ are Noetherian what implications follow for the ring C?

Question 15.1 is inspired by the constructions and examples featured in earlier chapters. By Corollary 5.15, if the base ring R for Inclusion Construction 5.3 is a Noetherian ring and R^* is the x-adic completion of R, then B/xB and $B[1/x]$ are Noetherian, where x is a nonzero nonunit of R and B is the Approximation Domain of Equation 5.4.5.

Remarks 15.2 and Proposition 15.3 give some straightforward answers in general to Question 15.1.

REMARKS 15.2. Let x be an element of a ring C such that C/xC and $C[1/x]$ are Noetherian. Then:

(1) $\operatorname{Spec} C$ is Noetherian.
(2) Every ideal of C that contains x is finitely generated. In particular, every $P \in \operatorname{Spec} C$ such that $x \in P$ is finitely generated.
(3) For every $P \in \operatorname{Spec} C$ such that $x \notin P$, the localization C_P is Noetherian.
(4) For every $P \in \operatorname{Spec} C$, the maximal ideal PC_P of C_P is finitely generated.
(5) There exist examples C where C is not Noetherian.
(6) There are examples with C/xC and $C[1/x]$ Noetherian and universally catenary, but C is not Noetherian and not catenary.

PROOF. Item 1 follows from Proposition 5.20. For item 2, let I be an ideal of C with $x \in I$. Since C/xC is Noetherian, there exist finitely many elements $a_1, \dots, a_n$ in I that generate the image of I in C/xC. It follows that $I = (x, a_1, \dots, a_n)C$ is finitely generated. If $P \in \operatorname{Spec} C$ and $x \notin P$, then C_P is a localization of the Noetherian ring $C[1/x]$. Therefore C_P is Noetherian. This establishes item 3.

For item 4, let $P \in \operatorname{Spec} C$. Either $x \in P$, in which case P is finitely generated by item 2 and *a fortiori* PC_P is finitely generated, or $x \notin P$, in which case C_P is Noetherian by item 3 and PC_P is finitely generated.

By Theorem 14.3, the rings B of Examples 14.1 are non-Noetherian local UFDs of dimension 3 for which the element x of B satisfies the desired properties of x in C. This establishes item 5.

By Remark 14.4, the rings B of Examples 14.1 are not catenary, even though $B[1/x]$ and B/xB are universally catenary. □

PROPOSITION 15.3. *Let x be an element of a UFD C.*

(1) *If the ring C/xC is Noetherian, then C_Q is Noetherian, for every prime ideal Q of C with $x \in Q$ and* $\operatorname{ht} Q \leq 2$.
(2) *If the rings C/xC and $C[1/x]$ are both Noetherian, then C_Q is Noetherian, for every prime ideal Q of C with* $\operatorname{ht} Q \leq 2$.

PROOF. For item 1, if $\operatorname{ht} Q = 1$, then C_Q is a DVR. Hence C_Q is Noetherian for Q of height-one. If Q has height two, then the prime ideals of C_Q are QC_Q, $\{Q_\alpha C_Q\}$ and (0), where the Q_α range over all height-one prime ideals of C contained in Q. Since C is a UFD, every such Q_α is principal. Thus every prime ideal of C_Q is finitely generated, and so C_Q is Noetherian by Cohen's Theorem 2.25.

For item 2, if $x \notin Q$, then C_Q is Noetherian by Remark 15.2.3. If $x \in Q$, then item 1 implies that C_Q is Noetherian. □

We use the setting from Insider Construction 10.7, but we restrict to a Noetherian base ring R and just one f, and the elements of $\underline{\tau}$ are algebraically independent in R^* over R.

SETTING 15.4. *Basic setup of Insider Construction* 10.7 *for Chapter* 15. Let R be a Noetherian domain with field of fractions $\mathcal{Q}(R) = K$. Let $x \in R$ be a nonzero nonunit of R, and let R^* be the x-adic completion of R. Let $\underline{\tau} = \{\tau_1, \ldots, \tau_s\} \subseteq xR^*$ be a set of elements that are algebraically independent over the base ring R. Assume that $K(\underline{\tau}) \subseteq \mathcal{Q}(R^*)$, the total ring of fractions of R^*, and that

$$\Psi^* : T := R[\underline{\tau}] = R[\tau_1, \ldots, \tau_s] \hookrightarrow R^*[1/x]$$

is flat.

Let $f \in (\underline{\tau})R[\underline{\tau}] \subseteq xR^*$ be transcendental over R. Since $f \in (\underline{\tau})R[\underline{\tau}]$, f has zero constant term as a polynomial in the τ_i. Define Intersection Domains

$$A := K(f) \cap R^*, \qquad \text{and } D := K(\tau) \cap R^*.$$

Let B be the Approximation Domain over R constructed using f as in Section 5.2; that is, B is the Approximation Domain of Inclusion Construction 5.3 associated to R and f. [1]

With Setting 15.4, Noetherian Flatness Theorem 6.3.1 implies D is also the Approximation Domain of Inclusion Construction 5.3 associated to $\underline{\tau}$, as in Definition 5.7. Thus D is a Noetherian Limit Intersection Domain, as defined in Definition 6.4.

Corollary 5.15 implies that B/xB and $B[1/x]$ are Noetherian. The non-flat locus of the extension $B \hookrightarrow R^*[1/x]$ gives more information about Noetherian properties of B such as that certain homomorphic images of B and certain localizations of B at prime ideals are Noetherian; see Local FlatnessTheorem 6.13, and Theorem 6.15.

[1] The examples of this chapter are Approximation Domains.

15.2. More properties of Insider Constructions

This section concerns properties of the constructed rings in the generality of Setting 15.4. These apply to Example 15.10 and examples described in Chapter 16.

REMARK 15.5. The fact that $R[\tau_1, \dots, \tau_s] \hookrightarrow R^*[1/x]$ is flat in in Setting 15.4 and that then non-flat locus of $\varphi : S = R[f] \hookrightarrow T$ is closed imply that the non-flat locus of $\psi : S = R[f] \hookrightarrow R^*[1/x]$ is closed; that is, the non-flat locus of ψ is determined by an ideal of $R^*[1/x]$. Let L be the ideal of R generated by the nonconstant coefficients of f. By Corollary 7.29, the ideal LT determines the non-flat locus of $\varphi : S = R[f] \hookrightarrow T$. Then, by Theorem 10.12, parts 1 and 2, the ideal $LR^*[1/x]$ determines the non-flat locus of both maps below:

$$B \hookrightarrow R^*[1/x] \text{ and } \psi : S := R[f] \hookrightarrow R^*[1/x].$$

Proposition 15.6 shows that an appropriate choice of f ensures that certain prime ideals of B are not finitely generated.

PROPOSITION 15.6. *Assume Setting 15.4. Let P be a prime ideal of R such that $x \notin P$ and $PB \neq B$. If the polynomial f has the form $f = p_1\tau_1 + \cdots + p_s\tau_s$, where each $p_i \in P$, and if $Q \in \operatorname{Spec} B$ satisfies:*

(i) $\mathfrak{q} := Q \cap R \subseteq P$, *and* (ii) $\{f_n\}_{n\in\mathbb{N}} \subseteq Q$.

Then

$$B = \bigcup_{n=1}^{\infty} B_n \implies Q = \bigcup_{n\in\mathbb{N}} (Q \cap B_n) \implies Q = \bigcup_{n\in\mathbb{N}} (Q \cap B_n)B.$$

Moreover $f_{n+1} \notin (Q \cap B_n)B$, Q is not finitely generated, and $Q = (\mathfrak{q}, \{f_n\}_{n\in\mathbb{N}})B$.

PROOF. For each $i \in \{1, 2, \dots, s\}$ and each $n \in \mathbb{N}$, let τ_{in} be the nth-endpiece of τ_i, as defined in Equation 5.4.3. The expression for f implies that for each $n \in \mathbb{N}$

$$f_n := p_1\tau_{1n} + \cdots + p_s\tau_{sn} \in Q \cap B_n.$$

By Endpiece Recursion Relation 5.5, $\tau_{in} = x\tau_{i,n+1} + c_{i,n+1}x$, with $c_{i,n+1} \in R$; it follows that

$$f_n = xf_{n+1} + p_1c_{1,n+1}x + \cdots + p_sc_{s,n+1}x = xf_{n+1} + xp', \tag{15.6.1}$$

where each $c_{i,n+1} \in R$ and $p' \in P$.

For each $n > 0$, $Q_n := Q \cap B_n$ is a prime ideal of B_n and is a localization of $Q \cap U_n = Q \cap (R[f_n]) = (\mathfrak{q}, f_n)R[f_n]$, a prime ideal of the polynomial ring $U_n = R[f_n]$ in the variable f_n. Thus $f_n \in Q \cap (R[f_n])$ implies $Q \cap B_n = (\mathfrak{q}, f_n)B_n$.

Suppose $f_{n+1} \in (Q \cap B_n)B = (\mathfrak{q}, f_n)B \subseteq (P, f_n)B$. Then $f_{n+1} = pa + f_nb$, where $a, b \in B$ and $p \in \mathfrak{q} \subseteq P$. Substituting $pa + f_nb$ for f_{n+1} in Equation 15.6.1 yields

$$f_n = xpa + xf_nb + xp' \implies (1 - xb)f_n = xpa + xp'.$$

By Proposition 5.17, the element $1-xb$ is a unit of B. Thus $f_n = xa'p + xp''$ for new elements $a' \in B$ and $p'' \in PB$. By Proposition 5.17.2.e, $f_n \in PB \cap U_n = PU_n$. This contradicts the fact that U_n is the polynomial ring $R[f_n]$. Thus $f_{n+1} \notin (Q \cap B_n)B$.

Since $Q = \bigcup_{n=1}^{\infty}(Q \cap B_n)B$, it follows that $Q = (\mathfrak{q}, \{f_n\}_{n\in\mathbb{N}})B$, and Q is not finitely generated. □

Proposition 15.7 describes prime ideals of B having height equal to the dimension of R.

PROPOSITION 15.7. *Assume Setting* 15.4. *Also assume that* $(R, \mathfrak{m}_R)$ *is a Noetherian local domain of dimension* $n > 0$. *Let* P *be a non-maximal prime ideal of* B *such that* $\operatorname{ht} P = n$, *and let* L *be the ideal of* R *generated by the nonconstant coefficients of* f. *Then*

(1) *The maximal ideal* $\mathfrak{m}_B$ *of* B *is* $\mathfrak{m}_R B$, $\dim B = n+1$, *and* $\dim(B/P) = 1$.
(2) *There exists a prime ideal* $P^* \in \operatorname{Spec} R^*$ *such that*
  (a) $PR^* \subseteq P^* \subsetneq \mathfrak{m}_{R^*}$, *and* $P = PR^* \cap B = P^* \cap B$.
  (b) $\operatorname{ht} P^* \le n-1$ *and* $LR^* \subseteq P^*$.
(3) $LR^* \cap B \subseteq P$, *and* $Q^* \cap B \subseteq P$, *for some prime ideal* Q^* *of* R^* *that is minimal over* LR^*.

PROOF. For item 1, by Proposition 5.17.5, B is local with maximal ideal $\mathfrak{m}_B = \mathfrak{m}_R B$. Since $B = \bigcup_{n=0}^{\infty} R[f_n]_{(\mathfrak{m}_R, f_n)}$ is the Approximation Domain of Insider Construction 10.7, Proposition 5.19.2 implies that $\dim B \le n+1$. Since P is a non-maximal prime ideal of height n, $\dim B = n+1$.

For item 2, $\operatorname{ht} \mathfrak{m}_B = n+1$. We define P^* in each of the two cases below:

Case i: $x \in P$. By Theorem 5.14.2, $R/xR = B/xB = R^*/xR^*$. Then the ideal $P^* := PR^*$ is non-maximal in R^*, and $P^* \cap B \neq \mathfrak{m}_B$.

Case ii: $x \notin P$. Let P^* be a prime ideal of R^* minimal over PR^* as in Proposition 5.17.4. Then $x \notin P^*$ and $x \notin P^* \cap B$ imply that P^* is not maximal and $P^* \cap B \neq \mathfrak{m}_B$.

In either case, $P \subseteq PR^* \cap B \subseteq P^* \cap B \subsetneq \mathfrak{m}_B$. Then $P = PR^* \cap B = P^* \cap B$, since $\operatorname{ht} P = n$ and $\operatorname{ht} \mathfrak{m}_B = n+1$. Since P^* is a non-maximal prime ideal of R^*, $\operatorname{ht} P^* < \dim R^* = n = \operatorname{ht} P$. By Remark 2.37.10, the extension $B_P \hookrightarrow R^*_{P^*}$ is not flat.

In Case i, $B_P \hookrightarrow R^*_{P^*}$ not flat implies that $(L, x)R^* \cap B \subseteq P$, by Theorem 6.13. Thus item 2 holds in Case i.

In Case ii, $R^*[1/x]_{P^* R^*[1/x]} = R^*_{P^*}$, since $x \notin P^*$. By Theorem 10.12.3, the nonflat locus of the extension $B \hookrightarrow R^*[1/x]$ is $LR^*[1/x]$. Thus $LR^*[1/x] \subseteq P^* R^*[1/x]$. It follows that $LR^* \subseteq P^*$ and $LR^* \cap B \subseteq P^* \cap B = P$, and so item 2 holds.

For item 3, choose a prime ideal Q^* of R^* such that $LR^* \subseteq Q^* \subseteq P^*$ and Q^* is minimal over LR^*. Then $Q^* \cap B \subseteq P^* \cap B = P$. □

Proposition 15.8 relates the constructed ring B to a ring of lower dimension. It is used in Proposition 15.14 to show that a homomorphic image of Example 15.10 is isomorphic to the three-dimensional ring B of Example 14.12.

PROPOSITION 15.8. *Assume Setting* 15.4 *and also assume that* $(R, \mathfrak{m})$ *is a regular local ring of dimension at least* 3, *that* $x, w \in R$ *are part of a regular system of parameters of* R, *and that* R^* *is the* x*-adic completion of* R.

(i) *Let* L *be the ideal of* R *generated by the coefficients of* f.
(ii) *Let* $\pi : R^* \to R^*/wR^*$ *be the natural homomorphism, and let* $\overline{}$ *denote the image under* π *in the RLR* $R^*/wR^* = \overline{R^*}$. *Assume that* $\overline{\tau_1}, \dots, \overline{\tau_s}$ *are algebraically independent over* R/wR.
(iii) *Let* B' *be the Approximation Domain formed by taking* $\overline{R}$ *as the base ring and defining the endpieces* $\overline{f}_n$ *of* $\overline{f}$ *analogously to Equation 5.4.1. That is,* B' *is defined by setting*

$$U'_n = \overline{R}[\overline{f}_n], \; B'_n = (U'_n)_{\mathfrak{n}'_n}, \; U' = \bigcup_{n=1}^{\infty} U'_n, \; \textit{and} \; B' = \bigcup_{n=1}^{\infty} B'_n,$$

where $\mathfrak{n}'_n = (\overline{\mathfrak{m}_R}, \overline{f_n})U'_n$.
If $L \not\subseteq wR^*[1/x]$, *then* $B' = B/wB = \overline{B}$ *and* $\overline{B}$ *is a UFD.*

PROOF. Since w is part of a regular system of parameters, w is a prime element of R. By Proposition 5.17.2, wB is a prime ideal of B; see Exercise 1. By Theorem 10.12.3, the non-flat locus of $B \hookrightarrow R^*[1/x]$ is defined by $LR^*[1/x]$. Since L is not contained in $wR^*[1/x]$, we have that

$$B \hookrightarrow R^*[1/x]_{wR^*[1/x]}$$

is flat, and therefore $B_{wR^*[1/x] \cap B} \hookrightarrow R^*[1/x]_{wR^*[1/x]}$ is faithfully flat. From this it follows that $wR^*[1/x] \cap B = wB$. Hence $\overline{B} = B/(wR^* \cap B) = B/wB$. Since

$$\overline{R}/x\overline{R} = \overline{B}/x\overline{B} = \overline{R^*}/x\overline{R^*},$$

the integral domain $\overline{R^*}$ is the $(\overline{x})$-adic completion of $\overline{R}$. Since the coefficients of f generate L and are not all contained in wR^*, the element $\overline{f}$ of the integral domain $\overline{B}$ is transcendental over $\overline{R}$. Since the endpieces $\overline{f}_n$ differ from $\overline{f}$ by elements of $\overline{R}$, they are also transcendental over $\overline{R}$.

Recall that $U_0[1/x] = U_n[1/x] = U[1/x]$ by Construction Properties Theorem 5.14.4 and Remark 5.16.3, and thus $wU \cap U_n = wU_n$, for each $n \in \mathbb{N}$. Then $wB \cap B_n = wB_n$, since B_n is a localization of U_n. Since $wR^* \cap B = wB$, it follows that $wR^* \cap B_n = wB_n$. Thus

$$\overline{R} \subseteq \overline{B_n} = B_n/wB_n \subseteq \overline{B} = B/wB \subseteq \overline{R^*} = R^*/wR^*.$$

Therefore $\overline{B} = \bigcup_{n=0}^{\infty} \overline{B_n}$. Then $B' = \overline{B}$, since $B'_n = \overline{B}_n$. By Theorem 5.24, the ring $\overline{B}$ is a UFD. □

REMARK 15.9. In the notation of Proposition 15.8, if R is the localized polynomial ring $R = k[x, y_1, \ldots, y_n]_{(x,y_1,\ldots,y_n)}$ and $w \in (y_1, \ldots, y_n)k[y_1, \ldots, y_n]$, then the assumption that $\overline{\tau_1}, \ldots, \overline{\tau_s}$ are algebraically independent over R/wR holds.

15.3. A four-dimensional non-Noetherian domain

We present a four-dimensional Approximation Domain B constructed using Setting 15.4; we identify Noetherian properties of B.

EXAMPLE 15.10. Let k be a field, let x, y and z be indeterminates over k. We use Setting 15.4 with $r = 1 = s$ (one τ and one f) to construct a four-dimensional Approximation Domain B such that

$$R := k[x, y, z]_{(x,y,z)} \subseteq B \subseteq R^* := k[y, z]_{(y,z)}[[x]].$$

With $\tau \in xk[[x]]$ transcendental over $k(x)$ and $D := k(x, y, z, \tau) \cap k[y, z]_{(y,z)}[[x]]$, the Intersection Domain D is a Local Prototype as in Definition 4.28. In addition, D is a three-dimensional RLR that is a directed union of four-dimensional RLRs.

We construct B inside D using Insider Construction 10.7. Define

$$f := yz\tau \quad \text{and} \quad A := k(x, y, z, f) \cap R^*.$$

The domain $A = D$, since $k(x, y, z, f) = k(x, y, z, \tau)$. For each integer $n \geq 0$, let τ_n be the nth endpiece of τ as in Equation 5.4.1. Then the nth endpiece of f is

$f_n = yz\tau_n$. By Remarks 5.16.3, the integral domains U_n, B_n, U and Approximation Domain B can be chosen as follows:

$$(15.10.2)\qquad \begin{aligned} U_n &:= k[x,y,z,f_n] & B_n &:= k[x,y,z,f_n]_{(x,y,z,f_n)} \\ U &:= \bigcup_{n=0}^{\infty} U_n \quad\text{and} & B &:= \bigcup_{n=0}^{\infty} B_n. \end{aligned}$$

Theorem 15.11 lists properties of the ring B that we prove in this section.

THEOREM 15.11. *Assume the setting of Example 15.10. Then:*

(1) *The ring B is a four-dimensional non-Noetherian local UFD with maximal ideal $\mathfrak{m}_B = (x,y,z)B$, and the $\mathfrak{m}_B$-adic completion of B is the three-dimensional RLR $k[[x,y,z]]$.*
(2) *The ring $B[1/x]$ is a Noetherian regular UFD and the ring B/xB is a two-dimensional RLR. For a prime ideal P of B, we have*

$$B_P \text{ is an RLR} \iff B_P \text{ is Noetherian} \iff (yz,x)R^* \cap B \not\subseteq P.$$

If $\operatorname{ht} P \le 2$, then B_P is an RLR.
(3) *The height-one prime ideals yB and zB are not contracted from R^*.*
(4) *If w is a prime element of B such that $w \notin yR^* \cup zR^*$, then $wR^* \cap B = wB$, and wB is the contraction of a height-one prime ideal of R^*.*
(5) *The prime spectrum of B is Noetherian.*
(6) *Concerning the finite generation of ideals of B:*
   (a) *Every height-one prime ideal is principal.*
   (b) *The ideals $Q_1 = (y,\{f_n\})B = yR^* \cap B$, $Q_2 = (z,\{f_n\})B = zR^* \cap B$ and $Q_3 = (y,z,\{f_n\})B = (y,z)R^* \cap B$ are prime and are not finitely generated, and $\operatorname{ht} Q_1 = \operatorname{ht} Q_2 = 2$ and $\operatorname{ht} Q_3 = 3$.*
   (c) *The prime ideals $(x,y)B$ and $(x,z)B$ have height three.*
   (d) *If P is a height-two prime ideal of B that contains an element of the form $y + g(z,x)$ or $z + h(x,y)$, where $0 \ne g(z,x) \in (x,z)k[x,z]$ and $0 \ne h(x,y) \in (x,y)k[x,y]$, then P is generated by two elements.*
   (e) *If $\mathfrak{a}$ is an ideal of B that contains $x + yzg(y,z)$, for some polynomial $g(y,z) \in k[y,z]$, then $\mathfrak{a}$ is finitely generated.*
   (f) *B has infinitely many height-three prime ideals that are not finitely generated, such as $Q_{i,\alpha} = (y - \alpha x^i, z, \{f_n\})B$, where $i \in \mathbb{N}$ and $\alpha \in k$.*
(7) *Every maximal ideal of $B[1/x]$ has height 2 or 3, and $B[1/x]$ has infinitely many maximal ideals of height three and infinitely many maximal ideals of height two.*
(8) *$B_{(x,y)B}$ and $B_{(x,z)B}$ are three-dimensional non-Noetherian local UFDs.*

We use Proposition 15.12 in the proof of Theorem 15.11.

PROPOSITION 15.12. *Assume the notation of Example 15.10. Then:*

(1) *The non-flat locus of the maps $\alpha : R[f] \hookrightarrow R^*[1/x]$ and $\psi : B \hookrightarrow R^*[1/x]$ is defined by the ideal $(yz)R^*[1/x]$.*
(2) *For every height-one prime ideal $\mathfrak{p}$ of R^* except yR^* and zR^*, we have $\operatorname{ht}(\mathfrak{p} \cap B) \le 1$.*
(3) *If w is a prime element of B such that $w \notin yR^* \cup zR^*$, then $wR^* \cap B = wB$.*

PROOF. Item 1 follows from Theorem 10.12, parts 2 and 3; see Remark 15.5.

By Theorem 5.14.1, item 2 holds if $\mathfrak{p} = xR^*$. Let $\mathfrak{p}$ be a height-one prime of R^* such that $\mathfrak{p} \notin \{xR^*, yR^*, zR^*\}$. Then $xyz \notin \mathfrak{p}$. Since $\mathfrak{p}R^*[1/x]$ does not contain $(yz)R^*[1/x]$ and $R^*[1/x] \subseteq (R^*)_{\mathfrak{p}}$, the map $B_{\mathfrak{p}\cap B} \hookrightarrow (R^*)_{\mathfrak{p}}$ is faithfully flat. Thus $\text{ht}(\mathfrak{p} \cap B) \leq 1$. This establishes item 2.

Let $\mathfrak{p}$ be a height-one prime ideal of R^* that contains wR^*. Then $\mathfrak{p} \neq yR^*$ or zR^*. By item 2, $\mathfrak{p} \cap B$ has height at most one. We have that the height-one prime ideal $\mathfrak{p} \cap B \supseteq wR^* \cap B \supseteq wB$, a height-one prime ideal. Thus item 3 follows. □

PROOF. (Of Theorem 15.11) For item 1, the map $\psi : B \hookrightarrow R^*[1/x]$ is not flat by Proposition 15.121, and so the ring B of Example 15.10 is non-Noetherian by Noetherian Flatness Theorem 6.3.1. The maximal ideal of B is $\mathfrak{m}_B = (x, y, z)B$, since the maximal ideal of each U_n is $(x, y, z, f_n)U_n$ and $f_n \in (x, y, z)U_{n+1}$. Also $\dim B \leq 4$ by Proposition 5.19. Since the chain of prime ideals

$$(0) \subsetneq yB \subsetneq (y, \{f_n\})B \subsetneq (y, z, \{f_n\})B \subsetneq \mathfrak{m}_B = (x, y, z)B$$

has strict containments—consider the elements y, f, z, x, we have $\dim B = 4$. The ring B is a UFD by Theorem 5.24. By Construction Properties Theorem 5.14.3, the (x)-adic completion of B is R^*, and so the $\mathfrak{m}_B$-adic completion of B is $k[[x, y, z]]$. Thus item 1 holds.

For item 2, Theorem 5.24 implies that the ring $B[1/x]$ is a Noetherian regular UFD. By Construction Properties Theorem 5.14.2, we have $R/xR = B/xB$, and so B/xB is a two-dimensional RLR. The non-flat locus of ψ is $yzR^*[1/x]$, the ideal generated by the coefficients of f, by Theorem 10.12.2. By Theorem 10.12.8, the domain B_P is Noetherian if and only if $(yz, x)R \not\subseteq P$. Combined with the fact that B/xB is a regular local ring, this implies that B_P is a regular local ring. Proposition 15.3 implies the last statement of item 2.

For item 3, yB and zB are not contracted from R^*, since

$$yB \subsetneq (y, f)B \subseteq yR^* \cap B \text{ and } zB \subsetneq (z, f)B \subseteq zR^* \cap B.$$

By Proposition 15.12, item 4 holds. Item 5 follows from Corollary 5.21.

For item 6a, since B is a UFD, every height-one prime ideal is principal. For items 6b and 6c, we have the following containments, most of which are proper:

$$(15.12.a) \qquad yB \subsetneq Q_1 \subseteq yR^* \cap B \subsetneq (x, y)B \subsetneq (x, y, z)B.$$

The containment $yR^* \cap B \subseteq (x, y)B$ follows from $(x, y)B/xB = (x, y)R^*/xR^*$. The sequence of elements f_n, x, z yields the asserted proper containments. The ideals $yB, (x, y)B$, and $(x, y, z)B$ are prime ideals by Proposition 5.17.2.b. Since $\text{ht}((x, y, z)B) = 4$, we have $\text{ht}(yR^* \cap B) = 2$ and $\text{ht}((x, y)B) = 3$. Also $yR^* \cap B$ is clearly a prime ideal that contains Q_1. By Proposition 15.6 with $Q_1 = yR^* \cap B$,

$$f_{n+1} \notin (yR^* \cap B_n)B; \quad yR^* \cap B = \cup_{n=1}^{\infty}(yR^* \cap B_n)B = (y, \{f_n\})B = Q_1;$$

and $yR^* \cap B$ is not finitely generated. Thus Q_1 is a prime ideal, $\text{ht}\, Q_1 = 2$ and Q_1 is not finitely generated. Similarly, Q_2 is a prime ideal, $Q_2 = zR^* \cap B$, $\text{ht}\, Q_2 = 2$, Q_2 is not finitely generated and $\text{ht}(x, z)B = 3$. Proposition 15.6 implies that $f_{n+1} \notin (y, z)R^* \cap B$, $Q_3 = (y, z)R^* \cap B$ and Q_3 is a prime ideal that is not finitely generated. Since $Q_1 \subsetneq Q_3 \subsetneq (x, y, z)B$, we have $\text{ht}\, Q_3 = 3$.

For item 6d, observe that the maximal ideal of $R = k[x,y,z]_{(x,y,z)}$ is generated by the regular system of parameters x, y, z. Since $\{x, w = y + g(x,z), z\}$ and $\{x, y, v = z + h(x,y)\}$ also generate the maximal ideal of R, they are regular systems of parameters. Since $y \notin wB$ and $z \notin vB$, the condition of Proposition 15.8 that the ideal $L = yzR$ generated by the coefficients of f is not contained in either wB or vB is satisfied. Thus, by Proposition 15.8, B/wB and B/vB are UFDs, and so every height-one prime ideal of B/wB or of B/vB is principal. It follows that every height-two prime ideal of B that contains such an element w or v is finitely generated.

For item 6e, assume $\mathfrak{a}$ is an ideal of B that contains a prime element of the form $x + yzg(y,z)$. Let $\mathfrak{p} = (x + yzg(y,z))B$. Then $\mathfrak{p}$ is a prime ideal and $\mathfrak{p} \subseteq \mathfrak{a}$. Item 4 above implies the prime ideal $\mathfrak{p}$ Since $g(y,z) \neq 0$, x is a regular element on $R^*/\mathfrak{p}R^*$. Also the ideal $\mathfrak{p}$ is contracted from R^* and so $\mathfrak{p}R^* \cap B = \mathfrak{p}$. By Theorem 6.15.3, the ring $B/\mathfrak{p}$ is Noetherian. Therefore $\mathfrak{a}$ is finitely generated.

For item 6f, the ideals $Q_{i,\alpha} = (y - \alpha x^i, z, \{f_n\})B$ are prime ideals, since each $Q_{i,\alpha} \cap B_n$ is a prime ideal. By Proposition 15.6, using $P = (y - \alpha x^i, z)R$ as the prime ideal of R in Proposition 15.6, we see that each $Q_{i,\alpha}$ is not finitely generated. If $i > j$ or $i = j$ and $\alpha \neq \beta \in k$, then

$$\alpha x^i - \beta x^j \in Q_{i,\alpha} \implies \alpha x^{i-j} - \beta \in Q_{i,\alpha}.$$

If $i > j$ and $\beta \neq 0$, this would imply that $Q_{i,\alpha}$ contains a unit; if $i \geq j$ and $\beta = 0$ but $\alpha \neq 0$, then $Q_{i,\alpha}$ contains x or a unit. if $i = j$ and $\alpha \neq \beta$, then $Q_{i,\alpha}$ contains a unit. Since each $Q_{i,\alpha}$ is a proper prime ideal, there are infinitely many distinct prime ideals of the form $Q_{i,\alpha}$.

For item 7, Theorem 5.18.2 with $n = 3$ and $s = 1$ implies $2 \leq \operatorname{ht} P \leq 3$, for every prime ideal of B maximal with respect to $x \notin P$. Equivalently, every maximal ideal of $B[1/x]$ has height 2 or 3. By item 6f, there are infinitely many height-three maximal ideals of $B[1/x]$, such as the $Q_{i,\alpha}B[1/x]$.

We claim, for each $i \in \mathbb{N}$, the ideal $Q_i := (y - z, y - x^i)B[1/x]$ is a height-two ideal of B maximal with respect to $x \notin Q_i$, and so Q_i is a maximal ideal of $B[1/x]$ of height 2. To see this, observe that $\mathfrak{q}_i := (y - z, y - x^i)R \in \operatorname{Spec} R$ and $\operatorname{ht} \mathfrak{q}_i = 2$. By Proposition 5.17.2, $Q_i = \mathfrak{q}_i B \in \operatorname{Spec} B$. Also $(y - z)B \subsetneq Q_i \subsetneq \mathfrak{m}_B$, since $(y - z)B_n \subsetneq \mathfrak{q}_i B_n \subsetneq \mathfrak{m}_{B_n}$, for every $n \in \mathbb{N}$, and so $2 \leq \operatorname{ht} Q_i \leq 3$. Suppose that $Q_i \subseteq P$, for some $P \in \operatorname{Spec} B$ with $\operatorname{ht} P = 3$. Then, by Proposition 15.7.3, $yz \in P$ and so y or $z \in P$. In either case, $y - z \in Q_i \subseteq P \implies (y, z, x)B = \mathfrak{m}_B \subseteq P$, a contradiction to $\operatorname{ht} P = 3 < \operatorname{ht} \mathfrak{m}_B = 4$. Thus Q_i is not contained in any height-three prime ideal of B. Hence $\operatorname{ht} Q_i = 2$ and Q_i is maximal with respect to not containing x, as desired for the claim.

To complete item 7, for $i < j \in \mathbb{N}$, we show that $Q_i \neq Q_j$. Suppose not; then $x^{j-i}(x^i - 1) = x^j - x^i = (y - x^i) - (y - x^j) \in Q_i \cap R = \mathfrak{q}_i$. This implies $x^i \in \mathfrak{q}_i$, a contradiction. Thus $B[1/x]$ contains infinitely many height-two maximal ideals.

For item 8 with $P = (x,y)B$, by Local Flatness Theorem 6.13, B_P is not Noetherian, since $(yz, x)R = (yz, x)R^* \cap R \subseteq P = (x,y)B$. The rest of item 8 follows from items 2, 6b and 6c. Similarly, $B_{(x,z)B}$ is a non-Noetherian local UFD. □

We display a partial picture of Spec(B) and make comments about the diagram. Here β ranges over the elements of k.

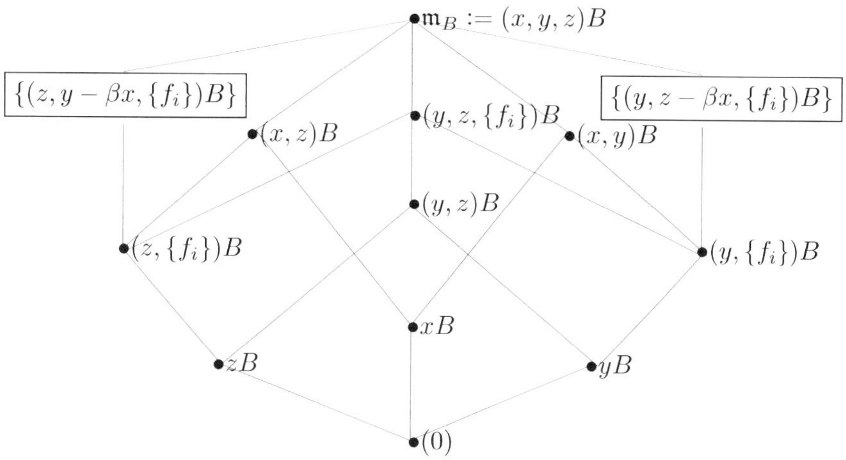

Diagram 15.12.0

Comments regarding Diagram 15.12.0. A line going from a box at one level to a bullet at a higher level indicates that every prime ideal in the lower level box is contained in the prime ideal in the higher level box. The links connecting every pair of bullets shown are saturated—no other prime ideals are between the two endpoints of each link. If one or more prime ideals are between two prime ideals, say $P_1 \subsetneq P_2 \subsetneq P_3 \in \operatorname{Spec} B$, and B_{P_3} is Noetherian, then there are infinitely many prime ideals (not shown) between P_1 and P_3. For example, there are infinitely many prime ideals between (0) and $(y,z)B$ (e.g. $(y - \beta z)B$ for every $\beta \in k[x]$); these are not shown in the picture. On the other hand, the only prime ideal between yB and $(x,y)B$ is the prime ideal $yR^* \cap B$, by Proposition 15.13.8. If g is a prime element of B such that $yz \notin \sqrt{(x,g)B}$, then the height-one prime ideal gB of B is contained in a height-two prime P of B that contains x; see Proposition 15.13.6. The prime ideals of B that contain $\{f_i\}$ but not x are not finitely generated by Proposition 15.6.

Other properties of $\operatorname{Spec} B$ are given in Proposition 15.13.

PROPOSITION 15.13. *Let B be the ring constructed in Example* 15.10. *Then:*

(1) $\operatorname{Spec} B$ *has no maximal saturated chain of length 2.*
(2) *Let P be a prime ideal of B of height three. Then*
   (a) *P contains $yR^* \cap B$ or $zR^* \cap B$.*
   (b) *$P = (y, g(z,x))R^* \cap B$ or $P = (z, h(y,x)R^* \cap B$, for some irreducible polynomial-power series $g(z,x) \in k[z]_{(z)}[[x]]$ and $h(y,x)$ in $k[y]_{(y)}[[x]]$.*
   (c) *$(x,y)B$ and $(x,z)B$ are the only height-three prime ideals of B that contain x.*
(3) *Let $g(z,x) \in k[z]_{(z)}[[x]]$ be an irreducible polynomial-power series such that $g(z,x)R^* \cap B \neq 0$. Then $P = (y, g(z,x))R^* \cap B$ is a height-three prime ideal of B. If $h(y,x)$ is irreducible in $k[y]_{(y)}[[x]]$ and $h(y,x)R^* \cap B \neq 0$, then $P = (z, h(y,x))R^* \cap B$ is a height-three prime ideal of B.*
(4) *Let P be a prime ideal of B of height three, $P \neq (x,y)B$ and $P \neq (x,z)B$. Then (a) P is not finitely generated, and (b) B_P is an RLR.*

(5) *Every prime ideal P of B such that $yz \notin P$ has height ≤ 2.*
(6) *If $g \in B$ and $yz \notin \sqrt{(g,x)B}$, then $(g,x)B$ is contained in a height-two prime ideal of B.*
(7) *If $g \in B$, $yz \in \sqrt{(g,x)B}$, and $P \in \operatorname{Spec} B$ is minimal over $(g,x)B$, then $P = (x,y)B$ or $P = (x,z)B$; thus every prime ideal of B minimal over $(g,x)B$ has height 3.*
(8) *There is a unique prime ideal N of B such that $yB \subsetneq N \subsetneq (x,y)B$, namely $N = yR^* \cap B$. Similarly, $zR^* \cap B$ is the unique prime ideal between zB and $(x,z)B$.*

PROOF. For item 1, if $(0) = P_0 \subsetneq P_1 \subsetneq \dots \subsetneq P_h$ is a maximal chain then $P_h = \mathfrak{m}_B = \mathfrak{m}_R B$ and $h > 0$. If $h = 2$ and $x \notin P_1$, then the ideal $P_1 B[1/x]$ is maximal, a contradiction to Theorem 15.11.7. If $x \in P_1$, then the extension $P_1 = xB \subsetneq \mathfrak{m}_B$ is saturated, a contradiction to $\dim(B/xB) = \dim(R/xR) = 2$.

For item 2, Proposition 15.7.3 implies that $P = P^* \cap B$, for some non-maximal prime ideal P^* of R^* with $yzR^* \subseteq P^*$. Then $\operatorname{ht} P^* \leq 2$ and $y \in P^*$ or $z \in P^*$. Since the proof is similar for z, assume $yR^* \subseteq P^*$, and so $yR^* \cap B \subseteq P$. By Theorem 15.11.6b, $\operatorname{ht}(yR^* \cap B) = 2$. Then $\operatorname{ht} P = 3$ implies $yR^* \cap B \subsetneq P$ and $yR^* \subsetneq P^*$, and $\operatorname{ht}(yR^*) = 1$ implies $\operatorname{ht} P^* = 2$. Since $R^* = k[y,z]_{(y,z)}[[x]]$ is a UFD, so is $R^*/yR^* = k[z]_{(z)}[[x]]$. Hence the height-one prime ideal P^*/yR^* of R^*/yR^* is principal, generated by an irreducible element of the form $g(z,x)$ in the ring $k[z]_{(z)}[[x]]$. That is, $P^* = (y, g(z,x))R^*$ and so $P = (y, g(z,x))R^* \cap B$. Thus items 2a and 2b hold. Item 2c follows from item 2a.

For item 3, clearly $P^* = (y, g(z,x))k[y,z]_{(y,z)}[[x]]$ is a height-two prime ideal of $R^* = k[y,z]_{(y,z)}[[x]]$, and $g(z,x) \in B \setminus (yR^* \cap B)$. Let $P = (y, g(z,x))R^* \cap B$. Thus $(yR^* \cap B) \subsetneq P \subsetneq \mathfrak{m}_B$, and so P is a height-three prime ideal of B. Similarly the other part of item 3 holds.

For item 4.a, by item 2, every prime ideal of height three contains $\{f_n\}_{n=1}^{\infty}$. By Proposition 15.6, every such prime ideal that avoids x is not finitely generated. Since $(x,y)B$ and $(x,z)B$ are the only height-three prime ideals that contain x, item 4.a holds. Item 4.b holds by Theorem 15.11.2 if P is a height-three prime ideal that does not contain x. Since $B_{(x,y)}B$ contains the non-finitely generated prime ideal $(y, \{f_n\}_{n=1}^{\infty})$, $B_{(x,y)}B$ is not Noetherian. Similarly $B_{(x,z)}B$ is not Noetherian.

For item 5, if $yz \notin P$, then P is not maximal, and, by item 2, $\operatorname{ht} P \neq 3$. Thus $\operatorname{ht} P \leq 2$.

For item 6, first suppose $g \in xB$. Then $\sqrt{(g,x)B} = xB \subsetneq P = (x, y+z)B$, a prime ideal since $(x, y+z)B_n$ is a prime ideal for every n. Also $yz \notin P$ and $xB \subsetneq P$, and so $\operatorname{ht} P = 2$ by item 5.

Now suppose that $g \notin xB$ and $yz \notin \sqrt{(g,x)B}$. Since B/xB is Noetherian, $\sqrt{(g,x)B}$ has a primary decomposition, and so there exists a prime ideal P such that $xB \subsetneq (g,x)B \subseteq P$ and $yz \notin P$. By item 5, $\operatorname{ht} P = 2$.

For item 7, let $P \in \operatorname{Spec} B$ be minimal over $\sqrt{(g,x)B}$. Then $yz \in \sqrt{(g,x)B}$ implies $yz \in P$. Also B/xB is Noetherian and so $\operatorname{ht}_{B/xB}(P/xB) \leq 1$, by Krull Altitude Theorem 2.23. Thus $P \neq \mathfrak{m}_B$, and so $\operatorname{ht} P \leq 3$. Since $yz \in P$, either y or z is in P; say $y \in P$. Then $(x,y)B \subseteq P$ and $\operatorname{ht}((x,y)B) = 3$ imply that $P = (x,y)B$.

For item 8, assume that $yB \subsetneq N \subsetneq (x,y)B$. Then $x \notin N$ and so there exists $N^* \in \operatorname{Spec} R^*$ with $NR^* \subseteq N^*$ and $x \notin N^*$ by Proposition 5.17.4. Then $y \in N$

implies $yR^* \subseteq N^*$. Thus there are chains of prime ideals of B

$$(0) \neq yB \subsetneq N \subseteq N^* \cap B \subsetneq (x,y)B, \text{ and } yB \subseteq yR^* \cap B \subseteq N^* \cap B.$$

Since $\operatorname{ht}(x,y)B = 3$, we have $N^* \cap B = N$. Also $\operatorname{ht}(N^* \cap B) = 2 = \operatorname{ht}(yR^* \cap B)$ by Theorem 15.11.6b, and so $N = yR^* \cap B$. □

PROPOSITION 15.14. *The homomorphic image $B/(y-z)B$ of the ring B of Example 15.10 is isomorphic to the three-dimensional ring of Example 14.12.*

PROOF. In order to distinguish and compare the two constructed rings, let the base ring R_C denote the base ring for Example 14.12 and let C denote the ring of Example 14.12 over R_C; let B be the ring of Example 15.10 with base ring R. Thus,

$$R_C := k[x,y]_{(x,y)}, \qquad R_C^* = k[y]_{(y)}[[x]], \quad C = \bigcup_{n=0}^{\infty} k[x,y,y^2\tau_n]_{(y,z)}$$

$$R := k[x,y,z]_{(x,y,z)}, \qquad R^* = k[y,z]_{(y,z)}[[x]], \quad B = \bigcup_{n=0}^{\infty} k[x,y,z,yz\tau_n]_{(x,y,z,yz\tau_n)}.$$

Let π be the natural map $B \to \overline{B} = B/(y-z)B$. Since π is the identity on $k[[x]]$, the element τ is in both R_C^* and R^*. Then $\pi(y) = \pi(z) = y$ and $\pi(f) = \pi(yz\tau) = y^2\tau$. Thus $\overline{R} \cong k[x,y]_{(x,y)} = R_C$. Here $L = yzR$, and so $L \not\subseteq (y-z)R^*[1/x]$. By Proposition 15.8 with $w = y - z \in R$, we have $B' = B/(y-z)B$, where B' is the approximation domain over $\overline{R} = R/(y-z)R$ using the element $\overline{f}$, which corresponds to $y^2\tau$ as an element of $k[y]_{(y)}[[x]]$. Thus $\overline{B} \cong C$. □

REMARK 15.15. Let B be the non-Noetherian local UFD of dimension four constructed in Example 15.10, and let I be an ideal of B. By Proposition 5.17.5, IR^* is $\mathfrak{m}_{R^*}$-primary $\iff$ I is $\mathfrak{m}_B$-primary,

We do not know the answers to Questions 15.16.

QUESTIONS 15.16. For the ring B constructed in Example 15.10,

(1) Does every prime ideal of B that is not finitely generated contain the set $\{f_i\}_{i=1}^{\infty}$?
(2) Are the prime ideals

$$Q_1 := (y, f_1, \ldots, f_i, \ldots)B \text{ and } Q_2 := (z, f_1, \ldots, f_i, \ldots)B$$

the only height-two prime ideals of B that are not finitely generated?

REMARK 15.17. The analysis in this chapter of Example 15.10 implies that a "Yes" answer to Question 15.16.2 yields a "Yes" answer to Question 15.16.1. By Proposition 15.13.2, every prime ideal of height three contains the set $\{f_i\}_{i=0}^{\infty}$, and the prime ideals of height three other than $(x,y)B$ and $(x,z)B$ are not finitely generated. The prime ideals of height one and height four are known to be finitely generated since B is a local UFD. Thus Question 15.16.1 boils down to whether every prime ideal of B of height two that is not finitely generated contains the set $\{f_i\}_{i=0}^{\infty}$. For some computations related to the two questions, see Exercise 2.

Exercises

(1) Let $(R, \mathfrak{m})$ be a regular local ring of dimension d, and let $a_1, \ldots a_d$ be a regular system of parameters for R.[2] If $\{b_1, \ldots, b_e\}$ is a subset of $\{a_1, \ldots a_d\}$, prove that $R/(b_1, \ldots, b_e)R$ is an RLR, and conclude that $(b_1, \ldots, b_e)R$ is a prime ideal of R.

(2) Assume the notation of Example 15.10, and write $\tau = \sum_{i=1}^{\infty} c_i x^i$, where $c_i \in k$. Let $n \in \mathbb{N}_0$ and $r \in \mathbb{N}$, and let P be a prime ideal of B that contains f_n and f_{n+r}, but $x \notin P$. Prove:
   (a) $x^r f_{n+r} - f_n = ayzx \in P$, where $a \in k[x]$.
   (b) If $a = 0$, then all the coefficients $c_{n+r}, c_{n+r-1}, \cdots, c_{n+1}$ are 0.
   (c) If $a \neq 0$, then $y \in P$ or $z \in P$.
   (d) With the additional assumption that $c_i \neq 0$, for some i with $n < i \leq n+r$, the two prime ideals given in Question 15.16.2 are the only prime ideals of height two that contain f_n and f_{n+r}, but that do not contain x.

   SUGGESTION. For part a, use Endpiece Recursion Relation 5.5, and $f_n = yz\tau_n$. For part c, show $a = ux^t$, for some $u \in k[x]$ and some power $t \in \mathbb{N}_0$; also use x is in the Jacobson radical of B. For part d, show $f_i \in P$, for every $i \in \mathbb{N}$.

(3) Let R be an integral domain and let $\tau_1, \ldots, \tau_s$ be algebraically independent over R. Let $f \in R[\tau_1, \ldots, \tau_s]$. If $f \notin R$, prove that f is transcendental over R.

---

[2]See the description of "Regular local ring" just above Remarks 2.10.

CHAPTER 16

Non-Noetherian examples in higher dimension

In this chapter we extend the construction of Chapter 14 to obtain local domains of dimension > 3 that are not Noetherian, but are close to being Noetherian. For every $m, d \in \mathbb{N}$, there exist non-catenary non-Noetherian local unique factorization domains B of dimension $d+2$ such that:

(i) B has exactly m prime ideals of height $d+1$;

(ii) These m prime ideals are not finitely generated;

(iii) The localization of B at every non-maximal prime ideal of B is Noetherian.

These examples use Insider Construction 10.7 of Section 10.2 and results from Chapter 15.

Section 16.1 contains Theorem 16.2; its proof establishes the properties of the ring B discussed above. In Section 16.2 we analyze a four-dimensional case of the example in more detail. Section 16.3 relates the examples of this chapter to the $D+M$ construction, finite conductor domains and coherence.

16.1. Higher dimensional non-Noetherian examples

The main examples in this chapter are similar to Examples 14.1 and Examples 10.15. Setting 16.1 describes the rings of the examples and gives the basic set-up for this section. Theorem 16.2 shows that these rings have many properties similar to those of Examples 14.1 and 15.10.

SETTING 16.1. Let $d, m \in \mathbb{N}$, let $x, y_1, \dots, y_d$ be indeterminates over a field k, and let $\underline{y} := \{y_1, \dots, y_d\}$. Define

$$R := k[x, \underline{y}]_{(x,\underline{y})}, \quad \mathfrak{m}_R := (x, \underline{y})R, \quad R^* = k[\underline{y}]_{(\underline{y})k[\underline{y}]}[[x]].$$

Then R^* is the x-adic completion of R. Let $\underline{\tau} := \tau_1, \dots, \tau_d \in xk[[x]]$ be algebraically independent elements over $k(x)$. Let $p_1, \dots, p_m$ be m non-associate nonzero prime elements of $R \setminus (x, y_2, \dots, y_d)R$ such that with $\mathfrak{q}_i := (p_i, y_2, \dots, y_d)R$, then $R/\mathfrak{q}_i$ is a DVR. It follows that $R^*/\mathfrak{q}_i R^*$ is a DVR. For example, let $p_i = y_1 - x^i$. Set $q = p_1 p_2 \cdots p_m$, and consider the element

$$f := q\tau_1 + y_2\tau_2 + \dots + y_d\tau_d \in xk[[x]].$$

Let f_n denote the nth-endpiece of f, defined analogously to τ_n in Equation 5.4.1. Define B to be the Approximation Domain corresponding to f, a nested union of localized polynomial rings of dimension $d+2$:

$$B := \bigcup_{n=1}^{\infty} B_n, \quad \text{where} \quad B_n = k[x, \underline{y}, f_n]_{(x,\underline{y},f_n)}.$$

The Local Prototype $D := k(x, \underline{y}, \underline{\tau}) \cap R^*$ contains both B and $A := k(x, \underline{y}, f) \cap R^*$.

THEOREM 16.2. *Assume Setting 16.1 and define* $Q_i = (p_i, y_2, \dots, y_d)R^* \cap B$, *for every* i *with* $1 \le i \le m$. *Then:*

(1) *The integral domain* B *is a non-Noetherian local UFD of dimension* $d+2$ *with maximal ideal* $\mathfrak{m}_B = \mathfrak{m}_R B$.
(2) *The Intersection Domain* A *equals its Approximation Domain* B.
(3) *The* $\mathfrak{m}_B$*-adic completion of* B *is* $k[[x, \underline{y}]]$, *the formal power series ring, a regular local domain of dimension* $d+1$.
(4) *The ring* $B[1/x]$ *is a Noetherian regular UFD, the ring* B/xB *is an RLR of dimension* d, *and, for every non-maximal prime ideal* P *of* B, *the ring* B_P *is an RLR.*
(5) *The ring* B *has exactly* m *prime ideals of height* $d+1$, *namely* $Q_1, \dots, Q_m$.
(6) *Each* Q_i *is a nested union* $Q_i = \bigcup_{n=1}^{\infty} Q_{in}$, *where* $Q_{in} = (p_i, y_2, \dots, y_d, f_n)$, *and each* Q_i *is not finitely generated.*
(7) *The local ring* $(B, \mathfrak{m}_B)$ *birationally dominates a localized polynomial ring in* $d+2$ *variables over the field* k.
(8) *Every saturated chain of prime ideals of* B *has length either* $d+1$ *or* $d+2$, *and there exist saturated chains of prime ideals of lengths both* $d+1$ *and* $d+2$. *Thus* B *is not catenary, and every maximal ideal of* $B[1/x]$ *other than the* $Q_i B[1/x]$ *has height* d.
(9) *Each height-one prime ideal of* B *is the contraction of a height-one prime ideal of* R^*.
(10) $\operatorname{Spec} B$ *is Noetherian.*

PROOF. Let $L := (q, y_2, \dots, y_d)R$, the ideal generated by the coefficients of f.

For item 1, the ideal $LR^*[1/x]$ defines the non-flat locus of $\beta : B \hookrightarrow R^*[1/x]$, by Theorem 10.12.2, and, by Theorem 10.12.4, B is not Noetherian. By Proposition 5.17.5, B is local with maximal ideal $\mathfrak{m}_B = \mathfrak{m}_R B$. By Theorem 5.24, B is a UFD. By Proposition 5.19, $\dim B \le d+2$.

For the rest of item 1 and part of item 5, since $(p_i, y_2, \dots, y_d)R^*$ is a prime ideal of R^*, the ideal Q_i is prime. By Proposition 5.17.2, the ideals $p_i B$ and $(p_i, y_2, \dots, y_j)B$ are prime, for every j with $2 \le j \le d$. The inclusions in the chain of prime ideals

$$(0) \subset p_i B \subset (p_i, y_2)B \subset \cdots \subset (p_i, y_2, \dots, y_d)B \subset Q_i \subset \mathfrak{m}_B$$

are strict because the contractions to B_n are strict for each n; to verify this, consider the list $p_i, y_2, y_3, \dots, y_d, f, x$, and use that $f \in Q_i \setminus (p_i, y_2, \dots, y_d)B_n$, for each n. Thus $\dim B = d+2$, each Q_i has height $d+1$, and $(p_i, y_2, \dots, y_d)B$ has height d, for each i. This proves item 1 and part of item 5.

For item 2, the non-flat locus of the extension $R[f] \hookrightarrow R[\underline{\tau}]$ is $LR[\underline{\tau}]$, by Theorem 10.12.1, and $\operatorname{ht}(LR^*[1/x]) > 1$. Since $R[\tau]$ is a UFD, Proposition 9.19 implies equality of the approximation and intersection domains B and A corresponding to the element f of R^*. This completes item 2.

By Construction Properties Theorem 5.14.3, the (x)-adic completion of B is R^*. Hence the $\mathfrak{m}_B$-adic completion of B is the same as the $\mathfrak{m}_R$-adic completion of R, that is $\widehat{B} = k[[x, y_1, \dots, y_d]]$. This proves item 3.

For item 4, by Theorem 5.24, the ring $B[1/x]$ is a Noetherian regular UFD. Then $R/xR = B/xB$ implies B/xB is a d-dimensional RLR. For every non-maximal prime ideal P of B, $(x, q, y_2, \dots, y_d)R = (L, x)R^* \cap R \not\subseteq P$, and so, by Theorem 10.12.8, B_P is Noetherian. For the last part of item 4, if $x \notin P$, then B_P is a localization of the Noetherian regular ring $B[1/x]$ and so B_P is an RLR. If $x \in P$, let $\overline{P}$ be the image of P in B/xB. Then $(B/xB)_{\overline{P}} = B_P/xB_P$ is a localization of an RLR and so is an RLR. Thus PB_P/xB_P is generated by h elements, where $h = \operatorname{ht}(PB_P/xB_P)$. Hence PB_P is generated by $h+1$ elements, if x is added to preimages of the generators for PB_P/xB_P. Since $\operatorname{ht} P = \operatorname{ht}(PB_P) \geq h+1$, and B_P is Noetherian, we have that B_P is an RLR. Thus item 4 holds.

For the remainder of item 5, if P is a height-$(d+1)$ prime ideal of B, then Proposition 15.7 implies that $P = P^* \cap B$, for a non-maximal prime ideal P^* of R^* such that $(q, y_2, \dots, y_d)R^* = LR^* \subseteq P^*$, and so $(p_i, y_2, \dots, y_d)R^* \subseteq P^*$, for some i with $1 \leq i \leq m$. It follows that $Q_i = (p_i, y_2, \dots, y_d)R^* \cap B \subseteq P^* \cap B = P$. Then $\operatorname{ht} Q_i = \operatorname{ht} P$ implies $Q_i = P$. Thus item 5 holds.

Item 6 holds by Proposition 15.6. Item 7 follows from the construction of B, since $B_n \subseteq B \subseteq \mathcal{Q}(B_n)$.

For item 8, there always exists a saturated chain of prime ideals of B between (0) and $\mathfrak{m}_B$ that contains the height-one prime xB, and $B/xB = R^*/xR^*$ implies that this chain has length equal to $\dim R^* = d+1$. Since $\dim B = d+2$, there also exists a saturated chain of prime ideals in B of length $d+2$. Hence B is not catenary.

Suppose that

$$(0) \subsetneq P_1 \subsetneq P_2 \subsetneq \dots \subsetneq P \subsetneq \mathfrak{m}_B \tag{16.2.0}$$

is a saturated chain of prime ideals of B, and so $\dim(B/P) = 1$. Thus P is a non-maximal prime ideal of B, and so, by item 4, B_P is an RLR. It follows that B_P is catenary; see Remark 3.27. Hence $h = \operatorname{ht} P = \operatorname{ht} PB_P$ is the number of links from (0) to P in *every* saturated chain of the form Equation 16.2.0.

There are two cases:

Case i: If $x \notin P$, then Theorem 5.18 implies that $\operatorname{ht} P$ is d or $d+1$. Therefore every chain from (0) to P has length d or $d+1$.

Case ii: If $x \in P$, then $\operatorname{ht}(B/xB) = \operatorname{ht}(R/xR) = d$, and there exists a saturated chain of prime ideals of B of the form

$$(0) \subsetneq xB \subsetneq \dots \subsetneq P \subsetneq \mathfrak{m}_B.$$

Here the non-zero prime ideals correspond to prime ideals of B/xB. Since xB has height one, since $\dim(B/xB) = \dim(R/xR) = d$, and since B_P is catenary, it follows that $\operatorname{ht} P = \operatorname{ht}_{B/xB}(P/xB) + 1 = d$ and every saturated chain from (0) to P has length d.

Thus in either case every saturated chain of prime ideals in B has length $d+1$ or $d+2$.

For the last part of item 8, suppose that $P \in \operatorname{Spec} B$ has $\operatorname{ht} P = h$ and that $PB[1/x]$ is a maximal ideal, but $P \neq Q_i$ for every i. Then P is maximal with respect to $x \notin P$ and $h \neq d+1$, by items 1 and 5. By Theorem 5.18.1, $\dim(B/P) = 1$, and so there is a saturated chain of length $h+1$ of the form $(0) \subsetneq \dots \subsetneq P \subsetneq \mathfrak{m}_B$ in $\operatorname{Spec} B$. Since $h \neq d+1$, it follows that $h = d$. This completes the proof of item 8.

For item 9, since R^* is a Krull domain and $B = A = \mathcal{Q}(B) \cap R^*$, it follows that B is a Krull domain and each height-one prime of B is the contraction of a height-one prime of R^*. Item 10 holds by Corollary 5.20. □

QUESTION 16.3. For the ring B constructed in Theorem 16.2, are the prime ideals

$$Q_i := (p_i, y_2, \dots, y_d, f_1, \dots, f_i, \dots)B$$

the only prime ideals of B that are not finitely generated?

16.2. A 4-dimensional prime spectrum

In Example 16.4, we study in more detail a four-dimensional special case of Examples 10.15.

EXAMPLE 16.4. Set $R := k[x, y, z]_{(x,y,z)}$, where x, y and z are indeterminates over a field k and let R^* denote the x-adic completion $k[y, z]_{(y,z)}[[x]]$ of R. and let $\mathfrak{m}_R$ and $\mathfrak{m}_{R^*}$ denote the maximal ideals of R and R^*, respectively. Let τ and σ in $xk[[x]]$ be algebraically independent over $k(x)$. Say

$$\tau := \sum_{n=1}^{\infty} c_n x^n \quad \text{and} \quad \sigma := \sum_{n=1}^{\infty} d_n x^n,$$

where the c_n and d_n are in k.

Let D be the Local Prototype $D := R^* \cap k(x, y, z, \tau, \sigma)$. By Local Prototype Theorem 10.6.1 the domain D is Noetherian and equals the approximation domain associated to τ, σ. In addition, D is a three-dimensional RLR that is a directed union of 5-dimensional RLRs, and the extension $T := R[\tau, \sigma] \hookrightarrow R^*[1/x]$ is flat.

Define $f := y\tau + z\sigma$ and $A := R^* \cap k(x, y, z, f)$, that is, A is the intersection domain associated with f. This notation fits that of Theorem 16.2, with $d = 2$, $m = 1$, $y_1 = y$, $y_2 = z$, $\tau_1 = \tau$, $\tau_2 = \sigma$, and the coefficient q of $\tau_1 = \tau$ in the expression for f is just one prime element $p_1 = y$ of $R \setminus (x, z)$.

For each integer $n \geq 0$, let τ_n and σ_n be the nth endpieces of τ and σ as in Equation 5.4.1. Then the nth endpiece of f is $f_n = y\tau_n + z\sigma_n$. As in Equation 5.5.1,

$$\tau_n = x\tau_{n+1} + xc_{n+1} \quad \text{and} \quad \sigma_n = x\sigma_{n+1} + xd_{n+1},$$

where c_{n+1} and d_{n+1} are in the field k. Therefore

$$\begin{aligned} f_n = y\tau_n + z\sigma_n &= yx\tau_{n+1} + yxc_{n+1} + zx\sigma_{n+1} + zxd_{n+1} \\ &= xf_{n+1} + yxc_{n+1} + zxd_{n+1}. \end{aligned} \tag{16.4.1}$$

The approximation domains U_n, B_n, U and B for A are as follows:

$$\begin{aligned} &\text{For } n \geq 0, \quad U_n := k[x, y, z, f_n] \qquad B_n := k[x, y, z, f_n]_{(x,y,z,f_n)} \\ &U := \bigcup_{n=0}^{\infty} U_n \quad \text{and} \qquad B := \bigcup_{n=0}^{\infty} B_n. \end{aligned} \tag{16.4.2}$$

By Theorem 16.2, B is a four-dimensional local non-Noetherian UFD with exactly one prime ideal of height three, namely $Q = (y, z)R^* \cap B$, Q is not finitely generated, and $B = A$. The remaining properties of Theorem 16.2, adjusted for this set-up, also hold for B.

We establish other properties of Example 16.4 in Propositions 16.5, 16.6 and 16.7.

PROPOSITION 16.5. *Assume the notation of Example* 16.4. *The homomorphic image B/zB of the ring B is isomorphic to the three-dimensional ring called B in Example* 14.9.

PROOF. For this proof, let B denote the Approximation Domain formed with respect to $f = y\tau + z\sigma$, as in Example 16.4. Revise the notation for Example 14.9:

$$R_E := k[x,y]_{(x,y)}, \quad g = y\tau, \quad g_n = y\tau_n, \quad E = \bigcup R_E[g_n]_{(x,y,g_n)}.$$

Thus E is the Approximation Domain constructed for g over R_E and called B in Example 14.9.

We show that the ring $B/zB \cong E$: By Proposition 15.8 and Remark 15.9 with $w = z$, the ring $B' = B/zB$, where B' is the Approximation Domain over $\overline{R} = R/zR$ using the element $\overline{f}$, transcendental over $\overline{R}$. Let $\psi_0 : \overline{R} \to R_E$ denote the k-isomorphism defined by $\overline{x} \mapsto x$ and $\overline{y} \mapsto y$. Then, as in the proof of Proposition 15.8, $\overline{R^*}$ is the $(\overline{x})$-adic completion of $\overline{R}$. Thus ψ_0 extends to an isomorphism $\psi : \overline{R^*} \to (R_E)^*$ that agrees with ψ_0 on $\overline{R}$ and such that $\psi(\overline{\tau}) = \tau$. Then $\psi(\overline{f}) = \psi(\overline{y} \cdot \overline{\tau} + \overline{z} \cdot \overline{\sigma}) = y\tau$, which is the transcendental element g used in the construction of E. Thus ψ is an isomorphism from $\overline{B} = B/zB$ to E, the ring constructed in Example 14.9. □

PROPOSITION 16.6. *In Example* 16.4, *let* $Q := (y,z)R^* \cap B$. *Then*

(1) *There exist infinitely many height-two prime ideals of* B *not contained in* Q *and each of these prime ideals is contracted from* R^*.
(2) *For certain height-one prime ideals* $\mathfrak{p}$ *contained in* Q, *there exist infinitely many height-two prime ideals between* $\mathfrak{p}$ *and* Q *that are contracted from* R^*, *and infinitely many that are not contracted from* R^*. *Hence the map* $\operatorname{Spec} R^* \to \operatorname{Spec} B$ *is not surjective.*

PROOF. For item 1, since $x \notin Q$ and $B/xB \cong R/xR$ is a Noetherian ring of dimension two, there are infinitely many height-two prime ideals of B containing xB; see Exercise 6 of Chapter 2. This proves there are infinitely many height-two prime ideals of B not contained in Q. If P is a height-two prime ideal of B not contained in Q, then $\operatorname{ht}(\mathfrak{m}_B/P) = 1$, by Theorem 16.2.5, and so, by Proposition 5.17.5, P is contracted from R^*. This completes the proof of item 1.

For item 2 we show that $\mathfrak{p} = zB$ has the stated properties. By Proposition 16.5, the ring B/zB is isomorphic to the ring called B in Example 14.9. For convenience we relabel the ring of Example 14.9 as B'. By Theorem 14.3, B' has exactly one non-finitely generated prime ideal, which we label Q', and $\operatorname{ht} Q' = 2$. It follows that $Q/zB = Q'$. By Discussion 14.10, there are infinitely many height-one prime ideals contained in Q' of Type II (that is, prime ideals that are contracted from R^*/zR^*) and infinitely many height-one prime ideals contained in Q' of Type III (that is, prime ideals that are not contracted from R^*/zR^*). The preimages in B of these prime ideals are height-two prime ideals of B that are contained in Q and contain zB. It follows that there are infinitely many contracted from R^* and there are infinitely many not contracted from R^*, as desired for item 2. □

Proposition 16.7 concerns the non-flat locus and the height-one prime ideals of the ring B of Example 16.4.

PROPOSITION 16.7. *Assume the notation of Example* 16.4. *Then:*

(1) *The ideal* $F_1 := (y,z)R^*[1/x]$ *defines the non-flat locus of the map* $B \hookrightarrow R^*[1/x]$.
(2) *If* $\mathfrak{p}$ *is a height-one prime ideal of* R^*, *then* $\operatorname{ht}(\mathfrak{p} \cap B) \leq 1$.
(3) *If* w *is a prime element of* B, *then* $wR^* \cap B = wB$.

PROOF. Item 1 follows from Theorem 10.12.2.

Item 2 is clear if $\mathfrak{p} = xR^*$. Let $\mathfrak{p}$ be a height-one prime of R^* other than xR^*. Since $\mathfrak{p}$ does not contain $(y,z)R^*$ and does not contain x, the map $B_{\mathfrak{p}\cap B} \hookrightarrow (R^*)_{\mathfrak{p}}$ is faithfully flat. Thus $\text{ht}(\mathfrak{p} \cap B) \le 1$. This establishes item 2.

Item 3 is clear if $wB = xB$. Assume that $wB \ne xB$ and let $\mathfrak{p}$ be a height-one prime ideal of R^* that contains wR^*. Then $\mathfrak{p}R^*[1/x] \cap R^* = \mathfrak{p}$, and by item 2, $\mathfrak{p} \cap B$ has height at most one. Now $wB \subseteq wR^* \cap B \subseteq \mathfrak{p} \cap B$. Thus item 3 follows. □

We display a partial picture of Spec(B) and make comments about the diagram.

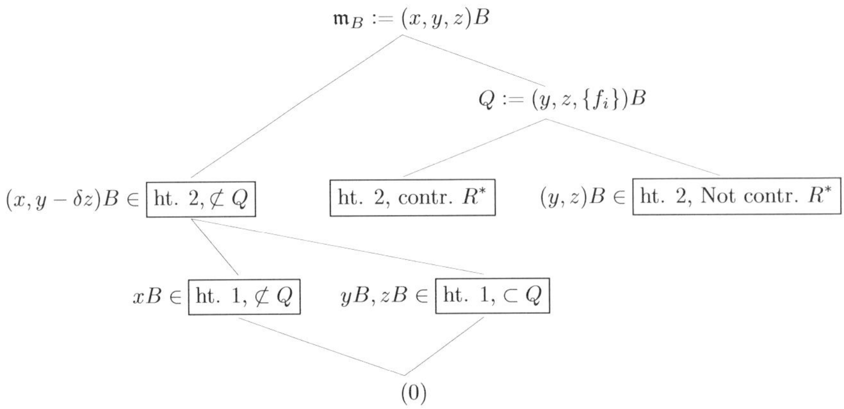

Diagram 16.6.0

Comments on Diagram 16.6.0. The abbreviation "contr. R^*" means "contracted from R^*". A line going from a box at one level to a box at a higher level indicates that every prime ideal in the lower level box is contained in at least one prime ideal in the higher level box. By Theorem 16.2.8, B has no maximal saturated chain of length 2. For $P \in \operatorname{Spec} B$, $x \in P \implies P \not\subseteq Q$. The two lines connecting levels 1 and 2 of the diagram are justified by Remarks 16.8.2: For every pair of prime elements g and h of B with $g \in Q$ and $h \notin Q$, there is a height-two prime ideal P of B that contains both gB and hB, but P is not contained in Q. There are no lines connecting the lower level righthand box to higher boxes that are contained in Q because we are uncertain about what inclusion relations exist for these prime ideals. We discuss this situation in Remarks 16.11.

REMARKS 16.8. Assume the notation of Example 16.4.

(1) By Theorem 16.6.8, the localization $B[1/x]$ has a unique maximal ideal $QB[1/x] = (y,z,f)B[1/x]$ of height three and all the other maximal ideals have height two. There are infinitely many maximal ideals of $B[1/x]$ of height two, such as $(y, z-\beta x^i)B[1/x]$, for nonzero $\beta \in k$ and $i \in \mathbb{N}$, because every $P_{\beta,i} := (y, z-\beta x^i)B$ is a height-two prime ideal, by Proposition 5.17.2, and P is not contained in Q,

(2) Let $p, q \in \mathfrak{m}_B$ be prime elements of B such that $q \in Q$ and $p \notin Q$. We claim that there exists a height-two prime ideal P of B such that $(p,q)B \subseteq P$. Let P be a prime ideal of B that is minimal over $(p,q)B$. Since $p \notin Q$ and $qB \subseteq Q$, it follows that $P \ne Q$ and $1 < \text{ht}\, P$. Also $\text{ht}\, P \ne 3$. If $\text{ht}\, P = 2$, for some prime ideal $P \in \operatorname{Spec} B$, there is nothing to prove. Suppose that every prime ideal of B that is minimal over $(p,q)B$ has height 4. Then $(p,q)B$ is primary for

$\mathfrak{m}_B$. By Proposition 5.17.5.b, $(p,q)R^*$ is $\mathfrak{m}_{R^*}$-primary. But R^* is Noetherian and $\operatorname{ht}\mathfrak{m}_{R^*} = 3$, and so this would contradict Krull Altitude Theorem 2.23. Hence there is a prime ideal P containing $(p,q)B$ having height two.

(3) Define

$$C_n := \frac{B_n}{(y,z)B_n} \quad \text{and} \quad C := \frac{B}{(y,z)B}.$$

Let $P = (y,z)R^*[1/x]$. Since $x \notin P$ and $B[1/x]$ is a localization of $S = R[f]$, it follows that $S_{P\cap S} = B_{P\cap B}$. Similarly $S_{P\cap S} = (B_n)_{P\cap B_n}$. It follows that $(y,z)B \cap B_n = (y,z)B_n$. Therefore $C = \bigcup_{n=0}^{\infty} C_n$.

We show that C is a rank 2 valuation domain with principal maximal ideal generated by the image of x. For each positive integer n, let $g_n \in C_n$ denote the image in C_n of the element $f_n \in B_n$ and let x denote the image of x. Then $C_n = k[x, g_n]_{(x,g_n)}$ is a 2-dimensional RLR. By Equation 16.4.1, $f_n = xf_{n+1} + x(c_n y + d_n z)$. It follows that $g_n = xg_{n+1}$ for each $n \in \mathbb{N}$. Thus C is an infinite directed union of local quadratric transforms of 2-dimensional regular local rings. Hence C is a valuation domain of dimension at most 2 by [**2**, Remark 2, p. 332 and Lemma 12, p.337]. By items 2 and 4 of Proposition 16.6, $\dim C \geq 2$, and therefore C is a valuation domain of rank 2. The maximal ideal of C is xC.

By Proposition 16.5, $B/zB \cong E$, where E is the ring B of Example 14.9. By an argument similar to that of Proposition 15.8 and by Proposition 16.5, we see that the above ring C is isomorphic to E/yE.

QUESTION 16.9. Let B be the ring constructed in Example 16.4: Is Q the only prime ideal of B that is not finitely generated?

Proposition 16.6 implies that the only possible non-finitely generated prime ideals of B other than Q have height two. We do not know whether every height-two prime ideal of B is finitely generated. We show in Theorem 16.10 that certain of the height-two prime ideals of B are finitely generated.

THEOREM 16.10. *Assume the notation of Example 16.4. Let w be a prime element of B. Then*

(1) *B/wB is Noetherian if and only if $w \notin Q$.*
(2) *Every ideal of B that is not contained in Q is finitely generated.*
(3) *If $w \in (y,z)k[x,y,z]$ and w has a nontrivial degree one term in y or z, then Q/wB is the unique non-finitely generated prime ideal of B/wB, and so Q is the unique non-finitely generated prime ideal of B that contains w.*

PROOF. For item 1, if $w \in Q$, then B/wB is not Noetherian since Q is not finitely generated. Assume $w \notin Q$. Since B/xB is known to be Noetherian, we may assume that $wB \neq xB$. By Proposition 16.7.1, $QR^*[1/x] = (y,z)R^*[1/x]$ defines the non-flat locus of $\varphi : B \to R^*[1/x]$. Since $wR^*[1/x] + (y,z)R^*[1/x] = R^*[1/x]$, Theorem 6.15 with $I = wB$ and $z = x$ implies that B/wB is Noetherian.

For item 2, we use that every non-finitely generated ideal is contained in an ideal maximal with respect to not being finitely generated and the latter ideal is prime. Thus it suffices to show every prime ideal P not contained in Q is finitely generated. If $P \not\subseteq Q$, then, since B is a UFD, there exists a prime element $w \in P \setminus Q$. By item 1, B/wB is Noetherian, and so P is finitely generated.

For item 3, w is a prime element of R and of R^*. Let $^-$ denote image under the canonical map $\pi : R^* \to R^*/wR^*$. We may assume that w is linear in z, that the coefficient of z is 1 and therefore that $w = z - yg(x,y)$, where $g(x,y) \in k[x,y]$. Thus $\overline{R} \cong k[x,y]_{(x,y)}$. By Proposition 15.8, $\overline{B}$ is the approximation domain over $\overline{R}$ with respect to the transcendental element

$$\overline{f} = \overline{y} \cdot \overline{\tau} + \overline{z} \cdot \overline{\sigma} = \overline{y} \cdot \overline{\tau} + \overline{y} \cdot \overline{g(x,y)} \cdot \overline{\sigma}.$$

The setting of Theorem 5.24 applies with B replaced by $\overline{B}$, the underlying ring R replaced by $\overline{R}$, and $z = \overline{x}$. Thus the ring $\overline{B}$ is a UFD, and so every height-one prime ideal of $\overline{B}$ is principal. Since $w \in Q$ and Q is not finitely generated, it follows that $\operatorname{ht}(\overline{Q}) = 2$ and that $\overline{Q}$ is the unique non-finitely generated prime ideal of $\overline{B}$. Hence the theorem holds. □

REMARKS 16.11. It follows from Proposition 5.17.5 that every height-two prime ideal of B that is not contained in Q is contracted from a prime ideal of R^*. As we state in Proposition 16.6.2, there are infinitely many height-two prime ideals of B that are contained in Q and are contracted from R^* and there are infinitely many height-two prime ideals of B that are contained in Q and are *not* contracted from R^*. Indeed, among the height-two prime ideals between zB and Q, there are infinitely many contracted from R^* and infinitely many that are not contracted from R^*, by Proposition 16.5. A similar statement holds with zB replaced by yB.

Since B_Q is a three-dimensional regular local ring, for each height-one prime ideal $\mathfrak{p}$ of B with $\mathfrak{p} \subset Q$, the set

$$\mathcal{S}_{\mathfrak{p}} = \{P \in \operatorname{Spec} B \mid \mathfrak{p} \subset P \subset Q \text{ and } \operatorname{ht} P = 2\}$$

is infinite. The infinite set $\mathcal{S}_{\mathfrak{p}}$ is the disjoint union of the sets $\mathcal{S}_{\mathfrak{p}c}$ and $\mathcal{S}_{\mathfrak{p}n}$, where the elements of $\mathcal{S}_{\mathfrak{p}c}$ are contracted from R^* and the elements of $\mathcal{S}_{\mathfrak{p}n}$ are not contracted from R^*.

We do not know whether there exists a height-one prime ideal $\mathfrak{p}$ contained in Q having the property that one of the sets $\mathcal{S}_{\mathfrak{p}c}$ or $\mathcal{S}_{\mathfrak{p}n}$ is empty. Furthermore if one of these sets is empty, which one is empty? If there are some such height-one prime ideals $\mathfrak{p}$ with one of the sets $\mathcal{S}_{\mathfrak{p}c}$ or $\mathcal{S}_{\mathfrak{p}n}$ empty, which height-one primes are they? It would be interesting to know the answers to these questions.

16.3. $D + M$ constructions, coherence, finite conductor domains

Related to Example 16.4, Evan Houston raised the question: "How does this example compare to a ring constructed using the three-dimensional ring of Example 14.9 and applying the popular "$D + M$" technique of multiplicative ideal theory?" In this section, we compare the two constructions in Example 16.13, and we also give connections to other concepts.

REMARK 16.12. The "$D+M$" construction involves an integral domain D and a prime ideal M of an extension domain E of D such that $D \cap M = (0)$. Then $D + M = \{a + b \mid a \in D,\ b \in M\}$. Moreover, for $a, a' \in D$ and $b, b' \in M$, if $a + b = a' + b'$, then $a = a'$ and $b = b'$. For more information on the $D + M$ construction, see for example the work of Gilmer in [**52**, p. 95], [**54**] or the paper of Brewer and Rutter [**25**].

Gilmer in [**53**, p. 583] remarks that the first use of the $D + M$ construction seems to be by Krull to give an example of a one-dimensional integrally closed local domain that is not a valuation domain [**106**, p. 670]. An earlier related construction by Prüfer gives an example of an integrally closed domain that is not completely integrally closed [**150**, p. 19]. Seidenberg uses the $D+M$ construction in his study of the dimension theory of polynomial rings [**166**].

Since D embeds in E/M, the ring $D+M$ is a pullback as in the paper of Gabelli and Houston [**55**] or the book of Leuschke and R. Wiegand [**110**, p. 42]. That is, we have the following commutative diagram:

$$\begin{array}{ccc} R := D+M & \hookrightarrow & E \\ \downarrow & & \downarrow \\ D = R/M & \hookrightarrow & E/M \end{array}$$

Exercise 14.4 includes properties of $D+M$ constructions and gives an outline for an example.

In Example 16.13, we consider a "$D+M$" construction that contrasts nicely with Example 16.4.

EXAMPLE 16.13. Let $(B, \mathfrak{m}_B)$ be the ring of Example 14.9. Thus k is a coefficient field of B and $B = k + \mathfrak{m}_B$. Assume the field k is the field of fractions of a DVR V, and let t be a generator of the maximal ideal of V. Define

$$C \ := \ V \ + \ \mathfrak{m}_B \ = \ \{\, a \ + \ b \mid a \ \in \ V, \ b \ \in \ \mathfrak{m}_B \,\}.$$

The integral domain C has the following properties:

(1) The maximal ideal $\mathfrak{m}_B$ of B is also a prime ideal of C, and $C/\mathfrak{m}_B \cong V$.
(2) C has a unique maximal ideal $\mathfrak{m}_C$; moreover, $\mathfrak{m}_C = tC$.
(3) $\mathfrak{m}_B = \bigcap_{n=1}^{\infty} t^n C$, and $B = C_{\mathfrak{m}_B} = C[1/t]$.
(4) Each $P \in \operatorname{Spec} C$ with $P \neq \mathfrak{m}_C$ is contained in $\mathfrak{m}_B$, and $P \in \operatorname{Spec} B$.
(5) $\dim C = 4$ and C has a unique prime ideal of height h, for $h = 2, 3$ or 4.
(6) $\mathfrak{m}_C$ is the only nonzero prime ideal of C that is finitely generated. Indeed, every nonzero proper ideal of B is an ideal of C that is not finitely generated.

Thus C is a non-Noetherian non-catenary four-dimensional local domain.

PROOF. Since C is a subring of B, $\mathfrak{m}_B \cap V = (0)$ and $V\mathfrak{m}_B = \mathfrak{m}_B$, item 1 holds. We have $C/(tV + \mathfrak{m}_B) = V/tV$. Thus $tV + \mathfrak{m}_B$ is a maximal ideal of C. Let $b \in \mathfrak{m}_B$. Since $1 + b$ is a unit of the local ring B, we have

$$\frac{1}{1+b} \ = \ 1 - \frac{b}{1+b} \quad \text{and} \quad \frac{b}{1+b} \ \in \ \mathfrak{m}_B.$$

Hence $1+b$ is a unit of C. Let $a+b \in C \setminus (tV + \mathfrak{m}_B)$, where $a \in V \setminus tV$ and $b \in \mathfrak{m}_B$. Then a is a unit of V and thus a unit of C. Moreover, $a^{-1}(a+b) = 1 + a^{-1}b$ and $a^{-1}b \in \mathfrak{m}_B$. Therefore $a+b$ is a unit of C. We conclude that $\mathfrak{m}_C := tV + \mathfrak{m}_B$ is the unique maximal ideal of C. Since t is a unit of B, we have $\mathfrak{m}_B = t\mathfrak{m}_B$. Hence $\mathfrak{m}_C = tV + \mathfrak{m}_B = tC$. This proves item 2.

For item 3, since t is a unit of B, we have $\mathfrak{m}_B = t^n\mathfrak{m}_B \subseteq t^n C$ for all $n \in \mathbb{N}$. Thus $\mathfrak{m}_B \subseteq \bigcap_{n=1}^{\infty} t^n C$. If $a + b \in \bigcap_{n=1}^{\infty} t^n C$ with $a \in V$ and $b \in \mathfrak{m}_B$, then

$$b \ \in \ \bigcap_{n=1}^{\infty} t^n C \ \implies a \ \in \ (\bigcap_{n=1}^{\infty} t^n C) \cap V \ = \ \bigcap_{n=1}^{\infty} t^n V \ = \ (0).$$

Hence $\mathfrak{m}_B = \bigcap_{n=1}^{\infty} t^n C$. Again using $t\mathfrak{m}_B = \mathfrak{m}_B$, we obtain

$$C[1/t] \ = \ V[1/t] \ + \ \mathfrak{m}_B \ = \ k \ + \ \mathfrak{m}_B \ = \ B.$$

Since $t \notin \mathfrak{m}_B$, we have $B = C[1/t] \subseteq C_{\mathfrak{m}_B} \subseteq B_{\mathfrak{m}_B} = B$. This proves item 3.

Item 4 follows from part c of Exercise 14.4 and from Exercise 16.1. Item 5 now follows from item 4 and the structure of Spec B.

For item 6, let J be a nonzero proper ideal of B. Since t is a unit of B, we have $J = tJ$. This implies by Nakayama's Lemma that J as an ideal of C is not finitely generated; see [**25**, Lemma 1]. Thus item 6 follows from item 4.

By item 6, C is non-Noetherian. Since $(0) \subsetneq xB \subsetneq \mathfrak{m}_B \subsetneq tC$ is a saturated chain of prime ideals of C of length 3, and $(0) \subsetneq yB \subsetneq Q \subsetneq \mathfrak{m}_B \subsetneq tC$ is a saturated chain of prime ideals of C of length 4, the ring C is not catenary. □

DEFINITIONS 16.14. An integral domain R is said to be a *finite conductor domain* if for elements a, b in the field of fractions of R the R-module $aR \cap bR$ is finitely generated. This concept is considered in the paper of McAdam [**123**].

A ring R is said to be *coherent* if every finitely generated ideal of R is finitely presented. By a theorem of Chase [**32**, Theorem 2.2], this condition is equivalent to each of the following:

(1) For each finitely generated ideal I and element a of R, the ideal $(I :_R a) = \{b \in R \mid ba \in I\}$ is finitely generated.
(2) For each $a \in R$ the ideal $(0 :_R a) = \{b \in R \mid ba = 0\}$ is finitely generated, and the intersection of two finitely generated ideals of R is again finitely generated.

A coherent integral domain is a finite conductor domain. Examples of finite conductor domains that are not coherent are given by Glaz in [**58**, Example 4.4] and by Olberding and Saydam in [**145**, Proposition 3.7]. On the other hand, by a result of Brewer and Rutter [**25**, Prop. 2], the ring of Example 16.13 is not a finite conductor domain and thus is not coherent.

REMARK 16.15. A theorem due to Rotthaus and Sega implies that the approximation domains B in the setting of Theorems 14.3, 15.11 and 16.2 are coherent and regular [**161**].[1] That is, they are coherent and every finitely generated submodule of a free module over B has a finite free resolution. For the ring $B = \bigcup_{n=1}^{\infty} B_n$ of these constructions, it is stated in [**161**] that $B_n[1/x] = B_{n+k}[1/x] = B[1/x]$ and that B_{n+k} is generated over B_n by a single element for all positive integers n and k. This is not correct for the local rings B_n. However, if instead of asserting these statements for the localized polynomial rings B_n and their union B of the construction, one makes the statements for the underlying polynomial rings U_n and their union U defined in Equation 5.4.5, or those defined in Examples 14.1, then one does have that $U_n[1/x] = U_{n+k}[1/x] = U[1/x]$ and that U_{n+k} is generated over U_n by a single element for all positive integers n and k; see Theorem 5.14.4 and Equation 5.4.4. Thus the Rotthaus-Sega argument yields that U and its localization B are coherent. Hence the rings of Examples 14.1, 15.10, and 16.4 are coherent.

Bruce Olberding pointed out to us that every non-finitely generated prime ideal P of either the ring B of Example 14.1 or of Example 16.4 is not an associated prime of a finitely generated ideal I of B; that is, $P \neq (I : a)$, where $a \in B$ and I is a

[1]The result of Rotthaus and Sega is more general than the settings of Chapters 14-16: Let $r, n \in \mathbb{N}$, let $x, y_1 \dots, y_r$ be indeterminates over a field k, let $\tau_1, \dots, \tau_n$ be algebraically independent elements of $xk[[x]]$, and let $R := k[x, y_1 \dots, y_r]_{(x, y_1 \dots, y_r)}$. They show every approximation domain constructed with Insider Construction 10.7 that uses just one polynomial $f \in R[\tau_1, \dots, \tau_n]$ (so that $m = 1$, for the m of Insider Construction 10.7) is coherent regular.

finitely generated ideal of B. By Definition 16.14.1, if B is coherent and $P = (I : a)$, where $a \in B$ and I is a finitely generated ideal of B, then P is also finitely generated.

Exercises

(1) As in Remarks 16.12, let B be an integral domain of the form $B = K + M$, where M is a maximal ideal of B and K is a field. Let D be an integral domain with field of fractions K, and let $C = D + M$, a $D + M$ construction as in Example 16.13. If $P \in \operatorname{Spec} C$ and $P \subseteq M$, prove that $P \in \operatorname{Spec} B$.

Suggestion. Notice that $B = (D \setminus \{0\})^{-1}C$. Show that $PB = P$.

(2) Let K denote the field of fractions of the integral domain B of Example 14.9, let t be an indeterminate over K and let V denote the DVR $K[t]_{(t)}$. Let M denote the maximal ideal of V. Thus $V = K + M$. Define $C := B + M$. Show that the integral domain C has the following properties:
(a) $\mathfrak{m}_B C$ is the unique maximal ideal of C, and is generated by two elements.
(b) For every nonzero element $a \in \mathfrak{m}_B$, we have $M \subset aC$.
(c) M is the unique prime ideal of C of height one; moreover M is not finitely generated as an ideal of C.
(d) $\dim C = 4$ and C has a unique prime ideal of height h, for $h = 1, 3$ or 4.
(e) For each $P \in \operatorname{Spec} C$ with $\operatorname{ht} P \geq 2$, the ring C_P is not Noetherian.
(f) C has precisely two prime ideals that are not finitely generated.
(g) C is non-catenary.

(3) Let $C = V + \mathfrak{m}_B$ be as in Example 16.13. Assume that V has a coefficient field L, and that L is the field of fractions of a DVR V_1. Define $C_1 := V_1 + tC$. Let s be a generator of V_1. Show that the integral domain C_1 has the following properties:
(a) The maximal ideal $\mathfrak{m}_C$ of C is also a prime ideal of C_1, and $C_1/\mathfrak{m}_C \cong V_1$.
(b) The principal ideal sC_1 is the unique maximal ideal of C_1.
(c) $\mathfrak{m}_C = \bigcap_{n=1}^{\infty} s^n C_1$, and $C = C_1[1/s]$.
(d) Each $P \in \operatorname{Spec} C_1$ with $P \neq sC_1$ is contained in $\mathfrak{m}_C$; thus $P \in \operatorname{Spec} C$.
(e) $\dim C_1 = 5$.
(f) C_1 has a unique prime ideal of height h for $h = 2, 3, 4$, or 5.
(g) The maximal ideal of C_1 is the only nonzero prime ideal of C_1 that is finitely generated. Indeed, every nonzero proper ideal of C is an ideal of C_1 that is not finitely generated.
(h) C_1 is non-catenary.

Comment. For item h, exhibit two saturated chains of prime ideals from (0) to sC_1 of different lengths.

CHAPTER 17

The Homomorphic Image Construction

This chapter contains a description and analysis of Homomorphic Image Construction 17.2, a more complex construction than Inclusion Construction 5.3. Construction 17.2 leads to more sophisticated examples. In both constructions, R is an integral domain and R^* is the ideal-adic completion of a principal ideal of R.

- *Inclusion Construction* 5.3 defines an Intersection Domain A inside R^*; thus $A = A_{\text{inc}} := R^* \cap L$ for a subfield L of the total quotient ring of R^*.
- *Homomorphic Image Construction* 17.2 yields an Intersection Domain A inside a homomorphic image R^*/I of R^*; here $A = A_{\text{hom}} := (R^*/I) \cap K$, where I is an ideal of R^* such that $P \cap R = (0)$ for each associated prime P of I, and K is the field of fractions of R.

Homomorphic Image Construction 17.2 is defined in Section 17.1. Section 17.2 contains the construction of an Approximation Domain for Construction 17.2. Construction Properties Theorem 17.11, Noetherian Flatness Theorem 17.13 and Weak Flatness Theorem 17.19 are proved in Sections 17.3-17.5; they are Homomorphic Image versions of theorems from earlier chapters for Inclusion Construction 5.3.

Theorem 17.17 relates Homomorphic Image Construction 17.2 for an ideal I to the resulting construction if a power of I replaces I. In Section 17.6, Inclusion Construction 5.3 is identified with a special case of Homomorphic Image Construction 17.2, in such a way that Approximation Domains for Inclusion Construction 5.3 correspond to Approximation Domains fitting the Homomorphic Image format of Section 17.2. A connection between the two constructions is established in Section 17.7. Homomorphic Image Construction 17.2 is used with the Local Prototypes of Definition 4.28 for the construction of non-catenary examples in Chapter 18.

17.1. Two construction methods and a picture

Setting 17.1 for Homomorphic Image Construction 17.2 is the same as Setting 5.1; we repeat it here for convenience. This facilitates comparison with Inclusion Construction 5.3.

SETTING 17.1. Let R be an integral domain with field of fractions $K := \mathcal{Q}(R)$. Assume $x \in R$ is a nonzero nonunit such that $\bigcap_{n\geq 1} x^n R = (0)$, the x-adic completion R^* is Noetherian, and x is a regular element of R^*.

CONSTRUCTION 17.2. *Homomorphic Image Construction*: With R, x and R^* as in Setting 17.1, let I be an ideal of R^* such that $P \cap R = (0)$ for each $P \in \operatorname{Spec} R^*$ that is associated to I. Define the *Intersection Domain* $A = A_{\text{hom}} := K \cap (R^*/I)$. The ring A_{hom} is a subring of a homomorphic image of R^* and is a birational extension of R.

NOTE 17.3. The condition in Construction 17.2, that $P \cap R = (0)$ for every prime ideal P of R^* that is associated to I, implies that the field of fractions K of R embeds in the total quotient ring $\mathcal{Q}(R^*/I)$ of R^*/I. To see this, observe that the canonical map $R \to R^*/I$ is an injection and that regular elements of R remain regular as elements of R^*/I. In this connection see Exercise 1 of this chapter.

We summarize Inclusion Construction 5.3, relabeled as Construction 17.4, for easy reference and comparison to Homomorphic Image Construction 17.2.

CONSTRUCTION 17.4. (Inclusion Construction 5.3): Assume Setting 17.1. Let $\tau_1, \dots, \tau_s \in xR^*$ be algebraically independent elements over R such that $K(\tau_1, \dots, \tau_s) \subseteq \mathcal{Q}(R^*)$. This inclusion implies that nonzero elements of $K[\tau_1, \dots, \tau_s]$ are units in $\mathcal{Q}(R^*)$ and therefore are regular elements of R^*. The *Intersection Domain* $A = A_{\text{inc}} := K(\tau_1, \dots, \tau_s) \cap R^*$.

In Construction 17.2, the Intersection Domain A_{hom} is an integral domain that is birational over R and is a subring of a homomorphic image of a power series extension of R. The Intersection Domain A_{inc} associated with Inclusion Construction 17.4 is an integral domain that is not algebraic over R and is a subring of a power series extension of R.

PICTURE 17.5. The diagram below shows the relationships among these rings.

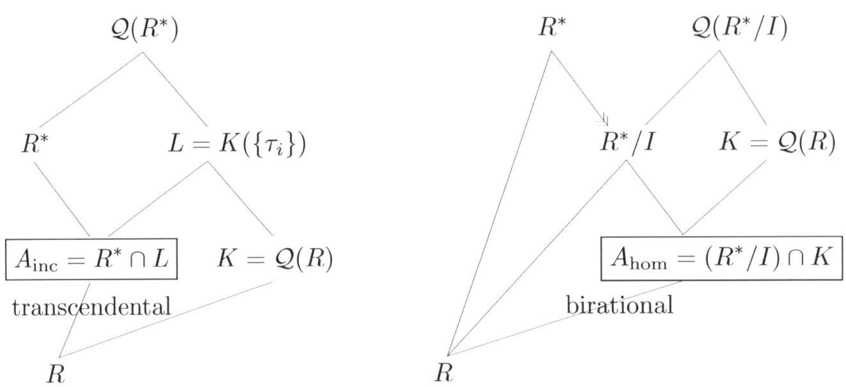

(17.4) $A := L \cap R^*$ (17.2) $A := K \cap (R^*/I)$

REMARKS 17.6. Homomorphic Image Construction 17.2 is widely applicable. If a Noetherian local domain R is essentially finitely generated over a field, then there often exist ideals I in the completion $\widehat{R}$ of R, or in an ideal-adic completion R^* of R, such that the intersection domain $\mathcal{Q}(R) \cap (\widehat{R}/I)$, or $\mathcal{Q}(R) \cap (R^*/I)$, is a Noetherian local domain that birationally dominates R; see Theorem 4.2. Construction 17.2 may be used to describe Example 4.15 of Nagata, Christel's Example 4.17, and other examples given by Brodmann and Rotthaus, Heitmann, Ogoma and Weston, [**26**], [**27**], [**95**], [**146**], [**147**], and [**183**].

While Inclusion Construction 17.4 is simpler, Homomorphic Image Construction 17.2 has more flexibility and yields examples that are not possible with Construction 17.4. Construction 17.4 is not sufficient to obtain certain types of rings such as Ogoma's celebrated example [**146**] of a normal non-catenary Noetherian local domain. As Theorem 6.27 shows, the universally catenary property holds for every Noetherian ring constructed using Inclusion Construction 17.4 over a Noetherian universally catenary local domain R.

Remark 17.16 and Example 18.15 show that examples constructed with Homomorphic Image Construction 17.2 may result in a non-catenary Noetherian local domain even if the base domain is universally catenary, Noetherian and local. Example 18.15 is a Noetherian local domain with geometrically regular formal fibers that is not universally catenary.

17.2. Approximations for Homomorphic Image Construction

The approximation methods in this chapter describe a subring B inside the constructed Intersection Domain A of Construction 17.2. This subring is useful for describing A.

The Approximation Domain B for Construction 17.2 is a nested union of birational extensions of R that are essentially finitely generated R-algebras. As with the Approximation Domain for Inclusion Construction 17.4 from Definition 5.7, we approach A using a sequence of "approximation rings" over R. We use the *frontpieces* of the power series involved, rather than the endpieces that are used for the approximations in Inclusion Construction 17.4.

A goal of these computations is to prove Noetherian Flatness Theorem 17.13 for Homomorphic Image Construction 17.2.

FRONTPIECE NOTATION 17.7. Let R be an integral domain with field of fractions $K := \mathcal{Q}(R)$. Let $x \in R$ be a nonzero nonunit such that $\bigcap_{n\geq 1} x^n R = (0)$, the x-adic completion R^* is Noetherian, and x is a regular element of R^*. Let I be an ideal of R^* such that $P \cap R = (0)$, for each $P \in \operatorname{Spec} R^*$ that is associated to R^*/I. As in Construction 17.2, define $A = A_{\text{hom}} := K \cap (R^*/I)$.

Since $I \subset R^*$, each $\gamma \in I$ has an expansion as a power series in x over R,

$$\gamma := \sum_{i=0}^{\infty} a_i x^i, \quad \text{where} \quad a_i \in R.$$

For each positive integer n we define the nth *frontpiece* γ_n of γ with respect to this expansion:

$$\gamma_n := \sum_{j=0}^{n} \frac{a_j x^j}{x^n}.$$

Thus, if $I := (\sigma_1, \dots, \sigma_t)R^*$, then for each σ_i

$$\sigma_i := \sum_{j=0}^{\infty} a_{ij} x^j, \qquad \text{where the} \quad a_{ij} \in R,$$

and the nth frontpiece σ_{in} of σ_i is

$$\sigma_{in} := \sum_{j=0}^{n} \frac{a_{ij} x^j}{x^n} \in K. \tag{17.7.1}$$

For Homomorphic Image Construction 17.2, we obtain approximating rings as follows: We define

$$U_n := R[\sigma_{1n}, \dots, \sigma_{tn}], \qquad \text{and } B_n := (1 + xU_n)^{-1} U_n. \tag{17.7.2}$$

The rings U_n and B_n are subrings of K. Proposition 17.9 implies that they may also be considered to be subrings of R^*/I. First we show in Proposition 17.8 that the approximating rings U_n and B_n are nested.

PROPOSITION 17.8. *With the setting of Frontpiece Notation 17.7, for each integer $n \geq 0$ and for each integer i with $1 \leq i \leq t$:*

(1) $\sigma_{in} = -xa_{i,n+1} + x\sigma_{i,n+1}$.
(2) $(x, \sigma_i)R^* = (x, a_{i0})R^*$ *and hence* $(x, I)R^* = (x, a_{10}, \dots, a_{t0})R^*$.
(3) $(x, \sigma_i)R^* = (x, x^n\sigma_{in})R^*$ *and hence* $(x, I)R^* = (x, x^n\sigma_{1n}, \dots, x^n\sigma_{tn})R^*$.

Thus $R \subseteq U_0$, $U_n \subseteq U_{n+1}$ and $B_n \subseteq B_{n+1}$, for each positive integer n.

PROOF. For item 1, Definition 17.7.1 implies $\sigma_{i,n+1} := \sum_{j=0}^{n+1} \frac{a_{ij}x^j}{x^{n+1}}$. Thus

$$x\sigma_{i,n+1} = \sum_{j=0}^{n+1} \frac{a_{ij}x^{j+1}}{x^{n+1}} = \sum_{j=0}^{n} \frac{a_{ij}x^j}{x^n} + xa_{i,n+1} = \sigma_{in} + xa_{i,n+1}.$$

For item 2, by definition

$$\sigma_i := \sum_{j=0}^{\infty} a_{ij}x^j = a_{i0} + x\Big(\sum_{j=1}^{\infty} a_{ij}x^{j-1}\Big).$$

For item 3, the following equation in R^* holds:

$$\sigma_i = \sum_{j=0}^{\infty} a_{ij}x^j = x^n\sigma_{in} + x^{n+1}\Big(\sum_{j=n+1}^{\infty} a_{ij}x^{j-n-1}\Big),$$

since $x^n\sigma_{in} \in R$. The asserted inclusions follow from this equation. □

Proposition 17.9 shows that the frontpieces of each of the power series σ_i generating the ideal I are elements of R^*/I, and of K, and thus of the Intersection Domain A of Homomorphic Image Construction 17.2. Even though they appear different, Proposition 17.9 also shows that the nth frontpiece in Frontpiece Notation 17.7 of each σ_i is, modulo the ideal I, the same as the negative of the nth endpiece of σ_i from Endpiece Notation 5.4. Equivalently, if the nth frontpiece and the nth endpiece of σ_i are both considered as elements of R^*, then their images in R^*/I are equal.

PROPOSITION 17.9. *Assume the setting of Frontpiece Notation 17.7 and let n be a positive integer. As an element of R^*/I, the nth frontpiece σ_{in} is the image in R^*/I of the negative of the nth endpiece of σ_i defined in Endpiece Notation 5.4, that is, for $\sigma_i := \sum_{j=0}^{\infty} a_{ij}x^j$, where each $a_{ij} \in R$,*

$$\sigma_{in} = -\sum_{j=n+1}^{\infty} \frac{a_{ij}x^j}{x^n} = -\sum_{j=n+1}^{\infty} a_{ij}x^{j-n} \qquad (\text{mod } I).$$

Itfollows that $\sigma_{in} \in K \cap (R^*/I)$, and so U_n and B_n are subrings of A and of R^*/I.

PROOF. Let π denote the natural homomorphism from R^* onto R^*/I. Since the restriction of π to R is the identity map on R, $x^n\pi(\sigma_{in}) = \pi(x^n\sigma_{in}) = x^n\sigma_{in}$;

also $\pi(\sigma_i) = 0$, and so

$$\sigma_i = x^n\sigma_{in} + \sum_{j=n+1}^{\infty} a_{ij}x^j \implies x^n\sigma_{in} = \sigma_i - \sum_{j=n+1}^{\infty} a_{ij}x^j$$
$$\implies x^n\pi(\sigma_{in}) = \pi(\sigma_i) - \pi(\sum_{j=n+1}^{\infty} a_{ij}x^j)$$
$$\implies x^n\pi(\sigma_{in}) = -x^n\pi(\sum_{j=n+1}^{\infty} a_{ij}x^{j-n}).$$

Hence $\pi(\sigma_{in}) = -\pi(\sum_{j=n+1}^{\infty} a_{ij}x^{j-n})$, since x is a regular element of R^*/I. Thus $\sigma_{in} = -\sum_{j=n+1}^{\infty} a_{ij}x^{j-n} \pmod I$. □

DEFINITION 17.10. Assume the setting of Frontpiece Notation 17.7. We define the nested union U, the *Approximation Domain* B and the *Intersection Domain* A:

$$(17.10.1)\quad U := \bigcup_{n=1}^{\infty} U_n, \qquad B := \bigcup_{n=1}^{\infty} B_n = (1+xU)^{-1}U, \qquad A := K \cap (R^*/I).$$

By Remark 3.3.1, the element x is in the Jacobson radical of R^*. By Proposition 17.9, $B \subseteq A$. Construction 17.2 is said to be *limit-intersecting* if $B = A$.

17.3. Basic properties of the Approximation Domains

Construction Properties Theorem 17.11 (Homomorphic Image Version) relates to the analysis of Homomorphic Image Construction 17.2.[1] The proof uses Lemma 5.12 to establish relationships among rings that arise in Homomorphic Image Construction 17.2 and the approximations in Section 17.2.

CONSTRUCTION PROPERTIES THEOREM 17.11. (Homomorphic Image Version) *Let R be an integral domain with field of fractions K. Let $x \in R$ be a nonzero nonunit such that $\bigcap_{n\in\mathbb{N}} x^nR = (0)$, the x-adic completion R^* is Noetherian, and x is a regular element of R^*. Let $I = (\sigma_1, \dots, \sigma_t)R^*$ be an ideal of R^* such that $P \cap R = (0)$ for each $P \in \operatorname{Spec} R^*$ that is associated to R^*/I. Assume the notation of Definition 17.10. Then, for each positive integer n:*

(1) *The ideals of R containing x^n are in one-to-one inclusion preserving correspondence with the ideals of R^* containing x^n. In particular, $(I, x)R^* = (a_{10}, \dots, a_{t0}, x)R^*$, and $(I, x)R^* \cap R = (a_{10}, \dots, a_{t0}, x)R^* \cap R = (a_{10}, \dots, a_{t0}, x)R$.*
(2) *The ideal $(a_{10}, \dots, a_{t0}, x)R$ equals $(x(R^*/I)) \cap R$ under the identification of R as a subring of R^*/I, and the element x is in the Jacobson radical of both R^*/I and B.*
(3) $(x^n(R^*/I)) \cap A = x^nA$, $(x^n(R^*/I)) \cap U = x^nU$, $(x^n(R^*/I)) \cap B = x^nB$.
(4) $U/x^nU = B/x^nB = A/x^nA = R^*/(x^nR^* + I) = R/((x^nR^* + I) \cap R)$. *The rings A, U and B all have x-adic completion R^*/I, that is, $A^* = U^* = B^* = R^*/I$.*
(5) $R[1/x] = U[1/x]$, $U = R[1/x] \cap B = R[1/x] \cap A = R[1/x] \cap (R^*/I)$. *Also:*

[1] When there is no confusion, we omit "Homomorphic Image Version". We also may write "Inclusion Version" for Theorem 5.14 if it seems helpful to distinguish the two versions.

(a) *$B[1/x]$ is a localization of R and of U.*
(b) *The integral domains R, U, B and A all have the same field of fractions K; that is, U, B and A are birational extensions of R.*
(c) *If P is a prime ideal of B such that $x \notin P$, then $R_{P\cap R} = B_P = U_{P\cap U}$.*

(6) *If R is local with maximal ideal $\mathfrak{m}_R$, then*
- *R^*, A, and B are local with maximal ideals $\mathfrak{m}_{R^*} = \mathfrak{m}_R R^*, \mathfrak{m}_A := \mathfrak{m}_R A$ and $\mathfrak{m}_B = \mathfrak{m}_R B$ respectively.*
- *$\widehat{A} = \widehat{U} = \widehat{B} = \widehat{R}/I\widehat{R} = \widehat{(R/I)}$.*

PROOF. The first assertion of item 1 follows because R/x^nR is canonically isomorphic to R^*/x^nR^*. The next assertion of item 1 is Proposition 17.8.2. If $\gamma = \sum_{i=1}^t \sigma_i\beta_i + x\tau \in (I,x)R^* \cap R$, where $\tau, \beta_i \in R^*$, then each $\beta_i = b_i + x\beta_i'$, where $b_i \in R$, $\beta_i' \in R^*$. Thus $\gamma - \sum_{i=1}^t a_{i0}b_i \in xR^* \cap R = xR$, and so $\gamma \in (a_{10}, \dots, a_{t0}, x)R$. Since $x(R^*/I) = (x,I)R^*/I$, it follows that $(a_{10}, \dots, a_{t0}, x)R \subseteq x(R^*/I) \cap R$.

The reverse inclusion in item 2 follows from $(I,x)R^* = (a_{10}, \dots, a_{t0}, x)R^*$. For the last part of item 2, since $x \in \mathcal{J}(R^*)$, the element $1 + ax$ is outside every maximal ideal of R^*, for every $a \in R^*$. Thus $x \in \mathcal{J}(R^*/I)$. By the definition of B in Equation 17.10.1, $x \in \mathcal{J}(B)$.

The first assertion of item 3 follows from the definition of A as $(R^*/I) \cap K$. To see that $x(R^*/I) \cap U \subseteq xU$, let $g \in x(R^*/I) \cap U$. Then $g \in U_n$, for some n, implies $g = r_0 + g_0$, where $r_0 \in R$, $g_0 \in (\sigma_{1n}, \dots, \sigma_{tn})U_n$. Also $\sigma_{in} = -xa_{i,n+1} + x\sigma_{i,n+1}$, and so $g_0 \in xU_{n+1} \subseteq x(R^*/I)$. Now $r_0 \in (x, \sigma_1, \dots, \sigma_t)R^* = (I,x)R^*$. Thus by item 1, $r_0 \in (a_{10}, \dots, a_{t0}, x)R$. Also each $a_{i0} = x\sigma_{i1} - xa_{i1} \in xU$. Thus $r_0 \in xU$, as desired. This proves that $x^n(R^*/I) \cap U = x^nU$. Since $B = (1+xU)^{-1}U$, we also have $x^n(R^*/I) \cap B = x^nB$. Thus item 3 holds.

For item 4, we show that, with $S = A$ and $T = R^*/I$, condition 1 of the four equivalent conditions in Lemma 5.12 holds: that is, (i) $xA = x(R^*/I) \cap A$ and (ii) $A/xA = (R^*/I)/(x(R^*/I))$. Part i holds by item 3. Part i implies that the kernel of the composition ψ of the maps shown

$$\psi : A \overset{\subseteq}{\hookrightarrow} (R^*/I) \to (R^*/I)/(x(R^*/I)) = R^*/(x,I)R^*$$

is xA. By Definitions 3.1, $R + xR^* = R^*$, and so condition 4(ii) of Lemma 5.12

$$R + (x(R^*/I)) \subseteq A + (x(R^*/I)) \implies A + (x(R^*/I)) = R^*/I.$$

Thus the map ψ is surjective, and so part ii holds. Also Item 4 for the ring A follows from statements 2 and 3 of the four equivalent statements in Lemma 5.12. The proofs of item 4 for the rings U and B are similar.

For item 5, if $g \in U$, then $g \in U_n$, for some n. By Equation 17.7.1, each $\sigma_{in} \in R[1/x]$, and so $g \in R[1/x]$. Thus $U[1/x] = R[1/x]$. By item 3,

$$xB \cap U = ((xR^*/I) \cap B) \cap U = x(R^*/I) \cap U = xU;$$

similarly $xA \cap U = xU$. Item 4 and (i) $\implies$ (iv) of Lemma 5.12 with $S = U$ and $T = B$, $T = A$, or $T = R^*/I$ imply that $U = U[1/x] \cap B = R[1/x] \cap B$, $U = R[1/x] \cap A$, and $U = R[1/x] \cap R^*/I$. For statement a of item 5, $B = (1+xU)^{-1}U$ implies $B[1/x] = (1+xU)^{-1}U[1/x]$ is a localization of $U[1/x] = R[1/x]$, and of U and R. For statement b, the integral domains R, U, B and A all have the same field of fractions since each is contained in A, and $A \subseteq K$. For statement c of item 5, since $1/x \in B_P$, the ring B_P is a localization of R and of U. Thus $B_P = R_{P\cap R} = U_{P\cap U}$; see Exercise 5.1.

For item 6, notice that $x \in \mathfrak{m}_R$. By item 2, $x \in \mathcal{J}(B)$ and $x \in \mathcal{J}(R^*)$, that is, x is in every maximal ideal of B and of R^*/I. Also R^* is local with maximal ideal $\mathfrak{m}_{R^*} = \mathfrak{m}_R R^*$. By item 4, $B/xB = A/xA = R^*/(xR^* + I) = R/((xR^* + I) \cap R)$, it follows that B and A are local with maximal ideals $\mathfrak{m}_R B$ and $\mathfrak{m}_R A$. For the rest of item 6 above, Fact 3.2 implies that $\widehat{R^*} = \widehat{R}$. By item 4, $A^* = U^* = B^* = R^*/I$. Thus the $\mathfrak{m}_A$-adic completion of A, the $\mathfrak{m}_U$-adic completion of U and $\mathfrak{m}_B$-adic completion of B are equal to the completion of R^*/I, which is $\widehat{R}/I$. □

REMARK 17.12. Theorem 17.11 implies the statements below.

(1) The definitions in Definition 17.10.1 of B and U are independent of
  (a) the choice of generators for I, and
  (b) the representation of the generators of I as power series in x,
  by Theorem 17.11.5.
(2) The rings $U = R[1/x] \cap (R^*/I)$ and $B = (1 + xU)^{-1}U$ are uniquely determined by x and the ideal I of R^* by Theorem 17.11.5.
(3) If $b \in B$ is a unit of A, then b is already a unit of B. This follows by Theorem 17.11.4, since x is in the Jacobson radical of B.
(4) The diagram below displays the relationships among these rings.

$$\begin{array}{ccccccccc}
\mathcal{Q}(R) & = & \mathcal{Q}(U) & = & \mathcal{Q}(B) & = & \mathcal{Q}(A) & \xrightarrow{\subseteq} & \mathcal{Q}(R^*/I) \\
\uparrow & & \uparrow & & \uparrow & & \uparrow & & \uparrow \\
R[1/x] & = & U[1/x] & \xrightarrow{\subseteq} & B[1/x] & \xrightarrow{\subseteq} & A[1/x] & \xrightarrow{\subseteq} & (R^*/I)[1/x] \\
\uparrow & & \uparrow & & \uparrow & & \uparrow & & \uparrow \\
R & \xrightarrow{\subseteq} & U = \cup U_n & \xrightarrow{\subseteq} & B & \xrightarrow{\subseteq} & A & \xrightarrow{\subseteq} & R^*/I.
\end{array}$$

17.4. Noetherian flatness for homomorphic images

Noetherian Flatness Theorem 17.13 (Homomorphic Image Version) gives precise conditions for the Approximation Domain B of Homomorphic Image Construction 17.2 to be Noetherian.

NOETHERIAN FLATNESS THEOREM 17.13. (Homomorphic Image Version) *Let R be an integral domain with field of fractions K. Let $x \in R$ be a nonzero nonunit such that $\bigcap_{n \geq 1} x^n R = (0)$, the x-adic completion R^* is Noetherian, and x is a regular element of R^*. Let I be an ideal of R^* such that $\mathfrak{p} \cap R = (0)$ for each $\mathfrak{p} \in \operatorname{Ass}(R^*/I)$. As in Frontpiece Notation 17.7.2 and Definition 17.10.1, let*

$$U := \bigcup_{n=1}^{\infty} U_n, \quad B := \bigcup_{n=1}^{\infty} B_n = (1 + xU)^{-1}U, \quad \text{and} \quad A := K \cap (R^*/I).$$

Then

(1) *The following statements are equivalent:*
  (a) *The extension $\psi : R \hookrightarrow (R^*/I)[1/x]$ is flat.*
  (b) *The ring B is Noetherian.*
  (c) *The extension $B \hookrightarrow R^*/I$ is faithfully flat.*
  (d) *The ring $A := K \cap (R^*/I)$ is Noetherian and $A = B$.*
  (e) *The ring A is Noetherian and A is a localization of a subring of $R[1/x]$.*

(2) *If R is assumed to be Noetherian, then statements a-e are equivalent to the ring U being Noetherian.*

PROOF. For item 1, (a) $\implies$ (b), if ψ is flat, then, by factoring ψ through $U[1/x] = R[1/x] \hookrightarrow (R^*/I)[1/x]$, it follows that $U \hookrightarrow (R^*/I)[1/x]$ is flat. By Lemma 6.2.3, with $S = U$ and $T = R^*/I$, the ring B is Noetherian.

For (b) $\implies$ (c), $B^* = R^*/I$ is flat over B, by Theorem 17.11.4 and Remark 3.3.3. By Proposition 5.17.1, $x \in \mathcal{J}(B)$, and so, by Remark 3.3.4, $B^* = R^*/I$ is faithfully flat over B.

For (c) $\implies$ (d), again Theorem 17.11.4 yields $B^* = R^*/I$, and so B^* is faithfully flat over B. Then

$$B = \mathcal{Q}(B) \cap (R^*/I) = \mathcal{Q}(A) \cap (R^*/I) = K \cap (R^*/I) = A$$

by Remark 2.37.9 and Theorem 5.14.2. By Remark 2.37.8, A is Noetherian.

The implication (d) $\implies$ (e) holds since $B = A$ is a localization of U and U is a subring of $R[1/x] = U[1/x]$, by Theorem 5.14.5.

For (e) $\implies$ (a): since A is a localization of a subring D of $R[1/x]$, write $A := \Gamma^{-1}D$, where Γ is a multiplicatively closed subset of D. Now

$$R \subseteq A = \Gamma^{-1}D \subseteq \Gamma^{-1}R[1/x] = \Gamma^{-1}A[1/x] = A[1/x].$$

Since A is Noetherian, $A \hookrightarrow A^* = R^*/I$ is flat by Remark 3.3.2. Therefore $A[1/x] \hookrightarrow (R^*/I)[1/x]$ is flat, and so

$$R \subseteq R[1/x] \subseteq \Gamma^{-1}R[1/x] = A[1/x] \hookrightarrow (R^*/I)[1/x]$$

is a composition of flat extensions. It follows that $R \hookrightarrow (R^*/I)[1/x]$ is flat.

For item 2, R Noetherian implies $R[1/x]$ is Noetherian. Assume condition d holds. Then the composite embedding.

$$U \hookrightarrow B = A \hookrightarrow B^* = A^* = R^*/I$$

is flat because B is a localization of U and A is Noetherian; see Remark 3.3.3. By Remark 3.3.4 again, A^* is faithfully flat over A. Thus Lemma 6.2, parts 1 and 3, with $S = U$ and $T = R^*/I$ yields that $S[1/x] = U[1/x] = R[1/x]$ is Noetherian, and hence U is Noetherian by Lemma 6.2.4.

If U is Noetherian, then the localization B of U is Noetherian, and as above $B = A$. Hence A is a localization of U, a subring of $R[1/x]$. Thus condition e holds and the proof is complete. □

COROLLARY 17.14. *Let R, I and x be as in Noetherian Flatness Theorem* 17.13 *(Homomorphic Image version). If* $\dim(R^*/I) = 1$, *then* $\varphi : R \hookrightarrow W := (R^*/I)[1/x]$ *is flat and therefore the equivalent conditions of Theorem* 17.13 *all hold.*

PROOF. By Construction Properties Theorem 17.11.2, x is in the Jacobson radical of R^*/I. Thus $\dim(R^*/I) = 1$ implies that $\dim W = 0$. The hypothesis on the ideal I implies that every prime ideal P of W contracts to (0) in R. Hence

$$\varphi_P : R_{P \cap R} = R_{(0)} = K \hookrightarrow W_P.$$

Thus W_P is a K-module and so a vector space over K. By Remark 2.37.2, φ_P is flat. Since flatness is a local property by Remark 2.371, the map φ is flat. □

REMARKS 17.15. Let R, I, x, A and B be as in Noetherian Flatness Theorem 17.13:

(1) We show in Section 17.6 that the Intersection Domain and Approximation Domain of Inclusion Construction 17.4 are isomorphic to the domains constructed in Homomorphic Image Construction 17.2 with a different base ring. Thus, by Remark 6.10, there are examples using Construction 17.2 such that the Intersection Domain A is Noetherian, but the Approximation Domain $B \neq A$, and other examples where $A = B$ is non-Noetherian.

(2) A necessary and sufficient condition that $A = B$ is that A is a localization of $R[1/x] \cap A$. Indeed, Theorem 17.11.5 implies that $R[1/x] \cap A = U$ and, by Definition 17.10.1, $B = (1 + xU)^{-1}U$. Therefore the condition is sufficient. On the other hand, if $A = \Gamma^{-1}U$, where Γ is a multiplicatively closed subset of U, then by Remark 17.12.3, each $y \in \Gamma$ is a unit of B, and so $\Gamma^{-1}U \subseteq B$ and $A = B$. See also Theorem 17.19 for more discussion of when $A = B$,

(3) Section 10.1 describes a family of Prototype examples where the conditions of the Inclusion version of Noetherian Flatness Theorem 6.3 hold. Under the identifications of Diagram 17.20.1 below, these examples become examples where $R \hookrightarrow (R^*/I)[1/x]$ is flat for Homomorphic Image Construction 17.2; see Remark 17.25.

REMARK 17.16. By Theorem 6.27, the universal catenary property is preserved by Inclusion Construction 5.3. In contrast, consider the constructed domains A and B of Homomorphic Image Construction 17.2, for $(R, \mathfrak{m})$ a universally catenary Noetherian local domain, $x \in \mathfrak{m}$ an appropriate nonzero element and I an ideal of the x-adic completion R^* of R. Then A and B are local and

$$A^* = B^* = R^*/I, \quad \text{and so} \quad \widehat{A} = \widehat{B} = \widehat{R}/I\widehat{R},$$

by Construction Properties Theorem 17.11.4. Even if $A = B$ and is Noetherian as in Noetherian Flatness Theorem 17.13, it is not necessarily true that $\widehat{R}/I\widehat{R}$ is equidimensional. In Example 18.15, with base ring R a localized polynomial ring in 3 variables over a field, so that R is certainly universally catenary, we construct a Noetherian local domain A that is not universally catenary by using Homomorphic Image Construction 17.2.

Theorem 17.17 extends the range of applications of Homomorphic Image Construction 17.2.

THEOREM 17.17. *Assume Setting 17.1 and Construction 17.2. Thus R is an integral domain with field of fractions K, x is a nonzero nonunit of R, and R^*, the x-adic completion of R, is Noetherian. The ideal I of R^* has the property that $\mathfrak{p} \cap R = (0)$, for each $\mathfrak{p} \in \operatorname{Ass}(R^*/I)$. Assume in addition that I is generated by a regular sequence of R^*. If $R \hookrightarrow (R^*/I)[1/x]$ is flat, then, for each $n \in \mathbb{N}$:*

(1) $\operatorname{Ass}(R^*/I^n) = \operatorname{Ass}(R^*/I)$,
(2) *R canonically embeds in R^*/I^n, and*
(3) *$R \hookrightarrow (R^*/I^n)[1/x]$ is flat.*

PROOF. Let $I = (\sigma_1, \dots, \sigma_r)R^*$, where $\sigma_1, \dots, \sigma_r$ is a regular sequence in R^*. Then the sequence $\sigma_1, \dots, \sigma_r$ is quasi-regular in the sense of [**122**, Theorem 16.2, page 125]; that is, the associated graded ring of R^* with respect to I, which is the direct sum $R^*/I \oplus I/I^2 \oplus \dots$, is a polynomial ring in r variables over R^*/I. For each positive integer n, the component I^n/I^{n+1} is a free (R^*/I)-module generated by the monomials of total degree n in these variables. Thus $\text{Ass}(I^n/I^{n+1}) = \text{Ass}(R^*/I)$; that is, a prime ideal P of R^* annihilates a nonzero element of R^*/I if and only if P annihilates a nonzero element of I^n/I^{n+1}.

For item 1 we proceed by induction: assume $\text{Ass}(R^*/I^n) = \text{Ass}(R^*/I)$ and $n \in \mathbb{N}$. Consider the exact sequence

$$0 \to I^n/I^{n+1} \hookrightarrow R^*/I^{n+1} \to R^*/I^n \to 0. \tag{17.17.0}$$

Then $\text{Ass}(R^*/I) = \text{Ass}(I^n/I^{n+1}) \subseteq \text{Ass}(R^*/I^{n+1})$. Also

$$\text{Ass}(R^*/I^{n+1}) \subseteq \text{Ass}(I^n/I^{n+1}) \cup \text{Ass}(R^*/I^n) = \text{Ass}(R^*/I)$$

by [**122**, Theorem 6.3, p. 38], and so it follows that $\text{Ass}(R^*/I^{n+1}) = \text{Ass}(R^*/I)$. Thus R canonically embeds in R^*/I^n for each $n \in \mathbb{N}$.

That $R \hookrightarrow (R^*/I^n)[1/x]$ is flat for every $n \in \mathbb{N}$ now follows by induction on n and by considering the exact sequence obtained by tensoring over R the short exact sequence (17.17.0) with $R[1/x]$. □

EXAMPLE 17.18. Let $R = k[x, y]$ be the polynomial ring in the variables x and y over a field k and let $R^* = k[y][[x]]$ be the x-adic completion of R. Fix an element $\tau \in xk[[x]]$ such that x and τ are algebraically independent over k, and define the $k[[x]]$-algebra homomorphism $\theta : k[y][[x]] \to k[[x]]$, by setting $\theta(y) = \tau$. Then $\ker(\theta) = (y - \tau)R^*$. Set $I := (y - \tau)R^*$. Notice that $\theta(R) = k[x, \tau] \cong R$ because x and τ are algebraically independent over k. Hence $I \cap R = (0)$. Also I is a prime ideal generated by a regular element of R^*, and $(I, x)R^* = (y, x)R^*$ is a maximal ideal of R^*. Corollary 17.14 and Theorem 17.17 imply that for each positive integer n, the Intersection Domain $A_n := (R^*/I^n) \cap k(x, y)$ is a one-dimensional Noetherian local domain having x-adic completion R^*/I^n. Since x generates an ideal primary for the unique maximal ideal of R^*/I^n, the ring R^*/I^n is also the completion of A_n with respect to the powers of the unique maximal ideal $\mathfrak{n}_n$ of A_n. Since R^*/I is a DVR, Remark 2.1 implies that A_1 is a DVR. For $n > 1$, the completion of A_n has nonzero nilpotent elements, and hence the integral closure of A_n is not a finitely generated A_n-module by Remarks 3.19. The inclusion $I^{n+1} \subsetneq I^n$ and the fact that A_n has completion R^*/I^n imply that $A_{n+1} \subsetneq A_n$ for each $n \in \mathbb{N}$; see Exercise 2 of this chapter. Hence the rings A_n form a strictly descending chain

$$A_1 \supset A_2 \supset \dots \supset A_n \supset \cdots$$

of one-dimensional local birational extensions of $R = k[x, y]$.

17.5. Weak Flatness for Homomorphic Image Construction 17.2

Theorem 17.19 is a version of Weak Flatness Theorem 9.9 that applies to Homomorphic Image Construction 17.2.

WEAK FLATNESS THEOREM 17.19. (Homomorphic Image Version) *Assume Setting* 17.1 *and Construction* 17.2. *Thus R is an integral domain with field of fractions K, x is a nonzero nonunit of R and R^*, the x-adic completion of R, is*

Noetherian. The ideal I of R^* has the property that $\mathfrak{p} \cap R = (0)$ for each $\mathfrak{p} \in \operatorname{Ass}(R^*/I)$. Let the rings A and B be as defined in Section 17.10. Assume in addition that R and B are Krull domains. Then

(1) *If the extension $R \hookrightarrow (R^*/I)[1/x]$ is weakly flat, then $A = B$, that is, the construction is limit-intersecting as in Definition 17.10.*
(2) *If R^*/I is a Krull domain, then the following statements are equivalent:*
  (a) *$A = B$.*
  (b) *$R \hookrightarrow (R^*/I)[1/x]$ is weakly flat.*
  (c) *The extension $B \hookrightarrow (R^*/I)[1/x]$ is weakly flat.*
  (d) *The extension $B \hookrightarrow R^*/I$ is weakly flat.*

PROOF. Theorem 17.11.3 implies that each height-one prime of B containing xB is contracted from R^*/I. Using Frontpiece Notation 17.7, Definition 17.10 and Theorem 17.11 yields that $B[1/x]$ is a localization of $R[1/x] = U[1/x]$. Since $R \hookrightarrow (R^*/I)[1/x]$ is weakly flat, it follows that $B \hookrightarrow (R^*/I)[1/x]$ is weakly flat by Remark 9.5.b. Therefore $B \hookrightarrow R^*/I$ is weakly flat. By Proposition 9.3.1, we have $B = \mathcal{Q}(B) \cap (R^*/I) = A$. This proves item 1.

For item 2, since R^*/I is a normal integral domain, $A = (R^*/I) \cap \mathcal{Q}(R)$ is a Krull domain. As noted in the proof of item 1, Theorem 17.11 implies that each height-one prime of B containing xB is contracted from R^*/I and $B[1/x]$ is a localization of $R[1/x] = U[1/x]$. It follows that (b), (c) and (d) are equivalent. By Proposition 9.3.3, (a) $\implies$ (d), and by Proposition 9.3.1, (d) $\implies$ (a). $\square$

17.6. Inclusion Constructions are Homomorphic Image Constructions

For this section we revise the setting so that R denotes the base ring for Inclusion Construction 17.4.

SETTING 17.20. Let R, x, and R^* be as in Setting 17.1. As in Construction 17.4, let $\tau_1, \ldots, \tau_s \in xR^*$ be algebraically independent elements over R such that $K(\tau_1, \ldots, \tau_s) \subseteq \mathcal{Q}(R^*)$. We define A to be the Intersection Domain $A = A_{\text{inc}} := K(\tau_1, \ldots, \tau_s) \cap R^*$, a subring of R^* that is not algebraic over R. Thus

$$\tau_i := \sum_{j=1}^{\infty} r_{ij} x^j \quad \text{where} \quad r_{ij} \in R.$$

Let $t_1, \ldots, t_s$ be indeterminates over R, define $S := R[t_1, \ldots, t_s]$, let S^* be the x-adic completion of S, and let I denote the ideal $(t_1 - \tau_1, \ldots, t_s - \tau_s)S^*$. Notice that $S^*/I \cong R^*$ implies that $P \cap S = (0)$ for each prime ideal $P \in \operatorname{Ass}(S^*/I)$. Thus we are in the setting of Homomorphic Image Construction 17.2 where we define the Intersection Domain $D := A_{\text{hom}} := K(t_1, \ldots, t_s) \cap (S^*/I)$. Let $\sigma_i := t_i - \tau_i$, for each i with $1 \le i \le s$. For each $n \in \mathbb{N}_0$ and each i with $1 \le i \le s$, the element τ_{in} of R^* is the nth endpiece of τ_i and the element $\sigma_{in} \in S^*/I$ is the nth frontpiece of σ_i.

17.6.1. Matching up Intersection Domains. Consider Diagram 17.20.1, where λ is the R-algebra isomorphism of S into R^* that maps $t_i \to \tau_i$ for $i = 1, \ldots, s$. Here $D := A_{\text{hom}} := \mathcal{Q}(S) \cap (S^*/I)$; that is, A_{hom} is the Intersection Domain of Construction 17.2, if R and R^* there are replaced by S and S^*. The map λ naturally extends to a homomorphism of S^* onto R^*, and the ideal I is the kernel of this extension. Thus there is an induced isomorphism of S^*/I onto R^* that we also label λ.

(17.20.1)

$$\begin{array}{ccccccc} & & S := R[t_1,\dots,t_s] & \longrightarrow & D := K(t_1,\dots,t_s)\cap(S^*/I) & \longrightarrow & S^*/I \\ & & \lambda\big\downarrow & & \lambda\big\downarrow & & \lambda\big\downarrow \\ R & \longrightarrow & S' := R[\tau_1,\dots,\tau_s] & \longrightarrow & A := \mathcal{Q}(R)(\tau_1,\dots,\tau_s)\cap R^* & \longrightarrow & R^*. \end{array}$$

Then λ maps D isomorphically onto A via $\lambda(t_i) = \tau_i$, for every i with $1 \le i \le s$.

PROPOSITION 17.21. *Inclusion Construction* 5.3 *is a special case, up to isomorphism, of Homomorphic Image Construction* 17.2. *That is, under the identifications of Diagram* 17.20.*1, the Intersection Domain of Inclusion Construction* 17.2 *fits the description of the Intersection Domain of Homomorphic Image Construction* 17.2.

PROOF. Since λ maps $D = A_{\text{hom}}$ isomorphically onto $A = A_{\text{inc}}$, Construction 17.2 includes Construction 5.3, up to isomorphism, as a special case. □

17.6.2. Matching up Approximation Domains. By Proposition 17.22, the identifications of Diagram 17.20.1 transform the Approximation Domain for Inclusion Construction 5.3 into the Approximation Domain of Homomorphic Image Construction 17.2. That is, the formula given in Equation 5.4.5 of Section 5.2 using endpieces becomes the formula given in Definition 17.10 defined on S and S^*/I using frontpieces.

PROPOSITION 17.22. *Assume Setting* 17.20. *As in Frontpiece Notation* 17.7, *define σ_{in} to be the nth frontpiece for σ_i in S^*/I. Denote by V_n, C_n, V, C the rings constructed in Frontpiece Notation* 17.7 *and Equation* 17.10.1 *with S as the base ring, as shown in Equations* 17.22.*1. Define U_n, B_n, U, B using Endpiece Notation* 5.4 *and Equations* 5.4.*4 and* 5.4.5 *over R. Thus*

$$\begin{aligned} V_n &:= S[\sigma_{1n},\dots,\sigma_{sn}] = R[t_1,\dots,t_s][\sigma_{1n},\dots,\sigma_{sn}], \\ C_n &:= (1+xV_n)^{-1}V_n, \\ V &:= \bigcup_{n=1}^{\infty} V_n, \qquad C := \bigcup_{n=1}^{\infty} C_n = (1+xV)^{-1}V \\ U &:= \bigcup_{n=1}^{\infty} U_n, \ \textit{where } U_n = R[\tau_{1n},\dots,\tau_{sn}] \quad \textit{and} \\ B &:= \bigcup_{n=1}^{\infty} B_n = (1+xU)^{-1}U, \ \textit{where } B_n = (1+xU_n)^{-1}U_n. \end{aligned} \tag{17.22.1}$$

Then the R-algebra isomorphism λ has the following properties:

$$\lambda(D) = A, \quad \lambda(\sigma_{in}) = \tau_{in}, \quad \lambda(V_n) = U_n, \quad \lambda(C_n) = B_n, \quad \lambda(V) = U, \quad \lambda(C) = B,$$

for all i with $1 \le i \le s$ and all $n \in \mathbb{N}$.

PROOF. Let r_{ij} be elements of R such that

$$\tau_i := \sum_{j=1}^{\infty} r_{ij}x^j, \qquad \tau_{in} := \sum_{j=n+1}^{\infty} r_{ij}x^{j-n}$$

$$\sigma_i := t_i - \tau_i = t_i - \sum_{j=1}^{\infty} r_{ij}x^j, \qquad \sigma_{in} := \frac{t_i - \sum_{j=1}^{n} r_{ij}x^j}{x^n}$$

$$\Longrightarrow \lambda(\sigma_{in}) = \frac{\tau_i - \sum_{j=1}^{n} r_{ij}x^j}{x^n}$$

$$= \tau_{in}.$$

The remaining statements of Proposition 17.22 now follow. □

REMARK 17.23. With Setting 17.20, Proposition 17.22 implies that each V_n is a polynomial ring over R in the variables $\sigma_{1n}, \dots, \sigma_{sn}$, since each U_n is a polynomial ring over R in the variables $\tau_{1n}, \dots \tau_{sn}$. Thus

$$V_n := S[\sigma_{1n}, \dots, \sigma_{sn}] = R[t_1, \dots, t_s][\sigma_{1n}, \dots, \sigma_{sn}] \overset{\lambda}{\cong} R[\sigma_{1n}, \dots, \sigma_{sn}],$$

where λ is defined as in Diagram 17.20.1; that is, $\lambda(t_i) = \tau_i$, for each i.

17.6.3. Making an Inclusion Prototype a homomorphic image. The identifications of Diagram 17.20.1 to the Prototypes and Local Prototypes of Definitions 10.3.and 4.28 give them the form of Homomorphic Image Construction 17.2.

These Prototypes are used to produce Homomorphic Image examples of noncatenary Noetherian local domains in Chapter 18.

SETTING 17.24. Let x be an indeterminate over a field k. Let r be a nonnegative integer and s a positive integer. Assume that $\tau_1, \dots, \tau_s \in xk[[x]]$ are algebraically independent over $k(x)$ and let $y_1, \dots, y_r$ and $t_1, \dots, t_s$ be additional indeterminates. We define the following rings:
(17.24.a)

$$R := k[x, y_1, \dots, y_r], \; R^* = k[y_1, \dots, y_r][[x]], \; V = k(x, \tau_1, \dots, \tau_s) \cap k[[x]].$$

Notice that R^* is the x-adic completion of R and V is a DVR by Remark 2.1.

By Prototype Theorem 10.2, the Prototype D of Definition 10.3 satisfies:

$$\begin{aligned} D := A_{\text{incl}} &:= k(x, y_1, \dots, y_r, \tau_1, \dots \tau_s) \cap R^* \\ &= (1 + xV[y_1, \dots, y_r])^{-1}V[y_1, \dots, y_r] \qquad (17.24.b) \\ &= B_{\text{incl}} := (1 + xU_{\text{incl}})^{-1}U_{\text{incl}}, \end{aligned}$$

where $U_{\text{incl}} := \bigcup_{n\in\mathbb{N}} R[\tau_{1n}, \dots \tau_{sn}]$, each τ_{in} is the nth endpiece of τ_i and each $\tau_{in} \in R^*$, for $1 \le i \le s$. By Construction Properties Theorem 5.14.3, the ring R^* is the x-adic completion of each of the rings $A_{\text{incl}}, B_{\text{incl}}$ and U_{incl}.

Set $S := k[x, y_1, \dots, y_r, t_1, \dots, t_s]$, let S^* be the x-adic completion of S, and let $\sigma_i := t_i - \tau_i$, for each i. Define

$$I := (t_1 - \tau_1, \dots, t_s - \tau_s)S^* = (\sigma_1, \dots, \sigma_s)S^*,$$

By Definition 17.10,
(17.24.c)

$$A_{\text{hom}} := k(x, y_1, \dots, y_r, t_1, \dots, t_s) \cap (S^*/I), \quad B_{\text{hom}} := (1 + xU_{\text{hom}})^{-1}U_{\text{hom}},$$

where $U_{\text{hom}} := \bigcup_{n\in\mathbb{N}} S[\sigma_{1n}, \dots, \sigma_{sn}]$, each σ_{in} is the nth frontpiece of σ_i and each $\sigma_{in} \in \mathcal{Q}(S) \cap (S^*/I)$, for $1 \le i \le s$, by Proposition 17.9. Then $I := (\sigma_1, \dots, \sigma_s)S^*$ is a prime ideal of S^*, and $S^*/I \cong k[y_1, \dots, y_r][[x]] = R^*$. The fact that $\tau_1, \dots, \tau_s$ are algebraically independent over $k(x)$ implies that $I \cap S = (0)$. By Construction Properties Theorem 17.11.4, the ring S^*/I is the x-adic completion of each of the rings $A_{\text{hom}}, B_{\text{hom}}$ and U_{hom}.

REMARK 17.25. Prototypes constructed using Inclusion Construction are isomorphic by Propositions 17.21 and 17.22 to their "translations" into Homomorphic Image Constructions: Assume Setting 17.24. Under the identifications given in Diagram 17.20.1, $A_{\text{hom}} = B_{\text{hom}}$ is isomorphic to $A_{\text{incl}} = B_{\text{incl}}$. Thus B_{hom} is Noetherian. Moreover:

(1) B_{hom} is isomorphic to a directed union of localizations of polynomial rings in $r + s + 1$ variables over k.
(2) $B_{\text{hom}} = A_{\text{hom}}$ is Noetherian of dimension $r + 1$. Let V be the DVR $k(x, \tau_1, \dots, \tau_s) \cap k[[x]]$. Then A is isomorphic to a localization of the polynomial ring $V[y_1, \dots, y_r]$. Thus A_{hom} is a regular Noetherian integral domain. That is, every localization of A at a prime ideal of A is an RLR.
(3) The canonical map $\alpha : S \hookrightarrow (S^*/I)[1/x]$ is flat.
(4) If k has characteristic zero, then $B_{\text{hom}} = A_{\text{hom}}$ is excellent.

Remark 17.25 implies that there exist Prototypes with a specific format that is easy to use.

PROPOSITION 17.26. *Let r be a nonnegative integer and s a positive integer, and let $x, y_1, \dots, y_r$ and $t_1, \dots, t_s$ be indeterminates over a field k. Assume the elements $\tau_1, \dots, \tau_s$ of $xk[[x]]$ are algebraically independent over $k(x)$.*

(1) *Set $S := k[x, y_1, \dots, y_r, t_1, \dots, t_s]$, and let S^* be the x-adic completion of S. Set $I := (t_1 - \tau_1, \dots, t_s - \tau_s)S^*$. Then*
$$A_{hom} := k(x, y_1, \dots, y_r, t_1, \dots, t_s) \cap (S^*/I)$$
is a Noetherian domain.
(2) (Local version) *Set $S := k[x, y_1, \dots, y_r, t_1, \dots, t_s]_{(x, y_1, \dots, y_r, t_1, \dots, t_s)}$. Let S^* be the x-adic completion of S. Set $I := (t_1 - \tau_1, \dots, t_s - \tau_s)S^*$. Then*
$$A_{hom} := k(x, y_1, \dots, y_r, t_1, \dots, t_s) \cap (S^*/I)$$
is a local Noetherian domain.
(3) *Let $A = A_{hom}$ be as in item 2, and let $V = k(x, \tau_1, \dots, \tau_s) \cap k[[x]]$. Then:*
   (a) *$A_{hom} = B_{hom} \cong V[y_1, \dots, y_r]_{(x, y_1, \dots, y_r)}$. Thus A is an RLR.*
   (b) *If V is excellent, then A is excellent.*

EXAMPLE 17.27. Assume Setting 17.24. Then $t_1 - \tau_1, \dots, t_s - \tau_s$ is a regular sequence in S^*. Let $I = (t_1 - \tau_1, \dots, t_s - \tau_s)S^*$. Theorem 17.17 implies that $S \hookrightarrow (S^*/I^n)[1/x]$ is flat for every positive integer n. Using I^n in place of I, Proposition 17.26.2 implies the existence for every r and n in $\mathbb{N}$ of a Noetherian local domain A having dimension $r + 1$ such that the x-adic completion A^* of A has nilradical $\mathfrak{n}$ with $\mathfrak{n}^{n-1} \neq (0)$.

Here are specific examples where Remark 17.25 and Proposition 17.26 apply. Example 17.28 shows that the dimension of U can be greater than the dimension of B_{hom}.

EXAMPLES 17.28. (1) Let the notation be as in Proposition 17.26.1, and let $S := k[x, t_1, \dots, t_s]$, that is, there are no y variables. Let S^* denote the x-adic completion of S. Then $I = (t_1 - \tau_i, \dots, t_s - \tau_s)S^*$, and, by Proposition 17.26.1,

$$\begin{aligned} V := (S^*/I) \cap \mathcal{Q}(S) &= (k[t_1, \dots, t_s][[x]]/(t_1 - \tau_i, \dots, t_s - \tau_s)) \cap k(x, t_1, \dots, t_s) \\ &= A_{\text{hom}} = B_{\text{hom}} = k(x, \tau_1, \dots, \tau_s) \cap k[[x]]. \end{aligned}$$

The DVR V is also obtained by localizing $U = U_{\text{hom}} = \bigcup_{n \in \mathbb{N}} S[\sigma_{1n}, \dots, \sigma_{sn}]$ at the prime ideal xU; each σ_{in} is the nth frontpiece of $\sigma_i = t_i - \tau_i$. In this example $S[1/x] = U[1/x]$ has dimension $s + 1$ and so $\dim U = s + 1$, while

$$\dim(S^*/I) = \dim A_{\text{hom}} = \dim B_{\text{hom}} = 1.$$

(2) Essentially the same example as in item 1 can be obtained by using Theorem 10.2 as follows. Let $R = k[x]$. Then $R^* = k[[x]]$, and

$$A_{\text{incl}} = k(x, \tau_1, \dots, \tau_s) \cap k[[x]] \quad \text{and} \quad A_{\text{incl}} = B_{\text{incl}},$$

by Theorem 10.2. In this case U_{incl} is a directed union of polynomial rings in $s + 1$ variables over k,

$$U_{\text{incl}} = \bigcup_{n=1}^{\infty} k[x][\tau_{1n}, \dots, \tau_{sn}],$$

where the τ_{in} are the nth endpieces of the τ_i as in Section 5.2. By Proposition 17.9, the endpieces are related to the frontpieces of the homomorphic image construction.

(3) Assume the notation of Proposition 17.25.2. Set $S = k[x, t_1, \dots, t_s]_{(x, t_1, \dots, t_s)}$, a regular local domain of dimension $s + 1$. This gives a modification of Example 17.28.1. In this case $S[1/x] = U[1/x]$ has dimension s, while we still have $S^*/I \cong k[[x]]$. Thus $\dim(S^*/I) = 1 = \dim A_{\text{hom}} = \dim B_{\text{hom}}$, whereas $\dim U = s + 1$.

One can also obtain a local version of Example 17.28.2 using the inclusion construction with $R = k[x]_{(x)}$ and applying Theorem 10.2. Then $R^* = k[[x]]$.

With $S := k[x, t_1, \dots, t_s]$ or $S = k[x, t_1, \dots, t_s]_{(x, t_1, \dots, t_s)}$ as in Example 17.28.1 or 17.28.3, the domains B_n constructed from S as in Section 17.2 are RLRs of dimension $s + 1$ dominated by $k[[x]]$ and having k as a coefficient field. In either case, since $(S^*/I)[1/x]$ is a field, the extension $S \hookrightarrow (S^*/I)[1/x]$ is flat. Thus by Theorem 17.13 the family $\{B_n\}_{n \in \mathbb{N}}$ is a directed union of RLRs of dimension $s + 1$ whose union B is Noetherian, and is in fact a DVR.

(4) As in Proposition 17.26.2, with $r = 1$ and $y_1 = y$, the ring

$$S = k[x, y, t_1, \dots, t_s]_{(x, y, t_1, \dots, t_s)}.$$

Then $S^*/I \cong k[y]_{(y)}[[x]]$. By Remark 17.25, the extension $S \hookrightarrow (S^*/I)[1/x]$ is flat. Let $V = k[[x]] \cap k(x, \tau_1, \dots, \tau_s)$. Then V is a DVR and

$$(S^*/I) \cap \mathcal{Q}(S) \cong V[y]_{(x,y)}$$

is a 2-dimensional regular local domain that is the directed union of RLRs of dimension $s + 2$.

17.7. Connecting the two constructions

In Setting 17.29, we repeat the setting for Inclusion Construction 5.3. Proposition 17.30 concerns the same set of algebraically independent elements $\tau_1, \dots, \tau_n$ over the same base ring R in terms of Homomorphic Image Construction 17.2.

SETTING 17.29. Let R be an integral domain with field of fractions K and let x be a nonzero nonunit of R. Assume that the x-adic completion R^* of R, is Noetherian, $\underline{\tau} = \tau_1, \dots, \tau_n \in R^*$ are algebraically independent over R, and the Intersection Domain of Inclusion Construction 5.3 is $A := K(\underline{\tau}) \cap R^*$. Also U and the Approximation Domain B are given by

$$U_j := R[\tau_{1j}, \dots, \tau_{nj}] \qquad B_j := (1 + xU_j)^{-1}U_j.$$

$$U := \bigcup_{j=1}^{\infty} U_j = \bigcup_{j=1}^{\infty} R[\tau_{1j}, \dots, \tau_{nj}] \quad \text{and} \quad B := \bigcup_{j=1}^{\infty} B_n,$$

where each τ_{ij} is an jth endpiece of τ_i.

PROPOSITION 17.30. *Assume Setting* 17.29, *and assume that* $P \cap R = (0)$, *for each* $P \in R^*$ *that is associated to the ideal* $I = (\tau_1, \dots, \tau_n)R^*$. *Let* V *and* E *be the Approximation Domains for Homomorphic Image Construction* 17.2 *corresponding to* I *defined by*

$$V_j := R[\tau'_{1j}, \dots, \tau'_{nj}], \qquad \text{and } E_j := (1 + xV_j)^{-1}V_j.$$

$$V := \bigcup_{j=1}^{\infty} V_j, \qquad E := \bigcup_{j=1}^{\infty} E_j = (1 + xV)^{-1}V,$$

where the τ'_{ij} *are the* j^{th} *frontpieces of* τ_i *as defined in Frontpiece Notation* 17.7, *for each* $i, j \in \mathbb{N}$. *Then the canonical map* $\varphi : R^* \to R^*/I$ *restricts to a homomorphism that surjectively maps* U *onto* V *and* B *onto* E.

PROOF. The map φ is the identity map on R, and Proposition 17.9 implies φ maps the negatives of endpiece generators of each U_j to the corresponding frontpieces that generate V_j. Therefore $\varphi(U) = V$, and so $\varphi(B) = E$. □

COROLLARY 17.31. *Assume Setting* 17.29 *and assume that* $P \cap R = (0)$, *for each* $P \in R^*$ *that is associated to the ideal* $I = (\tau_1, \dots, \tau_n)R^*$. *Let* B *be the Approximation Domain defined over* R *using Inclusion Construction* 5.3 *associated to elements* $\tau_1, \dots, \tau_n$ *of* R^* *algebraically independent over* R, *and let* E *be the Approximation Domain defined over* R *associated to the ideal* $I = (\tau_1, \dots, \tau_n)R^*$ *using Homomorphic Image Construction* 17.2. *Then:*

(1) *If* B *is Noetherian, so is* E.
(2) *If* $R[\underline{\tau}] \hookrightarrow R^*[1/x]$ *is flat, then so is the map* $R \hookrightarrow (R^*/I)[1/x]$.
(3) *If* $B \hookrightarrow R^*$ *is faithfully flat, then so is the map* $E \hookrightarrow R^*/I$.
(4) *If* B *is excellent, then* E *is excellent.*
(5) *If* B *is catenary, resp. universally catenary, then* E *is catenary, resp. universally catenary.*

PROOF. The first item holds since the homomorphic image of a Noetherian domain is also Noetherian. Items 2 and 3 follow since the "If" statements are equivalent to B Noetherian, and the "Then" statements are equivalent to E is Noetherian, by Noetherian Flatness Theorems 6.3.1 and 17.13.1. Items 4 and 5 follow from the fact that the properties of excellence, catenary and universally catenary are preserved under homomorphic image; see Remarks 3.48. □

EXAMPLE 17.32. Let $R = k[x, y]$, R^* and $\tau \in xk[[x]]$ be as in Example 17.18. Let K be the field of fractions of R, and set $f = y - \tau$. Let $A = K(f) \cap R^*$

be the intersection domain of Inclusion Construction 5.3 as in Setting 17.29. Then $I = fR^*$ is a prime ideal of R^* and $I \cap R = (0)$. The hypotheses of Proposition 17.30 are satisfied with f replacing the τ of Proposition 17.30, and $\varphi : R^* \to R^*/I$ maps $A = B$ surjectively onto $E = (R^*/I) \cap K$. In this example, A is a Noetherian regular UFD with $\dim A = 2$ and E is a DVR birationally dominating the two-dimensional RLR $R_{(x,y)R}$. Since E also has residue field k, the dimension formula [**122**, page 119] implies that E is not essentially finitely generated over R.

17.7.1. Insider Methods for Homomorphic Image Construction 17.2. Theorem 17.35 gives machinery for using Insider Construction 10.7 with the Homomorphic Image Construction 17.2 to construct additional examples. Theorem 17.35 yields examples that do not fit Inclusion Construction 5.3. This machinery is used in Chapter 19 for the construction of Ogoma's Example 19.13.

This is a two-step process:

(1) Use Inclusion Construction 5.3 with elements $\tau_1, \dots, \tau_n$ in an x-adic completion R^* of a Noetherian domain R such that the result is a Prototype D, or more generally D is a Noetherian Limit Intersection Domain. [2]

(2) Determine an appropriate ideal I of D such that IR^* is a proper ideal of R^* and $P \cap R = (0)$, for each associated prime P of IR^*.

In the application of Theorem 17.35 to Insider Construction 5.3, the ideal I is chosen to be $(f_1, \dots, f_m)D$, where the f_i are elements of $R[\tau_1, \dots, \tau_n]$ that are algebraically independent over R; see Corollary 17.39. Then Theorem 17.35 gives criteria for $\varphi : R \hookrightarrow (R^*/IR^*)[1/x]$ to be flat. If φ is flat, then $C = \mathcal{Q}(R) \cap (R^*/IR^*)$ is a Noetherian Intersection Domain that is equal to its Approximation Domain.

In some cases the Intersection Domain $\mathcal{Q}(R) \cap (R^*/IR^*)$ has additional properties such as being excellent.

SETTING 17.33. Let $x \in R$ be a nonzero nonunit of a Noetherian domain R, and let R^* be the x-adic completion of R. Let $\underline{\tau} = \{\tau_1, \dots, \tau_n\}$ be a set of elements of R^* that are algebraically independent over $K = \mathcal{Q}(R)$. Assume that nonzero elements of $R[\underline{\tau}]$ are regular elements in R^*. Let $D := K(\underline{\tau}) \cap R^*$ be the Intersection Domain of Inclusion Construction 5.3.

Assume that I is a proper ideal of R^* and $P \cap R = (0)$, for each associated prime P of R^*/I. Let $C = K \cap (R^*/I)$ be the Intersection Domain of Homomorphic Image Construction 17.2 and let E be the Approximation Domain of Definition 17.10 corresponding to the ideal I of R^*.

REMARK 17.34. Assume Setting 17.33. Then:

(1) By Noetherian Flatness Theorem 6.3.1, the extension $R[\tau_1, \dots, \tau_n] \hookrightarrow R^*[1/x]$ is flat $\iff$

$$D = K(\tau_1, \dots, \tau_n) \cap R^*$$

is a Noetherian Limit Intersection Domain.

(2) By Noetherian Flatness Theorem 17.13.1, the map

$$\varphi : R \hookrightarrow (R^*/I)[1/x]$$

is flat $\iff$ $C = K \cap (R^*/I)$ is a Noetherian Limit Intersection Domain, that is, $C = E$.

---

[2] See Definition 6.4.

Theorem 17.35 uses Insider Construction 10.7 and Homomorphic Image Construction 17.2 to obtain conditions that imply the flatness of $\varphi : R \hookrightarrow (R^*/I)[1/x]$.

THEOREM 17.35. *Assume Setting 17.33 and that* $R[\underline{\tau}] \hookrightarrow R^*[1/x]$ *is flat.* [3] *Let* $L = I \cap R[\underline{\tau}]$, *and consider the following extensions:*

$$R \overset{\beta}{\hookrightarrow} R[\underline{\tau}]/L \overset{\theta}{\hookrightarrow} (D/(I \cap D))[1/x] \overset{\alpha}{\hookrightarrow} (R^*/I)[1/x].$$

Let $\psi := \theta \circ \beta : R \hookrightarrow (D/(I \cap D))[1/x]$ *and*

$$\varphi := \alpha \circ \theta \circ \beta = \alpha \circ \psi : R \hookrightarrow (R^*/I)[1/x].$$

Then:

(1) α *is faithfully flat and* θ *is flat.*
(2) *If* β *is flat, then* ψ *is flat.*
(3) ψ *is flat* $\iff$ $\varphi : R \hookrightarrow (R^*/I)[1/x]$ *is flat.*
(4) *If* β *is flat or* ψ *is flat, then* C *is a Noetherian Limit Intersection Domain.*

PROOF. For item 1, since D is an Approximation Domain, Construction Properties Theorem 5.14.4 implies $D[1/x]$ is a localization of $R[\underline{\tau}]$ and the x-adic completion of D is R^*. Noetherian Flatness Theorem 6.3.1 and D is a Noetherian Limit Intersection Domain imply the extension $D \hookrightarrow R^*$ is a faithfully flat extension. It follows that $I \cap D = (I \cap D)R^* \cap D$. Also D Noetherian implies $D/(I \cap D)$ is Noetherian; hence $\alpha' : D/(I \cap D) \hookrightarrow R^*/I$ is faithfully flat. By Fact 2.38,

$$\alpha : (D/(I \cap D))[1/x] \hookrightarrow (R^*/I)[1/x]$$

is faithfully flat since adjoining $1/x$ is the same as tensoring with $R[1/x]$.

Since $D[1/x]$ is a localization of $R[\underline{\tau}]$, it follows that

$$(D/(I \cap D))[1/x] = D[1/x]/(I \cap D)D[1/x])$$

is a localization of $R[\underline{\tau}]/L$. Thus $\theta : R[\underline{\tau}]/L \hookrightarrow (D/(I \cap D))[1/x]$ is flat, and so item 1 holds.

Flatness of β and transitivity of flatness imply flatness of $\psi = \theta \circ \beta$. This proves items 2. By Remarks 2.37.13 and 2.37.15, the map ψ is flat if and only if $\varphi = \alpha \circ \psi$ is flat. This proves item 3.

For item 4, β is flat or ψ is flat implies that φ is flat, by items 2 and 3. By Remark 17.34.2, φ is flat implies $C = K \cap (R^*/I)$ is a Noetherian Limit Intersection Domain, and so item 4 holds. □

COROLLARY 17.36. *Assume the notation of Theorem 17.35. Consider the following extension where* φ *is the composite map:*

$$\varphi : R \overset{\beta}{\hookrightarrow} R[\underline{\tau}]/L \overset{\theta}{\hookrightarrow} (D/(I \cap D))[1/x] \overset{\alpha}{\hookrightarrow} (R^*/I)[1/x]. \tag{17.36.e1}$$

(1) *Let* $\mathfrak{q}^* \in \operatorname{Spec} R^*$ *with* $x \notin \mathfrak{q}^*$ *and* $I \subseteq \mathfrak{q}^*$. *Then the following statements are equivalent:*
   (a) $\beta_{(\mathfrak{q}^* \cap R[\underline{\tau}])} : R \hookrightarrow R[\underline{\tau}]_{(\mathfrak{q}^* \cap R[\underline{\tau}])}/L_{(\mathfrak{q}^* \cap R[\underline{\tau}])}$ *is flat.*
   (b) $\psi_{(\mathfrak{q}^* \cap D)} = \theta_{(\mathfrak{q}^* \cap D)} \circ \beta_{(\mathfrak{q}^* \cap R[\underline{\tau}])} : R \hookrightarrow D_{(\mathfrak{q}^* \cap D)}/(I \cap D)D_{(\mathfrak{q}^* \cap D)}$ *is flat.*
   (c) *The composite extension* $\varphi_{\mathfrak{q}^*} : R \hookrightarrow R^*_{\mathfrak{q}^*}/IR^*_{\mathfrak{q}^*}$ *is flat.*
   (d) $\beta_x : R \hookrightarrow (R[\underline{\tau}]/L)[1/x]$ *is flat.*

[3]It follows that D is a Noetherian Limit Intersection Domain by Remark 17.34.1.

(2) *The non-flat loci of the maps in Equation* 17.36.*e1 are closed and defined by ideals as follows:*
   (a) *The non-flat locus of* $\beta : R \hookrightarrow R[\tau_1, \dots, \tau_n]/L$ *is closed and defined by an ideal* F *of* $(R[\tau_1, \dots, \tau_n]/L)$.
   (b) *The non-flat loci of* φ, ψ, *and* β_x *are also defined by* F.

PROOF. For each $\mathfrak{q}^* \in \operatorname{Spec} R^*$ with $x \notin \mathfrak{q}^*$ and $I \subseteq \mathfrak{q}^*$, define the following extensions involving localizations of the extension of Equation 17.36.e1 with $\varphi_{\mathfrak{q}^*}$ as the composite map:
(17.36.e2)

$$\varphi_{\mathfrak{q}^*} : R \overset{\beta_{(\mathfrak{q}^* \cap R[\underline{\tau}])}}{\hookrightarrow} R[\underline{\tau}]_{(\mathfrak{q}^* \cap R[\underline{\tau}])} \overset{\theta_{(\mathfrak{q}^* \cap D)}}{\hookrightarrow} D_{(\mathfrak{q}^* \cap D)}/(I \cap D)D_{(\mathfrak{q}^* \cap D)} \overset{\alpha_{\mathfrak{q}^*}}{\hookrightarrow} R^*_{\mathfrak{q}^*}/IR^*_{\mathfrak{q}^*}.$$

For part 1, Theorem 17.35 implies that the local maps $\alpha_{\mathfrak{q}^*}, \theta_{\mathfrak{q}^* \cap D}$ and therefore also $\alpha_{\mathfrak{q}^*} \circ \theta_{\mathfrak{q}^* \cap D}$, for $\mathfrak{q}^* \in \operatorname{Spec} R^*$, are faithfully flat. Equation 17.36.e2, Remarks 2.37.13 and 2.37.15, and the faithful flatness of $\theta_{\mathfrak{q}^* \cap D}$ together imply

$$\beta_{(\mathfrak{q}^* \cap R[\underline{\tau}])} \text{ is flat} \iff \psi_{\mathfrak{q}^* \cap D} \text{ is flat}.$$

Thus statements a and b are equivalent. Similarly $\alpha_{\mathfrak{q}^*}$ faithfully flat implies that

$$\psi_{\mathfrak{q}^* \cap D} \text{ is flat} \iff \varphi_{\mathfrak{q}^*} \text{ is flat},$$

and so statements b and c are equivalent. Statement a is equivalent to statement d, since $(R[\tau_1, \dots, \tau_n]/L)[1/x]$ is a localization of $R[\tau_1, \dots, \tau_n]/L$.

For statement a of part 2, the extension $\beta : R \hookrightarrow (R[\tau_1, \dots, \tau_n]/L)[1/x]$ has finite type. By Theorem 2.42, the non-flat locus of β is closed and defined by

$$F := \cap\{\mathfrak{p}/L \mid \mathfrak{p} \in \operatorname{Spec}(R[\underline{\tau}]), x \notin \mathfrak{p}, L \subseteq \mathfrak{p} \text{ and } \beta_{\mathfrak{p}} \text{ is not flat}\},$$

where $\beta_{\mathfrak{p}} : R \hookrightarrow (R[\underline{\tau}]/L)_{(\mathfrak{p}/L)} = R[\underline{\tau}]_{\mathfrak{p}}/L_{\mathfrak{p}}$. Thus statement a of part 2 holds.

Statement b of part 2 now follows from Proposition 2.43 and Equation 17.36.e1, since α and $\alpha \circ \varphi$ are flat. □

PROPOSITION 17.37. *Assume Setting* 17.33, *and also assume that* R *is an excellent local domain. Let* $\psi = \theta \circ \beta : R \hookrightarrow (D/(I \cap D))[1/x]$, *as in Equation* 17.36.*e1. Then:*

(1) *If* ψ *is flat and* I *is a radical ideal, then* C *is analytically unramified.*
(2) *If* ψ *is smooth, then the formal fibers of* C *are geometrically regular.*
(3) *If* ψ *is smooth and* $C^* = R^*/I$ *is equidimensional, then* C *is excellent.*

PROOF. For item 1, since ψ is flat, C is a Noetherian Limit Intersection Domain by Theorem 17.35.4. Since R is excellent, the rings R^* and $C^* = R^*/I$ are excellent by Remarks 3.48. Since I is a radical ideal, C^* is reduced. Since C^* is excellent, C^* is Nagata, and so condition 2 of the Rees Finite Integral Closure Theorem 3.21 holds. By the equivalence of conditions 1 and 2 in Theorem 3.21, C^* is analytically unramified. In other words, $\widehat{C^*} = \widehat{C}$ is reduced. Hence C is analytically unramified.

For item 2, let $\mathfrak{p} \in \operatorname{Spec} C$. First assume $x \notin \mathfrak{p}$. Then $C[1/x] \subseteq C_{\mathfrak{p}}$ implies that $C_{\mathfrak{p}} = R_{(\mathfrak{p} \cap R)}$, by Construction Properties Theorem 17.11.5a. Let $\mathfrak{q} \in \operatorname{Spec} \widehat{C}$ be such that $\mathfrak{q} \cap C = \mathfrak{p}$. By localizing Equation 17.36.e1, it follows that:

$$C_{\mathfrak{p}} = R_{(\mathfrak{p} \cap R)} \overset{\psi_{(\mathfrak{q} \cap (D/(I \cap D)))}}{\hookrightarrow} D/(I \cap D)_{(\mathfrak{q} \cap (D/(I \cap D)))} \overset{\alpha_{(\mathfrak{q} \cap C^*)}}{\hookrightarrow} C^*_{(\mathfrak{q} \cap C^*)} \overset{\delta}{\hookrightarrow} \widehat{C}_{\mathfrak{q}},$$

where δ is the inclusion map. Also $(D/(I \cap D))_{(\mathfrak{q} \cap (D/(I \cap D)))}$ is a further localization of $(D/(I \cap D))[1/x]$, which is a localization of $R[\underline{\tau}]/(I \cap (R[\underline{\tau}])$ by Construction

Properties Theorem 5.14.4, and hence is excellent. Thus $(D/(I\cap D))_{(\mathfrak{q}\cap(D/(I\cap D))}$ is excellent, and $\widehat{(D/(I\cap D))}_{(\mathfrak{q}\cap(D/(I\cap D))} = \widehat{C}_{\mathfrak{q}}$. Thus the map $\gamma = \delta\circ\alpha_{(\mathfrak{q}\cap C^*)}$ shown in the equation below is faithfully flat with geometrically regular fibers:

$$C_{\mathfrak{p}} \overset{\psi_{(\mathfrak{q}\cap(D/(I\cap D))}}{\hookrightarrow} (D/(I\cap D))_{(\mathfrak{q}\cap(D/(I\cap D))} \overset{\gamma}{\hookrightarrow} \widehat{C}_{\mathfrak{q}}$$

Also $\psi_{(\mathfrak{q}\cap(D/(I\cap D))}$ is smooth and hence regular. Thus $\gamma\circ\psi_{(\mathfrak{q}\cap(D/(I\cap D))}$ is regular, and so the formal fiber over $\mathfrak{p}$ is geometrically regular.

Assume $x\in\mathfrak{p}\in\operatorname{Spec}C$, and again let $\mathfrak{q}\in\operatorname{Spec}\widehat{C}$ be such that $\mathfrak{q}\cap C=\mathfrak{p}$. Then $C/xC = C^*/xC^*$; see Remarks 3.3.6. Thus $\mathfrak{p}/xC = \mathfrak{p}^*/xC^*$, for some prime ideal $\mathfrak{p}^*$ of $C^* = R^*/I$. Therefore

$$\kappa(\mathfrak{p}) = \mathcal{Q}(C/\mathfrak{p}) = \mathcal{Q}(C^*/\mathfrak{p}^*) = \kappa(\mathfrak{p}^*), \text{ and } \mathfrak{p}C^* = \mathfrak{p}^*.$$

Since C^* is excellent, the formal fiber $\widehat{C^*}_{\mathfrak{q}}/\mathfrak{p}^*\widehat{C^*}_{\mathfrak{q}} = \widehat{C}_{\mathfrak{q}}/\mathfrak{p}\widehat{C}_{\mathfrak{q}}$ is an RLR. For every finite extension field F of $\kappa(\mathfrak{p}) = \kappa(\mathfrak{p}^*)$, the ring $(\widehat{C}_{\mathfrak{q}}/\mathfrak{p}\widehat{C}_{\mathfrak{q}})\otimes_C F = (\widehat{C}_{\mathfrak{q}}/\mathfrak{p}^*\widehat{C})\otimes_{C^*} F$ is a Noetherian regular ring. Thus the formal fiber over $\mathfrak{p}$ is geometrically regular. This completes the proof of item 2.

For item 3, by Definition 8.22 and item 2, it suffices to show that C is universally catenary. Let $d=\dim C$, then $d=\dim C^* = \dim\widehat{C}$ by Remark 3.3.4. Since C^* is equidimensional, $\dim(C^*/\mathfrak{p}^*) = d$ for every minimal prime ideal $\mathfrak{p}^*$ of C^*. Since C^* is excellent and hence universally catenary, Ratliff's Equidimension Theorem 3.26 implies the completion $\widehat{C}/\mathfrak{p}^*\widehat{C}$ of $C^*/\mathfrak{p}^*$ is equidimensional. Hence $\dim(\widehat{C}/\widehat{\mathfrak{p}}) = \dim(C^*/\mathfrak{p}^*) = d$ for every prime ideal $\widehat{\mathfrak{p}}$ of $\widehat{C}$ that is minimal over $\mathfrak{p}^*\widehat{C}$.

Let $\widehat{\mathfrak{p}}$ be a minimal prime ideal of $\widehat{C}$. Since $\widehat{C}$ is flat over C^*, $\widehat{\mathfrak{p}}\cap C^* = \mathfrak{p}^*$ is a minimal prime of C^*; see Remark 2.37.10. It follows that $\dim(\widehat{C}/\widehat{\mathfrak{p}}) = d$. Hence $\widehat{C}$ is equidimensional. Therefore C is universally catenary. □

Corollary 17.39 yields a procedure for constructions of examples using Prototype Theorem 10.2 and Insider Construction 10.7 together with Homomorphic Image Construction 17.2. The first part of the setting for Corollary 17.39 is the same as for Corollary 10.10, an application of Insider Construction 10.7.

SETTING 17.38. Let x and $\underline{y} = \{y_1,\dots,y_r\}$ be indeterminates over a field k, let $R = k[x,\underline{y}]_{(x,\underline{y})}$ or $R = k[x,\underline{y}]$, and let R^* be the x-adic completion of R. Let $\underline{\tau} = \tau_1,\dots,\tau_n$ be elements of $xk[[x]]\subseteq R^*$ that are algebraically independent over $k[x]$, and let $D = k(x,\underline{y},\underline{\tau})\cap R^*$ be the Protoype of Definition 4.28 or Equation 10.1.b. Let $\underline{f} = f_1,\dots,f_m\in R[\underline{\tau}]$ be algebraically independent over R.

Assume $P\cap R = (0)$, for every $P\in\operatorname{Ass}(R^*/(\underline{f})R^*)$. Apply Homomorphic Image Construction 17.2: let $C = k(x,\underline{y})\cap(R^*/(\underline{f})R^*)$ be the Intersection Domain corresponding to $I := (\underline{f})R^*$; let E be the Approximation Domain of Definition 17.10 corresponding to the ideal I of R^*. Let $L = I\cap R[\underline{\tau}]$, and let β,θ and α be the inclusion maps shown

$$R \overset{\beta}{\hookrightarrow} \frac{R[\underline{\tau}]}{L} \overset{\theta}{\hookrightarrow} (\frac{D}{I\cap D})[\frac{1}{x}] \overset{\alpha}{\hookrightarrow} (\frac{R^*}{I})[\frac{1}{x}],$$

COROLLARY 17.39. *Assume Setting* 17.38. *Then:*

(1) *If either of the maps*

$$\psi = \theta\circ\beta : R \hookrightarrow (\frac{D}{I\cap D})[\frac{1}{x}] \quad or \quad \beta : R\hookrightarrow (\frac{R[\underline{\tau}]}{I\cap R[\underline{\tau}]})[\frac{1}{x}],$$

is flat, then $C = k(x, \underline{y}) \cap (R^*/IR^*)$ *is a Noetherian Limit Intersectin Domain.*

(2) *Assume* $R = k[x, \underline{y}]_{(x,\underline{y})}$ *and the ideal* I *is equidimensional. If* ψ *is smooth, then* C *is excellent.*

PROOF. Item 1 follows from Theorem 17.35.4, and item 2 follows from Proposition 17.37.3. □

Exercises

(1) Let A be an integral domain and let $A \hookrightarrow B$ be an injective map to an extension ring B. For an ideal I of B, prove that the following are equivalent:
   (i) The induced map $A \to B/I$ is injective, and each nonzero element of A is regular on B/I.
   (ii) The field of fractions $\mathcal{Q}(A)$ of A naturally embeds in the total quotient ring $\mathcal{Q}(B/I)$ of B/I.

   If B is Noetherian, prove that conditions (i) and (ii) are also equivalent to the following condition:
   (iii) For each prime ideal P of B that is associated to I we have $P \cap A = (0)$.

(2) Let A be an integral domain and let $A \hookrightarrow B$ be an injective map to an extension ring B. Let I be an ideal of B such that $I \cap A = (0)$ and every nonzero element of A is a regular element on B/I. Let $C := \mathcal{Q}(A) \cap (B/I)$.
   (i) Prove that $C = \{a/b \mid a, b \in A,\ b \neq 0 \text{ and } a \in I + bB\}$.
   (ii) Assume that $J \subseteq I$ is an ideal of B such that every nonzero element of A is a regular element on B/J. Let $D := \mathcal{Q}(A) \cap (B/J)$. Prove that $D \subseteq C$.

   SUGGESTION. Item ii is immediate from item i. To see item i, observe that $bC = b(B/I) \cap \mathcal{Q}(A)$, and $a \in bC \iff a \in b(B/I) \iff a \in I + bB$.

(3) For the strictly descending chain of one-dimensional local domains

$$A_1 \supset A_2 \supset \cdots \supset A_n \supset \cdots$$

   given in Example 17.18 of birational extensions of $R = k[x, y]$, prove that $D := \bigcap_{n=1}^{\infty} A_n = R_{(x,y)R}$.

   SUGGESTION. Since $\mathfrak{n}_n \cap R = (x, y)R$, we have $R_{(x,y)R} \subset A_n$ for each $n \in \mathbb{N}$. By Exercise 2, the ring A_n may be described as

$$A_n = \{\ a/b \mid a, b \in R,\ b \neq 0 \text{ and } a \in I^n + bR^*\ \}.$$

   Show that $a \in I^n + bR^*$ for all $n \in \mathbb{N}$ if and only if $a/b \in R_{(x,y)R}$.

(4) Assume the setting of Frontpiece Notation 17.7 and Definition 17.10. If J is a proper ideal of B, prove that JB^* is a proper ideal of B^*, where B^* is the x-adic completion of B.

(5) Assume the setting of Frontpiece Notation 17.7, and let W denote the set of elements of R^* that are regular on R^*/I. Prove that the natural homomorphism $\pi : R^* \to R^*/I$ extends to a homomorphism $\overline{\pi} : W^{-1}R^* \to W^{-1}(R^*/I)$.

(6) Describe Example 17.28.4 in terms of Inclusion Construction 5.3. In particular, determine the appropriate base ring R for this construction.

CHAPTER 18

Catenary local rings with geometrically normal formal fibers

This chapter concerns the catenary property for a Noetherian local ring $(R, \mathfrak{m})$ having geometrically normal formal fibers.[1] Recall that a ring R is *catenary* if, for every pair of comparable prime ideals $P \subset Q$ of R, every saturated chain of prime ideals from P to Q has the same length. The ring R is *universally catenary* if every finitely generated R-algebra is catenary. From Definition 3.39, the ring R has *geometrically normal*, respectively, *geometrically regular, formal fibers* if, for each prime P of R and for each finite algebraic extension k' of the field $k(P) := R_P/PR_P$, the ring $\widehat{R} \otimes_R k'$ is normal, respectively, regular. By Remark 3.42, regular fibers are normal.

By Corollary 18.6.1, if the ring $(R, \mathfrak{m})$ has geometrically normal formal fibers, then the Henselization R^h of R is universally catenary.[2] Relations among the catenary and universally catenary properties of R and the fibers of the map $R \hookrightarrow R^h$ are given in Section 18.2. For each integer $n \geq 2$, Example 18.20 of Section 18.5 provides an example of a catenary Noetherian local integral domain of dimension n that has geometrically regular formal fibers and is not universally catenary. A more detailed summary of this chapter is given in Section 18.1.

We thank M. Brodmann and R. Sharp for raising a question on catenary and universally catenary rings that motivated our work in this chapter.

18.1. History, terminology and summary

Krull proves in [**107**] that every integral domain that is a finitely generated algebra over a field is catenary. Cohen proves in [**35**] that every complete Noetherian local ring is catenary. These results motivated the question of whether every Noetherian ring (or equivalently every Noetherian local integral domain) is catenary. Nagata answers this question by giving an example of a family of non-catenary Noetherian local domains in [**134**]; see also [**137**, Example 2, pages 203-205]. Each domain in this family is not integrally closed and has the property that its integral closure is catenary and Noetherian.

These examples of Nagata motivated the question of whether the integral closure of a Noetherian local domain is catenary. Work on this question continued for

[1]The material in this chapter comes from a paper we wrote that is included in a volume dedicated to Shreeram S. Abhyankar in celebration of his seventieth birthday. In his mathematical work Ram has opened up many avenues. In this chapter we are pursuing one of these related to power series and completions.

[2]The terms "Henselization" and "Henselian" are defined in Remarks 3.32.1 and Definition 3.30.

over 20 years with Ratliff being a leading researcher in this area, [**151**], [**152**]. The following terminology is used in [**137**, page 122] and [**152**, page 5]:

DEFINITION 18.1. A ring R satisfies the *chain condition for prime ideals* if for every pair $\mathfrak{p} \subsetneq \mathfrak{q}$ of prime ideals in R and every integral extension R' of $R_{\mathfrak{q}}/\mathfrak{p}R_{\mathfrak{q}}$ every maximal chain of prime ideals in R' has length equal to $\dim R'$

Ratliff in [**152**, (3.3.2), page 23] states the chain conjecture as follows:

THE CHAIN CONJECTURE 18.2. *Let A' be the derived normal ring of a Noetherian integral domain A. Then A' satisfies the chain condition for prime ideals.*

In 1980, T. Ogoma resolved this question by establishing the existence of a 3-dimensional Nagata local domain that is integrally closed but not catenary [**146**]. Heitmann in [**98**] later gives an alternate presentation of Ogoma's Example. We present a version of Ogoma's Example in Example 19.13.

Heitmann in [**96**] obtains the following notable characterization of the complete Noetherian local rings that are the completion of a UFD. He proves that every complete Noetherian local ring $(T, \mathfrak{n})$ that has depth at least two [3] and has the property that no element in the prime subring of T is a zerodivisor on T is the completion of a Noetherian local UFD. Let x, y, z, w be indeterminates over a field k, and let $T := k[[x, y, z, w]]/(xy, xz)$. Heitmann uses his result to establish the existence of a 3-dimensional Noetherian local UFD $(R, \mathfrak{m})$ having completion T. It follows that R is catenary but not universally catenary [**96**, Theorem 9].

Section 18.2 includes conditions for a Noetherian local ring $(R, \mathfrak{m})$ to be universally catenary. Theorem 18.8 asserts that R is universally catenary if and only if the set Γ_R is empty, where

$$\Gamma_R := \{P \in \operatorname{Spec}(R^h) \mid \dim(R^h/P) < \dim(R/(P \cap R))\}.$$

Moreover the subset Γ_R of $\operatorname{Spec} R^h$ is *stable under generalization* in the sense that, if $Q \in \Gamma_R$ and $P \in \operatorname{Spec} R^h$ is such that $P \subseteq Q$, then $P \in \Gamma_R$. Thus Γ_R satisfies a "strong" Going-down property.

Theorem 18.9 states that a Noetherian local domain R having geometrically normal formal fibers is catenary but is not universally catenary if and only if the set Γ_R is nonempty and $\dim(R^h/P) = 1$ for each prime ideal P in Γ_R. In this case, Γ_R is a subset of the minimal primes of R^h. Since R^h is Noetherian, Γ_R is finite. Thus, as observed in Corollary 18.10, if R is catenary but not universally catenary, then there exists a minimal prime $\widehat{q}$ of the $\mathfrak{m}$-adic completion $\widehat{R}$ of R such that $\dim(\widehat{R}/\widehat{q}) = 1$. If R is catenary, each minimal prime $\widehat{q}$ of $\widehat{R}$ such that $\dim(\widehat{R}/\widehat{q}) \neq \dim(R)$ must have $\dim(\widehat{R}/\widehat{q}) = 1$.

Theorem 18.12 gives conditions such that the flatness and Noetherian properties for the integral domains associated with ideals $I_1, \dots, I_n$ of an ideal-adic completion R^* in Homomorphic Image Construction 17.2 transfer to the integral domain associated with their intersection $I = I_1 \cap \cdots \cap I_n$. Similarly, Theorem 18.14 contains conditions so that geometrically regular formal fibers for the constructed ring of ideals $I_1, \dots, I_n$ transfer to rings constructed using the intersection $I = I_1 \cap \cdots \cap I_n$ have geometrically regular formal fibers. In Section 18.5, Theorem 18.12 is applied to produce Noetherian local domains that are not universally catenary. Section 18.6 concerns the depths of the constructed rings.

[3] See Definition 3.35.

18.2. Geometrically normal formal fibers and the catenary property

Throughout this section $(R, \mathfrak{m})$ is a Noetherian local ring with $\mathfrak{m}$-adic completion $\widehat{R}$. The ring R is *formally equidimensional*, or in other terminology *quasi-unmixed*, provided $\dim(\widehat{R}/\widehat{q}) = \dim \widehat{R}$ for every minimal prime $\widehat{q}$ of $\widehat{R}$. Ratliff's Equidimension Theorem 3.26, that R is universal catenary if and only if R is formally equidimensional, is crucial for our work. We use Theorem 3.26 to prove:

THEOREM 18.3. *Let $(R, \mathfrak{m})$ be a Henselian Noetherian local ring having geometrically normal formal fibers. Then:*

(1) *For each prime ideal P of R, the extension $P\widehat{R}$ to the $\mathfrak{m}$-adic completion of R is also a prime ideal.*
(2) *The ring R is universally catenary.*

PROOF. Item 2 follows from item 1 and Theorem 3.26. In order to prove item 1, observe that the completion of R/P is $\widehat{R}/P\widehat{R}$, and R/P is a Noetherian Henselian local integral domain having geometrically normal formal fibers. Pass from R to R/P; then for item 1 it suffices to prove: If R is a Henselian Noetherian local integral domain having geometrically normal formal fibers, then the completion $\widehat{R}$ of R is an integral domain.

For this, assume that R as above is an integral domain. By Definition 3.39, geometrically normal formal fibers are geometrically reduced. Let U be the nonzero elements of R; then the ring $U^{-1}\widehat{R}$ is reduced. Every element of U is a regular element of $\widehat{R}$ by the flatness of $\widehat{R}$ over R, and so $U^{-1}\widehat{R}$ has the same total quotient ring as $\widehat{R}$. Thus $\widehat{R}$ is reduced. By Theorem 8.19, R is Nagata, and so the integral closure $\overline{R}$ of R is a finitely generated R-module by Definition 2.20. Moreover, since R is Henselian, $\overline{R}$ is local; see Remark 3.32.5. Since $\overline{R}$ is an integrally closed integral domain, $\overline{R}$ is normal. The completion $\widehat{\overline{R}}$ of $\overline{R}$ is $\widehat{R} \otimes_R \overline{R}$ by [**137**, (17.8)].

The assumption that the formal fibers of R are normal implies that the formal fibers of $\overline{R}$ are normal: To see this, let $\overline{P} \in \operatorname{Spec} \overline{R}$ and let $P = \overline{P} \cap R$. Since $\overline{R}$ is a finite R-module, $k(\overline{P}) = \overline{R}_{\overline{P}}/\overline{P}\overline{R}_{\overline{P}}$ is a finite $k(P)$-module, where $k(P) = R_P/PR_P$. Thus $k(\overline{P})$ is a finite field extension of $k(P)$. Since R has generically normal formal fibers,

$$\widehat{R} \otimes_R k(P) \otimes_{k(P)} k(\overline{P}) = \widehat{R} \otimes_R k(\overline{P}) = \widehat{R} \otimes_R \overline{R} \otimes_{\overline{R}} k(\overline{P}) = \widehat{\overline{R}} \otimes_{\overline{R}} k(\overline{P})$$

is a normal ring. That is, for each $\overline{P} \in \operatorname{Spec} \overline{R}$, the fiber ring of the map $\varphi : \overline{R} \to \widehat{\overline{R}}$ over $\overline{P}$ is normal. Since $\overline{R}$ is a normal ring and φ is a flat local homomorphism with normal fibers, it follows that $\widehat{\overline{R}}$ is normal by Theorem 3.33.3. Since $\widehat{\overline{R}}$ is local, $\widehat{\overline{R}}$ is an integral domain, by Remark 2.3. Also $\widehat{R}$ is a flat R-module, and so $\widehat{R} = \widehat{R} \otimes_R R$ is a subring of $\widehat{\overline{R}} = \widehat{R} \otimes_R \overline{R}$. Therefore $\widehat{R}$ is an integral domain, as desired for the completion of the proof of Theorem 18.3. □

REMARK 18.4. Let $(R, \mathfrak{m})$ be a Noetherian local domain. An interesting result proved by Nagata establishes the existence of a one-to-one correspondence between the minimal primes of the Henselization R^h of R and the maximal ideals of the integral closure $\overline{R}$ of R; see Remarks 3.32.9. Moreover, if a maximal ideal $\overline{\mathfrak{m}}$ of $\overline{R}$ corresponds to a minimal prime $\mathfrak{q}$ of R^h, then the integral closure of the Henselian local domain $R^h/\mathfrak{q}$ is the Henselization of $\overline{R}_{\overline{\mathfrak{m}}}$; see [**137**, Ex. 2, page 188], [**131**]. Therefore $\operatorname{ht}(\overline{\mathfrak{m}}) = \dim(R^h/\mathfrak{q})$.

REMARK 18.5. Let $(R, \mathfrak{m})$ be a Noetherian local ring, let $\widehat{R}$ denote the $\mathfrak{m}$-adic completion of R, and let R^h denote the Henselization of R. The canonical map $f : R \hookrightarrow R^h$ is a regular map with zero-dimensional fibers by Remarks 8.28.2, and $\widehat{R}$ is also the completion of R^h with respect to its unique maximal ideal $\mathfrak{m}^h = \mathfrak{m}R^h$ by Remarks 3.32.1. The canonical maps are displayed below:

$$R \overset{f}{\hookrightarrow} R^h \overset{g}{\hookrightarrow} \widehat{R},$$

Then $g \circ f : R \hookrightarrow \widehat{R}$ has (geometrically) normal fibers if and only if $g : R^h \hookrightarrow \widehat{R}$ has (geometrically) normal fibers.

To see this, let P be a prime ideal of R and let $U = R \setminus P$. Then

$$PR^h = P_1 \cap \cdots \cap P_n$$

where the P_i are the minimal prime ideals of PR^h. Since $\widehat{R}$ is faithfully flat over R^h, finite intersections distribute over this extension, and so $P\widehat{R} = \cap_{i=1}^n (P_i\widehat{R})$. Let $S = U^{-1}(\widehat{R}/P\widehat{R})$ denote the fiber over P in $\widehat{R}$ and let $q_i = P_iS$. The ideals $q_1, \ldots, q_n$ of S intersect in (0) and are pairwise comaximal because for $i \neq j$, $(P_i + P_j) \cap U \neq \emptyset$. Therefore $S \cong \prod_{i=1}^n (S/q_i)$. By Remark 2.3, a Noetherian ring is normal if and only if it is a finite product of normal Noetherian domains. Thus the fiber over P in $\widehat{R}$ is normal if and only if the fiber over each of the P_i in $\widehat{R}$ is normal.

To see $g \circ f : R \hookrightarrow \widehat{R}$ has geometrically normal fibers if and only if $g : R^h \hookrightarrow \widehat{R}$ has geometrically normal fibers, observe that a geometrically normal morphism is an example of a *P-morphism*, in the sense of Grothendieck in [**62**, Vol. 24, Example (7.3.8) (vii′), p. 194]. Thus, since f is a regular morphism and is faithfully flat, it follows by [**62**, Vol. 24, (7.3.4), (P_I) and (P_{II}), p. 193] that g has geometrically normal fibers if and only if $g \circ f$ has geometrically normal fibers.

COROLLARY 18.6. *Let R be a Noetherian local domain having geometrically normal formal fibers. Then*

(1) *The Henselization R^h of R is universally catenary.*
(2) *If the integral closure $\overline{R}$ of R is again local, then R is universally catenary.*

In particular, if R is a normal Noetherian local domain having geometrically normal formal fibers, then R is universally catenary.

PROOF. For item 1, the Henselization R^h of R is a Noetherian local ring having geometrically normal formal fibers by Remark 18.5, and so Theorem 18.3 implies that R^h is universally catenary. For item 2, if the integral closure of R is local, then, by Remark 18.4, the Henselization R^h has a unique minimal prime. Since R^h is universally catenary, the completion $\widehat{R}$ is equidimensional by Ratliff's Equidimensional Theorem 3.25, and hence R is universally catenary. □

Theorem 18.7 relates the catenary property of R to the height of maximal ideals in the integral closure of R.

THEOREM 18.7. *Let $(R, \mathfrak{m})$ be a Noetherian local domain of dimension d and let $\overline{R}$ denote the integral closure of R. If $\overline{R}$ contains a maximal ideal $\overline{\mathfrak{m}}$ such that $\operatorname{ht}(\overline{\mathfrak{m}}) = r \notin \{1, d\}$, then R has a saturated chain of prime ideals of length $\leq r$. Hence R is not catenary in this case.*

PROOF. Since $\overline{R}$ has only finitely many maximal ideals [**137**, (33.10)], there exists $b \in \overline{\mathfrak{m}}$ such that b is in no other maximal ideal of $\overline{R}$. Let $R' = R[b] \subseteq \overline{R}$ and let $\mathfrak{m}' = \overline{\mathfrak{m}} \cap R'$. Notice that $\overline{\mathfrak{m}}$ is the unique prime ideal of $\overline{R}$ that contains $\mathfrak{m}'$. We show that $\operatorname{ht} \mathfrak{m}' = r$: Let $S = R' \setminus \mathfrak{m}'$. The extension $R'_{\mathfrak{m}'} \hookrightarrow S^{-1}\overline{R}$ is integral. Since $b \in \mathfrak{m}'$ and the only maximal ideal of $\overline{R}$ that contains b is $\overline{\mathfrak{m}}$, the ring $S^{-1}\overline{R}$ is local with maximal ideal $\overline{\mathfrak{m}}(S^{-1}\overline{R})$. Since $S \subseteq \overline{R} \setminus \overline{\mathfrak{m}}$ and $S^{-1}\overline{R}$ is integrally closed, we have $S^{-1}\overline{R} = \overline{R}_{\overline{\mathfrak{m}}}$. Thus $R'_{\mathfrak{m}'} \hookrightarrow \overline{R}_{\overline{\mathfrak{m}}}$ is integral, and so $\dim(R'_{\mathfrak{m}'}) = \dim(R_{\overline{\mathfrak{m}}})$, by the Going-up Theorem [**122**, Theorem 9.4]. Therefore $\operatorname{ht} \mathfrak{m}' = r$.

Since R' is a finitely generated R-module and is birational over R, there exists a nonzero element $a \in \mathfrak{m}$ such that $aR' \subseteq R$. It follows that $R[1/a] = R'[1/a]$. The maximal ideals of $R[1/a]$ have the form $PR[1/a]$, where $P \in \operatorname{Spec} R$ is maximal with respect to not containing a. For $P \in \operatorname{Spec} R$ such that $PR[1/a]$ is maximal in $R[1/a]$, there are no prime ideals strictly between P and $\mathfrak{m}$ by Theorem 2.23. If $\operatorname{ht} P = h$, then there exists a saturated chain $(0) \subsetneq \cdots \subsetneq P \subsetneq \mathfrak{m}$ of prime ideals of R of length $h+1$. Thus, to show R is not catenary, it suffices to establish the existence of a maximal ideal of $R[1/a]$ having height different from $d-1$. Since $R[1/a] = R'[1/a]$, the maximal ideals of $R[1/a]$ correspond to the prime ideals P' in R' maximal with respect to not containing a. Since $\operatorname{ht} \mathfrak{m}' > 1$, there exists $c \in \mathfrak{m}'$ such that c is not in any minimal prime of aR' nor in any maximal ideal of R' other than $\mathfrak{m}'$. Hence there exist prime ideals of R' containing c and not containing a. Let $P' \in \operatorname{Spec}(R')$ be maximal with respect to $c \in P'$ and $a \notin P'$. Then $P' \subset \mathfrak{m}'$, so $\operatorname{ht} P' \leq r - 1 < d - 1$. It follows that there exists a saturated chain of prime ideals of R of length $\leq r$, and hence R is not catenary. □

THEOREM 18.8. *Let $(R, \mathfrak{m})$ be a Noetherian local integral domain having geometrically normal formal fibers and let R^h denote the Henselization of R. Consider the set*

$$\Gamma_R := \{P \in \operatorname{Spec}(R^h) \mid \dim(R^h/P) < \dim(R/(P \cap R))\}.$$

Then the following statements hold.

(1) *For $\mathfrak{p} \in \operatorname{Spec}(R)$, the ring $R/\mathfrak{p}$ is not universally catenary if and only if there exists $P \in \Gamma_R$ such that $\mathfrak{p} = P \cap R$.*
(2) *The set Γ_R is empty if and only if R is universally catenary.*
(3) *If $Q \in \Gamma_R$, then each prime ideal P of R^h such that $P \subseteq Q$ is also in Γ_R, that is, the subset Γ_R of $\operatorname{Spec} R^h$ is stable under generalization.*
(4) *If $\mathfrak{p} \subset \mathfrak{q}$ are prime ideals in R and if there exists $Q \in \Gamma_R$ with $Q \cap R = \mathfrak{q}$, then there also exists $P \in \Gamma_R$ with $P \cap R = \mathfrak{p}$ and $P \subseteq Q$.*

PROOF. The map of $R/\mathfrak{p}$ to its $\mathfrak{m}$-adic completion $\widehat{R}/\mathfrak{p}\widehat{R}$ factors through the Henselization $R^h/\mathfrak{p}R^h$. Since $R \hookrightarrow \widehat{R}$ has geometrically normal fibers, so does the map $R^h \hookrightarrow \widehat{R}$ by Remark 18.5. Theorem 18.3 implies that each prime ideal P of R^h extends to a prime ideal $P\widehat{R}$. Therefore, by Theorem 3.26, the ring $R/\mathfrak{p}$ is universally catenary if and only if $R^h/\mathfrak{p}R^h$ is equidimensional if and only if there does not exist $P \in \Gamma_R$ with $P \cap R = \mathfrak{p}$. This proves items 1 and 2.

For item 3, let $P \in \operatorname{Spec} R^h$ be such that $P \subset Q$, and let $\operatorname{ht}(Q/P) = n$. Since the fibers of the map $R \hookrightarrow R^h$ are zero-dimensional, the contraction to R of an ascending chain of primes

$$P = P_0 \subsetneq P_1 \subsetneq \cdots \subsetneq P_n = Q$$

of R^h is a strictly ascending chain of primes from $\mathfrak{p} := P \cap R$ to $\mathfrak{q} := Q \cap R$. Hence $\text{ht}(\mathfrak{q}/\mathfrak{p}) \geq n$. Since R^h is catenary, we have

$$\dim(R^h/P) \;=\; n + \dim(R^h/Q) \;<\; n + \dim(R/\mathfrak{q}) \;\leq\; \dim(R/\mathfrak{p}),$$

where the strict inequality is because $Q \in \Gamma_R$. Therefore $P \in \Gamma_R$.

It remains to prove item 4. The extension $R \hookrightarrow R^h$ is fairhfully flat, and so the extension satisfies the Going-down property, by Remark 2.37.10. Thus there exists a prime ideal P of R^h such that $P \subseteq Q$ and $P \cap R = \mathfrak{p}$. By item 3, $P \in \Gamma_R$. □

Recall that the *dimension* of a prime ideal $\mathfrak{p}$ of a ring R refers to the Krull dimension of the factor ring, that is, the dimension of $\mathfrak{p}$ is $\dim(R/\mathfrak{p})$.

THEOREM 18.9. *Let $(R, \mathfrak{m})$ be a Noetherian local integral domain having geometrically normal formal fibers and let Γ_R be defined as in Theorem* 18.8. *The ring R is catenary but not universally catenary if and only if*

(i) *the set Γ_R is nonempty, and*
(ii) $\dim(R^h/P) = 1$, *for each prime ideal $P \in \Gamma_R$.*

If these conditions hold, then each $P \in \Gamma_R$ is a minimal prime of R^h, and Γ_R is a finite nonempty open subset of $\operatorname{Spec} R^h$.

PROOF. Assume that R is catenary but not universally catenary. By Theorem 18.8, the set Γ_R is nonempty and there exist minimal primes P of R^h such that $\dim(R^h/P) < \dim(R^h)$. By Remark 18.4, if a maximal ideal $\overline{\mathfrak{m}}$ of $\overline{R}$ corresponds to a minimal prime P of R^h, then $\text{ht}(\overline{\mathfrak{m}}) = \dim(R^h/P)$. Since R is catenary, Theorem 18.7 implies that the height of each maximal ideal of the integral closure $\overline{R}$ of R is either one or $\dim(R)$. Therefore $\dim(R^h/P) = 1$ for each minimal prime P of R^h for which $\dim(R^h/P) \neq \dim(R^h)$. Item 4 of Theorem 18.8 implies each $P \in \Gamma_R$ is a minimal prime of R^h and $\dim(R^h/P) = 1$.

For the converse, assume that Γ_R is nonempty and each prime ideal $W \in \Gamma_R$ has dimension one. Then R is not universally catenary by item 2 of Theorem 18.8. By item 3 of Theorem 18.8, if $W \in \Gamma_R$ and $V \in \text{Spec}(R^h)$ with $V \subsetneq W$, then $V \in \Gamma_R$. But then $\dim(R^h/W) = 1 = \dim(R^h/V)$ is a contradiction. Therefore every element of Γ_R is a minimal prime ideal of R^h; by item 4 of Theorem 18.8 every element of Γ_R lies over (0) in R.

To show R is catenary, it suffices to show for each nonzero non-maximal prime ideal $\mathfrak{p}$ of R that $\text{ht}(\mathfrak{p}) + \dim(R/\mathfrak{p}) = \dim(R)$ [**122**, Theorem 31.4]. Let P be a minimal prime ideal of $\mathfrak{p}R^h$ in R^h. Since R^h is flat over R with zero-dimensional fibers, $\text{ht}(\mathfrak{p}) = \text{ht}(P)$. Thus P is nonzero and non-maximal. Let Q be a minimal prime of R^h with $Q \subseteq P$. Then $Q \cap R = (0)$. We show $Q \notin \Gamma_R$: If $Q \in \Gamma_R$, then $\dim(R^h/Q) = 1$ by assumption. Thus $0 \neq \dim(R^h/P \leq \dim(R^h/Q) = 1$, and so $Q = P$. But $P \cap R = \mathfrak{p} \neq 0$ and $Q \cap R = (0)$, a contradiction. Thus $Q \notin \Gamma_R$, and so $\dim(R^h/Q) = \dim(R^h)$. Since R^h is catenary, $\text{ht}(P) + \dim(R^h/P) = \dim(R^h)$. Also $P \notin \Gamma_R$, since $P \cap R \neq (0)$. Therefore $\dim(R/\mathfrak{p}) = \dim(R^h/P)$, and so $\text{ht}(\mathfrak{p}) + \dim(R/\mathfrak{p}) = \dim(R)$. Thus R is catenary. □

COROLLARY 18.10. *If R has geometrically normal formal fibers and is catenary but not universally catenary, then there exist minimal prime ideals q of the $\mathfrak{m}$-adic completion $\widehat{R}$ of R such that* $\dim(\widehat{R}/q) = 1$.

PROOF. By Theorem 18.9, each prime ideal $Q \in \Gamma_R$ has dimension one and is a minimal prime of R^h. Moreover, $Q\widehat{R} := q$ is a minimal prime of $\widehat{R}$. Since $\dim(R^h/Q) = 1$, we have $\dim(\widehat{R}/q) = 1$. □

18.3. Flatness for the intersection of finitely many ideals

Assume the setting and notation of Homomorphic Image Construction 17.2 and Noetherian Flatness Theorem 17.13:

SETTING AND NOTATION 18.11. Let R be an integral domain with field of fractions $K := \mathcal{Q}(R)$. Let $x \in R$ be a nonzero nonunit such that $\bigcap_{n\geq 1} x^n R = (0)$, the x-adic completion R^* is Noetherian, and x is a regular element of R^*. Let I be an ideal of R^* having the property that $\mathfrak{p} \cap R = (0)$ for each $\mathfrak{p} \in \operatorname{Ass}(R^*/I)$. As in Frontpiece Notation 17.7.2 and Definition 17.10.1, let

$$U := \bigcup_{n=1}^{\infty} U_n, \quad B := \bigcup_{n=1}^{\infty} B_n = (1+xU)^{-1}U, \quad \text{and} \quad A := K \cap (R^*/I).$$

As shown in Noetherian Flatness Theorem 17.13, flatness of a certain map is equivalent to $B = A$ and B is Noetherian, for the ring B of Setting 18.11. Theorem 18.12 gives conditions for this flatness and the Noetherian property to transfer to an integral domain associated with an intersection of ideals.

THEOREM 18.12. *Assume Setting 18.11 for each of n ideals of R^*; thus R is an integral domain with field of fractions $K := \mathcal{Q}(R)$, the element $x \in R$ be a nonzero nonunit such that $\bigcap_{n\geq 1} x^n R = (0)$, the x-adic completion R^* is Noetherian, and x is a regular element of R^*, and $I_1, \ldots, I_n$ are ideals of R^* such that, for each $i \in \{1, \ldots, n\}$, each associated prime of R^*/I_i intersects R in (0). Also assume the map $R \hookrightarrow (R^*/I_i)[1/x]$ is flat for each i and that the localizations at x of the I_i are pairwise comaximal; that is, for all $i \neq j$, $(I_i + I_j)R^*[1/x] = R^*[1/x]$. Let $I := I_1 \cap \cdots \cap I_n$, $A := K \cap (R^*/I)$ and, for $i \in \{1, 2, \ldots n\}$, let $A_i := K \cap (R^*/I_i)$. Then*

1. *Each associated prime of R^*/I intersects R in (0).*
2. *The map $R \hookrightarrow (R^*/I)[1/x]$ is flat, and so the ring A is Noetherian and is equal to its associated approximation ring B. The x-adic completion A^* of A is R^*/I, and the x-adic completion A_i^* of A_i is R^*/I_i, for $i \in \{1, \ldots, n\}$.*
3. *The ring $A^*[1/x] \cong A_1^*[1/x] \times \cdots \times A_n^*[1/x]$. If $Q \in \operatorname{Spec}(A^*)$ and $x \notin Q$, then A_Q^* is a localization of precisely one of the A_i^*.*
4. *$A \subseteq A_1 \cap \cdots \cap A_n$ and $\cap_{i=1}^n A_i[1/x] \subseteq A_P$ for each $P \in \operatorname{Spec} A$ with $x \notin P$. Thus $A[1/x] = \cap_{i=1}^n A_i[1/x]$.*

PROOF. By Construction Properties Theorem 17.11.4, the x-adic completion A_i^* of A_i is R^*/I_i. Since $\operatorname{Ass}(R^*/I) \subseteq \bigcup_{i=1}^n \operatorname{Ass}(R^*/I_i)$, the condition on associated primes of Noetherian Flatness Theorem 17.13 holds for the ideal I, so item 1 holds.

For item 2, the natural R-algebra homomorphism $\pi : R^* \to \bigoplus_{i=1}^n (R^*/I_i)$ has kernel I. Since $(I_i + I_j)R^*[1/x] = R^*[1/x]$, for all $i \neq j$, the localization of π at x is onto. Thus $(R^*/I)[1/x] \cong \bigoplus_{i=1}^n (R^*/I_i)[1/x] = \bigoplus_{i=1}^n (A_i^*)[1/x]$ is flat over R. Therefore A is Noetherian and is equal to its associated approximation ring B, by Noetherian Flatness Theorem 17.13, and $A^* = R^*/I$ is the x-adic completion of A, by Theorem 17.11.4.

For item 3, if $Q \in \operatorname{Spec}(A^*)$ and $x \notin Q$, then A^*_Q is a localization of

$$A^*[1/x] \cong A^*_1[1/x] \oplus \cdots \oplus A^*_n[1/x].$$

Every prime ideal of $\bigoplus_{i=1}^n A^*_i[1/x]$ has the form $Q_iA^*_i[1/x] \oplus \bigoplus_{j\neq i} A^*_j[1/x]$, where $Q_i \in \operatorname{Spec}(A^*_i)$ for a unique $i \in \{1,\ldots,n\}$. It follows that A^*_Q is a localization of A^*_i for precisely this i. That is, $A^*_Q = (A^*_i)_{Q_i}$.

Since R^*/I_i is a homomorphic image of R^*/I, the ring $A \subseteq A_i$, for each i; see Exercise 2 of Chapter 17. Let $P \in \operatorname{Spec} A$ with $x \notin P$. Since $A^* = R^*/I$ is faithfully flat over A, there exists $P^* \in \operatorname{Spec}(A^*)$ with $P^* \cap A = P$. Then $x \notin P^*$ implies $A^*_{P^*} = (A^*_i)_{P^*_i}$, where $P^*_i \in \operatorname{Spec}(A^*_i)$ for some $i \in \{1,\ldots,n\}$. Let $P_i = P^*_i \cap A_i$. Since $A_P \hookrightarrow A^*_{P^*}$ and $(A_i)_{P_i} \hookrightarrow (A^*_i)_{P^*_i}$ are faithfully flat,

$$A_P = A^*_{P^*} \cap K = (A^*_i)_{P^*_i} \cap K = (A_i)_{P_i} \supseteq (A_i)[1/x],$$

by Remark 2.37.9. It follows that $\bigcap_{i=1}^n A_i[1/x] \subseteq A_P$. Thus

$$\bigcap_{i=1}^n A_i[1/x] \;\subseteq\; \bigcap\{A_P \mid P \in \operatorname{Spec} A \text{ and } x \notin P\} \;=\; A[1/x].$$

Since $A[1/x] \subseteq A_i[1/x]$, for each i, we have $A[1/x] = \bigcap_{i=1}^n A_i[1/x]$. □

18.4. Regular maps and geometrically regular formal fibers

Proposition 18.13 shows that certain regularity conditions on the base ring R and the extension $R \hookrightarrow R^*/I$ in Noetherian Flatness Theorem 17.13 (Homomorphic Image Version) yield geometrically regular formal fibers for the constructed ring A.

PROPOSITION 18.13. *Let R, x, R^*, A, B and I be as in Setting and Notation 18.11. Assume that the map $\psi_P : R_{P\cap R} \hookrightarrow (R^*/I)_P$ is regular, for each $P \in \operatorname{Spec}(R^*/I)$ with $x \notin P$. Then $A = B$ and moreover:*

1. *A is Noetherian and the map $A \longrightarrow A^* = R^*/I$ is regular.*
2. *If R is Noetherian semilocal with geometrically regular formal fibers and x is in the Jacobson radical of R, then A has geometrically regular formal fibers.*

PROOF. By Remark 2.37.1, flatness is a local property. Since regularity of a map includes flatness, the map $\psi_x : R \hookrightarrow (R^*/I)[1/x]$ is flat. By Theorem 17.13, the intersection ring A is Noetherian with x-adic completion $A^* = R^*/I$. Hence $A \longrightarrow A^*$ is flat.

Let $Q \in \operatorname{Spec}(A)$, let $\mathfrak{q} = Q \cap R$, let $k(Q)$ denote the field of fractions of A/Q, and let $A^*_{QA^*} = (A \setminus Q)^{-1}A^*$.

CASE 1. $x \in Q$. Then $R/\mathfrak{q} = A/Q = A^*/QA^*$. By Equation 3.29.0,

$$A^* \otimes_A k(Q) \;=\; \frac{A^*_{QA^*}}{QA^*_{QA^*}} \;=\; \frac{A_Q}{QA_Q} \;=\; k(Q).$$

Thus regularity holds in this case.

CASE 2. $x \notin Q$. Let L be a finite algebraic field extension of $k(Q)$. We show the ring $A^* \otimes_A L$ is regular. There is a natural embedding $A^* \otimes_A k(Q) \hookrightarrow A^* \otimes_A L$. Let $W \in \operatorname{Spec}(A^* \otimes_A L)$ and let $W' = W \cap (A^* \otimes_A k(Q))$. We have maps

$$\operatorname{Spec}(A^* \otimes_A k(Q)) \overset{\theta,\cong}{\to} \operatorname{Spec}\Big(\frac{A^*_{QA^*}}{QA^*_{QA^*}}\Big) \quad\text{and}\quad \operatorname{Spec}\Big(\frac{A^*_{QA^*}}{QA^*_{QA^*}}\Big) \overset{\rho}{\to} \operatorname{Spec} A^*,$$

since $A^*_{QA^*}/QA^*_{QA^*} = A^* \otimes_A k(Q)$ by Equation 3.29.0, and $A^* \to A^*_{QA^*}/QA^*_{QA^*}$. Let P be the prime ideal $P := \rho(\theta(W')) \in \operatorname{Spec}(A^*)$; then $P \cap A = Q$.

By assumption the map

$$R_{\mathfrak{q}} \hookrightarrow (R^*/I)_P = A^*_P$$

is regular. Since $x \notin Q$, it follows that $R_{\mathfrak{q}} = U_{Q\cap U} = A_Q$ and that $k(\mathfrak{q}) = k(Q)$. Thus the ring $A^*_P \otimes_{A_Q} L$ is regular. Therefore the localization$(A^* \otimes_A L)_W$ of $A^*_P \otimes_{A_Q} L$ is also regular.

For item 2, use a theorem of Rotthaus [**157**, (3.2), p. 179]: If R is a Noetherian semilocal ring with geometrically regular formal fibers and I_0 is an ideal of R contained in the Jacobson radical of R, then the I_0-adic completion of R also has geometrically regular formal fibers; see also [**122**, Remark 2, p. 260]. Thus R^* has geometrically regular formal fibers. Since the formal fibers of R^*/I are a subset of the formal fibers of R^*, the map $A^* = R^*/I \longrightarrow \widehat{A} = \widehat{(R^*/I)}$ is regular. By item 1, the map $A \to A^*$ is regular. The composition of two regular maps is regular [**122**, Thm. 32.1 (i)]. Therefore A has geometrically regular formal fibers, that is, the map $A \longrightarrow \widehat{A}$ is regular. □

Theorem 18.14 describes conditions such that the property of regularity of formal fibers for a Noetherian domain $A = B$ of Setting 18.11 transfers to an integral domain associated with an intersection of ideals.

THEOREM 18.14. *Let n be a positive integer, let R be a Noetherian integral domain with field of fractions K, let x be a nonzero nonunit of R, and let R^* denote the x-adic completion of R. Let $I_1, \ldots, I_n$ be ideals of R^* and let $I := I_1 \cap \cdots \cap I_n$. Assume that*

(1) *For each associated prime ideal P of each R^*/I_i, we have $P \cap R = (0)$.*
(2) *R is semilocal with geometrically regular formal fibers and x is in the Jacobson radical of R.*
(3) *Each $(R^*/I_i)[1/x]$ is a flat R-module and, for each $i \neq j$, the ideals $I_i R^*[1/x]$ and $I_j R^*[1/x]$ are comaximal in $R^*[1/x]$.*
(4) *For $i = 1, \ldots, n$, $A_i := K \cap (R^*/I_i)$ has geometrically regular formal fibers.*

Then $A := K \cap (R^*/I)$ is Noetherian and is equal to its approximation domain B, and A has geometrically regular formal fibers.

PROOF. Since R has geometrically regular formal fibers, it suffices to show, for $W \in \operatorname{Spec}(R^*/I)$ with $x \notin W$ and $W_0 := W \cap R$, that $R_{W_0} \longrightarrow (R^*/I)_W$ is regular, by Proposition 18.13.2. As in Theorem 18.12,

$$(R^*/I)[1/x] \;=\; (R^*/I_1)[1/x] \;\oplus \cdots \oplus\; (R^*/I_n)[1/x].$$

It follows that $(R^*/I)_W$ is a localization of R^*/I_i for some $i \in \{1, \ldots, n\}$. If $(R^*/I)_W = (R^*/I_i)_{W_i}$, where $W_i \in \operatorname{Spec}(R^*/I_i)$, then $R_{W_0} = (A_i)_{W_i \cap A_i}$ and $(A_i)_{W_i \cap A_i} \longrightarrow (R^*/I_i)_{W_i}$ is regular. Thus $R_{W_0} \longrightarrow (R^*/I)_W$ is regular. □

18.5. Examples that are not universally catenary

This section includes non-excellent examples obtained using Inclusion Construction Prototypes that are adjusted to the terminology of Homomorphic Image Construction 17.2 as in Remark 17.25.

The ring A of Example 18.15 is a two-dimensional Noetherian local domain that birationally dominates a three-dimensional RLR. The ring A has geometrically regular formal fibers, and is not universally catenary. This example is obtained via an intersection of two ideals.

EXAMPLE 18.15. Let k be a field of characteristic zero, and let x, y and z be indeterminates over k. Let $R = k[x,y,z]_{(x,y,z)}$, let K denote the field of fractions of R, and let $\tau_1, \tau_2, \tau_3 \in xk[[x]]$ be algebraically independent over $K = k(x,y,z)$. Let $R^* = k[y,z]_{(y,z)}[[x]]$, the x-adic completion of R. Apply Proposition 17.26.2, where the ideal I is the height-two prime ideal $Q := (z-\tau_1, y-\tau_2)R^*$, and then where I is $P := (z-\tau_3)R^*$, a height-one prime ideal. Then $A_1 := K \cap (R^*/P)$ and $A_2 := K \cap (R^*/Q)$ are Noetherian domains of the form described in Remark 17.25. Thus $(R^*/P)[1/x]$ and $(R^*/Q)[1/x]$ are both flat over R. Here $R^*/P \cong k[y]_{(y)}[[x]]$ and $R^*/Q \cong k[[x]]$. The ring $V := k[[x]] \cap k(x, \tau_3)$ is a DVR, and the Intersection Domain $A_1 \cong V[y]_{(x,y)}$ is a two-dimensional regular local domain that is a directed union of three-dimensional RLRs. The Intersection Domain A_2 is a DVR. By Remark 17.25.1d and the characteristic zero assumption, the intersection rings A_1 and A_2 are excellent.

Since $\tau_1, \tau_3 \in xk[[x]]$, the ideal $(z-\tau_1, z-\tau_3)R^*$ has radical $(x,z)R^*$. Hence the ideal $P+Q$ is primary for the maximal ideal $(x,y,z)R^*$, and so, in particular, P is not contained in Q. Let I be the ideal $P \cap Q$ of R^*; the representation $I = P \cap Q$ is irredundant and $\operatorname{Ass}(R^*/I) = \{P, Q\}$. Since $P \cap R = Q \cap R = (0)$, the ring R injects into R^*/I. Let $A := K \cap (R^*/I)$.

By Theorem 18.12.1, the inclusion $R \hookrightarrow (R^*/I)[1/x]$ is flat, the ring A is Noetherian, A equals its Approximation Domain B and A is a localization of a subring of $R[1/x]$. The map $A \hookrightarrow \widehat{A}$ of A into its completion factors through the map $A \hookrightarrow A^* = R^*/I$. Since R^*/I has minimal primes P/I and Q/I with $\dim R^*/P = 2$ and $\dim R^*/Q = 1$, and since $\widehat{A}$ is faithfully flat over $A^* = R^*/I$, the ring $\widehat{A}$ is not equidimensional. It follows that A is not universally catenary by Ratliff's Equidimension Theorem 3.25. By Remark 3.27, every homomorphic image of a regular local ring, or even of a Cohen-Macaulay local ring, is universally catenary; thus A is not a homomorphic image of a regular local ring.

Finally we show that the ring A of Example 18.15 has geometrically regular formal fibers; that is, the map $\phi : A \hookrightarrow \widehat{A}$ is regular. By the definition of R and the observations above, $A = B$ and A_1 and A_2 are excellent. Thus the hypotheses of Theorem 18.14 are satisfied, and so A has geometrically regular formal fibers.

REMARKS 18.16. The completion $\widehat{A}$ of the ring A of Example 18.15 has two minimal primes, one of dimension one and one of dimension two. As we observe above, A is not universally catenary by Ratliff's Equidimension Theorem 3.26. Another example of a Noetherian local domain that is not universally catenary but has geometrically regular formal fibers is given by Grothendieck in [**62**, (18.7.7), page 144] using a gluing construction; also see Greco's article[**61**, (1.1)]. We obtain rings similar to the ring A of Example 18.15 that have any finite number of minimal prime ideals and that are not universally catenary in Examples 18.18-18.20.

NOTES 18.17. We outline the general procedure used for the remaining examples of this section and give some justification here. Let $n \in \mathbb{N}$ and let R be a localized polynomial ring R over a field in $n+1$ variables, where x is one of the variables. Use Proposition 17.26.2 to obtain, for each i with $1 \le i \le n$, a suitable ideal I_i of the x-adic completion R^* of R and an integral domain A_i inside R^* associated to I_i so that the I_i and the A_i fit the hypotheses of Theorem 18.12, and so that the ring A of Theorem 18.12 associated to the intersection $I = \bigcap_{i=1}^n I_i$ has the desired properties. By Construction Properties Theorem 17.11.4, the x-adic completion A_i^* of A_i is R^*/I_i.

If $\operatorname{char} k = 0$, the rings A_i are excellent by Remark 17.25.1d. Thus the A_i have generically regular formal fibers if $\operatorname{char} k = 0$. By Theorem 18.14, A has geometrically regular formal fibers. On the other hand, If k is a perfect field with $\operatorname{char} k \neq 0$, it follows from Remark 10.5 that each A_i is not a Nagata ring, and is not excellent.

We construct in Example 18.18 a two-dimensional Noetherian local domain having geometrically regular formal fibers such that the completion has any desired finite number of minimal primes of dimensions one and two.

EXAMPLE 18.18. Let r and s be positive integers and let R be the localized polynomial ring in three variables $R := k[x, y, z]_{(x,y,z)}$, where k is a field of characteristic zero and the field of fractions of R is $K := k(x, y, z)$. Then the x-adic completion of R is $R^* := k[y, z]_{(y,z)}[[x]]$. Let $\tau_1, \dots, \tau_r, \beta_1, \beta_2, \dots, \beta_s, \gamma \in xk[[x]]$ be algebraically independent power series over $k(x)$. Define, as in Proposition 17.26.2,

$$Q_i := (z - \tau_i, y - \gamma)R^* \quad \text{and} \quad P_j := (z - \beta_j)R^*,$$

for $i \in \{1, \dots, r\}$ and $j \in \{1, \dots, s\}$. Apply Theorem 18.12 with $I_i = Q_i$, for $1 \le i \le r$, and $I_{r+j} = P_j$, for $1 \le j \le s$. Then $\{I_\ell \mid 1 \le \ell \le r + s\}$ satisfies the comaximality condition of Theorem 18.12 at the localization at x. As in Notes 18.17, Remark 17.25 implies that each map $R \hookrightarrow (R^*/I_\ell)[1/x]$ is flat and each $A_\ell := K \cap (R^*/I_\ell)$ is excellent. Let $I := I_1 \cap \cdots \cap I_{r+s}$ and $A := K \cap (R^*/I)$. By Theorem 18.12, the map $R \hookrightarrow (R^*/I)[1/x]$ is flat and A is Noetherian. Since $I = \bigcap_{1 \le \ell \le r+s} I_\ell$ and $\widehat{R}$ is the completion of R^*, we have $I\widehat{R} = \bigcap_{1 \le \ell \le r+s}(I_\ell \widehat{R})$, by Remark 2.37.11. Since each R^*/I_ℓ is a regular local ring, the extension $I_\ell \widehat{R}$ is a prime ideal. Then

$$\widehat{A} = \widehat{A^*} = \widehat{R^*}/I\widehat{R^*} = \widehat{R}/I\widehat{R} = \widehat{R}/(\cap_{1 \le \ell \le r+s} I_\ell \widehat{R}).$$

Thus the minimal primes of $\widehat{A}$ all have the form $\mathfrak{p}_\ell := I_\ell \widehat{A}$.

For J an ideal of R* containing I, let $\bar{J}$ denote the image of J in R^*/I. Then, for each i with $1 \le i \le r$, $\dim((R^*/I)/\bar{Q}_i) = \dim(R^*/Q_i) = 1$ and, for each j with $1 \le j \le s$, $\dim((R^*/I)/\bar{P}_j) = 2$. Thus A^* contains r minimal primes of dimension one and s minimal primes of dimension two. Since A^* modulo each of its minimal primes is a regular local ring, the completion $\widehat{A}$ of A also has precisely r minimal primes of dimension one and s minimal primes of dimension two.

We show that the stated properties hold for the integral domain A. The format of the general Homomorphic Image Construction 17.2 and the details of the construction of the integral domain A imply that A birationally dominates the 3-dimensional regular local domain R and that A is birationally dominated by each of the A_i.

By the definition of R and the observations given in Proposition 18.13, the hypotheses of Theorem 18.14 are satisfied. Theorem 18.14 implies that A has geometrically regular formal fibers. Since $\dim(A) = 2$, A is catenary, but not universally catenary.

Example 18.19 shows, for every integer $n \geq 2$, that there is a Noetherian local domain $(A, \mathfrak{m})$ of dimension n that has geometrically regular formal fibers and is catenary but not universally catenary.

EXAMPLE 18.19. Let $R = k[x, y_1, \dots, y_n]_{(x,y_1,\dots,y_n)}$ be a localized polynomial ring of dimension $n+1$ where k is a field of characteristic zero. Let $\sigma, \tau_1, \dots, \tau_n$ be $n+1$ elements of $xk[[x]]$ that are algebraically independent over $k(x)$ and consider the ideals

$$I_1 = (y_1 - \sigma)R^* \quad \text{and} \quad I_2 = (y_1 - \tau_1, \dots, y_n - \tau_n)R^*.$$

of the ring $R^* = k[y_1, \dots, y_n]_{(y_1,\dots,y_n)}[[x]]$. Then the ring

$$A = k(x, y_1, \dots, y_n) \cap (R^*/(I_1 \cap I_2))$$

is the desired example. As in Notes 18.17, each ring $k(x, y_1, \dots, y_n) \cap R^*/I_i$ is excellent. This implies that A is Noetherian with geometrically regular fibers. By an argument similar to that of Example 18.18, the completion $\widehat{A}$ of A has two minimal primes, $I_1\widehat{A}$ having dimension n and $I_2\widehat{A}$ having dimension one. Therefore the Henselization A^h has precisely two minimal prime ideals. Label these two prime ideals P and Q, where $P\widehat{A} = I_1\widehat{A}$ and $Q\widehat{A} = I_2\widehat{A}$. Thus $\dim(A^h/P) = n$ and $\dim(A^h/Q) = 1$. By Theorem 18.9, A is catenary but not universally catenary. By Theorem 18.14, A has geometrically regular formal fibers.

Example 18.20 is a construction, for each positive integer t and specified nonnegative integers $n_1, \dots, n_t$ with $n_1 \geq 1$, of a t-dimensional Noetherian local domain A with geometrically regular formal fibers. The domain A birationally dominates a $(t+1)$-dimensional regular local domain, and the completion $\widehat{A}$ of A contains exactly n_r minimal primes $\mathfrak{p}_{rj}$ of dimension $t+1-r$, for each r with $1 \leq r \leq t$. Moreover, each $\widehat{A}/\mathfrak{p}_{rj}$ is a regular local ring of dimension $t+1-r$. If $n_i > 0$ for some $i \neq 1$, then A is not universally catenary and is not a homomorphic image of a regular local domain. It follows from Remark 18.4 that the derived normal ring $\overline{A}$ of A has exactly n_r maximal ideals of height $t+1-r$ for each r with $1 \leq r \leq t$.

EXAMPLE 18.20. Let t be a positive integer and let n_r be a nonnegative integer for each r with $1 \leq r \leq t$. Assume that $n_1 \geq 1$. We construct a t-dimensional Noetherian local domain A that has geometrically regular formal fibers such that $\widehat{A}$ has exactly n_r minimal primes of dimension $t+1-r$ for each r. Let $x, y_1 \dots, y_t$ be indeterminates over a field k of characteristic zero.

Let $R = k[x, y_1, \dots, y_t]_{(x,y_1,\dots,y_t)}$, let $R^* = k[y_1, \dots, y_t]_{(y_1,\dots,y_t)}[[x]]$ denote the x-adic completion of R, and let K denote the field of fractions of R. For every r with $1 \leq r \leq t$ and $n_r = 0$, define $P_{r0} := R^*$.

For every $r, j, i \in \mathbb{N}$ such that $1 \leq r \leq t$, $1 \leq j \leq n_r$ and $1 \leq i \leq r$, choose elements $\{\tau_{rji}\}$ of $xk[[x]]$ so that the set $\bigcup\{\tau_{rji}\}$ is algebraically independent over $k(x)$. For every r, j with $1 \leq r \leq t$ and $1 \leq j \leq n_r$, let $P_{rj} := (y_1 - \tau_{rj1}, \dots, y_r - \tau_{rjr})$, a prime ideal of height r in R^*. If $n_r > 0$, then R^*/P_{rj} is a regular local ring of dimension $t+1-r$. Theorem 17.11.4 implies that $A_{rj} := K \cap (R^*/P_{rj})$ has

x-adic completion R^*/P_{rj}, and Theorem 17.11.4 implies that A_{rj} is local. Proposition 17.26.2 applies to this situation, with R, $y_1, \dots, y_r$, $y_{r+1}, \dots y_t$, and P_{rj} taking the place of S, $t_1, \dots, t_s$, $y_1, \dots, y_r$, and I respectively in Setting 17.24. Thus, by Remark 17.25, the extension $R \hookrightarrow (R^*/P_{rj})[1/x]$ is flat and A_{rj} is an RLR of dimension $t+1-r$ that has x-adic completion R^*/P_{rj}.

Let $I := \bigcap P_{rj}$ be the intersection of all the prime ideals P_{rj}. Since the elements τ_{rji} of $xk[[x]]$ are algebraically independent over $k(x)$, the sum $P_{rj}+P_{mi}$ has radical $(x, y_1, \dots, y_m)R^*$, for every r, m, i, j with $r \leq m \leq t$ and $(r,j) \neq (m,i)$. Thus $(P_{rj} + P_{mi})R^*[1/x] = R^*[1/x]$, the representation of I as the intersection of the P_{rj} is irredundant, and $\operatorname{Ass}(R^*/I) = \{P_{rj} \,|\, 1 \leq r \leq t, 1 \leq j \leq n_r\}$. Since each $P_{rj} \cap R = (0)$, we have $R \hookrightarrow R^*/I$, and the intersection domain $A := K \cap (R^*/I)$ is well defined. Moreover the x-adic completion A^* of A is R^*/I by Construction Properties Theorem 17.11.4.

By Theorem 18.12.2, the map $R \hookrightarrow (R^*/I)[1/x]$ is flat, A is Noetherian and A is a localization of a subring of $R[1/x]$. Since $I = \bigcap P_{rj}$ and $\widehat{R}$ is the completion of R^*, we have $I\widehat{R} = \bigcap P_{rj}\widehat{R}$ by Remark 2.37.11. Since R^*/P_{rj} is a regular local ring, the extension $P_{rj}\widehat{R}$ is a prime ideal. Then

$$\widehat{A} = \widehat{A^*} = \widehat{R^*}/I\widehat{R^*} = \widehat{R}/I\widehat{R} = \widehat{R}/(\cap P_{rj}\widehat{R^*}).$$

Thus the minimal primes of $\widehat{A}$ all have the form $\mathfrak{p}_{rj} := P_{rj}\widehat{A}$. Since R^*/P_{rj} is a regular local ring of dimension $t+1-r$, each $\widehat{A}/\mathfrak{p}_{rj}$ is a regular local ring of dimension $t+1-r$. The ring A birationally dominates the $(t+1)$-dimensional regular local domain R. By Theorem 18.14, A has geometrically regular formal fibers.

REMARKS 18.21. (1) Examples 18.18 and 18.19 are special cases of Example 18.20. By Theorem 18.9, the ring A constructed in Example 18.20 is catenary if and only if each minimal prime of $\widehat{A}$ has dimension either one or t. By taking $n_r = 0$ for $r \notin \{1, t\}$ in Example 18.20, we obtain additional examples of catenary Noetherian local domains A of dimension t having geometrically regular formal fibers for which the completion $\widehat{A}$ has precisely n_t minimal primes of dimension one and n_1 minimal primes of dimension t; thus A is not universally catenary.

(2) Let $(A, \mathfrak{n})$ be a Noetherian local domain constructed as in Example 18.20, let A^h denote the Henselization of A, and let A^* denote the x-adic completion of A. Since each minimal prime of $\widehat{A}$ is the extension of a minimal prime of A^h and also the extension of a minimal prime of A^*, the minimal primes of A^h and A^* are in a natural one-to-one correspondence. Let P be the minimal prime of A^h corresponding to a minimal prime $\mathfrak{p}$ of A^*. Since the minimal primes of A^* extend to pairwise comaximal prime ideals of $A^*[1/x]$, for each prime ideal $Q \supset P$ of A^h with $x \notin Q$, the prime ideal P is the unique minimal prime of A^h contained in Q. Let $\mathfrak{q} := Q \cap A$. Then $\operatorname{ht} \mathfrak{q} = \operatorname{ht} Q$, and either $\dim(A/\mathfrak{q}) > \dim(A^h/Q)$ or else every saturated chain of prime ideals of A containing $\mathfrak{q}$ has length less than $\dim A$.

In connection with Remark 18.21.2, we ask:

QUESTION 18.22. Let $(A, \mathfrak{n})$ be a Noetherian local domain constructed as in Example 18.20. If A is not catenary, what can be said about the cardinality of the set

$$\Gamma_A := \{P \in \operatorname{Spec}(A^h) \mid \dim(A^h/P) < \dim(A/(P \cap A))\}?$$

Is the set Γ_A ever infinite?

18.6. The depth of the constructed rings

We thank Lucho Avramov for suggesting we consider the depth of the rings constructed in Example 18.20; "depth" is defined in Definition 3.35.

REMARK 18.23. The catenary rings that arise from the construction in Example 18.20 all have depth one. However, Example 18.20 can be used to construct, for each integer $t \geq 3$ and integer d with $2 \leq d \leq t-1$, an example of a non-catenary Noetherian local domain A of dimension t and depth d having geometrically regular formal fibers. The x-adic completion A^* of A has precisely two minimal primes, one of dimension t and one of dimension d. To establish the existence of such an example, with notation as in Example 18.20, we set $m = t - d + 1$ and take $n_r = 0$ for $r \notin \{1, m\}$ and $n_1 = n_m = 1$. Let

$$P_1 := P_{11} = (y_1 - \tau_{111})R^* \text{ and } P_m := P_{m1} = (y_1 - \tau_{m11}, \ldots, y_m - \tau_{m1m})R^*.$$

Consider $A^* = R^*/(P_1 \cap P_m)$ and the short exact sequence

$$0 \longrightarrow \frac{P_1}{P_1 \cap P_m} \longrightarrow \frac{R^*}{P_1 \cap P_m} \longrightarrow \frac{R^*}{P_1} \longrightarrow 0.$$

Since P_1 is principal and not contained in P_m, we have $P_1 \cap P_m = P_1 P_m$ and $P_1/(P_1 \cap P_m) \cong R^*/P_m$. It follows that $\operatorname{depth} A^* = \operatorname{depth}(R^*/P_m) = d$; [**103**, page 103, ex 14] or [**29**, Prop. 1.2.9, page 11]. Since the local ring A and its x-adic completion have the same completion $\widehat{A}$ with respect to their maximal ideals, we have $\operatorname{depth} A = \operatorname{depth} \widehat{A} = \operatorname{depth} A^*$ [**122**, Theorem 17.5]. By Remark 18.4, the derived normal ring $\overline{A}$ of A has precisely two maximal ideals one, of height t and one of height d.

Exercises

(1) Let $(R, \mathfrak{m})$ be a three-dimensional Noetherian local domain such that each height-one prime ideal of R is the radical of a principal ideal. Prove that R is catenary.

(2) Let $(R, \mathfrak{m})$ be a catenary Noetherian local domain having geometrically normal formal fibers. If R is not universally catenary, prove that R has depth one.

SUGGESTION. Use Theorem 18.9 and the following theorem:

THEOREM 18.24. [**122**, Theorem 17.2] *Let $(R, \mathfrak{m})$ be a Noetherian local ring and $M \neq (0)$ a finite R-module. Then* $\operatorname{depth} M \leq \dim R/\mathfrak{p}$, *for every prime ideal* $\mathfrak{p}$ *of R associated to M.*

(3) Let R be an integral domain with field of fractions K and let R' be a subring of K that contains R. If $P \in \operatorname{Spec} R$ is such that $R' \subseteq R_P$, prove that there exists a unique prime ideal $P' \in \operatorname{Spec} R'$ such that $P' \cap R = P$.

(4) For the rings A and A^* of Example 18.15, prove that A^* is universally catenary.

CHAPTER 19

An Ogoma-like example

In this chapter we present an example that has the same properties as Ogoma's famous example of a normal Noetherian local domain that is not catenary. The example is constructed using Homomorphic Image Construction 17.2.

If a Noetherian local ring A dominates an excellent local subring R and A has the same completion as R, then A has Cohen-Macaulay formal fibers; see Corollary 19.2. This applies to examples obtained by Inclusion Construction 5.3, but not to examples obtained by Homomorphic Image Construction 17.2. Construction 17.2 is useful for creating examples where the formal fibers are not Cohen-Macaulay.

Section 19.1 features integral domains B and A arising from Inclusion Construction 5.3 and an integral domain C obtained with Homomorphic Image Construction 17.2. Theorems 19.8 and 19.9 show that A and B are non-Noetherian and $B \subsetneq A$. By Theorem 19.11, C is a 2-dimensional Noetherian local domain that is a homomorphic image of B and the generic formal fiber of C is not Cohen-Macaulay.

Section 19.2 features Example 19.13, which is similar to Ogoma's Example 19.4. Theorem 19.15 establishes that Example 19.13 has the properties of Ogoma's Example.

19.1. Cohen-Macaulay formal fibers

Proposition 19.1 is analogous to [**122**, Theorem 32.1(ii)]. The distinction is that we are considering regular fibers rather than geometrically regular fibers. The proof given in [**122**, Theorem 32.1(ii)] applies to establish Proposition 19.1.

PROPOSITION 19.1. *Suppose R, S, and T are Noetherian commutative rings and suppose we have maps $R \to S$ and $S \to T$ and the composite map $R \to T$. Assume*

(i) *$R \to T$ is flat with regular fibers,*
(ii) *$S \to T$ is faithfully flat.*

Then $R \to S$ is flat with regular fibers.

Theorem 7.4 and Proposition 19.1 imply the following result concerning Cohen-Macaulay formal fibers.

COROLLARY 19.2. *Let B be a Noetherian local ring that dominates an excellent local subring R and has the same completion as R. Then B has Cohen-Macaulay formal fibers. Thus every Noetherian ring A obtained via Inclusion Construction* 5.3 *with an excellent local domain R as the base ring has Cohen-Macaulay formal fibers.*

PROOF. Consider the inclusion maps $R \hookrightarrow B \hookrightarrow \widehat{B} = \widehat{R}$. Since R is excellent, the map $R \hookrightarrow \widehat{R}$ is faithfully flat with regular fibers. Since B is Noetherian, the map $B \hookrightarrow \widehat{B}$ is faithfully flat. By Proposition 19.1, the map $R \hookrightarrow B$ is flat with regular fibers. Theorem 7.4 implies that B has Cohen-Macaulay formal fibers. The second statement follows from Construction Properties Theorem 5.14 and Proposition 5.17.5.a. □

DISCUSSION 19.3. Let $(A, \mathfrak{n})$ be a Noetherian local domain that has a coefficient field k and has the property that the field of fractions L of A is finitely generated over k. Corollary 4.3 implies that there exists a local subring $(R, \mathfrak{m})$ of A such that A birationally dominates R. The ring R is essentially finitely generated over k, and there exists an ideal I of the completion $\widehat{R}$ of R such that $A = L \cap (\widehat{R}/I)$ and $\widehat{A} = \widehat{R}/I$. Depending on properties of the ideal I, the generic formal fiber of A is or is not Cohen-Macaulay.

Corollary 19.2 is related to Ogoma's famous example, described in Example 19.4, of a Nagata local domain of dimension three whose generic formal fiber is not equidimensional.

EXAMPLE 19.4. [**146**] (Features of Ogoma's construction) Let k be a countable field of infinite but countable transcendence degree over the field $\mathbb{Q}$ of rational numbers, let x, y, z, w be variables over k, and let $R = k[x, y, z, w]_{(x,y,z,w)}$ be the localized polynomial ring with maximal ideal $\mathfrak{m} = (x, y, z, w)$ and $\mathfrak{m}$-adic completion $\widehat{R} = k[[x, y, z, w]]$. By a clever enumeration of the prime elements in R, Ogoma constructs a "multi-adic" completion R^{multi} inside $\widehat{R}$ and three power series f, g, h in R^{multi} such that the following five conditions hold for f, g, h and the ideals $I := (fg, fh)\widehat{R}$ and $\widehat{P} := (f, g, h)\widehat{R}$ of $\widehat{R}$: [1]

(a) f, g, h are algebraically independent over $K := k(x, y, z, w)$.
(b) f, g, h are part of a regular system of parameters for $\widehat{R} = k[[x, y, z, w]]$.
(c) $\widehat{P} \cap R = (0)$, i.e., $\widehat{P}$ is in the generic formal fiber of R.
(d) $C := K \cap (\widehat{R}/I)$ is a Nagata local domain[2] with completion $\widehat{C} = \widehat{R}/I$.
(e) $\widehat{C} = \widehat{R}/I$ has a minimal prime ideal $f\widehat{R}/I$ of dimension 3 and a minimal prime ideal $(g, h)\widehat{R}/I$ of dimension 2. Thus C fails to be formally equidimensional. By Ratliff's Equidimension Theorem 3.26, C is not universally catenary, and so C is a counterexample to Chain Conjecture 18.2.

REMARKS 19.5. (Cohen-Macaulay formal fibers) Luchezar Avramov pointed out to us the following facts about formal fibers.

(1) Every homomorphic image R/I of a regular local ring R has the property that every local ring in a formal fiber of R/I has the form $\widehat{R}_{\mathfrak{q}}/\mathfrak{p}\widehat{R}_{\mathfrak{q}}$, where $\mathfrak{q}$ is a prime ideal of $\widehat{R}$ with $I \subseteq \mathfrak{q}$ and $\mathfrak{p} = \mathfrak{q} \cap R$ is a prime ideal of R with $I \subseteq \mathfrak{p}$. Then $\widehat{R}_{\mathfrak{q}}$ is a regular local ring and the inclusion map $R \hookrightarrow \widehat{R}_{\mathfrak{q}}$ is flat and factors through the inclusion map $R_{\mathfrak{p}} \hookrightarrow \widehat{R}_{\mathfrak{q}}$ that is faithfully flat. Since $R_{\mathfrak{p}}$ is regular, $\mathfrak{p}R_{\mathfrak{p}}$ is generated by $\operatorname{ht} \mathfrak{p}$ elements. Since the map $R_{\mathfrak{p}} \hookrightarrow \widehat{R}_{\mathfrak{q}}$ is faithfully flat, $\operatorname{ht} \mathfrak{p} = \operatorname{ht} \mathfrak{p}\widehat{R}_{\mathfrak{q}}$. Therefore $\widehat{R}_{\mathfrak{q}}/\mathfrak{p}\widehat{R}_{\mathfrak{q}}$ is a homomorphic image of the regular local ring $\widehat{R}_{\mathfrak{q}}$ modulo the complete intersection ideal $\mathfrak{p}\widehat{R}_{\mathfrak{q}}$. It follows that $\widehat{R}_{\mathfrak{q}}/\mathfrak{p}\widehat{R}_{\mathfrak{q}}$ is Cohen-Macaulay [**122**, Theorem 17.4, p. 135].

---

[1] We have changed the elements to f, g, h instead of g, h, ℓ, the notation used by Ogoma.

[2] Ogoma [**146**, page 158] actually constructs C as a directed union of birational extensions of R. He proves that C is Noetherian and that $\widehat{C} = \widehat{R}/I$. It follows that $C = K \cap (\widehat{R}/I)$. Heitmann observes in [**98**] that C is already normal.

(2) Every homomorphic image R/I of a Cohen-Macaulay local ring R has Cohen-Macaulay formal fibers. Let $\mathfrak{p}$ be a prime ideal of R with $I \subseteq \mathfrak{p}$ and let $\mathfrak{q} \in \operatorname{Spec} \widehat{R}$ with $\mathfrak{q} \cap R = \mathfrak{p}$. Since R is Cohen-Macaulay, $\widehat{R}$ is Cohen-Macaulay [**122**, Theorem 17.5]. It follows that $R_{\mathfrak{p}} \hookrightarrow \widehat{R}_q$ is a faithfully flat local map of Cohen-Macaulay local rings, and $\widehat{R}_q/\mathfrak{p}\widehat{R}_q$ is Cohen-Macaulay by [**122**, Corollary, page 181].

It is interesting that while regular local rings need not have regular formal fibers, they do have Cohen-Macaulay formal fibers.

In Example 19.6, the two forms of the basic construction technique (Inclusion and Homomorphic Image Constructions 5.3 and 17.2) are used to obtain three rings A, B and C. The ring B maps surjectively onto C, while A does not.

EXAMPLE 19.6. Let x, y, z be variables over a field k and let R be the localized polynomial ring $R = k[x,y,z]_{(x,y,z)}$. Let $\tau_1, \tau_2 \in xk[[x]]$ be formal power series in x that are algebraically independent over $k(x)$. Consider the discrete valuation domain

$$V := k(x, \tau_1, \tau_2) \cap k[[x]] = \bigcup_{n=1}^{\infty} k[x, \tau_{1n}, \tau_{2n}]_{(x,\tau_{1n},\tau_{2n})},$$

where τ_{1n}, τ_{2n} are the endpieces of τ_1 and τ_2, for $n \in \mathbb{N}_0$, as in Notation 5.4. The equality holds by Remark 4.20. Thus V is a nested union of localized polynomial rings in 3 variables over k. Let D be the Local Prototype of Definition 4.28 associated to $(\tau_1, \tau_2, x, y, z)$:

$$(19.6.1)\quad D := V[y,z]_{(x,y,z)} = U_{((x,y,z)D)\cap U}, \quad \text{where} \quad U := \bigcup_{n=1}^{\infty} k[x,y,z,\tau_{1n},\tau_{2n}].$$

By Prototype Theorems 10.2 and 10.6 and Proposition 5.17.5a, the ring D is a 3-dimensional regular local ring; D is a localization of a nested union of polynomial rings in 5 variables; $D = k(x,y,z,\tau_1,\tau_2) \cap R^*$, where $R^* = k[y,z]_{(y,z)}[[x]]$ is the x-adic completion of R; D has maximal ideal $\mathfrak{m}_D = (x,y,z)D$ and $\mathfrak{m}_D$-adic completion $\widehat{R} = k[[x,y,z]]$; and D dominates the localized polynomial ring R.

Consider the following elements of R^*:

$$(19.6.2)\quad s := y + \tau_1, \quad t := z + \tau_2, \quad \rho := s^2 = (y+\tau_1)^2 \quad \text{and} \quad \sigma := st = (y+\tau_1)(z+\tau_2).$$

The elements s and t are algebraically independent over $k(x,y,z)$ as are the elements ρ and σ. Let ρ_n and σ_n denote the endpieces of ρ and σ as defined in Endpiece Notation 5.4. Define the ideals $I := (\rho,\sigma)R^*$, $P_1 := sR^*$ and $P_2 := (s,t)R^*$.

NOTES 19.7. (1) $\operatorname{ht} I = 1$, $I = P_1P_2$, and $P_1 \cap R = P_2 \cap R = (0)$.
(2) P_1 and P_2 are the associated prime ideals of I in R^*.
(3) The inclusion $\alpha : R \hookrightarrow R^*/I$ extends to $\alpha_{\mathcal{Q}} : k(x,y,z) \hookrightarrow \mathcal{Q}(R^*/I)$, where $k(x,y,z)$ is the field of fractions of R, and $\mathcal{Q}(R^*/I)$ is the total quotient ring of R^*/I.
(4) $\sigma^2/\rho = (z+\tau_2)^2 = t^2$ and $D \cap IR^* = (\sigma,\rho)D = ID$.
(5) The inclusion $\alpha : R \hookrightarrow R^*/I$ factors into

$$R \overset{\beta}{\hookrightarrow} D/ID \overset{\delta}{\hookrightarrow} R^*/I.$$

(6) The image of t^2 in D/ID is transcendental over R, considered in the extension $R \hookrightarrow D/ID$.

PROOF. Items 1-5 follow from Equation 19.6.2 and the other definitions. For item 6, since $D = k(x, y, z, s, t) \cap R^*$, and s and t are algebraically independent over $k(x, y, z)$, they are also algebraically independent over R. Since the image of t is transcendental over R as an element of D/sD and $ID = (s^2, st)D \subseteq sD$, it follows that the image of t^2 is transcendental over R as an element of D/ID. □

To continue the notation for Example 19.6: Define rings A and C by:

$$A := k(x, y, z, \rho, \sigma) \cap R^* \qquad C := k(x, y, z) \ \cap \ (R^*/I).$$

In analogy with the rings D and U of Equation 19.6.1, there exist rings $W \subseteq U$ and $B \subseteq D$ defined as follows:

$$W := \bigcup_{n=1}^{\infty} k[x, y, z, \rho_n, \sigma_n], \quad B := \bigcup_{n=1}^{\infty} R[\rho_n, \sigma_n]_{(x,y,z,\rho_n,\sigma_n)} = W_{(x,y,z)W},$$

where ρ_n, σ_n are the nth endpieces of ρ, σ, respectively, for each $n \in \mathbb{N}_0$. Also there is a nested union Approximation Domain E associated to C from Equation 17.10.1:

$$E = \bigcup_{n=1}^{\infty} R[\bar{\rho_n}, \bar{\sigma_n}]_{(x,y,z,\bar{\rho_n},\bar{\sigma_n})},$$

where $\bar{\rho_n}, \bar{\sigma_n}$ are the nth frontpieces of ρ and σ. By Proposition 5.17 and Theorem 17.11.6, A, B and C are local domains with maximal ideals $(x, y, z)B$, $(x, y, z)A$ and $(x, y, z)B$, $(x, y, z)C$, respectively. Also $B \subseteq A$ with A birationally dominating B.

Theorems 19.8 and 19.9 show A and B are non-Noetherian and $B \subsetneq A$. Theorem 19.11 shows that C is a Noetherian local domain with completion $\widehat{C} = \widehat{R}/I\widehat{R}$ such that C has a non-Cohen-Macaulay formal fiber.

THEOREM 19.8. *Assume the notation of Example 19.6. The local integral domains $B \subseteq A$ both have x-adic completions R^* and have completions $\widehat{R}$ with respect to the powers of their maximal ideals. Let $Q := I \cap B$. Then:*

1. *B is a UFD,*
2. *$Q = P_1 \cap B = P_2 \cap B$,*
3. $\operatorname{ht}(P_1 \cap B) > \operatorname{ht}(P_1) = 1$,
4. *B fails to have Cohen-Macaulay formal fibers, and*
5. *B is non-Noetherian.*

PROOF. It follows from Construction Properties Theorem 5.14 that R^* is the x-adic completion of both A and B, and $\widehat{R}$ is the completion of both A and B.

For item 1, Theorem 5.24.1 implies B is a UFD.

For item 2, it suffices to show $I \cap W = P_1 \cap W = P_2 \cap W$. It is clear that $I \cap W \subseteq P_1 \cap W \subseteq P_2 \cap W$. Let $v \in P_2 \cap W$; say $v \in P_2 \cap k[x, y, z, \rho_n, \sigma_n]$, for $n \in \mathbb{N}$. Let $\sigma = \sum_{i=1}^{\infty} c_i x^i$, where each $c_i \in k[x, y, z]$. Then Equation 5.4.1 implies

$$\sigma_n = \sum_{i=n+1}^{\infty} c_i x^{i-n} \implies x^n \sigma_n = \sum_{i=n+1}^{\infty} c_i x^i = \sigma - \sum_{i=1}^{n} c_i x^i \implies x^n \sigma_n \in k[x, y, z, \sigma].$$

Similarly $\rho_n \in k[x, y, z, \rho, \sigma]$. Since v is a polynomial in σ_n, ρ_n, it follows that $x^t v \in k[x, y, z, \rho, \sigma]$ for some $t \in \mathbb{N}$; say

$$x^t v = \sum b_{ij} \rho^i \sigma^j, \quad \text{where } b_{ij} \in k[x, y, z], \quad \text{for all } i, j \in \mathbb{N}.$$

Then $b_{00} = 0$, since $\rho, \sigma \in P_2$ and $P_2 \cap k[x,y,z] = (0)$. This implies that $v \in P_1$. Thus $P_1 \cap B = P_2 \cap B$, and so part of item 2 holds.

For item 3, Theorem 5.14 implies $B[1/x]$ is a localization of the ring $W_0[1/x]$. The ideal $J = (\rho, \sigma)W_0$ is a prime ideal of height 2 and $x \notin J$. Then $B_{P_1 \cap B} = (W_0)_J$ and $x \notin P_1 \cap B$. Therefore $\operatorname{ht}(P_1 \cap B) = 2$. Since $P_1 = sR^*$ has height one, this proves item 3. Moreover $J[1/x] = (\rho, \sigma)W_0[1/x] = (\rho, \sigma)W_n[1/x]$ is a prime ideal of height 2 in $W_0[1/x] = W_n[1/x]$, for every $n \in \mathbb{N}_0$, and $(\rho, \sigma)W_n \subseteq I \cap W_n$. Thus $I \cap W$ contains and is contained in the height-two prime ideal $JW = P_1 \cap W$ of W, and so $Q = I \cap W = P_1 \cap W$. Hence item 2 holds.

Item 3 implies item 5, since $\operatorname{ht}(P_1 \cap B) > \operatorname{ht} P_1$ implies that $B \to \widehat{R}$ fails to satisfy the Going-down property, so $\widehat{R}$ is not flat over B and B is not Noetherian.

For item 4, as above, $Qk[[x,y,z]]_{P_2} = (\rho, \sigma)_{P_2} = I_{P_2}$. Thus $\widehat{R}_{P_2}/I\widehat{R}_{P_2}$ is a formal fiber of B. Since $k[y,z]_{(y,z)}[[x]]/I = k[s,t]_{(s,t)}[[x]]/(s^2, st)$, we see that $P_2\widehat{R}/I\widehat{R} = (s,t)k[[x,y,z]]/(s^2, st)k[[x,y,z]]$ is an embedded associated prime of the ring $\widehat{R}/I\widehat{R}$. Hence $(\widehat{R}/I\widehat{R})_{P_2\widehat{R}}$ is not Cohen-Macaulay and the embedding $B \longrightarrow k[[x,y,z]]$ fails to have Cohen-Macaulay formal fibers. [3] □

THEOREM 19.9. *Assume the notation of Example 19.6. Then:*

(1) *A is a local Krull domain with maximal ideal $(x,y,z)A$ and completion $\widehat{R}$,*
(2) *$P_1 \cap A \subsetneq P_2 \cap A$, so $B \subsetneq A$,*
(3) *A is non-Noetherian.*

PROOF. For item 1, it follows from Construction Properties Theorem 5.14 that $(x,y,z)A$ is the maximal ideal of A. By definition, A is the intersection of a field with the Krull domain R^*; thus A is a Krull domain.

For item 2, let $Q_i := P_i \cap A$, for $i = 1, 2$. Then $\sigma^2/\rho = t^2 \in (Q_2 \setminus B) \setminus Q_1$, by Note 19.7.4.

For item 3, assume A is Noetherian. Then A is a regular local ring, since the maximal ideal is $(x,y,z)A$, and so the embedding $A \hookrightarrow R^* = k[y,z]_{(y,z)}[[x]]$ is faithfully flat. In particular, A is a UFD and the ideal $Q_1 = sR^* \cap A = P_1 \cap A$ is a prime ideal of height one in A. Thus Q_1 is principal; write $Q_1 = vA$. By Equation 19.6.2, $\rho = s^2 \in Q_1$ and $\sigma = st \in Q_1$. Write $\rho = s^2 = bv$ and $\sigma = st = cv$ for some $b, c \in A$. Then

$$tbv = s^2 t = scv \implies tb = sc$$

Since $s = y + \tau_1$ and $t = z + \tau_2$ are non-associate prime elements of the UFD D, it follows that $b = sd$ and $c = td$ for some $d \in D$. Thus $s^2 = bv = sdv \implies s = dv$, where $d \in D$. Since D is a Prototype (and so Noetherian), the extension $D \hookrightarrow R^*$ is faithfully flat. Therefore

$$vA = Q_1 = sR^* \cap A \subseteq sR^* \cap D = sD \implies v = sa,$$

for some $a \in D$. Since also $s = dv$, where $d \in D$, it follows that a is a unit in $D \subseteq k[y,z]_{(y,z)}[[x]]$. Write:

(19.9.a) $\quad v = sa = h(\rho, \sigma)/g(\rho, \sigma), \quad$ where $\quad h(\rho, \sigma),\ g(\rho, \sigma) \in k[x,y,z][\rho, \sigma].$

Now $a \in D = U_{((x,y,z)D) \cap U}$, so $a = g_1/g_2$, where $g_1, g_2 \in k[x,y,z,\tau_{1n},\tau_{2n}]$, for some $n \in \mathbb{N}$, and g_2 as a unit in $R^* = k[y,z]_{(y,z)}[[x]]$ has nonzero constant term. There exists $m \in \mathbb{N}$ such that $x^m g_1 := f_1$ and $x^m g_2 := f_2$ are in the polynomial ring

[3] This also implies that B is non-Noetherian by Corollary 19.2.

$k[x,y,z,\tau_1,\tau_2] = k[x,y,z][s,t]$. Regard $f_2(s,t)$ as a polynomial in s and t with coefficients in $k[x,y,z]$. Then $f_2 k[y,z]_{(y,z)}[[x]] = x^m k[y,z]_{(y,z)}[[x]] = x^m k[s,t]_{(s,t)}[[x]]$. Since g_2 is a unit in R^*, $f_2 \notin (s,t)k[s,t]_{(s,t)}[[x]]$. It follows that the constant term of $f_2(s,t) \in k[x,y,z][s,t]$ is a nonzero element of $k[x,y,z]$. Since

$$a = \frac{x^m g_1}{x^m g_2} = \frac{f_1}{f_2}, \tag{19.9.b}$$

and a is a unit of D, the constant term of $f_1(s,t) \in k[x,y,z][s,t]$ is also nonzero. Equations 19.9.a and 19.9.b together yield

$$s f_1(s,t) g(s^2, st) = f_2(s,t) h(s^2, st). \tag{19.9.c}$$

The term of lowest total degree in s and t on the left hand side of Equation 19.9.c has odd degree, while the term of lowest total degree in s and t on the right hand side has even degree, a contradiction. Therefore the assumption that A is Noetherian leads to a contradiction. We conclude that A is not Noetherian. □

REMARKS 19.10. (i) Although A is not Noetherian, the proof of Theorem 19.9 does not rule out the possibility that A is a UFD. The proof does show that if A is a UFD, then $\operatorname{ht}(P_1 \cap A) > \operatorname{ht}(P_1)$. It would be interesting to know whether the non-flat map $A \to \widehat{A} = \widehat{R}$ has the property that $\operatorname{ht}(\widehat{Q} \cap A) \le \operatorname{ht}(\widehat{Q})$, for each $\widehat{Q} \in \operatorname{Spec} \widehat{R}$. It would also be interesting to know the dimension of A.
(ii) We observe the close connection of the integral domains $A \subseteq D$ of Example 19.6. The extension of fields $\mathcal{Q}(A) \subseteq \mathcal{Q}(D)$ has degree two and $A = \mathcal{Q}(A) \cap D$, yet A is non-Noetherian, while D is Noetherian.

THEOREM 19.11. *Assume the notation of Example* 19.6. *Then:*

1. $C^* = R^*/I$ *and* $\widehat{C} = \widehat{R}/I\widehat{R}$.
2. C *is a two-dimensional Noetherian local domain.*
3. $C = E$.
4. *The generic formal fiber of* C *is not Cohen-Macaulay.*

PROOF. Item 1 follows from Construction Properties Theorem 17.11. Hence, if C is Noetherian, then $\dim C = \dim \widehat{C} = \dim(\widehat{R}/I\widehat{R}) = 2$. To show that C is a Noetherian Limit Intersection Domain, by the Noetherian Flatness Theorem 17.13, it suffices to show that the canonical map φ is flat, where:

$$\begin{aligned} R = k[x,y,z]_{(x,y,z)} \overset{\varphi}{\hookrightarrow} (R^*/I)[1/x] &= (k[y,z]_{(y,z)}[x]]/I)[1/x] \\ &= (k[s,t]_{(s,t)}[[x]]/(s^2,st)k[s,t]_{(s,t)}[[x]])[1/x], \end{aligned}$$

since $k[[x]][s,t]_{(s,t)} = k[[x]][y,z]_{(y,z)} \subseteq k[y,z]_{(y,z)}[[x]] = k[[x]][y,z]_{(y,z)}[[x]]$. Thus it suffices to show for every prime ideal Q^* of R^* with $x \notin Q^*$ that the map

$$\varphi_{Q^*} : R \hookrightarrow R^*_{Q^*}/IR^*_{Q^*} = (R^*/I)_{Q^*}$$

is flat, where $I = P_1 P_2 \subseteq Q^*$.

If $Q^* = P_2 = (s,t)R^*$, then φ_{Q^*} is flat, since $P_2 \cap R = (0)$ implies $R_{P_2 \cap R}$ is a field.

If $Q^* \neq P_2$, then $P_2 R^*_{Q^*} = R^*_{Q^*}$, because $\operatorname{ht} P_2 = 2$ and

$$\dim {R^*}_{Q^*} \le \dim R^*[1/x] = 2.$$

Hence $IR^*_{Q^*} = P_1 R^*_{Q^*} = sR^*_{Q^*}$. Thus it suffices to show

$$\varphi_{Q^*} : R \longrightarrow R^*_{Q^*}/sR^*_{Q^*} = (R^*/sR^*)_{Q^*}$$

is flat. To see that φ_{Q^*} is flat, since $R \subseteq D_{Q^* \cap D} \subseteq R^*_{Q^*}$ and $sR^* \cap R = (0)$, the map φ_{Q^*} factors through a homomorphic image of $D = V[y,z]_{(x,y,z)}$. That is, φ_{Q^*} is the composition of the following maps:

$$R \xrightarrow{\ \gamma\ } (D/sD)_{D\cap Q^*} \xrightarrow{\ \psi_{Q^*}\ } (R^*/sR^*)_{Q^*}.$$

By Theorem 17.35.1, the map φ_{Q^*} is flat $\iff$ γ is flat. To show that γ is flat, observe that the ring $(D/sD)_{D\cap Q^*}$ is a localization of $(D/sD)[1/x]$, since $x \notin Q^*$. By Construction Properties Theorem 5.14.4, it follows that $(D/sD)_{D\cap Q^*}$ is a localization of the polynomial ring:

$$k[x,y,z,\tau_1,\tau_2]/sk[x,y,z,\tau_1,\tau_2] = k[x,y,z,s,t]/sk[x,y,z,s,t],$$

which is clearly flat over R. This implies that γ is flat, and so C is Noetherian and $C = E$. Thus items 2 and 3 hold.

For item 4, $P_2\widehat{R}/I\widehat{R} = \mathfrak{p}$ is an embedded associated prime of (0) of $\widehat{C}$, and so $\widehat{C}_{\mathfrak{p}}$ is not Cohen-Macaulay. Since $\mathfrak{p} \cap C = (0)$, the generic formal fiber of C is not Cohen-Macaulay. □

PROPOSITION 19.12. *The canonical map $B \hookrightarrow R^*/I$ factors through C, and $B/Q \cong C$, where $Q = I \cap B = s\widehat{R} \cap B$. On the other hand, the canonical map $A \hookrightarrow R^*/I$ fails to factor through C.*

PROOF. By Proposition 17.30, the restriction of the canonical map $R^* \to R^*/I$ is a map $\varphi : B \twoheadrightarrow C$ such that the following diagram commutes:

(19.12.1)
$$\begin{array}{ccc} B & \longrightarrow & R^* \\ {\scriptstyle\varphi}\big\downarrow & & \big\downarrow \\ C & \longrightarrow & R^*/I. \end{array}$$

Thus $C \cong B/Q$ is a homomorphic image of B.

Suppose that the canonical map $\zeta : A \hookrightarrow R^*/I$ factors through C. Then there is a map $\psi : A \to C$ such that $\varphi = i_2 \circ \psi$, that is:

$$R \overset{i_1}{\hookrightarrow} A \overset{\psi}{\to} C \overset{i_2}{\hookrightarrow} R^*/I,$$

where i_1 and i_2 are inclusion maps and $\psi\ |_R$ is inclusion. Thus $\ker(\psi) = \ker(\zeta)$. As in Notes 19.7.5, the map ζ factors through D as $\delta \circ \gamma$, since $I \cap D = (\rho, \sigma)D$:

$$R \overset{i_1}{\hookrightarrow} A \xrightarrow{\ \gamma\ } D/(\rho,\sigma)D \xrightarrow{\ \delta\ } R^*/I,$$

where δ is inclusion and so injective. Then $\ker(\gamma) = \ker(\zeta) = \ker(\psi)$, Thus $A/\ker(\psi)$ embeds in D/ID. By Notes 19.7.4 and 19.7.6, the image under the map γ of the element $(\sigma^2/\rho) = t^2$ of A in $D/(\rho,\sigma)D$ is transcendental over R. Hence $A/\ker(\gamma)$ is transcendental over R. Since C is a birational extension of R, the map $A \longrightarrow R^*/I$ fails to factor through C. □

19.2. Constructing the Ogoma-like example

Example 19.13 provides a ring having the features of Ogoma's Example outlined in Example 19.4. The setting is similar to that of Example 19.6 with one more variable.

EXAMPLE 19.13. (Ogoma-like example) Let x, y, z, w be variables over a field k of characteristic 0, let $R = k[x, y, z, w]_{(x,y,z,w)}$ be the localized polynomial ring, let $\mathfrak{m} = (x, y, z, w)R$ and let $R^* = k[y, z, w]_{(y,z,w)}[[x]]$ be the x-adic completion of R. Let σ, τ, ρ be elements of $xk[[x]]$ that are algebraically independent over $k(x)$. Set

$$f := y + \sigma, \quad g := z + \tau, \quad h := w + \rho;$$
$$D := k(x, y, z, w, \sigma, \tau, \rho) \cap R^*, \quad I = (fg, fh)R[f, g, h] \text{ and } T := R[f, g, h]/I.$$

Thus D is a Local Prototype over a field of characteristic zero, and so D is an excellent local domain, by Theorem 10.6.2. If P is an associated prime ideal of IR^* in R^*, then $fg, fh \in P \implies fR^* \subseteq P$ or $(g, h)R^* \subseteq P$; that is, $P = fR^*$ or $P = (g, h)R^*$, since fR^* and $(g, h)R^*$ are prime ideals. Thus $P \cap R = (0)$, for every associated prime ideal of IR^*. Therefore Setting 17.1 and the conditions of Homomorphic Image Construction 17.2 hold for the ideal IR^*.

Define C and E to be the Intersection Domain of Homomorphic Image Construction 17.2 and the associated Approximation Domain as in Definition 17.10:

$$C := k(x, y, z, w) \cap (R^*/IR^*),$$
$$E := \bigcup_{n=1}^{\infty} k[x, y, z, w, (fg)_n, (fh)_n]_{(x,y,z,w,\sigma,\tau,\rho,(fg)_n,(fh)_n)},$$

where the $(fg)_n, (fh)_n$ are the nth frontpieces of fg, fh, respectively.

Proposition 19.14 records properties of the rings in Example 19.13.

PROPOSITION 19.14. *Assume notation as in Example* 19.13. *Then:*

(1) $R[f, g, h] = R[\sigma, \tau, \rho]$.
(2) $\theta : R[\sigma, \tau, \rho] \hookrightarrow R^*[1/x]$ *is flat, and* D *is an excellent local domain.*
(3) $IR^* \cap R[f, g, h] = IR^* \cap R[\sigma, \tau, \rho] = IR[\sigma, \tau, \rho] = I$.
(4) $\varphi = \lambda \circ \psi$, *where*

$$R \overset{\psi}{\hookrightarrow} T = R[f, g, h]/I \overset{\lambda}{\hookrightarrow} (R^*/IR^*)[1/x]$$

and the map λ *is flat.*

(5) D/ID *is excellent with* x*-adic completion*

$$(D/ID)^* = R^*/IR^*.$$

PROOF. Item 1 follows from the definitions of f, g and h. Item 2 holds by Prototype Theorem 10.6.

For item 3, by item 2, $R[f, g, h][1/x] \hookrightarrow R^*[1/x]$ is faithfully flat. Since the elements f, g, h generate maximal ideals of $R^*[1/x]$ and $R[f, g, h][1/x]$, and $R[f, g, h]_{(f,g,h)R[f,g,h]} = R[f, g, h][1/x]_{(f,g,h)R[f,g,h][1/x]}$,

$$R[f, g, h]_{(f,g,h)R[f,g,h]} \hookrightarrow R^*[1/x]_{(f,g,h)R^*[1/x]} = R^*_{(f,g,h)R^*}$$

is faithfully flat, and so item 3 follows.

Item 4 follows from items 3 and 2.

For item 5, D is excellent and so D/ID is excellent. Since R^* is the x-adic completion of D and $D/ID \hookrightarrow R^*/IR^*$, item 5 follows. □

THEOREM 19.15. *Assume the setting of Example* 19.13. *Then* C *is a 3-dimensional normal Nagata integral domain that is not catenary. Therefore* C *is a counterexample to Chain Conjecture* 18.2.

The proof of Theorem 19.15 is given in several steps. The first step is to show that $C = E$ is Noetherian. For this, we use the Elkik ideal H of T as an R-algebra; see Equation 7.13.b and Theorem 7.15.4.

PROPOSITION 19.16. *Assume notation as in Example* 19.13, *and let H be the Elkik ideal of T as an R-algebra. Then*

(1) *$H = (f, g, h)T$.*
(2) *The map $\varphi : R \hookrightarrow (R^*/IR^*)[1/x]$ is flat.*
(3) *$C = E$ is Noetherian with x-adic completion $E^* = R^*/IR^*$.*

PROOF. For item 1, the Jacobian matrix associated to the R-algebra T of fg, fh with respect to the indeterminates f, g, h over R is $M := \begin{bmatrix} g & f & 0 \\ h & 0 & f \end{bmatrix}$; see Definition 7.13.2. The 2×2 minors of M are:

$$\det \begin{bmatrix} g & f \\ h & 0 \end{bmatrix} = -fh, \quad \det \begin{bmatrix} g & 0 \\ h & f \end{bmatrix} = fg, \quad \det \begin{bmatrix} f & 0 \\ 0 & f \end{bmatrix} = f^2.$$

Since fg, fh generate I, the colon ideal $[(fg, fh) :_{R[f,g,h]} I]$ from Definition 7.13.3 is all of $R[f, g, h]$. It follows that the images in T of the 2×2 minors of M are elements in H. In particular, $f^2T \subseteq H$. Since H is a radical ideal, $fT \subseteq H$.

In addition, the Elkik ideal H contains the images in T of

$$\Delta(fg) \cdot [fg :_{R[f,g,h]} I] \text{ and } \Delta(fh) \cdot [fh :_{R[f,g,h]} I],$$

where $\Delta(fg)$ is the ideal of T generated by the 1×1 minors of the Jacobian matrix $\mathcal{J}((fg); f, g, h)$ of fg with respect to the indeterminates f, g, h, and $\Delta(fh)$ is the ideal of T generated by the 1×1 minors of the Jacobian matrix $\mathcal{J}((fh); f, g, h)$ of fh with respect to the indeterminates f, g, h. It is straightforward to see that $\mathcal{J}((fg); f, g, h) = [g, f, 0]$ and $\mathcal{J}((fh); f, g, h) = [h, 0, f]$. Thus

$$\Delta(fg) = (g, f)T \quad \text{and} \quad \Delta(fh) = (h, f)T.$$

Also $[fg :_{R[f,g,h]} I] = gR[f, g, h]$ and $[fh :_{R[f,g,h]} I] = hR[f, g, h]$. Thus

$$g^2T \subseteq \Delta(fg) \cdot [fg :_{R[f,g,h]} I] \quad \text{and} \quad h^2T \subseteq \Delta(fh) \cdot [fh :_{R[f,g,h]} I],$$

and both of these ideals are in H. Since H is a radical ideal, $(g, h)T \subseteq H$.

Let $P := (f, g, h)T$. The arguments above show that $P \subseteq H$. Since T_P has two minimal prime ideals, T_P is not an integral domain and so is not an RLR. Hence the map $\psi_P : R \to T_P$ is not smooth. Since $H \not\subseteq P$ implies that ψ_P is smooth, it follows that $P \subseteq H$, and so $H = P$. This establishes item 1.

For item 2, to show φ is flat, it suffices to show $\varphi_Q : R \to (R^*/IR^*)[1/x]_Q$ is flat for each $Q \in \operatorname{Spec}(R^*/IR^*)[1/x]$. Proposition 19.14.4 gives $\varphi = \lambda \circ \psi$, where

$$R \stackrel{\psi}{\hookrightarrow} T = R[f, g, h]/I \stackrel{\lambda}{\hookrightarrow} (R^*/IR^*)[1/x].$$

Since λ is flat, it suffices to show $\psi_{Q \cap T} : R \to T_{Q \cap T}$ is flat, for each $Q \in \operatorname{Spec}(R^*/IR^*)[1/x]$.

If $H \not\subseteq Q \cap T$, then $\psi_{Q \cap T}$ is smooth, and hence flat. Assume $H \subseteq Q \cap T$. Then $(f, g, h)R^*/IR^* \subseteq Q$. Since $(f, g, h)R^*[1/x]$ is a maximal ideal of $R^*[1/x]$, it follows that $(f, g, h)(R^*/IR^*)[1/x]$ is a maximal ideal. Thus $Q = (f, g, h)(R^*/IR^*)[1/x]$ and $Q \cap T = H$. To complete the proof, it suffices to prove that $\psi_H : R \hookrightarrow T_H$ is flat. Since $(f, g, h)T \cap R = (0)$, the map ψ_H factors

$$R \stackrel{\alpha}{\hookrightarrow} R_{(0)} = \mathcal{Q}(R) \stackrel{\beta}{\hookrightarrow} T_H.$$

The map β is an extension of a field; hence β is flat. The map α is a localization and so is flat. Thus ψ_H is flat. This proves that φ is flat.

For item 3, Noetherian Flatness Theorem 17.13.1(a $\iff$ d) implies $C = E$ is Noetherian, and Construction Properties Theorem 17.11.4 implies $C^* = R^*/IR^*$. $\square$

PROPOSITION 19.17. *Assume notation as in Example 19.13. Let $\widehat{C}$ denote the completion of C with respect to its maximal ideal $\mathfrak{m}_C$, let $\mathfrak{q}_1 := (x, f)\widehat{C}$ and let $\mathfrak{q}_2 := (x, g, h)\widehat{C}$. Also, set $\mathfrak{p}_1 = \mathfrak{q}_1 \cap C$ and $\mathfrak{p}_2 = \mathfrak{q}_2 \cap C$. Then*

(1) *$\widehat{C}$ is not equidimensional. Thus C is not universally catenary.*
(2) *The ideals $\mathfrak{q}_1$, $\mathfrak{q}_2$ are height 1 prime ideals of $\widehat{C}$ with $x\widehat{C} = \mathfrak{q}_1 \cap \mathfrak{q}_2$.*
(3) *The ideals $\mathfrak{p}_1$, $\mathfrak{p}_2$ are height 1 prime ideals of C with $xC = \mathfrak{p}_1 \cap \mathfrak{p}_2$.*
(4) *$\widehat{C}_{\mathfrak{q}_1}$, $\widehat{C}_{\mathfrak{q}_2}$, $C_{\mathfrak{p}_1}$ and $C_{\mathfrak{p}_2}$ are DVRs with maximal ideal generated by x.*
(5) *There exist saturated chains of prime ideals in C of length 2 and 3. Hence C is not catenary.*
(6) *The formal fibers of C are reduced.*
(7) *Assume $Q \in \operatorname{Spec}\widehat{C}$, $x \notin Q$, $H \nsubseteq Q$ and $P := Q \cap R \neq (0)$. Then the formal fiber of P is a Noetherian regular ring.*
(8) *C is a Nagata integral domain.*

PROOF. For item 1, the $\mathfrak{m}$-adic completion $\widehat{R}$ of R is the 4-dimensional RLR $k[[x, y, z, w]]$. By Theorem 17.11.6, $\mathfrak{m}_C = \mathfrak{m}C$. Since $\widehat{C} = \widehat{C^*} = \widehat{R^*/IR^*}$, it follows that $\widehat{C} = \widehat{R}/I\widehat{R}$, by [**122**, Theorem 8.11]. Since $f = y+\sigma$, $g = z+\tau$, and $h = w+\rho$ are part of a regular system of parameters for $\widehat{R}$, the ideals $f\widehat{R}$ and $(g, h)\widehat{R}$ are prime ideals with $\operatorname{ht}(f\widehat{R}) = 1$, and $\operatorname{ht}(g, h)\widehat{R} = 2$, and their intersection is equal to their product. Thus

$$I\widehat{R} = (fg, fh)\widehat{R} = (f\widehat{R})((g, h)\widehat{R}) = (f\widehat{R}) \cap ((g, h)\widehat{R}).$$

It follows that $\widehat{C} = \widehat{R}/(f\widehat{R} \cap (g, h)\widehat{R})$, has two minimal prime ideals $f\widehat{C}$ and $(g, h)\widehat{C}$, $\dim(\widehat{C}/f\widehat{C}) = 3$, and $\dim(\widehat{C}/(g, h)\widehat{C}) = 2$. Thus $\dim \widehat{C} = 3$ and $\widehat{C}$ is not equidimensional.

The second statement of item 1 follows by Ratliff's Equidimension Theorem 3.25.

Items 2, 3, 4, and 5 follow from the fact that $\mathfrak{q}_1$ and $\mathfrak{q}_2$ are the images in $\widehat{C} = \widehat{R}/I\widehat{R}$ of the prime ideals $(x, f)\widehat{R}$ and $(x, g, h)\widehat{R}$ and $\widehat{C}$ is faithfully flat over C. In more detail, we have:

$$\widehat{C} = \frac{\widehat{R}}{(fg, fh)\widehat{R}} = \frac{k[[x, y, z, w]]}{(fg, fh)k[[x, y, z, w]]}.$$

Since $\mathfrak{q}_1 = (x, f)\widehat{C} = (x, y)\widehat{C}$, we have $\mathfrak{p}_1 = (x, y)C$ and $\dim(C/\mathfrak{p}_1) = 2$. Hence

$$(0) \subsetneq \mathfrak{p}_1 \subsetneq (x, y, z)C \subsetneq \mathfrak{m}_C$$

is a saturated chain of prime ideals in C of length 3.

Since $\mathfrak{q}_2 = (x, g, h)\widehat{C} = (x, z, w)\widehat{C}$, we have $\mathfrak{p}_2 = (x, z, w)C$ and $\dim(C/\mathfrak{p}_2) = 1$. Hence

$$(0) \subsetneq \mathfrak{p}_2 \subsetneq \mathfrak{m}_C$$

is a saturated chain of prime ideals in C of length 2.

For item 6, for every $P \in \operatorname{Spec} C$, the formal fiber at P is $(C \setminus P)^{-1}(\widehat{C}/P\widehat{C})$. It follows that the formal fiber at P is reduced if and only if $P\widehat{C}$ is a radical ideal.

Assume $P \in \operatorname{Spec} C$ and consider the following cases:

Case i: $x \in P$. Then $C/xC = R^*/(xR^* + IR^*)$, by Construction Properties Theorem 17.11.3. Hence $P/xC = P^*/(xR^* + IR^*)$, for a prime ideal P^* of R^* with $(x, I)R^* \subseteq P^*$. Then $P^*/IR^* = PR^*/IR^*$ is a prime ideal of R^*/IR^*. Since R^*/IR^* is an excellent ring, R^*/IR^* is Nagata; see Remarks 3.48. By Theorem 8.19, the formal fiber in $\widehat{C} = \widehat{R}/I\widehat{R}$ over the prime ideal PR^*/IR^* is reduced. It follows that the formal fiber of $\widehat{C} = \widehat{R}/I\widehat{R}$ over P is reduced.

Case ii: $x \notin P$. Let $Q \in \operatorname{Spec}(\widehat{C}) = \operatorname{Spec}(\widehat{R}/I\widehat{R})$ with $Q \cap C = P$. By Construction Properties Theorem 17.11.5c, it follows that $C_P = R_{P\cap R}$. By Construction Properties Theorem 5.14.4, $D[1/x]$ is a localization of $R[\sigma, \tau, \rho] = R[f, g, h]$ and so $(D/ID)[1/x]$ is a localization of T. Thus $(D/ID)_{Q\cap(D/ID)} = T_{Q\cap T}$. Consider the maps

$$(19.17.1) \qquad C_P = R_{P\cap R} \hookrightarrow T_{Q\cap T} = (D/ID)_{Q\cap(D/ID)} \hookrightarrow (R^*/IR^*)_{Q\cap(R^*/IR^*)}.$$

Case iia: Assume in addition that $H \subseteq Q$. Then $Q\cap(R^*/IR^*)[1/x]$ is a maximal ideal of $(R^*/IR^*)[1/x]$ and so is equal to $H(R^*/IR^*)[1/x]$. Thus $H\widehat{C} = Q$ and $Q \cap T = H$. Hence $P \cap R = Q \cap R = H \cap R = (0)$. As above, $C_P = R_{P\cap R} = R_{(0)}$. Since the field of fractions of C is the same as the field of fractions of R, the prime ideal $P = (0)$. By the proof of item 1, $(0)\widehat{C} = (f\widehat{C}) \cap ((g, h)\widehat{C})$ is an intersection of two prime ideals. Therefore the formal fiber of C over (0) is reduced.

Case iia′: Assume $x \notin Q$, $H \nsubseteq Q$ and $Q \cap R = (0)$. Then the computation from case iia shows the formal fiber of C over (0) is reduced.

Thus the final case for item 2 of Proposition 19.17 is Case iib.

Case iib: Assume $x \notin Q$, $H \nsubseteq Q$ and $P := Q \cap R \neq (0)$.

Consider the maps

$$C_P = R_{P\cap R} \overset{\alpha}{\hookrightarrow} T_{Q\cap T} = (D/ID)_{Q\cap(D/ID)} \overset{\beta}{\hookrightarrow} \widehat{C}_Q$$

Since $Q \cap T$ does not contain the Elkik ideal H, α is smooth. Since D/ID is an excellent local ring with completion $\widehat{C}$, it follows that the extension $D/ID \hookrightarrow \widehat{C}$ and so also the extension β is faithfully flat and regular. The composition of regular maps is regular. Thus $\beta \circ \alpha$ is regular. Therefore the formal fiber of P is a Noetherian regular ring.

This completes the proof of item 6 of Proposition 19.17 and also proves item 7.

For item 8 of Proposition 19.17, the formal fibers of C are reduced, by item 6. Since the characteristic of k is 0, the formal fibers of C are also geometrically reduced; see Remark 3.40. By Theorem 8.19, C is Nagata. □

Proposition 19.18 gives an additional important property of the Ogoma-like example.

PROPOSITION 19.18. *Assume notation as in Example* 19.13. *Then C is integrally closed.*

PROOF. By item 2 of Remark 2.3, or [**137**, (12.9), p. 41] or [**103**, Theorem 54, p. 35], to prove that C is integrally closed, it suffices to show:

(a) $C_{\mathfrak{p}}$ is a DVR for each $\mathfrak{p} \in \operatorname{Spec} C$ with $\operatorname{ht} \mathfrak{p} = 1$, and

(b) If $\mathfrak{p}$ is an associated prime of a nonzero principal ideal aC, then $\operatorname{ht} \mathfrak{p} = 1$.

Let $\mathfrak{p} \in \operatorname{Spec} C$ with $\operatorname{ht} \mathfrak{p} = 1$. If $x \in \mathfrak{p}$, then item 4 of Proposition 19.17 implies that $C_{\mathfrak{p}}$ is a DVR. If $x \notin \mathfrak{p}$, then Theorem 17.11.5c implies that $C_{\mathfrak{p}} = R_{\mathfrak{p}\cap R}$. Since R is an RLR, $C_{\mathfrak{p}}$ is a DVR. This establishes item a.

If $\mathfrak{p}$ is an associated prime of a nonzero principal ideal aC, then by [**137**, (12.5)] or [**103**, Theorem 121], $\mathfrak{p}$ is an associated prime of bC for every nonzero element $b \in \mathfrak{p}$. If $x \in \mathfrak{p}$, then $\mathfrak{p}$ an associated prime of xC implies $\operatorname{ht}\mathfrak{p} = 1$. If $x \notin \mathfrak{p}$, then Theorem 17.11.5c implies that $C_{\mathfrak{p}} = R_{\mathfrak{p}\cap R}$. Hence $C_{\mathfrak{p}}$ is an RLR and $\mathfrak{p}$ an associated prime of a principal ideal implies $\operatorname{ht}\mathfrak{p} = 1$. □

REMARK 19.19. M. C. Kang has shown that every Noetherian normal integral domain A of dimension 3 is *almost Cohen-Macaulay*; see [**102**, Ex.2]. Almost Cohen-Macaulay is defined by $\operatorname{grade}(P, R) = \operatorname{grade}(PR_P, R_P)$, for every $P \in \operatorname{Spec} A$; equivalently, $\operatorname{depth}_P A = \operatorname{depth}_{PA_P} A_P$, for every $P \in \operatorname{Spec} A$. Thus, as Cristodor Ionescu observes in [**101**, Example 2.9], Example 19.13 provides a commutative Noetherian ring that is almost Cohen-Macaulay, but is not catenary.

CHAPTER 20

Multi-ideal-adic completions of Noetherian rings

In this chapter we consider a variation of the usual ideal-adic completion of a Noetherian ring R.[1] Instead of successive powers of a fixed ideal I, we use a *multi-adic* filtration formed from a more general descending sequence $\{I_n\}_{n=0}^{\infty}$ of ideals. We develop the mechanics of a *multi-ideal-adic* completion R^* of R, abbreviated *multi-adic* completion. With additional hypotheses on the ideals of the filtration, we show that R^* is Noetherian. In the case where R is local, we prove that R^* is excellent, or Henselian or universally catenary if R has the stated property.

20.1. Ideal filtrations and completions

Let R be a commutative ring with identity. A *filtration* on R is a descending sequence $\{I_n\}_{n=0}^{\infty}$ of ideals of R. As stated in Chapter 3, associated to a filtration there is a well-defined completion

$$R^* = \varprojlim_n R/I_n,$$

and a canonical homomorphism $\psi : R \to R^*$. If $\bigcap_{n=0}^{\infty} I_n = (0)$, then ψ is injective and R may be regarded as a subring of R^*. A filtration $\{I_n\}_{n=0}^{\infty}$ is said to be *multiplicative* if $I_0 = R$ and $I_n I_m \subseteq I_{n+m}$, for all $m \geq 0$, $n \geq 0$. A. well-known example of a multiplicative filtration on R is the I-adic filtration $\{I^n\}_{n=0}^{\infty}$, where I is a fixed ideal of R.

In this chapter we consider filtrations of ideals of R that are *not* multiplicative, and examine the completions associated to these filtrations. We assume the ring R is Noetherian. Instead of successive powers of a fixed ideal I, we use a filtration formed from a more general descending sequence $\{I_n\}_{n=0}^{\infty}$ of ideals. We require that, for each $n > 0$, the nth ideal I_n is contained in the nth power of the Jacobson radical of R, and that $I_{nk} \subseteq I_n^k$ for all $k, n \geq 0$. We call the associated completion a *multi-adic* completion, and denote it by R^*. The basics of the multi-adic construction and the relationship between this completion and certain ideal-adic completions are considered in Section 20.2. In Sections 20.3 and 20.4, we prove that the multi-adic completion R^* with respect to such ideals $\{I_n\}$ has the properties stated above.

The process of passing to completion gives an analytic flavor to algebra. Often we view completions in terms of power series, or in terms of coherent sequences as in [**15**, pages 103-104]. Sometimes results are established by demonstrating for each n that they hold at the nth stage in the inverse limit.

Multi-adic completions are interesting from another point of view. Many examples in commutative algebra can be considered as subrings of R^*/J, where R^* is

[1] The material in this chapter is adapted from our paper [**89**] dedicated to Melvin Hochster on the occasion of his 65th birthday. Hochster's brilliant work has had a tremendous impact on commutative algebra.

a multi-adic completion of a localized polynomial ring R over a countable ground field and J is an ideal of R^*. In particular, certain counterexamples of Brodmann and Rotthaus, Heitmann, Nishimura, Ogoma, Rotthaus and Weston can be interpreted in this way, see [**26**], [**27**], [**98**], [**140**], [**143**], [**146**], [**147**], [**156**], [**157**], [**183**]. For many of these examples, a particular enumeration, $\{p_1, p_2, \dots\}$, of countably many non-associate prime elements is chosen and the ideals I_n are defined to be $I_n := (p_1 p_2 \dots p_n)^n$. The Noetherian property in these examples is a trivial consequence of the fact that every ideal of R^* that contains a power of one of the ideals I_n is extended from R. An advantage of R^* over the I_n-adic completion $\widehat{R}_n$ is that an ideal of R^* is more likely to be extended from R than is an ideal of $\widehat{R}_n$.

20.2. Basic mechanics for the multi-adic completion

SETTING 20.1. Let R be a Noetherian ring with Jacobson radical $\mathcal{J}$, and let $\mathbb{N}$ denote the set of positive integers. For each $n \in \mathbb{N}$, let $\mathfrak{q}_n$ be an ideal of R. Assume that the sequence $\{\mathfrak{q}_n\}$ is descending, that is $\mathfrak{q}_{n+1} \subseteq \mathfrak{q}_n$, and that $\mathfrak{q}_n \subseteq \mathcal{J}^n$, for each $n \in \mathbb{N}$. Also assume, for each pair of integers $k, n \in \mathbb{N}$, that $\mathfrak{q}_{nk} \subseteq \mathfrak{q}_n^k$.

Let $\mathcal{F} = \{\mathfrak{q}_k\}_{k \geq 0}$ be a filtration

$$R = \mathfrak{q}_0 \supseteq \mathfrak{q}_1 \supseteq \cdots \supseteq \mathfrak{q}_k \supseteq \mathfrak{q}_{k+1} \supseteq \cdots$$

of R satisfying the conditions in the previous paragraph and let

$$R^* := \varprojlim_k R/\mathfrak{q}_k \tag{20.1.1}$$

denote the completion of R with respect to $\mathcal{F}$.

Let $\widehat{R} := \varprojlim_k R/\mathcal{J}^k$ denote the completion of R with respect to the powers of the Jacobson radical $\mathcal{J}$ of R, and, for each $n \in \mathbb{N}$, let

$$\widehat{R}^{\mathfrak{q}_n} := \varprojlim_k R/\mathfrak{q}_n^k \tag{20.1.2}$$

denote the completion of R with respect to the powers of $\mathfrak{q}_n$. [2]

REMARK 20.2. Assume Setting 20.1. Then:

(1) If each ideal $\mathfrak{q}_j = b_j R$, for some $b_j \in R$, then every element γ of R^* may be expressed as an infinite sum $\gamma = \sum_{j=0}^{\infty} a_j b_j$, where each $a_j \in R$.

(2) If R is countable, then R^* has infinite transcendence degree over R.

To see the first item, let $\gamma \in R^*$. Then

$$\gamma = \{a_1, a_2, \dots\} \in \varprojlim_t R/\mathfrak{q}_t = R^*,$$

where $a_1 \in R/\mathfrak{q}_1$, $a_2 \in R/\mathfrak{q}_2$ and $a_2 + \mathfrak{q}_1/\mathfrak{q}_2 = a_1$ in $R/\mathfrak{q}_1$, $\cdots$, and so forth, is a coherent sequence as in [**15**, pp. 103-104]. This leads to an expression for γ as

$$\gamma = \sum_{j=0}^{\infty} a_j b_j, \tag{20.2.a}$$

where each $a_j \in R$. The justification is analogous to that of Remarks 3.5.1.

For item 2, again use that the elements γ of R^* are in one-to-one correspondence with coherent sequences $\{a_n\}_{n=1}^{\infty}$. For every n, there are at least two distinct choices

[2]In this chapter we allow the use of "$\widehat{\ }$" with annotations to indicate completions other than a completion with respect to powers of the maximal ideal of a local ring

for a_n in the nth position of such a sequence. Thus there are at least $2^{\aleph_0}$ coherent sequences. $\square$

REMARK 20.3. Assume notation as in Setting 20.1. For each fixed $n \in \mathbb{N}$,

$$R^* = \varprojlim_k R/\mathfrak{q}_k = \varprojlim_k R/\mathfrak{q}_{nk},$$

where $k \in \mathbb{N}$ varies. This holds because the limit of a subsequence is the same as the limit of the original sequence.

We establish in Proposition 20.4 canonical inclusion relations among $\widehat{R}^{\mathcal{J}}$ and the completions defined in Equations 20.1.1 and 20.1.2.

PROPOSITION 20.4. *Let the notation be as in Setting 20.1. For each $n \in \mathbb{N}$, we have canonical inclusions*

$$R \subseteq R^* \subseteq \widehat{R}^{\mathfrak{q}_n} \subseteq \widehat{R}^{\mathfrak{q}_{n-1}} \subseteq \cdots \subseteq \widehat{R}^{\mathfrak{q}_1} \subseteq \widehat{R}.$$

PROOF. The inclusion $R \subseteq R^*$ is clear since the intersection of the ideals $\mathfrak{q}_k$ is zero. For the inclusion $R^* \subseteq \widehat{R}^{\mathfrak{q}_n}$, by Remark 20.3, $R^* = \varprojlim_k R/\mathfrak{q}_{nk}$. Notice that

$$\mathfrak{q}_{nk} \subseteq \mathfrak{q}_n^k \subseteq \mathfrak{q}_{n-1}^k \subseteq \cdots \subseteq \mathcal{J}^k.$$

$\square$

To complete the proof of Proposition 20.4, we state and prove a general result about completions with respect to ideal filtrations (see also [**144**, Section 9.5]). We define the respective completions using coherent sequences as in [**15**, pages 103-104].

LEMMA 20.5. *Let R be a Noetherian ring with Jacobson radical $\mathcal{J}$ and let $\{H_k\}_{k\in\mathbb{N}}$, $\{I_k\}_{k\in\mathbb{N}}$ and $\{L_k\}_{k\in\mathbb{N}}$ be descending sequences of ideals of R such that, for each $k \in \mathbb{N}$, we have inclusions*

$$L_k \subseteq I_k \subseteq H_k \subseteq \mathcal{J}^k.$$

We denote the families of natural surjections arising from these inclusions as:

$$\delta_k : R/L_k \to R/I_k, \quad \lambda_k : R/I_k \to R/H_k \quad \text{and} \quad \theta_k : R/H_k \to R/\mathcal{J}^k,$$

and the completions with respect to these families as:

$$\widehat{R}^L = \varprojlim_k R/L_k, \quad \widehat{R}^I := \varprojlim_k R/I_k \quad \widehat{R}^H := \varprojlim_k R/H_k \quad \text{and} \quad \widehat{R} := \varprojlim_k R/\mathcal{J}^k.$$

Then

(1) *These families of surjections induce canonical injective maps Δ, Λ and Θ among the completions as shown in the diagram below.*
(2) *For each positive integer k we have a commutative diagram as displayed below, where the vertical maps are the natural surjections.*

$$\begin{array}{ccccccc}
R/L_k & \xrightarrow{\delta_k} & R/I_k & \xrightarrow{\lambda_k} & R/H_k & \xrightarrow{\theta_k} & R/\mathcal{J}^k \\
\uparrow & & \uparrow & & \uparrow & & \uparrow \\
\widehat{R}^L & \xrightarrow{\Delta} & \widehat{R}^I & \xrightarrow{\Lambda} & \widehat{R}^H & \xrightarrow{\Theta} & \widehat{R}
\end{array}$$

(3) *The composition* $\Lambda \cdot \Delta$ *is the canonical map induced by the natural surjections* $\lambda_k \cdot \delta_k : R/L_k \to R/H_k$. *Similarly, the other compositions in the bottom row are the canonical maps induced by the appropriate natural surjections.*

PROOF. By the universal property of inverse limits,[3] in each case there is a unique homomorphism of the completions. For example, the family of homomorphisms $\{\delta_k\}_{k\in\mathbb{N}}$ induces a unique homomorphism

$$\widehat{R}^{L} \xrightarrow{\ \Delta\ } \widehat{R}^{I}.$$

To define Δ, let $x = \{x_k\}_{k\in\mathbb{N}} \in \widehat{R}^{L}$ be a coherent sequence, where each $x_k \in R/L_k$. Then $\delta_k(x_k) \in R/I_k$ and we define $\Delta(x) := \{\delta_k(x_k)\}_{k\in\mathbb{N}} \in \widehat{R}^{I}$.

To show the maps on the completions are injective, consider for example the map Δ. Suppose $x = \{x_k\}_{k\in\mathbb{N}} \in \varprojlim_k R/L_k$ with $\Delta(x) = 0$. Then $\delta_k(x_k) = 0$ in R/I_k, that is, $x_k \in I_k R/L_k$, for every $k \in \mathbb{N}$. For $v \in \mathbb{N}$, consider the following commutative diagram:

$$\begin{array}{ccc} R/L_k & \xrightarrow{\ \delta_k\ } & R/I_k \\ {\scriptstyle \beta_{k,kv}}\uparrow & & \uparrow{\scriptstyle \alpha_{k,kv}} \\ R/L_{kv} & \xrightarrow{\ \delta_{kv}\ } & R/I_{kv} \end{array} \tag{20.1}$$

where $\beta_{k,kv}$ and $\alpha_{k,kv}$ are the canonical surjections associated with the inverse limits. We have $x_{kv} \in I_{kv}R/L_{kv}$. Therefore

$$x_k = \beta_{k,kv}(x_{kv}) \in I_{kv}(R/L_k) \subseteq \mathcal{J}^{kv}(R/L_k),$$

for every $v \in \mathbb{N}$. Since $\mathcal{J}(R/L_k)$ is contained in the Jacobson radical of R/L_k and R/L_k is Noetherian, we have

$$\bigcap_{v\in\mathbb{N}} \mathcal{J}^{kv}(R/L_k) = (0).$$

Therefore $x_k = 0$ for each $k \in \mathbb{N}$, and so Δ is injective. The remaining assertions are clear. □

LEMMA 20.6. *With* R^* *and* $\widehat{R}^{\mathfrak{q}_n}$ *as in Setting* 20.1, *we have*

$$R^* = \bigcap_{n\in\mathbb{N}} \widehat{R}^{\mathfrak{q}_n}.$$

PROOF. The inclusion "$\subseteq$" is shown in Proposition 20.4. For the reverse inclusion, fix positive integers n and k, and let $L_\ell = \mathfrak{q}_{nk\ell}$, $I_\ell = \mathfrak{q}_{nk}^{\ell}$ and $H_\ell = \mathfrak{q}_n^{\ell}$ for each $\ell \in \mathbb{N}$. Then $L_\ell \subseteq I_\ell \subseteq H_\ell \subseteq \mathcal{J}^{\ell}$, as in Lemma 20.5 and

$$\widehat{R}^{L} := \varprojlim_{\ell} R/\mathfrak{q}_{nk\ell} = R^*, \quad \widehat{R}^{I} := \varprojlim_{\ell} R/\mathfrak{q}_{nk}^{\ell} = \widehat{R_{nk}}^{\mathfrak{q}_{nk}}, \quad \widehat{R}^{II} := \varprojlim_{\ell} R/\mathfrak{q}_n^{\ell} = \widehat{R}^{\mathfrak{q}_n}.$$

[3]See, for example,[**122**, pages 271-272].

As above, $\widehat{R} := \varprojlim_{\ell} R/\mathcal{J}^\ell$. Thus the diagram in Lemma 20.5.2 becomes

$$\begin{array}{ccccccc}
R/\mathfrak{q}_{nk\ell} & \xrightarrow{\delta_\ell} & R/\mathfrak{q}_{nk}^\ell & \xrightarrow{\lambda_\ell} & R/\mathfrak{q}_n^\ell & \xrightarrow{\theta_\ell} & R/\mathcal{J}^\ell \\
\uparrow & & \uparrow & & \uparrow & & \uparrow \\
\widehat{R}^L := R^* & \xrightarrow{\Delta} & \widehat{R}^I := \widehat{R}^{\mathfrak{q}_{nk}} & \xrightarrow{\Lambda} & \widehat{R}^{II} := \widehat{R}^{\mathfrak{q}_n} & \xrightarrow{\Theta} & \widehat{R}.
\end{array}$$

Consider the commutative diagram below where the maps named Δ, Λ and Θ are renamed $\Delta_{nk}, \Lambda_{nk,n}$ and Θ_n to identify the subscripts of the completions that are used for the maps. Let $\Omega_n = \Lambda_{nk} \circ \Delta_{nk}$ and $\Gamma := \Theta_n \circ \Lambda_{nk} \circ \Delta_{nk}$. Then $\Omega_n : R^* \to \widehat{R}^{\mathfrak{q}_n}$ is the map on the inverse limit induced by the maps $\lambda_\ell \circ \delta_\ell$ and $\Gamma : R^* \to \widehat{R}$ is the map on the inverse limit induced by the maps $\theta_\ell \circ \lambda_\ell \circ \delta_\ell$.

$$(20.6.1) \qquad \begin{array}{ccc}
\widehat{R} & \xleftarrow{\Theta_n} & \widehat{R}^{\mathfrak{q}_n} \\
\Gamma \uparrow & \overset{\Omega_n}{\nearrow} & \uparrow \Lambda_{nk,n} \\
R^* & \xrightarrow{\Delta_{nk}} & \widehat{R}^{\mathfrak{q}_{nk}}
\end{array}$$

Let $\widehat{y} \in \bigcap_{n \in \mathbb{N}} \widehat{R}^{\mathfrak{q}_n}$. We show there is an element $\xi \in R^*$ such that $\Gamma(\xi) = \Theta_n(\widehat{y})$. This is sufficient to ensure that $\widehat{y} \in R^*$, since the maps Θ_n are injective and Diagram 20.6.1 is commutative.

First, we define ξ: For each $t \in \mathbb{N}$,

$$\widehat{y} = \{y_{1,t}, y_{2,t}, \dots, \} \in \varprojlim_{\ell} R/\mathfrak{q}_t^\ell \;=\; \widehat{R}^{\mathfrak{q}_t},$$

where $y_{1,t} \in R/\mathfrak{q}_t$, $y_{2,t} \in R/\mathfrak{q}_t^2$ and $(y_{2,t} + \mathfrak{q}_t)/\mathfrak{q}_t^2 = y_{1,t}$ in $R/\mathfrak{q}_t$, $\cdots$, is a coherent sequence as in [**15**, pp. 103-104]. Choose $z_t \in R$ so that $z_t + \mathfrak{q}_t = y_{1,t}$. Then $\widehat{y} - z_t \in \mathfrak{q}_t \widehat{R}^{\mathfrak{q}_t}$. If s and t are positive integers with $s \geq t$, then $\mathfrak{q}_s \subseteq \mathfrak{q}_t$. Therefore $z_t - z_s \in \mathfrak{q}_t \widehat{R}^{\mathfrak{q}_t} \cap R = \mathfrak{q}_t R$. Thus $\xi := \{z_t\}_{t \in \mathbb{N}} \in R^*$. It follows that, for all $t \in \mathbb{N}$, $\widehat{y} - z_t \in \mathfrak{q}_t \widehat{R}^{\mathfrak{q}_t} \subseteq \mathcal{J}^t \widehat{R}$. Hence $\Gamma(\xi) = \Theta_n(\widehat{y})$. This completes the proof of Lemma 20.6. $\square$

The following special case of Setting 20.1 is used by Brodmann, Heitmann, Nishimura, Ogoma, Rotthaus, and Weston for the construction of numerous examples.

SETTING 20.7. Let R be a Noetherian ring with Jacobson radical $\mathcal{J}$. For each $i \in \mathbb{N}$, let $p_i \in \mathcal{J}$ be a non-zero-divisor (that is, a regular element) on R.

For each $n \in \mathbb{N}$, let $q_n = (p_1 \cdots p_n)^n$. Let $\mathcal{F}_0 = \{(q_k)\}_{k \geq 0}$ be the filtration

$$R \supseteq (q_1) \supseteq \cdots \supseteq (q_k) \supseteq (q_{k+1}) \supseteq \cdots$$

of R and define $R^* := \varprojlim_{k} R/(q_k)$ to be the completion of R with respect to $\mathcal{F}_0$.

REMARK 20.8. In Setting 20.7, assume further that $R = K[x_1, \dots, x_n]_{(x_1,\dots,x_n)}$, a localized polynomial ring over a countable field K, and that $\{p_1, p_2, \dots\}$ is an enumeration of all the prime elements (up to associates) in R. As in 20.7, let $R^* := \varprojlim_{n} R/(q_n)$, where each $q_n = (p_1 \cdots p_n)^n$.

The ring R^* is often useful for the construction of Noetherian local rings with a bad locus (regular, Cohen-Macaulay, normal). In particular, Brodmann, Heitmann, Nishimura, Ogoma, Rotthaus, and Weston make use of special subrings of this multi-adic completion R^* for their examples. The first such example was constructed by Rotthaus in [**156**]. In this paper, Rotthaus obtains a regular local Nagata ring A that contains a prime element ω so that the singular locus of the quotient ring $A/(\omega)$ is not closed. This ring A is situated between the localized polynomial ring R and its $*$-completion R^*; thus, in general R^* is bigger than R. In the Rotthaus example, the singular locus of $(A/(\omega))^*$ is defined by a height one prime ideal Q that intersects $A/(\omega)$ in (0). Since all ideals $Q + (p_n)$ are extended from $A/(\omega)$, the singular locus of $A/(\omega)$ is not closed.

REMARK 20.9. For R and R^* as in Remark 20.8, the ring R^* is also the "ideal-completion", or "R-completion of R. This completion is defined and used in the paper of Zelinsky [**194**], the work of Matlis [**118**] and [**119**], and the book of Fuchs and Salce[**50**]. The *ideal-topology*, or *R-topology* on an integral domain R is the linear topology defined by letting the nonzero ideals of R be a subbase for the open neighborhoods of 0. The nonzero principal ideals of R also define a subbase for the open neighborhoods of 0. Recent work on ideal completions has been done by Tchamna in [**179**]. In particular, Tchamna observes in [**179**, Theorem 4.1] that the ideal-completion of a countable Noetherian local domain is also a multi-ideal-adic completion.

20.3. Preserving Noetherian under multi-adic completion

THEOREM 20.10. *Let the notation be as in Setting* 20.1. *Then the ring* R^* *defined in Equation* 20.1.*1 is Noetherian.*

PROOF. It suffices to show each ideal I of R^* is finitely generated. Since $\widehat{R}$ is Noetherian, there exist $f_1, \dots, f_s \in I$ such that $I\widehat{R} = (f_1, \dots, f_s)\widehat{R}$. Since $\widehat{R}^{\mathfrak{q}_n} \hookrightarrow \widehat{R}$ is faithfully flat, $I\widehat{R}^{\mathfrak{q}_n} = I\widehat{R} \cap \widehat{R}^{\mathfrak{q}_n} = (f_1, \dots, f_s)\widehat{R}^{\mathfrak{q}_n}$, for each $n \in \mathbb{N}$.

Let $f \in I \subseteq R^*$. Then $f \in I\widehat{R}^{\mathfrak{q}_1}$, and so

$$f \;=\; \sum_{i=1}^{s} \widehat{b}_{i0} f_i,$$

where $\widehat{b}_{i0} \in \widehat{R}^{\mathfrak{q}_1}$. Consider R as "$\mathfrak{q}_0$", and so $\widehat{b}_{i0} \in \mathfrak{q}_0 \widehat{R}^{\mathfrak{q}_1}$. Since $\widehat{R_1}/\mathfrak{q}_1\widehat{R}^{\mathfrak{q}_1} \cong R/\mathfrak{q}_1$, for all i with $1 \le i \le s$, we have $\widehat{b}_{i0} = a_{i0} + \widehat{c}_{i1}$, where $a_{i0} \in R = \mathfrak{q}_0 R$ and $\widehat{c}_{i1} \in \mathfrak{q}_1\widehat{R}^{\mathfrak{q}_1}$. Then

$$f \;=\; \sum_{i=1}^{s} a_{i0} f_i \;+\; \sum_{i=1}^{s} \widehat{c}_{i1} f_i.$$

Notice that

$$\widehat{d_1} \;:=\; \sum_{i=1}^{s} \widehat{c}_{i1} f_i \in (\mathfrak{q}_1 I)\widehat{R}^{\mathfrak{q}_1} \cap R^* \subseteq \widehat{R}^{\mathfrak{q}_2}.$$

By the faithful flatness of the extension $\widehat{R}^{\mathfrak{q}_2} \hookrightarrow \widehat{R}^{\mathfrak{q}_1}$, we see $\widehat{d_1} \in (\mathfrak{q}_1 I)\widehat{R}^{\mathfrak{q}_2}$, and therefore there exist $\widehat{b}_{i1} \in \mathfrak{q}_1\widehat{R}^{\mathfrak{q}-2}$ with

$$\widehat{d_1} = \sum_{i=1}^{s} \widehat{b}_{i1} f_i.$$

As before, using that $\widehat{R}^{\mathfrak{q}_2}/\mathfrak{q}_2\widehat{R}^{\mathfrak{q}_2} \cong R/\mathfrak{q}_2$, we can write $\widehat{b}_{i1} = a_{i1} + \widehat{c}_{i2}$, where $a_{i1} \in R$ and $\widehat{c}_{i2} \in \mathfrak{q}_2\widehat{R}^{\mathfrak{q}_2}$. This implies that $a_{i1} \in \mathfrak{q}_1\widehat{R}^{\mathfrak{q}_2} \cap R = \mathfrak{q}_1$. We have:

$$f = \sum_{i=1}^{s}(a_{i0} + a_{i1})f_i + \sum_{i=1}^{s}\widehat{c}_{i2}f_i.$$

Now set

$$\widehat{d}_2 := \sum_{i=1}^{s}\widehat{c}_{i2}f_i.$$

Then $\widehat{d}_2 \in (\mathfrak{q}_2 I)\widehat{R}^{\mathfrak{q}_2} \cap R^* \subseteq \widehat{R}^{\mathfrak{q}_3}$ and, since the extension $\widehat{R}^{\mathfrak{q}_3} \hookrightarrow \widehat{R}^{\mathfrak{q}_2}$ is faithfully flat, we have $\widehat{d}_2 \in (\mathfrak{q}_2 I)\widehat{R}^{\mathfrak{q}_3}$. We repeat the process. By a simple induction argument,

$$f = \sum_{i=1}^{s}(a_{i0} + a_{i1} + a_{i2} + \ldots)f_i,$$

where $a_{ij} \in \mathfrak{q}_j$ and $a_{i0} + a_{i1} + a_{i2} + \ldots \in R^*$. Thus $f \in (f_1, \ldots, f_s)R^*$. Hence I is finitely generated and R^* is Noetherian. □

COROLLARY 20.11. *Assume Setting* 20.1. *Then the inclusion map* $R \hookrightarrow R^*$ *is faithfully flat.*

PROOF. Let $R \overset{\alpha}{\hookrightarrow} R^* \overset{\beta}{\hookrightarrow} \widehat{R^*} = \widehat{R}$ with α and β the inclusion maps. By Remark 4.4, $\beta : R^* \hookrightarrow \widehat{R^*} = \widehat{R}$ and $\beta \circ \alpha : R \hookrightarrow \widehat{R^*}$ are both faithfully flat. By Remark 2.37.14, α is faithfully flat. □

We use Proposition 20.12 in the next section on preserving excellence.

PROPOSITION 20.12. *Assume notation as in Setting* 20.1, *and let the ring* R^* *be defined as in Equation* 20.1.1. *If* M *is a finitely generated* R^**-module, then*

$$M \cong \varprojlim_k (M/\mathfrak{q}_k M),$$

that is, M *is* $*$*-complete.*

PROOF. If $F = (R^*)^n$ is a finitely generated free R^*-module, then one can see directly that

$$F \cong \varprojlim_k F/\mathfrak{q}_k F,$$

and so F is $*$-complete.

Let M be a finitely generated R^*-module. Consider an exact sequence:

$$0 \longrightarrow N \longrightarrow F \longrightarrow M \longrightarrow 0,$$

where F is a finitely generated free R^*-module. This induces an exact sequence:

$$0 \longrightarrow \tilde{N} \longrightarrow F^* \longrightarrow M^* \longrightarrow 0,$$

where $\tilde{N}$ is the completion of N with respect to the induced filtration $\{\mathfrak{q}_k F \cap N\}_{k \geq 0}$; see [**15**, (10.3)].

This gives a commutative diagram:

$$\begin{array}{ccccccccc} 0 & \longrightarrow & N & \longrightarrow & F & \longrightarrow & M & \longrightarrow & 0 \\ & & \downarrow & & \cong\downarrow & & \gamma\downarrow & & \\ 0 & \longrightarrow & \tilde{N} & \longrightarrow & F^* & \longrightarrow & M^* & \longrightarrow & 0 \end{array}$$

where γ is the canonical map $\gamma : M \longrightarrow M^*$. The diagram shows that γ is surjective. We have

$$\bigcap_{k=1}^{\infty}(\mathfrak{q}_k M) \subseteq \bigcap_{k=1}^{\infty} J^k M = (0),$$

where the last equality is by [**15**, (10.19)]. Therefore γ is also injective. $\square$

REMARK 20.13. Let the notation be as in Setting 20.1, and let B be a finite R^*-algebra. Let $\widehat{B}^{\mathfrak{q}_n} \cong B \otimes_{R^*} \widehat{R}^{\mathfrak{q}_n}$ denote the $\mathfrak{q}_n$-adic completion of B. By Proposition 20.4, and Corollary 20.11, we have a sequence of inclusions:

$$B \hookrightarrow \cdots \hookrightarrow \widehat{B}^{\mathfrak{q}_{n+1}} \hookrightarrow \widehat{B}^{\mathfrak{q}_n} \hookrightarrow \ldots \hookrightarrow \widehat{B}^{\mathfrak{q}_1} \hookrightarrow \widehat{B}^{\mathcal{J}B},$$

where $\widehat{B}^{\mathcal{J}B}$ denotes the completion of B with respect to $\mathcal{J}B$. Let $\mathcal{J}_0$ denote the Jacobson radical of B. Since every maximal ideal of B lies over a maximal ideal of R^*, we have $\mathcal{J}B \subseteq \mathcal{J}_0$.

THEOREM 20.14. *Assume the notation of Setting* 20.1 *and Remark* 20.13. *Thus B is a finite R^*-algebra. Let $\widehat{I}$ be an ideal of $\widehat{B}^{\mathcal{J}B}$, let $I := \widehat{I} \cap B$, and, for each $n \in \mathbb{N}$, let $I_n := \widehat{I} \cap \widehat{B}^{\mathfrak{q}_n}$. If $\widehat{I} = I_n\widehat{B}^{\mathcal{J}B}$, for all n, then $\widehat{I} = I\widehat{B}^{\mathcal{J}B}$.*

PROOF. By replacing B by B/I, we may assume that $(0) = I = \widehat{I} \cap B$. To prove the theorem, it suffices to show that $\widehat{I} = 0$.

For each $n \in \mathbb{N}$, we define ideals $\mathfrak{c}_n$ of $\widehat{B}^{\mathfrak{q}_n}$ and $\mathfrak{a}_n$ of B:

$$\mathfrak{c}_n := I_n + \mathfrak{q}_n \widehat{B}^{\mathfrak{q}_n}, \qquad \mathfrak{a}_n := \mathfrak{c}_n \cap B.$$

Since $B/\mathfrak{q}_n B = \widehat{B}^{\mathfrak{q}_n}/\mathfrak{q}_n \widehat{B}^{\mathfrak{q}_n}$, the ideals of B containing $\mathfrak{q}_n$ are in one-to-one inclusion-preserving correspondence with the ideals of $\widehat{B}^{\mathfrak{q}_n}$ containing $\mathfrak{q}_n \widehat{B}^{\mathfrak{q}_n}$, and so

$$\text{(20.14.1)} \qquad \mathfrak{a}_n \widehat{B}^{\mathfrak{q}_n} = \mathfrak{c}_n, \quad \mathfrak{a}_{n+1} \widehat{B}^{\mathfrak{q}_n} = \mathfrak{a}_{n+1} \widehat{B}^{\mathfrak{q}_{n+1}} \widehat{B}^{\mathfrak{q}_n} = \mathfrak{c}_{n+1} \widehat{B}^{\mathfrak{q}_n}.$$

Since $\widehat{B}^{\mathcal{J}B}$ is faithfully flat over $\widehat{B}^{\mathfrak{q}_n}$ and $\widehat{I}$ is extended,

$$\text{(20.14.2)} \qquad I_{n+1} \widehat{B}^{\mathfrak{q}_n} = (I_{n+1}\widehat{B}^{\mathcal{J}B}) \cap \widehat{B}^{\mathfrak{q}_n} = \widehat{I} \cap \widehat{B}^{\mathfrak{q}_n} = I_n.$$

Thus Equations 20.14.1 and 20.14.2 and $\mathfrak{q}_{n+1} \widehat{B}^{\mathfrak{q}_n} \subseteq \mathfrak{q}_n \widehat{B}^{\mathfrak{q}_n}$ imply that:

$$\begin{aligned}\mathfrak{a}_n \widehat{B}^{\mathfrak{q}_n} &= \mathfrak{c}_n = I_n + \mathfrak{q}_n \widehat{B}^{\mathfrak{q}_n} = I_{n+1} \widehat{B}^{\mathfrak{q}_n} + \mathfrak{q}_n \widehat{B}^{\mathfrak{q}_n} = \mathfrak{c}_{n+1} \widehat{B}^{\mathfrak{q}_n} + \mathfrak{q}_n \widehat{B}^{\mathfrak{q}_n} \\ &= \mathfrak{a}_{n+1} \widehat{B}^{\mathfrak{q}_n} + \mathfrak{q}_n \widehat{B}^{\mathfrak{q}_n},\end{aligned}$$

for all $n \in \mathbb{N}$. Since $\widehat{B}^{\mathfrak{q}_n}$ is faithfully flat over B, the equation above implies that

$$\text{(20.14.3)} \qquad \mathfrak{a}_{n+1} + \mathfrak{q}_n B = (\mathfrak{a}_{n+1} \widehat{B}^{\mathfrak{q}_n} + \mathfrak{q}_n \widehat{B}^{\mathfrak{q}_n}) \cap B = \mathfrak{a}_n \widehat{B}^{\mathfrak{q}_n} \cap B = \mathfrak{a}_n.$$

Thus also

$$\text{(20.14.4)} \qquad \mathfrak{a}_n\widehat{B}^{\mathcal{J}B} \subseteq \mathfrak{a}_{n+1}\widehat{B}^{\mathcal{J}B} + \mathfrak{q}_n\widehat{B}^{\mathcal{J}B} \subseteq I_{n+1}\widehat{B}^{\mathcal{J}B} + \mathfrak{q}_n\widehat{B}^{\mathcal{J}B} = \widehat{I} + \mathfrak{q}_n\widehat{B}^{\mathcal{J}B}.$$

Now $\mathfrak{q}_n \subseteq \mathcal{J}^n\widehat{B}^{\mathcal{J}B}$, and $\mathcal{J} \subseteq \mathcal{J}_0$, the Jacobson radical of B. Then

$$\bigcap_{n\in\mathbb{N}}(\mathfrak{a}_n\widehat{B}^{\mathcal{J}B}) \subseteq \bigcap_{n\in\mathbb{N}}(\widehat{I} + \mathfrak{q}_n\widehat{B}^{\mathcal{J}B}) \subseteq \bigcap_{n\in\mathbb{N}}(\widehat{I} + \mathcal{J}^n\widehat{B}^{\mathcal{J}B}) = \widehat{I},$$

by applying Equation 20.14.4. Since $\widehat{I} \cap B = (0)$, we have

$$0 = \widehat{I} \cap B \supseteq \Big(\bigcap_{n\in\mathbb{N}} (\mathfrak{a}_n \widehat{B}^{\mathcal{J}B})\Big) \cap B \supseteq \bigcap_{n\in\mathbb{N}} ((\mathfrak{a}_n \widehat{B}^{\mathcal{J}B}) \cap B) = \bigcap_{n\in\mathbb{N}} \mathfrak{a}_n,$$

where the last equality is because $\widehat{B}^{\mathcal{J}B}$ is faithfully flat over B. Thus $\bigcap_{n\in\mathbb{N}} \mathfrak{a}_n = (0)$.

CLAIM. $\widehat{I} = (0)$.

PROOF OF CLAIM. Suppose $\widehat{I} \neq 0$. Then there exists $d \in \mathbb{N}$ so that $\widehat{I} \not\subseteq \mathcal{J}_0^d \widehat{B}^{\mathcal{J}B}$. By hypothesis, $\widehat{I} = I_d \widehat{B}^{\mathcal{J}B}$, and so $I_d \widehat{B}^{\mathcal{J}B} \not\subseteq \mathcal{J}_0^d \widehat{B}$. Since $\widehat{B}^{\mathcal{J}B}$ is faithfully flat over $\widehat{B}^{\mathfrak{q}_d}$, we have $I_d \not\subseteq \mathcal{J}_0^d \widehat{B}^{\mathfrak{q}_d}$. By Equation 20.14.1,

$$\mathfrak{a}_d \widehat{B}^{\mathfrak{q}_d} = \mathfrak{c}_d = I_d + \mathfrak{q}_d \widehat{B}^{\mathfrak{q}_d} \not\subseteq \mathcal{J}_0^d \widehat{B}^{\mathfrak{q}_d},$$

and so there exists an element $y_d \in \mathfrak{a}_d$ with $y_d \notin \mathcal{J}_0^d$.

By Equation 20.14.3, $\mathfrak{a}_{d+1} + \mathfrak{q}_d B = \mathfrak{a}_d$. Hence there exists $y_{d+1} \in \mathfrak{a}_{d+1}$ and $q_d \in \mathfrak{q}_d B$ so that $y_{d+1} + q_d = y_d$. Recursively construct sequences of elements $y_{n+1} \in \mathfrak{a}_{n+1}$ and $q_n \in \mathfrak{q}_n B$ such that $y_{n+1} + q_n = y_n$, for each $n \geq d$.

The sequence $\xi = (y_n + \mathfrak{q}_n B) \in \varprojlim_n B/\mathfrak{q}_n B = B$ corresponds to a nonzero element $y \in B$ such that, for every $n \geq d$, there exists an element $g_n \in \mathfrak{q}_n B$ with $y = y_n + g_n$. This shows that $y \in \mathfrak{a}_n$, for all $n \geq d$, and $y \notin \mathcal{J}_0^d \widehat{B}^{\mathcal{J}B}$. Therefore $\bigcap_{n\in\mathbb{N}} \mathfrak{a}_n \neq 0$, a contradiction. Thus $\widehat{I} = (0)$.

□

20.4. Preserving excellence or Henselian under multi-adic completion

The first four results of this section concern preservation of excellence.

THEOREM 20.15. *Assume notation as in Setting 20.1, and let the ring R^* be defined as in Equation 20.1.1. If $(R, \mathfrak{m})$ is an excellent local ring, then R^* is excellent.*

The following result is crucial to the proof of Theorem 20.15.

LEMMA 20.16. [**122**, Theorem 32.5, page 259] *Let A be a semilocal Noetherian ring. Assume, for every finite A-algebra C that is an integral domain and every maximal ideal $\mathfrak{a}$ of C, that the local domain $B = C_{\mathfrak{a}}$ has the following property: $(\widehat{B})_Q$ is a regular local ring for every prime ideal Q of $\widehat{B}$ such that $Q \cap B = (0)$. Then A is a G-ring, that is, $A_{\mathfrak{p}} \hookrightarrow \widehat{A_{\mathfrak{p}}}$ is regular for every prime ideal $\mathfrak{p}$ of A.*

We use Proposition 20.17 in the proof of Theorem 20.15.

PROPOSITION 20.17. *Assume Setting 20.1. Let R be a Noetherian semilocal ring with geometrically regular formal fibers. Then R^* has geometrically regular formal fibers.*

PROOF. Let C be an integral domain that is a finite R^*-algebra, let $\mathfrak{a}$ be a maximal ideal of C and let $B = C_{\mathfrak{a}}$. Then C is semilocal; let $\mathfrak{a}, \mathfrak{a}_2, \dots, \mathfrak{a}_t$ be the maximal ideals of C. By Theorem 3.14, $\widehat{C} = \widehat{B} \times \widehat{C_{\mathfrak{a}_2}} \times \cdot \times \widehat{C_{\mathfrak{a}_t}}$. By Fact 3.2, $\widehat{R^*} = \widehat{R}$. It follows that $\widehat{B}$ is a local integral domain that is a finite $\widehat{R}$-algebra. Let $P \in \operatorname{Sing}(\widehat{B})$, that is, $\widehat{B}_P$ is not a regular local ring. By Lemma 20.16, to prove that R^* has geometrically regular formal fibers, it suffices to prove that $P \cap B \neq (0)$.

The Noetherian complete semilocal ring $\widehat{R}$ has the property J-2 in the sense of Matsumura, that is, for every finite $\widehat{R}$-algebra, such as $\widehat{B} = B \otimes_{R^*} \widehat{R^*}$, the subset $\operatorname{Reg}(\operatorname{Spec}(\widehat{B}))$, of primes where the localization of $\widehat{B}$ is regular, is an open subset in the Zariski topology; see [**120**, pp. 246–249]. Thus there is a reduced ideal $\widehat{I}$ in $\widehat{B}$ that defines the singular locus; that is, $\operatorname{Sing}(\widehat{B}) = \mathcal{V}(\widehat{I})$.

If $\widehat{I} = (0)$, then the nilradical $\sqrt{(0)} = (0)$ in $\widehat{B}$ and so $\widehat{B}$ is a reduced ring. Then the localization $\widehat{B}_Q$ is a field, for every minimal prime Q of $\widehat{B}$, a contradiction to $Q \in \operatorname{Sing}(\widehat{B}$. We conclude that $\widehat{I} \neq (0)$.

For every $n \in \mathbb{N}$, $\widehat{B}^{\mathfrak{q}_n} \cong \widehat{R}^{\mathfrak{q}_n} \otimes_{R^*} B$ is a finite $\widehat{R}^{\mathfrak{q}_n}$-algebra. Since $\widehat{R}^{\mathfrak{q}_n}$ has geometrically regular formal fibers so has $\widehat{B}^{\mathfrak{q}_n}$; see [**157**]. This implies that $\widehat{I}$ is extended from $\widehat{B}^{\mathfrak{q}_n}$, for every $n \in \mathbb{N}$; see Exercise 3. By Theorem 20.14, $\widehat{I}$ is extended from B, and so $\widehat{I} = I\widehat{B}$, where $0 \neq I := \widehat{I} \cap B$. Since $\widehat{I} \subseteq P$, we have $(0) \neq I \subseteq P \cap B$. This completes the proof of Proposition 20.17. □

PROOF OF THEOREM 20.15. It remains to show that R^* is universally catenary. The extensions $R \hookrightarrow R^* \hookrightarrow \widehat{R}$ are injective local homomorphisms, R^* is Noetherian, and $\widehat{R^*} = \widehat{R}$. Thus Proposition 20.18 below implies that R^* is universally catenary. □

PROPOSITION 20.18. *Let $(A, \mathfrak{m})$ be a Noetherian local universally catenary ring and let $(B, \mathfrak{n})$ be a Noetherian local subring of the $\mathfrak{m}$-adic completion $\widehat{A}$ of A with $A \subseteq B \subseteq \widehat{A}$ and $\widehat{B} = \widehat{A}$, where $\widehat{B}$ is the $\mathfrak{n}$-adic completion of B. Then B is universally catenary.*

PROOF. By [**122**, Theorem 31.7], it suffices to show for $P \in \operatorname{Spec}(B)$ that $\widehat{A}/P\widehat{A}$ is equidimensional. We may assume that $P \cap A = (0)$, and hence that A is a domain.

Let Q and W in $\operatorname{Spec}(\widehat{A})$ be minimal primes over $P\widehat{A}$.

CLAIM. $\dim(\widehat{A}/Q) = \dim(\widehat{A}/W)$.

PROOF OF CLAIM. Since B is Noetherian, the canonical morphisms $B_P \longrightarrow \widehat{A}_Q$ and $B_P \longrightarrow \widehat{A}_W$ are flat. By [**122**, Theorem 15.1],

$$\dim(\widehat{A}_Q) = \dim(B_P) + \dim(\widehat{A}_Q/P\widehat{A}_Q), \quad \dim(\widehat{A}_W) = \dim(B_P) + \dim(\widehat{A}_W/\widehat{P A_W}).$$

Since Q and W are minimal over $P\widehat{A}$, it follows that:

$$\dim(\widehat{A}_Q) = \dim(\widehat{A}_W) = \dim(B_P).$$

Let $q \subseteq Q$ and $w \subseteq W$ be minimal primes of $\widehat{A}$ so that:

$$\dim(\widehat{A}_Q) = \dim(\widehat{A}_Q/q\widehat{A}_Q) \quad \text{and} \quad \dim(\widehat{A}_W) = \dim(\widehat{A}_W/w\widehat{A}_W).$$

Since we have reduced to the case where A is a universally catenary domain, its completion $\widehat{A}$ is equidimensional and therefore:

$$\dim(\widehat{A}/q) = \dim(\widehat{A}/w).$$

Since a complete local ring is catenary [**122**, Theorem 29.4], we have:

$$\dim(\widehat{A}/q) = \dim(\widehat{A}_Q/q\widehat{A}_Q) + \dim(\widehat{A}/Q),$$
$$\dim(\widehat{A}/w) = \dim(\widehat{A}_W/w\widehat{A}_W) + \dim(\widehat{A}/W).$$

Since $\dim(\widehat{A}/q) = \dim(\widehat{A}/w)$ and $\dim(\widehat{A}_Q) = \dim(\widehat{A}_W)$, it follows that

$$\dim(\widehat{A}/Q) = \dim(\widehat{A}/W).$$

This completes the proof of Proposition 20.18.

□

REMARK 20.19. Let R be a universally catenary Noetherian local ring. Proposition 20.18 implies that *every* Noetherian local subring B of $\widehat{R}$ with $R \subseteq B$ and $\widehat{B} = \widehat{R}$ is universally catenary. Hence, for each ideal I of R, the I-adic completion of R is universally catenary. Also R^* as in Setting 20.1 is universally catenary. Proposition 20.18 also implies that the Henselization of R is universally catenary. Seydi shows that the I-adic completions of universally catenary rings are universally catenary in [**167**]. Proposition 20.18 establishes this result for a larger class of rings.

PROPOSITION 20.20. *With notation as in Setting* 20.1, *let* $(R, \mathfrak{m}, k)$ *be a Noetherian local ring. If* R *is Henselian, then* R^* *is Henselian.*

PROOF. Assume that R is Henselian. It is well known that every ideal-adic completion of R is Henselian, see [**156**, p.6]. Thus $\widehat{R}^{\mathfrak{q}_n}$ is Henselian for all $n \in \mathbb{N}$. Let $\mathfrak{n}$ denote the nilradical of $\widehat{R}$. Then $\mathfrak{n} \cap R^*$ is the nilradical of R^*, and to prove R^* is Henselian, it suffices to prove that $R' := R^*/(\mathfrak{n} \cap R^*)$ is Henselian [**137**, (43.15)]. To prove R' is Henselian, by [**156**, Prop. 3, page 76], it suffices to show:

If $f \in R'[x]$ is a monic polynomial and its image $\bar{f} \in k[x]$ has a simple root, then f has a root in R'.

Let $f \in R'[x]$ be a monic polynomial such that $\bar{f} \in k[x]$ has a simple root. Since $\widehat{R}^{\mathfrak{q}_n}/(\mathfrak{n} \cap \widehat{R}^{\mathfrak{q}_n})$ is Henselian, for each $n \in \mathbb{N}$, there exists $\widehat{\alpha}_n \in \widehat{R}^{\mathfrak{q}_n}/(\mathfrak{n} \cap \widehat{R}^{\mathfrak{q}_n})$ with $f(\widehat{\alpha}_n) = 0$. Since f is monic and $\widehat{R}/\mathfrak{n}$ is reduced, f has only finitely many roots in $\widehat{R}/\mathfrak{n}$. Thus there is an α so that $\alpha = \widehat{\alpha}_n$, for infinitely many $n \in \mathbb{N}$. By Lemma 20.14, $R^* = \bigcap_{n \in \mathbb{N}} \widehat{R}^{\mathfrak{q}_n}$. Hence

$$R' = R^*/(\mathfrak{n} \cap R^*) = \bigcap_{n \in \mathbb{N}} \widehat{R}^{\mathfrak{q}_n}/(\mathfrak{n} \cap \widehat{R}^{\mathfrak{q}_n}),$$

and so there exists $\alpha \in R'$ such that $f(\alpha) = 0$. □

Exercises

(1) Let R be a Noetherian semilocal ring, let S be an R-algebra, and let $\varphi : R \to S$ be the canonical R-algebra homomorphism of R into S. Assume that S is a finite R-algebra, and therefore that $\varphi(R) \hookrightarrow S$ is a finite integral extension.
   (a) Prove that a prime ideal P of S is maximal in S if and only if $P \cap \varphi(R)$ is maximal in $\varphi(R)$.
   (b) Let $\mathfrak{m}$ be a maximal ideal of $\varphi(R)$. Prove that there exists at least one and at most finitely many prime ideals of S lying over $\mathfrak{m}$.
   (c) Prove that S is a Noetherian semilocal ring.
   (d) Let $\mathcal{J}(S)$ and $\mathcal{J}(\varphi(R))$ denote the Jacobson radicals of S and $\varphi(R)$. Prove that there exists a positive integer n such that $\mathcal{J}(S)^n \subseteq \mathcal{J}(\varphi(R))S$.
   (e) Prove that $\widehat{S}$ is also the $\mathcal{J}(\varphi(R))S$-adic completion of S.

(2) Let R denote the ring and $\{\mathfrak{q}_n\}$ the family of ideals given in Remark 20.8. Consider the linear topology obtained by letting the ideals $\mathfrak{q}_n$ be a subbase for the open neighborhoods of 0. Prove the ideals $\mathfrak{q}_n$ are also a subbase for the ideal-topology on R.

(3) Let R and S be Noetherian local rings and $\varphi : R \hookrightarrow S$ a faithfully flat local homomorphism. Assume that φ has regular fibers and that $\operatorname{Sing} S = \mathcal{V}(I)$ for a reduced ideal I of S. Then $\operatorname{Sing} R = \mathcal{V}(I \cap R)$.

SUGGESTION. Apply Theorem 3.33; see also [**122**, Theorem 23.9]. Since S is Noetherian, I has finitely many minimal primes $\mathfrak{q}_1, \dots, \mathfrak{q}_n$ and $I = \cap_{i=1}^n \mathfrak{q}_i$. Let $\mathfrak{p}_i = \mathfrak{q}_i \cap R$. Then $I \cap R = \cap_{i=1}^n \mathfrak{p}_i$. Let $\mathfrak{p}$ be a prime ideal of R. Since S is faithfully flat over R, there exists a prime ideal $\mathfrak{q}$ of S such that $\mathfrak{q} \cap R = \mathfrak{p}$. By Theorem 3.33, $S_{\mathfrak{q}}$ is regular if and only if $R_{\mathfrak{p}}$ is regular. Since $S_{\mathfrak{q}}$ is not regular if and only if $I \subseteq \mathfrak{q}$, it follows that $R_{\mathfrak{p}}$ is not regular if and only if $I \cap R \subseteq \mathfrak{p}$.

CHAPTER 21

Noetherian flatness and multi–adic constructions

In this chapter we define a construction analogous to Inclusion Constructions 5.3 using the multi-adic completion of Chapter 20 in place of the x-adic completion.

The multi-adic version of the inclusion construction is in Sections 21.2.

21.1. Flatness lemmas

This section contains expanded versions of two flatness lemmas that were crucial for the results obtained for Inclusion Construction 5.3. Fortunately, it is rather straightforward to extend Lemma 5.12 and Lemma 6.2 to the versions given in Lemma 21.2 and Lemma 21.3. These versions are useful for results concerning the multi-adic completion.

SETTING 21.1. Let $S \hookrightarrow T$ be an extension of commutative rings and let Δ be a multiplicatively closed subset of S such that $\Delta \subseteq \{$ nonzerodivisors of $T\}$.

LEMMA 21.2. *Assume Setting* 21.1. *The following statements are equivalent:*

(1) *For every* $x \in \Delta$, (i) $xS = xT \cap S$ *and* (ii) $\frac{S}{xS} = \frac{T}{xT}$.
(2) *For every* $x \in \Delta$, (i) $S = S[\frac{1}{x}] \cap T$ *and* (ii) $T[\frac{1}{x}] = S[\frac{1}{x}] + T$.
(3) *For every nonunit* $x \in \Delta$, *the* x*-adic completion of* S *equals the* x*-adic completion of* T.
(4) (i) $S = \Delta^{-1}S \cap T$, *and* (ii) $\Delta^{-1}T = \Delta^{-1}S + T$.

PROOF. Item 1 is equivalent to item 2 and to item 3 by Lemma 5.12. To show item 1i implies item 4i, let $\frac{a}{b} \in \Delta^{-1}S \cap T$, for some $a \in S$ and $b \in \Delta$. Then $a \in bT \cap S = bS$ by item 1i. Thus $\frac{a}{b} \in S$. Since $S \subseteq \Delta^{-1}S \cap T$ is obvious, item 4i holds.

To show item 1ii implies item 4ii, let $\frac{t}{b} \in \Delta^{-1}T$, where $t \in T$ and $b \in \Delta$. Item 1ii implies that $T = S + bT$. It follows that $t = s + bt'$, for some $s \in S$ and $t' \in T$. Thus

$$\frac{t}{b} = \frac{s}{b} + t' \in \Delta^{-1}S + T.$$

Therefore item 4ii holds.

To show item 4i implies item 2i, observe that $S \subseteq S[1/x] \cap T$. For the other direction, let $t = \frac{s}{x^n}$, for $t \in T$, $s \in S$ and $n \in \mathbb{N}_0$. Then $t = \frac{s}{x^n} \in \Delta^{-1}S \cap T = S$.

For item 4 implies item 2ii: the inclusion $S[\frac{1}{x}] + T \subseteq T[\frac{1}{x}]$ is clear. Let $t \in T$ and $n \in \mathbb{N}_0$. Then $\frac{t}{x^n} \in \Delta^{-1}T = \Delta^{-1}S + T$, and so there exists $b \in \Delta$, $s \in S$ and $t' \in T$ such that $\frac{t}{x^n} = \frac{s}{b} + t'$ holds. Thus $\frac{x^n s}{b} \in T \cap \Delta^{-1}S = S$, and so $\frac{s}{b} \in S[1/x]$ as desired. $\square$

Lemma 21.3 extends parts of Lemma 6.2 to a multi-adic setting.

LEMMA 21.3. *Assume Setting* 21.1 *and the equivalent conditions of Lemma* 21.2 *hold. Let* $X \subset S$ *be a set of generators of* Δ, *let* $\mathcal{W} := \{1 + xS \mid x \in X\}$ *and let* $C = \mathcal{W}^{-1}S$. *Let* $\mathcal{J}(S)$ *denote the Jacobson radical of* S. *Then:*

(1) $\Delta^{-1}T$ *is flat over* $S \iff T$ *is flat over* S.
(2) *If* T *is flat over* S, *then* $D = \mathcal{W}^{-1}T$ *is faithfully flat over* C.
(3) *If* T *is Noetherian and* T *is flat over* S, *then* C *is Noetherian.*
(4) *Assume that* $X \subseteq \mathcal{J}(S)$. *If* T *is Noetherian and* T *is flat over* S, *then* S *is Noetherian.*
(4′) *Assume that* $y \in X$ *is such that* $X \setminus \{y\} \subseteq \mathcal{J}(S)$.
  (a) *If* C *is Noetherian and* $S[1/y]$ *is Noetherian, then* S *is Noetherian.*
  (b) *If* T *is Noetherian,* T *is flat over* S, *and* $S[1/y]$ *is Noetherian, then* S *is Noetherian.*

PROOF. If T is flat over S, then, by transitivity of flatness, $\Delta^{-1}T$ is flat over S. For the converse, Lemma 21.2 implies that $S = \Delta^{-1}S \cap T$ and $\Delta^{-1}T = \Delta^{-1}S + T$. Thus the following sequence is exact.

$$0 \to S = \Delta^{-1}S \cap T \xrightarrow{\quad\alpha\quad} \Delta^{-1}S \oplus T \xrightarrow{\quad\beta\quad} \Delta^{-1}T = \Delta^{-1}S + T \to 0,$$

where $\alpha(b) = (b, -b)$ for all $b \in S$ and $\beta(c, d) = c + d$ for all $c \in \Delta^{-1}S,\ d \in T$. Since the two end terms are flat S-modules, the middle term $\Delta^{-1}S \oplus T$ is also S-flat by Remark 2.37.12. By Definition 2.36, a direct summand of a flat S-module is S-flat. Hence T is S-flat. Thus item 1 holds.

For item 2, if T is flat over S, then D is flat over C. For every $x \in \Delta$, $1 + xC$ consists entirely of units of C, and xC is contained in the Jacobson radical of C. Moreover C/xC is a localization of $S/xS = T/xT$ with

$$C/xC = \mathcal{W}^{-1}(S/xS) = \mathcal{W}^{-1}(T/xT) = D/xD.$$

Thus every maximal ideal of C is contained in a maximal ideal of D. It follows that D is faithfully flat over C.

For item 3, D is Noetherian since D is a localization of T. By item 2, D is faithfully flat over C. Hence C is Noetherian.

For item 4, since $X \subseteq \mathcal{J}(S)$, $\mathcal{W}$ consists of units of S. Thus $\mathcal{W}^{-1}S = S$, and S is Noetherian by item 3.

For item 4′, item 3 implies it suffices to prove statement a of item 4′. For this, let $\mathcal{V}_y = \{1 + yS\}$ and let $\mathcal{W}_1 = \{1 + xS \mid x \in X \setminus \{y\}\}$. Since $X \setminus \{y\} \subseteq \mathcal{J}(S)$, it follows that $\mathcal{W}_1$ consists of units of S. Thus ${\mathcal{W}_1}^{-1}S = S$. Then $\mathcal{W}^{-1}S = {\mathcal{V}_y}^{-1}{\mathcal{W}_1}^{-1}S = {\mathcal{V}_y}^{-1}S$ is Noetherian. By Lemma 6.2, S is Noetherian. □

21.2. Multi-adic inclusion constructions

SETTING 21.4. Let $(R, \mathfrak{m})$ be a Noetherian local domain, let $\{p_i\}_{i \in \mathbb{N}}$ be an infinite sequence of elements of $\mathfrak{m} \setminus \{0\}$ that determine distinct principal ideals of R, and let Δ be the multiplicatively closed subset of R generated by the $\{p_i\}$. For every $n \in \mathbb{N}$, set $q_n := p_1 \cdot \ldots \cdot p_n \in \mathfrak{m}^n$, and $I_n := q_n^n R$. Let

$$R^* := \varprojlim_n R/I_n,$$

the multi-adic completion of R with respect to the nested sequence of principal ideals $\{I_n\}$, as defined in Equation 20.1.1.

Definitions 21.5 give the notation and conditions for the Multi-adic Inclusion Construction.

DEFINITIONS 21.5. Assume Setting 21.4.

(1) By Remark 20.2.1, every $\gamma \in R^*$ has a "power series" expression

$$\gamma = \sum_{j=0}^{\infty} c_j q_j^j, \tag{21.5.a}$$

where each $c_j \in R$.

(2) Let $\gamma \in R^*$ be algebraically independent over $\mathcal{Q}(R)$; for every $n \in \mathbb{N}_0$, define the nth *endpiece* γ_n of γ by

$$\gamma_n = \sum_{j=n+1}^{\infty} c_j q_j^j \cdot \frac{1}{q_n^n}.$$

(3) Let $\tau_1, \dots, \tau_s \in R^*$ be algebraically independent over $\mathcal{Q}(R)$. Assume that the elements of $R[\tau_1, \dots, \tau_s]$ are regular in R^*. As in Inclusion Construction 5.3, define the *Multi-adic Inclusion Intersection Domain*

$$A = \mathcal{Q}(R)(\tau_1, \dots, \tau_s) \cap R^*.$$

(4) With $\tau_1, \dots, \tau_s \in R^*$ as in item 3, write each $\tau_i = \sum_{j=0}^{\infty} a_{ij} q_j^j$, where each $a_{ij} \in R$. As in item 2, for every $n \in \mathbb{N}_0$, the nth *endpiece* τ_{in} of τ_i is

$$\tau_{in} = \frac{1}{q_n^n} \sum_{j=n+1}^{\infty} a_{ij} q_j^j.$$

(5) Define integral domains U and B associated to the Multi-adic Construction:

$$U_n = R[\tau_{1n}, \dots, \tau_{sn}]; \quad U = \bigcup U_n; \quad B_n = (U_n)_{(\mathfrak{m}, \tau_{1n}, \dots, \tau_{sn})}; \quad B = \bigcup B_n.$$

Then $B = U_{(\mathfrak{m}, \tau_1, \dots, \tau_s)}$ is called the *Multi-adic Inclusion Approximation Domain.*

REMARK 21.6. The rings U_n, B_n, U and B are independent of choice of expressions for the τ_i. This statement follows by an argument analogous to that of Proposition 5.9, since $q_n^n R = R \cap q_n^n R^*$.

REMARK 21.7. For each $n \in \mathbb{N}$, the following relation holds for the nth and $(n+1)^{\text{st}}$ endpieces of an element $\gamma \in R^*$. Let $\gamma = \sum_{j=0}^{\infty} a_j q_j^j$, where each $a_j \in R$. For this, write

$$\begin{aligned}
\gamma_n = (\sum_{j=n+1}^{\infty} a_i q_j^j) \cdot \frac{1}{q_n^n} &= (a_{n+1} q_{n+1}^{n+1}) \cdot \frac{1}{q_n^n} + (\sum_{j=n+2}^{\infty} a_i q_j^j) \cdot (\frac{1}{q_n^n}) \\
&= (a_{n+1} q_{n+1}^{n+1}) \cdot \frac{1}{q_n^n} + (\gamma_{n+1}) \cdot (q_{n+1}^{n+1}) \cdot (\frac{1}{q_n^n}).
\end{aligned}$$

Lemma 21.8 is an adaptation of Lemma 5.13. The proof is similar to the proof of Lemma 5.13, but with the multi-adic notation.

LEMMA 21.8. *Assume Setting 21.4 and the notation of Definition 21.5. Then:*

(1) *For every $\eta \in U$ and every $t \in \mathbb{N}$, there exist elements $g_t \in R$ and $\delta_t \in U$ such that $\eta = g_t + q_t^t \delta_t$.*
(2) *For each $t \in \mathbb{N}$, $q_t^t R^* \cap U = q_t^t U$.*

Theorem 21.9 gives properties of Multi-adic Inclusion Construction 21.5 that correspond to parts of Construction Properties Theorem 5.14.

THEOREM 21.9. *Assume Setting* 21.4 *and the notation of Definition* 21.5. *Then, for every* $n \in \mathbb{N}$*:*

(1) $q_n^n R^* \cap A = q_n^n A$, $\quad q_n^n R^* \cap B = q_n^n B \quad$ *and* $\quad q_n^n R^* \cap U = q_n^n U$.
(2) $R/q_n^n R = U/q_n^n U = B/q_n^n B = A/q_n^n A = R^*/q_n^n R^*$, *and these rings are all Noetherian.*
(3) *The multi-adic completions of the rings* U, B *and* A *are all equal to* R^*, *that is,* $R^* = U^* = B^* = A^*$.
(4) $\Delta^{-1}B$ *is a localization of* $R[\tau_1, \ldots, \tau_s]$.

PROOF. The proofs of items 1 and 2 are similar to those of Theorem 5.14, except they use Lemma 21.8 in place of Lemma 5.13.

Item 3 follows from items 1 and 2 and the definition of R^*.

For item 4, observe that the denominators of the endpieces in the construction are all contained in Δ. Thus

$$R[\tau_1, \ldots, \tau_s] \subseteq U \subseteq \Delta^{-1}U \subseteq \Delta^{-1}R[\tau_1, \ldots, \tau_s].$$

Since B is a localization of U, the ring $\Delta^{-1}B$ is a localization of $R[\tau_1, \ldots, \tau_s]$. □

Since the p_i are regular on R^* and hence on B, Theorem 21.9 implies Proposition 21.10:

PROPOSITION 21.10. *Assume Setting* 21.4 *and the notation of Definition* 21.5. *Then, for every* $x \in \Delta$*:*

(1) x *is regular on* R^*.
(2) $xB = xR^* \cap B$.
(3) $B/xB = R^*/xR^*$.

COROLLARY 21.11. *Assume Setting* 21.4 *and notation as in Definition* 21.5. *Then* $\Delta^{-1}R^*$ *is flat over* B *if and only if* R^* *is flat over* B.

PROOF. In view of Proposition 21.10, Lemmas 21.2 and 21.3 apply with $S = B$ and $T = R^*$. □

Corollary 21.11 leads to a Noetherian Flatness Theorem 21.12 for Multi-adic Inclusion Construction 21.5:

NOETHERIAN FLATNESS THEOREM 21.12. *Assume Setting* 21.4 *and the notation of Definitions* 21.5. *The following statements are equivalent:*

(1) $B \overset{\alpha}{\hookrightarrow} \Delta^{-1}R^*$ *is flat.*
(2) $B \hookrightarrow R^*$ *is flat.*
(3) B *is Noetherian.*
(4) $R[\tau_1, \ldots, \tau_s] \hookrightarrow \Delta^{-1}R^*$ *is flat.*

PROOF. By Corollary 21.11 Item 1 $\iff$ item 2.

For item 1 $\implies$ item 3, as in Section 21.1, define $\mathcal{W} = < \bigcup_{x\in\Delta}(1+xB) >$. Since B is a local domain, $\mathcal{W}$ consists of units of B and $\mathcal{W}^{-1}B = B$. Thus Lemma 21.3.3 applies with $C = B$ and $T = R^*$.

Item 3 $\implies$ item 2 by Theorem 21.9.3 and Corollary 20.11. Thus items 1,2 and 3 are equivalent.

For item 1 $\Longrightarrow$ item 4, observe that item 1 implies $\Delta^{-1}B \hookrightarrow \Delta^{-1}R^*$ is flat. By Theorem 21.9.4, $\Delta^{-1}B$ is a localization of $R[\tau_1,\dots,\tau_s]$. Thus the composition $R[\tau_1,\dots,\tau_s] \hookrightarrow \Delta^{-1}B \hookrightarrow \Delta^{-1}R^*$ is flat.

To show item 4 $\Longrightarrow$ item 1, let $Q \in \operatorname{Spec}(\Delta^{-1}R^*)$, let $Q' = Q \cap R^*$, let $\mathfrak{q} = Q \cap B$, and let $\mathfrak{p} = Q \cap R[\tau_1,\dots,\tau_s]$. Consider the extensions

$$(21.12.0) \qquad R[\tau_1,\dots,\tau_s]_{\mathfrak{p}} \hookrightarrow B_{\mathfrak{q}} \overset{\alpha_Q}{\hookrightarrow} (\Delta^{-1}R^*)_Q.$$

By Theorem 21.9.4, $\Delta^{-1}B$ is a localization of $R[\tau_1,\dots,\tau_s]$. Since $\Delta \cap Q' = \emptyset$, it follows that $\Delta \cap \mathfrak{q} = \emptyset$, that $\Delta^{-1}B \subset B_{\mathfrak{q}}$, and hence that the ring $B_{\mathfrak{q}}$ is a further localization of $R[\tau_1,\dots,\tau_s]$. Now $R[\tau_1,\dots,\tau_s] \subseteq R[\tau_1,\dots,\tau_s]_{\mathfrak{p}} \subseteq B_{\mathfrak{q}}$ implies that $B_{\mathfrak{q}}$ is a localization of $R[\tau_1,\dots,\tau_s]_{\mathfrak{p}}$ that dominates $R[\tau_1,\dots,\tau_s]_{\mathfrak{p}}$. Since both are local rings, they are equal. Hence Equation 21.12.0 becomes

$$(21.12.1) \qquad R[\tau_1,\dots,\tau_s]_{\mathfrak{p}} = B_{\mathfrak{q}} \overset{\alpha_Q}{\hookrightarrow} (\Delta^{-1}R^*)_Q.$$

By hypothesis, $R[\tau_1,\dots,\tau_s] \hookrightarrow \Delta^{-1}R^*$ is flat. Thus α_Q is flat for every prime ideal $Q \in \operatorname{Spec}(\Delta^{-1}R^*)$, and so α is flat, as desired. $\square$

CHAPTER 22

Idealwise algebraic independence

Let $(R, \mathfrak{m})$ be an excellent normal local domain with field of fractions K and completion $(\widehat{R}, \widehat{\mathfrak{m}})$. In this chapter, we consider the case where Inclusion Construction 5.3 yields a localized polynomial ring over the original ring R. That is, for elements $\tau_1, \ldots, \tau_n \in \widehat{\mathfrak{m}}$ that are algebraically independent over K, the intersection domain $A = K(\tau_1, \ldots, \tau_n) \cap \widehat{R} = R[\tau_1, \ldots, \tau_n]_{(\mathfrak{m}, \tau_1, \ldots, \tau_n)}$. In this case the elements $\tau_1, \ldots, \tau_n$ are said to be *idealwise independent*. We analyze idealwise independence in some depth, find parallels with other concepts in this book, and present results related to this situation.

If R is countable with $\dim(R) > 1$, we show in Theorem 6.4.5 the existence of an infinite sequence of elements $\tau_1, \tau_2, \ldots$ of $\widehat{\mathfrak{m}}$ such that $\tau_1, \ldots, \tau_n$ are idealwise independent over R for each positive integer n. Then the subfield $K(\tau_1, \tau_2, \ldots)$ of $\mathcal{Q}(\widehat{R})$ has the property that the intersection domain $A = K(\tau_1, \tau_2, \ldots) \cap \widehat{R}$ is a localized polynomial ring in infinitely many variables over R. In particular, this intersection domain A is not Noetherian.

22.1. Idealwise independence, weakly flat and PDE extensions

We use the following setting throughout this chapter and Chapter 23.

SETTING AND NOTATION 22.1. Let $(R, \mathfrak{m})$ be an excellent normal local domain with field of fractions K and completion $(\widehat{R}, \widehat{\mathfrak{m}})$. By Theorem 8.23, $\widehat{R}$ is also a normal domain. Let $t_1, \ldots, t_n, \ldots$ be indeterminates over R, and assume that $\tau_1, \tau_2, \ldots, \tau_n, \ldots \in \widehat{\mathfrak{m}}$ are algebraically independent over K. For each integer $n \geq 0$ and ∞, we consider the following localized polynomial rings:

$S_n := R[t_1, \ldots, t_n]_{(\mathfrak{m}, t_1, \ldots, t_n)}$,

$R_n := R[\tau_1, \ldots, \tau_n]_{(\mathfrak{m}, \tau_1, \ldots, \tau_n)}$,

$S_\infty := R[t_1, \ldots, t_n, \ldots]_{(\mathfrak{m}, t_1, \ldots, t_n, \ldots)}$ and

$R_\infty := R[\tau_1, \ldots, \tau_n, \ldots]_{(\mathfrak{m}, \tau_1, \ldots, \tau_n, \ldots)}$.

For $n = 0$, we define $R_0 = R = S_0$. Of course, S_n is R-isomorphic to R_n and S_∞ is R-isomorphic to R_∞ with respect to the R-algebra homomorphism taking $t_i \to \tau_i$ for each i. When working with a particular n or ∞, we sometimes define S to be R_n or R_∞.

The completion $\widehat{S_n}$ of S_n is $\widehat{R}[[t_1, \ldots, t_n]]$. Let $\lambda : \widehat{S_n} \to \widehat{R}$ be the $\widehat{R}$-algebra surjection with $\mathfrak{p} := \ker(\lambda) = (t_1 - \tau_1, \ldots, t_n - \tau_n)\widehat{S_n}$, and let λ_1 be the restriction of λ to S_n. The following diagram commutes.

$$\begin{array}{ccccc} & & S_n = R[t_1, \ldots, t_n]_{(\mathfrak{m}, t_1, \ldots, t_n)} & \xrightarrow{\subseteq} & \widehat{S_n} = \widehat{R}[[t_1, \ldots, t_n]] \\ & & \Big\downarrow{\scriptstyle \lambda_1, \cong} & & \Big\downarrow{\scriptstyle \lambda} \\ R & \xrightarrow{\subseteq} & R_n = R[\tau_1, \ldots, \tau_n]_{(\mathfrak{m}, \tau_1, \ldots, \tau_n)} & \xrightarrow{\subseteq} & \widehat{R}. \end{array} \qquad (22.1.0)$$

The map λ_1 takes $t_i \to \tau_i$, and λ_1 is an R-algebra isomorphism because $\mathfrak{p} \cap S_n = (0)$. Moreover

$$\widehat{R} = \frac{\widehat{S_n}}{(t_1 - \tau_1, \dots, t_n - \tau_n)\widehat{S_n}}. \tag{22.1.1}$$

The central definition of this chapter is the following:

DEFINITION 22.2. Let $(R, \mathfrak{m})$ and $\tau_1, \dots, \tau_n \in \widehat{\mathfrak{m}}$ be as in Setting 22.1. We say that $\tau_1, \dots, \tau_n$ are *idealwise independent over* R if

$$\widehat{R} \cap K(\tau_1, \dots, \tau_n) = R_n.$$

Similarly, an infinite sequence $\{\tau_i\}_{i=1}^{\infty}$ in $\widehat{\mathfrak{m}}$ as in Setting 22.1 is *idealwise independent over* R if $\widehat{R} \cap K(\{\tau_i\}_{i=1}^{\infty}) = R_{\infty}$.

REMARKS 22.3. Assume Setting and Notation 22.1.

(1) A subset of an idealwise independent set $\{\tau_1, \dots, \tau_n\}$ over R is also idealwise independent over R. For example, to see that $\tau_1, \dots, \tau_m$ are idealwise independent over R for $m \leq n$, let K denote the field of fractions of R and observe that

$$\begin{aligned} \widehat{R} \cap K(\tau_1, \dots, \tau_m) &= \widehat{R} \cap K(\tau_1, \dots, \tau_n) \cap K(\tau_1, \dots, \tau_m) \\ &= R[\tau_1, \dots, \tau_n]_{(\mathfrak{m}, \tau_1, \dots, \tau_n)} \cap K(\tau_1, \dots, \tau_m) = R[\tau_1, \dots, \tau_m]_{(\mathfrak{m}, \tau_1, \dots, \tau_m)}. \end{aligned}$$

(2) Idealwise independence is a strong property of the elements $\tau_1, \dots, \tau_n$ and of the embedding map $\varphi : R_n \hookrightarrow \widehat{R}$. It is often difficult to compute $\widehat{R} \cap L$ for an intermediate field L between the field K and the field of fractions of $\widehat{R}$. In order for $\widehat{R} \cap L$ to be the localized polynomial ring R_n, there can be no new quotients in $\widehat{R}$ other than those in $\varphi(R_n)$; that is, if $f/g \in \widehat{R}$ and $f, g \in R_n$, then $f/g \in R_n$. This does not happen, for example, if one of the τ_i is in the completion of R with respect to a principal ideal; in particular, if $\dim(R) = 1$, then there do not exist idealwise independent elements over R.

The following example, considered in Chapter 4, illustrates Remark 22.3.2. This is Example 4.11; other details are given in Remarks 4.12.

EXAMPLE 22.4. Let $R = \mathbb{Q}[x, y]_{(x,y)}$, the localized ring of polynomials in two variables over the rational numbers. The elements $\tau_1 = e^x - 1, \tau_2 = e^y - 1$, and $e^x - e^y = \tau_1 - \tau_2$ of $\widehat{R} = \mathbb{Q}[[x, y]]$ belong to completions of R with respect to principal ideals (and so are not idealwise independent). If $R_2 = \mathbb{Q}[x, y, \tau_1, \tau_2]_{(x,y,\tau_1,\tau_2)}$ and L is the field of fractions of R_2, then the elements $(e^x - 1)/x, (e^y - 1)/y$, and $(e^x - e^y)/(x - y)$ are certainly in $L \cap \widehat{R}$ but not in R_2. Valebrega's Theorem 4.9 implies that $L \cap \widehat{R}$ is a two-dimensional regular local ring with completion $\widehat{R}$.

Recall the concepts PDE, weakly flat and height-one preserving from Definitions 2.14 in Chapter 2 and 9.1 in Chapter 9. We state the definitions again here for Krull domains.

DEFINITIONS 22.5. Let $S \hookrightarrow T$ be an extension of Krull domains.

- T is a PDE extension of S if for every height-one prime ideal Q in T, the height of $Q \cap S$ is at most one.
- T is a *height-one preserving* extension of S if for every height-one prime ideal P of S with $PT \neq T$ there exists a height-one prime ideal Q of T with $PT \subseteq Q$.

- T is *weakly flat* over S if every height-one prime ideal P of S with $PT \neq T$ satisfies $PT \cap S = P$.

We summarize the results of this chapter.

SUMMARY 22.6. Let $(R, \mathfrak{m})$ be an excellent normal local domain of dimension d with field of fractions K and completion $(\widehat{R}, \widehat{\mathfrak{m}})$. In Section 22.1 we consider idealwise independent elements as defined in Definition 22.2. We show in Theorem 22.11 that $\tau_1, \dots, \tau_n \in \widehat{\mathfrak{m}}$ are idealwise independent over R if and only if the extension $R[\tau_1, \dots, \tau_n] \hookrightarrow \widehat{R}$ is weakly flat in the sense of Definition 22.5. If R has the additional property that every height-one prime ideal of R is the radical of a principal ideal, we show Section 22.1 that a sufficient condition for $\tau_1, \dots, \tau_n$ to be idealwise independent over R is that the extension $R[\tau_1, \dots, \tau_n] \hookrightarrow \widehat{R}$ satisfies PDE ("pas d'éclatement", or in English "no blowing up"), defined in Definitions 22.5; see Theorem 22.1.2. Diagram 22.9.0, at the end of Section 22.1 displays the relationships among these concepts and some others, for extensions of Krull domains.

In Sections 22.2 and 22.3 we present two methods for obtaining idealwise independent elements over a countable ring R. The method in Section 22.2 is to find elements $\tau_1, \dots, \tau_n \in \widehat{\mathfrak{m}}$ so that (1) $\tau_1, \dots, \tau_n$ are algebraically independent over R, and (2) for every prime ideal P of $R_n = R[\tau_1, \dots, \tau_n]_{(\mathfrak{m}, \tau_1, \dots, \tau_n)}$ with $\dim(R_n/P) = n$, the ideal $P\widehat{R}$ is $\widehat{\mathfrak{m}}$-primary. In this case, we say that $\tau_1, \dots, \tau_n$ are *primarily independent* over R. If R is countable and $\dim(R) > 2$, we show in Theorem 22.31 the existence over R of idealwise independent elements that fail to be primarily independent.

The main theorem of this chapter is Theorem 22.20: For every countable excellent normal local domain R of dimension at least two, there exists an infinite sequence $\tau_1, \tau_2, \dots$ of elements of $\widehat{\mathfrak{m}}$ that are primarily independent over R. It follows that $A = K(\tau_1, \tau_2, \dots) \cap \widehat{R}$ is an infinite-dimensional non-Noetherian local domain. Thus, for the example $R = k[x, y]_{(x,y)}$ with k a countable field, there exists for every positive integer n and $n = \infty$, an extension $A_n = L_n \cap \widehat{R}$ of R such that $\dim(A_n) = \dim(R) + n$. In particular, the canonical surjection $\widehat{A_n} \to \widehat{R}$ has a nonzero kernel.

In Section 22.3 we define $\tau \in \widehat{\mathfrak{m}}$ to be *residually algebraically independent* over R if τ is algebraically independent over R and, for each height-one prime ideal P of $\widehat{R}$ such that $P \cap R \neq 0$, the image of τ in $\widehat{R}/P$ is algebraically independent over $R/(P \cap R)$. We extend the concept of residual algebraic independence to a finite or infinite number of elements $\tau_1, \dots, \tau_n, \dots \in \widehat{\mathfrak{m}}$; Theorem 22.27 shows the equivalence of residual algebraic independence to the extension $R[\tau_1, \dots, \tau_n] \hookrightarrow \widehat{R}$ satisfying PDE.

Theorem 22.30.1 shows that that primary independence implies residual algebraic independence and that primary independence implies idealwise independence. If every height-one prime ideal of R is the radical of a principal ideal, Theorem 22.30.3 shows that residual algebraic independence implies idealwise independence.

For R of dimension two, Theorem 22.30.2 shows that primary independence is equivalent to residual algebraic independence. Hence residual algebraic independence implies idealwise independence if $\dim R = 2$. As remarked above, if R has dimension greater than two, then primary independence is stronger than residual algebraic independence. Theorems 22.35 and 22.37 show the existence of idealwise independent elements that fail to be residually algebraically independent.

The following diagram summarizes some relationships among the independence concepts for one element τ of $\widehat{\mathfrak{m}}$, over a local normal excellent domain $(R, \mathfrak{m})$. In the diagram we use "ind." and "resid." to abbreviate "independent" and "residually algebraic".

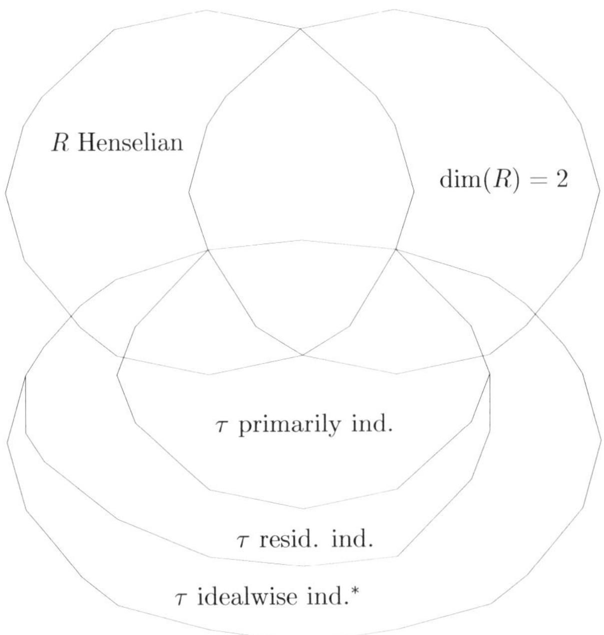

* In order to conclude that the idealwise independent set contains the residually algebraically independent set for $\dim R > 2$, we assume that every height-one prime ideal of R is the radical of a principal ideal.

Diagram 23.25 of Section 23.4.0 displays many more relationships among the independence concepts and other related properties.

In the remainder of this section we discuss some properties of extensions of Krull domains related to idealwise independence. A diagram near the end of this section displays the relationships among these properties.

REMARK 22.7. Let $S \hookrightarrow T$ be an extension of Krull domains. If S is a UFD, or more generally, if every height-one prime ideal of S is the radical of a principal ideal, then T is a height-one preserving extension of S. This is clear from the fact that every minimal prime of a principal ideal in a Krull domain has height one.

REMARK 22.8. Let $(R, \mathfrak{m})$ and $\tau_1, \ldots, \tau_n \in \widehat{\mathfrak{m}}$ be as in Setting 22.1. Also assume that each height-one prime ideal of R is the radical of a principal ideal. This property is preserved in a polynomial ring extension, by Fact 2.19 and Proposition 2.17. Thus Remark 22.7 implies that the embedding

$$\varphi : R_n = R[\tau_1, \ldots, \tau_n]_{(\mathfrak{m}, \tau_1, \ldots, \tau_n)} \hookrightarrow \widehat{R}$$

is a height-one preserving extension.

Corollary 22.9 is immediate from Remark 22.8 and Proposition 9.16.

COROLLARY 22.9. *Let* $(R, \mathfrak{m})$ *and* $\tau_1, \ldots, \tau_n \in \widehat{\mathfrak{m}}$ *be as in Setting* 22.1. *Assume that each height-one prime ideal of* R *is the radical of a principal ideal. Let* $R_n = R[\tau_1, \ldots, \tau_n]_{(\mathfrak{m}, \tau_1, \ldots, \tau_n)}$. *If* $R_n \hookrightarrow \widehat{R}$ *satisfies PDE, then* $\widehat{R}$ *is weakly flat over* R_n.

Let $S \hookrightarrow T$ be an extension of Krull domains, and let F be the field of fractions of S. Throughout the diagram "Q" denotes a prime ideal $Q \in \operatorname{Spec}(T)$ with $\operatorname{ht}(Q) = 1$, and "P" denotes $P \in \operatorname{Spec}(S)$ with $\operatorname{ht}(P) = 1$. Diagram 22.9.0. illustrates the relationships among the terms in Definitions 22.5 using the results (9.16), (9.4), (9.6), and (9.11):

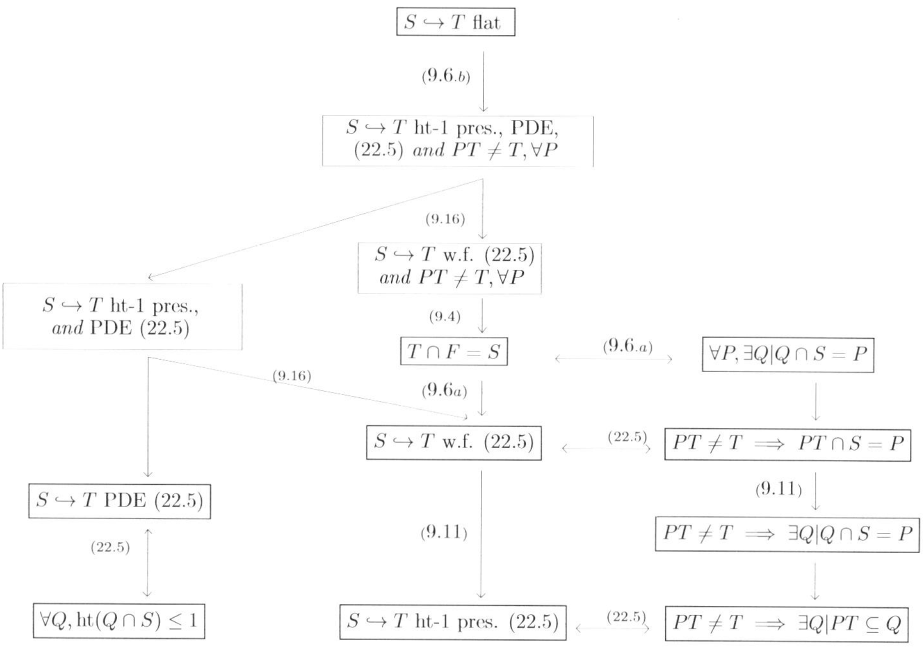

Diagram 22.9.0. The relationships among properties of an extension $S \hookrightarrow T$ of Krull domains.

REMARK 22.10. Let $S \hookrightarrow T$ be an extension of Krull domains. If $PT \neq T$ for every height-one prime ideal P of S, then Corollary 9.4.2 states that $S \hookrightarrow T$ is weakly flat if and only if $S = \mathcal{Q}(S) \cap T$. If S and T are local Krull domains with T dominating S, then $PT \neq T$ for each height one prime ideal of S. Therefore, for $R_n = R[\tau_1, \ldots, \tau_n]_{(\mathfrak{m}, \tau_1, \ldots, \tau_n)}$ and $\widehat{R}$ as in Setting 22.1, $R_n \hookrightarrow \widehat{R}$ is weakly flat if and only if $R_n = \mathcal{Q}(R_n) \cap \widehat{R}$.

THEOREM 22.11. *Let* $(R, \mathfrak{m})$ *be an excellent normal local domain with* $\mathfrak{m}$*-adic completion* $(\widehat{R}, \widehat{\mathfrak{m}})$ *and let* $\tau_1, \ldots, \tau_n \in \widehat{\mathfrak{m}}$ *be algebraically independent elements over* R. *Then:*

(1) $\tau_1, \ldots, \tau_n$ *are idealwise independent over* $R \iff R[\tau_1, \ldots, \tau_n] \hookrightarrow \widehat{R}$ *is weakly flat.*

(2) *If* $R[\tau_1,\dots,\tau_n] \hookrightarrow \widehat{R}$ *satisfies PDE and each height-one prime ideal of* R *is the radical of a principal ideal, then* $R[\tau_1,\dots,\tau_n] \hookrightarrow \widehat{R}$ *is weakly flat.*

In view of Remark 9.6.*b, these assertions also hold with* $R[\tau_1,\dots,\tau_n]$ *replaced by its localization* $R[\tau_1,\dots,\tau_n]_{(\mathfrak{m},\tau_1,\dots,\tau_n)}$.

PROOF. Item 1 is Remark 22.10, and item 2 is Corollary 22.9. □

In order to demonstrate idealwise independence we develop in the next two sections the concepts of primary independence and residual algebraic independence. Primary independence implies idealwise independence. If we assume that every height-one prime ideal of the base ring R is the radical of a principal ideal, then residual algebraic independence implies idealwise independence.

22.2. Primarily independent elements

In this section we introduce primary independence, a concept that implies idealwise independence; see Proposition 22.15. Let R be a countable excellent normal local domain of dimension at least two. Theorem 22.20 shows that there are infinitely many primarily independent elements over R.

DEFINITION 22.12. Let $(R,\mathfrak{m})$ be an excellent normal local domain. Elements $\tau_1,\dots,\tau_n \in \widehat{\mathfrak{m}}$ that are algebraically independent over R are called *primarily independent over* R, if the ideal $P\widehat{R}$ is $\widehat{\mathfrak{m}}$-primary, for every prime ideal P of $R_n = R[\tau_1,\dots,\tau_n]_{(\mathfrak{m},\tau_1,\dots,\tau_n)}$ such that $\dim(R_n/P) \le n$. A countably infinite sequence $\{\tau_i\}_{i=1}^{\infty}$ of elements of $\widehat{\mathfrak{m}}$ is *primarily independent over* R if $\tau_1,\dots,\tau_n$ are primarily independent over R for each n.

REMARKS 22.13. (1) By Diagram 22.1.1, primary independence of $\tau_1,\dots,\tau_n$ as defined in Definition 22.12 is equivalent to the statement that for every prime ideal P of R_n with $\dim(R_n/P) \le n$, the ideal $\lambda^{-1}(P\widehat{R}) = P\widehat{S_n} + \ker(\lambda)$ is primary for the maximal ideal of $\widehat{S_n}$.

(2) A subset of a primarily independent set is again primarily independent. For example, if $\tau_1,\dots,\tau_n$ are primarily independent over R, to see that $\tau_1,\dots,\tau_{n-1}$ are primarily independent, let P be a prime ideal of R_{n-1} with $\dim(R_{n-1}/P) \le n-1$. Then PR_n is a prime ideal of R_n with $\dim(R_n/PR_n) \le n$, and so $P\widehat{R}$ is primary for the maximal ideal of $\widehat{R}$.

(3) Every prime ideal P' of R_n such that $\dim(R_n/P') \le n$ contains a prime ideal P such that $\dim(R_n/P) = n$. Hence $\tau_1,\dots,\tau_n$ are primarily independent over R if and only if $P\widehat{R}$ is $\widehat{\mathfrak{m}}$-primary for every prime ideal P of R_n such that $\dim(R_n/P) = n$.

LEMMA 22.14. *Assume Setting* 22.1 *with* $(R,\mathfrak{m})$ *an excellent normal local domain of dimension at least 2. Let* $R_n = R[\tau_1,\dots,\tau_n]_{(\mathfrak{m},\tau_1,\dots,\tau_n)}$, *where* n *is a positive integer and* $\tau_1,\dots,\tau_n$ *are primarily independent over* R. *Let* P *be a prime ideal of* $S_n = R[t_1,\dots,t_n]_{(\mathfrak{m},t_1,\dots,t_n)}$ *such that* $\dim(S_n/P) \ge n+1$ *and let* $\mathfrak{p} = \lambda_1(P)$ *be the corresponding ideal of* R_n. *Then* $\dim(R_n/\mathfrak{p}) = \dim(S_n/P) \ge n+1$, *and*

(1) *The ideal* $\mathfrak{p}\widehat{R}$ *is not* $\widehat{\mathfrak{m}}$*-primary, and*

(2) $\mathfrak{p}\widehat{R} \cap R_n = \mathfrak{p}$.

PROOF. For item 1, if $\dim(R_n/\mathfrak{p}) \geq n+1$ and if $\mathfrak{p}\widehat{R}$ is primary for $\widehat{\mathfrak{m}}$, then Diagram 22.1.1 shows that $\mathfrak{p}\widehat{R} = P\widehat{S_n} + \ker(\lambda)$ is primary for the maximal ideal of $\widehat{S_n}$. Hence the maximal ideal of $\widehat{S_n}/P\widehat{S_n}$ is the radical of an n-generated ideal. Also $\widehat{S_n}/P\widehat{S_n} \cong \widehat{(S_n/P)}$, the completion of S_n/P, and $\dim(S_n/P) \geq n+1$ implies that $\dim(\widehat{S_n/P}) \geq n+1$. This is a contradiction by Theorem 2.23.

For item 2, if $\dim(R_n/\mathfrak{p}) = n+1$, and $\mathfrak{p} \subsetneq (\mathfrak{p}\widehat{R} \cap R_n)$, then $\dim(\frac{R_n}{(\mathfrak{p}\widehat{R}\cap R_n)}) \leq n$. Thus $\mathfrak{p}\widehat{R} = (\mathfrak{p}\widehat{R} \cap R_n)\widehat{R}$ is primary for $\widehat{\mathfrak{m}}$, a contradiction to item 1. Therefore $\mathfrak{p}\widehat{R} \cap R_n = \mathfrak{p}$ for each $\mathfrak{p}$ such that $\dim(R_n/\mathfrak{p}) = n+1$.

Assume that $\dim(R_n/\mathfrak{p}) > n+1$ and let

$$\mathcal{A} \;:=\; \{\mathfrak{q} \in \operatorname{Spec} R_n \mid \mathfrak{p} \subset \mathfrak{q} \text{ and } \dim(R_n/\mathfrak{q}) \;=\; n+1\}.$$

Proposition 3.28 implies that $\mathfrak{p} = \bigcap_{\mathfrak{q}\in\mathcal{A}} \mathfrak{q}$. Since $\mathfrak{q}\widehat{R} \cap R_n = \mathfrak{q}$, for each prime ideal $\mathfrak{q} \in \mathcal{A}$, it follows that

$$\mathfrak{p} \subseteq \mathfrak{p}\widehat{R} \cap R_n = (\bigcap_{\mathfrak{q}\in\mathcal{A}} \mathfrak{q})\widehat{R} \cap R_n \subseteq \bigcap_{\mathfrak{q}\in\mathcal{A}} (\mathfrak{q}\widehat{R} \cap R_n) \subseteq \bigcap_{\mathfrak{q}\in\mathcal{A}} \mathfrak{q} = \mathfrak{p}.$$

□

PROPOSITION 22.15. *Let $(R, \mathfrak{m})$ be an excellent normal local domain of dimension at least 2.*

(1) *Let n be a positive integer, and let $R_n = R[\tau_1, \dots, \tau_n]_{(\mathfrak{m},\tau_1,\dots,\tau_n)}$, where $\tau_1, \dots, \tau_n$ are primarily independent over R. Then $R_n = L \cap \widehat{R}$, where L is the field of fractions of R_n. Thus $\tau_1, \dots, \tau_n$ are idealwise independent elements of $\widehat{R}$ over R.*
(2) *If $\{\tau_i\}_{i=1}^{\infty}$ is a countably infinite sequence of primarily independent elements of $\widehat{\mathfrak{m}}$ over R, then $\{\tau_i\}_{i=1}^{\infty}$ are idealwise independent over R.*

PROOF. Since item 2 is a consequence of item 1, it suffices to prove item 1. Let $\mathfrak{p}$ be a height-one prime ideal of R_n. Then $\dim(R_n/\mathfrak{p}) \geq n+1$, since R_n is catenary and $\dim R \geq 2$. Lemma 22.14.2 implies that $\mathfrak{p}\widehat{R} \cap R_n = \mathfrak{p}$. Therefore $\widehat{R}$ is weakly flat over R_n. Hence by Theorem 22.11.1, we have $R_n = L \cap \widehat{R}$. □

PROPOSITION 22.16. *Let $(R, \mathfrak{m})$, $\tau_1, \dots, \tau_n \in \widehat{\mathfrak{m}}$ and λ be as in Setting 22.1. Thus λ restricts to an isomorphism*

$$\lambda_1 : S_n \;=\; R[t_1, \dots, t_n]_{(\mathfrak{m},t_1,\dots,t_n)} \;\to\; R_n \;=\; R[\tau_1, \dots, \tau_n]_{(\mathfrak{m},\tau_1,\dots,\tau_n)},$$

where $t_1, \dots, t_n$ are indeterminates over R. Then the following are equivalent:

(1) *For each prime ideal P of S_n such that $\dim(S_n/P) \geq n$ and each prime ideal $\widehat{P}$ of $\widehat{S_n}$ minimal over $P\widehat{S_n}$, the images of $t_1 - \tau_1, \dots, t_n - \tau_n$ in $\widehat{S_n}/\widehat{P}$ generate an ideal of height n in $\widehat{S_n}/\widehat{P}$.*
(2) *For each prime ideal P of S_n with $\dim(S_n/P) \geq n$ and each nonnegative integer $i \leq n$, the element $t_i - \tau_i$ is outside every prime ideal $\widehat{Q}$ of $\widehat{S}_n$ minimal over $(P, t_1 - \tau_1, \dots, t_{i-1} - \tau_{i-1})\widehat{S}_n$.*
(3) *For each prime ideal P of S_n such that $\dim(S_n/P) = n$, the images of $t_1 - \tau_1, \dots, t_n - \tau_n$ in $\widehat{S_n}/P\widehat{S_n}$ generate an ideal primary for the maximal ideal of $\widehat{S}_n/P\widehat{S}_n$.*
(4) *The elements $\tau_1, \dots, \tau_n$ are primarily independent over R.*

PROOF. Observe that items 1 and 2 are equivalent, and that item 1 and 2 imply item 3.

For the equivalence of items 3 and 4, Equation 22.1.1 implies

$$(22.16.0) \qquad \frac{S_n}{P} = \frac{R_n}{\lambda_1(P)}, \quad \text{and} \quad \frac{\widehat{S_n}}{(P, t_1 - \tau_1, \ldots, t_n - \tau_n)\widehat{S_n}} \cong \frac{\widehat{R}}{\lambda_1(P)\widehat{R}},$$

for every $P \in \operatorname{Spec} S_n$. Item 4 is equivalent to the ideal $\mathfrak{p}\widehat{R}$ being $\widehat{\mathfrak{m}}$-primary, for every $\mathfrak{p} \in R_n$ with $\dim(R_n/\mathfrak{p}) = n$, by Definition 22.12 and Remark 22.13.3. Let $\mathfrak{p} \in R_n$ and $P \in \operatorname{Spec} S_n$ be such that $\lambda_1(P) = \mathfrak{p}$. By Equation 22.16.0,

$$\begin{aligned} &\dim(R_n/\mathfrak{p}) = n \quad \text{and} \quad \mathfrak{p}\widehat{R} \text{ is } \widehat{\mathfrak{m}}\text{-primary} \\ \iff &\dim(S_n/P) = n \quad \text{and} \quad (P, t_1 - \tau_1, \ldots, t_n - \tau_n)(\widehat{S_n}) \text{ is } \widehat{\mathfrak{m}}\text{-primary} \\ \iff &\dim(S_n/P) = n \quad \text{and} \quad (t_1 - \tau_1, \ldots, t_n - \tau_n)\Big(\frac{\widehat{S_n}}{P\widehat{S_n}}\Big) \text{ is } \Big(\frac{\widehat{\mathfrak{m}}\widehat{S_n}}{P\widehat{S_n}}\Big)\text{-primary.} \end{aligned}$$

Hence item 3 is equivalent to item 4.

It remains to show that item 3 implies item 1. For this, let P be a prime ideal of S_n such that $\dim(S_n/P) = n + h$, where $h \geq 0$. There exist $s_1, \ldots, s_h \in S_n$ so that if $I = (P, s_1, \ldots, s_h)S_n$, then for each minimal prime Q of I we have $\dim(S_n/Q) = n$. Item 3 implies that the images of $t_1 - \tau_1, \ldots, t_n - \tau_n$ in $\widehat{S_n}/Q\widehat{S_n}$ generate an ideal primary for the maximal ideal of $\widehat{S_n}/Q\widehat{S_n}$. It follows that the images of $t_1 - \tau_1, \ldots, t_n - \tau_n$ in $\widehat{S_n}/I\widehat{S_n}$ generate an ideal primary for the maximal ideal of $\widehat{S_n}/I\widehat{S_n}$, and therefore that the images of $s_1, \ldots, s_h, t_1 - \tau_1, \ldots, t_n - \tau_n$ in $\widehat{S_n}/P\widehat{S_n}$ are a system of parameters for the $(n+h)$-dimensional local ring $\widehat{S_n}/P\widehat{S_n}$. Let $\widehat{P}$ be a minimal prime ideal of $P\widehat{S_n}$. By Ratliff's equidimension Theorem 3.26, $\dim(\widehat{S_n}/\widehat{P}) = n + h$. Since the images of $s_1, \ldots, s_h, t_1 - \tau_1, \ldots, t_n - \tau_n$ in the complete local domain $\widehat{S_n}/\widehat{P}$ are a system of parameters, it follows that the images of $t_1 - \tau_1, \ldots, t_n - \tau_n$ in $\widehat{S_n}/\widehat{P}$ generate an ideal of height n in $\widehat{S_n}/\widehat{P}$. Therefore item 1 holds. □

COROLLARY 22.17. *With the notation of Setting 22.1 and Proposition 22.16, assume that $\tau_1, \ldots, \tau_n$ are primarily independent over R.*

(1) *Let I be an ideal of S_n such that $\dim(S_n/I) = n$. It follows that the ideal $(I, t_1 - \tau_1, \ldots, t_n - \tau_n)\widehat{S_n}$ is primary to the maximal ideal of $\widehat{S_n}$.*
(2) *Let $P \in \operatorname{Spec} S_n$ be a prime ideal with $\dim(S_n/P) > n$. Then the ideal $\widehat{W} = (P, t_1 - \tau_1, \ldots, t_n - \tau_n)\widehat{S_n}$ has $\operatorname{ht}(\widehat{W}) = \operatorname{ht}(P) + n$ and $\widehat{W} \cap S_n = P$.*

PROOF. For item 1, let $P \in \operatorname{Spec}(S_n)$ be minimal over I. Then $\dim(S_n/P) \leq n$. By Remark 22.13.1, $(P, t_1 - \tau_1, \ldots, t_n - \tau_n)\widehat{S_n}$ is primary to the maximal ideal of $\widehat{S_n}$. Exercise 1 completes the proof of item 1.

For item 2, by Proposition 22.16.1, $\operatorname{ht}(\widehat{W}) = \operatorname{ht}(P) + n$. Let λ_1 be the restriction to S_n of the canonical homomorphism $\lambda : \widehat{S_n} \to \widehat{R}$ from Setting 22.1. Then $\lambda_1 : S_n \overset{\cong}{\to} R_n$. Thus $\dim(R_n/\lambda_1(P)) > n$, and so $\lambda_1(P)\widehat{R} \cap R_n = \lambda_1(P)$, by Lemma 22.14.2. Now

$$\widehat{W} \cap S_n = \lambda^{-1}(\lambda_1(P)\widehat{R}) \cap \lambda_1^{-1}(R_n) = \lambda_1^{-1}(\lambda_1(P)\widehat{R} \cap R_n) = \lambda_1^{-1}(\lambda_1(P)) = P.$$

□

To establish the existence of primarily independent elements, we use the following prime avoidance lemma over a complete local ring. (This is similar to [**30**, Lemma 3],[**186**, Lemma 10], [**171**] and [**110**, Lemma 14.2].) We also use this result in two constructions given in Section 22.3.

LEMMA 22.18. *Let $(T, \mathfrak{n})$ be a complete Noetherian local ring of dimension at least 2, and let $t \in \mathfrak{n} \setminus \mathfrak{n}^2$. Assume that I is an ideal of T containing t, and that $\mathcal{U}$ is a countable set of prime ideals of T each of which fails to contain I. Then there exists an element $a \in I \cap \mathfrak{n}^2$ such that $t - a \notin \bigcup\{Q \mid Q \in \mathcal{U}\}$.*

PROOF. Let $\{P_i\}_{i=1}^{\infty}$ be an enumeration of the prime ideals of $\mathcal{U}$. We may assume that there are no containment relations among the prime ideals of $\mathcal{U}$. Choose $f_1 \in \mathfrak{n}^2 \cap I$ so that $t - f_1 \notin P_1$. Then choose $f_2 \in P_1 \cap \mathfrak{n}^3 \cap I$ so that $t - f_1 - f_2 \notin P_2$. Note that $f_2 \in P_1$ implies $t - f_1 - f_2 \notin P_1$. Successively, by induction, choose

$$f_n \in P_1 \cap P_2 \cap \cdots \cap P_{n-1} \cap \mathfrak{n}^{n+1} \cap I$$

so that $t - f_1 - \ \dots \ - f_n \notin \bigcup_{i=1}^{n} P_i$ for each positive integer n. Then we have a Cauchy sequence $\{f_1 + \cdots + f_n\}_{n=1}^{\infty}$ in T that converges to an element $a \in \mathfrak{n}^2$. Now

$$t - a = (t - f_1 - \ \dots \ - f_n) + (f_{n+1} + \dots),$$

where $(t - f_1 - \ \dots \ - f_n) \notin P_n, (f_{n+1} + \dots) \in P_n$. Therefore $t - a \notin P_n$, for all n, and $t - a \in I$. $\square$

REMARK 22.19. Let $A \hookrightarrow B$ be an extension of Krull domains. If α is a nonzero nonunit of B such that $\alpha \notin Q$, for each height-one prime Q of B such that $Q \cap A \neq (0)$, then $\alpha B \cap A = (0)$. In particular, such an element α is algebraically independent over A.

THEOREM 22.20. *Let $(R, \mathfrak{m})$ be a countable excellent normal local domain of dimension at least 2, and let $(\widehat{R}, \widehat{\mathfrak{m}})$ be the completion of R. Then:*

(1) *There exists $\tau \in \widehat{\mathfrak{m}}$ that is primarily independent over R.*
(2) *Let $n \in \mathbb{N}$. If $\tau_1, \dots, \tau_{n-1} \in \widehat{\mathfrak{m}}$ are primarily independent over R, then there exists $\tau_n \in \widehat{\mathfrak{m}}$ such that $\tau_1, \dots, \tau_{n-1}, \tau_n$ are primarily independent over R.*
(3) *There exists an infinite sequence $\tau_1, \dots, \tau_n, \dots \in \widehat{\mathfrak{m}}$ of elements that are primarily independent over R.*

PROOF. Item 2 implies item 1 and item 3. To prove item 2, let $t_1, \dots, t_n$ be indeterminates over R, and let the notation be as in Setting 22.1. Thus we have $S_{n-1} \cong R_{n-1}$, under the R-algebra isomorphism taking $t_i \to \tau_i$. Let $\widehat{\mathfrak{n}}$ denote the maximal ideal of $\widehat{S_n}$. We show the existence of $a \in \widehat{\mathfrak{n}}^2$ such that:

(i) There is an $\widehat{R}$-algebra surjection : $\lambda_a : \widehat{S_n} \to \widehat{R}$ with $\ker \lambda_a = (t_1 - \tau_1, \dots, t_{n-1} - \tau_{n-1}, t_n - a)\widehat{S_n}$.
(ii) The elements $\tau_1, \dots, \tau_{n-1}$, together with the image τ_n of t_n under the map λ_a are primarily independent over R.

To see items i and ii: Let $\widehat{I} = (t_1 - \tau_1, \dots, t_{n-1} - \tau_{n-1})\widehat{S_{n-1}}$. Since S_n is countable and Noetherian, enumerate

$$\{P_j\}_{j=1}^{\infty} = \{\mathfrak{p} \in \operatorname{Spec} S_n \mid \dim(S_n/P_j) \geq n\}.$$

Let $\mathcal{U} = \{\widehat{\mathfrak{p}} \in \operatorname{Spec} \widehat{S_n} \mid \widehat{\mathfrak{p}}$ is minimal over $(P_j, \widehat{I})\widehat{S_n}$, for some $P_j\}$. Then $\mathcal{U}$ is countable and $\widehat{\mathfrak{n}} \notin \mathcal{U}$ since $(P_j, \widehat{I})\widehat{S_n}$ is generated by $n - 1$ elements over $P_j\widehat{S_n}$ and

$\dim(\widehat{S_n}/P_j\widehat{S_n}) \geq n$. By Lemma 22.18 with the ideal I of Lemma 22.18 taken to be $\widehat{\mathfrak{n}}$, there exists an element $a \in \widehat{\mathfrak{n}}^2$ so that $t_n - a$ is not in $\widehat{Q}$, for every prime ideal $\widehat{Q} \in \mathcal{U}$.

Then

$$\widehat{S_n} = \widehat{S_{n-1}}[[t_n]] = \widehat{S_{n-1}}[[t_n - a]], \qquad \frac{\widehat{S_n}}{(t_n - a)\widehat{S_n}} = \widehat{S_{n-1}}, \text{ and } \frac{\widehat{S_{n-1}}}{\widehat{I}} \cong \widehat{R}.$$

Let λ_a be the composition

$$\widehat{S_n} \to \frac{\widehat{S_n}}{(t_n - a)\widehat{S_n}} = \widehat{S_{n-1}} \to \frac{\widehat{S_{n-1}}}{\widehat{I}} \to \widehat{R} = \frac{\widehat{S_n}}{(\widehat{I}, t_n - a)\widehat{S_n}}.$$

Define $\lambda_a(t_n) = \tau_n$, an element in $\widehat{R}$.

The $\widehat{R}$-algebra surjection $\lambda_a : \widehat{S_n} \to \widehat{R}$ has $\ker(\lambda_a) = (\widehat{I}, t_n - a)\widehat{S_n}$. The kernel of λ_a is also equal to $(\widehat{I}, t_n - \tau_n)\widehat{S_n}$. Therefore the setting will be as in Diagram 22.1.0 after we establish Claim 22.21.

CLAIM 22.21. $(\widehat{I}, t_n - \tau_n)\widehat{S_n} \cap S_n = (0)$.

PROOF. (of Claim 22.21) Since $\tau_1, \dots, \tau_{n-1}$ are algebraically independent over R, we have $\widehat{I} \cap S_{n-1} = (0)$. Let $R'_n = R_{n-1}[t_n]_{(\max(R_{n-1}), t_n))}$. Consider the diagram:

$$\begin{array}{ccc} S_n = S_{n-1}[t_n]_{(\max(S_{n-1}), t_n)} & \xrightarrow{\subset} & \widehat{S_n} = \widehat{S_{n-1}}[[t_n]] \\ \cong \downarrow & & \lambda' \downarrow \\ R'_n = R_{n-1}[t_n]_{(\max(R_{n-1}), t_n)} & \xrightarrow{\subset} & \widehat{R}[[t_n]] \cong (\widehat{S_{n-1}}/\widehat{I})[[t_n]], \end{array}$$

where λ' restricted to $\widehat{S_{n-1}}$ is the canonical projection : $\widehat{S_{n-1}} \to \widehat{S_{n-1}}/\widehat{I} \cong \widehat{R}$, and $\lambda'(t_n) = t_n$.

For $\widehat{Q}$ a prime ideal of $\widehat{S_n}$, we have $\widehat{Q} \in \mathcal{U} \iff \lambda'(\widehat{Q}) = \widehat{P}$, where $\widehat{P}$ is a prime ideal of $\widehat{R}[[t_n]] \cong (\widehat{S_{n-1}}/\widehat{I})[[t_n]]$ minimal over $\lambda'(P_j)\widehat{R}[[t_n]]$ for some prime ideal P_j of S_n such that $\dim(S_n/P_j) \leq n$. Since $t_n - a$ is outside every $\widehat{Q} \in \mathcal{U}$, $t_n - \lambda'(a) = \lambda'(t_n - a)$ is outside every prime ideal $\widehat{P}$ of $\widehat{R}[[t_n]]$, such that $\widehat{P}$ is minimal over $\lambda'(P_j)\widehat{R}[[t_n]]$. Since S_n is catenary and $\dim(S_n) = n + \dim(R)$,

$$\{P_j\}_{j=1}^{\infty} = \{\mathfrak{p} \in \operatorname{Spec} S_n \mid \dim(S_n/\mathfrak{p}) \geq n\} = \{\mathfrak{p} \in \operatorname{Spec} S_n \mid \operatorname{ht}(\mathfrak{p}) \leq \dim(R)\}.$$

Suppose $\widehat{P}$ is a height-one prime ideal of $\widehat{R}[[t_n]]$ such that $P := \widehat{P} \cap R'_n \neq (0)$. Then $\widehat{P}$ is a minimal prime ideal of $P\widehat{R}[[t_n]]$. Also $P = \lambda'(Q)$, where Q is a height-one prime ideal of S_n and $\dim(S_n/Q) = n + \dim(R) - 1 \geq n$. Therefore $Q \in \{P_j\}_{j=1}^{\infty}$. By choice of a, we have $t_n - \lambda'(a) \notin \widehat{P}$. By Remark 22.19, $(t_n - \lambda'(a))\widehat{R}[[t_n]]) \cap R'_n = (0)$. Hence $(\widehat{I}, t_n - \tau_n)\widehat{S_n} \cap S_n = (0)$. □

CLAIM 22.22. Let P be a prime ideal of S_n such that $\dim(S_n/P) = n$. Then the ideal $(P, \widehat{I}, t_n - \tau_n)\widehat{S_n}$ is $\widehat{\mathfrak{n}}$-primary.

PROOF. (of Claim 22.22) Let $Q = P \cap S_{n-1}$. Either $QS_n = P$, or $QS_n \subsetneq P$. If $QS_n = P$, then $\dim(S_{n-1}/Q) = n-1$ and the primary independence of $\tau_1, \dots, \tau_{n-1}$ implies that $(Q, \widehat{I})\widehat{S_{n-1}}$ is primary for the maximal ideal of $\widehat{S_{n-1}}$. Therefore $(Q, \widehat{I}, t_n - \tau_n)\widehat{S_n} = (P, \widehat{I}, t_n - \tau_n)\widehat{S_n}$ is $\widehat{\mathfrak{n}}$-primary in this case. On the other hand, if $QS_n \subsetneq P$, then $\dim(S_{n-1}/Q) = n$. Let $\widehat{Q}'$ be a minimal prime ideal of

$(Q,\widehat{I})\widehat{S_{n-1}}$. By Proposition 22.16, $\dim(\widehat{S_{n-1}}/\widehat{Q}') = 1$, and hence $\dim(\widehat{S_n}/\widehat{Q}'\widehat{S_n}) = 2$. The primary independence of $\tau_1, \dots \tau_{n-1}$ implies that $\widehat{Q}' \cap S_{n-1} = Q$. Therefore $\widehat{Q}'\widehat{S_{n-1}}[[t_n]] \cap S_n = QS_n \subsetneq P$, so P is not contained in $\widehat{Q}'\widehat{S_n}$. Therefore $\dim(\widehat{S_n}/(P,\widehat{I})\widehat{S_n})) = 1$ and our choice of a implies that $(P,\widehat{I}, t_n - \tau_n)\widehat{S_n}$ is $\widehat{\mathfrak{n}}$-primary. □

This completes the proof of Theorem 22.20. □

COROLLARY 22.23. *Let $(R,\mathfrak{m})$ be a countable excellent normal local domain of dimension at least 2, and let K denote the field of fractions of R. Then there exist $\tau_1, \dots, \tau_n, \dots \in \widehat{\mathfrak{m}}$ such that $A = K(\tau_1, \tau_2, \dots) \cap \widehat{R}$ is an infinite-dimensional non-Noetherian local domain. In particular, for k a countable field, the localized polynomial ring $R = k[x,y]_{(x,y)}$ has such extensions inside $\widehat{R} = k[[x,y]]$.*

PROOF. By Theorem 22.20.3, there exist $\tau_1, \dots, \tau_n, \dots \in \widehat{\mathfrak{m}}$ that are primarily independent over R. By Proposition 22.15.2, primarily independent elements are idealwise independent. It follows that $A = K(\tau_1, \tau_2, \dots) \cap \widehat{R}$ is an infinite-dimensional local domain. In particular, A is not Noetherian. □

22.3. Residually algebraically independent elements

We introduce in this section a third concept, that of residual algebraic independence. Residual algebraic independence is weaker than primary independence. In Theorem 22.31 we show that over every countable normal excellent local domain $(R,\mathfrak{m})$ of dimension at least three there exists an element residually algebraically independent over R that is not primarily independent over R. In Theorems 22.35 and 22.37 we show the existence of idealwise independent elements that fail to be residually algebraically independent.

DEFINITION 22.24. Let $(\widehat{R},\widehat{\mathfrak{m}})$ be a complete normal Noetherian local domain and let A be a Krull subdomain of $\widehat{R}$ such that $A \hookrightarrow \widehat{R}$ satisfies PDE.

(1) An element $\tau \in \widehat{\mathfrak{m}}$ is *residually algebraically independent with respect to $\widehat{R}$ over A*, if τ is algebraically independent over A and, for each height-one prime $\widehat{P}$ of $\widehat{R}$ such that $\widehat{P} \cap A \neq (0)$, the image of τ in $\widehat{R}/\widehat{P}$ is algebraically independent over the integral domain $A/(\widehat{P} \cap A)$.
(2) Elements $\tau_1, \dots \tau_n \in \widehat{\mathfrak{m}}$ are said to be *residually algebraically independent* over A, if τ_{i+1} is residually algebraically independent over $A[\tau_1, \dots, \tau_i]$, for each $0 \le i < n$.
(3) An infinite sequence $\{\tau_i\}_{i=1}^{\infty}$ of elements of $\widehat{\mathfrak{m}}$ is *residually algebraically independent* over A, if $\tau_1, \dots \tau_n$ are residually algebraically independent over A, for each positive integer n.

REMARK 22.25. Let $\tau_1, \dots, \tau_n \in \widehat{\mathfrak{m}}$ be algebraically independent over R. If the τ_i are residually algebraically independent over R, as in Condition 2 of Definition 22.24, then the τ_i satisfy:

(2′) For each height-one prime ideal $\widehat{P}$ of $\widehat{R}$ with $\widehat{P} \cap R \neq 0$, the images of $\tau_1, \dots, \tau_n$ in $\widehat{R}/\widehat{P}$ are algebraically independent over $R/(\widehat{P} \cap R)$.

The proof that Condition 2 implies Condition 2′ is left to the reader in Exercise 2. Construction 22.32 shows that Condition 2′ does not imply Condition 2.

Proposition 22.26 relates residual algebraic independence for τ over A to the PDE property of Definition 22.5 for $A[\tau] \hookrightarrow \widehat{R}$. By Corollary 9.14, for an extension of Krull domains, the PDE property is equivalent to the LF_1 property of Definition 9.1.3.

PROPOSITION 22.26. *Let $(R, \mathfrak{m})$ and $\tau \in \widehat{\mathfrak{m}}$ be as in Setting 22.1. Let A be a Krull subdomain of $\widehat{R}$ such that τ is algebraically independent over A and $A \hookrightarrow \widehat{R}$ satisfies PDE. Then τ is residually algebraically independent with respect to $\widehat{R}$ over $A \iff A[\tau] \hookrightarrow \widehat{R}$ satisfies PDE.*

PROOF. Assume $A[\tau] \hookrightarrow \widehat{R}$ does not satisfy PDE. Then there exists a prime ideal $\widehat{P}$ of $\widehat{R}$ of height one such that $\operatorname{ht}(\widehat{P} \cap A[\tau]) \geq 2$. Then $\widehat{P} \cap A \neq 0$, and $\operatorname{ht}(\widehat{P} \cap A) = 1$, since PDE holds for $A \hookrightarrow \widehat{R}$. Thus, with $\mathfrak{p} = \widehat{P} \cap A$, we have $\mathfrak{p}A[\tau] \subsetneq \widehat{P} \cap A[\tau]$; that is, there exists $f(\tau) \in (\widehat{P} \cap A[\tau]) \setminus \mathfrak{p}A[\tau]$, or equivalently there is a nonzero polynomial $\bar{f}(x) \in (A/\mathfrak{p})[x]$ so that $\bar{f}(\bar{\tau}) = \bar{0}$ in $A[\tau]/(\widehat{P} \cap A[\tau])$, where $\bar{\tau}$ denotes the image of τ in $\widehat{R}/\widehat{P}$. This means that $\bar{\tau}$ is algebraic over $A/(\widehat{P} \cap A)$. Hence τ is not residually algebraically independent with respect to $\widehat{R}$ over A.

For the converse, assume that $A[\tau] \hookrightarrow \widehat{R}$ satisfies PDE and let $\widehat{P}$ be a height-one prime ideal of $\widehat{R}$ such that $\widehat{P} \cap A = \mathfrak{p} \neq 0$. Since $A[\tau] \hookrightarrow \widehat{R}$ satisfies PDE, $\widehat{P} \cap A[\tau] = \mathfrak{p}A[\tau]$ and $A[\tau]/(\mathfrak{p}A[\tau])$ canonically embeds in $\widehat{R}/\widehat{P}$. Hence the image of τ in $A[\tau]/\mathfrak{p}A[\tau]$ is algebraically independent over $A/\mathfrak{p}$. It follows that τ is residually algebraically independent with respect to $\widehat{R}$ over A. □

THEOREM 22.27. *Let $(R, \mathfrak{m})$ be an excellent normal local domain with completion $(\widehat{R}, \widehat{\mathfrak{m}})$ and let $\tau_1, \ldots, \tau_n \in \widehat{\mathfrak{m}}$ be algebraically independent over R. The following statements are equivalent:*

(1) *The elements $\tau_1, \ldots, \tau_n$ are residually algebraically independent with respect to $\widehat{R}$ over R.*
(2) *For each integer i with $1 \leq i \leq n$, if $\widehat{P}$ is a height-one prime ideal of $\widehat{R}$ such that $\widehat{P} \cap R[\tau_1, \ldots, \tau_{i-1}] \neq 0$, then $\operatorname{ht}(\widehat{P} \cap R[\tau_1, \ldots, \tau_i]) = 1$.*
(3) *$R[\tau_1, \ldots, \tau_n] \hookrightarrow \widehat{R}$ satisfies PDE.*

If each height-one prime ideal of R is the radical of a principal ideal, then these equivalent conditions imply the map $R[\tau_1, \ldots, \tau_n] \hookrightarrow \widehat{R}$ is weakly flat.

PROOF. The equivalence of the three items follows from Proposition 22.26. The last sentence follows from Theorem 22.11) □

The equivalence of items 1 and 3 of Theorem 22.27 implies:

COROLLARY 22.28. *Assume the notation of Theorem 22.27. If $\tau_1, \ldots, \tau_n$ are residually algebraically independent with respect to $\widehat{R}$ over R, then every permutation of $\tau_1, \ldots, \tau_n$ is residually algebraically independent with respect to $\widehat{R}$ over R.*

PROPOSITION 22.29. *Assume Setting 22.1. Also assume that $\dim R \geq 2$, and that $\{\tau_i\}_{i=1}^m \subseteq \widehat{\mathfrak{m}}$ is primarily independent over R, where m is either a positive integer or $m = \infty$. Let n be an integer with $0 \leq n \leq m$ and let $\widehat{P}$ be a height-one prime ideal of $\widehat{R}$ such that $\mathfrak{p} := \widehat{P} \cap R_n \neq 0$. Let $\widehat{W} = (\mathfrak{p}, t_1 - \tau_1, \ldots, t_n - \tau_n)\widehat{S_n}$. Then:*

(1) $\operatorname{ht} \mathfrak{p} = 1$.

(2) $\lambda^{-1}(\widehat{P}) = \widehat{W}$.

(3) *For every integer* j *with* $n \leq j \leq m$, $\widehat{P} \cap R_j = \mathfrak{p}R_j$.

(4) *For every integer* j *with* $0 \leq j \leq m$, $\operatorname{ht}(\widehat{P} \cap R_j) \leq 1$.

PROOF. It follows from Diagram 22.1.0 that $\lambda(\widehat{W}) = \mathfrak{p}\widehat{R} \subseteq \widehat{P}$. By Corollary 22.17.2, $\operatorname{ht}(\widehat{W}) = \operatorname{ht}(\mathfrak{p}) + n$. Also, $\widehat{W} \subseteq (\widehat{P}, t_1 - \tau_1, \ldots, t_n - \tau_n) = \lambda^{-1}(\widehat{P})$ and thus

$$1 + n \;\leq\; \operatorname{ht}(\mathfrak{p}) + n \;=\; \operatorname{ht}(\widehat{W}) \;\leq\; \operatorname{ht}(\lambda^{-1}(\widehat{P})) \;\leq\; \operatorname{ht}(\widehat{P}) + n \;=\; 1 + n.$$

Therefore $\operatorname{ht}(\mathfrak{p}) = 1$, $\lambda^{-1}(\widehat{P}) = \widehat{W}$, and $\lambda(\widehat{W}) = \widehat{P}$. This proves items 1 and 2. If $n \leq j \leq m$, then $\operatorname{ht}(\widehat{P} \cap R_j) = 1$, by item 1. Since $\operatorname{ht}\mathfrak{p} = 1$, the ideal $\mathfrak{p}R_j$ is a height-one prime ideal, and $\mathfrak{p}R_j \subseteq \widehat{P} \cap R_j$. Thus $\mathfrak{p}R_j = \widehat{P} \cap R_j$. This proves item 3. If $\widehat{P} \cap R_j = (0)$, then $\operatorname{ht}(\widehat{P} \cap R_j) = 0$. If $\widehat{P} \cap R_j \neq (0)$, then $\operatorname{ht}(\widehat{P} \cap R_j) = 1$, by item 1. Thus item 4 holds. $\square$

THEOREM 22.30. *Let* $(R, \mathfrak{m})$ *and* $\{\tau_i\}_{i=1}^m \subseteq \widehat{\mathfrak{m}}$ *be as in Setting 22.1, where* $\dim R \geq 2$ *and* m *is either a positive integer or* $m = \infty$.

(1) *If* $\{\tau_i\}_{i=1}^m$ *is primarily independent over* R, *then* $\{\tau_i\}_{i=1}^m$ *is residually algebraically independent over* R.

(2) *If* $\dim R = 2$, *then* $\{\tau_i\}_{i=1}^m$ *is primarily independent over* R *if and only if* $\{\tau_i\}_{i=1}^m$ *is residually algebraically independent over* R.

(3) *If each height-one prime ideal of* R *is the radical of a principal ideal and* $\{\tau_i\}_{i=1}^m$ *is residually algebraically independent over* R, *then* $\{\tau_i\}_{i=1}^m$ *is idealwise independent over* R.

PROOF. Item 1 follows from Proposition 22.29, and the equivalence of items 1 and 3 of Theorem 22.27. To prove item 2, assume that $\dim R = 2$ and $n \leq m$ is a positive integer such that $\tau_1, \ldots, \tau_n$ are residually algebraically independent over R. Let $R_n = R[\tau_1, \ldots, \tau_n]_{(\mathfrak{m}, \tau_1, \ldots, \tau_n)}$. By Theorem 22.27, $R_n \hookrightarrow \widehat{R}$ satisfies PDE. Let $\mathfrak{p}$ be a prime ideal of R_n such that $\dim(R_n/\mathfrak{p}) \leq n$. Since $\dim R_n = n + 2$ and R_n is catenary, it follows that $\operatorname{ht}\mathfrak{p} \geq 2$. To show $\tau_1, \ldots, \tau_n$ are primarily independent over R, it suffices to show that $\mathfrak{p}\widehat{R}$ is $\widehat{\mathfrak{m}}$-primary. Since $\dim(\widehat{R}) = 2$, $\mathfrak{p}\widehat{R}$ is $\widehat{\mathfrak{m}}$-primary is equivalent to $\mathfrak{p}$ is not contained in a height-one prime ideal of $\widehat{R}$. This last statement holds since $R_n \hookrightarrow \widehat{R}$ satisfies PDE.

Item 3 follows from Theorems 22.27 and 22.11. $\square$

THEOREM 22.31. *Let* $(R, \mathfrak{m})$ *be a countable excellent normal local domain of dimension* d *and let* $(\widehat{R}, \widehat{\mathfrak{m}})$ *be the completion of* R. *If* $d \geq 3$, *then there exists an element* $\tau \in \widehat{\mathfrak{m}}$ *that is residually algebraically independent over* R, *but not primarily independent over* R.

PROOF. The proof uses techniques similar to those in the proof of Theorem 22.20. Let t be an indeterminate over R and set $S_1 = R[t]_{(\mathfrak{m},t)}$. Thus $\widehat{S_1} = \widehat{R}[[t]]$. Let $\widehat{\mathfrak{n}_1}$ denote the maximal ideal of $\widehat{S_1}$. Then

$$S_1 \;=\; R[t]_{(\mathfrak{m},t)} \;\hookrightarrow\; \widehat{S_1} = \widehat{R}[[t]]$$

is the top line of a diagram similar to Diagram 22.1.0 for $n = 1$. We seek an appropriate $\widehat{R}$-algebra homomorphism $\lambda : \widehat{S_1} \longrightarrow \widehat{R}$ to complete the diagram. For this, let $\mathfrak{p}_0$ be a prime ideal of R with $\operatorname{ht}\mathfrak{p}_0 = d - 1$, and let $\widehat{P_0}$ be a minimal prime ideal of $\mathfrak{p}_0\widehat{R}$. Then $\operatorname{ht}\widehat{P_0} = d-1$. Define $\widehat{Q_0} := (\widehat{P_0},t)\widehat{S_1}$ and $Q_0 := \widehat{Q_0}\cap S_1 = (\mathfrak{p}_0,t)S_1$. Then $\operatorname{ht}\widehat{Q_0} = d = \operatorname{ht} Q_0$. Let

$$\mathcal{U} = \{\widehat{Q} \in \operatorname{Spec}\widehat{S_1} \mid \operatorname{ht}\widehat{Q} \le d,\ \operatorname{ht}(\widehat{Q}\cap S_1) = \operatorname{ht}\widehat{Q} \text{ and } \widehat{Q}\neq\widehat{Q_0}\ \}.$$

Then $\mathcal{U}$ is countable, since $\operatorname{Spec} S_1$ is countable, and each prime ideal of $\mathcal{U}$ is one of the finitely many prime ideals of $\operatorname{Spec}\widehat{S_1}$ minimal over $\mathfrak{q}\widehat{S_1}$, for some $\mathfrak{q}\in\operatorname{Spec} S_1$. Apply Lemma 22.18 with $T = \widehat{S_1}$, $\mathfrak{n} = \widehat{\mathfrak{n}_1}$ and $I = \widehat{Q_0}$, to obtain an element $a\in\widehat{Q_0}\,\cap\,\widehat{\mathfrak{n}_1}^2$ so that $t-a\in\widehat{Q}_0$, but $t-a$ is not in any prime ideal in $\mathcal{U}$. Since $a\in\widehat{\mathfrak{n}_1}^2$, it follows that $\widehat{R}[[t]] = \widehat{R}[[t-a]]$. Define λ to be the natural surjection

$$\lambda : \widehat{S_1} \longrightarrow \widehat{S_1}/(t-a)\widehat{S_1} = \widehat{R},$$

and define $\tau := \lambda(t) = \lambda(a)$.

By Remark 22.19, $(t-a)\widehat{S_1}\cap S_1 = (0)$, and so the map λ restricted to S_1 is an isomorphism from S_1 onto $R_1 := R[\tau]_{(\mathfrak{m},\tau)}$. The prime ideal $\lambda(Q_0)$ in $R_1 = R[\tau]_{(\mathfrak{m},\tau)}$ has $\operatorname{ht}(\lambda(Q_0)) = d$. Thus $\dim(R_1/\lambda(Q_0)) = 1$. Since Diagram 22.1.0 is commutative, $\lambda(Q_0)\widehat{R}\subseteq\lambda(\widehat{Q_0})$. Since $(t-\tau)\widehat{S_1} = (t-a)\widehat{S_1}\subseteq\widehat{Q_0}$, the prime ideal $\lambda(\widehat{Q_0})$ has height $d-1$. Therefore $\lambda(Q_0)\widehat{R}$ is not $\widehat{\mathfrak{m}}$-primary. Hence τ is not primarily independent.

To prove that τ is residually algebraically independent over R, by Theorem 22.27, it suffices to show the extension $R_1 = R[\tau]_{(\mathfrak{m},\tau)} \hookrightarrow \widehat{R}$ satisfies PDE.

Let $\widehat{P}$ be a height-one prime ideal of $\widehat{R}$. If $\widehat{P}\cap R = (0)$, then $\operatorname{ht}(\widehat{P}\cap R_1)\le 1$. It remains to consider the case where $\widehat{P}\cap R\neq 0$. Then $\operatorname{ht}(\widehat{P}\cap R) = 1$, and so $\mathfrak{p} := \widehat{P}\cap R_1$ has $\operatorname{ht}\mathfrak{p}\le 2$.

Let $\widehat{Q_2} := \lambda^{-1}(\widehat{P})$ in $\widehat{S_1}$. Then $\operatorname{ht}(\widehat{Q_2}) = 2$, since $\widehat{Q_2} = (\widehat{P},t-a)\widehat{S_1}$.

Suppose that $\operatorname{ht}\mathfrak{p} = 2$. Then, under the R-isomorphism of S_1 to R_1 taking t to τ, $\mathfrak{p}$ corresponds to a height-two prime ideal P of S_1. Since $\widehat{S_1}$ is flat over S_1, $\operatorname{ht}(\widehat{Q_2}\cap S_1)\le 2$. Hence $P\subseteq\widehat{Q_2}\cap S_1$ implies $P = \widehat{Q_2}\cap S_1$. The following diagram illustrates this situation:

$$\begin{array}{ccc}
P = \widehat{Q}_2\cap S_1 \text{ (ht 2)} & \xrightarrow{\ \subseteq\ } & \widehat{Q}_2 = \lambda^{-1}(\widehat{P}) = (\widehat{P},(t-a))\widehat{S_1} \text{ (ht 2)} \\
\cong\big\downarrow & & \lambda\big\downarrow \\
\mathfrak{p} = \widehat{P}\cap R_1 \text{ (ht 2)} & \xrightarrow{\ \subseteq\ } & \widehat{P} \text{ (ht 1 in } \widehat{R}).
\end{array}$$

Then $\operatorname{ht}P = \operatorname{ht}\widehat{Q_2} = 2 < d = \operatorname{ht}\widehat{Q_0}$ implies $\widehat{Q_2}\in\mathcal{U}$, a contradiction to $t-a\in\widehat{Q_2}$.

Thus $\operatorname{ht}(\widehat{P}\cap R_1) = 1$, so τ is residually algebraically independent over R. □

Construction 22.32 shows that condition 2 of Definition 22.24 is stronger than Condition 2′ of Remark 22.25.

CONSTRUCTION 22.32. Let $(R,\mathfrak{m})$ be a countable excellent local unique factorization domain (UFD) of dimension two and let $(\widehat{R},\widehat{\mathfrak{m}})$ be the completion of R; for example, $R = \mathbb{Q}[x,y]_{(x,y)}$ and $\widehat{R} = \mathbb{Q}[[x,y]]$. As in Theorem 22.20, construct $\tau_1\in\widehat{\mathfrak{m}}$ primarily independent over R. By Theorem 22.30.1, $\tau_1\in\widehat{\mathfrak{m}}$ is residually algebraic over R. Let t_1,t_2 be variables over R and let $S_2 := R[t_1,t_2]_{(\mathfrak{m},t_1,t_2)}$. Thus $\widehat{S_2} = \widehat{R}[[t_1,t_2]]$. Let $\widehat{\mathfrak{n}}$ denote the maximal ideal of $\widehat{S_2}$. Consider the ideal

$I := (t_1, t_2, t_1 - \tau_1)\widehat{S_2}$ and define

$$\mathcal{U} = \{\widehat{Q} \in \operatorname{Spec}(\widehat{S_2}) \mid I \not\subseteq \widehat{Q}\ ,\ \operatorname{ht}\widehat{Q} = \operatorname{ht}(\widehat{Q} \cap S_2),\ \text{and}\ \ \widehat{Q}\ \text{is minimal over}$$
$$(P, t_1 - \tau_1)\widehat{S_2},\ \text{for some}\ P \in \operatorname{Spec} S_2\ \text{with}\ \operatorname{ht} P \leq 2\}.$$

If a prime ideal P of S_2 occurs as $\widehat{Q} \cap S_2$, for some $\widehat{Q} \in \mathcal{U}$, then $(t_1, t_2)S_2 \not\subseteq P$, since otherwise

$$I = (t_1, t_2, t_1 - \tau_1)\widehat{S_2} \subseteq (P, t_1 - \tau_1)\widehat{S_2} \subseteq \widehat{Q},$$

a contradiction. Thus $t_1, t_2 \notin P$, for every prime ideal P of S_2 in the description of $\mathcal{U}$. By Lemma 22.18, there exists $a \in \widehat{\mathfrak{n}}^2 \cap I$ so that $t_2 - a \notin \bigcup\{\widehat{Q} : \widehat{Q} \in \mathcal{U}\}$. Then $\widehat{S_2} = \widehat{R}[[t_1, t_2]] = \widehat{R}[[t_1 - \tau_1, t_2 - a]]$. Let λ denote the canonical R-algebra surjection

$$\lambda :\ \widehat{S_2} \longrightarrow \widehat{S_2}/(t_1 - \tau_1, t_2 - a)\widehat{S_2} = \widehat{R},$$

with $\operatorname{ht}(\ker(\lambda)) = 2$. Define $\tau_2 := \lambda(t_2)$.

CLAIM 22.33. The element τ_2 is not residually algebraically independent over $R[\tau_1]$; thus τ_1, τ_2 do not satisfy item 2 of Definition 22.24.

PROOF. (of Claim 22.33) Let $\widehat{W}$ be a prime ideal of $\widehat{S_2}$ that is minimal over $I = (t_1, t_2, t_1 - \tau_1)\widehat{S_2}$. Then $\operatorname{ht}\widehat{W} \leq 3$, and $t_2 - a \in I \subseteq \widehat{W}$, since $t_2 \in I$ and $a \in I$. Thus $\ker(\lambda) \subseteq \widehat{W}$. Let $\widehat{P} = \lambda(\widehat{W}) \subset \widehat{R}$. Then $\operatorname{ht}\widehat{P} \leq 1$. Since $0 \neq \tau_1 = \lambda(t_1) \in \widehat{P}$, $\operatorname{ht}\widehat{P} = 1$. Since τ_1 is residually algebraically independent over R, the extension $R[\tau_1] \hookrightarrow \widehat{R}$ satisfies PDE by Proposition 22.26. Therefore $\operatorname{ht}(\widehat{P} \cap R[\tau_1]) \leq 1$. But $\tau_1 \in \widehat{P} \cap R[\tau_1]$, and so $\operatorname{ht}(\widehat{P} \cap R[\tau_1]) = 1$ and $\widehat{P} \cap R = (0)$. Also $\tau_2 = \lambda(t_2) \in \widehat{P}$; thus $\tau_1, \tau_2 \in \widehat{P} \cap R[\tau_1, \tau_2]$, and so $\operatorname{ht}(\widehat{P} \cap R[\tau_1, \tau_2]) \geq 2$. Thus $R[\tau_1, \tau_2] \hookrightarrow \widehat{R}$ does not satisfy PDE. By Proposition 22.26, τ_2 is not residually algebraically independent over $R[\tau_1]$. □

CLAIM 22.34. For each height-one prime ideal $\widehat{P}$ of $\widehat{R}$ with $\widehat{P} \cap R \neq 0$, the images of τ_1 and τ_2 in $\widehat{R}/\widehat{P}$ are algebraically independent over $R/(\widehat{P} \cap R)$. That is, τ_1, τ_2 satisfy item $2'$ of Remark 22.25.

PROOF. (of Claim 22.34) Suppose $\widehat{P}$ is a height-one prime ideal of $\widehat{R}$ with $\widehat{P} \cap R \neq (0)$ and let $\widehat{Q} = \lambda^{-1}(\widehat{P})$. Then $\operatorname{ht}(\widehat{Q}) = 3$ and $\operatorname{ht}(\widehat{P} \cap R) = 1$. Set $R_1 := R[\tau_1]_{(\mathfrak{m}, \tau_1)}$ and $R_2 := R[\tau_1, \tau_2]_{(\mathfrak{m}, \tau_1, \tau_2)}$. By Proposition 22.26 and the residual algebraic independence of τ_1 over R, we have $\operatorname{ht}(\widehat{P} \cap R_1) = 1$, and so $\operatorname{ht}(\widehat{P} \cap R_2) \leq 2$. If $\operatorname{ht}(\widehat{P} \cap R_2) = 1$, we are done by Proposition 22.26. Suppose $\operatorname{ht}(\widehat{P} \cap R_2) = 2$. The following diagram illustrates this situation:

$$\begin{array}{ccccccccc}
 & & \widehat{Q} \cap S_1 & \xrightarrow{\subseteq} & \widehat{Q} \cap S_2 8 > \subseteq >> \widehat{Q} = \lambda^{-1}(\widehat{P}) & \xrightarrow{\subseteq} & \widehat{S_2} \\
 & & \cong\big\downarrow & & \cong\big\downarrow & & \lambda\big\downarrow & & \lambda\big\downarrow \\
\widehat{P} \cap R & \xrightarrow{\subseteq} & \widehat{P} \cap R_1 & \xrightarrow{\subseteq} & \widehat{P} \cap R_2 & \xrightarrow{\subseteq} & \widehat{P} & \xrightarrow{\subseteq} & \widehat{R}.
\end{array}$$

Thus $\widehat{Q} \cap S_2 = P$ is a prime ideal of height 2, and $\operatorname{ht}(\widehat{Q} \cap S_1) = 1$. Also $P \neq (t_1, t_2)S_2$ because $(t_1, t_2)S_2 \cap R = (0)$. But this means that $\widehat{Q} \in \mathcal{U}$ since $\widehat{Q}$ is minimal over $(P, t_1 - \tau_1)\widehat{S_2}$ where P is a prime ideal of S_2 with $\dim(S_2/P) = 2$ and $P \neq (t_1, t_2)S_2$, a contradiction to the choice of a. Thus item $2'$ holds. □

Theorem 22.35 gives a method to obtain an idealwise independent element that fails to be residually algebraically independent.

THEOREM 22.35. *Let $(R, \mathfrak{m})$ be a countable excellent local UFD of dimension at least two. Assume there exists a height-one prime P of R such that P is contained in at least two distinct height-one prime ideals $\widehat{P}$ and $\widehat{Q}$ of $\widehat{R}$. Also assume that $\widehat{P}$ is not the radical of a principal ideal in $\widehat{R}$. Then there exists $\tau \in \mathfrak{m}\widehat{R}$ that is idealwise independent but not residually algebraically independent over R.*

PROOF. Let t be an indeterminate over R and $S_1 = R[t]_{(\mathfrak{m},t)}$. Then $\widehat{S_1} = \widehat{R}[[t]]$. Let $\widehat{\mathfrak{n}_1}$ denote the maximal ideal of $\widehat{S_1}$. By Lemma 22.18 with $I = (\widehat{P}, t)\widehat{S_1}$ and

$$\mathcal{U} = \{\mathfrak{p} \in \operatorname{Spec}(\widehat{S_1}) \mid \mathfrak{p} \neq I,\ \operatorname{ht}(\mathfrak{p}) \leq 2, \text{ and } \mathfrak{p} \text{ minimal over } \mathfrak{p} \cap S_1\},$$

there exists $a \in (\widehat{P}, t)\widehat{S_1} \cap \widehat{\mathfrak{n}_1}^2$ such that

$$t - a \in (\widehat{P}, t)\widehat{S_1} \setminus (\bigcup \{\mathfrak{p} \mid \mathfrak{p} \in \mathcal{U}\}).$$

Thus, if $t - a \in \mathfrak{p}$, for some prime ideal $\mathfrak{p} \neq (\widehat{P}, t)\widehat{S_1}$ of $\widehat{S_1}$ with $\operatorname{ht}(\mathfrak{p}) \leq 2$, then $\operatorname{ht}(\mathfrak{p}) > \operatorname{ht}(\mathfrak{p} \cap S_1)$. Furthermore the choice of $t - a$ ensures that each height-one prime ideal $\widehat{\mathfrak{q}}$ other than $\widehat{P}$ of $\widehat{R}$ has the property that $\operatorname{ht}(\widehat{\mathfrak{q}} \cap R_1) \leq 1$.

Let λ be the $\widehat{R}$-algebra surjection $\widehat{S_1} \to \widehat{R}$ with kernel $(t - a)\widehat{S_1}$. By Remark 22.19, $(t-a)\widehat{S_1} \cap S_1 = (0)$. Define $\tau := \lambda(t) = \lambda(a)$ in $\mathfrak{m}\widehat{R}$. Then the restriction of λ to S_1 maps S_1 isomorphically onto $R_1 := R[\tau]_{(\mathfrak{m},\tau)}$, and so τ is algebraically independent over R. Since $a \in \mathfrak{n}_1^2$, it follows that $\widehat{S_1} = \widehat{R}[[t-a]]$. Also $t - a \in (\widehat{P}, t)\widehat{S_1}$ implies $(\widehat{P}, t - a)\widehat{S_1} \subseteq (\widehat{P}, t)\widehat{S_1}$. Then $\operatorname{ht}((\widehat{P}, t - a)\widehat{R}[[t - a]]) = 2 = \operatorname{ht}((\widehat{P}, t)\widehat{R}[[t]])$ implies that $(\widehat{P}, t - a)\widehat{S_1} = (\widehat{P}, t)\widehat{S_1}$. Thus $\tau = \lambda(t) \in \lambda((\widehat{P}, t - a)\widehat{S_1}) \subseteq \widehat{P}$; that is, the image of τ in $\widehat{R}/\widehat{P}$ is 0. Therefore τ is not residually algebraically independent over R.

To see that τ is idealwise independent over R, it suffices to show $R_1 \hookrightarrow \widehat{R}$ is weakly flat, by Theorem 22.11.1. Since $R_1 \hookrightarrow \widehat{R}$ is a local map, this is equivalent to showing that each height-one prime ideal of R_1 is the contraction of a height-one prime ideal of $\widehat{R}$. For $\mathfrak{q}$ a height-one prime ideal of R_1 and $\lambda_1 : S_1 \to R_1$ the restriction of λ, let $\mathfrak{q}_1 := \lambda_1^{-1}(\mathfrak{q})$ denote the corresponding height-one prime ideal of S_1. Consider separately the two cases: (i) $\mathfrak{q}_1 \not\subseteq (\widehat{P}, t)\widehat{S_1}$, and (ii) $\mathfrak{q}_1 \subseteq (\widehat{P}, t)\widehat{S_1}$.

In the first case, let $\mathbf{w}_1$ be a height-two prime ideal of $\widehat{S_1}$ containing the height-two ideal $(\mathfrak{q}_1, t - a)\widehat{S_1}$. Since $\mathfrak{q}_1 \not\subseteq (\widehat{P}, t)\widehat{S_1}$, the choice of $t - a$ implies $\mathbf{w}_1 \cap S_1$ has height at most one. Therefore $\mathbf{w}_1 \cap S_1 = \mathfrak{q}_1$. Let $\mathbf{w} = \lambda(\mathbf{w}_1)$. Then $\mathbf{w}$ is a height-one prime ideal of $\widehat{R}$ and $\mathbf{w} \cap R_1 = \mathfrak{q}$. Thus $\mathfrak{q}$ is the contraction of a height-one prime ideal of $\widehat{R}$, for every height-one prime ideal $\mathfrak{q}$ of R_1 such that $\mathfrak{q}_1 := \lambda_1^{-1}(\mathfrak{q})$ is not contained in $(\widehat{P}, t)\widehat{S_1}$.

For the second case, assume $\mathfrak{q}_1 \subseteq (\widehat{P}, t)\widehat{S_1}$. Equivalently, $\mathfrak{q}$ is a height-one prime ideal of R_1 such that $\mathfrak{q} \subseteq (P, \tau)R_1$, since $\lambda((\widehat{P}, t)\widehat{S_1}) \cap R_1 = (P, \tau)R_1$. First consider the prime ideal $\mathfrak{q} = PR_1$, which is contained in $\widehat{Q} \cap R_1$. By the hypothesis and the choice of a,

$$\operatorname{ht}(\widehat{Q} \cap R_1) \leq \operatorname{ht} Q = 1 \implies PR_1 = \widehat{Q} \cap R_1.$$

Thus PR_1 is the contraction of a height-one prime ideal of $\widehat{R}$.

Finally, let $\mathfrak{q}$ be a height-one prime ideal of R_1 such that $\mathfrak{q} \subseteq (P,\tau)R_1$ and $\mathfrak{q} \neq PR_1$. Since R is a UFD, R_1 is a UFD and $\mathfrak{q} = fR_1$ for an element $f \in \mathfrak{q}$. Since $\widehat{P}$ is not the radical of a principal ideal, there exists a height-one prime ideal $\widehat{\mathfrak{q}} \neq \widehat{P}$ of $\widehat{R}$ such that $f \in \widehat{\mathfrak{q}}$. Since $\operatorname{ht}(\widehat{\mathfrak{q}} \cap R_1) \leq 1$, we have $\widehat{\mathfrak{q}} \cap R_1 = fR_1 = \mathfrak{q}$. Therefore τ is idealwise independent over R. $\square$

EXAMPLE 22.36. Let k be the algebraic closure of the field $\mathbb{Q}$ and $z^2 = x^3 + y^7$. Then $R = k[x,y,z]_{(x,y,z)}$ is a countable excellent local UFD having a height-one prime ideal P satisfying the conditions in Theorem 22.35. That R is a UFD is shown in [**162**, page 32]. Since $z - xy$ is an irreducible element of R, the ideal $P = (z - xy)R$ is a height-one prime ideal of R. It is observed in [**67**, pages 300-301] that in the completion $\widehat{R}$ of R there exist distinct height-one prime ideals $\widehat{P}$ and $\widehat{Q}$ lying over P. Moreover, the blowup of $\widehat{P}$ has a unique exceptional prime divisor and this exceptional prime divisor is not the unique exceptional prime divisor on the blowup of an $\widehat{\mathfrak{m}}$-primary ideal. Therefore $\widehat{P}$ is not the radical of a principal ideal of $\widehat{R}$.

Theorem 22.37 gives an alternative method to obtain idealwise independent elements that are not residually algebraically independent.

THEOREM 22.37. *Let $(R, \mathfrak{m})$ be a countable excellent local UFD of dimension at least two. Assume there exists a height-one prime ideal P_0 of R such that P_0 is contained in at least two distinct height-one prime ideals $\widehat{P}$ and $\widehat{Q}$ of $\widehat{R}$. Also assume that the Henselization $(R^h, \mathfrak{m}^h)$ of R is a UFD. Then there exists $\tau \in \mathfrak{m}\widehat{R}$ that is idealwise independent but not residually algebraically independent over R.*

PROOF. Since R is excellent, $P := \widehat{P} \cap R^h$ and $Q := \widehat{Q} \cap R^h$ are distinct height-one prime ideals of R^h with $P\widehat{R} = \widehat{P}$, and $Q\widehat{R} = \widehat{Q}$. Let $x \in R^h$ be such that $xR^h = P$. Theorem 22.20 implies there exists $y \in \mathfrak{m}\widehat{R}$ that is primarily independent and hence residually algebraically independent over R^h.

We show that $\tau = xy$ is idealwise independent but not residually algebraically independent over R. Since x is nonzero and algebraic over R, xy is algebraically independent over R. Let $R_1 = R[xy]_{(\mathfrak{m},xy)}$. Then $\widehat{P} \cap R_1 = x\widehat{R} \cap R_1 \supseteq (P_0, xy)R_1$ implies that $\operatorname{ht}(\widehat{P} \cap R_1) \geq 2$. Since R_1 is a UFD, Theorem 22.27 implies xy is not residually algebraically independent over R.

Since y is idealwise independent over R^h, every height-one prime ideal of the polynomial ring $R^h[y]$ contained in the maximal ideal $\mathfrak{n} = (\mathfrak{m}^h, y)R^h[y]$ is the contraction of a height-one prime ideal of $\widehat{R}$. To show xy is idealwise independent over R, it suffices to show every prime element $w \in (\mathfrak{m}, xy)R[xy]$ is such that $wR[xy]$ is the contraction of a height-one prime ideal of $R^h[y]$ contained in $\mathfrak{n}$. If $w \notin (P, xy)R^h[xy]$, then the constant term of w as a polynomial in $R^h[xy]$ is in $\mathfrak{m}^h \setminus P$. Thus $w \in \mathfrak{n}$ and $w \notin xR^h[y]$. Since $R^h[xy][1/x] = R^h[y][1/x]$ and $xR^h[y] \cap R^h[xy] = (x, xy)R^h[xy]$, it follows that there is a prime factor u of w in $R^h[xy]$ such that $u \in \mathfrak{n} \setminus xR^h[y]$. Then $uR^h[y]$ is a height-one prime ideal of $R^h[y]$ and $uR^h[y] \cap R^h[xy] = uR^h[xy]$. Since $R^h[xy]$ is faithfully flat over $R[xy]$, it follows that $uR^h[y] \cap R[xy] = wR[xy]$.

We have $QR^h[xy] = QR^h[y] \cap R^h[xy]$ and $QR^h[xy] \cap R[xy] = P_0R[xy]$. Thus it remains to show, for a prime element $w \in (\mathfrak{m}, xy)R[xy]$ such that $w \in (P, xy)R^h[xy]$ and $wR[xy] \neq P_0R[xy]$, that $wR[xy]$ is the contraction of a height-one prime ideal of $R^h[y]$ contained in $\mathfrak{n}$. Since $(P, xy)R^h[xy] \cap R[xy] = (P_0, xy)R[xy]$, it follows that w is a nonconstant polynomial in $R[xy]$ and the constant term w_0 of w is in P_0. In the polynomial ring $R^h[y]$ we have $w = x^n v$, where $v \notin xR^h[y]$. If v_0 denotes the constant term of v as a polynomial in $R^h[y]$, then $x^n v_0 = w_0 \in P_0 \subseteq R$ implies $x^n v_0 \in Q \subseteq R^h$. Since $x \in R^h \setminus Q$, we must have $v_0 \in Q$ and hence $v \in \mathfrak{n}$. Also $v \notin xR^h[y]$ implies there is a height-one prime ideal $\mathbf{v}$ of $R^h[y]$ with $v \in \mathbf{v}$ and $x \notin \mathbf{v}$. Then, since $R^h[y]_{\mathbf{v}}$ is a localization of $R^h[xy]$, $\mathbf{v} \cap R^h[xy]$ is a height-one prime ideal of $R^h[xy]$ that is contained in $(\mathfrak{m}^h, xy)R^h[xy]$. It follows that $\mathbf{v} \cap R^h[xy] = wR^h[xy]$, which completes the proof of Theorem 22.37. □

Example 4.13 is a specific example fitting the hypothesis of Theorem 22.37. In more generality, we have:

EXAMPLE 22.38. Let $R = k[s,t]_{(s,t)}$ be a localized polynomial ring in two variables s and t over a countable field k where k has characteristic not equal to 2. Let $P_0 = (s^2 - t^2 - t^3)R$. Then P_0 is a height-one prime ideal of R and $P_0\widehat{R} = (s^2 - t^2 - t^3)k[[s,t]]$ is the product of two distinct height-one prime ideals of $\widehat{R}$.

REMARK 22.39. Let $(R, \mathfrak{m})$ be excellent normal local domain and let $(\widehat{R}, \widehat{\mathfrak{m}})$ be its completion. Assume that $\tau \in \widehat{\mathfrak{m}}$ is algebraically independent over R. By Theorem 22.11, the extension $R[\tau] \hookrightarrow \widehat{R}$ is weakly flat if and only if τ is idealwise independent over R. By Theorem 22.27, this extension satisfies PDE (or equivalently LF_1) if and only if τ is residually algebraically independent over R. Thus Examples 22.36 and 22.38 give extensions of Krull domains $R[\tau] \hookrightarrow \widehat{R}$, that are weakly flat, but do not satisfy PDE. In fact, in these examples the ring $R[\tau]$ is a 3-dimensional excellent UFD.

Exercises

(1) Assume Setting 22.1. Let I be an ideal of R such that $(P, t_1 - \tau_1, \ldots, t_n - \tau_n)\widehat{S_n}$ is primary to the maximal ideal of $\widehat{S_n}$, for every $P \in \operatorname{Spec} S_n$ with P minimal over I. Prove that $(I, t_1 - \tau_1, \ldots, t_n - \tau_n)\widehat{S_n}$ is primary to the maximal ideal of $\widehat{S_n}$.

(2) Prove Remark 22.25: Let $\tau_1, \ldots, \tau_n \in \widehat{\mathfrak{m}}$ be algebraically independent over R, and let $\widehat{P}$ be a height-one prime ideal of $\widehat{R}$ such that $\widehat{P} \cap R[\tau_1, \ldots, \tau_{i-1}] \neq (0)$. Show that the images of the τ_i in $\widehat{R}/\widehat{P}$ are algebraically independent over $R[\tau_1, \ldots, \tau_{i-1}]/(\widehat{P} \cap R[\tau_1, \ldots, \tau_{i-1}])$.

(3) As in Remark 22.19, let $A \hookrightarrow B$ be an extension of Krull domains, and let α be a nonzero nonunit of B such that $\alpha \notin Q$ for each height-one prime Q of B such that $Q \cap A \neq (0)$.
   (a) Prove that $\alpha B \cap A = (0)$ as asserted in Remark 22.19.
   (b) Prove that α is algebraically independent over A.

   SUGGESTION. For part a, see Remarks 2.12.

(4) Let $R = k[s,t]_{(s,t)}$ and the field k be as in Example 22.38.
   (a) Prove as asserted in Example 22.38 that $(s^2 - t^2 - t^3)R$ is a prime ideal.
   (b) Prove that $s^2 - t^2 - t^3$ factors in the power series ring $k[[s,t]]$ as the product of two non-associate prime elements.

CHAPTER 23

Idealwise algebraic independence II

This chapter relates the three concepts of independence from Chapter 22 to flatness conditions of extensions of Krull domains. Implications among the concepts are given, as well as some conclusions concerning their equivalence in special situations. We also investigate their stability under change of base ring.

Setting 22.1 from Chapter 22 is used in this chapter. Thus $(R, \mathfrak{m})$ is an excellent normal local domain with field of fractions K and completion $(\widehat{R}, \widehat{\mathfrak{m}})$, and $t_1, \ldots, t_n$ are indeterminates over R. The elements $\tau_1, \ldots, \tau_n \in \widehat{\mathfrak{m}}$ are algebraically independent over R, and we have embeddings:

$$R \hookrightarrow S = R_n = R[\tau_1, \ldots, \tau_n]_{(\mathfrak{m}, \tau_1, \ldots, \tau_n)} \stackrel{\varphi}{\hookrightarrow} \widehat{R}$$

Using this setting and other terminology of Chapter 22, we summarize the results of this chapter.

SUMMARY 23.1. In Section 23.1 we describe the three concepts of idealwise independence, residual algebraic independence, and primary independence defined in Definitions 22.2, 22.24, and 22.12 in terms of certain flatness conditions on the embedding

$$\varphi : R[\tau_1, \ldots, \tau_n]_{(\mathfrak{m}, \tau_1, \ldots, \tau_n)} \hookrightarrow \widehat{R}.$$

In Section 23.2 we investigate the stability of these independence concepts under base change, composition and polynomial extension. We prove in Corollary 23.18 the existence of uncountable excellent normal local domains R such that $\widehat{R}$ contains infinite sets of primarily independent elements.

Corollary 23.20 of Section 23.3 states that residual algebraic independence or primary independence holds for elements over the original ring R if and only if the corresponding property holds over the Henselization R^h of R. Also idealwise independence descends from the Henselization to the ring R.

A diagram in Section 23.4 displays the relationships among the independence concepts and other related properties.

23.1. Primary independence and flatness

In this section we describe the concept of primary independence in terms of flatness of certain localizations of the canonical embedding of Setting 22.1

$$\varphi : R_n = R[\tau_1, \ldots, \tau_n]_{(\mathfrak{m}, \tau_1, \ldots, \tau_n)} \hookrightarrow \widehat{R}.$$

Chapter 22 contains flatness conditions for φ that are equivalent to idealwise independence and residual algebraic independence. Remark 23.2 summarizes these conditions.

REMARK 23.2. Let $(R, \mathfrak{m})$ and $\tau_1, \ldots, \tau_n \in \widehat{\mathfrak{m}}$ be as in Setting 22.1, and let $\varphi : R_n = R[\tau_1, \ldots, \tau_n]_{(\mathfrak{m}, \tau_1, \ldots, \tau_n)} \hookrightarrow \widehat{R}$ denote the canonical embedding. Then:

(1) $\tau_1, \ldots, \tau_n$ are idealwise independent over R if and only if the map $R_n \hookrightarrow \widehat{R}$ is weakly flat; see Definitions 22.2 and 22.5 and Theorem 22.11.

(2) The elements $\tau_1, \ldots, \tau_n$ are residually algebraically independent over R $\iff$ $\varphi : R_n = R[\tau_1, \ldots, \tau_n]_{(\mathfrak{m}, \tau_1, \ldots, \tau_n)} \longrightarrow \widehat{R}$ satisfies LF_1; see Definition 9.1.3, Corollary 9.14 and Theorem 22.27.

(3) If each height-one prime of R is the radical of a principal ideal and the elements $\tau_1, \ldots, \tau_n$ are residually algebraically independent over R, then the elements $\tau_1, \ldots, \tau_n$ are idealwise independent over R. See Theorem 22.30.3.

Lemma 23.3 is useful for obtaining a description of primary independence in terms of flatness of certain localizations of the embedding $\varphi : R_n \longrightarrow \widehat{R}$:

LEMMA 23.3. *Let $d \in \mathbb{N}$ and $n \in \mathbb{N}_0$, and let $(S, \mathfrak{m}) \hookrightarrow (T, \mathfrak{n})$ be a local embedding of catenary Noetherian local domains with $\dim T = d$ and $\dim S = d+n$. Assume the extension $S \hookrightarrow T$ satisfies:*

$$\operatorname{ht} P \geq d \implies PT \text{ is } \mathfrak{n}\text{-primary}, \quad \text{for every } P \in \operatorname{Spec} S. \quad \textit{Property } 23.3.0$$

Then, for every $Q \in \operatorname{Spec} T$ with $\operatorname{ht} Q \leq d-1$, we have $\operatorname{ht}(Q \cap S) \leq \operatorname{ht} Q$.

PROOF. If $Q \in \operatorname{Spec} T$ is such that $\operatorname{ht}(Q \cap S) \geq d$, then, by Property 23.3.0, $(Q \cap S)T$ is $\mathfrak{n}$-primary, and so $Q = \mathfrak{n}$ and $\operatorname{ht} Q = d$. Thus, for every $Q \in \operatorname{Spec} T$ with $\operatorname{ht} Q \leq d-1$, we have $\operatorname{ht}(Q \cap S) \leq d-1$. In particular, if $\operatorname{ht} Q = d-1$, then $\operatorname{ht}(Q \cap S) \leq \operatorname{ht} Q$.

Proceed by induction on $s \geq 1$ to show: $\operatorname{ht} Q = d-s \implies \operatorname{ht}(Q \cap S) \leq \operatorname{ht} Q$. Assume $s \geq 2$ and $\operatorname{ht}(P \cap S) \leq \operatorname{ht} P$, for every $P \in \operatorname{Spec} T$ with $d > \operatorname{ht} P \geq d-s+1$. Let $Q \in \operatorname{Spec} T$ with $\operatorname{ht} Q = d-s$. Suppose $\operatorname{ht}(Q \cap S) \geq d-s+1$; choose $b \in \mathfrak{m} \setminus Q$ and let $Q_1 \in \operatorname{Spec} T$ be minimal over $(b, Q)T$. Since T is catenary and Noetherian, we have $\operatorname{ht} Q_1 = d-s+1$. By the inductive hypothesis, $\operatorname{ht}(Q_1 \cap S) \leq d-s+1$. Since $b \in Q_1 \cap S$, the ideal $Q_1 \cap S$ properly contains $Q \cap S$. But this implies

$$d-s+1 \geq \operatorname{ht}(Q_1 \cap S) > \operatorname{ht}(Q \cap S) \geq d-s+1,$$

a contradiction. Thus $\operatorname{ht}(Q \cap S) \leq \operatorname{ht} Q$, for every $Q \in \operatorname{Spec} T$ with $\operatorname{ht} Q \leq d-1$. $\square$

Theorem 23.4 uses the LF_n notation of Definition 9.1.3 and uses Remark 9.2.

THEOREM 23.4. *Let $(R, \mathfrak{m})$ be an excellent normal local domain, and let the elements $\tau_1, \ldots, \tau_n \in \widehat{\mathfrak{m}}$ be as in Setting 22.1. Assume that $\dim R = d$. Then the elements $\tau_1, \ldots, \tau_n$ are primarily independent over R if and only if*

$$\varphi : R_n = R[\tau_1, \ldots, \tau_n]_{(\mathfrak{m}, \tau_1, \ldots, \tau_n)} \longrightarrow \widehat{R}$$

satisfies LF_{d-1}.

PROOF. To prove the $\implies$ direction: Since R_n is a localized polynomial ring over R, the map $R \hookrightarrow R_n$ has regular fibers. Since R is excellent, the map $R \hookrightarrow \widehat{R}$ has regular, hence Cohen-Macaulay, fibers. Consider the sequence

$$R \longrightarrow R_n \xrightarrow{\varphi} \widehat{R}.$$

To show φ satisfies LF_{d-1}, we show that $\varphi_{\widehat{Q}}$ is flat for every $\widehat{Q} \in \operatorname{Spec} \widehat{R}$ with $\operatorname{ht} \widehat{Q} \le d-1$. For this, by (2) $\implies$ (1) of Theorem 7.3, it suffices to show $\operatorname{ht}(\widehat{Q} \cap R_n) \le \operatorname{ht} \widehat{Q}$ for every $\widehat{Q} \in \operatorname{Spec} \widehat{R}$ with $\operatorname{ht} \widehat{Q} \le d-1$. This holds by Lemma 23.3, since primary independence implies Property 23.3.0.

For $\Longleftarrow$, let $P \in \operatorname{Spec} R_n$ be a prime ideal with $\dim(R_n/P) \le n$. Suppose that $P\widehat{R}$ is not $\widehat{\mathfrak{m}}$-primary and let $\widehat{Q} \supseteq P\widehat{R}$ be a minimal prime of $P\widehat{R}$. Then $\operatorname{ht}(\widehat{Q}) \le d-1$. Set $Q = \widehat{Q} \cap R_n$, then LF_{d-1} implies that the map

$$\varphi_{\widehat{Q}} : (R_n)_Q \longrightarrow \widehat{R}_{\widehat{Q}}$$

is faithfully flat. Hence by going-down (Remark 2.37.10), $\operatorname{ht} Q \le d-1$. But $P \subseteq Q$ and R_n is catenary, so $d-1 \ge \operatorname{ht} Q \ge \operatorname{ht} P \ge d$, a contradiction. We conclude that $\tau_1, \dots, \tau_n$ are primarily independent. $\square$

REMARK 23.5. Theorem 23.4 yields a different proof of statements (1) and (3) of Theorem 22.30, that primarily independent elements are residually algebraically independent and that in dimension two, the two concepts are equivalent. Consider again the basic Setting 22.1, with $d = \dim(R)$. Theorem 23.4 equates the LF_{d-1} condition on the extension $R_n = R[\tau_1, \dots, \tau_n]_{(\mathfrak{m}, \tau_1, \dots, \tau_n)} \longrightarrow \widehat{R}$ to the primary independence of the τ_i. Corollary 9.14 and Theorem 22.27 imply that the τ_i are residually algebraically independent if and only if the extension $R_n = R[\tau_1, \dots, \tau_n]_{(\mathfrak{m}, \tau_1, \dots, \tau_n)} \longrightarrow \widehat{R}$ satisfies LF_1. Clearly $\mathrm{LF}_i \implies \mathrm{LF}_{i-1}$, for $i > 1$, and if $d = \dim(R) = 2$, then $\mathrm{LF}_{d-1} = \mathrm{LF}_1$.

REMARK 23.6. In Setting 22.1, if $\tau_1, \dots, \tau_n$ are primarily independent over R and $\dim(R) = d$, then $\varphi : R_n \longrightarrow \widehat{R}$ satisfies LF_{d-1}, but *not* LF_d, that is, φ fails to be faithfully flat; for faithful flatness would imply going-down and hence that $\dim(R_n) \le d = \dim(\widehat{R})$.

EXAMPLE 23.7. By a modification of Example 9.17, it is possible to obtain, for each integer $d \ge 2$, an injective local map $\varphi : (A, \mathfrak{m}) \longrightarrow (B, \mathfrak{n})$ of normal Noetherian local domains with B essentially of finite type over A, $\varphi(\mathfrak{m})B = \mathfrak{n}$, and $\dim(B) = d$ such that φ satisfies LF_{d-1}, but fails to be faithfully flat over A. Let k be a field and let $x_1, \dots, x_d, y$ be indeterminates over k. Let A be the localization of $k[x_1, \dots, x_d, x_1y, \dots, x_dy]$ at the maximal ideal generated by $(x_1, \dots, x_d, x_1y, \dots, x_dy)$, and let B be the localization of $A[y]$ at the prime ideal $(x_1, \dots, x_d)A[y]$. Then A is an $d+1$-dimensional normal Noetherian local domain and B is an d-dimensional regular local domain birationally dominating A. For every non-maximal prime Q of B we have $B_Q = A_{Q \cap A}$. Hence $\varphi : A \longrightarrow B$ satisfies LF_{d-1}, but φ is not faithfully flat since $\dim(B) < \dim(A)$.

The local injective map $\varphi : (A, \mathfrak{m}) \longrightarrow (B, \mathfrak{n})$ of Example 23.7 is not height-one preserving. Remark 22.8 shows that if each height-one prime ideal of R is the radical of a principal ideal then the maps studied in this chapter are height-one preserving.

23.2. Composition, base change and polynomial extensions

In this section we investigate idealwise independence, residual algebraic independence, and primary independence under polynomial ring extensions and localizations of these polynomial extensions.

Proposition 23.8 implies that many of the properties of injective maps that we consider are stable under composition:

PROPOSITION 23.8. *Let $\varphi : A \longrightarrow B$ and $\psi : B \longrightarrow C$ be injective maps of commutative rings, and let $s \in \mathbb{N}$. That is,*

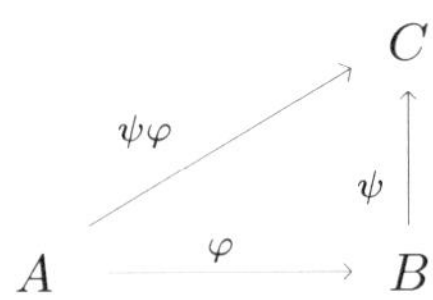

is a commutative diagram of commutative rings and injective maps. Then:

(1) *If φ and ψ satisfy LF_s, then $\psi\varphi$ satisfies LF_s.*
(2) *If C is Noetherian, ψ is faithfully flat and the composite map $\psi\varphi$ satisfies LF_s, then φ satisfies LF_s.*
(3) *Assume that A, B and C are Krull domains, and that $QC \neq C$, for each height-one prime Q of B. If φ and ψ are height-one preserving (respectively weakly flat), then $\psi\varphi$ is height-one preserving (respectively weakly flat).*

PROOF. The first item follows because a flat map satisfies going-down, see Remark 2.37.10. For item 2, since C is Noetherian and ψ is faithfully flat, B is Noetherian; see Remark 2.37.8. Let $Q \in \operatorname{Spec}(B)$ with $\operatorname{ht}(Q) = d \leq s$. We show $\varphi_Q : A_{Q\cap A} \longrightarrow B_Q$ is faithfully flat. By localization of B and C at $B \setminus Q$, we may assume that B is local with maximal ideal Q. Since C is faithfully flat over B, $QC \neq C$. Let $Q' \in \operatorname{Spec}(C)$ be a minimal prime of QC. Since C is Noetherian and B is local with maximal ideal Q, we have $\operatorname{ht}(Q') \leq d$ and $Q' \cap B = Q$. Since the composite map $\psi\varphi$ satisfies LF_s, the composite map

$$A_{Q'\cap A} = A_{Q\cap A} \xrightarrow{\varphi_Q} B_Q = B_{Q'\cap B} \xrightarrow{\psi_{Q'}} C_{Q'}$$

is faithfully flat. This and the faithful flatness of $\psi_{Q'} : B_{Q'\cap B} \longrightarrow C_{Q'}$ imply that φ_Q is faithfully flat [**120**, (4.B) page 27].

For item 3, let P be a height-one prime of A such that $PC \neq C$. Then $PB \neq B$ so if φ and ψ are height-one preserving then there exists a height-one prime Q of B such that $PB \subseteq Q$. By assumption, $QC \neq C$ (and ψ is height-one preserving), so there exists a height-one prime Q' of C such that $QC \subseteq Q'$. Hence $PC \subseteq Q'$.

If φ and ψ are weakly flat, then by Proposition 9.11 there exists a height-one prime Q of B such that $Q \cap A = P$. Again by assumption, $QC \neq C$; thus the weak flatness of ψ implies $QC \cap B = Q$. Now

$$P \subseteq PC \cap A \subseteq QC \cap A = QC \cap B \cap A = Q \cap A = P.$$

□

REMARKS 23.9. If in Proposition 23.8.3 the Krull domains B and C are local, but not necessarily Noetherian, and ψ is a local map, then clearly $QC \neq C$ for each height-one prime Q of B.

If a map λ of Krull domains is faithfully flat, then λ is height-one preserving, weakly flat and satisfies condition LF_k for every integer $k \in \mathbb{N}$. Thus if $\varphi : A \longrightarrow B$ and $\psi : B \longrightarrow C$ are injective maps of Krull domains such that one of φ or ψ is faithfully flat and the other is weakly flat (respectively height-one preserving or satisfies LF_k), then the composition $\psi\varphi$ is again weakly flat (respectively height-one preserving or satisfies LF_k). Moreover, if the map ψ is faithfully flat, we also obtain the following converse to Proposition 23.8.3:

PROPOSITION 23.10. *Let $\varphi : A \longrightarrow B$ and $\psi : B \longrightarrow C$ be injective maps of Krull domains. Assume that ψ is faithfully flat. If $\psi\varphi$ is height-one preserving (respectively weakly flat), then φ is height-one preserving (respectively weakly flat).*

PROOF. Let P be a height-one prime ideal of A such that $PB \neq B$. Since ψ is faithfully flat, $PC \neq C$; so if $\psi\varphi$ is height-one preserving, then there exists a height-one prime ideal Q' of C containing PC. Now $Q = Q' \cap B$ has height one by going-down for flat extensions, and $PB \subseteq Q' \cap B = Q$, so φ is height-one preserving. The proof of the weakly flat statement is similar, using Proposition 9.11. □

Next we consider a commutative square of commutative rings and injective maps:

$$\begin{array}{ccc} A' & \xrightarrow{\varphi'} & B' \\ {\scriptstyle\mu}\uparrow & & {\scriptstyle\nu}\uparrow \\ A & \xrightarrow{\varphi} & B \end{array}$$

PROPOSITION 23.11. *In the diagram above, assume that μ and ν are faithfully flat, and let $k \in \mathbb{N}$. Then:*

(1) *(Ascent) Assume that $B' = B \otimes_A A'$, or that B' is a localization of $B \otimes_A A'$. Let ν denote the canonical map associated with this tensor product. If $\varphi : A \longrightarrow B$ satisfies LF_k, then $\varphi' : A' \longrightarrow B'$ satisfies LF_k.*
(2) *(Descent) If B' is Noetherian and $\varphi' : A' \longrightarrow B'$ satisfies LF_k, then $\varphi : A \longrightarrow B$ satisfies LF_k.*
(3) *(Descent) Assume that the rings A, A', B and B' are Krull domains. If $\varphi' : A' \longrightarrow B'$ is height-one preserving (respectively weakly flat), then $\varphi : A \longrightarrow B$ is height-one preserving (respectively weakly flat).*

PROOF. For (1), assume that φ satisfies LF_k and let $Q' \in \operatorname{Spec}(B')$ with $\operatorname{ht}(Q') \leq k$. Put $Q = (\nu)^{-1}(Q')$, $P' = (\varphi')^{-1}(Q')$, and $P = \mu^{-1}(P') = \varphi^{-1}(Q)$ and consider the commutative diagrams:

$$\begin{array}{ccc} A' & \xrightarrow{\varphi'} & B' \\ {\scriptstyle\mu}\uparrow & & {\scriptstyle\nu}\uparrow \\ A & \xrightarrow{\varphi} & B \end{array} \qquad\qquad \begin{array}{ccc} A'_{P'} & \xrightarrow{\varphi'_{Q'}} & B'_{Q'} \\ {\scriptstyle\mu_{P'}}\uparrow & & {\scriptstyle\nu_{Q'}}\uparrow \\ A_P & \xrightarrow{\varphi_Q} & B_Q \end{array}$$

The flatness of ν implies that $\operatorname{ht}(Q) \leq k$ and so, by assumption, φ_Q is faithfully flat. The ring $B'_{Q'}$ is a localization of $B_Q \otimes_{A_P} A'_{P'}$, and so B_Q is faithfully flat over A_P implies $B'_{Q'}$ is faithfully flat over $A'_{P'}$.

For item 2, by Proposition 23.8.1, $\varphi'\mu = \nu\varphi$ satisfies LF_k. By Proposition 23.8.2, φ satisfies LF_k.

Item 3 follows immediately from the assumption that μ and ν are faithfully flat maps and hence going-down holds; see Remark 2.37.10. □

Next we examine the situation for polynomial extensions.

PROPOSITION 23.12. *Let $(R, \mathfrak{m})$ and $\{\tau_i\}_{i=1}^m \subseteq \widehat{\mathfrak{m}}$ be as in Setting 22.1, where m is either an integer or $m = \infty$, and the dimension of R is at least 2. Let z be an indeterminate over $\widehat{R}$. Then:*

(1) *$\{\tau_i\}_{i=1}^m$ is residually algebraically independent over R $\iff$ $\{\tau_i\}_{i=1}^m$ is residually algebraically independent over $R[z]_{(\mathfrak{m},z)}$.*
(2) *If $\{\tau_i\}_{i=1}^m$ is idealwise independent over $R[z]_{(\mathfrak{m},z)}$, then $\{\tau_i\}_{i=1}^m$ is idealwise independent over R.*

PROOF. Let $n \in \mathbb{N}$ be an integer with $n \leq m$. Set $R_n = R[\tau_1, \ldots, \tau_n]_{(\mathfrak{m},\tau_1,\ldots,\tau_n)}$. Let $\varphi : R_n \longrightarrow \widehat{R}$ and $\mu : R_n \longrightarrow R_n[z]$ be the inclusion maps. The following diagram commutes:

$$\begin{array}{ccccc} R_n[z]_{(\max(R_n),z)} & \xrightarrow{\varphi'} & R' = \widehat{R}[z]_{(\widehat{\mathfrak{m}},z)} & \xrightarrow{\psi} & \widehat{R'} = \widehat{R}[[z]] \\ {\scriptstyle\mu}\uparrow & & {\scriptstyle\mu'}\uparrow & & \\ R_n & \xrightarrow{\varphi} & \widehat{R}\,. & & \end{array}$$

The ring R' is a localization of the tensor product $\widehat{R} \otimes_{R_n} R_n[z]$ and Proposition 23.11 applies. Thus, for item 1, φ satisfies LF_1 if and only if φ' satisfies LF_1. Since the inclusion map ψ of $R' = \widehat{R}[z]_{(\widehat{\mathfrak{m}},z)}$ to its completion $\widehat{R}[[z]]$ is faithfully flat, Proposition 23.11 implies these statements are equivalent:

φ satisfies $LF_1 \iff \varphi'$ satisfies $LF_1 \iff \psi\varphi'$ satisfies LF_1.

For item 2, the τ_i are idealwise independent over $R[z]_{(\mathfrak{m},z)}$, then the map $\psi\varphi'$ is weakly flat by Remark 23.2. Thus φ' is weakly flat and the statement follows by Proposition 23.11. □

PROPOSITION 23.13. *Let $A \hookrightarrow B$ be an extension of Krull domains such that for each height-one prime $P \in \operatorname{Spec}(A)$ we have $PB \neq B$, and let Z be a (possibly uncountable) set of indeterminates over A. Then $A \hookrightarrow B$ is weakly flat if and only if $A[Z] \hookrightarrow B[Z]$ is weakly flat.*

PROOF. Let F denote the field of fractions of A. By Corollary 9.4, the extension $A \hookrightarrow B$ is weakly flat if and only if $F \cap B = A$. Thus the assertion follows from $F \cap B = A \iff F(Z) \cap B[Z] = A[Z]$. □

It would be interesting to know whether the converse of Proposition 23.12.2 is true. In this connection we have:

REMARKS 23.14. Let $\varphi : A \longrightarrow B$ be a weakly flat map of Krull domains, and let P be a height-one prime in A.

(1) Let Q be a minimal prime of the extended ideal PB. If the map $\varphi_Q : A \longrightarrow B_Q$ is weakly flat, then $\operatorname{ht} Q = 1$. To see this, observe that QB_Q is the unique minimal prime of PB_Q, so QB_Q is the radical of PB_Q. If φ_Q is weakly flat, then $PB_Q \cap A = P$ and hence $QB_Q \cap A = P$. It follows that $A_P \hookrightarrow B_Q$. Since A_P is a DVR and its maximal ideal PA_P extends to an ideal primary for the maximal ideal QB_Q of the Krull domain B_Q, we must have that B_Q is a DVR and hence $\operatorname{ht} Q = 1$.
(2) Thus if there exists a weakly flat map of Krull domains $\varphi : A \longrightarrow B$ and a minimal prime Q of PB such that $\operatorname{ht} Q > 1$, then the map $\varphi_Q : A \longrightarrow B_Q$ fails to be weakly flat,

(3) If P is the radical of a principal ideal, then each minimal prime of PB has height one.

QUESTION 23.15. Let $\varphi : A \longrightarrow B$ be a weakly flat map of Krull domains, and let P be a height-one prime in A, as in Remarks 23.14. Is it possible that the extended ideal PB has a minimal prime Q with $\operatorname{ht} Q > 1$?

REMARK 23.16. Primary independence never lifts to polynomial rings. With Setting 22.1 and $\tau \in \widehat{\mathfrak{m}}$, to see that $\tau \in \widehat{\mathfrak{m}}$ fails to be primarily independent over $R[z]_{(\mathfrak{m},z)}$, observe that $\mathfrak{m}R[z]_{(\mathfrak{m},z)}$ is a dimension-one prime ideal that extends to $\widehat{\mathfrak{m}}[[z]]$. The ideal $\widehat{\mathfrak{m}}[[z]]$ also has dimension one and is *not* $(\mathfrak{m}, z)$-primary in $\widehat{R}[[z]]$. Alternatively, in the language of locally flat maps, if the elements $\{\tau_i\}_{i=1}^m \subseteq \widehat{\mathfrak{m}}$ are primarily independent over R, then Proposition 23.11 implies that the map

$$\varphi' : R_n[z]_{(\max(R_n),z)} \longrightarrow \widehat{R}[[z]]$$

satisfies condition LF_{d-1}, where $d = \dim(R)$. For $\{\tau_i\}_{i=1}^m$ to be primarily independent over $R[z]_{(\mathfrak{m},z)}$, the map φ' has to satisfy LF_d, since $\dim R[z]_{(\mathfrak{m},z)} = d+1$. Proposition 23.11 forces $\varphi : R_n \longrightarrow \widehat{R}$ to satisfy condition LF_d and thus φ is flat, so that $n = 0$.

Theorem 23.17 gives a method to obtain residually algebraically independent and primarily independent elements over an uncountable excellent local domain. Theorem 23.17 uses the fact that, if A is a Noetherian ring and Z is a set of indeterminates over A, then the ring $A(Z)$ obtained by localizing the polynomial ring $A[Z]$ at the multiplicative system of polynomials whose coefficients generate the unit ideal of A is again a Noetherian ring [**55**, Theorem 6].

THEOREM 23.17. *Let $(R, \mathfrak{m})$ and $\{\tau_i\}_{i=1}^m \subset \widehat{\mathfrak{m}}$ be as in Setting 22.1, where m is either an integer or $m = \infty$, and $\dim(R) = d \geq 2$. Let Z be a set (possibly uncountable) of indeterminates over $\widehat{R}$ and let $R(Z) = R[Z]_{(\mathfrak{m}R[Z])}$. Then:*

(1) *$\{\tau_i\}_{i=1}^m$ is primarily independent over $R \iff \{\tau_i\}_{i=1}^m$ is primarily independent over $R(Z)$.*

(2) *$\{\tau_i\}_{i=1}^m$ is residually algebraically independent over $R \iff \{\tau_i\}_{i=1}^m$ is residually algebraically independent over $R(Z)$.*

(3) *If $\{\tau_i\}_{i=1}^m$ is idealwise independent over $R(Z)$, then $\{\tau_i\}_{i=1}^m$ is idealwise independent over R.*

PROOF. Let $n \in \mathbb{N}$ be an integer with $n \leq m$, put $R_n = R[\tau_1, \ldots, \tau_n]_{(\mathfrak{m},\tau_1,\ldots,\tau_n)}$ and let $\mathfrak{n}$ denote the maximal ideal of R_n. Let $\varphi : R_n \longrightarrow \widehat{R}$ and $\mu : R_n \longrightarrow R_n(Z) = R_n[Z]_{\mathfrak{n}R_n[Z]}$ be the inclusion maps. The following diagram commutes:

$$\begin{array}{ccccc}
R_n(Z) & \xrightarrow{\ \varphi'\ } & \widehat{R}(Z) & \xrightarrow{\ \psi\ } & \widehat{R(Z)} \\
\Big\uparrow{\scriptstyle \mu} & & \Big\uparrow{\scriptstyle \mu'} & & \\
R_n & \xrightarrow{\ \varphi\ } & \widehat{R} & & ,
\end{array}$$

where ψ is the inclusion map taking $\widehat{R}(Z)$ to its completion. Since R_n is a free R_n-module and $\widehat{R}$ is an R_n-module, $\widehat{R} \otimes_{R_n} R_n[Z] \cong \widehat{R}[Z]$. The ring $\widehat{R}(Z) = S^{-1}\widehat{R}[Z]$, where S is the multiplicatively closed set

$$S := \{f(Z) \in \widehat{R}[Z] \mid \text{ the coefficients of } f(Z) \text{ generate the unit ideal of } \widehat{R}\}.$$

Since $\widehat{R}$ is local with maximal ideal $\widehat{\mathfrak{m}}$, $S = \widehat{R}[Z] \setminus \widehat{\mathfrak{m}}\widehat{R}[Z]$, and so $\widehat{R}(Z) = \widehat{R}[Z]_{\widehat{\mathfrak{m}}\widehat{R}[Z]}$. Now Proposition 23.11 applies. Thus, for item 1, φ satisfies LF_{d-1} if and only if φ' satisfies LF_{d-1}. Similarly, for item 2, φ satisfies LF_1 if and only if φ' satisfies LF_1.

Since the inclusion map ψ is faithfully flat,

$$\varphi \text{ satisfies } LF_k \iff \varphi' \text{ satisfies } LF_k \iff \psi\varphi' \text{ satisfies } LF_k.$$

Items 1 and 2 hold since primary independence is equivalent to LF_{d-1} by Theorem 23.4, and residual algebraic independence is equivalent to LF_1 by Corollary 9.14.

For item 3, observe that $R(Z), \widehat{R}(Z)$ and $\widehat{R(Z)}$ are all excellent normal local domains, and so they are Krull domains. If the τ_i are idealwise independent over $R(Z)$, then the morphism $\psi\varphi'$ is weakly flat by Remark 23.2.1. By Proposition 23.10, φ' is weakly flat. Proposition 23.8.2 implies φ is also weakly flat. Item 3 follows by Remark 23.2.1. □

COROLLARY 23.18. *Let k be a countable field, let Z be an uncountable set of indeterminates over k and let x, y be additional indeterminates. Then the ring $R := k(Z)[x,y]_{(x,y)}$ is an uncountable excellent normal local domain of dimension two, and, for m a positive integer or $m = \infty$, there exist m primarily independent elements (and hence also residually algebraically and idealwise independent elements) over R.*

PROOF. Apply Proposition 22.15 and Theorems 22.20, 22.30 and 23.17. □

23.3. Passing to the Henselization

In this section we investigate idealwise independence, residual algebraic independence, and primary independence as we pass from R to the Henselization R^h of R. In particular, we show in Proposition 23.23 that for a single element $\tau \in \mathfrak{m}\widehat{R}$ the notions of idealwise independence and residual algebraic independence coincide if $R = R^h$. This implies that for every excellent normal local Henselian domain of dimension 2 all three concepts coincide for an element $\tau \in \widehat{\mathfrak{m}}$; that is, τ is idealwise independent $\iff$ τ is residually algebraically independent $\iff$ τ is primarily independent.

We use the commutative square of Proposition 23.11 and obtain the following result for Henselizations:

PROPOSITION 23.19. *Let $\varphi : (A, \mathfrak{m}) \hookrightarrow (B, \mathfrak{n})$ be an injective local map of normal Noetherian local domains, and let $\varphi^h : A^h \longrightarrow B^h$ denote the induced map of the Henselizations. Then:*

1. *For each k with $1 \le k \le \dim(B)$, φ satisfies LF_k $\iff$ φ^h satisfies LF_k. Thus φ satisfies PDE $\iff$ φ^h satisfies PDE.*
2. *(Descent) If φ^h is height-one preserving (respectively weakly flat), then φ is height-one preserving (respectively weakly flat).*

Using shorthand and diagrams, we show Proposition 23.19 schematically:

$$\boxed{\varphi \text{ is } LF_k} \iff \boxed{\varphi^h \text{ is } LF_k} \quad ; \quad \boxed{\varphi \text{ is PDE}} \iff \boxed{\varphi^h \text{ is PDE}}$$

$$\boxed{\varphi \text{ ht-1 pres}} \Longleftarrow \boxed{\varphi^h \text{ ht-1 pres}} \quad ; \quad \boxed{\varphi \text{ w.f.}} \Longleftarrow \boxed{\varphi^h \text{ w.f.}}.$$

PROOF. (of Proposition 23.19) Consider the commutative diagram:

$$\begin{array}{ccc} A^h & \xrightarrow{\varphi^h} & B^h \\ {\scriptstyle\mu}\uparrow & & {\scriptstyle\nu}\uparrow \\ A & \xrightarrow{\varphi} & B \end{array}$$

where μ and ν are the faithfully flat canonical injections; see Remarks 3.32.1 and 3.32.2. Since $\nu\circ\varphi$ is injective and A is normal, φ^h is injective by Remark 3.32.4. The statement for PDE in item 1 holds if the statement for LF_1 in item 1 holds, since Corollary 9.14 implies they are equivalent statements. Proposition 23.11.2 shows that if φ^h is LF_k, then φ is LF_k. Thus to prove item 1, it suffices to show that φ is LF_k implies φ^h is LF_k.

Assume that φ is LF_k. Let $Q' \in \mathrm{Spec}(B^h)$ with $\mathrm{ht}(Q') \le k$. Put $Q = Q' \cap B$, $P' = Q' \cap A^h$, and $P = P' \cap A$. We consider the localized diagram:

$$\begin{array}{ccc} A^h_{P'} & \xrightarrow{\varphi^h_{Q'}} & B^h_{Q'} \\ {\scriptstyle\mu_{P'}}\uparrow & & {\scriptstyle\nu_{Q'}}\uparrow \\ A_P & \xrightarrow{\varphi_Q} & B_Q \end{array}$$

The faithful flatness of ν implies $\mathrm{ht}(Q) \le k$. The LF_k condition on φ implies that φ_Q is flat. Since $\varphi^h_{Q'}$ and φ_Q are local maps of local rings, flatness of either map is equivalent to faithful flatness.

To show that $\varphi^h_{Q'} : A^h_{P'} \longrightarrow B^h_{Q'}$ is flat, apply Remark 7.2 with $M = B^h_{Q'}$ and $I = PA^h_{Q'}$. By Remark 7.2, $B^h_{Q'}$ is flat over $A^h_{P'}$ if and only if

(a) $PA^h_{Q'} \otimes_{A^h_{P'}} B^h_{Q'} \cong PA^h_{Q'}B^h_{Q'}$; that is, $PA^h_{Q'} \otimes_{A^h_{P'}} B^h_{Q'} \cong PB^h_{Q'}$, and

(b) $B^h_{Q'}/PA^h_{Q'}B^h_{Q'}$ is $A^h_{P'}/PA^h_{Q'}$-flat; that is, $B^h_{Q'}/PB^h_{Q'}$ is $A^h_{P'}/PA^h_{Q'}$-flat.

Note that P' is a minimal prime divisor of PA^h. By Remark 8.28.2, μ is a regular map. Therefore $(A^h/PA^h)_{P'} = (A^h/P')_{P'}$ is a field. Thus

$$\overline{\varphi^h_{Q'}} : (A^h/PA^h)_{P'} \longrightarrow (B^h/PB^h)_{Q'}$$

is faithfully flat and it remains to show that

$$PA^h_{P'} \otimes_{A^h_{P'}} B^h_{Q'} \cong PB^h_{Q'}.$$

This can be seen as follows:

$$\begin{aligned} PA^h_{P'} \otimes_{A^h_{P'}} B^h_{Q'} &\cong (P \otimes_{A_P} A^h_{P'}) \otimes_{A^h_{P'}} B^h_{Q'} && \text{by flatness of } \mu \\ &\cong P \otimes_{A_P} B^h_{Q'} \\ &\cong (P \otimes_{A_P} B_Q) \otimes_{B_Q} B^h_{Q'} \\ &\cong PB_Q \otimes_{B_Q} B^h_{Q'} && \text{by flatness of } \varphi_Q \\ &\cong PB^h_{Q'} && \text{by flatness of } \nu. \end{aligned}$$

This completes the proof of item 1.

Item 2 is proved in Proposition 23.11.3. □

COROLLARY 23.20. *Let* $(R, \mathfrak{m})$ *and* $\{\tau_i\}_{i=1}^m$ *be as in Setting 22.1, where* m *is either a positive integer or* $m = \infty$ *and* $\dim(R) = d \ge 2$. *Then the following diagram commutes*

$$
(23.20.0)\qquad
\begin{array}{ccc}
\widetilde{R_n} := R^h[\tau_1,\dots,\tau_n]_{(\mathfrak{m},\tau_1,\dots,\tau_n)} & \xrightarrow{\varphi'} & \widehat{R} \\
{\scriptstyle\mu}\uparrow & & \Vert \\
R_n := R[\tau_1,\dots,\tau_n]_{(\mathfrak{m},\tau_1,\dots,\tau_n)} & \xrightarrow{\varphi} & \widehat{R},
\end{array}
\text{ where}
$$

μ is the canonical faithfully flat injection from the proof of Proposition 23.19 and φ and φ' are the inclusion maps, and:

(1) *$\{\tau_i\}_{i=1}^m$ is primarily independent over R $\iff$ $\{\tau_i\}_{i=1}^m$ is primarily independent over R^h.*
(2) *$\{\tau_i\}_{i=1}^m$ is residually algebraically independent over R $\iff$ $\{\tau_i\}_{i=1}^m$ is residually algebraically independent over R^h.*
(3) (Descent) *If $\{\tau_i\}_{i=1}^m$ is idealwise independent over R^h, then $\{\tau_i\}_{i=1}^m$ is idealwise independent over R.*

PROOF. As in the proof of Proposition 23.19, Diagram 23.20.0 commutes. For items 1 and 2, it suffices to show the equivalence for every positive integer $n \le m$. Refer to Diagram 23.20.0. By Remark 8.28.5, the local rings $R_n = R[\tau_1,\dots,\tau_n]_{(\mathfrak{m},\tau_1,\dots,\tau_n)}$ and $\widetilde{R_n} = R^h[\tau_1,\dots,\tau_n]_{(\mathfrak{m},\tau_1,\dots,\tau_n)}$ have the same Henselization R_n^h. Also $R_n \subseteq \widetilde{R_n}$. By Theorem 23.4 and Proposition 23.19 we have:

$$
\begin{aligned}
&\tau_1,\dots,\tau_n \text{ are primarily (respectively residually algebraically)}\\
&\qquad\qquad \text{independent over } R \iff\\
&R_n \longrightarrow \widehat{R} \text{ satisfies } LF_{d-1} \text{ (respectively } LF_1) \iff\\
&R_n^h \longrightarrow \widehat{R} = \widehat{R^h} \text{ satisfies } LF_{d-1} \text{ (respectively } LF_1) \iff\\
&\widetilde{R}_n \longrightarrow \widehat{R} \text{ satisfies } LF_{d-1} \text{ (respectively } LF_1).
\end{aligned}
$$

The third statement on idealwise independence follows from Theorem 23.11.3 by considering Diagram 23.20.0. □

REMARK 23.21. The examples given in Theorems 22.35 and 22.37 show the converse to part 3 of Corollary 23.20 fails: weak flatness need not lift to the Henselization. With the notation of Proposition 23.19, if φ is weakly flat, then for every $P \in \operatorname{Spec}(A)$ of height one with $PB \ne B$ there exists by Proposition 9.11, $Q \in \operatorname{Spec}(B)$ of height one such that $P = Q \cap A$. In the Henselization A^h of A, the ideal PA^h is a finite intersection of height-one prime ideals P_i' of A^h by Remarks 8.28.2. Only one of the P_i' is contained in Q. Thus as in Theorems 22.35 and 22.37, one of the minimal prime divisors P_i' may fail the condition for weak flatness.

Let R be an excellent normal local domain with Henselization R^h. By Remark 8.28.9, R^h is an integral domain. Let K and K^h denote the fields of fractions of R and R^h respectively. Let L be an intermediate field with $K \subseteq L \subseteq K^h$. Since $K^h \cap \widehat{R} = R^h$, it follows that $T = L \cap \widehat{R} = L \cap R^h$. The intersection ring T is an étale extension of R, and T is an excellent reduced local ring with Henselization $T^h = R^h$, by [**159**, Corollary 1.5 and Proposition 1.9]. Since an étale extension of a normal domain is normal, the ring T is an excellent normal local domain.

The proof of corollary 23.22 uses that Henselian excellent normal local domains are algebraically closed in their completion; see Remark 8.28.5.

COROLLARY 23.22. *Let $(R, \mathfrak{m})$ and $\{\tau_i\}_{i=1}^m$ be as in Setting 22.1, where m denotes a positive integer or $m = \infty$. Let T be a Noetherian local domain dominating R and algebraic over R and dominated by $\widehat{R}$ with $\widehat{R} = \widehat{T}$. Then:*

(1) *$\{\tau_i\}_{i=1}^m$ is primarily independent over R $\iff$ $\{\tau_i\}_{i=1}^m$ is primarily independent over T.*
(2) *$\{\tau_i\}_{i=1}^m$ is residually algebraically independent over R $\iff$ $\{\tau_i\}_{i=1}^m$ is residually algebraically independent over T.*
(3) *If $\{\tau_i\}_{i=1}^m$ is idealwise independent over T, then $\{\tau_i\}_{i=1}^m$ is idealwise independent over R.*

PROOF. By Remark 8.28.5, $R \subseteq T \subseteq R^h = T^h$. Statements 1 and 2 follow from Corollary 23.20, parts 1 and 2.

For statement 3, use Remark 23.2.1: $\tau_1, \ldots, \tau_n$ are idealwise independent over R, respectively T, if and only if $R_n \hookrightarrow \widehat{R}$, respectively $T[\tau_1, \ldots, \tau_n]_{(-)} \hookrightarrow \widehat{T}$, is weakly flat. Suppose $\{\tau_i\}_{i=1}^m$ is idealwise independent over T. Then, for every $n \leq m$, $\{\tau_i\}_{i=1}^n$ is idealwise independent over T, and this implies

$$T[\tau_1, \ldots, \tau_n]_{(-)} \hookrightarrow \widehat{T} = \widehat{R}$$

is weakly flat. By Proposition 23.11.3, $R_n \hookrightarrow \widehat{R}$ is weakly flat, and so $\tau_1, \ldots, \tau_n$ are idealwise independent over R. Hence $\{\tau_i\}_{i=1}^m$ is idealwise independent over R. □

Theorem 22.30 states that, if R has the property that every height-one prime ideal is the radical of a principal ideal and $\tau \in \widehat{\mathfrak{m}}$ is residually algebraically independent over R, then τ is idealwise independent over R. Proposition 23.23 describes a situation in which idealwise independence implies residual algebraic independence.

PROPOSITION 23.23. *Let $(R, \mathfrak{m})$ and $\tau \in \widehat{\mathfrak{m}}$ be as in Setting 22.1. Suppose R has the property that, for each $P \in \operatorname{Spec}(R)$ with $\operatorname{ht}(P) = 1$, the ideal $P\widehat{R}$ is prime.*

(1) *If τ is idealwise independent over R, then τ is residually algebraically independent over R.*
(2) *If R has the additional property that every height-one prime ideal is the radical of a principal ideal, then τ is idealwise independent over R $\iff$ τ is residually algebraically independent over R.*

PROOF. For item 1, let $\widehat{P} \in \operatorname{Spec}(\widehat{R})$ be such that $\operatorname{ht}(\widehat{P}) = 1$ and $\widehat{P} \cap R \neq 0$. Then $\operatorname{ht}(\widehat{P} \cap R) = 1$ and $(\widehat{P} \cap R)R[\tau]$ is a prime ideal of $R[\tau]$ of height 1. Idealwise independence of τ implies that $(\widehat{P} \cap R)R[\tau] = (\widehat{P} \cap R)\widehat{R} \cap R[\tau]$. Since $(\widehat{P} \cap R)\widehat{R}$ is nonzero and prime, $\widehat{P} = (\widehat{P} \cap R)\widehat{R}$ and $\widehat{P} \cap R[\tau] = (\widehat{P} \cap R)R[\tau]$. Therefore $\operatorname{ht}(\widehat{P} \cap R[\tau]) = 1$ and Theorem 22.27 implies that τ is residually algebraically independent over R.

Item 1 implies item 2 by Theorem 22.30.3. □

REMARKS 23.24. (1) Let R be an excellent normal local domain. If R is Henselian, then R/P is Henselian for each height-one prime P of R. This implies that, for each $P \in \operatorname{Spec} R$ with $\operatorname{ht} P = 1$, the ideal $P\widehat{R}$ is prime, as in the hypothesis of Proposition 23.23. To see this, let P be a height-one prime of R such that $S := R/P$ is Henselian. Then the integral closure S' of S in its field of fractions is again local, in fact S' is an excellent normal local domain and so its completion

$\widehat{S'}$ is a normal domain by Theorem 8.23. Since $S \subseteq S'$ are finite R-modules and $\widehat{R}$ is a flat R-module,

$$\widehat{S} = S \otimes_R \widehat{R} \subseteq S' \otimes_R \widehat{R} = \widehat{S'},$$

by [**122**, Theorem 8.7, p.60]. Thus $\widehat{S} = \widehat{R}/P\widehat{R}$ is an integral domain, and so $P\widehat{R}$ is a prime ideal.

(2) There is an example in [**9**] of a normal Noetherian local domain R that is not Henselian but, for each prime ideal P of R of height-one, the domain R/P is Henselian.

(3) It is unclear whether Proposition 23.23 extends to more than one algebraically independent element $\tau \in \widehat{\mathfrak{m}}$, because the localized polynomial ring $R[\tau]_{(\mathfrak{m},\tau)}$ is not Henselian.

COROLLARY 23.25. *Let R be an excellent Henselian normal local domain of dimension 2, and assume the notation of Setting 22.1. Then:*

(1) *τ is residually algebraically independent over R $\iff$ τ is primarily independent over R.*
(2) *Either of these equivalent conditions implies τ is idealwise independent over R.*
(3) *If R has the additional property that every height-one prime ideal is the radical of a principal ideal, then the three conditions are equivalent.*

PROOF. This follows from Theorem 22.30, Proposition 22.15.1 and Proposition 23.23. □

23.4. Summary diagram for the independence concepts

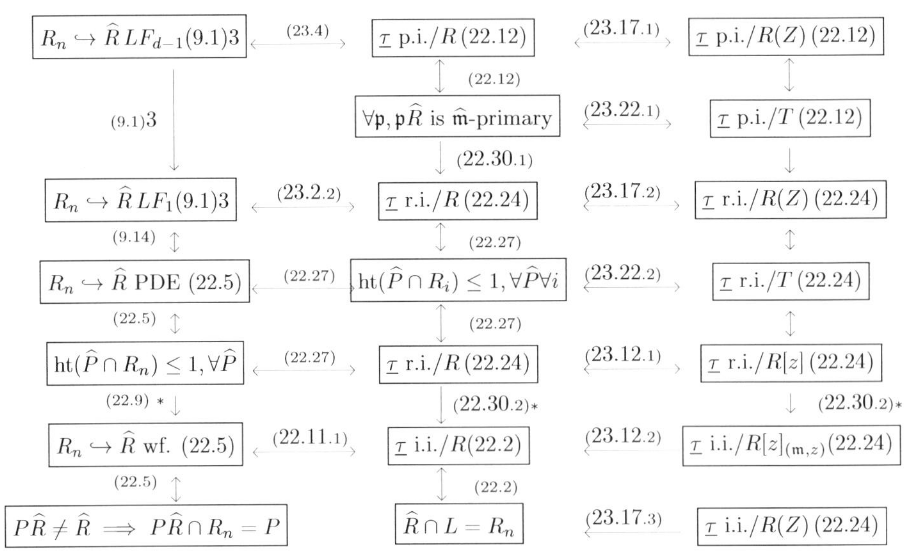

Diagram 23.25.0 Implications among the properties.

Diagram 23.25.0 uses Setting 22.1: $R, \mathfrak{m}, \underline{\tau} = \{\tau_1 \dots, \tau_n\}$, and $R_n = R[\underline{\tau}]_{(\mathfrak{m},\underline{\tau})}$. Let $d = \dim(R)$, $L =$ the field of fractions of R_n, let $\mathfrak{p}$ denote a prime ideal of R_n such that $\dim(R_n/\mathfrak{p}) \le d - 1$, P denotes a prime ideal of R_n with $\text{ht}(P) = 1$, $\widehat{P}$ in $\text{Spec}(\widehat{R})$ has $\text{ht}(\widehat{P}) = 1$, $R^h =$ the Henselization of R in $\widehat{R}$, T is a local Noetherian domain dominating and algebraic over R and dominated by $\widehat{R}$ with $\widehat{R} = \widehat{T}$, z is an indeterminate over the field of fractions of $\widehat{R}$ and Z is a possibly uncountable set of set of indeterminates over the field of fractions of $\widehat{R}$. Then we have the implications shown below.

We use the abbreviations "p. i.", "r. i." and "i. i." for "primarily independent", "residually algebraically independent" and "idealwise independent".

* We assume that every height-one prime ideal of R is a principal ideal in order to have the starred arrows.

NOTE 23.26. $R_n \hookrightarrow \widehat{R}$ is always height-one preserving by Proposition 22.8.

CHAPTER 24

Krull domains with Noetherian x-adic completions

This chapter contains applications of Inclusion Construction 5.3 for the case where the base ring R is a local Krull domain. With the setting and notation of Inclusion Construction 5.3 and the assumptions of Setting 24.2, R is a Krull domain and both the Intersection Domain A and Approximation Domain B of Definition 5.7 are Krull domains; see Theorem 9.7.

The result of applying Inclusion Construction 5.3 to a normal Noetherian integral domain R that satisfies the assumptions of Setting 24.2 may fail to be Noetherian.[1] Setting 24.2 yields an application to non-Noetherian Krull domains for which Construction 5.3 can be iterated.

The construction in Chapters 22 and 23 uses the entire $\mathfrak{m}$-adic completion of an excellent normal local domain $(R, \mathfrak{m})$, rather than a completion with respect to a principal ideal. Chapters 22 and 23 contain examples of subfields L of the field of fractions of $\widehat{R}$ such that the ring $A := L \cap \widehat{R}$ is a localized polynomial ring over R in finitely many or infinitely many variables. Thus A may be a non-Noetherian Krull domain, such as the example of Corollary 22.23.

Here, as in Inclusion Construction 5.3 and Chapters 4 to 10, we use completions with respect to a principal ideal. The base ring is a local Krull domain $(R, \mathfrak{m})$. We do not assume R is Noetherian, but we do assume in Setting 24.2 the existence of a nonzero nonunit x of R such that the x-adic completion R^* of R is an analytically normal Noetherian domain. Since R^* is a Krull domain, the ring $A := L \cap R^*$, for L a subfield of $\mathcal{Q}(R^*)$, is a Krull domain. Thus we can apply the results of Chapter 9 to the constructed ring A. This setting permits iterations of Inclusion Construction 5.3, as in Section 24.2.

The focus of this chapter is the *limit-intersecting* conditions of Definitions 24.6 that include the limit-intersecting condition given in Definition 5.10. By Theorem 9.9, these conditions imply that the ring A of Inclusion Construction 5.3 equals its Approximation Domain B from Definition 5.7. The two stronger forms of the limit-intersecting condition are useful for determining if A is Noetherian or excellent. Sections 24.3 and 24.4 contain several examples related to these concepts.

Concepts from earlier chapters are useful in this study, including various flatness conditions for extensions of Krull domains in Chapter 9. The following terms from Definitions 2.14 and 9.1 are restated for use in this chapter:

DEFINITIONS 24.1. Let $\varphi : S \hookrightarrow T$ be an extension of Krull domains.

- T is a *PDE* extension of S if for every height-one prime ideal Q in T, the height of $Q \cap S$ is at most one.

[1] See Chapters 14 and 16 for examples where the constructed domain is a non-Noetherian Krull domain.

- T is a *height-one preserving* extension of S if for every height-one prime ideal P of S with $PT \neq T$ there exists a height-one prime ideal Q of T with $PT \subseteq Q$.
- T is *weakly flat* over S if every height-one prime ideal P of S with $PT \neq T$ satisfies $PT \cap S = P$.
- Let $r \in \mathbb{N}$ be an integer with $1 \leq r \leq d = \dim(T)$ where d is an integer or $d = \infty$. Then φ is called *locally flat in height* r, abbreviated LF_r, if, for every prime ideal Q of T with $\text{ht}(Q) \leq r$, the induced map on the localizations $\varphi_Q : S_{Q\cap S} \longrightarrow T_Q$ is faithfully flat.

24.1. Applying Inclusion Construction 5.3

SETTING AND NOTATION 24.2. Let $(R, \mathfrak{m})$ be a local Krull domain with field of fractions F. Assume that x is a nonzero element of $\mathfrak{m}$ such that the x-adic completion $(R^*, \mathfrak{m}^*)$ of R is an analytically normal Noetherian local domain. By Exercise 24.1, we have $\bigcap_{i=1}^{\infty} x^n R = (0)$. Since the $\mathfrak{m}$-adic completion of R is the same as the $\mathfrak{m}^*$-adic completion of R^*, the $\mathfrak{m}$-adic completion $\widehat{R}$ of R is also a normal Noetherian local domain, Let F^* denote the field of fractions of R^*. Since R^* is Noetherian, $\widehat{R}$ is faithfully flat over R^* and $R^* = \widehat{R} \cap F^*$. Therefore $F \cap R^* = F \cap \widehat{R}$. Let d denote the dimension of the Noetherian domain R^*. It follows that d is also the dimension of $\widehat{R}$. [2]

Let $\tau_1, \ldots, \tau_s \in \mathfrak{m}^*$ be algebraically independent over F. The hypotheses of Inclusion Construction 5.3 are satisfied by R, x and the τ_i. As in Construction 5.3 and Equations 5.4.3 and 5.4.6, let τ_{in} denote the nth endpiece of τ_i, for each i, and

$$U_n := R[\tau_n, \ldots, \tau_{sn}],\ U := \varinjlim_{n\in\mathbb{N}} U_n,\ B := (1+xU)^{-1}U,\ A := F(\tau_1, \ldots, \tau_s) \cap R^*.$$

Remark 5.6.1 and Part 6 of Construction Properties Theorem 5.14 imply that B is local, that

$$B = \varinjlim_{n\in\mathbb{N}} B_n = \bigcup_{n=1}^{\infty} B_n, \text{ where } B_n := (U_n)_{(\mathfrak{m}, \tau_{1n}, \ldots, \tau_{sn})}, \text{for each } n \in \mathbb{N},$$

that each $B_n \subseteq B_{n+1}$ and that B_{n+1} dominates B_n. [3] By parts 3 and 4 of Construction Properties Theorem 5.14, the x-adic completions of A and of B are equal to R^*, and A birationally dominates B. Let $\widehat{R}[1/x]$ denote the localization of $\widehat{R}$ at the powers of x; similarly, let $R^*[1/x]$ denote the localization of R^* at the powers of x. The domains $\widehat{R}[1/x]$ and $R^*[1/x]$ have dimension $d - 1$.

We sometimes need the following assumption.

ASSUMPTION 24.3. $R = F \cap R^* = F \cap \widehat{R}$; equivalently, R^* and $\widehat{R}$ are weakly flat over R.

Corollary 9.4.ii implies the equivalence of the two statements in Assumption 24.3.

---

[2] If R is Noetherian, then d is also the dimension of R. However, if R is not Noetherian, then the dimension of R may be greater than d. This is illustrated by taking R to be the ring B of Example 12.7.

[3] The definition of B_n used in Remarks 5.6.1 and Equation 5.4.5 is different from that given here, but U_n is the same. It follows that, with Notation 24.2, $B_n \subseteq B_{n+1}$ and B_n is dominated by B_{n+1}.

REMARK 24.4. It is possible for $R \hookrightarrow R^*[1/x]$ to satisfy the conditions of Setting 24.2 but fail to satisfy Assumption 24.3. This is demonstrated by the iterative example of Section 12.1 as given in Theorem 12.3, with $R := B \neq A$; see Example 12.7. The Krull domain B of Example 12.7 with $B \neq A$ also illustrates that a directed union of normal Noetherian domains may be a non-Noetherian Krull domain.

Theorems 24.5 consists of applications of Construction Properties Theorem 5.14, Theorem 5.24, Noetherian Flatness Theorem 6.3, and Theorem 9.7.

THEOREM 24.5. *Assume Setting 24.2. Then the intermediate rings B_n, B and A have the following properties:*

(1) $x^nR^* \cap R = x^nR$, $\quad x^nR^* \cap A = x^nA$, $\quad x^nR^* \cap B = x^nB$ *and* $x^nR^* \cap U = x^nU$, *for each* $n \in \mathbb{N}$.
(2) $R/x^tR = U/x^tU = B/x^tB = A/x^tA = R^*/x^tR^*$, *for every positive integer* t.
(3) $A^* = B^* = R^*$.
(4) *For every* $n \in \mathbb{N}$, $B[1/x]$ *is a localization of* B_n, *i.e., for each* $n \in \mathbb{N}$, *there exists a multiplicatively closed subset* S_n *of* B_n *such that* $B[1/x] = S_n^{-1}B_n$.
(5) *The minimal prime ideals over* xR, xB, xA, *and* xR^* *are in one-to-one correspondence via extension and contraction of prime ideals.*
(6) B *and* A *are local rings, with* $B \subseteq A$ *and* A *dominating* B.
(7) *Every ideal of* R, B *or* A *that contains* x *is finitely generated by elements of* R. *In particular, the maximal ideal* $\mathfrak{m}$ *of* R *is finitely generated, and the maximal ideals of* B *and* A *are* $\mathfrak{m}B$ *and* $\mathfrak{m}A$.
(8) *If* B *is Noetherian, then* $B = A$.
(9) A *and* B *are local Krull domains.*

PROOF. Properties 1 - 4 are items 1 - 4 of Construction Properties Theorem 5.14. Property 5 follows from Properties 1 and 2. Property 6 follows from Proposition 5.17.5 and Remarks 5.6. Since R^* is Noetherian, property 7 follows from property 2. Property 8 is in Noetherian Flatness Theorem 6.3.1.

Property 9 follows from Theorem 9.7, property 6 above and Setting 24.2. □

24.2. Limit-intersecting elements

Let $(R, \mathfrak{m})$ be a Krull domain as in Setting 24.2. Remark 24.7.1 shows that each of the *limit-intersecting* properties of Definitions 24.6 implies $L \cap \widehat{R}$ is a directed union of localized polynomial ring extensions of R. In Note 24.15, we compare these limit-intersection concepts to the independence concepts of Definitions 22.2, 22.12, and 22.24.

DEFINITIONS 24.6. Assume Setting 24.2 and also assume Assumption 24.3.

(1) The elements $\tau_1, \ldots, \tau_s$ are said to be *limit-intersecting* in x over R provided $B = A$.
(2) The elements $\tau_1 \ldots, \tau_s$ are said to be *residually limit-intersecting* in x over R provided the inclusion map

$$B_0 := R[\tau_1, \ldots, \tau_s]_{(\mathfrak{m}, \tau_1, \ldots, \tau_s)} \longrightarrow R^*[1/x] \quad \text{is } LF_1. \qquad (24.6.2)$$

(3) The elements $\tau_1 \dots, \tau_s$ are said to be *primarily limit-intersecting* in x over R provided the inclusion map

$$B_0 := R[\tau_1, \dots, \tau_s]_{(\mathfrak{m}, \tau_1, \dots, \tau_s)} \longrightarrow R^*[1/x] \quad \text{is flat.} \qquad (24.6.3)$$

Since $R^*[1/x]$ and $\widehat{R}[1/x]$ have dimension $d - 1$, the condition LF_{d-1} is equivalent to primarily limit-intersecting, that is, to the flatness of the map $B_0 \longrightarrow R^*[1/x]$.

REMARKS 24.7. Assume Setting 24.2.

(1) The following statements are equivalent:

(a) $\tau_1, \cdots, \tau_s$ are limit-intersecting, that is, $A = B$, as in Definition 24.6.1

(b) $B_0 := R[\tau_1, \dots, \tau_s]_{(\mathfrak{m}, \tau_1, \dots, \tau_s)} \hookrightarrow R^*[1/x]$ is weakly flat.

(c) $B \hookrightarrow R^*[1/x]$ is weakly flat.

(d) $B \hookrightarrow R^*$ is weakly flat.

(e) $R[\tau_1, \dots, \tau_s] \hookrightarrow R^*$ is weakly flat.

This follows from Weak Flatness Theorem 9.9 and Proposition 9.8, since B is a Krull domain by Theorem 9.7.4.

(2) If an injective map of Krull domains is weakly flat, then it is height-one preserving by Corollary 9.4.ii. Thus any of the equivalent conditions of Remark 24.7.1 imply that $B \longrightarrow R^*$ is height-one preserving.

(3) If B is Noetherian, then, by Theorem 24.5.8, $A = B$, and so all the conclusions of Remark 24.7.1 hold.

(4) By Remark 22.8, if R is an excellent normal local domain such that every height-one prime ideal of R is the radical of a principal ideal, then the extension $B_0 \hookrightarrow R^*$ is height-one preserving. By Proposition 9.16, an extension of Krull domains that is height-one preserving and satisfies PDE is weakly flat.

(5) If $\{\tau_1, \dots, \tau_s\}$ are primarily limit-intersecting, then $\{\tau_1, \dots, \tau_s\}$ are residually limit-intersecting by Definition 24.6.

(6) If $d = 2$, then obviously $LF_1 = LF_{d-1}$. Hence in this case primarily limit-intersecting is equivalent to residually limit-intersecting.

(7) Since $\widehat{R}[1/x]$ is faithfully flat over $R^*[1/x]$, the statements obtained by replacing $R^*[1/x]$ by $\widehat{R}[1/x]$ in Definitions 24.6 parts 2 and 3 give definitions equivalent to those definitions; see Propositions 23.8 and 23.10 of Chapter 23, and Proposition 24.8.

(8) By Construction Properties Theorem 5.14, $R \hookrightarrow B_n$, for every n, and $R \hookrightarrow B$ are faithfully flat. Thus we have that

(8a) If residually limit-intersecting elements exist over R, then $R \hookrightarrow R^*[1/x]$ must be LF_1. This follows since $R \hookrightarrow B_0 \hookrightarrow R^*[1/x]_Q$ is the composition of a faithfully flat map followed by a flat map, for every height-one prime ideal $Q \in \operatorname{Spec}(R^*[1/x])$.

(8b) If primarily limit-intersecting elements exist over R, then $R \hookrightarrow R^*[1/x]$ must be flat.

(8c) If $B_0 \to R^*$ is weakly flat, then $R \to R^*$ is weakly flat; see Exercise 24.2.

(9) The examples of Remarks 9.12 and 24.4 show that in some situations R^* contains no limit-intersecting elements. Indeed, if R is complete with respect to some nonzero ideal I, and x is outside every minimal prime over I, then every element $\tau = \sum a_i x^i$ of R^* that is transcendental over R fails to be limit-intersecting in x. To see this, choose an element $z \in I$, z outside every minimal prime ideal of

xR; define $\sigma := \sum a_i z^i \in R$. Then $(\tau-\sigma)/(z-x) \in A$, but $(\tau-\sigma)/(z-x) \notin B$, and so $A \neq B$. Here $\tau - \sigma \in (z-x)A \cap R[\tau] \subseteq (z-x)R^* \cap R[\tau]$. Thus $B \hookrightarrow R^*$ is not weakly flat, and so, by item 4, $B \hookrightarrow R^*[1/x]$ is not weakly flat. By Remark 24.7.1, the element τ is not limit-intersecting. Moreover a minimal prime over $z-x$ in R^* intersects $R[\tau]$ in an ideal $\mathfrak{p}$ of height greater than one, because $\mathfrak{p}$ contains $z-x$ and $\tau - \sigma$.

PROPOSITION 24.8. *Assume Setting 24.2, and let k be a positive integer with $1 \le k \le d-1$. Then the following are equivalent:*

(1) *The inclusion map $\varphi : B_0 := R[\tau_1, \dots, \tau_s]_{(\mathfrak{m}, \tau_1, \dots, \tau_s)} \hookrightarrow R^*[1/x]$ is LF_k.*
(1′) *The canonical injection $\varphi_1 : B_0 := R[\tau_1, \dots, \tau_s]_{(\mathfrak{m}, \tau_1, \dots, \tau_s)} \longrightarrow \widehat{R}[1/x]$ is LF_k.*
(2) *The canonical injection $\varphi' : U_0 := R[\tau_1, \dots, \tau_s] \longrightarrow R^*[1/x]$ is LF_k.*
(2′) *The canonical injection $\varphi_1' : U_0 := R[\tau_1, \dots, \tau_s] \longrightarrow \widehat{R}[1/x]$ is LF_k.*
(3) *The canonical injection $\theta : B_n := R[\tau_{1n}, \dots, \tau_{sn}]_{(\mathfrak{m}, \tau_{1n}, \dots, \tau_{sn})} \longrightarrow R^*[1/x]$ is LF_k.*
(3′) *The canonical injection $\theta_1 : B_n := R[\tau_{1n}, \dots, \tau_{sn}]_{(\mathfrak{m}, \tau_{1n}, \dots, \tau_{sn})} \longrightarrow \widehat{R}[1/x]$ is LF_k.*
(4) *The canonical injection $\psi : B \longrightarrow R^*[1/x]$ is LF_k.*
(4′) *The canonical injection $\psi_1 : B \longrightarrow \widehat{R}[1/x]$ is LF_k.*

Each of these statements is also equivalent to LF_k of the corresponding inclusion map obtained by replacing B_0, B_n, U_0 and B by $B_0[1/x]$, $B_n[1/x]$, $U_0[1/x]$ and $B[1/x]$.

PROOF. We have:

$$U_0 \xrightarrow{\text{loc.}} B_0 \xrightarrow{\varphi} R^*[1/x] \xrightarrow{\text{f.f.}} \widehat{R}[1/x].$$

The injection $\varphi_1' : U_0 \longrightarrow \widehat{R}[1/x]$ factors as $\varphi' : U_0 \longrightarrow R^*[1/x]$ followed by the faithfully flat injection $R^*[1/x] \longrightarrow \widehat{R}[1/x]$. Therefore φ' is LF_k if and only if φ_1' is LF_k. The injection φ' factors through the localization $U_0 \longrightarrow B_0$ and so φ is LF_k if and only if φ' is LF_k.

For each $n \in \mathbb{N}$, B_n is a localization of U_n, and $B = (1+xU)^{-1}U$, by Theorem 5.14, parts 5.14 and 6. Thus

$$\begin{aligned} B[1/x] \longrightarrow R^*[1/x] \text{ is } LF_k &\iff U[1/x] \longrightarrow R^*[1/x] \text{ is } LF_k \\ \iff U_0[1/x] \longrightarrow R^*[1/x] \text{ is } LF_k &\iff B_n[1/x] \longrightarrow R^*[1/x] \text{ is } LF_k \\ &\iff B_0[1/x] \longrightarrow R^*[1/x] \text{ is } LF_k . \end{aligned}$$

Thus

$$\begin{aligned} \psi : B \longrightarrow R^*[1/x] \text{ is } LF_k &\iff U \longrightarrow R^*[1/x] \text{ is } LF_k \\ \iff \varphi' : U_0 \longrightarrow R^*[1/x] \text{ is } LF_k &\iff \theta : B_n \longrightarrow R^*[1/x] \text{ is } LF_k \\ &\iff \varphi : B_0 \longrightarrow R^*[1/x] \text{ is } LF_k . \end{aligned}$$ □

REMARKS 24.9. Assume Setting 24.2.

(1) Since B_0 is a localization of U_0, the elements $\tau_1, \dots, \tau_s$ are primarily limit-intersecting in x over R if and only if $U_0 = R[\tau_1, \dots, \tau_s] \hookrightarrow R^*[1/x]$ is flat; see Remark 9.2.

(2) If $(R, \mathfrak{m})$ is a one-dimensional local Krull domain, then R is a DVR, R^* is also a DVR, and $R^*[1/x]$ is flat over $U_0 = R[\tau_1, \dots, \tau_s]$. Therefore, in this situation, $\tau_1, \dots, \tau_s$ are primarily limit-intersecting in x over R if and only if $\tau_1, \dots, \tau_s$ are algebraically independent over R; see Corollary 6.6.

(3) Let $\tau_1, \dots, \tau_s \in k[[x]]$ be transcendental over $k(x)$, where k is a field and x is an indeterminate. Then $\tau_1, \dots, \tau_s$ are primarily limit-intersecting in x over $k[x]_{(x)}$ by item 2 above. If $x_1, \dots, x_m$ are additional indeterminates over $k(x)$, then, by Prototype Theorem 10.2 and Noetherian Flatness Theorem 6.3, the elements $\tau_1, \dots, \tau_s$ are primarily limit-intersecting in x over $k[x, x_1, \dots, x_m]_{(x, x_1, \dots, x_m)}$.

(4) If B is Noetherian, then $\tau_1, \dots, \tau_s$ are primarily limit-intersecting in x over R by Noetherian Flatness Theorem 6.3.

(5) By the equivalence of (1) and (3) of Proposition 24.8, the elements $\tau_1, \dots, \tau_s$ are primarily limit-intersecting in x over R if and only if the endpiece power series $\tau_{1n}, \dots, \tau_{sn}$ are primarily limit-intersecting in x over R.

(6) In view of Remark 9.2, and Proposition 24.8, we have $\tau_1, \dots, \tau_s$ are residually (respectively primarily) limit-intersecting in x over R if and only if the canonical map

$$S_0^{-1} B_0 = B[1/x] \longrightarrow R^*[1/x]$$

is LF_1 (respectively LF_{d-1} or equivalently flat). Here S_0 is the multiplicatively closed subset of B_0 from Theorem 24.5.4,

THEOREM 24.10. *Assume Setting 24.2 and Assumption 24.3. Thus $(R, \mathfrak{m})$ is a local Krull domain with field of fractions F, and $x \in \mathfrak{m}$ is such that the x-adic completion $(R^*, \mathfrak{m}^*)$ of R is an analytically normal Noetherian local domain and $R = R^* \cap F$. For elements $\tau_1, \dots, \tau_s \in \mathfrak{m}^*$ that are algebraically independent over R, the following are equivalent:*

(1) *The extension $R[\tau_1, \dots, \tau_s] \hookrightarrow R^*[1/x]$ is flat.*
(2) *The elements $\tau_1, \dots, \tau_s$ are primarily limit-intersecting in x over R.*
(3) *The extension $B \hookrightarrow R^*$ is faithfully flat.*
(4) *The intermediate rings A and B are equal and are Noetherian.*
(5) *The constructed ring B is Noetherian.*

If these equivalent conditions hold, then the Krull domain R is Noetherian.

PROOF. By Remark 24.9.1, items 1 and 2 are equivalent. Theorem 24.10 follows from Noetherian Flatness Theorem 6.3, parts 1 and 3. □

REMARK 24.11. Example 10.15 yields the existence of a three-dimensional regular local domain $R = k[x, y, z]_{(x,y,z)}$, over an arbitrary field k, and an element $f = y\tau_1 + z\tau_2$ in the x-adic completion of R such that f is residually limit-intersecting in x over R, but fails to be primarily limit-intersecting in x over R. In particular, the rings A and B constructed using f are equal, yet A and B are not Noetherian. The elements τ_1 and τ_2 are elements of $xk[[x]]$ that are algebraically independent over $k(x)$.

Proposition 24.12 gives criteria for an element τ in R^* to be residually limit-intersecting.

PROPOSITION 24.12. *Assume Setting 24.2 and $s = 1$. Then items 1 - 3 are equivalent:*

(1) *The element $\tau = \tau_1$ is residually limit-intersecting in x over R.*
(2) *If $\widehat{P}$ is a height-one prime ideal of $\widehat{R}$ such that $x \notin \widehat{P}$ and $\widehat{P} \cap R \neq (0)$, then $\operatorname{ht}(\widehat{P} \cap R[\tau]_{(\mathfrak{m},\tau)}) = 1$.*
(3) *$B \hookrightarrow R^*[1/x]$ is LF_1.*

If $(R, \mathfrak{m})$ is Noetherian, then item 4 is equivalent to items 1-3.

(4) *For every height-one prime ideal P of R such that $x \notin P$ and for every minimal prime divisor $\widehat{P}$ of $P\widehat{R}$ in $\widehat{R}$, the image $\bar{\tau}$ of τ in $\widehat{R}/\widehat{P}$ is algebraically independent over R/P.*

PROOF. By Proposition 24.8, item 1 is equivalent to item 3, $\psi : B \hookrightarrow R^*[1/x]$ being LF_1, and also item 1 is equivalent to $R[\tau] \hookrightarrow \widehat{R}[1/x]$ being LF_1.

For item 1 $\implies$ item 2, suppose item 2 fails; that is, there exists a prime ideal $\widehat{P}$ of $\widehat{R}$ of height one such that $x \notin \widehat{P}$, $\widehat{P} \cap R \neq (0)$, but $\operatorname{ht}(\widehat{P} \cap R[\tau]) \geq 2$. Let $\widehat{Q} := \widehat{P}\widehat{R}[1/x]$ and $Q := \widehat{Q} \cap R[\tau]_{(\mathfrak{m},\tau)}$. Then $\operatorname{ht} Q \geq 2$. By Definition 24.6.2 of residually limit-intersecting, the injective map $R[\tau]_{(\mathfrak{m},\tau)} \hookrightarrow \widehat{R}[1/x]$ is LF_1. Then the map $(R[\tau]_{(\mathfrak{m},\tau)})_Q \hookrightarrow (\widehat{R}[1/x])_{\widehat{Q}}$ is faithfully flat by Definition 24.1, a contradiction to $\operatorname{ht} Q \geq 2 > \operatorname{ht} \widehat{P} = \operatorname{ht} \widehat{Q}$.

For item 2 $\implies$ item 1, let $\widehat{Q} \in \operatorname{Spec}(\widehat{R}[1/x])$ have $\operatorname{ht} \widehat{Q} = 1$. Define

$$\widehat{P} := \widehat{Q}\widehat{R}[1/x] \cap \widehat{R} \in \operatorname{Spec} \widehat{R}, \qquad Q := \widehat{P} \cap (R[\tau]_{(\mathfrak{m},\tau)}) = \widehat{Q} \cap (R[\tau]_{(\mathfrak{m},\tau)}).$$

Case i: If $\widehat{Q} \cap R \neq (0)$, then item 2 implies that $\operatorname{ht} Q = 1$. Proposition 2.17 implies $R[\tau]_{(\mathfrak{m},\tau)}$ is a Krull domain. Since $\operatorname{ht} Q = 1$, $(R[\tau]_{(\mathfrak{m},\tau)})_Q$ is a DVR. The map $(R[\tau]_{(\mathfrak{m},\tau)})_Q \hookrightarrow (\widehat{R}[1/x])_{\widehat{Q}}$ is faithfully flat by Remark 2.39.2.
Case ii: If $\widehat{Q} \cap R = (0)$ but $Q \neq (0)$, then $\operatorname{ht} Q = 1$. By the same reasoning used for case i, the map $(R[\tau]_{(\mathfrak{m},\tau)})_Q \hookrightarrow (\widehat{R}[1/x])_{\widehat{Q}}$ is faithfully flat.
Case iii: If $\widehat{Q} \cap R = (0) = Q$, then $(R[\tau]_{(\mathfrak{m},\tau)})_Q = (R[\tau]_{(\mathfrak{m},\tau)})_{(0)}$ is a field, and so the map is again faithfully flat.

Thus in all cases $R[\tau] \hookrightarrow \widehat{R}[1/x]$ is LF_1, and so item 1 holds.

For item 2 $\iff$ item 4: Since R is Noetherian, the map $R \hookrightarrow \widehat{R}$ is flat, and Going-down holds; see Remark 2.37.10. Hence $\operatorname{ht} \widehat{P} \geq \operatorname{ht}(\widehat{P} \cap R)$, for each $\widehat{P} \in \operatorname{Spec} \widehat{R}$. If $P \in \operatorname{Spec} R$ and $\widehat{P} \in \operatorname{Spec} \widehat{R}$ is minimal over $P\widehat{R}$, then $P = \widehat{P} \cap R$, and the map $R_P \hookrightarrow \widehat{R}_{\widehat{P}}$ is faithfully flat with $P\widehat{R}_{\widehat{P}}$ primary for the maximal ideal of $\widehat{R}_{\widehat{P}}$. By Krull's Altitude Theorem 2.23, $\dim R_P \leq \dim \widehat{R}_{\widehat{P}}$. Therefore $\dim R_P = \dim \widehat{R}_{\widehat{P}}$. In particular, $\operatorname{ht} P = 1 \iff \operatorname{ht} \widehat{P} = 1$.

If $\widehat{P}$ is a height-one prime ideal of $\widehat{R}$ such that $P := \widehat{P} \cap R \neq 0$ and $x \notin \widehat{P}$, then P is a height-one prime ideal of R with $x \notin P$. Then $\operatorname{ht}(\widehat{P} \cap R[\tau]) = 1 \iff \widehat{P} \cap R[\tau] = PR[\tau]$ and $R[\tau]/(PR[\tau])$ canonically embeds in $\widehat{R}/\widehat{P}$. That is, the image of τ in $R[\tau]/PR[\tau]$ is algebraically independent over R/P. □

REMARK 24.13. Assume Setting 24.2. If R has the property that every height-one prime of R is the radical of a principal ideal, and τ is residually limit-intersecting in x over R, then the extension $B \hookrightarrow R^*[1/x]$ is height-one preserving by Remark 9.6.c, and hence weakly flat by Propositions 9.14, 9.16 and 24.12. Thus, with these assumptions, if τ is residually limit-intersecting, then τ is limit-intersecting.

Proposition 24.14 implies a transitive property of limit-intersecting elements.

PROPOSITION 24.14. *Assume Setting and Notation 24.2. Also assume $s > 1$. For every $j \in \{1, \dots s\}$, set $A(j) := F(\tau_1, \dots, \tau_j) \cap \widehat{R}$ and let $\mathfrak{m}(j)$ denote the maximal ideal of $A(j)$. Then the following statements are equivalent:*

(1) *$\tau_1, \dots, \tau_s$ are limit-intersecting in x over R.*
(2) *For every $j \in \{1, \dots, s\}$, the elements $\tau_1, \dots, \tau_j$ are limit-intersecting in x over R and the elements $\tau_{j+1}, \dots, \tau_s$ are limit-intersecting in x over $A(j)$.*
(3) *There exists a $j \in \{1, \dots, s\}$, such that the elements $\tau_1, \dots, \tau_j$ are limit-intersecting in x over R and the elements $\tau_{j+1}, \dots, \tau_s$ are limit-intersecting in x over $A(j)$.*

Moreover, the three adjusted statements that result if every instance of "limit-intersecting" in statements 1 through 3 is replaced by "residually limit-intersecting" are also equivalent, as are the three adjusted statements for "primarily limit-intersecting".

PROOF. Set $B(j) := \bigcup_{n=1}^{\infty} R[\tau_{1n}, \dots, \tau_{jn}]_{(\mathfrak{n}, \tau_{1n}, \dots, \tau_{jn})}$. That (2) implies (3) is clear, for any of the conditions—"limit-intersecting", "residually limit-intersecting" or "primarily limit-intersecting" .

For (3) $\implies$ (1), Remark 24.7.1 implies that $A(j) = B(j)$ under each of the conditions on $\tau_1, \dots, \tau_j$. By applying Definitions 24.6 of limit-intersecting, residually limit-intersecting, and primarily limit-intersecting in x over $A(j)$ to $\tau_{j+1}, \dots, \tau_s$ with Remark 24.7.8, we get the equivalence of the stated flatness properties for each of the maps

$$\begin{aligned}
\varphi_1 &: A(j)[\tau_{j+1}, \dots, \tau_s]_{(\mathfrak{m}(j), \tau_{j+1}, \dots, \tau_s)} && \longrightarrow A(j)^*[1/x] = R^*[1/x] \\
\varphi_2 &: A(j)[\tau_{j+1}, \dots, \tau_s]_{(\mathfrak{m}(j), \tau_{j+1}, \dots, \tau_s)}[1/x] && \longrightarrow R^*[1/x] \\
\varphi_3 &: B(j)[\tau_{j+1}, \dots, \tau_s]_{(\mathfrak{m}(j), \tau_{j+1}, \dots, \tau_s)}[1/x] && \longrightarrow R^*[1/x] \\
\varphi_4 &: R[\tau_1, \dots, \tau_s]_{(\mathfrak{n}, \tau_1, \dots, \tau_s)}[1/x] && \longrightarrow R^*[1/x] \\
\varphi_5 &: R[\tau_1, \dots, \tau_s]_{(\mathfrak{n}, \tau_1, \dots, \tau_s)} && \longrightarrow R^*[1/x].
\end{aligned}$$

Thus

$$\begin{aligned}
\psi : B \longrightarrow R^*[1/x] \text{ is } LF_k &\iff U \longrightarrow R^*[1/x] \text{ is } LF_k \\
&\iff \varphi' : U_0 \longrightarrow R^*[1/x] \text{ is } LF_k \\
&\iff \theta : B_n \longrightarrow R^*[1/x] \text{ is } LF_k \\
&\iff \varphi : B_0 \longrightarrow R^*[1/x] \text{ is } LF_k .
\end{aligned}$$

The respective flatness properties for φ_5 are equivalent to the conditions that $\tau_1, \dots, \tau_s$ be limit-intersecting, or residually limit-intersecting, or primarily limit-intersecting in x over R. Thus (3) $\implies$ (1) for each property.

For (1) $\implies$ (2), we go backwards: The statement of (1) for $\tau_1, \dots, \tau_s$ is equivalent to the respective flatness property for φ_5. This is equivalent to φ_4 and thus φ_3

having the respective flatness property. By Remark 24.7.1, $B(j)[\tau_{j+1}, \ldots, \tau_s]_{(-)} \longrightarrow R^*[1/x]$ has the appropriate flatness property. Also $B(j) \longrightarrow B(j)[\tau_{j+1}, \ldots, \tau_s]_{(-)}$ is flat, and so $B(j) \longrightarrow R^*[1/x]$ has the appropriate flatness property. Thus the $\tau_1, \ldots, \tau_j$ are limit-intersecting, or residually limit-intersecting or primarily limit-intersecting in x over R. Therefore $A(j) = B(j)$, and so $A(j) \longrightarrow R^*[1/x]$ has the appropriate flatness property. It follows that $\tau_{j+1}, \ldots, \tau_s$ are limit-intersecting, or residually limit-intersecting, or primarily limit-intersecting in x over $A(j)$, as desired. □

NOTE 24.15. The "limit-intersection" concepts of Definition 24.6 have names similar to the names used for "idealwise independence" concepts of Definitions 22.2, 22.12, and 22.24. The concepts apply in different settings. The similarities that exist among the terms is because the maps defining the concepts have similar properties, but in different settings.

(1) Consider the two settings and concepts: Setting 24.2 for the limit-intersection concepts has $(R, \mathfrak{m})$ a local Krull domain and $x \in \mathfrak{m}$ a nonzero element such that the x-adic completion R^* of R is an analytically normal Noetherian local domain. Limit-intersecting, residually limit-intersecting and primarily limit-intersecting involve a finite set $\underline{\tau}$ of elements in the maximal ideal $\mathfrak{m}^*$ of R^* that are algebraically independent over R. Then $R[\underline{\tau}]$ is a polynomial ring over R. The limit-intersecting properties involve the map $R[\underline{\tau}] \hookrightarrow R^*[1/x]$.

Setting 22.1 for the independence concepts has $(R, \mathfrak{m})$ an excellent normal local domain. Then $\widehat{R}$ is a normal Noetherian domain, by Theorem 8.23. Idealwise independent, residually algebraically independent and primarily independent involve a set $\underline{\tau}$ of elements in $\widehat{R}$ that are algebraically independent over R. The independence properties involve the map $R[\underline{\tau}]_{(\mathfrak{m},\underline{\tau})} \hookrightarrow \widehat{R}$. Since $(R, \mathfrak{m})$ Noetherian implies that $\widehat{R}$ is faithfully flat over R^*, if $R[\underline{\tau}] \hookrightarrow \widehat{R}$ is weakly flat, then $R[\underline{\tau}] \hookrightarrow R^*$ is weakly flat.

(2) Let $(R, \mathfrak{m})$ be an excellent normal local domain; thus R fits both settings.

(a) By Theorem 22.11.1, $\underline{\tau}$ is idealwise independent over R is equivalent to $R_n \hookrightarrow \widehat{R}$ is weakly flat, and therefore is equivalent to weak flatness of $R[\underline{\tau}] \hookrightarrow \widehat{R}$. By Remark 24.7.1, $\underline{\tau}$ is limit-intersecting in x over R is equivalent to weak flatness of $R[\underline{\tau}] \hookrightarrow R^*[1/x]$. Thus $\underline{\tau}$ idealwise independent over R implies that $\underline{\tau}$ is limit-intersecting in x over R.

(b) By Theorem 22.27, one element τ is residually algebraically independent over $R \iff R[\tau] \hookrightarrow \widehat{R}$ satisfies LF_1. By Definition 24.6, the element τ is residually limit-intersecting in x over $R \iff R[\tau] \hookrightarrow R^*[1/x]$ is LF_1. Thus residual algebraic independence of one element τ implies τ is limit-intersecting in x over R. Proposition 24.12 shows that τ residually limit-intersecting in x is equivalent to part of the condition for τ to be residually algebraically independent over R.

(c) Assume that $\dim R = d$. By Definition 24.6, $\underline{\tau}$ is primarily limit-intersecting if $R[\underline{\tau}] \hookrightarrow R^*[1/x]$ is flat, whereas, by Theorem 23.4, $\underline{\tau}$ is primarily independent $\iff R[\underline{\tau}] \hookrightarrow \widehat{R}$ satisfies LF_{d-1}. Since $\dim R^*[1/x] = d - 1$, the map $R[\underline{\tau}] \hookrightarrow R^*[1/x]$ is flat if it satisfies LF_{d-1}. Thus $\underline{\tau}$ primarily independent over R implies that $\underline{\tau}$ is primarily limit-intersecting in x over R.

24.3. Non-Noetherian intersection/approximation domains

Theorem 10.12 and Examples 10.15 yield examples where the constructed domains A and B are equal and are not Noetherian. Theorem 24.16 gives a specific example of an excellent regular local domain $(R, \mathfrak{m})$ of dimension three with $\mathfrak{m} = (x, y, z)R$ and $\widehat{R} = \mathbb{Q}[[x, y, z]]$ such that there exists an element $\tau \in yR^*$, where R^* is the (y)-adic completion of R, with τ limit-intersecting and residually limit-intersecting, but *not* primarily limit-intersecting in y over R. In this example, $B = A$ and B is non-Noetherian.

THEOREM 24.16. *Let $\mathbb{Q}[[x, y, z]]$ be the power series ring in three indeterminates x, y, z over the rational numbers $\mathbb{Q}$. Then $\mathbb{Q}[[x, y, z]]$ contains an excellent regular local three-dimensional domain $(R, \mathfrak{m})$, with maximal ideal $\mathfrak{m} = (x, y, z)R$ and an element τ in the (y)-adic completion R^* of R such that*

(24.16.1) *τ is residually limit-intersecting in y over R.*
(24.16.2) *τ is not primarily limit-intersecting in y over R.*
(24.16.3) *τ is limit-intersecting in y over R.*

In particular, the rings A and B constructed using τ as in Notation 24.2 are equal, yet A and B fail to be Noetherian.

PROOF. Define the following elements of $\mathbb{Q}[[x, y, z]]$:

$$\gamma := e^x - 1 \in x\mathbb{Q}[[x]], \qquad \delta := e^{x^2} - 1 \in x\mathbb{Q}[[x]],$$
$$\sigma := \gamma + z\delta \in \mathbb{Q}[z]_{(z)}[[x]] \qquad \text{and} \qquad \tau := e^y - 1 \in y\mathbb{Q}[[y]].$$

For each n, define the endpieces $\gamma_n, \delta_n, \sigma_n$ and τ_n as in Notation 5.4, considering γ, δ, σ as series in x and τ as a series in y. Thus, for example,

$$\gamma = \sum_{i=1}^{\infty} a_i x^i; \qquad \gamma_n = \sum_{i=n+1}^{\infty} a_i x^{i-n}, \qquad \text{and} \quad x^n\gamma_n + \sum_{i=1}^{n} a_i x^i = \gamma.$$

Here $a_i := 1/i!$. The δ_n, σ_n satisfy similar relations.

CLAIM 24.17. Define $V := \mathbb{Q}(x, \gamma, \delta) \cap \mathbb{Q}[[x]]$ and $D := \mathbb{Q}(x, z, \sigma) \cap \mathbb{Q}[z]_{(z)}[[x]]$. Then:

(1) The equalities (*1)-(*5) of Diagram 24.16.0 hold.
(2) V and $V[z]_{(x,z)}$ are excellent regular local rings, and
(3) The canonical local embedding $\psi : D \longrightarrow V[z]_{(x,z)}$ is a direct limit of the maps $\psi_n : \mathbb{Q}[x, z, \sigma_n]_{(x,z,\sigma_n)} \longrightarrow \mathbb{Q}[x, z, \gamma_n, \delta_n]_{(x,z,\gamma_n,\delta_n)}$, where $\psi_n(\sigma_n) = \gamma_n + z\delta_n$, and so is also faithfully flat.

PROOF. (of Claim 24.17) By Local Prototype Theorem 10.6,

$$(^*1): \quad V := \mathbb{Q}(x, \gamma, \delta) \cap \mathbb{Q}[[x]] = \varinjlim(\mathbb{Q}[x, \gamma_n, \delta_n]_{(x,\gamma_n,\delta_n)}) = \bigcup \mathbb{Q}[x, \gamma_n, \delta_n]_{(x,\gamma_n,\delta_n)}.$$

The equalities (*2) and (*3) follow from Theorem 10.6.1. Since V has characteristic zero, Theorem 10.6.2 implies V and $V[z]_{(x,z)}$ are excellent regular local rings.

To establish (*4) and (*5) and item 3, observe that for each positive integer n, the map

$$\psi_n : \mathbb{Q}[x, z, \sigma_n]_{(x,z,\sigma_n)} \longrightarrow \mathbb{Q}[x, z, \gamma_n, \delta_n]_{(x,z,\gamma_n,\delta_n)}$$

is faithfully flat. Thus the induced map on the direct limits:

$$\psi : D = \varinjlim \mathbb{Q}[x, z, \sigma_n]_{(x,z,\sigma_n)} \longrightarrow V[z]_{(x,z)} = \varinjlim \mathbb{Q}[x, z, \gamma_n, \delta_n]_{(x,z,\gamma_n,\delta_n)}$$

is also faithfully flat. The direct limits can also be expressed as directed unions, and the expression for $D[y]_{(x,y,z)}$ will be a similar directed union. $\square$

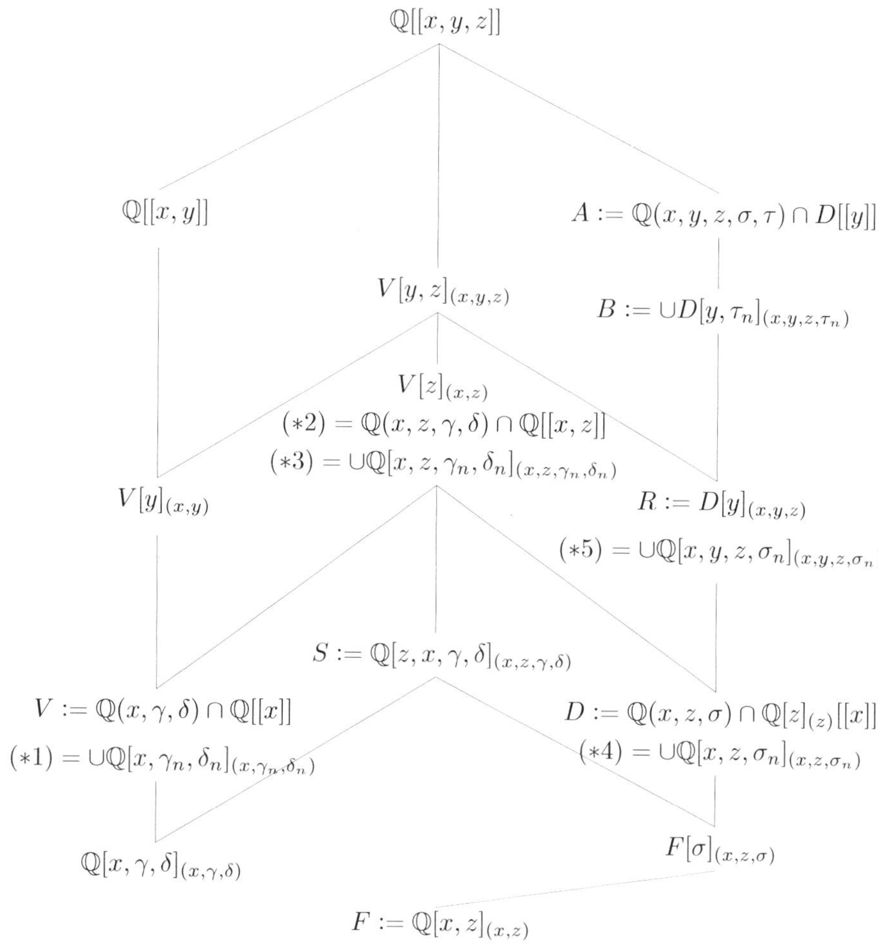

Diagram 24.16.0.

CLAIM 24.18. $D := \mathbb{Q}(x,z,\sigma) \cap \mathbb{Q}[z]_{(z)}[[x]]$ is excellent, and $R := D[y]_{(x,y,z)}$ is a three-dimensional excellent regular local domain with maximal ideal $\mathfrak{m} = (x,y,z)R$ and $\mathfrak{m}$-adic completion $\widehat{R} = \mathbb{Q}[[x,y,x]]$.

PROOF. (of Claim 24.18) By Theorem 4.9 of Valabrega, the ring

$$D := \mathbb{Q}(x,z,\sigma) \cap \mathbb{Q}[z]_{(z)}[[x]]$$

is a two-dimensional regular local domain and the completion $\widehat{D}$ of D with respect to the powers of its maximal ideal is canonically isomorphic to $\mathbb{Q}[[x,z]]$.

With an appropriate change of notation, Theorem 10.17 applies to prove Claim 24.18.

Let $F = \mathbb{Q}[x,z]_{(x,z)}$ and let F^* denote the x-adic completion of F. Consider the local injective map

$$F[\sigma]_{(x,z,\sigma)} \xrightarrow{\ \phi\ } F[\gamma,\delta]_{(x,z,\gamma,\delta)} := S.$$

Let $\phi_x : F[\sigma]_{(x,z,\sigma)} \longrightarrow S_x$ denote the composition of ϕ followed by the canonical map of S to S_x. We have the setting of Insider Construction 10.7 and Theorem 10.17, where F plays the role of R and $V[z]_{(x,z)}$ plays the role of B; here F is an excellent normal local domain.

By Theorem 10.17, to show D is excellent, it suffices to show that ϕ_x is a regular morphism. Let t_1 and t_2 be new variables over $\mathbb{Q}(x,y,z)$. The map ϕ_x may be identified as the inclusion map

$$\begin{array}{ccc} \mathbb{Q}[z,x,t_1+zt_2]_{(z,x,t_1+zt_2)} & \xrightarrow{\ \phi_x\ } & \mathbb{Q}[z,x,t_1,t_2]_{(z,x,t_1,t_2)}[1/x] \\ \downarrow{\scriptstyle \mu} & & \downarrow{\scriptstyle \nu} \\ \mathbb{Q}[z,x,\gamma+z\delta]_{(z,x,\gamma+z\delta)} & \xrightarrow{\ \phi_x\ } & \mathbb{Q}[z,x,\gamma,\delta]_{(z,x,\gamma,\delta)}[1/x] \end{array}$$

where μ and ν are the isomorphisms mapping $t_1 \to \gamma$ and $t_2 \to \delta$. Then ϕ_x is a regular morphism, since $\mathbb{Q}[z,x,t_1,t_2] = \mathbb{Q}[z,x,t_1+zt_2][t_2]$ is isomorphic to a polynomial ring in one variable over its subring $\mathbb{Q}[z,x,t_1+zt_2]$. By Theorem 10.17, D is excellent. This completes the proof of Claim 24.18. □

CLAIM 24.19. The element $\tau := e^y - 1$ is in the (y)-adic completion R^* of $R := D[y]_{(x,y,z)}$, but τ is not primarily limit-intersecting in y over R and the ring B constructed using τ is not Noetherian.

PROOF. (of Claim 24.19) Consider the height-two prime ideal $\widehat{P} := (z, y-x)\widehat{R}$ of $\widehat{R}$. Now $y \notin \widehat{P}$, so $\widehat{P}\widehat{R}_y$ is a height-two prime ideal of $\widehat{R}_y$. Moreover, the ideal $Q := \widehat{P} \cap R[\tau]_{(\mathfrak{m},\tau)}$ contains the element $\sigma - \tau$. Thus $\operatorname{ht}(Q) = 3$ and the canonical map $R[\tau]_{(\mathfrak{m},\tau)} \longrightarrow \widehat{R}_y$ is not flat. The Noetherian Flatness Theorem 6.3 implies that τ is not primarily limit-intersecting in y over R, and B is not Noetherian. □

For the completion of the proof of Theorem 24.16, it remains to show that τ is residually limit-intersecting in y over R. We first establish the following claim.

CLAIM 24.20. Let $\widehat{P}$ be a height-one prime ideal of $\widehat{R} = \mathbb{Q}[[x,y,z]]$, and suppose $\widehat{P} \cap \mathbb{Q}[[x,y]] \neq (0)$. Let $P_0 := \widehat{P} \cap R$. Then

(1) P_0 is a prime ideal of R of height at most one, and $P_0 = fR$, for some $f \in P_0$, and
(2) There exists $f_0 \in P_0 \cap \mathbb{Q}[x,y]$ such that $f_0R = fR = P_0$; that is, P_0 is extended from $\mathbb{Q}[x,y] \subseteq R$.

PROOF. We may assume P_0 is distinct from (0), xR, yR, and zR, since these are obviously extended. Since $\widehat{R}$ is faithfully flat over R and $P_0 \neq (0)$, P_0 has height one. Similarly $P_1 := \widehat{P} \cap V[y,z]_{(x,y,z)}$ has height at most one since $\widehat{R}$ is also the completion of $V[y,z]_{(x,y,z)}$. Also $P_1 \cap R = P_0$, and so P_1 is nonzero and hence has height one. The ring $R[1/x]$ is a localization of $\mathbb{Q}[x,y,z,\sigma][1/x]$, since $\sigma_n \in \mathbb{Q}[x,y,z,\sigma][1/x]$, for every $n \in \mathbb{N}$. Thus $\widehat{P} \cap \mathbb{Q}[x,y,z,\sigma]$ has height one and contains an element f that generates P_0, since R is a UFD.

Similarly, for every $n \in \mathbb{N}$, γ_n and δ_n are in $\mathbb{Q}[x,y,z,\gamma,\delta][1/x]$. Therefore the ring $V[y,z]_{(x,y,z)}[1/x]$ is a localization of $\mathbb{Q}[x,y,z,\gamma,\delta][1/x]$. Thus $\widehat{P} \cap \mathbb{Q}[x,y,z,\gamma,\delta]$

has height one and contains a generator g for P_1. Let $\widehat{h} \in \mathbb{Q}[[x,y]]$ be a generator of $\widehat{P} \cap \mathbb{Q}[[x,y]]$. Then $\widehat{h}\widehat{R} = \widehat{P}$. The following diagram illustrates this situation:

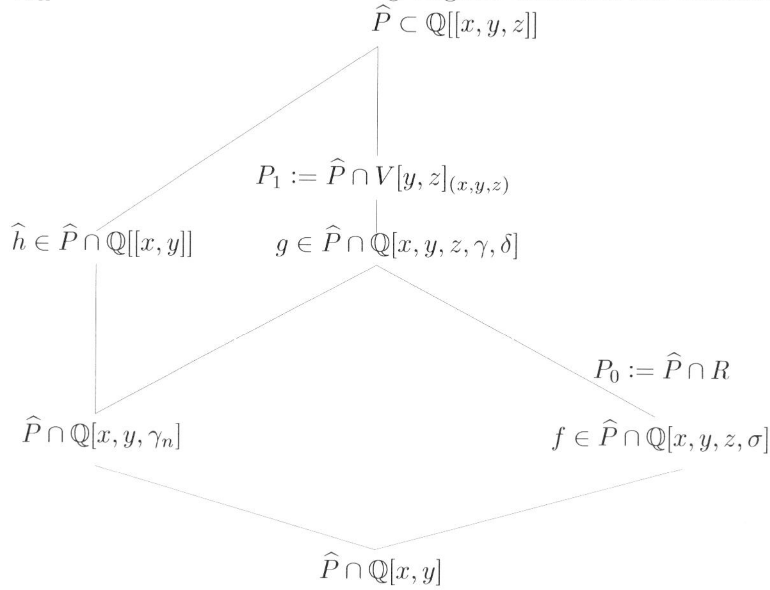

Picture for proof of (24.20)

Subclaim 1: Take g_0 to be the constant term of the generator g of $\widehat{P} \cap V[y,z]_{(x,y,z)}$ from above, with $g \in \widehat{P} \cap \mathbb{Q}[x,y,z,\gamma,\delta]$. Then $g_0 \in \widehat{P} \cap \mathbb{Q}[x,y,\gamma,\delta]$ and g_0 generates P_1.

PROOF. (of Subclaim 1): Write $g = g_0 + g_1 z + \cdots + g_r z^r$, where each $g_i \in \mathbb{Q}[x,y,\gamma,\delta]$. Since $g \in \widehat{P}$, we have $g = \widehat{h}(x,y)\phi(x,y,z)$, for some $\phi(x,y,z) \in \mathbb{Q}[[x,y,z]]$. Since g is irreducible and $P_1 \neq zV[y,z]_{(x,y,z)}$, we have $g_0 \neq (0)$.

Setting $z = 0$, we have $g_0 = g(0) = \widehat{h}(x,y)\phi(x,y,0) \in \mathbb{Q}[[x,y]]$. Thus $g_0 \in \widehat{h}\mathbb{Q}[[x,y]] \cap \mathbb{Q}[x,y,\gamma,\delta] \neq (0)$. Therefore $g_0 \in \mathbb{Q}[x,y,\gamma,\delta]$ and

$$g\mathbb{Q}[x,y,\gamma,\delta,z] = g_0\mathbb{Q}[x,y,\gamma,\delta].$$

Thus P_1 is extended from $\mathbb{Q}[x,y,\gamma,\delta]$. □

Subclaim 2: Write f as $f = f_0 + f_1 z + \cdots + f_r z^r$, where the $f_i \in \mathbb{Q}[x,y,\gamma,\delta]$. Then $f_0 \in \mathbb{Q}[x,y,\gamma]$.

PROOF. (of Subclaim 2) Since f is an element of $\mathbb{Q}[x,y,\sigma,z]$, write f as a polynomial

$$f = \sum a_{ij} z^i \sigma^j = \sum a_{ij} z^i (\gamma + z\delta)^j, \text{ where } a_{ij} \in \mathbb{Q}[x,y].$$

By setting $z = 0$, obtain $f_0 = f(0) = \sum a_{0j}(\gamma)^j \in \mathbb{Q}[x,y,\gamma]$. □

PROOF. Completion of proof of Claim 24.20. Since $f \in \widehat{P} \cap \mathbb{Q}[x,y,z,\gamma,\delta]$, it follows that $f = dg_0$, for g_0 from Subclaim 1 and some $d \in \mathbb{Q}[x,y,\gamma,\delta,z]$. Regarding d as a polynomial in z with coefficients in $\mathbb{Q}[x,y,\gamma,\delta]$ and setting $z = 0$, gives $f_0 = f(0) = d(0)g_0 \in \mathbb{Q}[x,y,\gamma,\delta]$. Thus f_0 is a multiple of g_0. Hence $g_0 \in \mathbb{Q}[x,y,\gamma]$, by Subclaim 2.

Now again use that $f = dg_0$ and set $z = 1$ to obtain that $d(1)g_0 = f(1) \in \mathbb{Q}[x, y, \gamma + \delta]$. This says that $f(1)$ is a multiple of the polynomial $g_0 \in \mathbb{Q}[x, y, \gamma]$. Since γ and δ are algebraically independent over $\mathbb{Q}[x, y]$, this implies $f(1)$ has degree 0 in $\gamma + \delta$ and g_0 has degree 0 in γ. Therefore $g_0 \in \mathbb{Q}[x, y]$, $d \in \mathbb{Q}$, $0 \neq f \in \mathbb{Q}[x, y]$, and $P_0 = fR$ is extended from $\mathbb{Q}[x, y]$. □

Claim 24.21 shows that τ is residually limit-intersecting by showing that τ satisfies part of the residual algebraic independence property.

CLAIM 24.21. (1) If $\widehat{P}$ is a height-one prime ideal of $\widehat{R}$ with $\operatorname{ht}(\widehat{P} \cap R) = 1$ and $y \notin \widehat{P}$, then the image $\bar{\tau}$ of τ in $\widehat{R}/\widehat{P}$ is algebraically independent over $R/(\widehat{P} \cap R)$.

(2) τ is residually limit-intersecting in y over R.

PROOF. (of Claim 24.21) By Proposition 24.12, item 1 $\implies$ item 2.

For item 1, let $P_0 := \widehat{P} \cap R$ and let $\pi : \mathbb{Q}[[x, y, z]] \longrightarrow \mathbb{Q}[[x, y, z]]/\widehat{P}$; we use $\bar{\ }$ to denote the image under π. If $\widehat{P} = x\widehat{R}$, then we have the commutative diagram:

$$\begin{array}{ccccc} R/P_0 & \longrightarrow & (R/P_0)[\bar{\tau}] & \longrightarrow & \mathbb{Q}[[x, y, z]]/\widehat{P} \\ \cong\uparrow & & \cong\uparrow & & \cong\uparrow \\ \mathbb{Q}[y, z]_{(y,z)} & \longrightarrow & \mathbb{Q}[y, z]_{(y,z)}[\tau] & \longrightarrow & \mathbb{Q}[[y, z]] \,. \end{array}$$

Since τ is transcendental over $\mathbb{Q}[y, z]$, $\bar{\tau}$ is algebraically independent over $R/(\widehat{P} \cap R)$ in this case. □

For the other height-one primes $\widehat{P}$ of $\widehat{R}$, we distinguish two cases:

CASE 1. $\widehat{P} \cap \mathbb{Q}[[x, y]] = (0)$.

Let $P_1 := V[y, z]_{(x,y,z)} \cap \widehat{P}$. We have the following commutative diagram of local injective morphisms:

$$\begin{array}{ccccc} R/P_0 & \longrightarrow & V[y, z]_{(x,y,z)}/P_1 & \longrightarrow & \mathbb{Q}[[x, y, z]]/\widehat{P} \\ & & \uparrow & & \uparrow \\ & & V[y]_{(x,y)} & \longrightarrow & \mathbb{Q}[[x, y]] \,, \end{array}$$

where $V[y, z]_{(x,y,z)}/P_1$ is algebraic over $V[y]_{(x,y)}$. Since $\tau \in \mathbb{Q}[[x, y]]$ is transcendental over $V[y]_{(x,y)}$, its image $\bar{\tau}$ in $\mathbb{Q}[[x, y, z]]/\widehat{P}$ is transcendental over $V[y, z]_{(x,y,z)}/P_1$ and thus is transcendental over R/P_0.

CASE 2. $\widehat{P} \cap \mathbb{Q}[[x, y]] \neq (0)$.

In this case, by Claim 24.20, the height-one prime $P_0 := \widehat{P} \cap R$ is extended from a prime ideal in $\mathbb{Q}[x, y]$. Let p be a prime element of $\mathbb{Q}[x, y]$ such that $(p) = \widehat{P} \cap \mathbb{Q}[x, y]$. We have the inclusions:

$$G := \mathbb{Q}[x, y]_{(x,y)}/(p) \hookrightarrow R/P_0 \hookrightarrow \widehat{R}/\widehat{P} \,,$$

where $R/P_0 = \varinjlim \mathbb{Q}[x, y, z, \sigma_n]_{(x,y,z,\sigma_n)}/(p)$ has transcendence degree ≤ 1 over $G[\bar{z}]_{(\bar{x},\bar{y},\bar{z})}$. It suffices to show that $\bar{\sigma}$ and $\bar{\tau}$ are algebraically independent over the field of fractions $\mathbb{Q}(\bar{x}, \bar{y}, \bar{z})$ of $G[\bar{z}]$.

Let G' be the integral closure of G in its field of fractions. Let $S := \widehat{R}/\widehat{P}$ and let S' denote the integral closure of S. Since S is complete, S' is local. Let $\mathfrak{m}$

denote the maximal ideal of S' and set $\mathfrak{n} := \mathfrak{m} \cap G'$. Define $H := G'_{\mathfrak{n}}$. Then H is a DVR dominated by the complete local ring S'. It follows that $\widehat{H}$, the $\mathfrak{n}H$-adic completion of H, is dominated by S'.

Now $H[\bar{z}]$ has transcendence degree at least one over $\mathbb{Q}[\bar{z}]$. Since $\widehat{P} \cap \mathbb{Q}[x, y] \neq 0$, the transcendence degree over $\mathbb{Q}$ of both $\mathbb{Q}[\bar{x}, \bar{y}]$ and H equals one. Thus $H[\bar{z}]$ has transcendence degree exactly one over $\mathbb{Q}(\bar{z})$. There exists an element $t \in H$ that is transcendental over $\mathbb{Q}[\bar{z}]$ and is such that t generates the maximal ideal of the DVR H. Then H is algebraic over $\mathbb{Q}[t]$ and H may be regarded as a subring of $\mathbb{C}[[t]]$, where $\mathbb{C}$ is the complex numbers. In order to show that $\bar{\sigma}$ and $\bar{\tau}$ are algebraically independent over $G[\bar{z}]$, it suffices to show that $\bar{\sigma}$ and $\bar{\tau}$ are algebraically independent over $H[\bar{z}]$ and thus it suffices to show that these elements are algebraically independent over $\mathbb{Q}(t, \bar{z})$. Thus it suffices to show that $\bar{\sigma}$ and $\bar{\tau}$ are algebraically independent over $\mathbb{C}(t, \bar{z})$.

We have the setup shown in the following diagram:

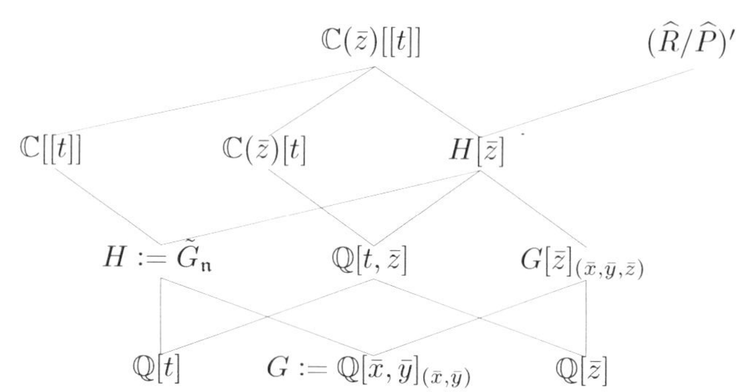

By [**18**, Corollary 1, p. 253], if $\bar{x}, \bar{x}^2, \bar{y} \in t\mathbb{C}[[t]]$ are linearly independent over $\mathbb{Q}$, then:

$$\operatorname{trdeg}_{\mathbb{C}(t)}(\mathbb{C}(t)(\bar{x}, \bar{x}^2, \bar{y}, e^{\bar{x}}, e^{\bar{x}^2}, e^{\bar{y}})) \geq 3;$$

see paragraph 2 of the proof of Theorem 12.18. Since $\bar{x}, \bar{x}^2$ and $\bar{y}$ are in H, these elements are algebraic over $\mathbb{Q}(t)$. Therefore if $\bar{x}, \bar{x}^2, \bar{y}$ are linearly independent over $\mathbb{Q}$, then the exponential functions $e^{\bar{x}}, e^{\bar{x}^2}, e^{\bar{y}}$ are algebraically independent over $\mathbb{Q}(t)$ and hence $\bar{\sigma}$ and $\bar{\tau}$ are algebraically independent over $G(\bar{z})$.

We observe that if $\bar{x}, \bar{x}^2, \bar{y} \in tH$ are linearly dependent over $\mathbb{Q}$, then there exist $a, b, c \in \mathbb{Q}$ such that

$$lp = ax + bx^2 + cy \quad \text{in} \quad \mathbb{Q}[x, y].$$

where $l \in \mathbb{Q}[x, y]$. Since $(p) \neq (x)$, we have $c \neq 0$. Hence we may assume $c = 1$ and $lp = y - ax - bx^2$ with $a, b \in \mathbb{Q}$. Since $y - ax - bx^2$ is irreducible in $\mathbb{Q}[x, y]$, we may assume $l = 1$. Also a and b cannot both be 0 since $y \notin \widehat{P}$. Thus if $\bar{x}, \bar{x}^2, \bar{y} \in tH$ are linearly dependent over $\mathbb{Q}$, then we may assume

$$p = y - ax - bx^2 \qquad \text{for some } a, b \in \mathbb{Q} \qquad \text{not both } 0.$$

It remains to show that $\bar{\sigma}$ and $\bar{\tau}$ are algebraically independent over $G(\bar{z})$ provided that $p = y - ax - bx^2$, that is

$$\bar{y} = a\bar{x} + b\bar{x}^2, \qquad \text{for } a, b \in \mathbb{Q}, \text{ not both } 0.$$

Suppose $h \in G[\bar{z}][u,v]$, where u, v are indeterminates and that $h(\bar{\sigma},\bar{\tau}) = 0$. This implies

$$h(e^{\bar{x}} + \bar{z}e^{\bar{x}^2}, e^{a\bar{x}+b\bar{x}^2}) = 0.$$

We have $e^{a\bar{x}} = (e^{\bar{x}})^a$ and $e^{b\bar{x}^2}$ are algebraic over $G(\bar{z}, e^{\bar{x}}, e^{\bar{x}^2})$ since a and b are rational. Since $p = y - ax - bx^2$, we have p is part of a regular system of parameters and the ideal $P = pQ[[x,y,z]]$ is a prime ideal. Hence we can treat $\bar{z}$ as a variable and set it equal to zero. By substituting $\bar{z} = 0$ we obtain an equation over G:

$$h(e^{\bar{x}}, e^{a\bar{x}+b\bar{x}^2}) = 0,$$

which implies that $b = 0$ since $\bar{x}$ and $\bar{x}^2$ are linearly independent over $\mathbb{Q}$. Now the only case to consider is the case where $p = y + ax$. The equation we obtain then is:

$$h(e^{\bar{x}} + \bar{z}e^{\bar{x}^2}, e^{a\bar{x}}) = 0,$$

which implies that h must be the zero polynomial, since $e^{\bar{x}^2}$ is transcendental over the algebraic closure of the field of fractions of $G[\bar{z}, e^{\bar{x}}]$. This completes the proof of Claim 24.21. □

Thus τ is residually limit-intersecting over R. Since R is a UFD, the element τ is limit-intersecting over R by Remark 24.13. This completes the proof of Theorem 24.16. □

REMARK 24.22. With notation as in Theorem 24.16, let u, v be indeterminates over $\mathbb{Q}[[x,y,z]]$. Then the height-one prime ideal $\widehat{Q} = (u - \tau)$ in $\mathbb{Q}[[x,y,z,u]]$ is in the generic formal fiber of the excellent regular local ring $R[u]_{(x,y,z,u)}$ and the intersection domain

$$B = K(u) \cap (\mathbb{Q}[[x,y,z,u]]/\widehat{Q}) \cong K(\tau) \cap \widehat{R},$$

where K is the fraction field of R, fails to be Noetherian. In a similar fashion this intersection ring $K(\tau) \cap \widehat{R}$ may be identified with the following ring: Let $\widehat{U} = (u - \tau, v - \sigma)$ be the height-two prime ideal in $\mathbb{Q}[[x,y,z,u,v]]$ that is in the generic formal fiber of the polynomial ring $\mathbb{Q}[x,y,z,u,v]_{(x,y,z,u,v)}$. Then:

$$\mathbb{Q}(x,y,z,u,v) \cap ((\mathbb{Q}[[x,y,z,u,v]])/\widehat{U}) \cong K(\tau) \cap \widehat{R}\,.$$

As shown in Theorem 24.16, this ring is not Noetherian. We do not know an example of a height-one prime ideal $\widehat{W}$ in the generic formal fiber of a polynomial ring T for which the intersection ring $A = \mathcal{Q}(T) \cap (\widehat{T}/\widehat{W})$ fails to be Noetherian. In Chapter 24, we present an example of such an intersection ring A whose completion is not equal to $\widehat{T}$. However in this example the ring A is still Noetherian.

24.4. A birational connection and additional examples

SETTING AND NOTATION 24.23. Let $(R, \mathfrak{m})$ be a Noetherian local domain with $\mathfrak{m}$-adic completion $\widehat{R}$. The field of fractions K of R canonically imbeds in the total quotient ring $\mathcal{Q}(\widehat{R})$ of $\widehat{R}$. A prime ideal $\mathfrak{p}$ of $\widehat{R}$ is in the *generic formal fiber* of R if $\mathfrak{p} \cap R = (0)$. The composite map $R \hookrightarrow \widehat{R} \to \widehat{R}/\mathfrak{p}$ is injective, and the intersection $K \cap (\widehat{R}/\mathfrak{p})$ is well defined.

In [**71**], Heinzer, Rotthaus and Sally describe an association between prime ideals in the generic formal fiber of R and birational extensions of R via Basic Construction Equation 1.2.0 For $\mathfrak{p}$ in the generic formal fiber of R, define

$$\varphi(\mathfrak{p}) := K \cap (\widehat{R}/\mathfrak{p}),$$

a birational local overring of R.

More generally, for an ideal $\mathfrak{a}$ of $\widehat{R}$ such that each associated prime ideal of $\mathfrak{a}$ is in the generic formal fiber of R, then the composite map $R \hookrightarrow \widehat{R} \to \widehat{R}/\mathfrak{a}$ is injective, and K canonically embeds in the total quotient ring of $\widehat{R}/\mathfrak{a}$, as in Construction 17.2. Hence the intersection domain

$$C := \varphi(\mathfrak{a}) := K \cap \widehat{R}/\mathfrak{a}$$

is well defined, and is a birational local overring of R.

REMARK 24.24. Assume Setting and Notation 24.23 where R is a localized polynomial ring. It is observed in [**71**, Theorem 2.5] that there exists a one-to-one correspondence between prime ideals $\widehat{\mathfrak{p}}$ of $\widehat{R}$ that are maximal in the generic formal fiber of R and DVRs C such that C birationally dominates R and $C/\mathfrak{m}C$ is a finitely generated R-module. The prime ideals maximal in the generic formal fiber of R have dimension 1 by Theorem 26.3.1. Example 24.25 demonstrates that this connection between the maximal ideals of the generic formal fiber of a localized polynomial ring R and certain birational extensions of R does not extend to prime ideals non-maximal in the generic formal fiber R.

EXAMPLE 24.25. For $S := \mathbb{Q}[x, y, z]_{(x,y,z)}$, the construction of Theorem 12.18 yields an example of a height-one prime ideal $\widehat{P}$ of $\widehat{S} = \mathbb{Q}[[x, y, z]]$ in the generic formal fiber of S. Then the canonical map $S \to \widehat{S}/\widehat{P}$ is injective. Moreover

$$\mathcal{Q}(S) \cap (\widehat{S}/\widehat{P}) = S.$$

PROOF. Let $\widehat{P} := (z - \tau)\widehat{S}$, where τ is as in Theorem 12.18. Then $\mathbb{Q}(x, y, z) \cap (\widehat{S}/\widehat{P})$ can be identified with the intersection $\mathbb{Q}(x, y, \tau) \cap \mathbb{Q}[[x, y]]$. Therefore

$$\mathbb{Q}(x, y, z) \cap (\widehat{S}/\widehat{P}) = S = \mathbb{Q}[x, y, z]_{(x,y,z)}.$$

□

With $S = \mathbb{Q}[x, y, z]_{(x,y,z)}$, every prime ideal of $\widehat{S} = \mathbb{Q}[[x, y, z]]$ that is maximal in the generic formal fiber of S has height 2 by Theorem 27.12; also see Remark 26.2.5. Thus the prime ideal $\widehat{P}$ is not maximal in the generic formal fiber of $S = \mathbb{Q}[x, y, z]_{(x,y,z)}$.

EXAMPLE 24.26. Again let $S = \mathbb{Q}[x, y, z]_{(x,y,z)}$. With a slight modification of Example 24.25, we exhibit a prime ideal $\widehat{P}$ in the generic formal fiber of S that *does* correspond to a nontrivial birational extension; that is, the intersection ring

$$A := \mathcal{Q}(S) \cap (\widehat{S}/\widehat{P})$$

is essentially finitely generated over S.

PROOF. Let τ be the element from Theorem 12.18. Let $\widehat{P} = (z - x\tau)\widehat{S}$. Since τ is transcendental over $\mathbb{Q}(x, y, z)$, the prime ideal $\widehat{P}$ is in the generic formal fiber of S. The ring S can be identified with a subring of $\widehat{S}/\widehat{P} \cong \mathbb{Q}[[x,y]]$ by replacing z by $x\tau$; thus $S = \mathbb{Q}[x,y,x\tau]_{(x,y,x\tau)}$. It follows that the Intersection Domain

$$\mathcal{Q}(S) \cap \mathbb{Q}[[x,y]] = \mathbb{Q}(x,y,\tau) \cap \mathbb{Q}[[x,y]] = \mathbb{Q}[x,y,\tau]_{(x,y,\tau)} = \mathbb{Q}[x,y,z/x]_{(x,y,z/x)}.$$

The ring $\mathbb{Q}[x,y,\tau]_{(x,y,\tau)}$ is then the essentially finitely generated birational extension of S defined as $S[z/x]_{(x,y,z/x)}$. □

EXAMPLE 24.27. Let $\sigma \in x\mathbb{Q}[[x]]$ and $\rho \in y\mathbb{Q}[[y]]$ be as in Theorem 12.18. If $D := \mathbb{Q}(x,\sigma) \cap \mathbb{Q}[[x]] = \bigcup_{n=1}^{\infty} \mathbb{Q}[x,\sigma_n]_{(x,\sigma_n)}$ and $T := D[y]_{(x,y)}$, and so T is regular local with completion $\widehat{T} = \mathbb{Q}[[x,y]]$, then the element ρ is primarily limit-intersecting in y over T.

PROOF. Since $\dim T = 2$, Remark 24.7.6 implies ρ is primarily limit-intersecting in y over T if $\varphi_y : T[\rho] \longrightarrow \mathbb{Q}[[x,y]][1/y]$ is LF_1. That is, the induced map $\varphi_{\widehat{P}} : T[\rho]_{\widehat{P} \cap T[\rho]} \longrightarrow \mathbb{Q}[[x,y]]_{\widehat{P}}$ is flat for every height-one prime ideal $\widehat{P}$ of $\mathbb{Q}[[x,y]]$ with $y \notin \widehat{P}$. It is equivalent to show for every height-one prime $\widehat{P}$ of $\mathbb{Q}[[x,y]]$ that $\widehat{P} \cap T[\rho]$ has height ≤ 1. If $\widehat{P} = (x)$, the statement is immediate, since ρ is algebraically independent over $\mathbb{Q}(y)$. Next we consider the case $\widehat{P} \cap Q[x,y,\sigma] = (0)$. Since $\mathbb{Q}(x,y,\sigma) = \mathbb{Q}(x,y,\sigma_n)$ for every positive integer n, $\widehat{P} \cap Q[x,y,\sigma] = (0)$ if and only if $\widehat{P} \cap Q[x,y,\sigma_n] = (0)$. Moreover, if this is true, then since the field of fractions of $T[\rho]$ has transcendence degree one over $\mathbb{Q}(x,y,\sigma)$, then $\widehat{P} \cap T[\rho]$ has height ≤ 1. The remaining case is where $P := \widehat{P} \cap Q[x,y,\sigma] \neq (0)$ and $xy \notin \widehat{P}$. By Proposition 6.3, $\bar{\rho}$ is transcendental over $\bar{T} = T/(\widehat{P} \cap T)$, and this is equivalent to $\operatorname{ht}(\widehat{P} \cap T[\tau]) = 1$, by Proposition 24.12. □

Still referring to ρ, σ, σ_n as in Theorem 12.18 and Example 24.27, and using that σ is primarily limit-intersecting in y over T, we have:

$$A := \mathcal{Q}(T)(\rho) \cap \mathbb{Q}[[x,y]] = \varinjlim T[\rho_n]_{(x,y,\rho_n)} = \varinjlim \mathbb{Q}[x,y,\sigma_n,\rho_n]_{(x,y,\sigma_n,\rho_n)}$$

where the endpieces ρ_n are defined as in Section 5.4; viz., $\rho := \sum_{n=1}^{\infty} b_i z^i$ and $\rho_n = \sum_{i=n+1}^{\infty} b_i z^{i-n}$. The philosophy here is that sufficient "independence" of the algebraically independent elements σ and ρ allows us to explicitly describe the intersection ring A.

The previous examples have been over localized polynomial rings, where we are free to exchange variables. The next example shows, over a different regular local domain, that an element in the completion with respect to one regular parameter x may be residually limit-intersecting with respect to x whereas the corresponding element in the completion with respect to another regular parameter y may be transcendental but fail to be residually limit-intersecting.

EXAMPLE 24.28. There exists a regular local ring R with $\widehat{R} = \mathbb{Q}[[x,y]]$ such that $\sigma = e^x - 1$ is residually limit-intersecting in x over R, whereas $\gamma = e^y - 1$ fails to be limit-intersecting in y over R.

PROOF. Let $\{\omega_i\}_{i \in I}$ be a transcendence basis of $\mathbb{Q}[[x]]$ over $\mathbb{Q}(x)$ such that:

$$\{e^{x^n}\}_{n \in \mathbb{N}} \subseteq \{\omega_i\}_{i \in I}.$$

Let D be the discrete valuation ring:

$$D = \mathbb{Q}(x, \{\omega_i\}_{i \in I, \omega_i \neq e^x}) \cap \mathbb{Q}[[x]].$$

Obviously, $\mathbb{Q}[[x]]$ has transcendence degree 1 over D. The set $\{e^x\}$ is a transcendence basis of $\mathbb{Q}[[x]]$ over D. Let $R = D[y]_{(x,y)}$.

By Remark 24.9.1, the element $\sigma = e^x - 1$ is primarily limit-intersecting and hence residually limit-intersecting in x over D. Moreover, by Remark 24.9.2, σ is also primarily and hence residually limit-intersecting over $R := D[y]_{(x,y)}$. However, the element $\gamma = e^y - 1$ is not residually limit-intersecting in y over R. To see this, consider the height-one prime ideal $P := (y - x^2)\mathbb{Q}[[x, y]]$. The prime ideal $W := P \cap R[\gamma]_{(x,y,\gamma)}$ contains the element $\gamma - e^{x^2} + 1 = e^y - e^{x^2}$. Therefore W has height greater than one and γ is not residually limit-intersecting in y over R. $\square$

Note that the intersection ring $\mathcal{Q}(R)(\gamma) \cap \mathbb{Q}[[x, y]]$ is a regular local ring with completion $\mathbb{Q}[[x, y]]$ by Theorem 4.9, a theorem of Valabrega.

Exercises

(1) Let A be a Krull domain and let x be a nonunit of A. Prove that $\bigcap x^n A = (0)$.

(2) Prove Remark 24.7.8c: With Notation 24.2, if $B_0 \hookrightarrow R^*$ is weakly flat, so is $R \hookrightarrow R^*$, and Assumption 24.3 holds.

SUGGESTION. Show that $U_0 \hookrightarrow R^*$ is weakly flat implies the result.

CHAPTER 25

Inclusion Constructions over excellent normal local domains

Let $(R, \mathfrak{m})$ be an excellent normal local domain. Let x be a nonzero element in $\mathfrak{m}$ and let R^* denote the x-adic completion of R. In this chapter we consider certain extension domains A inside R^* arising from Inclusion Construction 5.3. We use test criteria given in Theorem 7.3, Theorem 7.4 and Corollary 7.6, involving the heights of certain prime ideals to determine flatness for the map φ defined in Equation 25.1.0. These characterizations of flatness involve the condition that certain fibers are Cohen-Macaulay and other fibers are regular.

We give in Theorem 25.12 and Remarks 25.14 necessary and sufficient conditions for an element $\tau \in xR^*$ to be primarily limit-intersecting in x over R; see Remark 25.2. If R is countable, we prove in Theorem 25.19 the existence of an infinite sequence of elements of xR^* that are primarily limit-intersecting in x over R. Using this result we establish the existence of a normal Noetherian local domain B such that: B dominates R; B has x-adic completion R^*; and B contains a height-one prime ideal $\mathfrak{p}$ such that $R^*/\mathfrak{p}R^*$ is not reduced. Thus B is not a Nagata domain and hence is not excellent; see Remark 3.48.

25.1. Primarily limit-intersecting extensions and flatness

In this section, we consider properties of Inclusion Construction 5.3 under the assumptions of Setting 25.1.

SETTING 25.1. Let $(R, \mathfrak{m})$ be an excellent normal local domain and let x be a nonzero element in $\mathfrak{m}$. Let $(R^*, \mathfrak{m}^*)$ be the x-adic completion of R and let $(\widehat{R}, \widehat{\mathfrak{m}})$ be the $\mathfrak{m}$-adic completion of R. Thus R^* and $\widehat{R}$ are normal Noetherian local domains and $\widehat{R}$ is the $\mathfrak{m}^*$-adic completion of R^*. Let $\tau_1, \ldots, \tau_s$ be elements of xR^* that are algebraically independent over R, and set $U_0 = S := R[\tau_1, \ldots, \tau_s]$. The field of fractions L of S is a subfield of the field of fractions $\mathcal{Q}(R^*)$ of R^*. Define $A := L \cap R^*$.

REMARK 25.2. Noetherian Flatness Theorem 6.3 implies that $A = L \cap R^*$ is both Noetherian and a localization of a subring of $S[1/x]$ if and only if the extension φ is flat, where

$$(25.1.0) \qquad \varphi : S \longrightarrow R^*[1/x]$$

By Definition 24.6.3, the elements $\tau_1, \ldots, \tau_s$ are *primarily limit-intersecting* in x over R if and only if φ is flat.

THEOREM 25.3. *Assume notation as in Setting* 25.1. *That is,* $(R, \mathfrak{m})$ *is an excellent normal local domain,* x *is a nonzero element in* $\mathfrak{m}$*,* $(R^*, \mathfrak{m}^*)$ *is the* x*-adic* *completion of R, and the elements $\tau_1, \dots, \tau_s \in xR^*$ are algebraically independent over R. Then the following statements are equivalent:*

(1) $S := R[\tau_1, \dots, \tau_s] \hookrightarrow R^*[1/x]$ *is flat. Equivalently, $\tau_1, \dots, \tau_s$ are primarily limit-intersecting in x over R^*.*
(2) *For P a prime ideal of S and Q^* a prime ideal of R^* minimal over PR^*, if $x \notin Q^*$, then* $\operatorname{ht}(Q^*) = \operatorname{ht}(P)$.
(3) *If Q^* is a prime ideal of R^* with $x \notin Q^*$, then* $\operatorname{ht}(Q^*) \geq \operatorname{ht}(Q^* \cap S)$.

Moreover, if any of (1)-(3) hold, then $S \hookrightarrow R^*[1/x]$ has Cohen-Macaulay fibers.

PROOF. By Remark 25.2, we have the equivalence in item 1.

(1) $\Rightarrow$ (2): Let P be a prime ideal of S and let Q^* be a prime ideal of R^* that is minimal over PR^* and is such that $x \notin Q^*$. The assumption of item 1 implies flatness of the map:

$$\varphi_{Q^*} : S_{Q^* \cap S} \longrightarrow R^*_{Q^*}.$$

By Remark 2.37.10, we have $Q^* \cap S = P$, and by [**122**, Theorem 15.1], $\operatorname{ht} Q^* = \operatorname{ht} P$.

(2) $\Rightarrow$ (3): Let Q^* be a prime ideal of R^* with $x \notin Q^*$. Set $Q := Q^* \cap S$ and let $\mathbf{w}^*$ be a prime ideal of R^* that is minimal over QR^* and is contained in Q^*. Then $\operatorname{ht}(Q) = \operatorname{ht}(\mathbf{w}^*)$ by (2) since $x \notin \mathbf{w}^*$ and therefore $\operatorname{ht}(Q^*) \geq \operatorname{ht}(Q)$.

(3) $\Rightarrow$ (1): Let Q^* be a prime ideal of R^* with $x \notin Q^*$. Then for every prime ideal $\mathbf{w}^*$ of R^* contained in Q^*, we also have $x \notin \mathbf{w}^*$, and by (3), $\operatorname{ht}(\mathbf{w}^*) \geq \operatorname{ht}(\mathbf{w}^* \cap S)$. Therefore, by Theorem 7.4, $\varphi_{Q^*} : S_{Q^* \cap S} \longrightarrow R^*_{Q^*}$ is flat with Cohen-Macaulay fibers. □

With notation as in Setting 25.1, the map $R^* \hookrightarrow \widehat{R}$ is flat. Hence the corresponding statements in Theorem 25.3 with R^* replaced by $\widehat{R}$ also hold. We record this as

COROLLARY 25.4. *Assume notation as in Setting 25.1. Then the following statements are equivalent:*

(1) $S := R[\tau_1, \dots, \tau_s] \hookrightarrow \widehat{R}[1/x]$ *is flat.*
(2) *For P a prime ideal of S and $\widehat{Q}$ a prime ideal of $\widehat{R}$ minimal over $P\widehat{R}$, if $x \notin \widehat{Q}$, then* $\operatorname{ht}(\widehat{Q}) = \operatorname{ht}(P)$.
(3) *If $\widehat{Q}$ is a prime ideal of $\widehat{R}$ with $x \notin \widehat{Q}$, then* $\operatorname{ht}(\widehat{Q}) \geq \operatorname{ht}(\widehat{Q} \cap S)$.

Moreover, if any of (1)-(3) hold, then $S \hookrightarrow \widehat{R}[1/x]$ has Cohen-Macaulay fibers.

The following is another corollary to Theorem 25.3.

COROLLARY 25.5. *With the notation of Theorem 25.3, assume that $\widehat{R}[1/x]$ is flat over S. Let $P \in \operatorname{Spec} S$ with $\operatorname{ht} P \geq \dim R$. Then*

(1) *For every $\widehat{Q} \in \operatorname{Spec} \widehat{R}$ minimal over $P\widehat{R}$ we have $x \in \widehat{Q}$.*
(2) *Some power of x is in $P\widehat{R}$.*

PROOF. Clearly items 1 and 2 are equivalent. To prove these hold, suppose that $x \notin \widehat{Q}$. By Theorem 25.3.2, $\operatorname{ht}(P) = \operatorname{ht}(\widehat{Q})$. Since $\dim(R) = \dim(\widehat{R})$, we have $\operatorname{ht}(\widehat{Q}) \geq \dim(\widehat{R})$. But then $\operatorname{ht}(\widehat{Q}) = \dim(\widehat{R})$ and $\widehat{Q}$ is the maximal ideal of $\widehat{R}$. This contradicts the assumption that $x \notin \widehat{Q}$. We conclude that $x \in \widehat{Q}$. □

Theorem 25.3, together with results from Chapter 6, gives Corollary 25.6.

COROLLARY 25.6. *Assume notation as in Setting 25.1, and consider the following conditions:*

(1) *A is Noetherian and is a localization of a subring of $S[1/x]$.*
(2) *$S \hookrightarrow R^*[1/x]$ is flat.*
(3) *$S \hookrightarrow R^*[1/x]$ is flat with Cohen-Macaulay fibers.*
(4) *For every $Q^* \in \operatorname{Spec}(R^*)$ with $x \notin Q^*$, we have $\operatorname{ht}(Q^*) \geq \operatorname{ht}(Q^* \cap S)$.*
(5) *A is Noetherian.*
(6) *$A \hookrightarrow R^*$ is flat.*
(7) *$A \hookrightarrow R^*[1/x]$ is flat.*
(8) *$A \hookrightarrow R^*[1/x]$ is flat with Cohen-Macaulay fibers.*

Conditions (1)-(4) are equivalent, conditions (5)-(8) are equivalent and (1)-(4) imply (5)-(8).

PROOF. Item 1 is equivalent to item 2 by Noetherian Flatness Theorem 17.13, item 2 is equivalent to item 3 and item 7 is equivalent to item 8 by Theorem 7.4, and item 2 is equivalent to item 4 by Theorem 25.3.

It is obvious that item 1 implies item 5. By Construction Properties Theorem 5.14.3, the ring R^* is the x-adic completion of A, and so item 5 is equivalent to item 6. By Lemma 6.2).1, item 6 is equivalent to item 7. □

REMARKS 25.7. (1) With the notation of Corollary 25.6, if $\dim A = 2$, it follows that condition (7) of Corollary 25.6 holds. Since R^* is normal, so is A. Thus if $Q^* \in \operatorname{Spec} R^*$ with $x \notin Q^*$, then $A_{Q^* \cap A}$ is either a DVR or a field. The map $A \to R^*_{Q^*}$ factors as $A \to A_{Q^* \cap A} \to R^*_{Q^*}$. Since $R^*_{Q^*}$ is a torsionfree and hence flat $A_{Q^* \cap A}$-module, it follows that $A \to R^*_{Q^*}$ is flat. Therefore $A \hookrightarrow R^*[1/x]$ is flat and A is Noetherian.

(2) There exist examples where $\dim A = 2$ and conditions (5)-(8) of Corollary 25.6 hold, but yet conditions (1)-(4) fail to hold; see Theorem 12.3.

QUESTION 25.8. With the notation of Corollary 25.6, suppose for every prime ideal Q^* of R^* with $x \notin Q^*$ that $\operatorname{ht}(Q^*) \geq \operatorname{ht}(Q^* \cap A)$. Does it follow that R^* is flat over A or, equivalently, that A is Noetherian?

Theorem 25.3 also extends to give equivalences for the *locally flat in height k* property; see Definitions 24.1.

THEOREM 25.9. *Assume notation as in Setting 25.1. That is, $(R, \mathfrak{m})$ is an excellent normal local domain, x is a nonzero element in $\mathfrak{m}$, $(R^*, \mathfrak{m}^*)$ is the x-adic completion of R, and the elements $\tau_1, \ldots, \tau_s \in xR^*$ are algebraically independent over R. Then the following statements are equivalent:*

(1) *$S := R[\tau_1, \ldots, \tau_s] \hookrightarrow \widehat{R}[1/x]$ is LF_k.*
(2) *If P is a prime ideal of S and $\widehat{Q}$ is a prime ideal of $\widehat{R}$ minimal over $P\widehat{R}$ and if, moreover, $x \notin \widehat{Q}$ and $\operatorname{ht}(\widehat{Q}) \leq k$, then $\operatorname{ht}(\widehat{Q}) = \operatorname{ht}(P)$.*
(3) *If $\widehat{Q}$ is a prime ideal of $\widehat{R}$ with $x \notin \widehat{Q}$ and $\operatorname{ht}(\widehat{Q}) \leq k$, then $\operatorname{ht}(\widehat{Q}) \geq \operatorname{ht}(\widehat{Q} \cap S)$.*

PROOF. (1) $\Rightarrow$ (2): Let P be a prime ideal of S and let $\widehat{Q}$ be a prime ideal of $\widehat{R}$ that is minimal over $P\widehat{R}$ with $x \notin \widehat{Q}$ and $\operatorname{ht}(\widehat{Q}) \leq k$. The assumption of item 1 implies flatness for the map:

$$\varphi_{\widehat{Q}} : S_{\widehat{Q}\cap S} \longrightarrow \widehat{R}_{\widehat{Q}},$$

and we continue as in Theorem 25.3.

(2) $\Rightarrow$ (3): Let $\widehat{Q}$ be a prime ideal of $\widehat{R}$ with $x \notin \widehat{Q}$ and $\operatorname{ht}(\widehat{Q}) \le k$. Set $Q := \widehat{Q} \cap S$ and let $\widehat{W}$ be a prime ideal of $\widehat{R}$ that is minimal over $Q\widehat{R}$, and so that $\widehat{W} \subseteq \widehat{Q}$. Then $\operatorname{ht}(Q) = \operatorname{ht}(\widehat{W})$ by item 2 since $x \notin \widehat{W}$ and therefore $\operatorname{ht}(\widehat{Q}) \ge \operatorname{ht}(Q)$.

(3) $\Rightarrow$ (1): Let $\widehat{Q}$ be a prime ideal of $\widehat{R}$ with $x \notin \widehat{Q}$ and $\operatorname{ht}(\widehat{Q}) \le k$. Then for every prime ideal $\widehat{W}$ contained in $\widehat{Q}$, we also have $x \notin \widehat{W}$ and $\operatorname{ht}(\widehat{W}) \ge \operatorname{ht}(\widehat{W} \cap S)$, by item 3. To complete the proof it suffices to show that $\varphi_{\widehat{Q}} : S_{\widehat{Q}\cap S} \longrightarrow \widehat{R}_{\widehat{Q}}$ is flat, and this is a consequence of Theorem 7.4. □

25.2. Existence of primarily limit-intersecting extensions

In this section, we establish the existence of primary limit-intersecting elements over countable excellent normal local domains.To do this, we use the following prime avoidance lemma that is analogous to Lemma 22.18, but avoids the hypothesis of Lemma 22.18 that T is complete in its $\mathfrak{n}$-adic topology. See the articles [**30**], [**171**], [**186**] and the book [**110**, Lemma 14.2] for other prime avoidance results involving countably infinitely many prime ideals.

LEMMA 25.10. *Let $(T, \mathfrak{n})$ be a Noetherian local domain that is complete in the x-adic topology, where x is a nonzero element of $\mathfrak{n}$. Let $\mathcal{U}$ be a countable set of prime ideals of T such that $x \notin P$ for each $P \in \mathcal{U}$, and fix an arbitrary element $t \in \mathfrak{n} \setminus \mathfrak{n}^2$. Then there exists an element $a \in x^2T$ such that $t - a \notin \bigcup\{P : P \in \mathcal{U}\}$.*

PROOF. We may assume there are no inclusion relations among the $P \in \mathcal{U}$. We enumerate the prime ideals in $\mathcal{U}$ as $\{P_i\}_{i=1}^{\infty}$. We choose $b_2 \in T$ so that $t - b_2x^2 \notin P_1$ as follows: (i) if $t \in P_1$, let $b_2 = 1$. Since $x \notin P_1$, we have $t - x^2 \notin P_1$. (ii) if $t \notin P_1$, let b_2 be a nonzero element of P_1. Then $t - b_2x^2 \notin P_1$. Assume by induction that we have found $b_2, \dots, b_n$ in T such that

$$t - cx^2 \;:=\; t - b_2x^2 - \cdots - b_nx^n \;\notin\; P_1 \cup \cdots \cup P_{n-1}.$$

We choose $b_{n+1} \in T$ so that $t - cx^2 - b_{n+1}x^{n+1} \notin \bigcup_{i=1}^{n} P_i$ as follows: (i) if $t - cx^2 \in P_n$, let $b_{n+1} \in (\prod_{i=1}^{n-1} P_i) \setminus P_n$. (ii) if $t - cx^2 \notin P_n$, let b_{n+1} be any nonzero element in $\prod_{i=1}^{n} P_i$. Hence in either case there exists $b_{n+1} \in T$ so that

$$t - b_2x^2 - \cdots - b_{n+1}x^{n+1} \;\notin P_1 \cup \cdots \cup P_n.$$

Since T is complete in the x-adic topology, the Cauchy sequence

$$\{b_2x^2 + \cdots + b_nx^n\}_{n=2}^{\infty}$$

has a limit $a \in \mathfrak{n}^2$. Since T is Noetherian and local, every ideal of T is closed in the x-adic topology. Hence, for each integer $n \ge 2$, we have

$$t - a \;=\; (t - b_2x^2 - \cdots - b_nx^n) \;-\; (b_{n+1}x^{n+1} + \cdots),$$

where $t - b_2x^2 - \cdots - b_nx^n \notin P_{n-1}$ and $(b_{n+1}x^{n+1} + \cdots) \in P_{n-1}$. We conclude that $t - a \notin \bigcup_{i=1}^{\infty} P_i$. □

We use the following setting to describe necessary and sufficient conditions for an element to be primarily limit-intersecting.

SETTING 25.11. Let $(R, \mathfrak{m})$ be a d-dimensional excellent normal local domain with $d \geq 2$, let x be a nonzero element of $\mathfrak{m}$ and let R^* denote the x-adic completion of R. Let t be a variable over R, let $S := R[t]_{(\mathfrak{m},t)}$, and let S^* denote the I-adic completion of S, where $I := (x,t)S$. Then $S^* = R^*[[t]]$ is a $(d+1)$-dimensional normal Noetherian local domain with maximal ideal $\mathfrak{n}^* := (\mathfrak{m}, t)S^*$. For each element $a \in x^2S^*$, we have $S^* = R^*[[t]] = R^*[[t-a]]$. Let $\lambda_a : S^* \to R^*$ denote the canonical homomorphism $S^* \to S^*/(t-a)S^* = R^*$, and let $\tau_a = \lambda_a(t) = \lambda_a(a)$. Consider the set

$$\mathcal{U} := \{P^* \in \operatorname{Spec} S^* \mid \operatorname{ht}(P^* \cap S) = \operatorname{ht} P^*, \text{ and } x \notin P^* \}.$$

Since $S \hookrightarrow S^*$ is flat and thus satisfies the Going-down property, the set $\mathcal{U}$ can also be described as the set of all $P^* \in \operatorname{Spec} S^*$ such that $x \notin P^*$ and P^* is minimal over PS^* for some $P \in \operatorname{Spec} S$, see [**122**, Theorem 15.1]

THEOREM 25.12. *With the notation of Setting* 25.11, *the element* τ_a *is primarily limit-intersecting in* x *over* R *if and only if* $t - a \notin \bigcup\{P^* \mid P^* \in \mathcal{U}\}$.

PROOF. Consider the commutative diagram:

$$\begin{array}{ccccccc}
 & & S = R[t]_{(\mathfrak{m},t)} & \xrightarrow{\subseteq} & S^* = R^*[[t]] & \xrightarrow{\subseteq} & S^*[1/x] \\
 & & \lambda_0 \downarrow & & \lambda_a \downarrow & & \\
R & \xrightarrow{\subseteq} & R_1 = R[\tau_a]_{(\mathfrak{m},\tau_a)} & \longrightarrow & R^* & \xrightarrow{\subseteq} & R^*[1/x].
\end{array}$$

Diagram 25.12.0

The map λ_0 denotes the restriction of λ_a to S.

Assume that τ_a is primarily limit-intersecting in x over R. Then τ_a is algebraically independent over R and λ_0 is an isomorphism. If $t - a \in P^*$ for some $P^* \in \mathcal{U}$, we prove that $\varphi : R_1 \to R^*[1/x]$ is not flat. Let $Q^* := \lambda_a(P^*)$. We have $\operatorname{ht} Q^* = \operatorname{ht} P^* - 1$, and $x \notin P^*$ implies $x \notin Q^*$. Let $P := P^* \cap S$ and $Q := Q^* \cap R_1$. Commutativity of Diagram 25.12.0 and λ_0 an isomorphism imply that $\operatorname{ht} P = \operatorname{ht} Q$. Since $P^* \in \mathcal{U}$, we have $\operatorname{ht} P = \operatorname{ht} P^*$. It follows that $\operatorname{ht} Q > \operatorname{ht} Q^*$. This implies that $\varphi : R_1 \to R^*[1/x]$ is not flat.

For the converse, assume that $t - a \notin \bigcup\{P^* \mid P^* \in \mathcal{U}\}$. Since $a \in x^2S^*$ and S^* is complete in the (x,t)-adic topology, we have $S^* = R^*[[t]] = R^*[[t-a]]$. Thus

$$\mathfrak{p} := \ker(\lambda_a) = (t - \tau_a)S^* = (t-a)S^*$$

is a height-one prime ideal of S^*. Since $x \in R$ and $\mathfrak{p} \cap R = (0)$, we have $x \notin \mathfrak{p}$.

Since $t - a$ is outside every element of $\mathcal{U}$, we have $\mathfrak{p} \notin \mathcal{U}$. Since $\mathfrak{p}$ does not fit the condition of $\mathcal{U}$, we have $\operatorname{ht}(\mathfrak{p} \cap S) \neq \operatorname{ht} \mathfrak{p} = 1$, and so, by the faithful flatness of $S \hookrightarrow S^*$, $\mathfrak{p} \cap S = (0)$. Therefore the map $\lambda_0 : S \to R_1$ has trivial kernel, and so λ_0 is an isomorphism. Thus τ_a is algebraically independent over R.

Since R is excellent and R_1 is a localized polynomial ring over R, the hypotheses of Corollary 7.6 are satisfied. It follows that the element τ_a is primarily limit-intersecting in x over R provided that $\operatorname{ht}(Q_1^* \cap R_1) \leq \operatorname{ht} Q_1^*$ for every prime ideal $Q_1^* \in \operatorname{Spec}(R^*[1/x])$, or, equivalently, if for every $Q^* \in \operatorname{Spec} R^*$ with $x \notin Q^*$, we have $\operatorname{ht}(Q^* \cap R_1) \leq \operatorname{ht} Q^*$. Thus, to complete the proof of Theorem 25.12, it suffices to prove Claim 25.13. □

CLAIM 25.13. For every prime ideal $Q^* \in \operatorname{Spec} R^*$ with $x \notin Q^*$, we have

$$\operatorname{ht}(Q^* \cap R_1) \leq \operatorname{ht} Q^*.$$

PROOF. (of Claim 25.13) Since $\dim R^* = d$ and $x \notin Q^*$, we have $\operatorname{ht} Q^* = r \le d-1$. Since the map $R \hookrightarrow R^*$ is flat, we have $\operatorname{ht}(Q^* \cap R) \le \operatorname{ht} Q^* = r$. Suppose that $Q := Q^* \cap R_1$ has height at least $r+1$ in $\operatorname{Spec} R_1$. Since R_1 is a localized polynomial ring in one variable over R and $\operatorname{ht}(Q \cap R) \le r$, we have $\operatorname{ht}(Q) = r+1$. Let $P := \lambda_0^{-1}(Q) \in \operatorname{Spec} S$. Then $\operatorname{ht} P = r+1$ and $x \notin P$.

Let $P^* := \lambda_a^{-1}(Q^*)$. Since the prime ideals of S^* that contain $t-a$ and have height $r+1$ are in one-to-one correspondence with the prime ideals of R^* of height r, we have $\operatorname{ht} P^* = r+1$. By the commutativity of the diagram, we also have $x \notin P^*$ and $P \subseteq P^* \cap S$, and so

$$r+1 = \operatorname{ht} P \le \operatorname{ht}(P^* \cap S) \le \operatorname{ht} P^* = r+1,$$

where the last inequality holds because the map $S \hookrightarrow S^*$ is flat. It follows that $P = P^* \cap S$, and so $P^* \in \mathcal{U}$. This contradicts the fact that $t-a \notin P_1^*$ for each $P_1^* \in \mathcal{U}$. Thus we have $\operatorname{ht}(Q^* \cap R_1) \le r = \operatorname{ht} Q^*$, as asserted in Claim 25.13. This completes the proof of Theorem 25.12. □

Theorem 25.12 yields a necessary and sufficient condition for an element of R^* that is algebraically independent over R to be primarily limit-intersecting in x over R.

REMARKS 25.14. Assume notation as in Setting 25.11.

(1) For each $a \in x^2S^*$ as in Setting 25.11, we have $(t-a)S^* = (t-\tau_a)S^*$. Hence $t - a \notin \bigcup\{P^* \mid P^* \in \mathcal{U}\} \iff t-\tau_a \notin \bigcup\{P^* \mid P^* \in \mathcal{U}\}$.

(2) If $a \in R^*$, then the commutativity of Diagram 25.12.0 implies that $\tau_a = a$.

(3) For $\tau \in R^*$, we have $\tau = a_0 + a_1x + \tau'$, where a_0 and a_1 are in R and $\tau' \in x^2R^*$.
   (a) The rings $R[\tau]$ and $R[\tau']$ are equal. Hence τ is primarily limit-intersecting in x over R if and only if τ' is primarily limit-intersecting in x over R.
   (b) Assume $\tau \in R^*$ is algebraically independent over R. Then τ is primarily limit-intersecting in x over R if and only if $t-\tau' \notin \bigcup\{P^* \mid P^* \in \mathcal{U}\}$.

   Item 3b follows from Theorem 25.12 by setting $a = \tau'$ and applying item 3a and item 2.

We use Theorem 25.12 and Lemma 25.10 to prove Theorem 25.15.

THEOREM 25.15. *Let $(R, \mathfrak{m})$ be a countable excellent normal local domain with dimension $d \ge 2$, and let x be a nonzero element in $\mathfrak{m}$.Let R^* denote the x-adic completion of R. Then there exists an element $\tau \in xR^*$ that is primarily limit-intersecting in x over R.*

PROOF. As in Setting 25.11, let

$$\mathcal{U} := \{P^* \in \operatorname{Spec} S^* \mid \operatorname{ht}(P^* \cap S) = \operatorname{ht} P^*, \quad \text{and } x \notin P^* \}.$$

Since the ring S is countable and Noetherian, the set $\mathcal{U}$ is countable. Lemma 22.18 implies that there exists an element $a \in x^2S^*$ such that $t-a \notin \bigcup\{P^* \mid P^* \in \mathcal{U}\}$. By Theorem 25.12, the element τ_a is primarily limit-intersecting in x over R. □

To establish the existence of more than one primarily limit-intersecting element we use the following setting.

SETTING 25.16. Let $(R, \mathfrak{m})$ be a d-dimensional excellent normal local domain, let x be a nonzero element of $\mathfrak{m}$ and let R^* denote the x-adic completion of R. Let $t_1, \ldots, t_{n+1}$ be indeterminates over R, and let S_n and S_{n+1} denote the localized polynomial rings

$$S_n := R[t_1, \ldots, t_n]_{(\mathfrak{m}, t_1, \ldots, t_n)} \quad \text{and} \quad S_{n+1} := R[t_1, \ldots, t_{n+1}]_{(\mathfrak{m}, t_1, \ldots, t_{n+1})}.$$

Let S_n^* denote the I_n-adic completion of S_n, where $I_n := (x, t_1, \ldots, t_n)S_n$. Then $S_n^* = R^*[[t_1, \ldots, t_n]]$ is a $(d+n)$-dimensional normal Noetherian local domain with maximal ideal $\mathfrak{n}^* = (\mathfrak{m}, t_1, \ldots, t_n)S_n^*$. Assume that $\tau_1, \ldots, \tau_n \in xR^*$ are primarily limit-intersecting in x over R, and define $\lambda : S_n^* \to R^*$ to be the R^*-algebra homomorphism such that $\lambda(t_i) = \tau_i$, for $1 \leq i \leq n$.

Since $S_n^* = R^*[[t_1 - \tau_1, \ldots, t_n - \tau_n]]$, we have

$$\mathfrak{p}_n := \ker \lambda = (t_1 - \tau_1, \ldots, t_n - \tau_n)S_n^*.$$

Consider the commutative diagram:

$$\begin{array}{ccccccc}
 & & S_n = R[t_1, \ldots, t_n]_{(\mathfrak{m}, t_1, \ldots, t_n)} & \xrightarrow{\subseteq} & S_n^* = R^*[[t_1, \ldots, t_n]] & \xrightarrow{\subseteq} & S_n^*[1/x] \\
 & & \Big\downarrow {\scriptstyle \lambda_0,\ \cong} & & \Big\downarrow {\scriptstyle \lambda} & & \\
R & \xrightarrow{\subseteq} & R_n = R[\tau_1, \ldots, \tau_n]_{(\mathfrak{m}, \tau_1, \ldots, \tau_n)} & \xrightarrow{\varphi_0} & R^* & \xrightarrow{\alpha} & R^*[1/x].
\end{array}$$

Let S_{n+1}^* denote the I_{n+1}-adic completion of S_{n+1}, where

$$I_{n+1} := (x, t_1, \ldots, t_{n+1})S_{n+1}.$$

For each element $a \in x^2 S_{n+1}^*$, we have

$$S_{n+1}^* = S_n^*[[t_{n+1}]] = S_n^*[[t_{n+1} - a]]. \tag{25.16.1}$$

Let $\lambda_a : S^* \to R^*$ denote the composition

$$S_{n+1}^* = S_n^*[[t_{n+1}]] \longrightarrow \frac{S_n^*[[t_{n+1}]]}{(t_{n+1} - a)} = S_n^* \xrightarrow{\lambda} R^*,$$

and let $\tau_a := \lambda_a(t_{n+1}) = \lambda_a(a)$. We have $\ker \lambda_a = (\mathfrak{p}_n, t_{n+1} - a)S_{n+1}^*$. Consider the commutative diagram

$$\begin{array}{ccccccccc}
 & & S_n & \xrightarrow{\subseteq} & S_n^* & \xrightarrow{\subseteq} & S_{n+1}^* & \longrightarrow & S_{n+1}^*[1/x] \\
 & & \Big\downarrow {\scriptstyle \lambda_0,\ \cong} & & \Big\downarrow {\scriptstyle \lambda} & & \Big\downarrow {\scriptstyle \lambda_a} & & \Big\downarrow \\
R & \xrightarrow{\subseteq} & R_n & \xrightarrow{\varphi_0} & R^* & \xrightarrow{=} & R^* & \longrightarrow & R^*[1/x].
\end{array}$$

Diagram 25.16.2

Let

$$\mathcal{U} := \{P^* \in \operatorname{Spec} S_{n+1}^* \mid P^* \cap S_{n+1} = P,\ x \notin P \text{ and } P^* \text{ is minimal over } (P, \mathfrak{p}_n)S_{n+1}^*\}.$$

Notice that $x \notin P^*$ for each $P^* \in \mathcal{U}$, since $x \in R$ implies $\lambda_a(x) = x$.

THEOREM 25.17. *With the notation of Setting 25.16, the elements $\tau_1, \ldots, \tau_n, \tau_a$ are primarily limit-intersecting in x over R if and only if $t_{n+1} - a \notin \bigcup \{P^* \mid P^* \in \mathcal{U}\}$.*

PROOF. Assume that $\tau_1,\dots,\tau_n,\tau_a$ are primarily limit-intersecting in x over R. Then $\tau_1,\dots,\tau_n,\tau_a$ are algebraically independent over R. Consider the following commutative diagram:

$$\begin{array}{ccccc} & & S_{n+1}=R[t_1,\dots,t_{n+1}]_{(\mathfrak{m},t_1,\dots,t_{n+1})} & \xrightarrow{\subseteq} & S^*_{n+1}=R^*[[t_1,\dots,t_{n+1}]] \\ & & \lambda_1\downarrow & & \lambda_a\downarrow \\ R & \xrightarrow{\subseteq} & R_{n+1}=R[\tau_1,\dots,\tau_a]_{(\mathfrak{m},\tau_1,\dots,\tau_a)} & \longrightarrow & R^*. \end{array}$$

Diagram 25.17.0

The map λ_1 is the restriction of λ_a to S_{n+1}, and is an isomorphism since $\tau_1,\dots,\tau_n,\tau_a$ are algebraically independent over R.

If $t_{n+1}-a\in P^*$ for some $P^*\in\mathcal{U}$, we prove that $\varphi: R_{n+1}\to R^*[1/x]$ is not flat, a contradiction to our assumption that $\tau_1,\dots,\tau_n,\tau_a$ are primarily limit-intersecting. Since $P^*\in\mathcal{U}$, we have $\mathfrak{p}_n\subset P^*$. Then $t_{n+1}-a\in P^*$ implies $\ker\lambda_a\subset P^*$. Let $\lambda_a(P^*):=Q^*$. Then $\lambda_a^{-1}(Q^*)=P^*$ and $\operatorname{ht}P^*=n+1+\operatorname{ht}Q^*$. Since $P^*\in\mathcal{U}$, we have $x\notin P^*$. The commutativity of Diagram 25.17.0 implies that $x\notin Q^*$. Let $P:=P^*\cap S_{n+1}$ and let $Q:=Q^*\cap R_{n+1}$. Commutativity of Diagram 25.17.0 and λ_1 an isomorphism imply that $\operatorname{ht}P=\operatorname{ht}Q$. Since P^* is a minimal prime of $(P,\mathfrak{p}_n)S^*_{n+1}$ and $\mathfrak{p}_n$ is n-generated and S^*_{n+1} is Noetherian and catenary, we have $\operatorname{ht}P^*\le\operatorname{ht}P+n$. Hence $\operatorname{ht}P\ge\operatorname{ht}P^*-n$. Thus

$$\operatorname{ht}Q \;=\; \operatorname{ht}P \;\ge\; \operatorname{ht}P^*-n \;=\; \operatorname{ht}Q^*+n+1-n \;=\; \operatorname{ht}Q^*+1.$$

The fact that $\operatorname{ht}Q>\operatorname{ht}Q^*$ implies that the map $R_{n+1}\to R^*[1/x]$ is not flat.

For the converse, we have

Assumption 25.17.1: $\quad t_{n+1}-a\notin\bigcup\{\,P^*\mid P^*\in\mathcal{U}\,\}$.

Since $\lambda_a: S^*_{n+1}\to R^*$ is an extension of $\lambda: S^*_n\to R^*$ as in Diagram 25.16.2, we have $\ker\lambda_a\cap S_n=(0)$. Let $\mathfrak{p}:=(t_{n+1}-a)S^*_{n+1}=(t_{n+1}-\tau_a)S^*_{n+1}$. As in Equation 25.16.1, we have

$$S^*_{n+1} \;=\; R^*[[t_1,\dots,t_{n+1}]] \;=\; R^*[[t_1-\tau_1,\dots,t_n-\tau_n,t_{n+1}-a]].$$

Thus $P^*:=(\mathfrak{p}_n,\mathfrak{p})S^*_{n+1}$ is a prime ideal of height $n+1$ and $P^*\cap R^*=(0)$. It follows that $x\notin P^*$. We show that $P^*\cap S_{n+1}=(0)$. Assume that $P=P^*\cap S_{n+1}\neq(0)$. Since $\operatorname{ht}P^*=n+1$, P^* is minimal over $(P,\mathfrak{p}_n)S^*_{n+1}$, and so $P^*\in\mathcal{U}$, a contradiction to Assumption 25.17.1. Therefore $P^*\cap S_{n+1}=(0)$. It follows that $\mathfrak{p}\cap S_{n+1}=(0)$ since $\mathfrak{p}\subset P^*$. Thus $\ker\lambda_1=(0)$, and so λ_1 in Diagram 25.17.0 is an isomorphism. Therefore τ_a is algebraically independent over R_n.

Since R is excellent and R_{n+1} is a localized polynomial ring in $n+1$ variables over R, the hypotheses of Corollary 7.6 are satisfied. It follows that the elements $\tau_1,\dots,\tau_n,\tau_a$ are primarily limit-intersecting in x over R if for every $Q^*\in\operatorname{Spec}R^*$ with $x\notin Q^*$, we have $\operatorname{ht}(Q^*\cap R_{n+1})\le\operatorname{ht}Q^*$. Thus, to complete the proof of Theorem 25.17, it suffices to prove Claim 25.18. $\square$

CLAIM 25.18. Let $Q^*\in\operatorname{Spec}R^*$ with $x\notin Q^*$ and $\operatorname{ht}Q^*=r$. Then

$$\operatorname{ht}(Q^*\cap R_{n+1})\le\; r.$$

PROOF. (of Claim 25.18) Let $Q_1:=Q^*\cap R_{n+1}$ and let $Q_0:=Q^*\cap R_n$. Suppose $\operatorname{ht}Q_1>r$. Notice that $r<d$, since $d=\dim R^*$ and $x\notin Q^*$.

Since $\tau_1,\dots,\tau_n$ are primarily limit-intersecting in x over R, the extension

$$R_n:=R[\tau_1,\dots,\tau_n]_{(\mathfrak{m},\tau_1,\dots,\tau_n)}\hookrightarrow R^*[1/x]$$

from Diagram 25.16.2 is flat. Thus $\operatorname{ht} Q_0 \le r$ and $\operatorname{ht} Q_0 \le \operatorname{ht} L^*$ for every prime ideal L^* of R^* with $Q_0R^* \subseteq L^* \subseteq Q^*$. Since R_{n+1} is a localized polynomial ring in the indeterminate τ_a over R_n, we have that $\operatorname{ht} Q_1 \le \operatorname{ht} Q_0 + 1 = r + 1$. Thus $\operatorname{ht} Q_1 = r + 1$ and $\operatorname{ht} Q_0 = r$. It follows that Q^* is a minimal prime of Q_0R^*.

Let $h(\tau_a)$ be a polynomial in

$$(Q^* \cap R_n[\tau_a]) \setminus (Q^* \cap R_n)R_{n+1}.$$

It follows that $Q^* \cap R_{n+1} := Q_1$ is a minimal prime of the ideal $(Q^* \cap R_n, h(\tau_a))R_{n+1}$.

With notation from Diagram 25.16.2, define

$$P_0 := \lambda_0^{-1}(Q_0) \quad \text{and} \quad P_0^* := \lambda^{-1}(Q^*).$$

Since λ_0 is an isomorphism, P_0 is a prime ideal of S_n with $\operatorname{ht} P_0 = r$. Moreover, we have the following:

(1) $P_0^* \cap S_n = P_0$ (by commutativity in Diagram 25.16.2),
(2) $x \notin P_0^*$ (by item 1),
(3) P_0^* is a minimal prime of $(P_0, \mathfrak{p}_n)S_n^*$ (since $S_n^*/\mathfrak{p}_n = R^*$ in Diagram 25.16.2, and Q^* is a minimal prime of Q_0R^*),
(4) $\operatorname{ht} P_0^* = n+r$ (by the correspondence between prime ideals of S_n^* containing $\mathfrak{p}_n$ and prime ideals of R^*).

Consider the commutative diagram below with the left and right ends identified:

$$\begin{array}{ccccccccc}
S_{n+1}^* & \longleftarrow & S_n^* & \longleftarrow & S_n & \longrightarrow & S_{n+1} & \xrightarrow{\ \theta\ } & S_{n+1}^* \\
{\scriptstyle \lambda_a}\big\downarrow & & {\scriptstyle \lambda}\big\downarrow & & {\scriptstyle \lambda_0,\cong}\big\downarrow & & {\scriptstyle \lambda_1,\cong}\big\downarrow & & {\scriptstyle \lambda_a}\big\downarrow \\
R^* & \longleftarrow & R^* & \longleftarrow & R_n & \longrightarrow & R_{n+1} & \longrightarrow & R^*,
\end{array}$$

Diagram 25.18.0

where λ, λ_0 and λ_1 are as in Diagrams 25.16.2 and 25.17.0, and so λ_a restricted to S_n^* is λ. Let $h(t_{n+1}) = \lambda_1^{-1}(h(\tau_a))$ and set

$$P_1 := \lambda_1^{-1}(Q_1) \in \operatorname{Spec}(S_{n+1}), \quad \text{and} \quad P^* := \lambda_a^{-1}(Q^*) \in \operatorname{Spec}(S_{n+1}^*).$$

Then P_1 is a minimal prime of $(P_0, h(t_{n+1}))S_{n+1}$, since Q_1 is a minimal prime of $(Q_0, h(\tau_a))R_{n+1}$. Since $Q_1 \subseteq Q^*$, we have $h(t_{n+1}) \in P^*$ and $P_1S_{n+1}^* \subseteq P^*$ because $\lambda_a(h(t_{n+1})) = \lambda_1(h(t_{n+1})) = h(\tau_a) \in Q_1$ and $\lambda_a(P_1) = \lambda_1(P_1) = Q_1$. By the correspondence between prime ideals of S_{n+1}^* containing $\ker(\lambda_a) = \mathfrak{p}_{n+1}$ and prime ideals of R^*, we see

$$\operatorname{ht} P^* = \operatorname{ht} Q^* + n + 1 = r + n + 1.$$

Since $\lambda_a(P_0^*) \subseteq Q^*$, we have $P_0^* \subseteq P^*$, but $h(t_{n+1}) \notin P_0S_{n+1}$ implies $h(t_{n+1}) \notin P_0^*S_{n+1}^*$. Therefore

$$(P_0, \mathfrak{p}_n)S_{n+1}^* \subseteq P_0^*S_{n+1}^* \subsetneq (P_0^*, h(t_{n+1}))S_{n+1}^* \subseteq P^*.$$

By items 3 and 4 above, $\operatorname{ht} P_0^* = n+r$ and P_0^* is a minimal prime of $(P_0, \mathfrak{p}_n)S_n^*$. Since $\operatorname{ht} P^* = n + r + 1$, it follows that P^* is a minimal prime of $(P_0, h(t_{n+1}), \mathfrak{p}_n)S_{n+1}^*$. Since $(P_0, h(t_{n+1}), \mathfrak{p}_n)S_{n+1}^* \subseteq (P_1, \mathfrak{p}_n)S_{n+1}^* \subseteq P^*$, we have P^* is a minimal prime of $(P_1, \mathfrak{p}_n)S_{n+1}^*$. But then, by Assumption 25.17.1, $t_{n+1} - a \notin P^*$, a contradiction. This contradiction implies that $\operatorname{ht} Q_1 = r$. This completes the proof of Claim 25.18 and thus also the proof of Theorem 25.17. □

We use Theorem 25.15, Theorem 25.17 and Lemma 25.10 to prove in Theorem 25.19 the existence over a countable excellent normal local domain of dimension at least two of an infinite sequence of primarily limit-intersecting elements.

THEOREM 25.19. *Let R be a countable excellent normal local domain with dimension $d \geq 2$, let x be a nonzero element in the maximal ideal $\mathfrak{m}$ of R, and let R^* be the x-adic completion of R. Let n be a positive integer. Then*

(1) *If the elements $\tau_1, \dots, \tau_n \in xR^*$ are primarily limit-intersecting in x over R, then there exists an element $\tau_a \in xR^*$ such that $\tau_1, \dots, \tau_n, \tau_a$ are primarily limit-intersecting in x over R.*
(2) *There exists an infinite sequence $\tau_1, \dots, \tau_n, \dots \in xR^*$ of elements that are primarily limit-intersecting in x over R.*

PROOF. Since item 1 implies item 2, it suffices to prove item 1. Theorem 25.15 implies the existence of an element $\tau_1 \in xR^*$ that is primarily limit-intersecting in x over R. As in Setting 25.16, let

$$\mathcal{U} := \{P^* \in \operatorname{Spec} S^*_{n+1} \mid P^* \cap S_n = P \in \operatorname{Spec} S_n \text{ and } P^* \text{ is minimal over } (P, \mathfrak{p}_n)S^*_{n+1}\}.$$

Since the ring S_{n+1} is countable and Noetherian, the set $\mathcal{U}$ is countable. Lemma 22.18 implies that there exists an element $a \in x^2 S^*_{n+1}$ such that

$$t_{n+1} - a \notin \bigcup \{P^* \mid P^* \in \mathcal{U}\}.$$

By Theorem 25.17, the elements $\tau_1, \dots, \tau_n, \tau_a$ are primarily limit-intersecting in x over R. □

Using Theorem 25.15, we establish in Theorem 25.20, for every countable excellent normal local domain R of dimension $d \geq 2$, the existence of a primarily limit-intersecting element $\eta \in xR^*$ such that the constructed Noetherian domain

$$B = A = R^* \cap \mathcal{Q}(R[\eta])$$

is not a Nagata domain and hence is not excellent.

THEOREM 25.20. *Let R be a countable excellent normal local domain of dimension $d \geq 2$, let x be a nonzero element in the maximal ideal $\mathfrak{m}$ of R, and let R^* be the x-adic completion of R. There exists an element $\eta \in xR^*$ such that*

(1) *η is primarily limit-intersecting in x over R.*
(2) *The associated intersection domain $A := R^* \cap \mathcal{Q}(R[\eta])$ is equal to its approximation domain B.*
(3) *The ring A has a height-one prime ideal $\mathfrak{p}$ such that $R^*/\mathfrak{p}R^*$ is not reduced.*

Thus the integral domain $A = B$ associated to η is a normal Noetherian local domain that is not a Nagata domain and hence is not excellent.

PROOF. Since $\dim R \geq 2$, there exists $y \in \mathfrak{m}$ such that $\operatorname{ht}(x, y)R = 2$. By Theorem 25.15, there exists $\tau \in xR^*$ such that τ is primarily limit-intersecting in x over R. Hence the extension $R[\tau] \longrightarrow R^*[1/x]$ is flat. Let $n \in \mathbb{N}$ with $n \geq 2$, and let $\eta := (y + \tau)^n$. Since τ is algebraically independent over R, the element η is also algebraically independent over R. Moreover, the polynomial ring $R[\tau]$ is a free $R[\eta]$-module with $1, \tau, \dots, \tau^{n-1}$ as a free module basis. Hence the map $R[\eta] \longrightarrow R^*[1/x]$ is flat. It follows that η is primarily limit-intersecting in x over R. Therefore the intersection domain $A := R^* \cap \mathcal{Q}(R[\eta])$ is equal to its associated approximation domain B and is a normal Noetherian domain with x-adic completion R^*. Since η

is a prime element of the polynomial ring $R[\eta]$ and $B[1/x]$ is a localization of $R[\eta]$, it follows that $\mathfrak{p} := \eta B$ is a height-one prime ideal of B. Since $\tau \in R^*$, and $\eta = (y+\tau)^n$, the ring $R^*/\mathfrak{p}R^*$ contains nonzero nilpotent elements. Since a Nagata local domain is analytically unramified, it follows that the normal Noetherian domain B is not a Nagata ring, [**122**, page 264] or [**137**, (32.2)]. □

Let d be an integer with $d \geq 2$. In Examples 10.15 we give extensions that satisfy LF_{d-1} but do not satisfy LF_d; see Definition 24.1. These extensions are weakly flat but are not flat. In our setting these examples have the intersection domain A equal to its approximation domain B but A is not Noetherian In Theorem 25.21, we present a more general construction of examples with these properties.

THEOREM 25.21. *Let $(R, \mathfrak{m})$ be a countable excellent normal local domain. Assume that* $\dim R = d + 1 \geq 3$, *that $(x, y_1, \ldots, y_d)R$ is an $\mathfrak{m}$-primary ideal, and that R^* is the x-adic completion of R. Then there exists $f \in xR^*$ such that f is algebraically independent over R and the map $\varphi : R[f] \longrightarrow R^*[1/x]$ is weakly flat but not flat. Indeed, φ satisfies LF_{d-1}, but fails to satisfy LF_d. Thus the intersection domain $A := \mathcal{Q}(R[f]) \cap R^*$ is equal to its approximation domain B, but A is not Noetherian.*

PROOF. By Theorem 25.19, there exist elements $\tau_1, \ldots, \tau_d \in xR^*$ that are primarily limit-intersecting in x over R. Let

$$f := y_1\tau_1 + \cdots + y_d\tau_d.$$

Using that $\tau_1, \ldots, \tau_d$ are algebraically independent over R, we regard f as a polynomial in the polynomial ring $T := R[\tau_1, \ldots, \tau_d]$. Let $S := R[f]$. For $Q \in \operatorname{Spec} R^*[1/x]$ and $P := Q \cap T$, consider the composition φ_Q

$$S \longrightarrow T_P \longrightarrow R^*[1/x]_Q.$$

Since $\tau_1, \ldots, \tau_d$ are primarily limit-intersecting in x over R, the map $T \hookrightarrow R^*[1/x]$ is flat. Thus the map φ_Q is flat if and only if the map $S \longrightarrow T_P$ is flat. Let $\mathfrak{p} := P \cap R$.

Assume that P is a minimal prime of $(y_1, \ldots, y_d)T$. Then $\mathfrak{p}$ is a minimal prime of $(y_1, \ldots, y_d)R$. Since T is a polynomial ring over R, we have $P = \mathfrak{p}T$ and $\operatorname{ht}(\mathfrak{p}) = d = \operatorname{ht} P$. Notice that $(\mathfrak{p}, f)S = P \cap S$ and $\operatorname{ht}(\mathfrak{p}, f)S = d + 1$. Since a flat extension satisfies the Going-down property, the map $S \longrightarrow T_P$ is not flat. Hence φ does not satisfy LF_d.

Assume that $\operatorname{ht} P \leq d - 1$. Then $(y_1, \ldots, y_d)T$ is not contained in P. Hence $(y_1, \ldots, y_d)R$ is not contained in $\mathfrak{p}$. Consider the sequence

$$S = R[f] \hookrightarrow R_{\mathfrak{p}}[f] \xrightarrow{\psi} R_{\mathfrak{p}}[\tau_1, \ldots, \tau_d] \hookrightarrow T_P,$$

where the first and last injections are localizations. Since the nonconstant coefficients of f generate the unit ideal of $R_{\mathfrak{p}}$, the map ψ is flat; see Theorem 7.28. Thus φ satisfies LF_{d-1}.

We conclude that the intersection domain $A = R^* \cap \mathcal{Q}(R[f])$ is equal to its approximation domain B and is not Noetherian. □

Exercise

(1) Let $(R, \mathfrak{m})$ be a Noetherian local ring, let x be an element of $\mathfrak{m}$ and let R^* be the x-adic completion of R. Let S be the localized polynomial ring $R[t]_{(\mathfrak{m},t)}$

and let S^* denote the I-adic completion completion of S, where $I = (y,t)S$. Let a be an element of the ideal xR^*.

(a) Prove that R^* is complete in the a-adic topology on R^*, and that S^* is complete in the $(t-a)$-adic topology on S^*.
(b) Prove that S^* is the formal power series ring $R^*[[t]]$.
(c) Prove that $R^*[[t]] = R^*[[t-a]]$. Thus S^* is the formal power series ring in $t-a$ over R^*, as is used in the proof of Theorem 25.12.

COMMENT. Item a is a special case of Exercise 2 of [**122**, p. 63].

SUGGESTION. For item c, prove that every element of S^* has a unique expression as a power series in t over R^* and also a unique expression as a power series in $t-a$ over R^*.

CHAPTER 26

Weierstrass techniques for generic fiber rings

Let k be a field, let m and n be positive integers, and let $X = \{x_1, \dots, x_n\}$ and $Y = \{y_1, \dots, y_m\}$ be sets of independent variables over k. Define the rings A, B and C as follows:

$$(26.1.0) \qquad A := k[X]_{(X)}, \; B := k[[X]]\,[Y]_{(X,Y)} \quad \text{and} \quad C := k[Y]_{(Y)}[[X]].$$

That is, A is the usual localized polynomial ring in the variables of X. The rings B and C are "mixed polynomial-power series rings", over k using X, the power series variables, and Y, the polynomial variables, in two different ways: The ring B consists of polynomials in Y with coefficients in the power series ring $k[[X]]$, whereas C contains power series in the X variables over the localized polynomial ring $k[Y]_{[Y]}$. Thus C is complete in the (X)-adic topology, whereas B is not. The (X)-adic completion B^* of B is C.

The maps shown in the following sequences are local embeddings:

$$A := k[X]_{(X)} \hookrightarrow \widehat{A} := k[[X]], \quad \widehat{A} \hookrightarrow \widehat{B} = \widehat{C} = k[[X,Y]] \quad \text{and}$$
$$B := k[[X]]\,[Y]_{(X,Y)} \hookrightarrow C := k[Y]_{(Y)}[[X]] \hookrightarrow \widehat{B} = \widehat{C} = k[[X]]\,[[Y]].$$

There is a canonical inclusion map $B \hookrightarrow C$, and the ring C has infinite transcendence degree over B, even if $m = n = 1$. In Chapter 28, we consider this embedding and analyze the associated spectral map.

In this chapter, we develop techniques using the Weierstrass Preparation Theorem. These techniques are applied in Chapter 27 to describe the prime ideals maximal in generic fiber rings associated to the polynomial-power series rings $A, B,$ and C. In Chapter 27, we prove every prime ideal P in $k[[X]]$ that is maximal with respect to $P \cap A = (0)$ has $\operatorname{ht} P = n - 1$. For every prime ideal P of $k[[X]]\,[[Y]]$ such that P is maximal with respect to either $P \cap B = (0)$ or $P \cap C = (0)$, we prove $\operatorname{ht}(P) = n + m - 2$. In addition we prove each prime ideal P of $k[[X, Y]]$ that is maximal with respect to $P \cap k[[X]] = (0)$ has $\operatorname{ht} P = m$ or $n + m - 2$; see Theorem 26.3.

26.1. Terminology, background and results

NOTATION 26.1. Let $(R, \mathfrak{m})$ be a Noetherian local domain and let $\widehat{R}$ be the $\mathfrak{m}$-adic completion of R. The *formal fibers* of R are the fibers of the map $\operatorname{Spec} \widehat{R} \to \operatorname{Spec} R$. A prime ideal $\widehat{P}$ of $\widehat{R}$ is in the *generic formal fiber* of $R \iff \widehat{P} \cap R = (0)$. The *generic formal fiber ring* of R, denoted $\operatorname{Gff}(R)$, is the ring

$$\operatorname{Gff}(R) : = \; (R \setminus (0))^{-1}\widehat{R} \; = \; \widehat{R}[K],$$

where K denotes the field of fractions of R. Since $\operatorname{Gff}(R)$ is a localization of $\widehat{R}$, the prime ideals of $\operatorname{Gff}(R)$ are all extended from $\widehat{R}$. $\operatorname{Spec}\operatorname{Gff}(R)$ is precisely the set of prime ideals in the generic formal fiber of R.

For a prime ideal P of R, the formal fiber over P is $\operatorname{Spec}(\,(R_P/PR_P)\otimes_R \widehat{R}\,)$, or equivalently $\operatorname{Spec}((R\setminus P)^{-1}(\widehat{R}/P\widehat{R}))$; see Discussion 3.29 and Definition 3.44. Let $\operatorname{Gff}(R/P)$ denote the generic formal fiber ring of R/P. Since $\widehat{R}/P\widehat{R}$ is the completion of R/P, the formal fiber over P is $\operatorname{Spec}(\operatorname{Gff}(R/P))$.

Let $R \hookrightarrow S$ be an injective homomorphism of commutative rings. If R is an integral domain, the *generic fiber ring* of the map $R \hookrightarrow S$ is the localization $(R\setminus(0))^{-1}S$ of S.

The formal fibers encode important information about the structure of R. For example, R is excellent provided it is universally catenary and has geometrically regular formal fibers [**62**, (7.8.3), page 214]; see Definition 8.22.

Remarks 26.2 contains historical remarks regarding dimensions of generic formal fiber rings and heights of the maximal ideals of these rings:

REMARKS 26.2. (1) Let $(R,\mathfrak{m})$ be a Noetherian local domain. In [**121**] Matsumura remarks that, as the ring R gets closer to its $\mathfrak{m}$-adic completion $\widehat{R}$, it is natural to think that the dimension of the generic formal fiber ring $\operatorname{Gff}(R)$ gets smaller. He proves that the generic formal fiber ring of A has dimension $\dim A-1$, and the generic formal fiber rings of B and C have dimension $\dim B-2=\dim C-2$ in [**121**]. Matsumura speculates as to whether $\dim R-1, \dim R-2$ and 0 are the only possible values for $\dim(\operatorname{Gff}(R))$ in [**121**, p. 261].

(2) In answer to Matsumura's question, Rotthaus establishes the following result in [**158**]: For every pair t,n of nonnegative integers with $t<n$, there exists an excellent regular local ring R such that $\dim R=n$ and $\dim(\operatorname{Gff}(R))=t$.

(3) Let $(R,\mathfrak{m})$ be an n-dimensional universally catenary Noetherian local domain. Loepp and Rotthaus in [**114**] compare the dimension of the generic formal fiber ring of R with that of the localized polynomial ring $R[x]_{(\mathfrak{m},x)}$. Matsumura shows in [**121**] that the dimension of the generic formal fiber ring $\operatorname{Gff}(R[x]_{(\mathfrak{m},x)})$ is either n or $n-1$. Loepp and Rotthaus prove that $\dim(\operatorname{Gff}(R[x]_{(\mathfrak{m},x)}))=n$ implies that $\dim(\operatorname{Gff} R)=n-1$ [**114**, Theorem 2]. They show by example that in general the converse is not true, and they give sufficient conditions for the converse to hold.

(4) Let $(T,\mathfrak{m}_T)$ be a complete Noetherian local domain that contains a field of characteristic zero. Assume that $T/\mathfrak{m}_T$ has cardinality at least the cardinality of the real numbers. By adapting techniques developed by Heitmann in [**96**], in the articles [**112**] and [**113**], Loepp proves, among other things, for every prime ideal $\mathfrak{q}$ of T with $\mathfrak{q}\neq\mathfrak{m}_T$, there exists an excellent regular local ring R that has completion T and has generic formal fiber ring $\operatorname{Gff}(R)=T_{\mathfrak{q}}$. By varying the height of $\mathfrak{q}$, this yields examples where the dimension of the generic formal fiber ring is any integer t with $0\le t<\dim T$. Loepp also shows for these examples that, for each nonzero prime ideal $\mathfrak{p}$ of R, there exists a unique prime $\mathfrak{q}$ of T with $\mathfrak{q}\cap R=\mathfrak{p}$ and $\mathfrak{q}=\mathfrak{p}T$.

(5) Let $R,\mathfrak{m})$ be a countable Noetherian local domain. Heinzer, Rotthaus and Sally show in [**71**, Proposition 4.10, page 36] that:

(a) The generic formal fiber ring $\operatorname{Gff}(R)$ is a Jacobson ring in the sense that each prime ideal of $\operatorname{Gff}(R)$ is an intersection of maximal ideals of $\operatorname{Gff}(R)$.

(b) $\dim(\widehat{R}/P)=1$ for each prime ideal $P\in\operatorname{Spec}\widehat{R}$ that is maximal with respect to $P\cap R=(0)$.

(c) If $\widehat{R}$ is equidimensional of dimension n, then $\operatorname{ht} P = n-1$ for each prime ideal $P \in \operatorname{Spec} \widehat{R}$ that is maximal with respect to $P \cap R = (0)$.

(d) If $Q \in \operatorname{Spec} \widehat{R}$ with $\operatorname{ht} Q \geq 1$, then there exists a prime ideal $P \subset Q$ such that $P \cap R = (0)$ and $\operatorname{ht}(Q/P) = 1$.

If the field k is countable, it follows from this result that all ideals maximal in the generic formal fiber ring of A have the same height.

(6) In Matsumura's article [**121**] from item 1 above, he does not address the question of whether all ideals maximal in the generic formal fiber rings for A, B and C have the same height. In general, for an excellent regular local ring R it can happen that $\operatorname{Gff}(R)$ contains maximal ideals of different heights; see the article [**158**, Corollary 3.2] of Rotthaus.

(7) Charters and Loepp in [**31**, Theorem 3.1] extend Rotthaus's result mentioned in item 6: Let $(T, \mathfrak{m}_T)$ be a complete Noetherian local ring and let G be a nonempty subset of $\operatorname{Spec} T$ such that the number of maximal elements of G is finite. They prove there exists a Noetherian local domain R with completion T and with generic formal fiber exactly G if G satisfies the following conditions:

(a) $\mathfrak{m}_T \notin G$ and G contains the associated primes of T,

(b) If $\mathfrak{p} \subset \mathfrak{q}$ are in $\operatorname{Spec} T$ and $\mathfrak{q} \in G$, then $\mathfrak{p} \in G$, and

(c) Every $\mathfrak{q} \in G$ meets the prime subring of T in (0).

Charters and Loepp [**31**, Theorem 4.1] also show that, if T contains the ring of integers and, in addition to conditions a, b, and c,

(d) T is equidimensional, and

(e) $T_{\mathfrak{p}}$ is a regular local ring for each maximal element $\mathfrak{p}$ of G,

then there exists an excellent local domain R with completion T and with generic formal fiber exactly G; see [**31**, Theorem 4.1]. Since the maximal elements of the set G may be chosen to have different heights, this result provides many examples where the generic formal fiber ring contains maximal ideals of different heights.

The Weierstrass techniques developed in this chapter enable us to prove the following theorem in Chapter 27:

MAXIMAL GENERIC FIBERS THEOREM 26.3. *Let k be a field, let m and n be positive integers, and let $X = \{x_1, \ldots, x_n\}$ and $Y = \{y_1, \ldots, y_m\}$ be sets of independent variables over k. Then, for each of the rings $A := k[X]_{(X)}, B := k[[X]]\,[Y]_{(X,Y)}$ and $C := k[Y]_{(Y)}[[X]]$, every prime ideal maximal in the generic formal fiber ring has the same fixed height; more precisely*:

(1) *If P is a prime ideal of $\widehat{A}$ maximal with respect to $P \cap A = (0)$, then* $\operatorname{ht}(P) = n-1$.

(2) *If P is a prime ideal of $\widehat{B}$ maximal with respect to $P \cap B = (0)$, then* $\operatorname{ht}(P) = n+m-2$.

(3) *If P is a prime ideal of $\widehat{C}$ maximal with respect to $P \cap C = (0)$, then* $\operatorname{ht}(P) = n+m-2$.

(4) *There are at most two possible values for the height of a maximal ideal of the generic fiber ring $(\widehat{A} \setminus (0))^{-1}\widehat{C}$ of the inclusion map $\widehat{A} \hookrightarrow \widehat{C}$.*

  (a) *If $n \geq 2$ and P is a prime ideal of $\widehat{C}$ maximal with respect to $P \cap \widehat{A} = (0)$, then either* $\operatorname{ht} P = n+m-2$ *or* $\operatorname{ht} P = m$.

  (b) *If $n = 1$, then all ideals maximal in the generic fiber ring $(\widehat{A} \setminus (0))^{-1}\widehat{C}$ have height m.*

We were motivated to consider generic fiber rings for the embeddings displayed above because of questions related to Chapters 28 and 29 and ultimately because of the following question posed by Melvin Hochster and Yongwei Yao.

QUESTION 26.4. Let R be a complete Noetherian local domain. Can one describe or somehow classify the local maps of R to a complete Noetherian local domain S such that $U^{-1}S$ is a field, where $U = R \setminus (0)$, i.e., such that the generic fiber of $R \hookrightarrow S$ is trivial?

REMARK 26.5. By Cohen's structure theorems [**35**], [**137**, (31.6)], a complete Noetherian local domain R is a finite integral extension of a complete regular local domain R_0. If R has the same characteristic as its residue field, then R_0 is a formal power series ring over a field; see Remarks 3.19. The generic fiber of $R \hookrightarrow S$ is trivial if and only if the generic fiber of $R_0 \hookrightarrow S$ is trivial.

A local ring R is called equicharacteristic, if the ring and its residue field have the same characteristic; see Definition 3.18.1. If the equicharacteristic local ring has characteristic zero, then we say R is "equicharacteristic zero" or "of equal characteristic zero". Such a ring contains the field of rational numbers; see Exercise 26.1.

Thus, as Hochster and Yao remark, there is a natural way to construct such extensions in the case where the local ring R has characteristic zero and contains the rational numbers; consider

$$R = k[[x_1, \dots, x_n]] \hookrightarrow T = L[[x_1, \dots, x_n, y_1, \dots, y_m]] \to T/P = S, \tag{26.4.0}$$

where k is a subfield of L, the x_i, y_j are formal indeterminates, and P is a prime ideal of T maximal with respect to being disjoint from the image of $R \setminus \{0\}$. The prime ideals P that arise correspond to the maximal ideals of the generic fiber $(R\setminus(0))^{-1}T$. The composite extension $T \hookrightarrow S$ satisfies the condition of Question 26.4. If $k = \mathbb{Q}$, then, in Equation 26.4.0, R has characteristic zero and contains the rational numbers.

In Theorem 27.16, we answer Question 26.4 in the special case where the extension arises from the embedding in Sequence 26.4.0 with the field $L = k$. We prove in this case that the dimension of the extension ring S must be either 2 or n.

We introduce the following terminology for the condition of Question 26.4 with a more general setting:

DEFINITION 26.6. For R and S integral domains with R a subring of S, we say that S is a *trivial generic fiber* extension of R, or a *TGF* extension of R, if every nonzero prime ideal of S has nonzero intersection with R. If $R \overset{\varphi}{\hookrightarrow} S$, then φ is also called a *trivial generic fiber extension* or *TGF extension.*

As in Remark 26.5, every extension $R \hookrightarrow T$ from an integral domain R to a commutative ring T yields a TGF extension by considering a composition

$$R \hookrightarrow T \to T/P = S, \tag{26.6.0}$$

where $P \in \operatorname{Spec} T$ is maximal with respect to $P \cap R = (0)$. Thus the generic fiber ring and so also Theorem 26.3 give information regarding TGF extensions in the case where the smaller ring is a mixed polynomial-power series ring.

Theorem 26.3 is useful in the study of Sequence 26.4.0, because the map in Sequence 26.4.0 factors through:

$$R = k[[x_1, \dots, x_n]] \hookrightarrow k[[x_1, \dots, x_n]]\,[y_1, \dots, y_m] \hookrightarrow T = L[[x_1, \dots, x_n, y_1, \dots, y_m]].$$

The second extension of this sequence is TGF if $n = m = 1$ and $k = L$; see Exercise 1 of this chapter. We study TGF extensions in Chapters 28 and 29.

Section 26.2 contains implications of Weierstrass' Preparation Theorem to the prime ideals of power series rings. We first prove a technical proposition regarding a change of variables that provides a "nice" generating set for a given prime ideal P of a power series ring; then in Theorem 26.14 we prove that, in certain circumstances, a larger prime ideal can be found with the same contraction as P to a certain subring. In Section 27.3 weuse Valabrega's, Theorem 4.9, concerning subrings of a two-dimensional regular local domain.

In Sections 27.1 and 27.2, we prove parts 2 and 3 of Theorem 26.3 stated above. We apply Theorem 26.14 in Section 27.4 to prove part 1 of Theorem 26.3, and in Section 27.5 we prove part 4.

26.2. Variations on a theme by Weierstrass

The statement of Weierstrass Preparation Theorem 26.8 uses Definition 26.7.

DEFINITION 26.7. Let $(R, \mathfrak{m})$ be a local ring and let x be an indeterminate over R. A monic polynomial $F = x^s + a_{s-1}x^{s-1} + \cdots + a_0 \in R[x]$ is *distinguished* if $s > 0$ and $a_i \in \mathfrak{m}$ for $0 \le i \le s-1$.[1]

More generally, if $(R_0, \mathfrak{m}_0)$ is a local ring, Z is a finite set of indeterminates over R_0 and $R = R_0[Z]$, then a monic polynomial $f \in R[x]$ is *distinguished* if f is distinguished as a polynomial of $R_{(\mathfrak{m}_0, Z)}[x]$.

THEOREM 26.8. (Weierstrass) [**193**, Theorem 5, p. 139; Corollary 1, p. 145], [**23**, Proposition 6, p.510][2] *Let $(R, \mathfrak{m})$ be a formal power series ring in finitely many variables over a field, let $f \in R[[x]]$ be a formal power series and let $\overline{f}$ denote the image of f in $(R/\mathfrak{m})[[x]]$. Assume that $\operatorname{ord}\overline{f} = s > 0$.*[3] *There exists a unique ordered pair (u, F) such that u is a unit in $R[[x]]$ and $F \in R[x]$ is a distinguished monic polynomial of degree s such that $f = uF$.*

We often write "By Weierstrass" to mean "using Theorem 26.8 or Corollary 26.9".

COROLLARY 26.9. *Assume notation as in Theorem 26.8. Thus $f \in R[[x]]$ is such that $\overline{f} \in (R/\mathfrak{m}_R)[[x]]$ and $\operatorname{ord}\overline{f} = s > 0$. Then:*

1. *The ideal $fR[[x]] = FR[[x]]$ is extended from $R[x]$.*
2. *$R[[x]]/(f)$ is a free R-module of rank s.*
3. *Every $g \in R[[x]]$ has the form $g = qf + r$, where $q \in R[[x]]$ and $r \in R[x]$ is a polynomial with $\deg r \le s-1$.*
4. *If $f \in R[x]$ is a distinguished monic polynomial of degree s, then items 1-3 hold, and also $R[x]/fR[x] = R[[x]]/fR[[x]]$, with the identifications of the canonical map.*

Remarks 26.10 extends Corollary 26.9.

---

[1]Every polynomial that is "monic" in a variable x is assumed to have positive degree in x.

[2]The version in [**23**] assumes only that the local ring $(R, \mathfrak{m})$ is complete and Hausdorff in the $\mathfrak{m}$-adic topology.

[3]Here "$\operatorname{ord}\overline{f}$" refers to the order function with respect to the local ring $(R/\mathfrak{m})[[x]]$; see Definition 2.6.

REMARKS 26.10. Let $(R, \mathfrak{m})$ be a formal power series ring in finitely many variables over a field. Let $n \in \mathbb{N}$ and let x, y, and $Z = \{z_1, \dots, z_n\}$ be new variables over R. For $1 \leq i \leq n$, let $Z_i = \{z_1, \dots, z_i\}$.

(1) Let $g \in R[[x]]\,[Z, y]$ be monic as a polynomial in y of degree t, and let $f \in R[x]$ be a distinguished monic polynomial in x. Then there exists $g' \in R[x, Z, y]$ that is monic as a polynomial in y of degree t such that $(g', f)R[[x, y, Z]] = (g, f)R[[x, y, Z]]$.

If g is a distinguished monic polynomial in y as an element of $R[[x]]\,[Z, y]$, then g' is also distinguished in y as an element of $R[x, Z, y]$.

(2) Assume $n \geq 2$, and

$g_1 \in R[z_1]$ is a distinguished monic polynomial in z_1 of degree t_1,

$g_2 \in R[[z_1]]\,[z_2]$ is a distinguished monic polynomial in z_2 of degree t_2,

...

$g_n \in R[[Z_{n-1}]]\,[z_n]$ is a distinguished monic polynomial in z_n of degree t_n.

Then there exist monic polynomials $g_1' = g_1 \in R[z_1], g_2' \in R[Z_2], \dots, g_n' \in R[Z]$ such that

$g_1' \in R[z_1]$ is distinguished in z_1 of degree t_1,

$g_2' \in R[z_1, z_2]$ is distinguished in z_2 of degree t_2,

...

$g_n' \in R[Z_{n-1}, z_n]$ is distinguished in z_n of degree t_n,

and, for each i with $1 \leq i \leq n$,
$(g_1, g_2, \dots, g_i)R[[Z_i]] = (g_1', g_2', \dots, g_i')R[[Z_i]]$.

(3) Assume $g_1, g_2, \dots, g_n \in R[Z]$ are such that, for every $1 \leq i \leq n$, each $g_i \in R[Z_i]$ is distinguished monic in z_i. Then:
   (a) $g_1, \dots, g_n$ is a regular $R[Z]$-sequence,
   (b) The ring $R[Z]/(g_1, \dots, g_n)R[Z]$ is a finite free module extension of R, and $R[Z]/(g_1, \dots, g_n)R[Z] = R[[Z]]/(g_1, \dots, g_n)R[[Z]]$.
   (c) The (Z)-adic completion of $R[Z]$ is identical to the $(g_1, \dots, g_n)$-adic completion of $R[Z]$ and both equal $R[[Z]]$.
   (d) If I is an ideal of $R[[Z]]$ such that $g_1, \dots, g_n \in I$, then I is extended from $R[Z]$; that is, $I = (I \cap R[Z])R[[Z]]$.

PROOF. For item 1, consider the elements of $R[[x]]\,[Z, y]$ as polynomials in $Z \cup \{y\}$. Then $g = y^t + \sum_{j=1}^{e} a_j b_j$, where $e \in \mathbb{N}_0$, the $a_j \in R[[x]]$, and each b_j is a monomial in the variables $Z \cup \{y\}$ such that the power of y in each b_j is less than t.

By Weierstrass Corollary 26.9, each $a_j = q_j f + r_j$, where $q_j \in R[[x]]$, $r_j \in R[x]$, and $\deg_x r_j < \deg_x f$. Define $g' = y^t + \sum_{j=1}^{e} r_j b_j \in R[x, Z, y]$. Since the power of y in each b_j is less than t, g' is monic in y of degree t. Then

$$g = y^t + \sum_{j=1}^{e} a_j b_j = y^t + \sum_{j=1}^{e} (q_j f + r_j) b_j = y^t + f \sum_{j=1}^{e} q_j b_j + \sum_{j=1}^{e} r_j b_j = g' + f \sum_{j=1}^{e} q_j b_j.$$

Thus $(f, g)R[[x, Z, y]] = (f, g')R[[x, Z, y]]$.

The "distinguished" condition for g implies that the non-leading coefficients of g (in y) are in the maximal ideal $(\mathfrak{m}, x, Z)R[[x]]\,[Z]$. From the set-off equation above, $g - g' = f\sum_{j=1}^{e} q_j b_j$. Since $f \in R[x]$ is distinguished, $f \equiv x^s \pmod{m}$, and thus $f \in (\mathfrak{m}, x)R[x] \subseteq (\mathfrak{m}, x, Z)R[[x]]\,[Z, y]$. It follows that g' is distinguished. This completes item 1.

For item 2, take $g_1' = g_1$; the case $n = 2$ holds by item 1 with Z an empty set of variables. If $n > 2$, use induction. Assume, for each i with $2 \le i \le n$, there exist $g_i'' \in R[[z_1]][z_2, \dots z_i]$ distinguished monic in $R[[z_1]][z_2, \dots z_{i-1}, z_i]$ in z_i of degree t_i, and such that $(g_2'', \dots g_i'')R[[Z]] = (g_2, \dots, g_i)R[[Z]]$. Since $g_1 \in R[z_1]$ is distinguished monic in z_1, item 1 implies that there exist g_i' in $R[z_1][z_2, \dots z_i]$ distinguished monic (as an element of $R[[z_1]][z_2, \dots z_{i-1}, z_i]$) in z_i of degree t_i such that, for $i > 1$,

$$(g_1, g_i'')R[[Z_i]] = (g_1, g_i')R[[Z_i]].$$

It follows that $(g_1, g_2, \dots, g_i)R[[Z_i]] = (g_1, g_2', \dots, g_i')R[[Z_i]]$, for each i. This proves item 2.

For statement 3.a, g_1 monic in z_1 implies that g_1 is regular on $R[Z]$. For each $i > 1$, let $\overline{g_i}$ denote the image of g_i in $R[Z]/((g_1, \dots, g_{i-1})R[Z]$. Each $\overline{g_i}$ monic in z_i over

$$\frac{R[Z]}{(g_1, \dots, g_{i-1})R[Z]} = \frac{R[Z_{i-1}]}{(g_1, \dots, g_{i-1})}[Z \setminus Z_{i-1}]$$

implies $\overline{g_i}$ is regular on $R[Z]/(g_1, \dots, g_{i-1})R[Z]$. Thus $g_1, \dots, g_n$ is a regular sequence on $R[Z]$.

For statement 3.b, Weierstrass Corollary 26.9 implies the statement for $n = 1$. For $n > 1$, by induction, the ring $R_{n-1} := R[Z_{n-1}]/(g_1, \dots, g_{n-1})R[Z_{n-1}]$ is a finite free R-module. Since

$$\frac{R[Z]}{(g_1, \dots, g_n)R[Z]} = \frac{R_{n-1}[z_n]}{g_n R_{n-1}[z_n]}$$

is a finite free R_{n-1}-module, it follows that $R[Z]/(g_1, \dots, g_n)R[Z]$ is a finite free R-module. The second part of statement 3.b follows from Corollary 26.9.4.

For statement 3.c, the ring $R[Z]/(g_1, \dots g_n)$ is a finite free module over the complete ring R. Therefore $R[Z]/(g_1, \dots g_n)$ is complete. This implies that $R[[Z]]$ is the $(g_1, \dots g_n)$-adic completion and the Z-adic completion of $R[Z]$.

Statement 3.d holds by Fact 3.2.1.

This completes the proof of Remarks 26.10. □

We apply the Weierstrass Preparation Theorem 26.8 to examine the structure of a given prime ideal P in the power series ring $\widehat{A} = k[[X]]$, where $X = \{x_1, \dots, x_n\}$ is a set of n variables over the field k. Here $A = k[X]_{(X)}$ is the localized polynomial ring in these variables. Our procedure is to make a change of variables that yields a regular sequence in P of a nice form.

NOTATION 26.11. By a *change of variables*, we mean a finite sequence of *polynomial change of variables* of the type described below, where $X = \{x_1, \dots, x_n\}$ is a set of n variables over the field k. For example, with $e_i, f_i \in \mathbb{N}$, consider

$$x_1 \mapsto x_1 = z_1, (x_1 \text{ is fixed}) \qquad x_2 \mapsto x_2 + x_1^{e_2} = z_2, \qquad \dots,$$

$$x_{n-1} \mapsto x_{n-1} + x_1^{e_{n-1}} = z_{n-1}, \qquad x_n \mapsto x_n + x_1^{e_n} = z_n,$$

followed by:

$$z_1 \mapsto z_1 + z_n^{f_1} = t_1, \qquad z_2 \mapsto z_2 + z_n^{f_2} = t_2, \qquad \dots,$$

$$z_{n-1} \mapsto z_{n-1} + z_n^{f_{n-1}} = t_{n-1}, \qquad z_n \mapsto z_n = t_n, (z_n \text{ is fixed}).$$

Thus a change of variables defines an automorphism of $\widehat{A}$ that restricts to an automorphism of A.

THEOREM 26.12. *Let $R := k[[X]]$, where k is a field, and $X = \{x_1, \ldots, x_n\}$ is a finite set of indeterminates over k. Let $P \in \operatorname{Spec} R$ with $x_1 \notin P$ and $\operatorname{ht} P = r$, where $1 \le r \le n-1$. Then:*

(1) *There exists a change of variables*

$$x_1 \mapsto z_1 := x_1 \ (x_1 \text{ is fixed}), x_2 \mapsto z_2, \ \ldots, \ x_n \mapsto z_n,$$

and a regular sequence $f_1, \ldots, f_r \in P$ so that, with $Z_i = \{z_1, \ldots, z_i\}$ and for each i with $1 \le i \le n$, each f_i is distinguished monic in z_{n-i+1} as an element of

$$k[[Z_{n-r}]]\,[z_{n-r+1}, \ldots, z_{n-i}][z_{n-i+1}].$$

In particular

$$\begin{aligned}
&f_1 \in k[[Z_{n-r}]]\,[z_{n-r+1}, \ldots, z_{n-1}][z_n] && \textit{is monic in } z_n \\
&f_2 \in k[[Z_{n-r}]]\,[z_{n-r+1}, \ldots, z_{n-2}][z_{n-1}] && \textit{is monic in } z_{n-1}, \textit{ etc} \\
&\vdots \\
&f_r \in k[[Z_{n-r}]]\,[z_{n-r+1}] && \textit{is monic in } z_{n-r+1}.
\end{aligned}$$

(2) *More generally, if $f_1, \ldots, f_r \in P$ satisfy the properties of item 1 for some change of variables Z and $J := (f_1, \ldots, f_r)R$, then:*
   (a) *P is a minimal prime of J.*
   (b) *The ring R/J is a finite free module extension of $k[[Z_{n-r}]]$.*
   (c) *The (Z'_{n-r})-adic completion of $k[[Z_{n-r}]]\,[Z'_{n-r}]$ is identical to the $(f_1, \ldots f_r)$-adic completion and both equal $R = k[[X]] = k[[Z]]$.*
   (d) *If $P_1 := P \cap k[[Z_{n-r}]]\,[Z'_{n-r}]$, then $P_1R = P$, that is, P is extended from $k[[Z_{n-r}]]\,[Z'_{n-r}]$.*
   (e) *If I is an ideal of R such that $f_1, \ldots, f_r \in I$, then I is extended from $k[[Z_{n-r}]]\,[Z'_{n-r}]$; that is, $I = (I \cap k[[Z_{n-r}]]\,[Z'_{n-r}])R$.*
   (f) *$P \cap k[[Z_{n-r}]] = (0)$, and*

$$k[[Z_{n-r}]] \hookrightarrow k[[Z_{n-r}]]\,[Z'_{n-r}]/P_1 \cong k[[Z]]/P$$

   is a finite integral extension.

PROOF. For item 1, first find polynomials in one variable over power series rings:

CLAIM 26.13. There exists a change of variables

$$x_1 \mapsto z_1 := x_1 \ (x_1 \text{ is fixed}), x_2 \mapsto z_2, \ \ldots, \ x_n \mapsto z_n,$$

and a regular sequence $f_1, \ldots, f_r \in P$ so that, for every i with $1 \le i \le r$, and the polynomial $f_i \in k[[z_1, \ldots, z_{n-i}]]\,[z_{n-i+1}]$ is distinguished monic as a polynomial in z_{n-i+1}.

PROOF OF CLAIM 26.13. Since R is a unique factorization domain, there exists a nonzero prime element f in P. The power series f is therefore not a multiple of x_1, and so f must contain a monomial term $x_2^{i_2} \ldots x_n^{i_n}$ with a nonzero coefficient in k. This nonzero coefficient in k may be assumed to be 1. Given positive integers

$e_2, \dots, e_{n-1}$, there exists an automorphism $\sigma : R \to R$ defined by the change of variables:

$$x_1 \mapsto x_1 \qquad x_2 \mapsto t_2 := x_2 + x_n^{e_2} \quad \dots \quad x_{n-1} \mapsto t_{n-1} := x_{n-1} + x_n^{e_{n-1}} \qquad x_n \mapsto x_n.$$

If $n = 2$, $x_2 \mapsto x_2 = x_n$. For $n > 2$, assume that $e_2, \dots, e_{n-1} \in \mathbb{N}$ are chosen suitably so that f written as a power series in the variables $x_1, t_2, \dots, t_{n-1}, x_n$ contains a term $a_n x_n^s$, where s is a positive integer, and $a_n \in k$ is nonzero. Assume that the integer s is minimal among all integers i such that a term ax_n^i occurs in f with a nonzero coefficient $a \in k$; further assume that the coefficient $a_n = 1$.

Let $B_0 := k[[x_1, t_2, \dots, t_{n-1}]]\,[x_n]$. Weierstrass Theorem 26.8 implies

$$f = \varepsilon f_1 \in P,$$

where $f_1 \in B_0$ is a distinguished monic polynomial in x_n of degree s and ε is a unit in R. Since $f \in P$ is a prime element, $f_1 \in P$ is also a prime element. By Weierstrass Corollary 26.9, every element g of R can be written as:

$$g = f_1 h + q,$$

where $h \in k[[x_1, t_2, \dots, t_{n-1}, x_n]] = R$ and $q \in B_0$. Hence q is a polynomial in x_n of degree less than s. Let f_1 and $\alpha_1, \dots, \alpha_t$ be generators for P in R; write each $\alpha_i = f_1 h_i + q_i$, where each $h_i \in R$, and each $q_i \in B_0$ has degree in x_n less than s. Thus P is finitely generated by f_1 and the set $\{q_i\} \subset B_0$. It follows that P is extended from $P \cap B_0$.

This implies Claim 26.13 for $r = 1$, with f_1 as defined and $z_1 = x_1, z_2 = t_2, \dots,$ $z_{n-1} = t_{n-1}$, $z_n = x_n$.

For $r > 1$, use induction and $P_0 := P \cap k[[x_1, t_2, \dots, t_{n-1}]]$. For f_1 defined above, f_1 is distinguished monic in x_n over $k[[x_1, t_2, \dots, t_{n-1}]]$. Thus $f_1 \notin k[[x_1, t_2, \dots, t_{n-1}]]$. Since P is extended from $B_0 := k[[x_1, t_2, \dots, t_{n-1}]]\,[x_n]$ and $P \cap B_0$ has height r, it follows that $\operatorname{ht} P_0 = r - 1$. Since $x_1 \notin P$, we have $x_1 \notin P_0$. By induction there exists a change of variables

$$t_2 \mapsto z_2, \dots, t_{n-1} \mapsto z_{n-1}$$

of $k[[x_1, t_2, \dots, t_{n-1}]]$ and $f_2, \dots, f_r \in P_0$ so that, for every i with $2 \le i \le r$, the polynomial f_i is distinguished monic in $k[[x_1, z_2, \dots, z_{n-i}]]\,[z_{n-i+1}]$.

The polynomial f_1 is still distinguished monic in $k[[x_1, z_2, \dots, z_{n-1} = t_{n-1}]]\,[z_n]$. By Remark 26.10.3a, $f_1, \dots f_r$ is a regular sequence. This proves Claim 26.13. □

Claim 26.13 and Remark 26.10.2 complete the proof of item 1 of Theorem 26.12.

For item 2 of Theorem 26.12, apply [**103**, Theorem 132, p. 95]: "A proper ideal that contains a regular sequence of length r in a commutative ring has height at least r." It follows that P is a minimal prime ideal over $(f_1, \dots, f_r)k[[Z]]$. Thus Statement 2.a holds. Statements 2.b, 2.c, 2.d and 2.e of Theorem 26.12 follow from statements 3.b, 3.c, and 3.d of Remarks 26.10. For the proof of statement 2.f of Theorem 26.12, let $P_1 := P \cap k[[Z_{n-r}]]\,[Z'_{n-r}]$, and consider the maps

$$k[[Z_{n-r}]] \overset{\alpha}{\hookrightarrow} k[[Z_{n-r}]]\,[Z'_{n-r}]/(f_1, \dots, f_r) \to k[[Z_{n-r}]]\,[Z'_{n-r}]/P_1 \cong R/P.$$

By statement 2.b, the map α is free and hence flat. Since $\operatorname{ht} P_1 = \operatorname{ht} P = r$, the height of the prime ideal $P_1/(f_1, \dots, f_r)$ in $k[[Z_{n-r}]]\,[Z'_{n-r}]/(f_1, \dots, f_r)$ is 0. Flatness of α implies that $\operatorname{ht}(P \cap k[[Z_{n-r}]]) = 0$, that is, $P \cap k[[Z_{n-r}]] = (0)$. This completes the proof of Theorem 26.12. □

The following theorem is the technical heart of this chapter.

THEOREM 26.14. *Let k be a field and let y and $X = \{x_1, \dots, x_n\}$ be variables over k. Assume that V is a discrete valuation domain with completion $\widehat{V} = k[[y]]$ and that $k[y] \subseteq V \subseteq k[[y]]$ and that the field $k((y)) = k[[y]]\,[1/y]$ has uncountable transcendence degree over the quotient field $\mathcal{Q}(V)$ of V. Set $R_0 := V[[X]]$ and $R = \widehat{R_0} = k[[y, X]]$. Let $P \in \operatorname{Spec} R$ be such that:*

(i) $P \subseteq (X)R$ *(so $y \notin P$), and*
(ii) $\dim(R/P) > 2$.

Then there is a prime ideal $Q \in \operatorname{Spec} R$ such that

(1) $P \subset Q \subset (X)R$,
(2) $\dim(R/Q) = 2$, *and*
(3) $P \cap R_0 = Q \cap R_0$.

In particular, $P \cap k[[X]] = Q \cap k[[X]]$.

PROOF. Assume that P has height r. Since $\dim(R/P) > 2$, it follows that $0 \le r < n - 1$.

If $r > 0$, then Theorem 26.12 implies there exists a transformation $x_1 \mapsto z_1, \dots, x_n \mapsto z_n$ so that the variable y is fixed and there exist a regular sequence of distinguished monic polynomials $f_1, \dots, f_r \in P$ such that

$f_1 \in k[[y, z_1, \dots, z_{n-r}]]\,[z_{n-r+1}, \dots, z_n]$ is monic in z_n,
$f_2 \in k[[y, z_1, \dots, z_{n-r}]]\,[z_{n-r+1}, \dots, z_{n-1}]$ is monic in z_{n-1} etc,

$\vdots$

$f_r \in k[[y, z_1, \dots, z_{n-r}]]\,[z_{n-r+1}]$ is monic in z_{n-r+1},

and the other assertions of Theorem 26.12 hold, so the prime ideal P is minimal over $(f_1, \dots, f_r)R$. Let $Z_{n-r} = \{z_1, \dots, z_{n-r}\}$ and $Z'_{n-r} = \{z_{n-r+1}, \dots, z_{n-1}, z_n\}$. Define $D := k[[y, Z_{n-r}]]\,[Z'_{n-r}]_{(y,Z)}$ and $P_1 := P \cap D$, By Theorem 26.12, $P_1 R = P$.

If $r = 0$, then $P = (0)$. Put $z_i := x_i$, for each i with $1 \le i \le n$, set

$$Z = Z_n = X = \{z_1, \dots, z_n\} = Z_{n-r} \text{ and } Z'_{n-r} = \emptyset.$$

Define $D := k[[y, X]]$ and $P_1 = (0) = P$.

The following diagram shows these rings and ideals.

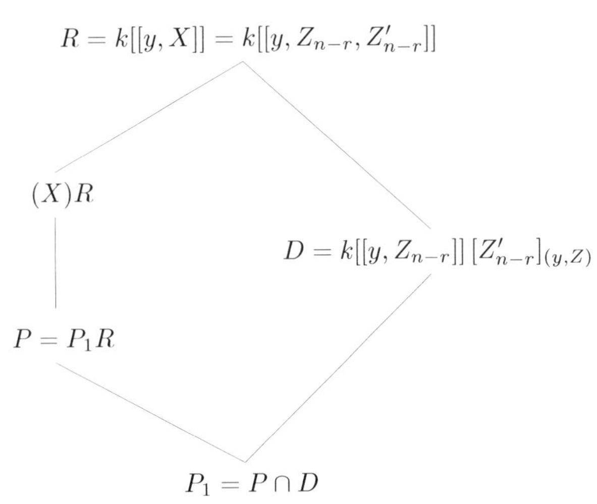

Note that $f_1, \dots, f_r \in P_1$. Let $g_1, \dots, g_s \in P_1$ be such that

$$P_1 = (f_1, \dots, f_r, g_1, \dots, g_s)D, \text{ then } P = P_1 R = (f_1, \dots, f_r, g_1, \dots, g_s)R.$$

For each $(i) := (i_1, \dots, i_n) \in \mathbb{N}^n$ and j, k with $1 \le j \le r$, $1 \le k \le s$, let $a_{j,(i)}$, $b_{k,(i)}$ denote the coefficients in $k[[y]]$ of the f_j, g_k, so that

$$f_j = \sum_{(i)\in\mathbb{N}^n} a_{j,(i)} z_1^{i_1} \dots z_n^{i_n}, \qquad g_k = \sum_{(i)\in\mathbb{N}^n} b_{k,(i)} z_1^{i_1} \dots z_n^{i_n} \in k[[y]]\,[[Z]].$$

Define

$$\Delta := \begin{cases} \{a_{j,(i)}, b_{k,(i)}\} \subseteq k[[y]], & \text{for } r > 0 \\ \emptyset, & \text{for } r = 0. \end{cases}$$

A key observation here is that in either case the set Δ is countable.

To continue the proof, we consider $S := \mathcal{Q}(V(\Delta)) \cap k[[y]]$, a discrete valuation domain, and its field of quotients $L := \mathcal{Q}(V(\Delta))$. Since Δ is a countable set, the field $k((y))$ has uncountable transcendence degree over L. Let $\gamma_2, \dots, \gamma_{n-r}$ be elements of $k[[y]]$ that are algebraically independent over L. We define $T := L(\gamma_2, \dots, \gamma_{n-r}) \cap k[[y]]$ and $E := \mathcal{Q}(T) = L(\gamma_2, \dots, \gamma_{n-r})$.

The diagram below shows the prime ideals P and P_1 and the containments among the relevant rings.

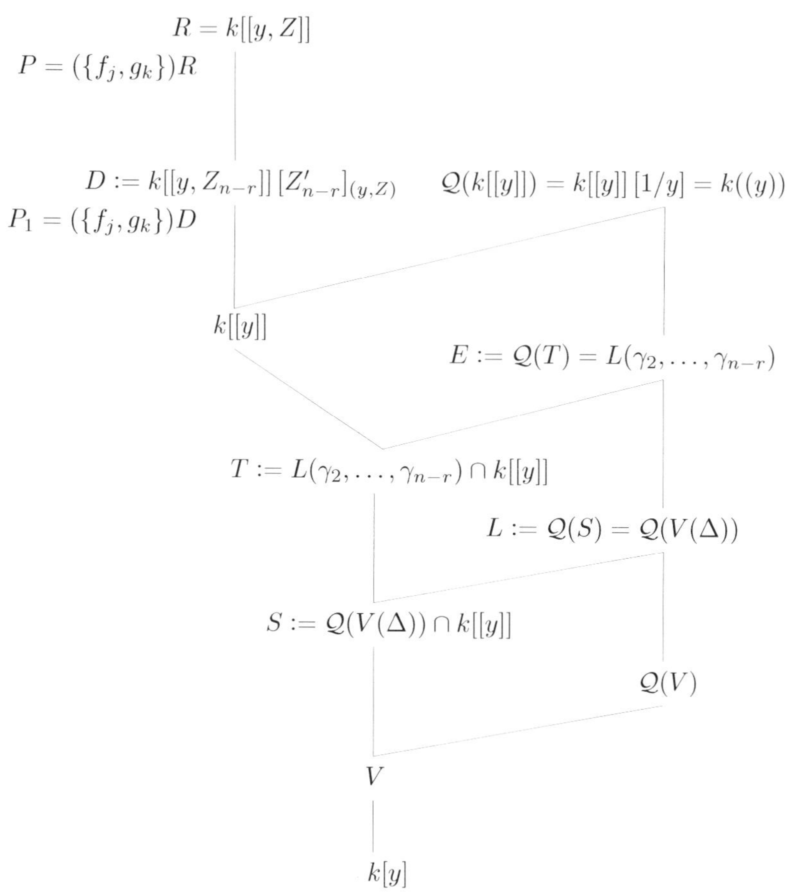

Let $P_2 := P \cap S[[Z_{n-r}]]\,[Z'_{n-r}]_{(y,Z)}$. Then $P_2R = P$, since

$$f_1, \ldots, f_r, g_1, \ldots, g_s \in S[[Z_{n-r}]]\,[Z'_{n-r}]_{(y,Z)},$$

and $P_2S[[Z]] = P \cap S[[Z]]$. Since $P \subseteq (x_1, \ldots, x_n)R = (Z)R$, there is a prime ideal $\widetilde{P}$ in $L[[Z]]$ that is minimal over $P_2L[[Z]]$. Since $L[[Z]]$ is flat over $S[[Z]]$,

$$\text{(26.14.i)} \qquad P \cap S[[Z]] = P_2S[[Z]] = \widetilde{P} \cap S[[Z]].$$

Note that $L[[X]] = L[[Z]]$ is the $(f_1, \ldots, f_r)$-adic (and the (Z'_{n-r})-adic) completion of $L[[Z_{n-r}]]\,[Z'_{n-r}]_{(Z)}$. In particular,

$$L[[Z_{n-r}]]\,[Z'_{n-r}]/(f_1, \ldots, f_r) = L[[Z_{n-r}]]\,[[Z'_{n-r}]]/(f_1, \ldots, f_r)$$

and this also holds with the field L replaced by its extension field E.

Since $L[[Z]]/\widetilde{P}$ is a homomorphic image of $L[[Z]]/(f_1, \ldots, f_r)$, it follows that $L[[Z]]/\widetilde{P}$ is integral (and finite) over $L[[Z_{n-r}]]$. This yields the commutative diagram:

$$\text{(26.14.0)} \qquad \begin{array}{ccccc} E[[Z_{n-r}]] & \overset{\varphi}{\hookrightarrow} & \dfrac{E[[Z_{n-r}]]\,[[Z'_{n-r}]]}{\widetilde{P}E[[Z]]} & = & \dfrac{E[[Z]]}{\widetilde{P}E[[Z]]} \\ \uparrow & & \uparrow & & \\ L[[Z_{n-r}]] & \overset{\psi}{\hookrightarrow} & \dfrac{L[[Z_{n-r}]]\,[[Z'_{n-r}]]}{\widetilde{P}} & = & \dfrac{L[[Z]]}{\widetilde{P}} \end{array}$$

with injective finite horizontal maps. Recall that E is the subfield of $k((y))$ obtained by adjoining $\gamma_2, \ldots, \gamma_{n-r}$ to the field L. Thus the vertical maps in Diagram 26.14.0 are faithfully flat.

Let $\mathfrak{q} := (z_2 - \gamma_2 z_1, \ldots, z_{n-r} - \gamma_{n-r} z_1)E[[Z_{n-r}]] \in \operatorname{Spec}(E[[Z_{n-r}]])$ and let $\widetilde{W}$ be a minimal prime of the ideal $(\widetilde{P}, \mathfrak{q})E[[Z]]$. Since

$$f_1, \ldots, f_r, z_2 - \gamma_2 z_1, \ldots, z_{n-r} - \gamma_{n-r} z_1$$

is a regular sequence in $T[[Z]]$, the prime ideal $W := \widetilde{W} \cap T[[Z]]$ has height $n-1$. Let $\widetilde{Q}$ be a minimal prime of $\widetilde{W}k((y))[[Z]]$ and let $Q := \widetilde{Q} \cap R$. Then

$$\text{(26.14.ii)} \qquad W = Q \cap T[[Z]] = \widetilde{W} \cap T[[Z]], \quad \text{and} \quad P \subset Q \subset (Z)R = (X)R.$$

Pictorially we have:

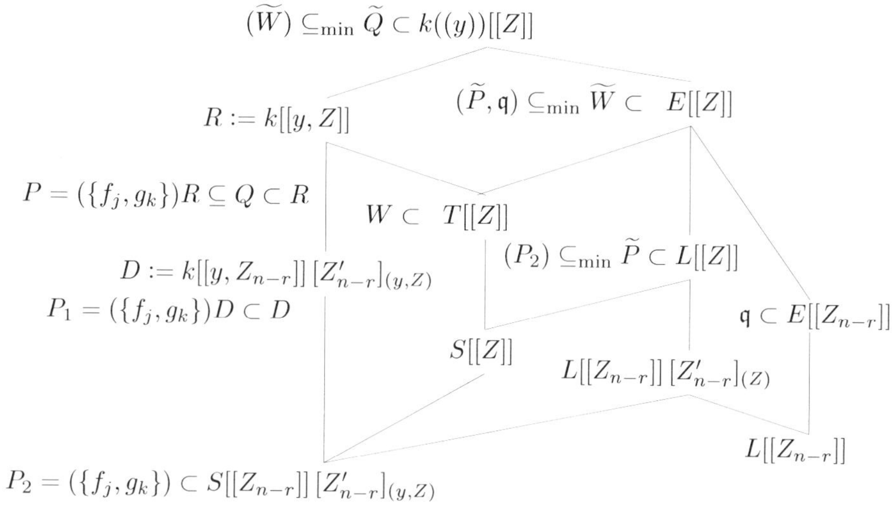

Notice that $\mathfrak{q}$ is a prime ideal of height $n - r - 1$. Since $k((y))[[Z]]$ is flat over $k[[y, Z]] = R$, it follows that $\operatorname{ht} Q = n - 1$ and $\dim(R/Q) = 2$. Thus $P_2 \subseteq W \cap S[[Z_{n-r}]]\,[Z'_{n-r}]_{(y,Z)}$.

CLAIM 26.15. $\mathfrak{q} \cap L[[Z_{n-r}]] = (0)$.

To show this we argue as in Matsumura [**121**, Proof of Theorem 2]: Suppose that

$$h = \sum_{m \in \mathbb{N}} h_m \in \mathfrak{q} \cap L[[z_1, \ldots, z_{n-r}]],$$

where $h_m \in L[z_1, \ldots, z_{n-r}]$ is a homogeneous polynomial of degree m:

$$h_m = \sum_{|(i)|=m} c_{(i)} z_1^{i_1} \ldots z_{n-r}^{i_{n-r}},$$

where $(i) := (i_1, \ldots, i_{n-r}) \in \mathbb{N}^{n-r}$, $|(i)| := i_1 + \cdots + i_{n-r}$ and $c_{(i)} \in L$. Consider the E-algebra homomorphism $\pi : E[[Z_{n-r}]] \to E[[z_1]]$ defined by $\pi(z_1) = z_1$ and $\pi(z_i) = \gamma_i z_1$ for $2 \le i \le n - r$. Then $\ker \pi = \mathfrak{q}$, and for each $m \in \mathbb{N}$:

$$\pi(h_m) = \pi\Big(\sum_{|(i)|=m} c_{(i)} z_1^{i_1} \ldots z_{n-r}^{i_{n-r}}\Big) = \sum_{|(i)|=m} c_{(i)} \gamma_2^{i_2} \ldots \gamma_{n-r}^{i_{n-r}} z_1^m$$

and

$$\pi(h) = \sum_{m \in \mathbb{N}} \pi(h_m) = \sum_{m \in \mathbb{N}} \sum_{|(i)|=m} c_{(i)} \gamma_2^{i_2} \ldots \gamma_{n-r}^{i_{n-r}} z_1^m.$$

Since $h \in \mathfrak{q}$, $\pi(h) = 0$. Since $\pi(h)$ is a power series in $E[[z_1]]$, each of its coefficients is zero, that is, for each $m \in \mathbb{N}$,

$$\sum_{|(i)|=m} c_{(i)} \gamma_2^{i_2} \ldots \gamma_{n-r}^{i_{n-r}} = 0.$$

Since the γ_i are algebraically independent over L, each $c_{(i)} = 0$. Therefore $h = 0$, and so $\mathfrak{q} \cap L[[Z_{n-r}]] = (0)$. This proves Claim 26.15.

CLAIM 26.16. $\mathfrak{q} = \widetilde{W} \cap E[[Z_{n-r}]]$.

To show Claim 26.16, consider the integral extension φ of Diagram 26.14.0:

$$E[[Z_{n-r}]] \overset{\varphi}{\hookrightarrow} \frac{E[[Z_{n-r}]]\,[[Z'_{n-r}]]}{\widetilde{P}E[[Z]]} = \frac{E[[Z]]}{PE[[Z]]}.$$

By definition, $\mathfrak{q} \in \operatorname{Spec}(E[[Z_{n-r}]])$ and $\widetilde{W}$ is a prime ideal of $E[[Z]]$ minimal over the ideal $(\mathfrak{q}, \widetilde{P})E[[Z]]$. Thus $\mathfrak{q} \subseteq \widetilde{W} \cap E[[Z_{n-r}]]$ and $\widetilde{W}E[[Z]]/PE[[[Z]]$ is a prime ideal of $E[[Z]]/PE[[[Z]]$ minimal over $\varphi(\mathfrak{q}) = \mathfrak{q}E[[Z]]/PE[[[Z]]$.

Suppose that $\mathfrak{q} \neq \widetilde{W} \cap E[[Z_{n-r}]]$; then the prime ideal $\varphi^{-1}(\widetilde{W}E[[Z]]/PE[[[Z]])$ properly contains $\mathfrak{q}$. Since φ is a flat extension, Remark 2.37.10 (Going-down) implies that there is a smaller prime ideal contained in $\varphi^{-1}(\widetilde{W}E[[Z]]/PE[[[Z]])$ that lies over $\mathfrak{q}$, a contradiction to $\widetilde{W}E[[Z]]/PE[[[Z]]$ being a minimal prime ideal over $\varphi(\mathfrak{q})$. This contradiction establishes that $\mathfrak{q} = \widetilde{W} \cap E[[Z_{n-r}]]$, as desired for Claim 26.16.

CLAIM 26.17. $\widetilde{P} = \widetilde{W} \cap L[[Z]]$.

For Claim 26.17, observe that Claims 26.15 and 26.16 imply

$$\widetilde{W} \cap L[[Z_{n-r}]] = \widetilde{W} \cap E[[Z_{n-r}]] \cap L[[Z_{n-r}]] = \mathfrak{q} \cap L[[Z_{n-r}]] = (0).$$

Consider the integral extension ψ of Diagram 26.14:

$$L[[Z_{n-r}]] \overset{\psi}{\hookrightarrow} L[[Z_{n-r}]]\,[[Z'_{n-r}]]/\widetilde{P} = L[[Z]]/\widetilde{P}.$$

Since $\widetilde{P} \in \operatorname{Spec}(L[[Z]])$, it follows that $\widetilde{P} \subseteq \widetilde{W} \cap L[[Z]]$.

Suppose that $\widetilde{P} \neq \widetilde{W} \cap L[[Z]]$. Then $(0) = \widetilde{P}/\widetilde{P} \subsetneq (\widetilde{W} \cap L[[Z]])/\widetilde{P}$ are prime ideals of $L[[Z]]/\widetilde{P}$ that both lie above the prime ideal (0). Since ψ is an integral extension of integral domains, it is not possible to have two distinct prime ideals in a chain that both lie above the same prime ideal (0) of $L[[Z_{n-r}]]$; see[**122**, Theorem 9.3]. This contradiction shows that $\widetilde{P} = \widetilde{W} \cap L[[Z]]$. Thus Claim 26.17 holds.

Consider the commutative diagram below, where all the maps are inclusions:

$$\begin{array}{ccc} R = k[[y,Z]] & \hookrightarrow & k((y))[[Z]] \\ \uparrow & & \uparrow \\ T[[Z]] & \hookrightarrow & E[[Z]] \\ \uparrow & & \uparrow \\ S[[Z]] & \hookrightarrow & L[[Z]]\,. \end{array} \tag{26.17.0}$$

Then $Q \cap T[[Z]] = W = \widetilde{W} \cap T[[Z]]$, by Equation 26.14.ii. Thus

$$Q \cap S[[Z]] = Q \cap T[[Z]] \cap S[[Z]] = \widetilde{W} \cap L[[Z]] \cap S[[Z]] = \widetilde{P} \cap S[[Z]],$$

using Claim 26.17. Since $P \cap S[[Z]] = \widetilde{P} \cap S[[Z]]$ by Equation 26.14.i, we conclude that $Q \cap S[[Z]] = P \cap S[[Z]]$ and therefore $Q \cap R_0 = P \cap R_0$. □

We record the following corollary.

COROLLARY 26.18. *Let k be a field, let $X = \{x_1, \dots, x_n\}$ and y be independent variables over k, and let $R = k[[y, X]]$. Assume $P \in \operatorname{Spec} R$ is such that:*

(i) $P \subseteq (x_1, \dots, x_n)R$ *and*
(ii) $\dim(R/P) > 2$.

Then there is a prime ideal $Q \in \operatorname{Spec} R$ so that

(1) $P \subset Q \subset (x_1, \dots, x_n)R$,
(2) $\dim(R/Q) = 2$, *and*
(3) $P \cap k[y]_{(y)}[[X]] = Q \cap k[y]_{(y)}[[X]]$.

In particular, $P \cap k[[x_1, \dots, x_n]] = Q \cap k[[x_1, \dots, x_n]]$.

PROOF. With notation as in Theorem 26.14, let $V = k[y]_{(y)}$. □

Exercises

(1) Prove that a local ring that has residue field of characteristic zero contains the field of rational numbers.
(2) Let $I = (f_1, \dots, f_r)R$ be a proper ideal of a ring R. If $(f_1, \dots, f_r)$ is a regular sequence on R of length r, prove that $\operatorname{ht} I \geq r$. This is a fact used in the proof of Theorem 26.12. A reference is given there to [**103**, Theorem 132] to confirm this fact.

CHAPTER 27

Generic fiber rings of mixed polynomial-power series rings

Our primary goal in this chapter is to prove Theorem 26.3. This theorem concerns the generic formal fiber rings (denoted Gff) of the rings $A := k[X]_{(X)}$, $B := k[[X]]\,[Y]_{(X,Y)}$ and $C := k[Y]_{(Y)}[[X]]$, where k is a field and X and Y are finite sets of indeterminates; see Equation 26.1.0 and Notation 26.1. Theorem 26.3 is proved using the techniques developed in Chapter 26.

Matsumura proves in [**121**] that the generic formal fiber ring of A has dimension one less than the dimension of A, and the generic formal fiber rings of B and C each have dimension equal to $\dim B - 2 = \dim C - 2$. Matsumura does not consider in [**121**] the question of whether for A, B or C every maximal ideal of the generic formal fiber ring has the same height.

For a local extension $R \hookrightarrow S$ of Noetherian local integral domains, Theorem 27.19 gives sufficient conditions in order that every maximal ideal of $\operatorname{Gff}(S)$ has height $h = \dim \operatorname{Gff}(R)$. Using Theorem 27.19, we prove in Theorem 27.22 that all prime ideals maximal in the generic formal fiber of a local domain essentially finitely generated over a field have the same height.

Sections 27.1 and 27.2 contain the proofs of parts 2 and 3 of Theorem 26.3. In Section 27.4 we prove part 1 of Theorem 26.3, using the results of Section 27.3 concerning subrings of power series rings in two variables. Section 27.5 contains the proof of part 4. Section 27.6 gives a more general result, Theorem 27.19, containing conditions that imply the dimension of the generic formal fiber ring of a Noetherian local ring R is equal to that of a Noetherian local extension ring S.

27.1. Weierstrass implications for the ring $B = k[[X]]\,[Y]_{(X,Y)}$

Remarks 27.1 are useful for Theorem 27.2 and Theorem 27.12.

REMARKS 27.1. Let n and m be positive integers, and let $X = \{x_1, \dots, x_n\}$ and $Y = \{y_1, \dots, y_m\}$ denote sets of variables over a field k. For each i with $1 \le i \le m$, let $Y_i := \{y_1, \dots, y_i\}$, and let $R_i = k[[X, y_i]]$. Let

$$B := k[[X]]\,[Y]_{(X,Y)} = k[[x_1, \dots, x_n]]\,[y_1, \dots, y_m]_{(x_1,\dots,x_n,y_1,\dots,y_m)} \text{ and}$$
$$C := k[Y]\,[[X]]_{(X,Y)} = k[y_1, \dots, y_m]\,[[x_1, \dots, x_n]]_{(x_1,\dots,x_n,y_1,\dots,y_m)}.$$

Let $P \in \operatorname{Spec}(k[[X, Y]])$. Then:

(1) The (X, Y)-adic completion of B and of C is $\widehat{B} = k[[X, Y]] = \widehat{C}$.

(2) Assume that $P \cap B = (0)$, $\dim(\widehat{B}/P) > 2$, and $P \subseteq (X, Y_{m-1})\widehat{B}$. Then there exists $Q \in \operatorname{Spec} \widehat{B}$ such that

$$P \subseteq Q \subseteq (X, Y_{m-1})\widehat{B}, \quad \operatorname{ht} Q = n + m - 2, \quad \text{and} \quad Q \cap B = (0).$$

(3) Assume that $P \cap C = (0)$, $\dim(\widehat{C}/P) > 2$, and $P \subseteq (X, Y_{m-1})\widehat{C}$. Then there exists $Q \in \operatorname{Spec} \widehat{C}$ such that

$$P \subseteq Q \subseteq (X, Y_{m-1})\widehat{C}, \quad \operatorname{ht} Q = n + m - 2, \quad \text{and} \quad Q \cap C = (0).$$

(4) Let $P_i := P \cap R_i$, for every i such that $1 \leq i \leq m$. Assume that $P_i \not\subseteq (X, Y_{i-1})k[[X, Y_i]]$, for some i. Then, for each i, there exists a prime element $g \in P_i \cap k[[X, Y_{i-1}]]\,[y_i]$ such that g is distinguished monic in y_i of positive degree as an element of $k[[X, Y_{i-1}]]\,[y_i]$, in the sense of Definition 26.7.

PROOF. Item 1 is clear. For items 2 and 3, let $R' = k[y_m]_{(y_m)}][[X, Y_{m-1}]]$. Corollary 26.18 with the ring $R := \widehat{B} = \widehat{C} = k[[X, Y]] = k[[y_m]]\,[[X, Y_{m-1}]]$ implies that there exists a prime ideal $Q \in \operatorname{Spec}(\widehat{C})$ such that

$$\dim(\widehat{C}/Q) = 2, \quad P \subseteq Q \subseteq (X, Y_{m-1})k[[X, Y]], \quad \text{and} \quad P \cap R' = Q \cap R'.$$

Since $k[[X, Y]]$ is catenary, $\operatorname{ht}(Q) = n + m - 2$. Since B and C both involve polynomials in y_m and power series in some of the other variables of $X \cup Y_{m-1}$, it follows that B and C are subrings of R'.

For item 2, $Q \cap B = Q \cap B \cap R' = P \cap R' \cap B = P \cap B = (0)$.

For item 3, $Q \cap C = Q \cap C \cap R' = P \cap R' \cap C = P \cap C = (0)$.

For item 4, by hypothesis $P_i \not\subseteq (X, Y_{i-1})k[[X, Y_i]]$. Since $P_i \subseteq (X, Y_i)k[[X, Y_i]]$, some element of P_i contains a term y_i^t, for some $t > 0$. Since $\widehat{C}$ is a UFD, there exists a nonzero prime element $f \in P_i$ that contains a term y_i^t, for some $t > 0$. Let t be the minimal positive power of y_i among such terms of f. We may assume the coefficient of y_i^t is 1. By Weierstrass Theorem 26.8, it follows that $f = \varepsilon g$, where $g \in k[[X, Y_{i-1}]]\,[y_i]$ is a nonzero distinguished monic polynomial in y_i of degree t and ε is a unit of $\widehat{C}$. Thus g is a prime element of $P_i \cap k[[X, Y_{i-1}]]\,[y_i]$ as desired. □

THEOREM 27.2. *Assume k is a field, X and Y are finite sets of indeterminates, $B = k[[X]]\,[Y]_{(X,Y)}$, $|X| = n$, and $|Y| = m$, where $m, n \in \mathbb{N}$. If Q is an ideal of $\widehat{B} = k[[X, Y]]$ maximal with the property that $Q \cap B = (0)$, then Q is a prime ideal of height $n + m - 2$.*

PROOF. Suppose first that Q is an ideal maximal with respect to the property that $Q \cap B = (0)$. Then clearly Q is prime. Matsumura shows in [**121**, Theorem 3] that the dimension of the generic formal fiber of B is at most $n + m - 2$. Therefore $\operatorname{ht} Q \leq n + m - 2$.

Assume $P \in \operatorname{Spec} \widehat{B}$ is an arbitrary prime ideal of height $r < n + m - 2$ with $P \cap B = (0)$. We construct a prime ideal $Q \in \operatorname{Spec} \widehat{B}$ with $P \subset Q$, $Q \cap B = (0)$, and $\operatorname{ht} Q = n + m - 2$. This will show that all prime ideals maximal in the generic fiber have height $n + m - 2$.

Use the following notation, for $1 \leq \ell < i \leq m$:

$$Y_i := \{y_1, y_2, \ldots, y_i\}; \; R_i := k[[X, Y_i]]; \; P_i := P \cap R_i; \; Y_{\ell,i} := \{y_\ell, y_{\ell+1}, \ldots, y_i\}.$$

In the case where $P \subseteq (X, Y_{m-1})\widehat{B} = (X, Y_{m-1})R_m$, there exists a Q as desired by Remark 27.1.2, and so Theorem 27.2 holds in this case.

Assume for the remainder of the proof that $P \not\subseteq (X, Y_{m-1})\widehat{B}$. For the construction of Q, we use Note 27.3.

NOTE 27.3. $P_1 \subseteq (X)k[[X, y_1]]$, and therefore $m \geq 2$.

PROOF. (of Note 27.3) If $P_1 \not\subseteq (X)k[[X, y_1]]$, then Remark 27.1.4 implies there exists a distinguished monic polynomial g such that $g \in P_1 \cap k[[X]][y_1] \subseteq P \cap B$, a contradiction to $P \cap B = (0)$.

Since $P_m = P \not\subseteq (X, Y_{m-1})\widehat{B}$, it follows that $m \geq 2$. □

The proof of Claim 27.4 completes the proof of Theorem 27.2:

CLAIM 27.4. Assume that $P \in \operatorname{Spec} \widehat{B}$, $\operatorname{ht} P = r < n + m - 2$, $P \cap B = (0)$, and $P \not\subseteq (X, Y_{m-1})\widehat{B}$. Then there exists a DVR V and a prime ideal such that

(i) $k[y_1]_{(y_1)} \subseteq V \subseteq k[[y_1]]$, and $k((y_1\beta)) = k[[y_1]][1/y_1]$ has uncountable transcendence degree over $\mathcal{Q}(V)$.
(ii) $P \subseteq V[[X]][[Y_{2,m}]]$.
(iii) $P \subset Q$, and $\operatorname{ht} Q = n + m - 2$.
(iv) $Q \cap V[[X]][Y_{2,m}] = P \cap V[[X]][Y_{2,m}]$.
(v) $Q \cap B = (0)$.

PROOF OF CLAIM 27.4. Let s be the smallest nonnegative integer with $P_{m-s} \subseteq (X, Y_{m-s-1})R_{m-s}$. By assumption, $P_m = P \not\subseteq (X, Y_{m-1})\widehat{B}$, and so $s > 0$ and $P \neq (0)$. By Note 27.3, $s = m - 1$ satisfies $P_{m-s} \subseteq (X, Y_{m-s-1})R_{m-s}$. Thus s exists and $0 < s \leq m - 1$.

The minimality of s implies $P_{m-i} \not\subseteq (X, Y_{m-i-1})R_{m-i}$, for every $0 \leq i \leq s-1$. By Remark 27.1.4, there exist prime polynomials $g_1, \ldots, g_s$ such that

$$g_1 \in P_m \cap k[[X, Y_{m-1}]][y_m], \ldots, g_i \in P_{m-i+1} \cap k[[X, Y_{m-i}]][y_{m-i-1}]. \\ \cdots g_s \in P_{m-s+1} \cap k[[X, Y_{m-s}]][y_{m-s-1}],$$

and each g_i is distinguished monic over $k[[X, Y_{m-i}]]$ in the variable y_{m-i+1}. By Remark 26.10.2, there exists a regular sequence

$$f_1 \in P_m, \ldots, f_i \in P_{m-i+1}, \ldots, f_s \in P_{m-s+1},$$

where each f_i is a distinguished monic element of $R_{m-s}[Y_{m-s+1,m-i}][y_{m-i+1}]$, monic in y_{m-i+1}. Each $f_i \in G := k[[X, Y_{m-s}]][Y_{m-s+1,m}] = R_{m-s}[Y_{m-s+1,m}]$.

By Theorem 26.12.2d, P is extended from G. Choose $h_1, \ldots, h_t \in G$ such that $P \cap G = (f_1, \ldots, f_s, h_1, \ldots, h_t)G$. Then $P = (f_1, \ldots, f_s, h_1, \ldots, h_t)\widehat{B}$. For integers d, ℓ with $1 \leq d \leq s$ and $1 \leq \ell \leq t$, express the f_d and h_ℓ in G as power series in $\widehat{B} = k[[y_1]][[y_2, \ldots, y_m]][[X]]$ with coefficients in $k[[y_1]]$:

$$f_d = \sum a_{d(i)(j)} y_2^{i_2} \cdots y_m^{i_m} x_1^{j_1} \cdots x_n^{j_n} \quad \text{and} \quad h_\ell = \sum b_{\ell(i)(j)} y_2^{i_2} \cdots y_m^{i_m} x_1^{j_1} \cdots x_n^{j_n},$$

where $(i) := (i_2, \ldots, i_m)$, $(j) := (j_1, \ldots, j_n)$, and $a_{d(i)(j)}, b_{\ell(i)(j)} \in k[[y_1]]$. The set $\Delta = \{a_{d(i)(j)}, b_{\ell(i)(j)}\}$ is countable.

For item i of Claim 27.4, define $V := k(y_1, \Delta) \cap k[[y_1]]$. Then V is a discrete valuation domain with completion $k[[y_1]]$. Since Δ is countable, the field of fractions $\mathcal{Q}(V)$ has countable transcendence degree over $k[y_1]_{(y_1)}$. By Fact 3.10, $k((y_1))$ has uncountable transcendence degree over $\mathcal{Q}(V)$. Thus item i of Claim 27.4 holds.

Item ii holds since $B \subseteq k[y_1]_{(y_1)}[[X, Y_{2,m}]] \subseteq V[[X, Y_{2,m}]]$.

Notice that $V[[X, Y_{2,m-s}]] \subseteq R_{m-s}$ and that $V[[X, Y_{2,m-s}]]$ has completion R_{m-s}. Set $\underline{f} := \{f_1, \ldots, f_s\}$. Then

$$\underline{f} \subset V[[X, Y_{2,m-s}]][Y_{m-s+1,m}] \subseteq G \text{ and } (\underline{f})G \cap R_{m-s} = (0).$$

Furthermore the extension

$$V[[X, Y_{2,m-s}]] \hookrightarrow \frac{V[[X, Y_{2,m-s}]]\,[Y_{m-s+1,m}]}{(\underline{f})V[[X, Y_{2,m-s}]]\,[Y_{m-s+1,m}]}$$

is finite and

$$P \cap (V[[X, Y_{2,m-s}]]) = P \cap R_{m-s} \cap (V[[X, Y_{2,m-s}]]) = P_{m-s} \cap (V[[X, Y_{2,m-s}]]).$$

Recall the choice of s minimal such that $P_{m-s} \subseteq (X, Y_{m-s-1})R_{m-s}$. Also R_{m-s} is the completion of $V[[X, Y_{2,m-s}]]$. It follows that

$$\begin{aligned} P_{m-s} \cap (V[[X, Y_{2,m-s}]]) &\subseteq (\,(X, Y_{m-s-1})R_{m-s}\,) \cap (V[[X, Y_{2,m-s}]]) \\ &\subseteq (X, Y_{m-s-1})(V[[X, Y_{2,m-s}]]). \end{aligned}$$

Consider the commutative diagram:

$$\begin{array}{ccc} R_{m-s} := k[[X, Y_{m-s}]] \hookrightarrow & \frac{R_{m-s}[[Y_{m-s+1,m}]]}{(\underline{f})R_{m-s}[[Y_{m-s+1,m}]]} = \frac{k[[X,Y]]}{(\underline{f})k[[X,Y]]} = \frac{\widehat{B}}{(\underline{f})\widehat{B}} \\ \uparrow & \uparrow \\ V[[X, Y_{2,m-s}]] \quad \hookrightarrow & \frac{V[[X,Y_{2,m-s}]]\,[Y_{m-s+1,m}]}{(\underline{f})V[[X,Y_{2,m-s}]]\,[Y_{m-s+1,m}]}\,. \end{array} \tag{27.4.0}$$

The horizontal maps are injective and finite and the vertical maps are completions.

The ring in the top right corner of Diagram 27.4.0 is $\widehat{B}/(\underline{f})\widehat{B}$. Let $\overline{P}$ denote the image of P in $\widehat{B}/(\underline{f})\widehat{B}$. Thus $\overline{P}$ lies over P_{m-s} in R_{m-s}. By Theorem 26.14, $P_{m-s} \subseteq (X, Y_{m-s-1})R_{m-s}$ implies there is a prime ideal Q_{m-s} of R_{m-s} such that

$$\begin{aligned} P_{m-s} \subseteq Q_{m-s} \subseteq (X, Y_{m-s-1})R_{m-s}, \quad &\dim(R_{m-s}/Q_{m-s}) = 2 \quad \text{and} \\ Q_{m-s} \cap (V[[X, Y_{2,m-s}]]) &= P_{m-s} \cap (V[[X, Y_{2,m-s}]]). \end{aligned}$$

(The rings of the diagram are catenary, so the height of Q_{m-s} is determined by its dimension.) Since the top row of the diagram is an integral extension, the "Going-up Theorem" holds. Therefore there is a prime ideal $\overline{Q}$ in $\widehat{B}/(f_1, \dots, f_s)\widehat{B}$ such that $\overline{Q}$ lies over Q_{m-s} with $\overline{P} \subseteq \overline{Q}$, [**122**, Theorem 9.4].

Let Q be the preimage in $\widehat{B} = k[[X, Y]]$ of $\overline{Q}$. Then Q has height

$$n + s - 2 + m - s = n + m - 2.$$

Thus item iii of Claim 27.4 holds. Moreover, it follows from Diagram 27.4.0 that Q and P have the same contraction to $V[[X, Y_{2,m}]]$.

For the proof of item v, since $B \subseteq V[[X, Y_{2,m-s}]]\,[Y_{m-s+1,m}]$, it follows that

$$\begin{aligned} Q \cap B &= Q \cap (V[[X, Y_{2,m-s}]]\,[Y_{m-s+1,m}]) \cap B \\ &= P \cap (V[[X, Y_{2,m-s}]]\,[Y_{m-s+1,m}]) \cap B = P \cap B = (0), \end{aligned}$$

as desired. This completes the proof of Claim 27.4. □

Theorem 27.2 is also proved. □

27.2. Weierstrass implications for the ring $C = k[Y]_{(Y)}[[X]]$

As before, k denotes a field, n and m are positive integers, and $X = \{x_1, \dots, x_n\}$ and $Y = \{y_1, \dots, y_m\}$ denote sets of variables over k. Consider the ring

$$C = k[y_1, \dots, y_m]_{(y_1,\dots,y_m)}[[x_1, \dots, x_n]] = k[Y]_{(Y)}[[X]].$$

The completion of C is $\widehat{C} = k[[Y, X]] = k[[X, Y]]$.

THEOREM 27.5. *Assume the notation above. If an ideal Q of $\widehat{C}$ is maximal with respect to the property that $Q \cap C = (0)$, then Q is a prime ideal and* $\operatorname{ht} Q = n+m-2$.

PROOF. As in Theorem 27.2, it is easy to see that an ideal Q maximal such that $Q \cap C = (0)$ is a prime ideal.

Let $B = k[[X]]\,[Y]_{(X,Y)} \subset C$. If $P \in \operatorname{Spec} \widehat{C} = \operatorname{Spec} \widehat{B}$ is such that $P \cap C = (0)$, then $P \cap B = (0)$, and so $\operatorname{ht} P \leq n+m-2$, by Theorem 27.2. Let P be a nonzero prime ideal of $\operatorname{Spec} \widehat{C}$ with $P \cap C = (0)$ and $\operatorname{ht} P = r < n+m-2$. It suffices to show there exists a prime ideal $Q \in \operatorname{Spec} \widehat{C}$ such that

$$P \subseteq Q, \;\; Q \cap C = (0), \;\; \operatorname{ht} Q = n+m-2.$$

In the notation of Remark 27.1, $Y_{m-1} = \{y_1, \ldots, y_{m-1}\}$. If $P \subseteq (X, Y_{m-1})\widehat{C}$, then Remark 27.1.3 implies the existence of $Q \in \operatorname{Spec} \widehat{C}$ with $\operatorname{ht} Q = n+m-2$ such that $P \subset Q$ and $Q \cap C = (0)$. Thus Theorem 27.5 holds if $P \subseteq (X, Y_{m-1})\widehat{C}$.

Therefore assume that P is not contained in $(X, Y_{m-1})\widehat{C}$ for the remainder of this proof. Then P is not contained in $(X)\widehat{C}$. Consider the ideal $J := (P, X)\widehat{C}$. Since C is complete in the XC-adic topology, if J is primary for the maximal ideal of $\widehat{C}$, then P is extended from C; see [**157**, Lemma 2]. Since $P \cap C = (0)$, J is not primary for the maximal ideal of $\widehat{C}$, and so $\operatorname{ht} J = n+s < n+m$, where $0 < s < m$.

Let $W \in \operatorname{Spec} \widehat{C}$ be a minimal prime ideal of J such that $\operatorname{ht} W = n+s$, and let $W_0 = W \cap k[[Y]]$. Then $W = (W_0, X)\widehat{C}$ and W_0 is a prime ideal of $k[[Y]]$ with $\operatorname{ht} W_0 = s$. By Proposition 26.12 applied to $k[[Y]]$ and the prime ideal $W_0 \in \operatorname{Spec} k[[Y]]$, there exists a change of variables $Y \mapsto Z$ with

$$y_1 \mapsto y_1 = z_1, \;\; y_2 \mapsto z_2, \;\; \ldots, \;\; y_m \mapsto z_m$$

and elements $f_1, \ldots, f_s \in W_0$, so that, with $Z_1 = \{z_1, \ldots, z_{m-s}\}$,

$$\begin{array}{ll} f_1 \in k[[Z_1]]\,[z_{m-s+1}, \ldots, z_m] & \text{is monic in } z_m \\ f_2 \in k[[Z_1]]\,[z_{m-s+1}, \ldots, z_{m-1}] & \text{is monic in } z_{m-1}, \text{ etc} \\ \vdots & \\ f_s \in k[[Z_1]]\,[z_{m-s+1}] & \text{is monic in } z_{m-s+1}. \end{array}$$

Then $z_1, \ldots, z_{m-s}, f_1, \ldots, f_s$ is a regular sequence in $k[[Z]] = k[[Y]]$. Let T be a set of $m-s$ additional variables: $T = \{t_{m-s+1}, \ldots, t_m\}$. Define the map:

$$\varphi : k[[Z_1, T]] \longrightarrow k[[z_1, \ldots, z_m]]$$

by $z_i \mapsto z_i$, for every $1 \leq i \leq m-s$, and $t_{m-j+1} \mapsto f_j$, for every j with $1 \leq j \leq s$. The embedding φ is finite (and free) and so is the extension ρ of φ to the power series ring in X over $k[[Z_1, T]]$:

$$\rho : k[[Z_1, T]]\,[[X]] \longrightarrow k[[z_1, \ldots, z_m]]\,[[X]] = \widehat{C}.$$

The contraction $\rho^{-1}(W) \in \operatorname{Spec} k[[Z_1, T, X]]$ of the prime ideal W of $\widehat{C}$ has height $n+s$, since $\operatorname{ht} W = n+s$. Moreover $\rho^{-1}(W)$ contains $(T, X)k[[Z_1, T, X]]$, a prime ideal of height $n+s$. Therefore $\rho^{-1}(W) = (T, X)k[[Z_1, T, X]]$. By construction, $P \subseteq W$, and so $\rho^{-1}(P) \subseteq (T, X)k[[Z_1, T, X]]$.

Next construct a DVR V that contains all the coefficients in $k[[z_1]]$ of the generators f_i of W_0 and of generators of P. As in the proof of Theorem 27.2, write expressions for the f_j as power series in $z_2, \ldots, z_m$ with coefficients in $k[[z_1]]$:

$$f_j = \sum a_{j(i)} z_2^{i_2} \cdots z_m^{i_m},$$

where $(i) := (i_2, \ldots, i_m), 1 \le j \le s$, and $a_{j(i)} \in k[[z_1]]$. Let $g_1, \ldots, g_e$ be a finite set of generators for P and express the g_d, where $1 \le d \le e$, as power series in $z_2, \ldots, z_m, x_1, \ldots, x_n$ with coefficients in $k[[z_1]]$:

$$g_d = \sum b_{d(i)(\ell)} z_2^{i_2} \cdots z_m^{i_m} x_1^{\ell_1} \cdots x_n^{\ell_n},$$

where $(i) := (i_2, \ldots, i_m)$, $(\ell) := (\ell_1, \ldots, \ell_n)$, and $b_{d(i)(\ell)} \in k[[z_1]]$. Let Δ denote the subset $\Delta = \{a_{j(i)}, b_{d(i)(\ell)}\}$ of $k[[z_1]]$. Define the discrete valuation domain:

$$V := k(z_1, \Delta) \cap k[[z_1]].$$

Since V is countably generated over $k(z_1)$, the field $k((z_1))$ has uncountable transcendence degree over $\mathcal{Q}(V) = k(z_1, \Delta)$. Moreover, by construction the ideal P is extended from $V[[z_2, \ldots, z_m]]\,[[X]]$.

Consider the embedding $\psi : V[[z_2, \ldots, z_{m-s}, T]] \longrightarrow V[[z_2, \ldots, z_m]]$. The map ψ is the restriction of φ above, so that ψ is the identity on V; $z_i \mapsto z_i$, for every i with $2 \le i \le m - s$; and $t_{m-j+1} \mapsto f_j$, for every j with $1 \le j \le s$.

Let σ be the extension of ψ to the power series rings:

$$\sigma : V[[z_2, \ldots, z_{m-s}, T]]\,[[X]] \longrightarrow V[[z_2, \ldots, z_m]]\,[[X]]$$

with $\sigma(x_i) = x_i$ for all i with $1 \le i \le n$.

Notice that the finite (free) embedding ρ defined above is the completion $\widehat{\sigma}$ of the map σ, that is, the extension of σ to the completions. Define

$$\lambda : V[[z_2, \ldots, z_{m-s}]]\,[T]_{(Z_1,T)}[[X]] \longrightarrow V[[z_2, \ldots, z_{m-s}]]\,[z_{m-s+1}, \ldots, z_m]_{(Z)}[[X]],$$

determined by $t_{m-j+1} \mapsto f_j$, for every j with $1 \le j \le s$, also a finite free embedding.

Consider the commutative diagram:

(27.5.0)

$$\begin{array}{ccc}
k[[Z_1, T, X]] & \xrightarrow{\widehat{\sigma}=\rho} & k[[Z, X]] = \widehat{C} \\
\uparrow & & \uparrow \\
V[[z_2, \ldots, z_{m-s}, T, X]] & \xrightarrow{\sigma} & V[[z_2, \ldots, z_m, X]] \\
\uparrow & & \uparrow \\
V[[z_2, \ldots, z_{m-s}]]\,[T]_{(Z_1,T)}[[X]] & \xrightarrow{\lambda} & V[[z_2, \ldots, z_{m-s}]]\,[z_{m-s+1}, \ldots, z_m]_{(Z)}[[X]],
\end{array}$$

where ρ and λ are finite embeddings.

Recall that $\rho^{-1}(P) \subseteq (T, X)k[[Z_1, T, X]]$. By Theorem 26.14, there exists a prime ideal Q_0 of the ring $k[[Z_1, T, X]]$ such that $\rho^{-1}(P) \subseteq Q_0$, $\operatorname{ht} Q_0 = n + m - 2$, and

$$Q_0 \cap V[[z_2, \ldots, z_{m-s}, T, X]] = \rho^{-1}(P) \cap V[[z_2, \ldots, z_{m-s}, T, X]]. \tag{27.5.1}$$

By the "Going-up theorem" [**122**, Theorem 9.4] used with the finite embedding ρ of Diagram 27.5.0, there is a prime ideal $Q \in \operatorname{Spec} \widehat{C}$ that lies over Q_0 and contains P. Moreover, Q also has height $n + m - 2$.

Since ρ is a finite free embedding,

$$P \cap k[[Z_1, T, X]] = \rho^{-1}(P) \subseteq Q \cap k[[Z_1, T, X]] = Q_0. \tag{27.5.2}$$

To complete the proof, we show $Q \cap C = (0)$. For this, take the intersections of the rings of Diagram 27.5.0 with P and Q and "chase" through the diagram. By Equations 27.5.1 and 27.5.2,

$$P \cap V[[z_2,\dots,z_{m-s},T,X]] = Q \cap V[[z_2,\dots,z_{m-s},T,X]],$$

at the left middle level of the diagram.

Let $\widetilde{P} = P \cap V[[z_2,\dots,z_{m-s}]]\,[z_{m-s+1},\dots,z_m]_{(Z)}[[X]]$. Then

$$\widetilde{P} \subseteq \widetilde{Q} = Q \cap V[[z_2,\dots,z_{m-s}]]\,[z_{m-s+1},\dots,z_m]_{(Z)}[[X]].$$

The prime ideals $\widetilde{P}$ and $\widetilde{Q}$ of $V[[z_2,\dots,z_{m-s}]]\,[z_{m-s+1},\dots,z_m]_{(Z)}[[X]]$ both lie over

$$P \cap V[[z_2,\dots,z_{m-s}]]\,[T]_{(Z_1,T)}[[X]] = Q \cap V[[z_2,\dots,z_{m-s}]]\,[T]_{(Z_1,T)}[[X]].$$

Since λ is a finite map, we conclude that $\widetilde{P} = \widetilde{Q}$.

Since $C \subseteq V[[z_2,\dots,z_{m-s}]]\,[z_{m-s+1},\dots,z_m]_{(Z)}[[X]]$, the intersection

$$\begin{aligned} Q \cap C &= Q \cap V[[z_2,\dots,z_m]]\,[z_{m-s+1},\dots,z_m]_{(Z)}[[X]] \cap C = \widetilde{Q} \cap C = \widetilde{P} \cap C \\ &= P \cap V[[z_2,\dots,z_m]]\,[z_{m-s+1},\dots,z_m]_{(Z)}[[X]] \cap C = P \cap C = (0). \end{aligned}$$

This completes the proof of Theorem 27.5. □

REMARK 27.6. With B and C as in Sections 27.1 and 27.2, we have

$$B = k[[X]]\,[Y]_{(X,Y)} \hookrightarrow k[Y]_{(Y)}[[X]] = C \quad \text{and} \quad \widehat{B} = k[[X,Y]] = \widehat{C}.$$

Thus for $P \in \operatorname{Spec} k[[X,Y]]$, if $P \cap C = (0)$, then $P \cap B = (0)$. By Theorems 27.2 and 27.5, each prime of $k[[X,Y]]$ maximal in the generic formal fiber of B or C has height $n+m-2$. Therefore each $P \in \operatorname{Spec} k[[X,Y]]$ maximal with respect to $P \cap C = (0)$ is also maximal with respect to $P \cap B = (0)$. However, if $n+m \geq 3$, the generic fiber of $B \hookrightarrow C$ is nonzero (see Propositions 28.25 and 28.27 of Chapter 28), and so there exist primes of $k[[X,Y]]$ maximal in the generic formal fiber of B that are not in the generic formal fiber of C.

27.3. Subrings of the power series ring $k[[z,t]]$

In this section we establish properties of certain subrings of the power series ring $k[[z,t]]$ that are useful in considering the generic formal fiber of localized polynomial rings over the field k.

NOTATION 27.7. Let k be a field and let z and t be independent variables over k. Consider countably many power series:

$$\alpha_i(z) = \sum_{j=0}^{\infty} a_{ij} z^j \in k[[z]]$$

with coefficients $a_{ij} \in k$. Let s be a positive integer and let $\omega_1,\dots,\omega_s \in k[[z,t]]$ be power series in z and t, say:

$$\omega_i = \sum_{j=0}^{\infty} \beta_{ij} t^j, \quad \text{where} \quad \beta_{ij}(z) = \sum_{\ell=0}^{\infty} b_{ij\ell} z^\ell \in k[[z]] \quad \text{and} \quad b_{ij\ell} \in k,$$

for each i with $1 \leq i \leq s$. Consider the subfield $k(z, \{\alpha_i\}, \{\beta_{ij}\})$ of $k((z))$ and the discrete rank-one valuation domain

$$V := k(z, \{\alpha_i\}, \{\beta_{ij}\}) \cap k[[z]]. \tag{27.7.0}$$

The completion of V is $\widehat{V} = k[[z]]$. Assume that $\omega_1, \ldots, \omega_r$ are algebraically independent over $\mathcal{Q}(V)(t)$ and that the elements $\omega_{r+1}, \ldots, \omega_s$ are algebraic over the field $\mathcal{Q}(V)(t, \{\omega_i\}_{i=1}^r)$. Notice that the set $\{\alpha_i\} \cup \{\beta_{ij}\}$ is countable, and that also the set of coefficients of the α_i and β_{ij}

$$\Delta := \{a_{ij}\} \cup \{b_{ij\ell}\}$$

is a countable subset of the field k. Let k_0 denote the prime subfield of k and let F denote the algebraic closure in k of the field $k_0(\Delta)$. The field F is countable and the power series $\alpha_i(z)$ and $\beta_{ij}(z)$ are in $F[[z]]$. Consider the subfield $F(z, \{\alpha_i\}, \{\beta_{ij}\})$ of $F((z))$ and the discrete rank-one valuation domain

$$V_0 := F(z, \{\alpha_i\}, \{\beta_{ij}\}) \cap F[[z]].$$

The completion of V_0 is $\widehat{V_0} = F[[z]]$. Since $\mathcal{Q}(V_0)(t) \subseteq \mathcal{Q}(V)(t)$, the elements $\omega_1, \ldots, \omega_r$ are algebraically independent over the field $\mathcal{Q}(V_0)(t)$. Let

$$\begin{aligned} E_0 :&= \mathcal{Q}(V_0)(t, \omega_1, \ldots, \omega_r), \qquad E := \mathcal{Q}(V)(t, \omega_1, \ldots, \omega_r), \qquad \text{and} \\ \widetilde{E} :&= \mathcal{Q}(V)(t, \omega_1, \ldots, \omega_s). \end{aligned}$$

Thus $E_0 \subseteq E \subseteq \widetilde{E}$, the field E_0 is a subfield of $\mathcal{Q}(V_0[[t]])$, and the fields E and $\widetilde{E}$ are subfields of $\mathcal{Q}(V[[t]])$ with $\widetilde{E}$ algebraic over E.

REMARK 27.8. Assume Notation 27.7. Define the integral domains:

$$\begin{aligned} D_0 :&= E_0 \cap V_0[[t]], \qquad D := E \cap V[[t]], \qquad \text{and} \\ \widetilde{D} :&= \widetilde{E} \cap V[[t]]. \end{aligned} \tag{27.8.0}$$

Thus $D_0 \subseteq D \subseteq \widetilde{D}$, and $\widetilde{D}$ is algebraic over D. A result of Valabrega, Theorem 4.9, implies that D_0, D and $\widetilde{D}$ are two-dimensional regular local rings with completions $\widehat{D_0} = F[[z,t]]$ and $\widehat{D} = \widehat{\widetilde{D}} = k[[z,t]]$. Moreover, $\mathcal{Q}(D_0) = E_0$ is a countable field, and $\mathcal{Q}(\widetilde{D}) = \widetilde{E}$ is algebraic over $\mathcal{Q}(D) = E$. If $\gamma \in zF[[z]]$ and $(t-\gamma)k[[z,t]] \cap D = (0)$, then $(t-\gamma)k[[z,t]] \cap \widetilde{D} = (0)$.

PROPOSITION 27.9. *Assume Notation 27.7. Define D_0 as in Equation 27.8.0. Then there exists a power series $\gamma \in zF[[z]]$ such that the prime ideal $(t-\gamma)F[[z,t]] \cap D_0 = (0)$, that is, $(t-\gamma)F[[z,t]]$ is in the generic formal fiber of D_0.*

PROOF. Since D_0 is countable there are only countably many prime ideals in D_0 and since D_0 is Noetherian there are only countably many prime ideals in $\widehat{D_0} = F[[z,t]]$ that lie over a nonzero prime of D_0. There are uncountably many primes in $F[[z,t]]$, which are generated by elements of the form $t-\sigma$ for some $\sigma \in zF[[z]]$. Thus there must exist an element $\gamma \in zF[[z]]$ with $(t-\gamma)F[[z,t]] \cap D_0 = (0)$. $\square$

For $\omega_i = \omega_i(t) = \sum_{j=0}^{\infty} \beta_{ij} t^j$ as in Notation 27.7 and γ an element of $zk[[z]]$, let $\omega_i(\gamma)$ denote the following power series in $k[[z]]$:

$$\omega_i(\gamma) := \sum_{j=0}^{\infty} \beta_{ij} \gamma^j \in k[[z]].$$

PROPOSITION 27.10. *Assume Notation 27.7. Let k' be a field with $\Delta \subset k' \subseteq k$, let V' be a DVR with $k'[z]_{(z)} \subseteq V' \subseteq k'[[z]]$, let $E' := \mathcal{Q}(V')(t, \omega_1, \ldots, \omega_r)$, and* *let* $D' = E' \cap V'[[t]]$*, similar to Equation 27.8.0. For an element* $\gamma \in zk'[[z]]$*, the following conditions are equivalent:*

(i) $(t-\gamma)k'[[z,t]] \cap D' = (0)$.

(ii) $\gamma, \omega_1(\gamma), \dots, \omega_r(\gamma)$ *are algebraically independent over* $\mathcal{Q}(V')$.

PROOF. (i) $\Rightarrow$ (ii): Assume by way of contradiction that $\{\gamma, \omega_1(\gamma), \dots, \omega_r(\gamma)\}$ is an algebraically dependent set over $\mathcal{Q}(V')$ and let $d_{(h)} \in V'$ be finitely many elements such that

$$\sum_{(h)} d_{(h)}\omega_1(\gamma)^{h_1}\dots\omega_r(\gamma)^{h_r}\gamma^{h_{r+1}} = 0$$

is a nontrivial equation of algebraic dependence for $\gamma, \omega_1(\gamma), \dots, \omega_r(\gamma)$, where each $(h) = (h_1 \dots, h_r, h_{r+1})$ is an $(r+1)$-tuple of nonnegative integers. It follows that

$$\sum_{(h)} d_{(h)}\omega_1^{h_1}\dots\omega_r^{h_r}t^{h_{r+1}} \in (t-\gamma)k'[[z,t]] \cap D' = (0).$$

Since $\omega_1, \dots, \omega_r$ are algebraically independent over $\mathcal{Q}(V')(t)$, we have $d_{(h)} = 0$ for all (h), a contradiction. This completes the proof that (i) $\Rightarrow$(ii).

(ii) $\Rightarrow$ (i): If $(t-\gamma)k'[[z,t]] \cap D' \neq (0)$, then there exists a nonzero element

$$\tau = \sum_{(h)} d_{(h)}\omega_1^{h_1}\dots\omega_r^{h_r}t^{h_{r+1}} \in (t-\gamma)k'[[z,t]] \cap V'[t, \omega_1, \dots, \omega_r],$$

with $d_{(h)} \in V$. But this implies that

$$\tau(\gamma) = \sum_{(h)} d_{(h)}\omega_1(\gamma)^{h_1}\dots\omega_r(\gamma)^{h_r}\gamma^{h_{r+1}} = 0.$$

Since $\gamma, \omega_1(\gamma), \dots, \omega_r(\gamma)$ are algebraically independent over $\mathcal{Q}(V')$, it follows that all the coefficients $d_{(h)} = 0$, a contradiction to the assumption that τ is nonzero. $\square$

Proposition 27.11 follows from Proposition 27.9. In the statement we include just the part of the setting necessary to apply Proposition 27.11 to the proof of Theorem 27.12, instead of the full setting and notation of Notation 27.7 and Proposition 27.9.

PROPOSITION 27.11. *Let* $\alpha_i(z) = \sum_{j=0}^{\infty} a_{ij}z^j \in k[[z]]$ *be countably many power series in a variable* z *over a field* k *with coefficients* $a_{ij} \in k$*, let* $\omega_1, \dots, \omega_s \in k[[z,t]]$ *be power series in* z *and in an additional variable* t*, and write:*

$$\omega_i = \sum_{j=0}^{\infty} \beta_{ij}t^j, \quad \textit{where} \quad \beta_{ij}(z) = \sum_{\ell=0}^{\infty} b_{ij\ell}z^\ell \in k[[z]] \quad \textit{and} \quad b_{ij\ell} \in k,$$

for each i *with* $1 \le i \le s$*. Let* $r \le s$ *be such that, with renumbering,* $\omega_1, \dots, \omega_r$ *is a maximal subset that is algebraically independent over* $\mathcal{Q}(V)(t)$*. Define*

$$V := k(z, \{\alpha_i\}, \{\beta_{ij}\}) \cap k[[z]], \quad E := \mathcal{Q}(V)(t, \omega_1, \dots, \omega_r), \quad D := E \cap V[[t]],$$
$$\widetilde{E} := \mathcal{Q}(V)(t, \omega_1, \dots, \omega_s), \quad \widetilde{D} := \widetilde{E} \cap V[[t]],$$

so that V *is a DVR with completion* $\widehat{V} = k[[z]]$*, and* $E \subseteq \widetilde{E} \subseteq \mathcal{Q}(V[[t]])$*. Then there exists a power series* $\gamma \in zk[z]]$ *such that*

(1) $(t-\gamma)k[[z,t]] \cap D = (0)$*, that is,* $(t-\gamma)k[[z,t]]$ *is in the generic formal fiber of* D.

(2) $(t-\gamma)k[[z,t]] \cap \widetilde{D} = (0)$, *that is*, $(t-\gamma)k[[z,t]]$ *is in the generic formal fiber of* $\widetilde{D}$.

PROOF. For the proof, use the (expanded) notation from Notation 27.7 and Remark 27.8. Recall that F is the algebraic closure in k of the field $k_0(\{a_{ij}\} \cup \{b_{ij\ell}\})$, where k_0 denotes the prime subfield of k. The field F is countable and the power series $\alpha_i(z)$ and $\beta_{ij}(z)$ are in $F[[z]]$. Also $V_0 := F(z, \{\alpha_i\}, \{\beta_{ij}\}) \cap F[[z]]$,

$$E_0 := \mathcal{Q}(V_0)(t, \omega_1, \dots, \omega_r) \subseteq E, \qquad \text{and} \qquad D_0 = E_0 \cap V_0[[t]] \subseteq D.$$

By Proposition 27.9, there exists $\gamma \in zF[[z]]$ with $(t-\gamma)F[[z,t]] \cap D_0 = (0)$. We show that $(t-\gamma)k[[z,t]] \cap D = (0)$.

Let $L := F(\{t_i\}_{i \in I})$, where $\{t_i\}_{i \in I}$ is a transcendence basis of k over F. Then $L \subseteq k$ and k is algebraic over L. The elements $\{\alpha_i\}, \{\beta_{ij}\}$ are contained in $F[[z]]$. Set

$$V_1 := L(z, \{\alpha_i\}, \{\beta_{ij}\}) \cap L[[z]] \quad \text{and} \quad D_1 := \mathcal{Q}(V_1)(t, \omega_1, \dots, \omega_r) \cap L[[z,t]].$$

Then V_1 is a discrete rank-one valuation domain with completion $L[[z]]$ and D_1 is a two-dimensional regular local domain with completion $\widehat{D_1} = L[[z,t]]$. Note that $\mathcal{Q}(V)$ and $\mathcal{Q}(D)$ are algebraic over $\mathcal{Q}(V_1)$ and $\mathcal{Q}(D_1)$, respectively.

To show $(t-\gamma)k[[z,t]] \cap D = (0)$, it suffices to prove that $(t-\gamma)k[[z,t]] \cap D_1 = (0)$, since $\mathcal{Q}(D)$ is algebraic over $\mathcal{Q}(V_1)$. For this, $(t-\gamma)k[[z,t]] \cap L[[z,t]] = (t-\gamma)L[[z,t]]$, and so it suffices to prove that $(t-\gamma)L[[z,t]] \cap D_1 = (0)$.

The commutative diagram

$$\begin{array}{ccc} F[[z]] & \xrightarrow{\{t_i\}\text{algebraically ind.}} & L[[z]] \\ \uparrow & & \uparrow \\ \mathcal{Q}(V_0) & \xrightarrow{\text{transcendence basis } \{t_i\}} & \mathcal{Q}(V_1) \end{array}$$

implies that the set $\{\gamma, \omega_1(\gamma), \dots, \omega_r(\gamma)\} \cup \{t_i\}$ is algebraically independent over $\mathcal{Q}(V_0)$. Therefore $\{\gamma, \omega_1(\gamma), \dots, \omega_r(\gamma)\}$ is algebraically independent over $\mathcal{Q}(V_1)$. By Proposition 27.10, $(t-\gamma)L[[z,t]] \cap D_1 = (0)$. Thus item 1 holds.

Item 2 follows from item 1, since $\widetilde{D}$ is algebraic over D. This completes the proof of Proposition 27.11. □

27.4. Weierstrass implications for the localized polynomial ring A

Let n be a positive integer, let $X = \{x_1, \dots, x_n\}$ be a set of n variables over a field k, and let $A := k[x_1, \dots, x_n]_{(x_1, \dots, x_n)} = k[X]_{(X)}$ denote the localized polynomial ring in these n variables over k. Then the completion of A is $\widehat{A} = k[[X]]$.

THEOREM 27.12. *For the localized polynomial ring* $A = k[X]_{(X)}$ *defined above, if* Q *is an ideal of* $\widehat{A}$ *maximal with respect to* $Q \cap A = (0)$, *then* Q *is a prime ideal of height* $n-1$.

PROOF. It is clear that Q as described in the statement is a prime ideal. Also the assertion holds for $n = 1$. Thus we assume $n \geq 2$. By Matsumura's result [**121**, Theorem 2], there exists a nonzero prime ideal $\mathfrak{p}$ in $k[[x_1, x_2]]$ such that $\mathfrak{p} \cap k[x_1, x_2]_{(x_1, x_2)} = (0)$. It follows that $\mathfrak{p}\widehat{A} \cap A = (0)$. Thus the generic formal fiber of A is nonzero, and Theorem 27.12 holds for $n = 2$.

Assume $n > 2$. Let $P \in \operatorname{Spec} \widehat{A}$ be a nonzero prime ideal with $P \cap A = (0)$ and $\operatorname{ht} P = r < n-1$. We construct $Q \in \operatorname{Spec} \widehat{A}$ of height $n-1$ with $P \subseteq Q$ and $Q \cap A = (0)$:

By Proposition 26.12, there exist a change of variables

$$x_1 \mapsto x_1 = z_1,\ x_2 \mapsto z_2,\ \dots,\ x_n \mapsto z_n$$

and polynomials in variables $z_{n-r+1}, \dots, z_n$ over $k[[z_1, \dots, z_{n-r}]]$ such that:

$$\begin{aligned} &f_1 \in k[[z_1, \dots, z_{n-r}]]\,[z_{n-r+1}, \dots, z_n] && \text{monic in } z_n \\ &f_2 \in k[[z_1, \dots, z_{n-r}]]\,[z_{n-r+1}, \dots, z_{n-1}] && \text{monic in } z_{n-1}, \text{ etc} \\ &\vdots \\ &f_r \in k[[z_1, \dots, z_{n-r}]]\,[z_{n-r+1}] && \text{monic in } z_{n-r+1}, \end{aligned}$$

$f_1, \dots, f_r$ is a regular sequence of $\widehat{A}$, P is a minimal prime of $(f_1, \dots, f_r)\widehat{A}$, and P is extended from

$$R := k[[z_1, \dots, z_{n-r}]]\,[z_{n-r+1}, \dots, z_n].$$

Let $P_0 := P \cap R$. By assumption, $\operatorname{ht} P = r$. Then $\operatorname{ht} P_0 = r$, since P_0 is also minimal over the regular sequence $f_1, \dots, f_r$ of length r; see [**103**, Theorem 132, p. 95]. By Proposition 26.12.2d, $P = P_0\widehat{A}$.

Extend $f_1, \dots, f_r$ to a set of generators of P_0, say:

$$P_0 = (f_1, \dots, f_r, g_1, \dots, g_s)R.$$

Using an argument similar to that in the proof of Theorem 26.14, write

$$f_j = \sum_{(i) \in \mathbb{N}^{n-1}} a_{j,(i)} z_2^{i_2} \cdots z_n^{i_n} \quad \text{and} \quad g_\ell = \sum_{(i) \in \mathbb{N}^{n-1}} b_{\ell,(i)} z_2^{i_2} \cdots z_n^{i_n},$$

where $(i) = (i_2, \dots, i_n) \in \mathbb{N}^{n-1}$ and $a_{j,(i)}, b_{\ell,(i)} \in k[[z_1]]$. Let

$$V_0 := k(z_1, \{a_{j,(i)}, b_{\ell,(i)}\}) \cap k[[z_1]].$$

Then V_0 is a discrete rank-one valuation domain with completion $k[[z_1]]$, and $k((z_1))$ has uncountable transcendence degree over the field of fractions $\mathcal{Q}(V_0)$ of V_0. Let $\gamma_3, \dots, \gamma_{n-r} \in k[[z_1]]$ be algebraically independent over $\mathcal{Q}(V_0)$ and define

$$\mathfrak{q} := (z_3 - \gamma_3 z_2, z_4 - \gamma_4 z_2, \dots, z_{n-r} - \gamma_{n-r} z_2)k[[z_1, \dots, z_{n-r}]].$$

Then $\mathfrak{q} \cap V_0[[z_2, \dots, z_{n-r}]] = (0)$, by an argument similar to that in Claim 26.15; see also [**121**]. Let $R_1 := V_0[[z_2, \dots, z_{n-r}]]\,[z_{n-r+1}, \dots, z_n]$, let $P_1 := P \cap R_1$ and consider the commutative diagram:

$$\begin{array}{ccc} k[[z_1, \dots, z_{n-r}]] & \xrightarrow{\ \beta\ } & R/P_0 \\ \uparrow & & \uparrow \\ V_0[[z_2, \dots, z_{n-r}]] & \xrightarrow{\ \alpha\ } & R_1/P_1. \end{array} \qquad (27.12.1)$$

The vertical maps of Diagram 27.12.1 are injections. By Proposition 26.12.2f, the horizontal map β of Diagram 27.12.1 is an injective finite integral extension. Since P_1 is a minimal prime ideal of the regular sequence $f_1, \dots, f_r$ on R_1, the argument of Proposition 26.12.2f shows that the map α is an injective finite integral extension.

CLAIM 27.13. Let W be a prime ideal of $k[[Z]]$ that is minimal over $(\mathfrak{q}, P)\widehat{A}$. Then:

(1) $\operatorname{ht} W = n - 2$.
(2) $W \cap k[[z_1, \dots, z_{n-r}]] = \mathfrak{q}$.
(3) $W \cap A = (0)$.
(4) $k[[z_1, z_2]] \cap W = (0)$.

PROOF. The prime ideal W is minimal over

$$I := (f_1, \dots, f_r, z_3 - \gamma_3 z_2, \dots, z_{n-r} - \gamma_{n-r} z_2)\widehat{A}.$$

For item 1, by Proposition 26.12, the elements $f_1, \dots, f_r$ form a regular sequence on $k[[Z]]$, and $k[[Z]]/(f_1, \dots, f_r)k[[Z]] = k[[Z_1]][Z_2]/(f_1, \dots, f_r)k[[Z_1]][Z_2]$ is a finite free module over $k[[Z_1]]$. Also $k[[Z_1]] = k[[Z_1']]$, where

$$Z_1' = \{z_1, z_2, z_3 - \gamma_3 z_2, \dots, z_{n-r} - \gamma_{n-r} z_2\}.$$

The elements of Z_1' are a regular system of parameters for the regular local ring $k[[Z_1]]$. Hence

$$z_3 - \gamma_3 z_2, \dots, z_{n-r} - \gamma_{n-r} z_2$$

is a regular sequence on the finite free module $k[[Z]]/(f_1, \dots, f_r)k[[Z]]$; see Proposition 26.12.2b. Thus $f_1, \dots, f_r, z_3 - \gamma_3 z_2, \dots, z_{n-r} - \gamma_{n-r} z_2$ is a regular sequence on $\widehat{A} = k[[Z]]$. By [**103**, Theorem 132, p. 95], every prime ideal minimal over I has the same height, namely the length of the regular sequence that generates I. Thus $\operatorname{ht} W = n - 2$.

For item 2, consider the prime ideal $(W \cap R)/P_0$ of R/P_0 in Diagram 27.12.1. Then P is extended from P_0 implies $W \cap R$ is minimal over $(\mathfrak{q}, P_0)$, and so the intersection $P_0 \cap k[[z_1, \dots, z_{n-r}]] = (0)$. It follows that

$$W \cap k[[z_1, \dots, z_{n-r}]] = (W \cap R) \cap k[[z_1, \dots, z_{n-r}]]$$
$$= ((W \cap R)/P_0) \cap k[[z_1, \dots, z_{n-r}]] = \mathfrak{q}.$$

For item 3, $\mathfrak{q} \cap V_0[[z_2, \dots, z_{n-r}]] = (0)$. That is, the prime ideal $(W \cap R_1)/P_1$ lies over (0) under the inclusion map α of Diagram 27.12.1. Thus $W \cap R_1 = P_1$. Now $W \cap A = W \cap R_1 \cap A = P_1 \cap A = (0)$, as desired.

For item 4, let $B := k[[z_1, \dots, z_{n-r}]] = k[[z_1, z_2, z_3 - \gamma_3 z_2, \dots, z_{n-r} - \gamma_{n-r} z_2]]$. Let $\sigma : B \to B/\mathfrak{q} \cong k[[z_1, z_2]]$ be the canonical map. Then $\sigma|_{k[[z_1, z_2]]}$, the restriction of σ to $k[[z_1, z_2]]$, is an isomorphism. Thus $k[[z_1, z_2]] \cap \ker \sigma = k[[z_1, z_2]] \cap \mathfrak{q} = (0)$. Since $W \cap k[[z_1, \dots, z_{n-r}]] = \mathfrak{q}$, by item 2, $k[[z_1, z_2]] \cap W = k[[z_1, z_2]] \cap \mathfrak{q} = (0)$.

This completes the proof of Claim 27.13. □

Now W satisfies $\operatorname{ht} W = n-2$, $W \cap A = (0)$, $x_1 (= z_1) \notin W$ and $P \subseteq W$. Apply Proposition 26.12 to the prime ideal W. For convenience, denote the new variables found in Proposition 26.12.1 using the prime ideal W by $z_1 = x_1, z_2, \dots, z_n$, let $Z = \{z_1, z_2, \dots, z_n\}$, and let $h_1, \dots, h_{n-2}$ denote the regular sequence in W such that

$$\begin{array}{lll} & h_1 \in k[[z_1, z_2]]\,[z_3, \dots, z_n] & \text{is monic in } z_n, \\ & h_2 \in k[[z_1, z_2]]\,[z_3, \dots, z_{n-1}] & \text{is monic in } z_{n-1}, \\ (27.13.0) & \vdots & \\ & h_{n-2} \in k[[z_1, z_2]]\,[z_3] & \text{is monic in } z_3. \end{array}$$

Define $W_0 = W \cap k[[z_1, z_2]]\,[z_3, \ldots, z_n]$. Then $W_0\widehat{A} = W$. For some $s \in \mathbb{N}_0$, let $\{h_j \mid n-1 \le j \le n-2+s\}$ be s additional polynomials in $k[[z_1, z_2]]\,[z_3, \ldots, z_n]$ so that $h_1, \ldots, h_{n-2}, h_{n-1}, \ldots, h_{n+s-2}$ generate W_0. Consider the coefficients in $k[[z_1]]$ of the h_j, for $1 \le j \le n+s-2$:

$$h_j = \sum_{(i)} c_{j(i)} z_2^{i_2} \ldots z_n^{i_n},$$

with $c_{j(i)} \in k[[z_1]]$. The set $\{c_{j(i)}\}$ is countable. Define

$$V := k(z_1, \{a_{j,(i)}, b_{\ell,(i)}, c_{j(i)}\}) \cap k[[z_1]].$$

Then V is a rank-one discrete valuation domain that is countably generated over $k[z_1]_{(z_1)}$ and W is extended from its contraction $W \cap (V[[z_2]]\,[z_3, \ldots, z_n])$.

Each h_i is a polynomial in $z_3, \ldots, z_n$ with coefficients in $V[[z_2]]$:

$$h_i = \sum \omega_{(i)} z_3^{i_3} \ldots z_n^{i_n},$$

with $\omega_{(i)} \in V[[z_2]] \subseteq k[[z_1, z_2]]$. By Valabrega's Theorem 4.9, the integral domain

$$D := \mathcal{Q}(V)(z_2, \{\omega_{(i)}\}) \cap k[[z_1, z_2]],$$

is a two-dimensional regular local domain with completion $\widehat{D} = k[[z_1, z_2]]$. Let $W_1 := W \cap D[z_3, \ldots, z_n]$. Then $W_1\widehat{A} = W_1 k[[z_1, \ldots, z_n]] = W$, and $D \subseteq k[[z_1, z_2]]$.

By Proposition 26.12.2f, $W \cap k[[z_1, z_2]] = (0)$. By Proposition 27.11 with $z_1 = z$ and $z_2 = t$, there exists a prime element $q \in k[[z_1, z_2]]$ with $qk[[z_1, z_2]] \cap D = (0)$. Since $W \cap k[[z_1, z_2]] = (0)$, it follows that $q \notin W$.

Let $Q \in \operatorname{Spec} \widehat{A}$ be a minimal prime of $(q, W)\widehat{A}$. Since $\operatorname{ht} W = n-2$ and $q \notin W$, $\operatorname{ht} Q = n-1$. Moreover, $P \subseteq W$ implies $P \subseteq Q$.

CLAIM 27.14. $Q \cap D[z_3, \ldots, z_n] = W_1$.

PROOF. (of Claim 27.14) Consider the commutative diagram:

$$\begin{array}{ccc} k[[z_1, z_2]] & \longrightarrow & k[[z_1, \ldots, z_n]]/W \\ \uparrow & & \uparrow \\ D & \xrightarrow{\sigma} & D[z_3, \ldots, z_n]/W_1\,. \end{array} \tag{27.14.0}$$

Since $W \cap k[[z_1, z_2]] = (0)$ and $W_1 \cap D \subseteq W \cap k[[z_1, z_2]] = (0)$, the horizontal maps of Diagram 27.14.0 are injective. Since W_1 contains the regular sequence $h_1, h_2, \ldots, h_r, \ldots, h_{n-2}$ of polynomials in $k[[z_1, z_2]][z_3, \ldots, z_n]$ from Equations 27.13.0; each monic in one of the variables of $\{z_3, \ldots, z_n\}$, it follows as in Proposition 26.12.2d that the map σ of Diagram 27.14.0 is a finite integral extension.

Consider the prime ideal $(Q \cap D[z_3, \ldots, z_n])/W_1$ of $D[z_3, \ldots, z_n]/W_1$. Since $W_1 \cap D = (0)$, the intersection of this prime ideal with D can be considered as

$$Q \cap D[z_3, \ldots, z_n] \cap D = Q \cap k[[z_1, z_2]] \cap D \subseteq qk[[z_1, z_2]] \cap D = (0),$$

where the last equality is by the choice of q. Thus $(Q \cap D[z_3, \ldots, z_n])/W_1$ lies above (0) in D. Since σ is a finite integral extension of integral domains, the ideal (0) is the only prime ideal of $D[z_3, \ldots, z_n])/W_1$ lying above (0). That is, $(Q \cap D[z_3, \ldots, z_n])/W_1 = (0)$. It follows that $Q \cap D[z_3, \ldots, z_n] = W_1$, and so Claim 27.14 holds. □

To complete the proof of Theorem 27.12, observe that $Q \cap D[z_3, \dots, z_n] = W_1$ implies $Q \cap A = (0)$. To see this:

$$z_1 \in V_0 := k(z_1, a_{j,(i)}, b_{\ell,(i)}) \cap k[[z_1]] \subseteq V := \mathcal{Q}(V_0)(\{c_{j(i)}\}) \cap k[[z_1]$$
$$\subseteq D := \mathcal{Q}(V)(z_2, \{\omega_{(i)}\}) \cap k[[z_1, z_2]],$$

where the $a_{j,(i)}, b_{\ell,(i)}, c_{j(i)} \in k[[z_1]]$. This implies that $z_1 \in D$. Also $z_2 \in D$. Thus

$$k[X] = k[Z] \subseteq D[z_3, \dots, z_n].$$

Therefore:

$$Q \cap D[z_3, \dots z_n] = W_1 \implies Q \cap k[Z] \subseteq W_1 \cap k[Z] \subseteq W \cap k[Z] = (0).$$

Thus $Q \cap A = Q \cap k[Z]_{(Z)} = (0)$. This completes the proof of Theorem 27.12. □

27.5. Generic fibers of power series ring extensions

In this section we apply the Weierstrass machinery from Section 26.2 to the generic fiber rings of power series extensions.

THEOREM 27.15. *Let $n \geq 2$ be an integer and let $y, x_1, \dots, x_n$ be variables over the field k. Let $X = \{x_1, \dots, x_n\}$ and let R_1 be the formal power series ring $k[[X]]$. Consider the extension $R_1 \hookrightarrow R_1[[y]] = R$. Let $U = R_1 \setminus (0)$. For $P \in \operatorname{Spec} R$ such that $P \cap U = \emptyset$, we have:*

1. *If $P \not\subseteq XR$, then $\dim R/P = n$, the ideal P is maximal in the generic fiber of the map $R_1 \hookrightarrow R$, equivalently P corresponds to a maximal ideal of the generic fiber ring $U^{-1}R$ as in Notation 26.1, and $R_1 \hookrightarrow R/P$ is finite. Moreover $P = gR$ is a principal ideal of R, where g is a monic polynomial of $R_1[y]$.*
2. *If $P \subseteq XR$, then there exists $Q \in \operatorname{Spec} R$ such that $P \subseteq Q$, $\dim R/Q = 2$ and Q extends to a maximal ideal of the generic fiber ring $U^{-1}R$.*

Assume $n > 2$. Then every prime ideal Q of R that extends to a maximal ideal of the generic fiber ring $U^{-1}R$ satisfies

$$\dim R/Q = \begin{cases} n & \text{if } R_1 \hookrightarrow R/Q \text{ is finite, or} \\ 2 & \text{if } Q \subset XR. \end{cases}$$

PROOF. Let $P \in \operatorname{Spec} R$ be such that $P \cap U = \emptyset$ or equivalently $P \cap R_1 = (0)$. Then R_1 embeds in R/P. If $\dim(R/P) \leq 1$, then the maximal ideal of R_1 generates an ideal primary for the maximal ideal of R/P. By Theorem 3.16, R/P is finite over R_1, and so $\dim R_1 = \dim(R/P)$, a contradiction. Thus $\dim(R/P) \geq 2$.

For item 1, if $P \not\subseteq XR$, then there exists a prime element $f \in P$ that contains a term y^s for some positive integer s. By Weierstrass, that is, by Theorem 26.8, it follows that $f = g\epsilon$, where $g \in k[[X]]\,[y]$ is a nonzero monic polynomial in y and ϵ is a unit of R. Then $fR = gR \subseteq P$ is a prime ideal and $R_1 \hookrightarrow R/gR$ is a finite integral extension. Since $P \cap R_1 = (0)$, it follows that $gR = P$, a principal ideal.

For item 2, if $P \subseteq XR$ and $\dim(R/P) > 2$, then Theorem 26.14 with $V = k[y]_{(y)}$ and $R_0 = V[[x]]$. implies there exists $Q \in \operatorname{Spec} R$ such that $\dim(R/Q) = 2$, $P \subset Q \subset XR$ and $P \cap R_1 = (0) = Q \cap R_1$, and so P does not extend to a maximal ideal of the generic fiber ring. Thus $Q \in \operatorname{Spec} R$ maximal in the generic fiber of $R_1 \hookrightarrow R$ implies that the dimension of $\dim(R/Q)$ is 2, or equivalently that $\operatorname{ht} Q = n - 1$. □

THEOREM 27.16. *Let n and m be positive integers, and let $X = \{x_1, \dots, x_n\}$ and $Y = \{y_1, \dots, y_m\}$ be sets of independent variables over the field k. Consider the formal power series rings $R_1 = k[[X]]$ and $R = k[[X, Y]]$ and the extension $R_1 \hookrightarrow R_1[[Y]] = R$. Let $U = R_1 \setminus (0)$. Let $Q \in \operatorname{Spec} R$ be maximal with respect to $Q \cap U = \emptyset$. If $n = 1$, then $\dim R/Q = 1$ and $R_1 \hookrightarrow R/Q$ is finite.*

If $n \geq 2$, there are two possibilities:

(1) *$R_1 \hookrightarrow R/Q$ is finite, in which case $\dim R/Q = \dim R_1 = n$, or*
(2) *$\dim R/Q = 2$.*

PROOF. First assume $n = 1$, and let $x = x_1$. Since Q is maximal with respect to $Q \cap U = \emptyset$, for each $P \in \operatorname{Spec} R$ with $Q \subsetneq P$ we have $P \cap U$ is nonempty and therefore $x \in P$. It follows that $\dim R/Q = 1$, for otherwise,

$$Q = \bigcap \{P \mid P \in \operatorname{Spec} R \text{ and } Q \subsetneq P\},$$

which implies $x \in Q$. By Theorem 3.16, $R_1 \hookrightarrow R/Q$ is finite.

It remains to consider the case where $n \geq 2$. We proceed by induction on m. Theorem 27.15 yields the assertion for $m = 1$. Suppose $Q \in \operatorname{Spec} R$ is maximal with respect to $Q \cap U = \emptyset$. As in the proof of Theorem 27.15, we have $\dim R/Q \geq 2$. If $Q \subseteq (X, y_1, \dots, y_{m-1})R$, then by Theorem 26.14 with $R_0 = k[y_m]_{(y_m)}[[X, y_1, \dots, y_{m-1}]]$, there exists $Q' \in \operatorname{Spec} R$ with $Q \subseteq Q'$, $\dim R/Q' = 2$, and $Q \cap R_0 = Q' \cap R_0$. Since $R_1 \subseteq R_0$, we have $Q' \cap U = \emptyset$. Since Q is maximal with respect to $Q \cap U = \emptyset$, we have $Q = Q'$, and so $\dim R/Q = 2$.

Otherwise, if $Q \not\subseteq (X, y_1, \dots, y_{m-1})R$, then there exists a prime element $f \in Q$ that contains a term y_m^s for some positive integer s. Let $R_2 = k[[X, y_1, \dots, y_{m-1}]]$. By Weierstrass, it follows that $f = g\epsilon$, where $g \in R_2[y_m]$ is a nonzero monic polynomial in y_m and ϵ is a unit of R. We have $fR = gR \subseteq Q$ is a prime ideal and $R_2 \hookrightarrow R/gR$ is a finite integral extension. Thus $R_2/(Q \cap R_2) \hookrightarrow R/Q$ is an integral extension. It follows that $Q \cap R_2$ is maximal in R_2 with respect to being disjoint from U. By induction $\dim R_2/(Q \cap R_2)$ is either n or 2. Since R/Q is integral over $R_2/(Q \cap R_2)$, $\dim R/Q$ is either n or 2. □

REMARK 27.17. Theorem 27.16 proves part 4 of Theorem 26.3. To see this, let $P \in \operatorname{Spec} k[[X, Y]]$ be maximal with $P \cap k[[X]] = (0)$. Since R is catenary, $\operatorname{ht} P + \dim R/P = n + m$. If $n = 1$, then Theorem 27.16 implies that $\dim(k[X, Y]/P) = n = 1$, and $\operatorname{ht} P = m$. If $n \geq 2$, the two cases are (i) $\operatorname{ht} P = m$ and (ii) $\operatorname{ht} P = n + m - 2$. These two cases are as in (a) and (b) of part 4 of Theorem 26.3.

Using the TGF terminology of Definition 26.6, Theorem 27.16 implies:

COROLLARY 27.18. *With the notation of Theorem 27.16, if $P \in \operatorname{Spec} R$ is such that $R_1 \hookrightarrow R/P =: S$ is a TGF extension, then $\dim S = \dim R_1 = n$ or $\dim S = 2$.*

27.6. Gff(R) and Gff(S) for S an extension domain of R

Theorem 27.19 is useful in considering properties of generic formal fiber rings.

THEOREM 27.19. *Let $\phi : (R, \mathfrak{m}) \hookrightarrow (S, \mathfrak{n})$ be an injective local map of Noetherian local integral domains. Consider the following properties:*

(1) *$\mathfrak{m}S$ is $\mathfrak{n}$-primary, and $S/\mathfrak{n}$ is finite algebraic over $R/\mathfrak{m}$.*
(2) *$R \hookrightarrow S$ is a TGF-extension and $\dim R = \dim S$; see Definition 26.6.*
(3) *R is analytically irreducible.*

(4) *R is analytically normal and S is universally catenary.*
(5) *All maximal ideals of* $\text{Gff}(R)$ *have the same height.*

If items 1, 2 and 3 hold, then $\dim \text{Gff}(R) = \dim \text{Gff}(S)$. *If, in addition, items 4 and 5 hold, then the maximal ideals of* $\text{Gff}(S)$ *all have height* $h = \dim \text{Gff}(R)$.

PROOF. Let $\widehat{R}$ and $\widehat{S}$ denote the $\mathfrak{m}$-adic completion of R and $\mathfrak{n}$-adic completion of S respectively, and let $\widehat{\phi} : \widehat{R} \to \widehat{S}$ be the natural extension of φ as defined above. Consider the commutative diagram

(27.19.a)
$$\begin{array}{ccc} \widehat{R} & \xrightarrow{\widehat{\phi}} & \widehat{S} \\ \uparrow & & \uparrow \\ R & \xrightarrow{\phi} & S, \end{array}$$

where the vertical maps are the natural inclusion maps to the completion. Assume items 1, 2 and 3 hold. Item 1 implies that $\widehat{S}$ is a finite $\widehat{R}$-module with respect to the map $\widehat{\phi}$ by [**122**, Theorem 8.4]. By item 2, we have $\dim \widehat{R} = \dim R = \dim S = \dim \widehat{S}$. Item 3 says that $\widehat{R}$ is an integral domain. It follows that the map $\widehat{\phi} : \widehat{R} \hookrightarrow \widehat{S}$ is injective. Let $Q \in \operatorname{Spec} \widehat{S}$ and let $P = Q \cap \widehat{R}$. Since $R \hookrightarrow S$ is a TGF-extension, by item 2, commutativity of Diagram 27.19.a implies that

$$Q \cap S = (0) \iff P \cap R = (0).$$

Therefore $\widehat{\phi}$ induces an injective finite map $\text{Gff}(R) \hookrightarrow \text{Gff}(S)$. We conclude that $\dim \text{Gff}(R) = \dim \text{Gff}(S)$.

Assume in addition that items 4 and 5 hold, and let $h = \dim \text{Gff}(R)$. The assumption that S is universally catenary implies that $\dim(\widehat{S}/\mathfrak{q}) = \dim S$ for each minimal prime $\mathfrak{q}$ of $\widehat{S}$; see [**122**, Theorem 31.7]. Since $\frac{\widehat{R}}{\mathfrak{q} \cap \widehat{R}} \hookrightarrow \frac{\widehat{S}}{\mathfrak{q}}$ is an integral extension, we have $\mathfrak{q} \cap \widehat{R} = (0)$. The assumption that $\widehat{R}$ is a normal domain implies that the going-down theorem holds for $\widehat{R} \hookrightarrow \widehat{S}/\mathfrak{q}$ by [**122**, Theorem 9.4(ii)]. Therefore for each $Q \in \operatorname{Spec} \widehat{S}$ we have $\operatorname{ht} Q = \operatorname{ht} P$, where $P = Q \cap \widehat{R}$. Hence if $\operatorname{ht} P = h$ for each $P \in \operatorname{Spec} \widehat{R}$ that is maximal with respect to $P \cap R = (0)$, then $\operatorname{ht} Q = h$ for each $Q \in \operatorname{Spec} \widehat{S}$ that is maximal with respect to $Q \cap S = (0)$. This completes the proof of Theorem 27.19. □

REMARK 27.20. We would like to thank Rodney Sharp and Roger Wiegand for their interest in Theorem 27.19. The hypotheses of Theorem 27.19 do not imply that S is a finite R-module, or even that S is essentially finitely generated over R. If $\phi : (R, \mathfrak{m}) \hookrightarrow (T, \mathfrak{n})$ is an extension of rank one discrete valuation rings (DVRs) such that $T/\mathfrak{n}$ is finite algebraic over $R/\mathfrak{m}$, then for every field F that contains R and is contained in the field of fractions of $\widehat{T}$, the ring $S := \widehat{T} \cap F$ is a DVR such that the extension $R \hookrightarrow S$ satisfies the hypotheses of Theorem 27.19.

As a specific example where S is essentially finite over R, but not a finite R-module, let $R = \mathbb{Z}_{5\mathbb{Z}}$, the integers localized at the prime ideal generated by 5, and let A be the integral closure of R in $\mathbb{Q}[i]$. Then A has two maximal ideals lying over $5R$, namely $(1+2i)A$ and $(1-2i)A$. Let $S = A_{(1+2i)A}$. Then the extension $R \hookrightarrow S$ satisfies the hypotheses of Theorem 27.19. Since S properly contains A, and every element in the field of fractions of A that is integral over R is contained in A, it follows that S is not finitely generated as an R-module. In Remark 27.31, we describe examples in higher dimension where S is not a finite R-module.

27.7. Formal fibers of prime ideals in polynomial rings

In this section we present a generalization of Theorem 27.12 and discuss related results concerning generic formal fibers of extensions of mixed polynomial-power series rings.

We were inspired to revisit and generalize Theorem 27.12 by Youngsu Kim. His interest in formal fibers and the material in [**86**] inspired us to consider the second question below; also see [**92**].

QUESTIONS 27.21. For $n \in \mathbb{N}$, let $x_1, \ldots, x_n$ be indeterminates over a field k and let $R = k[x_1, \ldots, x_n]_{(x_1,\ldots,x_n)}$ denote the localized polynomial ring with maximal ideal $\mathfrak{m} = (x_1, \ldots, x_n)R$. Let $\widehat{R}$ be the $\mathfrak{m}$-adic completion of R.

(1) For $P \in \operatorname{Spec} R$, what is the dimension of the generic formal fiber ring $\operatorname{Gff}(R/P)$?
(2) What heights are possible for maximal ideals of the ring $\operatorname{Gff}(R/P)$?

In connection with Question 27.21.1, for $P \in \operatorname{Spec} R$, the ring R/P is essentially finitely generated over a field and $\dim(\operatorname{Gff}(R/P)) = n - 1 - \operatorname{ht} P$ by a result of Matumura [**121**, Theorem 2 and Corollary, p. 263].

As a sharpening of Matsumura's result and of Theorem 27.12, we prove Theorem 27.22; see also Theorem 27.25. Thus the answer to Question 27.21.2 is that the height of *every* maximal ideal of $\operatorname{Gff}(R/P)$ is $n - 1 - \operatorname{ht} P$.

THEOREM 27.22. *Let S be a local domain essentially finitely generated over a field; thus $S = k[s_1, \ldots, s_r]_{\mathfrak{p}}$, where k is a field, $r \in \mathbb{N}$, the elements s_i are in S and $\mathfrak{p}$ is a prime ideal of the finitely generated k-algebra $k[s_1, \ldots, s_r]$. Let $\mathfrak{n} := \mathfrak{p}S$ and let $\widehat{S}$ denote the $\mathfrak{n}$-adic completion of S. Then every maximal ideal of* $\operatorname{Gff}(S)$ *has height* $\dim S - 1$. *Equivalently, if $Q \in \operatorname{Spec} \widehat{S}$ is maximal with respect to $Q \cap S = (0)$, then* $\operatorname{ht} Q = \dim S - 1$.

Theorem 27.22 is restated and proved in Theorem 27.25.

DISCUSSION 27.23. Let $\phi : (R, \mathfrak{m}) \hookrightarrow (S, \mathfrak{n})$ be an injective local map of the Noetherian local ring $(R, \mathfrak{m})$ into a Noetherian local ring $(S, \mathfrak{n})$. Let $\widehat{R} = \varprojlim_n R/\mathfrak{m}^n$ denote the $\mathfrak{m}$-adic completion of R and let $\widehat{S} = \varprojlim_n S/\mathfrak{n}^n$ denote the $\mathfrak{n}$-adic completion of S. For each $n \in \mathbb{N}$, we have $\mathfrak{m}^n \subseteq \mathfrak{n}^n \cap R$. Hence there exists a map

$$\phi_n : R/\mathfrak{m}^n \;\to\; R/(\mathfrak{n}^n \cap R) \;\hookrightarrow\; S/\mathfrak{n}^n, \qquad \text{for each } \; n \in \mathbb{N}.$$

The family of maps $\{\phi_n\}_{n \in \mathbb{N}}$ determines a unique map $\widehat{\phi} : \widehat{R} \to \widehat{S}$.

Since $\mathfrak{m}^n \subseteq \mathfrak{n}^n \cap R$, the $\mathfrak{m}$-adic topology on R is the subspace topology from Sif and only if for each positive integer n there exists a positive integer s_n such that $\mathfrak{n}^{s_n} \cap R \subseteq \mathfrak{m}^n$. Since $R/\mathfrak{m}^n$ is Artinian, the descending chain of ideals $\{\mathfrak{m}^n + (\mathfrak{n}^s \cap R)\}_{s \in \mathbb{N}}$ stabilizes. The ideal $\mathfrak{m}^n$ is closed in the $\mathfrak{m}$-adic topology, and it is closed in the subspace topology if and only if $\bigcap_{s \in \mathbb{N}} (\mathfrak{m}^n + (\mathfrak{n}^s \cap R)) = \mathfrak{m}^n$. Hence $\mathfrak{m}^n$ is closed in the subspace topology if and only if there exists a positive integer s_n such that $\mathfrak{n}^{s_n} \cap R \subseteq \mathfrak{m}^n$.

Thus the subspace topology from S is the same as the $\mathfrak{m}$-adic topology on R if and only if $\widehat{\varphi}$ is injective.

DISCUSSION 27.24. As in the statement of Theorem 27.22, let $S = k[z_1, \dots, z_r]_{\mathfrak{p}}$ be a local domain essentially finitely generated over a field k, so that $\mathfrak{p}$ is a maximal ideal of $k[z_1, \dots, z_r]$. We observe that S is a localization at a maximal ideal of an integral domain that is a finitely generated algebra over an extension field F of k.

To see this, let $A = k[x_1, \dots, x_r]$ be a polynomial ring in r variables over k, and let Q denote the kernel of the k-algebra homomorphism σ of A onto $k[z_1, \dots, z_r]$ defined by mapping $x_i \mapsto z_i$ for each i with $1 \le i \le r$. Using permutability of localization and residue class formation, there exists a maximal ideal $N \supseteq Q$ of A such that $S = A_N/QA_N$. Thus

$$Q := \ker \sigma \subseteq A := k[x_1, \dots, x_r] \xrightarrow{\sigma} k[z_1, \dots, z_r] \subseteq S := k[z_1, \dots z_r]_{\mathfrak{p}} = A_N/QA_N.$$

A version of Noether normalization as in [**120**, Theorem 24 (14.F) page 89] states that, if $\operatorname{ht} N = s$, then there exist elements $y_1, \dots, y_r$ in A such that A is integral over $B = k[y_1, \dots, y_r]$ and $N \cap B = (y_1, \dots, y_s)B$. It follows that $y_1, \dots, y_r$ are algebraically independent over k and A is a finitely generated B-module. Let F denote the field $k(y_{s+1}, \dots, y_r)$, and let U denote the multiplicatively closed set $k[y_{s+1}, \dots, y_r] \setminus (0)$. Then $U^{-1}B$ is the polynomial ring $F[y_1, \dots, y_s]$, and $U^{-1}A := C$ is a finitely generated $U^{-1}B$-module. Moreover NC is a maximal ideal of C such that

$$NC \cap U^{-1}B = (y_1, \dots, y_s)U^{-1}B = (y_1, \dots, y_s)F[y_1, \dots, y_s].$$

Hence $NC \cap U^{-1}B$ is a maximal ideal of $U^{-1}B$, and $(y_1, \dots, y_s)C$ is primary for the ideal NC of C. Thus $S = C_{NC}/QC_{NC}$ is a localization of the finitely generated F-algebra $D := C/QC$ at the maximal ideal NC/QC. The following commutative diagram displays this information:

$$\begin{array}{ccc}
Q \subseteq N \subseteq A := k[x_1, \dots, x_r] & \xrightarrow{\sigma} & U^{-1}A = C \subseteq S = C_{NC}/QC_{NC} \\
\text{integral}, \subseteq \big\uparrow & & \text{fin.gen}, \subseteq \big\uparrow \\
N \cap B = (y_1, \dots, y_s)B \subseteq B = k[y_1, \dots, y_r] & \xrightarrow{\subseteq} & U^{-1}B = F[y_1, \dots y_s].
\end{array}$$

Therefore S is a localization of an integral domain D at a maximal ideal of D and D is a finitely generated algebra over an extension field F of k.

THEOREM 27.25. *Let $A = k[s_1, \dots, s_r]$ be an integral domain that is a finitely generated algebra over a field k, let N be a maximal ideal of A, and let $Q \subset N$ be a prime ideal of A. Set $S = A_N/QA_N$ and $\mathfrak{n} = NS$. If $\dim S = d$, then every maximal ideal of the generic formal fiber ring $\operatorname{Gff}(S)$ has height $d - 1$.*

PROOF. Choose $x_1, \dots, x_d$ in $\mathfrak{n}$ such that $x_1, \dots, x_d$ are algebraically independent over k and $(x_1, \dots, x_d)S$ is $\mathfrak{n}$-primary. Set $R = k[x_1, \dots, x_d]_{(x_1, \dots, x_d)}$, a localized polynomial ring over k, and let $\mathfrak{m} = (x_1, \dots, x_d)R$. To prove Theorem 27.25, it suffices to show that the inclusion map $\phi : R \hookrightarrow S$ satisfies items 1 - 5 of Theorem 27.19. By construction ϕ is an injective local homomorphism and $\mathfrak{m}S$ is $\mathfrak{n}$-primary. Also $R/\mathfrak{m} = k$ and $S/\mathfrak{n} = A/N$ is a field that is a finitely generated k-algebra and hence a finite algebraic extension field of k; see [**122**, Theorem 5.2]. Therefore item 1 holds. Since $\dim S = d = \dim A/Q$, the field of fractions of S has transcendence degree d over the field k. Therefore S is algebraic over R. It follows that $R \hookrightarrow S$ is a TGF extension. Thus item 2 holds. Since R is a regular local ring, R is analytically irreducible and analytically normal. Since S is essentially finitely generated over a field, S is universally catenary. Therefore items 3 and 4 hold. Since R is a localized polynomial ring in d variables, Theorem 27.12 implies that every maximal ideal of $\mathrm{Gff}(R)$ has height $d-1$. By Theorem 27.19, every maximal ideal of $\mathrm{Gff}(S)$ has height $d-1$. $\square$

27.8. Other results on generic formal fibers

Theorems 27.2 and 27.6 give descriptions of the generic formal fiber ring of mixed polynomial-power series rings. We use Theorems 27.19, 27.2 and 27.6 to deduce Theorem 27.26.

THEOREM 27.26. *Let R be either $k[[X]]\,[Y]_{(X,Y)}$ or $k[Y]_{(Y)}[[X]]$, where m and n are positive integers and $X = \{x_1, \ldots, x_n\}$ and $Y = \{y_1, \ldots, y_m\}$ are sets of independent variables over a field k. Let $\mathfrak{m}$ denote the maximal ideal $(X,Y)R$ of R. Let $(S, \mathfrak{n})$ be a Noetherian local integral domain containing R such that:*

(1) *The injection $\varphi : (R, \mathfrak{m}) \hookrightarrow (S, \mathfrak{n})$ is a local map.*
(2) *$\mathfrak{m}S$ is $\mathfrak{n}$-primary, and $S/\mathfrak{n}$ is finite algebraic over $R/\mathfrak{m}$.*
(3) *$R \hookrightarrow S$ is a TGF-extension and $\dim R = \dim S$.*
(4) *S is universally catenary.*

Then every maximal ideal of the generic formal fiber ring $\mathrm{Gff}(S)$ has height $n+m-2$. Equivalently, if P is a prime ideal of $\widehat{S}$ maximal with respect to $P \cap S = (0)$, then $\mathrm{ht}(P) = n + m - 2$.

PROOF. We check that the conditions 1–5 of Theorem 27.19 are satisfied for R and S and the injection φ. Since the completion of R is $k[[X,Y]]$, R is analytically normal, and so also analytically irreducible. Items 1–4 of Theorem 27.26 ensure that the rest of conditions 1–4 of Theorem 27.19 hold. By Theorems 27.2 and 27.6, every maximal ideal of $\mathrm{Gff}(R)$ has height $n+m-2$, and so condition 5 of Theorem 27.19 holds. Thus we have every maximal ideal of $\mathrm{Gff}(S)$ has height $n+m-2$ by Theorem 27.19. $\square$

REMARK 27.27. Let k, X, Y, and R be as in Theorem 27.26. Let A be a finite integral extension domain of R and let S be the localization of A at a maximal ideal. As observed in the proof of Theorem 27.26, R is a local analytically normal integral domain. Since R is universally catenary, and since the universally catenary property is preserved under localizations and finitely generated algebras, it follows that S is universally catenary; that is, condition 4 of Theorem 27.26 holds We also have that conditions 1–3 of Theorem 27.26 hold. Thus the extension $R \hookrightarrow S$ satisfies the hypotheses of Theorem 27.26. Hence every maximal ideal of $\mathrm{Gff}(S)$ has height $n+m-2$.

Example 27.28 is an application of Theorem 27.26 and Remark 27.27.

EXAMPLE 27.28. Let k, X, Y, and R be as in Theorem 27.26. Let K denote the field of fractions of R, and let L be a finite algebraic extension field of K. Let A be the integral closure of R in L, and let S be a localization of A at a maximal ideal. The ring R is a Nagata ring by a result of Marot; see [**117**, Prop.3.5]. Therefore A is a finite integral extension of R and the conditions of Remark 27.27 apply to show that every maximal ideal of $\mathrm{Gff}(S)$ has height $n+m-2$.

DISCUSSION 27.29. Concerning prime ideals in an integral extension lying over a given prime ideal of an integral domain, McAdam in [**124**] introduces the following terminology.

DEFINITION 27.30. Let R be an integral domain and let T be the integral closure of R in an algebraic closure of the field of fractions of R. A prime ideal P of R is said to *n-split* if in T there are precisely n prime ideals (possibly $n = \infty$) that lie over P.

Heinzer and S. Wiegand show in [**94**, Theorem 1.1] that a nonzero prime ideal P of a Noetherian domain R is either 1-split or ∞-split, and if P is a nonzero 1-split prime ideal, then R is local with maximal ideal P. For a local domain R with maximal ideal P, Remark 3.32.5 implies that P is 1-split if and only if R is Henselian.

REMARK 27.31. With notation as in Example 27.28, since the sets X and Y are nonempty, the field K is a simple transcendental extension of a subfield. It follows that the regular local ring R is not Henselian; see the book of Berger, Kiehl, Kunz and Nastold [**21**, Satz 2.3.11, p. 60] and the paper of Schmidt [**164**]. As described in Discussion 27.29, there exists a finite algebraic field extension L/K such that the integral closure A of R in L has more than one maximal ideal. If $\mathfrak{n}$ is a maximal ideal of A, then $S = A_{\mathfrak{n}}$ is not a finite R-module, and gives an example $R \hookrightarrow S$ that satisfies the hypotheses of Theorem 27.19.

Exercise

(1) Let x and y be indeterminates over a field k. Let $R = k[[x]][y]$ and let $\tau \in yk[[y]]$ be such that y and τ are algebraically independent over k. Then we have the embedding $R = k[[x]][y] \hookrightarrow k[[x,y]] = \widehat{R}$. Let $\mathfrak{p} := (x - \tau)\widehat{R}$. Prove the following:
  (a) $\widehat{R}/\mathfrak{p} = k[[y]]$.
  (b) $\mathfrak{p} \cap k[x,y] = (0)$.
  (c) $\mathfrak{p} \cap R \neq (0)$.

  SUGGESTION. For item c, use Theorem 3.16.

CHAPTER 28

Mixed polynomial-power series rings and relations among their spectra

We are interested in the following sequence of two-dimensional nested mixed polynomial-power series rings:

$$(28.0.1) \qquad A := k[x,y] \hookrightarrow B := k[[y]]\,[x] \hookrightarrow C := k[x]\,[[y]] \hookrightarrow E := k[x,1/x]\,[[y]],$$

where k is a field and x and y are indeterminates over k. [1] That is, A is the usual polynomial ring in the two variables x and y over k, the ring B is all polynomials in the variable x with coefficients in the power series ring $k[[y]]$, the ring C is all power series in the variable y over the polynomial ring $k[x]$, and E is power series in the variable y over the ring $k[x,1/x]$. In Sequence 28.0.1 all the maps are flat; see Propositions 2.37.4 and 3.3.2. We also consider Sequence 28.0.2 consisting of embeddings between the rings C and E of Sequence 28.0.1:

$$(28.0.2) \qquad C \hookrightarrow D_1 := k[x]\,[[y/x]] \hookrightarrow \cdots \hookrightarrow D_n := k[x]\,[[y/x^n]] \hookrightarrow \cdots \hookrightarrow E.$$

With regard to Sequence 28.0.2, for n a positive integer, the map $C \hookrightarrow D_n$ is not flat, since $\operatorname{ht}(xD_n \cap C) = 2$ but $\operatorname{ht}(xD_n) = 1$; see Proposition 2.37.10. The map $D_n \hookrightarrow E$ is a localization followed by an ideal-adic completion of a Noetherian ring and therefore is flat. We discuss the spectra of the rings in Sequences 28.0.1 and 28.0.1, and we consider the maps induced on the spectra by the inclusion maps on the rings. For example, we determine whether there exist nonzero primes of one of the larger rings that intersect a smaller ring in zero.

28.1. Two motivations

We were led to consider these rings by questions that came up in two contexts.

The first motivation is a question about formal schemes that is discussed in the introduction to the paper [**14**] by Alonzo-Tarrio, Jeremias-Lopez and Lipman:

QUESTION 28.1. If a map between Noetherian formal schemes can be factored as a closed immersion followed by an open immersion, can this map also be factored as an open immersion followed by a closed immersion?[2]

Brian Conrad observed that an example to show the answer to Question 28.1 is "No" can be constructed for every triple $(R, x, \mathfrak{p})$ that satisfies the following three conditions; see [**14**]:

(28.1.1) R is an ideal-adic domain, that is, R is a Noetherian domain that is separated and complete with respect to the powers of a proper ideal I.

[1]The material in this chapter is adapted from our article [**87**] dedicated to Robert Gilmer, an outstanding algebraist, scholar and teacher.

[2]See Scheme Terminology 28.3 for a brief explanation of this terminology.

(28.1.2) x is a nonzero element of R such that the completion of $R[1/x]$ with respect to the powers of $IR[1/x]$, denoted $S := R_{\{x\}}$, is an integral domain.

(28.1.3) $\mathfrak{p}$ is a nonzero prime ideal of S that intersects R in (0).

The following example of such a triple $(R, x, \mathfrak{p})$ is described in [**14**]:

EXAMPLE 28.2. Let w, x, y, z be indeterminates over a field k. Let

$$R := k[w, x, z]\,[[y]] \quad \text{and} \quad S := k[w, x, 1/x, z]\,[[y]].$$

Notice that R is complete with respect to yR and S is complete with respect to yS. An indirect proof that there exist nonzero primes p of S for which $p \cap R = (0)$ is given in the paper [**14**] of Lipman, Alonzo-Tarrio and Jeremias-Lopez, using a result of Heinzer and Rotthaus [**70**, Theorem 1.12, p. 364]. A direct proof is given in [**87**, Proposition 4.9]. In Proposition 28.32 below we give a direct proof of a more general result due to Dumitrescu [**41**, Corollary 4].

In Scheme Terminology 28.3 we explain some of the terminology of formal schemes necessary for understanding Question 28.1; more details may be found in [**63**]. In Remark 28.4 we explain why a triple satisfying (28.1.1) to (28.1.3) yields examples that answer Question 28.1.

SCHEME TERMINOLOGY 28.3. Let R be a Noetherian integral domain and let K be its field of fractions. Let X denote the topological space $\operatorname{Spec} R$ with the Zariski topology defined in Section 2.1. We form a *sheaf*, denoted $\mathcal{O}$, on X by associating, to each open set U of X, the ring

$$\mathcal{O}(U) = \bigcap_{x \in U} R_{\mathfrak{p}_x},$$

where $\mathfrak{p}_x$ is the prime associated to the point $x \in U$; see [**168**, p. 235 and Theorem 1, p. 238]. For each pair $U \subseteq V$ of open subsets of X, there exists a natural inclusion map $\rho_U^V : \mathcal{O}(V) \hookrightarrow \mathcal{O}(U)$. The "ringed space" $(X, \mathcal{O})$ is identified with $\operatorname{Spec} R$ and is called an *affine scheme*; see [**168**, p. 242-3], [**63**, Definition I.10.1.2, p. 402]. Assume that $R = R^*$ is complete with respect to the I-adic topology, where I is a nonzero proper ideal of R (see Definition 3.1).Then the ringed space $(X, \mathcal{O})$ is denoted $\operatorname{Spf}(R)$ and is called the *formal spectrum* of R. It is also called a *Noetherian formal adic affine scheme*; see [**63**, I.10.1.7, p. 403]. An *immersion* is a morphism $f : Y \to X$ of schemes that factors as an isomorphism to a subscheme Z of X followed by a canonical injection $Z \to X$; see [**63**, (I.4.2.1)].

REMARK 28.4. Assume, in addition to R being a Noetherian integral domain complete with respect to the I-adic topology, that x is a nonzero element of R, that S is the completion of $R[1/x]$ with respect to the powers of $IR[1/x]$, and that $\mathfrak{p}$ is a prime ideal of S such that the triple $(R, x, \mathfrak{p})$ satisfies the three conditions 28.1.1 to 28.1.3.

The composition of the maps $R \to S \to S/\mathfrak{p}$ determines a map on formal spectra $\operatorname{Spf}(S/\mathfrak{p}) \to \operatorname{Spf}(S) \to \operatorname{Spf}(R)$ that is a closed immersion followed by an open immersion. This is because a surjection such as $S \to S/\mathfrak{p}$ of adic rings gives rise to a closed immersion $\operatorname{Spf}(S/\mathfrak{p}) \to \operatorname{Spf}(S)$ while a localization, such as that of R with respect to the powers of x, followed by the completion of $R[1/x]$ with respect to the powers of $IR[1/x]$ to obtain S gives rise to an open immersion $\operatorname{Spf}(S) \to \operatorname{Spf}(R)$ [**63**, I.10.14.4].

The map $\mathrm{Spf}(S/\mathfrak{p}) \to \mathrm{Spf}(R)$ cannot be factored as an open immersion followed by a closed one. This is because a closed immersion into $\mathrm{Spf}(R)$ corresponds to a surjective map of adic rings $R \to R/J$, where J is an ideal of R [**63**, page 441]. Thus if the map $\mathrm{Spf}(S/\mathfrak{p}) \to \mathrm{Spf}(R)$ factored as an open immersion followed by a closed one, we would have R-algebra homomorphisms from $R \to R/J \to S/\mathfrak{p}$, where $\mathrm{Spf}(S/\mathfrak{p}) \to \mathrm{Spf}(R/J)$ is an open immersion. Since $\mathfrak{p} \cap R = (0)$, we must have $J = (0)$. This implies $\mathrm{Spf}(S/\mathfrak{p}) \to \mathrm{Spf}(R)$ is an open immersion, that is, the composite map $\mathrm{Spf}(S/\mathfrak{p}) \to \mathrm{Spf}(S) \to \mathrm{Spf}(R)$, is an open immersion. But also $\mathrm{Spf}(S) \to \mathrm{Spf}(R)$ is an open immersion. It follows that $\mathrm{Spf}(S/\mathfrak{p}) \to \mathrm{Spf}(S)$ is both open and closed. Since S is an integral domain this implies $\mathrm{Spf}(S/\mathfrak{p}) \cong \mathrm{Spf}(S)$. Since $\mathfrak{p}$ is nonzero, this is a contradiction. Thus Example 28.2 shows that the answer to Question 28.1 is "No".

The second motivation for the material in this chapter comes from Question 26.4 of Melvin Hochster and Yongwei Yao "Can one describe or somehow classify the local maps $R \hookrightarrow S$ of complete local domains R and S such that every nonzero prime ideal of S has nonzero intersection with R?" The following example is a local map of the type described in the Hochster-Yao question.

EXAMPLE 28.5. Let x and y be indeterminates over a field k and consider the extension $R := k[[x,y]] \hookrightarrow S := k[[x]]\,[[y/x]]$.

To see this extension is TGF—the "trivial generic fiber" condition of Definition 26.6, it suffices to show $P \cap R \neq (0)$ for each $P \in \operatorname{Spec} S$ with $\operatorname{ht} P = 1$. This is clear if $x \in P$, while if $x \notin P$, then $k[[x]] \cap P = (0)$, and so $k[[x]] \hookrightarrow R/(P\cap R) \hookrightarrow S/P$ and S/P is finite over $k[[x]]$. Therefore $\dim R/(P \cap R) = 1$, and so $P \cap R \neq (0)$.

Definition 26.6 is related to Question 28.1. If a ring R and a nonzero element x of R satisfies conditions 28.1.1 and 28.1.2, then condition 28.1.3 simply says that the extension $R \hookrightarrow R_{\{x\}}$ is *not* TGF.

In some correspondence to Lipman regarding Question 28.1, Conrad asked: "Is there a nonzero prime ideal of $E := k[x,1/x]\,[[y]]$ that intersects $C = k[x]\,[[y]]$ in zero?" If there were such a prime ideal $\mathfrak{p}$, then the triple $(C, x, \mathfrak{p})$ would satisfy conditions 28.1.1 to 28.1.3. This would yield a two-dimensional example to show the answer to Question 28.1 is "No". Thus one can ask:

QUESTION 28.6. Let x and y be indeterminates over a field k. Is the extension $C := k[x]\,[[y]] \hookrightarrow E := k[x, 1/x]\,[[y]]$ TGF?

We show in Proposition 28.12.2 below that the answer to Question 28.6 is "Yes"; thus the triple $(C, x, \mathfrak{p})$ does not satisfy condition 28.1.3, although it does satisfy conditions 28.1.1 and 28.1.2. This is part of our analysis of the prime spectra of A, B, C, D_n and E, and the maps induced on these spectra by the inclusion maps on the rings.

REMARKS 28.7. (1) The extension $k[[x,y]] \hookrightarrow k[[x,y/x]]$ is, up to isomorphism, the same as the extension $k[[x,xy]] \hookrightarrow k[[x,y]]$.

(2) We show in Chapter 29 that the extension $R := k[[x,y,xz]] \hookrightarrow S := k[[x,y,z]]$ is not TGF. We also give more information about TGF extensions of local rings there.

(3) Takehiko Yasuda gives additional information on the TGF property in [**188**]. In particular, he shows that

$$\mathbb{C}[x,y][[z]] \hookrightarrow \mathbb{C}[x,x^{-1},y][[z]]$$

is not TGF, where $\mathbb{C}$ is the field of complex numbers [**188**, Theorem 2.7].

28.2. Trivial generic fiber (TGF) extensions and prime spectra

We record in Proposition 28.8 several basic facts about TGF extensions. We omit the proofs since they are straightforward.

PROPOSITION 28.8. *Let $R \hookrightarrow S$ and $S \hookrightarrow T$ be injective maps where R, S and T are integral domains.*

1. *If $R \hookrightarrow S$ and $S \hookrightarrow T$ are TGF extensions, then so is the composite map $R \hookrightarrow T$. Equivalently if the composite map $R \hookrightarrow T$ is not TGF, then at least one of the extensions $R \hookrightarrow S$ or $S \hookrightarrow T$ is not TGF.*
2. *If $R \hookrightarrow T$ is TGF, then $S \hookrightarrow T$ is TGF.*
3. *If the map $\operatorname{Spec} T \to \operatorname{Spec} S$ is surjective and $R \hookrightarrow T$ is TGF, then $R \hookrightarrow S$ is TGF.*

We use the following remark about prime ideals in a formal power series ring.

REMARKS 28.9. Let R be a commutative ring and let $R[[y]]$ denote the formal power series ring in the variable y over R. Then

1. Each maximal ideal of $R[[y]]$ is of the form $(\mathfrak{m}, y)R[[y]]$ where $\mathfrak{m}$ is a maximal ideal of R. Thus y is in every maximal ideal of $R[[y]]$.
2. If R is Noetherian with $\dim R[[y]] = n$ and $x_1, \dots, x_m$ are independent indeterminates over $R[[y]]$, then y is in every height $n+m$ maximal ideal of the polynomial ring $R[[y]]\,[x_1, \dots, x_m]$.

PROOF. Item 1 follows from Theorem 2.33. For item 2, let $\mathfrak{m}$ be a maximal ideal of $R[[y]]\,[x_1, \dots, x_m]$ with $\operatorname{ht}(\mathfrak{m}) = n+m$. By [**103**, Theorem 39], $\operatorname{ht}(\mathfrak{m} \cap R[[y]]) = n$; thus $\mathfrak{m} \cap R[[y]]$ is maximal in $R[[y]]$, and so, by item 1, $y \in \mathfrak{m}$. □

PROPOSITION 28.10. *Let n be a positive integer, let R be an n-dimensional Noetherian domain, let y be an indeterminate over R, and let $\mathfrak{q}$ be a prime ideal of height n in the power series ring $R[[y]]$. If $y \notin \mathfrak{q}$, then $\mathfrak{q}$ is contained in a unique maximal ideal of $R[[y]]$.*

PROOF. Since $R[[y]]$ has dimension $n+1$ and $y \notin \mathfrak{q}$, the ring $S := R[[y]]/\mathfrak{q}$ has dimension one. Moreover, S is complete with respect to the yS-adic topology [**122**, Theorem 8.7] and every maximal ideal of S is a minimal prime of the principal ideal yS. Hence S is a complete semilocal ring. Since S is also an integral domain, it must be local by [**122**, Theorem 8.15]. Therefore $\mathfrak{q}$ is contained in a unique maximal ideal of $R[[y]]$. □

In Section 28.3 we use the following corollary to Proposition 28.10.

COROLLARY 28.11. *Let R be a one-dimensional Noetherian domain and let $\mathfrak{q}$ be a height-one prime ideal of the power series ring $R[[y]]$. If $\mathfrak{q} \neq yR[[y]]$, then $\mathfrak{q}$ is contained in a unique maximal ideal of $R[[y]]$.*

PROPOSITION 28.12. *Consider the nested mixed polynomial-power series rings:*

$$\begin{aligned} A := k[x, y] &\hookrightarrow B := k[[y]]\,[x] \hookrightarrow C := k[x]\,[[y]] \\ &\hookrightarrow D_1 := k[x]\,[[y/x]] \hookrightarrow D_2 := k[x]\,[[y/x^2]] \hookrightarrow \cdots \\ &\hookrightarrow D_n := k[x]\,[[y/x^n]] \hookrightarrow \cdots \hookrightarrow E := k[x, 1/x]\,[[y]], \end{aligned}$$

where k is a field and x and y are indeterminates over k. Then

1. *If $S \in \{B, C, D_1, D_2, \cdots, D_n, \cdots, E\}$, then $A \hookrightarrow S$ is not TGF.*
2. *If $\{R, S\} \subset \{B, C, D_1, D_2, \cdots, D_n \cdots, E\}$ are such that $R \subseteq S$, then $R \hookrightarrow S$ is TGF.*
3. *Each of the proper associated maps on spectra fails to be surjective.*

PROOF. For item 1, let $\sigma(y) \in yk[[y]]$ be such that $\sigma(y)$ and y are algebraically independent over k. Then $(x - \sigma(y))S \cap A = (0)$, and so $A \hookrightarrow S$ is not TGF.

For item 2, observe that every maximal ideal of C, D_n or E is of height two with residue field finite algebraic over k. To show $R \hookrightarrow S$ is TGF, it suffices to show $\mathfrak{q} \cap R \neq (0)$ for each height-one prime ideal $\mathfrak{q}$ of S. This is clear if $y \in \mathfrak{q}$. If $y \notin \mathfrak{q}$, then $k[[y]] \cap \mathfrak{q} = (0)$, and so $k[[y]] \hookrightarrow R/(\mathfrak{q} \cap R) \hookrightarrow S/q$ are injections. By Corollary 28.11, $S/\mathfrak{q}$ is a one-dimensional local domain. Since the residue field of $S/\mathfrak{q}$ is finite algebraic over k, it follows that $S/\mathfrak{q}$ is finite over $k[[y]]$. Therefore $S/\mathfrak{q}$ is integral over $R/(\mathfrak{q} \cap R)$. Hence $\dim(R/(\mathfrak{q} \cap R) = 1$ and so $\mathfrak{q} \cap R \neq (0)$.

For item 3, observe that xD_n is a prime ideal of D_n and x is a unit of E. Thus $\operatorname{Spec} E \to \operatorname{Spec} D_n$ is not surjective. Now, considering $C = D_0$ and $n > 0$, we have $xD_n \cap D_{n-1} = (x, y/x^{n-1})D_{n-1}$. Therefore xD_{n-1} is not in the image of the map $\operatorname{Spec} D_n \to \operatorname{Spec} D_{n-1}$. The map from $\operatorname{Spec} C \to \operatorname{Spec} B$ is not onto, because $(1+xy)B$ is a prime ideal of B, but $1+xy$ is a unit in C. Similarly $\operatorname{Spec} B \to \operatorname{Spec} A$ is not onto, because $(1+y)A$ is a prime ideal of A, but $1+y$ is a unit in B. This completes the proof. □

QUESTION AND REMARKS 28.13. Which of the Spec maps of Proposition 28.12 are one-to-one and which are finite-to-one?

1. For $S \in \{B, C, D_1, D_2, \cdots, D_n, \cdots, E\}$, the generic fiber ring of the map $A \hookrightarrow S$ has infinitely many prime ideals and has dimension one. Every height-two maximal ideal of S contracts in A to a maximal ideal. Every maximal ideal of S containing y has height two. Also $yS \cap A = yA$ and the map $\operatorname{Spec} S/yS \to \operatorname{Spec} A/yA$ is one-to-one.
2. Suppose $R \hookrightarrow S$ is as in Proposition 28.12.2. Each height-two prime of S contracts in R to a height-two maximal ideal of R. Each height-one prime of R is the contraction of at most finitely many prime ideals of S and all of these prime ideals have height one. If $R \hookrightarrow S$ is flat, which is true if $S \in \{B, C, E\}$, then "going-down" holds for $R \hookrightarrow S$, and so, for P a height-one prime of S, we have $\operatorname{ht}(P \cap R) \leq 1$.
3. As mentioned in [**94**, Remark 1.5], C/P is Henselian for every nonzero prime ideal P of C other than yC.

28.3. Spectra for two-dimensional mixed polynomial-power series rings

Let x and y be indeterminates over a field k. We consider the prime spectra, as partially ordered sets, of the mixed polynomial-power series rings A, B, C, $D_1, D_2, \cdots, D_n, \cdots$ and E as given in Sequences 28.0.1 and 28.0.2 at the beginning of this chapter.

Even for k a countable field there are at least two non-order-isomorphic partially ordered sets that can be the prime spectrum of the polynomial ring $A := k[x, y]$. Let $\mathbb{Q}$ be the field of rational numbers, let F be a field contained in the algebraic closure of a finite field and let $\mathbb{Z}$ denote the ring of integers. Then, by [**184**] and [**185**], $\operatorname{Spec}\mathbb{Q}[x, y] \ncong \operatorname{Spec} F[x, y] \cong \operatorname{Spec}\mathbb{Z}[y]$.

The prime spectra of the rings B, C, $D_1, \cdots, D_n, \cdots$, and E of Sequences 28.0.1 and 28.0.2 are simpler since they involve power series in y. Remark 28.9.2 implies that y is in every maximal ideal of height two of each of these rings.

The partially ordered set $\operatorname{Spec} B = \operatorname{Spec}(k[[y]]\,[x])$ is similar to a prime ideal space studied by Heinzer and S. Wiegand in the countable case in [**94**] and then generalized by Shah to other cardinalities in [**169**].The ring $k[[y]]$ is uncountable, even if k is countable. It follows that $\operatorname{Spec} B$ is also uncountable. The partially ordered set $\operatorname{Spec} B$ can be described uniquely up to isomorphism by the axioms of [**169**] (similar to the CHP axioms of [**94**]), since $k[[y]]$ is Henselian and has cardinality at least equal to c, the cardinality of the real numbers $\mathbb{R}$. [3]

Theorem 28.14 characterizes $U := \operatorname{Spec} B$, for the ring B of Sequence 28.01, as a *Henselian affine* partially ordered set (where the "$\leq$" relation is "set containment").

THEOREM 28.14. [**94**, Theorem 2.7] [**169**, Theorem 2.4] *Let $B = k[[y]]\,[x]$ be as in Sequence 28.0.1, where k is a field, the cardinality of the set of maximal ideals of $k[x]$ is α and the cardinality of $k[[y]]$ is β. Then the partially ordered set $U := \operatorname{Spec} B$ under containment is called Henselian affine of type (β, α) and is characterized as a partially ordered set by the following axioms:*

(1) $|U| = \beta$.
(2) *U has a unique minimal element.*
(3) $\dim(U) = 2$ *and* $|\{$ *height-two elements of* $U\}| = \alpha$.
(4) *There exists a special height-one element $u \in U$ such that u is less than every height-two element of U, namely $u = (y)$, and the special element is unique.*
(5) *Every non-special height-one element of U is less than at most one height-two element.*
(6) *Every height-two element $t \in U$ is greater than exactly β height-one elements such that t is the unique height-two element above each. If $t_1, t_2 \in U$ are distinct height-two elements, then the special element from (4) is the unique height-one element less than both.*
(7) *There are exactly β height-one elements that are maximal.*

REMARK 28.15. (1) The axioms of Theorem 28.14 are redundant. We feel this redundancy helps in understanding the relationships among the prime ideals.

(2) The theorem applies to the spectrum of B by defining the unique minimal element to be the ideal (0) of B and the special height-one element to be the prime ideal yB. Every height-two maximal ideal $\mathfrak{m}$ of B has nonzero intersection with $k[[y]]$. Thus $\mathfrak{m}/yB$ is principal and so $\mathfrak{m} = (y, f(x))$, for some monic irreducible polynomial $f(x)$ of $k[x]$. Consider $\{f(x) + ay \,|\, a \in k[[y]]\}$. This set has cardinality β and each $f(x) + ay$ is contained in a nonempty finite set of height-one primes contained in $\mathfrak{m}$. If $\mathfrak{p}$ is a height-one prime contained in $\mathfrak{m}$ with $\mathfrak{p} \neq yB$, then

[3] Kearnes and Oman observe in [**105**] that some cardinality arguments are incomplete in the paper [**169**]. R. Wiegand and S. Wiegand show that Shah's results are correct in [**187**]. In Remarks 28.15.2 we give proofs of some items of Theorem 28.14.

$\mathfrak{p} \cap k[[y]] = (0)$, and so $\mathfrak{p}k((y))[x]$ is generated by a monic polynomial in $k((y))[x]$. But for $a, b \in k[[y]]$ with $a \neq b$, we have $(f(x) + ay, f(x) + by)k((y))[x] = k((y))[x]$. Therefore no height-one prime contained in $\mathfrak{m}$ contains both $f(x)+ay$ and $f(x)+by$. Since B is Noetherian and $|B| = \beta$ is an infinite cardinal, we conclude that the cardinality of the set of height-one prime ideals contained in $\mathfrak{m}$ is β. Examples of height-one maximals are $(1 + xyf(x, y))$, for various $f(x, y) \in k[[y]]\,[x]$. The set of height-one maximal ideals of B also has cardinality β.

(3) We observe that $\alpha = |k| \cdot \aleph_0$ and $\beta = |k|^{\aleph_0}$ in Theorem 28.14, where $\aleph_0 = |\mathbb{N}|$. The proof of this is Exercise 28.1.

(4) The axioms given in Theorem 28.14 *characterize* $\operatorname{Spec} B$ in the sense that every two partially ordered sets satisfying these axioms are order-isomorphic.

The picture of $\operatorname{Spec} B$ is shown below:

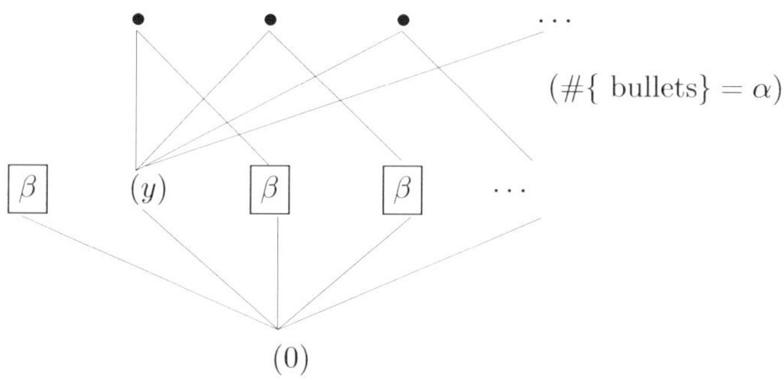

Spec $k[[y]]\,[x]$

In the diagram, β is the cardinality of $k[[y]]$, and α is the cardinality of the set of maximal ideals of $k[x]$ (and also the cardinality of the set of maximal ideals of $k[[y]]\,[x]$); the boxed β means there are cardinality β height-one primes in that position with respect to the partial ordering.

REMARK 28.16. In his paper [**180**], Peder Thompson shows that Theorem 28.14 relates to an example involving $\operatorname{cosupp}_R(R)$, the *cosupport* of a Noetherian ring R as a module over itself. The set $\operatorname{cosupp}_R(R)$ is precisely the set of prime ideals appearing in a minimal pure-injective resolution of R; see [**180**, Theorem 2.5, and the discussion on p. 9].[4]

Sather-Wagstaff and Wicklein ask in [**163**] whether the cosupport of a finitely generated module over a Noetherian ring need be a closed subset of $\operatorname{Spec} R$. Thompson considers the ring $T = k[[y]][x]$, as a module over itself in [**180**, Example 5.6].

[4] "Cosupport" is normally defined for derived categories of chain complexes. For a finitely generated module M over a Noetherian ring R, Sean Sather-Wagstaff and Richard Wicklein state in [**163**, Fact 4.2] that the *small co-support*, (the same "cosupport" used by Thompson), equals

$$\text{co-supp}_R(M) = \{\mathfrak{p} \in \operatorname{Spec} R \mid \operatorname{Ext}^i_R(k(\mathfrak{p}), M) \neq 0, \text{ for some } i \in \mathbb{N}_0\}$$

For a discussion of injective resolutions and $\operatorname{Ext}^i_R(k(\mathfrak{p}), M)$, see [**122**, Appendix B]. Definitions of "cosupport" in the literature vary. Thompson's definition of cosupport is introduced by Dave Benson, Srikanth Iyengar, and Henning Krause in their paper [**20**]. A motivation for this definition of cosupport is Amnon Neeman's paper [**139**].

By Theorem 28.14 above, T has uncountably many height-one maximal ideals. Every maximal ideal of T is in the cosupport of T by [**180**, Remark 4.8]. Thus, if the cosupport of T were closed, it would equal $\mathcal{V}(I)$ for some ideal I contained in every height-one maximal ideal, whereas

$$\bigcap\{\mathfrak{p} \in \operatorname{Spec} T \mid \mathfrak{p} \text{ is a height-one maximal ideal }\} = (0),$$

by Remark 2.12, since T is a UFD and thus a Krull domain. That is, the only possibility for I would be $I = (0)$.

On the other hand, the cosupport of $T/(x) = k[[y]]$ is $\{(y)\}$ by [**180**, Example 5.3], and so (0) is not in the cosupport of $k[[y]]$. Lemma 4.4 of [**180**], states that, if $f : R \to S$ is a finite map of Noetherian rings, if f^* is the spectral map $f^* : \operatorname{Spec} S \to \operatorname{Spec} R$, and if $\operatorname{cosupp}_S S$ is the cosupport of S over S, then $\operatorname{cosupp}_S S = (f^*)^{-1}(\operatorname{cosupp}_R R)$, where $\operatorname{cosupp}_R R$ is the cosupport of R over R. Thus (x) is not in the cosupport of T, and so the cosupport of T is not $\mathcal{V}((0))$. Therefore the cosupport of T is not closed. For more details, see [pT].

Next we consider $\operatorname{Spec} R[[y]]$, for R a Noetherian one-dimensional domain. We show in Theorem 28.19 below that $\operatorname{Spec} R[[y]]$ has the following picture:

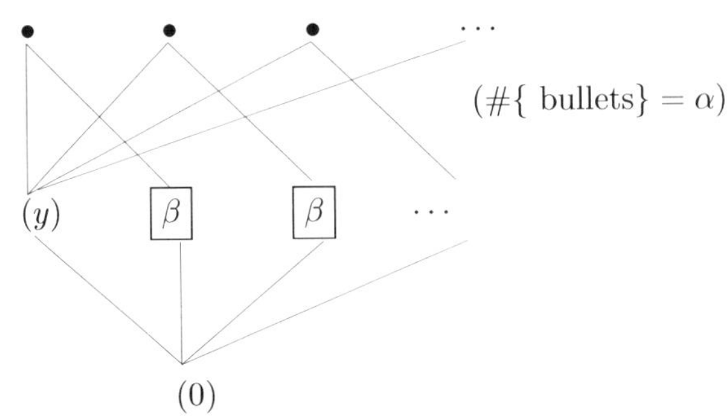

$\operatorname{Spec}(R[[y]])$

Here α is the cardinality of the set of maximal ideals of R (and also the cardinality of the set of maximal ideals of $R[[y]]$ by Remark 28.9.1) and β is the cardinality (uncountable) of $R[[y]]$; the boxed β (one for each maximal ideal of R) means that there are exactly β prime ideals in that position.

We give the following lemma and add some more arguments in order to justify the cardinalities that occur in the spectra of power series rings more precisely.

LEMMA 28.17. [**187**, Lemma 4.2] *Let T be a Noetherian domain, y an indeterminate and I a proper ideal of T. Let $\delta = |T|$ and $\gamma = |T/I|$. Then $\delta \leq \gamma^{\aleph_0}$, and $|T[[y]]| = \delta^{\aleph_0} = \gamma^{\aleph_0}$.*

PROOF. The first equality holds by Exercise 28.1. That $\delta^{\aleph_0} \geq \gamma^{\aleph_0}$ follows from $\gamma \leq \delta$. For the reverse inequality, $\bigcap_{n\geq 1} I^n = 0$ by the Krull Intersection Theorem [**122**, Theorem 8.10 (ii)]. Therefore there is a monomorphism

$$T \hookrightarrow \prod_{n\geq 1} T/I^n. \tag{28.17.0}$$

Now T/I^n has a finite filtration with factors I^{r-1}/I^r for each r with $1 \leq r \leq n$. Since I^{r-1}/I^r is a finitely generated (T/I)-module, $|I^{r-1}/I^r| \leq \gamma^{\aleph_0}$. Therefore $|T/I^n| \leq (\gamma^{\aleph_0})^n = \gamma^{\aleph_0}$, for each n. Thus $\delta \leq (\gamma^{\aleph_0})^{\aleph_0} = \gamma^{(\aleph_0^2)} = \gamma^{\aleph_0}$ by Equation 28.17.0. Finally, $\delta^{\aleph_0} \leq (\gamma^{\aleph_0})^{\aleph_0} = \gamma^{\aleph_0}$, and so $\delta^{\aleph_0} = \gamma^{\aleph_0}$. □

The following remarks, observed in the article [**187**] of R. Wiegand and S. Wiegand, are helpful for establishing the cardinaliies in Theorem 28.19.

REMARKS 28.18. Let $\aleph_0$ denote the cardinality of the set of natural numbers. Suppose that T is a commutative ring of cardinality δ, that $\mathfrak{m}$ is a maximal ideal of T and that γ is the cardinality of $T/\mathfrak{m}$. Then:

(1) The cardinality of $T[[y]]$ is $\delta^{\aleph_0}$, by Lemma 28.17 and Exercise 28.1. If T is Noetherian, then $T[[y]]$ is Noetherian, and so every prime ideal of $T[[y]]$ is finitely generated. Since the cardinality of the finite subsets of $T[[y]]$ is $\delta^{\aleph_0}$, it follows that $T[[y]]$ has at most $\delta^{\aleph_0}$ prime ideals.

(2) If T is Noetherian, then there are at least $\gamma^{\aleph_0}$ distinct height-one prime ideals (other than $(y)T[[y]]$) of $T[[y]]$ contained in $(\mathfrak{m}, y)T[[y]]$. To see this, choose a set $C = \{c_i \,|\, i \in I\}$ of elements of T so that $\{c_i + \mathfrak{m} \,|\, i \in I\}$ gives the distinct coset representatives for $T/\mathfrak{m}$. Thus there are γ elements of C, and for $c_i, c_j \in C$ with $c_i \neq c_j$, we have $c_i - c_j \notin \mathfrak{m}$. Now also let $a \in \mathfrak{m}, a \neq 0$. Consider the set

$$G = \{a + \sum_{n \in \mathbb{N}} d_n y^n \,|\, d_n \in C \,\forall n \in \mathbb{N}\}.$$

Each of the elements of G is in $(\mathfrak{m}, y)T[[y]] \setminus yT[[y]]$ and hence is contained in a height-one prime contained in $(\mathfrak{m}, y)T[[y]]$ distinct from $yT[[y]]$.

Moreover, $|G| = |C|^{\aleph_0} = \gamma^{\aleph_0}$. Let P be a height-one prime ideal of $T[[y]]$ contained in $(\mathfrak{m}, y)T[[y]]$ but such that $y \notin P$. If two distinct elements of G, say $f = a + \sum_{n \in \mathbb{N}} d_n y^n$ and $g = a + \sum_{n \in \mathbb{N}} e_n y^n$, with the $d_n, e_n \in C$, are both in P, then so is their difference; that is

$$f - g = \sum_{n \in \mathbb{N}} d_n y^n - \sum_{n \in \mathbb{N}} e_n y^n = \sum_{n \in \mathbb{N}} (d_n - e_n) y^n \in P.$$

Now let t be the smallest power of y so that $d_t \neq e_t$. Then $(f - g)/y^t \in P$, since P is prime and $y \notin P$, but the constant term, $d_t - e_t \notin \mathfrak{m}$, which contradicts the fact that $P \subseteq (\mathfrak{m}, y)T[[y]]$. Thus there must be at least $|C|^{\aleph_0} = \gamma^{\aleph_0}$ distinct height-one primes contained in $(\mathfrak{m}, y)T[[y]]$.

(3) If T is Noetherian, then there are exactly $\gamma^{\aleph_0} = \delta^{\aleph_0}$ distinct height-one prime ideals (other than $yT[[y]]$) of $T[[y]]$ contained in $(\mathfrak{m}, y)T[[y]]$. This follows from (1) and (2) and Lemma 28.17.

THEOREM 28.19. [**87**] [**187**] *Let R be a one-dimensional Noetherian domain with cardinality δ, let $\beta = \delta^{\aleph_0}$ and let α be the cardinality of the set of maximal ideals of R, where α may be finite. Let $U = \operatorname{Spec} R[[y]]$, where y is an indeterminate over R. Then U as a partially ordered set (where the "$\leq$" relation is "set containment") satisfies the following axioms:*

(1) $|U| = \beta$.
(2) *U has a unique minimal element, namely* (0).
(3) $\dim(U) = 2$ *and* $|\{$ *height-two elements of* $U \}| = \alpha$.
(4) *There exists a special height-one element $u \in U$ such that u is less than every height-two element of U, namely $u = (y)$. If $|\max(R)| > 1$, then the special element is unique.*

(5) *Every non-special height-one element of U is less than exactly one height-two element.*
(6) *Every height-two element $t \in U$ is greater than exactly β height-one elements that are less than only t. If $t_1, t_2 \in U$ are distinct height-two elements, then the special element from (4) is the unique height-one element less than both.*
(7) *There are no height-one maximal elements in U. Every maximal element has height two.*

The set U is characterized as a partially ordered set by the axioms 1-7. Every partially ordered set satisfying the axioms 1-7 is isomorphic to every other such partially ordered set.

PROOF. Item 1 follows from Remarks 28.18.1 and 28.18.3. Item 2 and the first part of item 3 are clear. The second part of item 3 follows immediately from Remark 28.9.1.

For items 4 and 5, suppose that P is a height-one prime of $R[[y]]$. If $P = yR[[y]]$, then P is contained in each maximal ideal of $R[[y]]$ by Remark 28.9.1, and so $yR[[y]]$ is the special element. If $y \notin P$, then, by Corollary 28.11, P is contained in a unique maximal ideal of $R[[y]]$.

For item 6, use Remarks 28.18.2 and 28.18.3.

All partially ordered sets satisfying the axioms of Theorem 28.14 are order-isomorphic, and the partially ordered set U of the present theorem satisfies the same axioms as in Theorem 28.14 except axiom (7) that involves height-one maximals. Since U has no height-one maximals, an order-isomorphism between two partially ordered sets as in Theorem 28.19 can be deduced by adding on height-one maximals and then deleting them. □

COROLLARY 28.20. *In the terminology of Sequences 28.0.1 and 28.0.2 at the beginning of this chapter, we have* $\operatorname{Spec} C \cong \operatorname{Spec} D_n \cong \operatorname{Spec} E$, *but* $\operatorname{Spec} B \not\cong \operatorname{Spec} C$.

PROOF. The rings C, D_n, and E are all formal power series rings in one variable over a one-dimensional Noetherian domain R, where R is either $k[x]$ or $k[x, 1/x]$. Thus the domain R satisfies the hypotheses of Theorem 28.19. Also the number of maximal ideals is the same for C, D_n, and E, because in each case, it is the same as the number of maximal ideals of R which is $|k[x]| = |k| \cdot \aleph_0$.

Thus in the picture of $R[[y]]$ shown above, for $R[[y]] = C, D_n$ or E, we have $\alpha = |k| \cdot \aleph_0$ and $\beta = |R[[y]]| = |R|^{\aleph_0}$, and so the spectra are isomorphic. The spectrum of B is not isomorphic to that of C, however, because B contains height-one maximal ideals, such as that generated by $1 + xy$, whereas C has no height-one maximal ideals. □

REMARK 28.21. As mentioned at the beginning of this section, it is shown in [**184**] and [**185**] that $\operatorname{Spec} \mathbb{Q}[x, y] \not\cong \operatorname{Spec} F[x, y] \cong \operatorname{Spec} \mathbb{Z}[y]$, where F is a field contained in the algebraic closure of a finite field. Corollary 28.22 shows that the spectra of power series extensions in y behave differently in that $\operatorname{Spec} \mathbb{Z}[[y]] \cong \operatorname{Spec} \mathbb{Q}[x]\,[[y]] \cong \operatorname{Spec} F[x]\,[[y]]$.

COROLLARY 28.22. *If $\mathbb{Z}$ is the ring of integers, $\mathbb{Q}$ is the rational numbers, F is a field contained in the algebraic closure of a finite field, and $\mathbb{R}$ is the real numbers,* *then*

$$\operatorname{Spec}\mathbb{Z}[[y]] \cong \operatorname{Spec}\mathbb{Q}[x]\,[[y]] \cong \operatorname{Spec} F[x]\,[[y]] \ncong \operatorname{Spec}\mathbb{R}[x]\,[[y]].$$

PROOF. The rings $\mathbb{Z}, \mathbb{Q}[x]$ and $F[x]$ are all countable with countably infinitely many maximal ideals. Thus if $R = \mathbb{Z}, \mathbb{Q}[x]$ or $F[x]$, then R satisfies the hypotheses of Theorem 28.19 with the cardinality conditions of parts (b) and (c). On the other hand, $\mathbb{R}[x]$ has uncountably many maximal ideals; thus $\mathbb{R}[x]\,[[y]]$ also has uncountably many maximal ideals. □

28.4. Higher dimensional mixed polynomial-power series rings

In analogy to Sequence 28.0.1, we display several embeddings involving three variables.

(28.4.0.1)

$$k[x,y,z] \overset{\alpha}{\hookrightarrow} k[[z]]\,[x,y] \overset{\beta}{\hookrightarrow} k[x]\,[[z]]\,[y] \overset{\gamma}{\hookrightarrow} k[x,y]\,[[z]] \overset{\delta}{\hookrightarrow} k[x]\,[[y,z]],$$

$$k[[z]]\,[x,y] \overset{\epsilon}{\hookrightarrow} k[[y,z]]\,[x] \overset{\zeta}{\hookrightarrow} k[x]\,[[y,z]] \overset{\eta}{\hookrightarrow} k[[x,y,z]],$$

where k is a field and x, y and z are indeterminates over k.

REMARKS 28.23. (1) By Proposition 28.12.2 every nonzero prime ideal of $C = k[x]\,[[y]]$ has nonzero intersection with $B = k[[y]]\,[x]$. In three or more variables, however, the analogous statements fail. We show below that the maps $\alpha, \beta, \gamma, \delta, \epsilon, \zeta, \eta$ in Sequence 28.4.0.1 fail to be TGF. Thus, by Proposition 28.8.2, no proper inclusion in Sequence 28.4.0.1 is TGF. The dimensions of the generic fiber rings of the maps in the diagram are either one or two.

(2) For those rings in Sequence 28.4.0.1 of form $R = S[[z]]$ (ending in a power series variable) where S is a ring, such as $R = k[x,y][[z]]$, we have some information concerning the prime spectra. By Proposition 28.10 every height-two prime ideal not containing z is contained in a unique maximal ideal. By [**137**, Theorem 15.1] the maximal ideals of $S[[z]]$ are of the form $(\mathfrak{m}, z)S[[z]]$, where $\mathfrak{m}$ is a maximal ideal of S, and thus the maximal ideals of $S[[z]]$ are in one-to-one correspondence with the maximal ideals of S. As in section 28.3, using Remarks 28.9, we see that maximal ideals of $\operatorname{Spec} k[[z]]\,[x,y]$ can have height two or three, that (z) is contained in every height-three prime ideal, and that every height-two prime ideal not containing (z) is contained in a unique maximal ideal.

(3) It follows by arguments analogous to that in Proposition 28.12.1, that α, δ, ϵ are not TGF. For α, let $\sigma(z) \in zk[[z]]$ be transcendental over $k(z)$; then $(x-\sigma)k[[z]]\,[x,y] \cap k[x,y,z] = (0)$. For δ and ϵ: let $\sigma(y) \in yk[[y]]$ be transcendental over $k(y)$; then $(x-\sigma)k[x]\,[[z,y]] \cap k[x]\,[[z]]\,[y] = (0)$, and $(x-\sigma)k[[y,z]]\,[x] \cap k[[z]]\,[x,y] = (0)$.

(4) By Main Theorem 26.3.4.a of Chapters 26 and 28 (proved in Theorem 27.5), η is not TGF and the dimension of the generic fiber ring of η is one.

In order to show in Proposition 28.25 below that the map β is not TGF, we first observe:

PROPOSITION 28.24. *The element* $\sigma = \sum_{n=1}^{\infty}(xz)^{n!} \in k[x]\,[[z]]$ *is transcendental over* $k[[z]]\,[x]$.

PROOF. Consider an expression

$$Z := a_\ell\sigma^\ell + a_{\ell-1}\sigma^{\ell-1} + \cdots + a_1\sigma + a_0,$$

where the $a_i \in k[[z]]\,[x]$ and $a_\ell \neq 0$. Let m be an integer greater than $\ell+1$ and greater than $\deg_x a_i$ for each i such that $0 \le i \le \ell$ and $a_i \neq 0$. Regard each $a_i\sigma^i$ as a power series in x with coefficients in $k[[z]]$.

For each i with $0 \le i \le \ell$, we have $i(m!) < (m+1)!$. It follows that the coefficient of $x^{i(m!)}$ in σ^i is nonzero, and the coefficient of x^j in σ^i is zero for every j with $i(m!) < j < (m+1)!$. Thus if $a_i \neq 0$ and $j = i(m!)+\deg_x a_i$, then the coefficient of x^j in $a_i\sigma^i$ is nonzero, while for j such that $i(m!) + \deg_x a_i < j < (m+1)!$, the coefficient of x^j in $a_i\sigma^i$ is zero. By our choice of m, for each i such that $0 \le i < \ell$ and $a_i \neq 0$, we have

$$(m+1)! > \ell(m!) + \deg_x a_\ell \ge i(m!) + m! > i(m!) + \deg_x a_i.$$

Thus in Z, regarded as a power series in x with coefficients in $k[[z]]$, the coefficient of x^j is nonzero for $j = \ell(m!) + \deg_x a_\ell$. Therefore $Z \neq 0$. We conclude that σ is transcendental over $k[[z]]\,[x]$. □

PROPOSITION 28.25. *In Sequence 28.4.0.1,* $k[[z]]\,[x,y] \overset{\beta}{\hookrightarrow} k[x]\,[[z]]\,[y]$ *is not a TGF-extension.*

PROOF. Fix an element $\sigma \in k[x]\,[[z]]$ that is transcendental over $k[[z]]\,[x]$. We define $\pi : k[x]\,[[z]]\,[y] \to k[x]\,[[z]]$ to be the identity map on $k[x]\,[[z]]$ and $\pi(y) = \sigma z$. Let $\mathfrak{q} = \ker \pi$. Then $y - \sigma z \in \mathfrak{q}$. If $h \in \mathfrak{q} \cap (k[[z]]\,[x,y])$, then

$$h = \sum_{j=0}^{s}\sum_{i=0}^{t}(\sum_{\ell\in\mathbb{N}} a_{ij\ell}z^\ell)x^iy^j, \text{ for some } s,t \in \mathbb{N} \text{ and } a_{ij\ell} \in k, \text{ and so}$$

$$0 = \pi(h) = \sum_{j=0}^{s}\sum_{i=0}^{t}(\sum_{\ell\in\mathbb{N}} a_{ij\ell}z^\ell)x^i(\sigma z)^j = \sum_{j=0}^{s}\sum_{i=0}^{t}(\sum_{\ell\in\mathbb{N}} a_{ij\ell}z^{\ell+j})x^i\sigma^j.$$

Since σ is transcendental over $k[[z]]\,[x]$, we have that x and σ are algebraically independent over $k((z))$. Thus each of the $a_{ij\ell} = 0$. Therefore $\mathfrak{q}\cap(k[[z]][x,y]) = (0)$, and so the embedding β is not TGF. □

The concept of "analytic independence" is useful in several arguments below.

DEFINITION AND REMARKS 28.26. Let I be an ideal of an integral domain A. Assume that A is complete and Hausdorff in the I-adic topology. Let B be a subring of A, let $a_1, \dots, a_n \in I$ and let $v_1, \dots, v_n$ be indeterminates over A. We say $a_1, \dots, a_n$ are *analytically independent* over B if the B-algebra homomorphism $\varphi : B[[v_1, \dots, v_n]] \to A$, where $\varphi(v_i) = a_i$ for each i, is injective.

(1) This definition of "analytically independent" is given in the book of Zariski and Samuel [**193**, page 258]. This use of the term applies to the work of Abhankar and Moh [**15**], and of Dumitrescu [**41**]. However, this definition does not agree with the use of the term "analytically independent" in Matsumura [**122**, page 107] and Swanson and Huneke [**175**, page 175].

(2) If, for example, a and b are elements of I, then we have power series expressions $\sum_{i=0}^\infty\sum_{j=0}^\infty c_{ij}a^ib^j$, where each $c_{ij} \in B$. If a and b are analytically independent over B, then the expression above is unique. Thus $\sum_{i=0}^\infty\sum_{j=0}^\infty c_{ij}a^ib^j = 0$ implies that every $c_{ij} = 0$.

PROPOSITION 28.27. *In Sequence 28.4.0.1, the extensions*

$$k[[y,z]]\,[x] \overset{\zeta}{\hookrightarrow} k[x]\,[[y,z]] \quad \text{and} \quad k[x]\,[[z]]\,[y] \overset{\gamma}{\hookrightarrow} k[x,y]\,[[z]]$$

are not TGF.

PROOF. For ζ, let $t = xy$ and let $\sigma \in k[[t]]$ be algebraically independent over $k(t)$. Define $\pi : k[x]\,[[y,z]] \to k[x]\,[[y]]$ as follows. For

$$f := \sum_{\ell=0}^{\infty} \sum_{m+n=\ell} f_{mn}(x) y^m z^n \in k[x]\,[[y,z]],$$

where $f_{mn}(x) \in k[x]$, define

$$\pi(f) := \sum_{\ell=0}^{\infty} \sum_{m+n=\ell} f_{mn}(x) y^m (\sigma y)^n \in k[x]\,[[y]].$$

In particular, $\pi(z) = \sigma y$. Let $\mathfrak{p} := \ker \pi$. Then $z - \sigma y \in \mathfrak{p}$, and so $\mathfrak{p} \neq (0)$. Let $h \in \mathfrak{p} \cap k[[y,z]]\,[x]$. We show $h = 0$. Now h is a polynomial with coefficients in $k[[y,z]]$, and we define $g \in k[[y,z]]\,[t]$, by, if $a_i(y,z) \in k[[y,z]]$ and

$$h := \sum_{i=0}^{r} a_i(y,z) x^i, \text{ then set } g := y^r h = \sum_{i=0}^{r} \Big(\sum_{\ell=0}^{\infty} \sum_{m+n=\ell} b_{imn} y^m z^n\Big) t^i.$$

The coefficients of g are in $k[[y,z]]$, since $y^r x^i = y^{r-i} t^i$. Thus

$$\begin{aligned} 0 = \pi(g) = \sum_{i=0}^{r} \Big(\sum_{\ell=0}^{\infty} \sum_{m+n=\ell} b_{imn} y^m (\sigma y)^n\Big) t^i &= \sum_{i=0}^{r} \Big(\sum_{\ell=0}^{\infty} \sum_{m+n=\ell} b_{imn} \sigma^n y^\ell\Big) t^i \\ &= \sum_{\ell=0}^{\infty} \Big(\sum_{m+n=\ell} \Big(\sum_{i=0}^{r} b_{imn} t^i\Big) \sigma^n\Big) y^\ell. \end{aligned}$$

The elements $t = xy$ and y of $k[x][[y,z]]$ are *analytically independent* over k in the sense of Definition 28.26; hence the coefficient of each y^ℓ (in $k[[t]]$) is 0. Since σ and t are algebraically independent over k, the coefficient of each σ^n is 0. It follows that each $b_{imn} = 0$, that $g = 0$ and hence that $h = 0$. Thus $\mathfrak{p} \cap k[[y,z]]\,[x] = (0)$, and so the extension ζ is not TGF.

To see that γ is not TGF, we switch variables in the proof for ζ, so that $t = yz$. Again choose $\sigma \in k[[t]]$ to be algebraically independent over $k(t)$. Define $\psi : k[x,y]\,[[z]] \to k[y]\,[[z]]$ by $\psi(x) = \sigma z$ and ψ is the identity on $k[y]\,[[z]]$. Then ψ can be extended to $\pi : k[y]\,[[x,z]] \to k[y]\,[[z]]$, which is similar to the π in the proof above. As above, set $\mathfrak{p} := \ker \pi$; then $\mathfrak{p} \cap k[[x,z]]\,[y] = (0)$. Thus $\mathfrak{p} \cap k[x]\,[[z]]\,[y] = (0)$ and γ is not TGF. □

PROPOSITION 28.28. *Let D be an integral domain and let x and t be indeterminates over D. Then $\sigma = \sum_{n=1}^{\infty} t^{n!} \in D[[x,t]]$ is algebraically independent over $D[[x,xt]]$.*

PROOF. Suppose that σ is algebraically dependent over $D[[x,xt]]$. Then there exists an equation

$$\gamma_\ell \sigma^\ell + \cdots + \gamma_i \sigma^i + \cdots + \gamma_1 \sigma + \gamma_0 = 0,$$

where each $\gamma_i \in D[[x, xt]]$, ℓ is a positive integer and $\gamma_\ell \neq 0$. This implies

$$\gamma := \gamma_\ell \sigma^\ell + \cdots + \gamma_i \sigma^i + \cdots + \gamma_1 \sigma = -\gamma_0$$

is an element of $D[[x, xt]]$. We obtain a contradiction by showing that $\gamma \notin D[[x, xt]]$.

For each i with $1 \leq i \leq \ell$, write

$$\gamma_i := \sum_{j=0}^{\infty} f_{ij}(x)(xt)^j \in D[[x, xt]],$$

where each $f_{ij}(x) \in D[[x]]$. Since $\gamma_\ell \neq 0$, there exists j such that $f_{\ell j}(x) \neq 0$. Let a_ℓ be the smallest such j, and let m_ℓ be the order of $f_{\ell a_\ell}(x)$, that is, $f_{\ell a_\ell}(x) = x^{m_\ell} g_\ell(x)$, where $g_\ell(0) \neq 0$. Let n be a positive integer such that

$$n \geq 2 + \max\{\ell, m_\ell, a_\ell\}.$$

Since $\ell < n$ and $1 \leq i \leq \ell$, we have

$$\begin{aligned}
\sigma^i = (t + \ldots + t^{n!} + \ldots)^i &= t(t + \ldots + t^{(n-1)!} + t^{n!} + \ldots)^{(i-1)} \\
&+ \ldots + t^{(n-1)!}(t + \ldots + t^{(n-1)!} + t^{n!} + \ldots)^{(i-1)} \\
&+ t^{n!}(t + \ldots + t^{(n-1)!} + t^{n!} + \ldots)^{(i-1)} + \\
&+ t^{(n+1)!}(t + t^2 + \ldots)^{(i-1)} + \ldots \\
&= \delta_i(t) + c_i t^{i(n!)} + t^{(n+1)!}\tau_i(t),
\end{aligned} \tag{28.28.0}$$

where $c_i = 1$ is a nonzero element of D, $\delta_i(t)$ is a polynomial in $D[t]$ of degree at most $(i-1)n! + (n-1)!$ and $\tau_i(t) \in D[[t]]$, for each i. We use the following two claims to complete the proof. □

CLAIM 28.29. The coefficient of $t^{\ell(n!)+a_\ell}$ in $\sigma^\ell \gamma_\ell = \sigma^\ell(\sum_{j=a_\ell}^{\infty} f_{\ell j}(x)(xt)^j)$ as a power series in $D[[x]]$ has order $m_\ell + a_\ell$, and hence, in particular, is nonzero.

PROOF. By the choice of n, $(n+1)! = (n+1)n! > \ell(n!) + n > \ell(n!) + a_\ell$. Hence, by the expression for σ^ℓ given in Equation 28.28.0, we see that all of the terms in $\sigma^\ell \gamma_\ell$ of the form $bt^{\ell(n!)+a_\ell}$, for some $b \in D[[x]]$, appear in the product

$$(\, t^{\ell(n!)} + \delta_\ell(t)\,)(x^{m_\ell} g_\ell(x)(xt)^{a_\ell} + \sum_{j=1+a_\ell}^{\ell(n!)+a_\ell} f_{\ell j}(x)(xt)^j).$$

One of the terms of the form $bt^{\ell(n!)+a_\ell}$ in this product is

$$(x^{m_\ell + a_\ell} g_\ell(x))t^{\ell(n!)+a_\ell} = (x^{m_\ell + a_\ell}(g_\ell(0) + xh_\ell(x)))t^{\ell(n!)+a_\ell},$$

where we write $g_\ell(x) = g_\ell(0) + xh_\ell(x)$ with $h_\ell(x) \in D[[x]]$. Since $g_\ell(0)$ is a nonzero element of D, $x^{m_\ell + a_\ell} g_\ell(x) \in D[[x]]$ has order $m_\ell + a_\ell$. The other terms in the product $\sigma^\ell \gamma_\ell$ that have the form $bt^{\ell(n!)+a_\ell}$, for some $b \in D[[x]]$, are in the product

$$(\delta_\ell(t))(\sum_{j=1+a_\ell}^{\ell(n!)+a_\ell} f_{\ell j}(x)(xt)^j) = \sum_{j=1+a_\ell}^{\ell(n!)+a_\ell} f_{\ell j}(x)(xt)^j \delta_\ell(t).$$

Since $\deg_t \delta_\ell \leq (\ell - 1)n! + (n-1)!$ and since, for each j with $f_{\ell j}(x) \neq 0$, we have $\deg_t f_{\ell j}(x)(xt)^j = j$, we see that each term $f_{\ell j}(x)(xt)^j \delta_\ell(t)$ has degree in t less than or equal to $j + (\ell - 1)n! + (n-1)!$. Thus each nonzero term in this product of the form $bt^{\ell(n!)+a_\ell}$ has $\ell(n!) + a_\ell \le j + (\ell-1)n! + (n-1)!$. That is,

$$\begin{aligned} j &\ge \ell(n!) + a_\ell - (\ell-1)(n!) - (n-1)! = a_\ell + n! - (n-1)! \\ &= a_\ell + (n-1)!(n-1) > a_\ell + m_\ell, \end{aligned}$$

since $n - 1 > m_\ell$. Moreover, for j such that $f_{\ell j}(x) \neq 0$, the order in x of $f_{\ell j}(x)(xt)^j$ is bigger than or equal to j. Thus the coefficient of $t^{\ell(n!)+a_\ell}$ in $\sigma^\ell\gamma_\ell$ as a power series in x has order $m_\ell + a_\ell$, as desired for Claim 28.29. □

CLAIM 28.30. For $i < \ell$, the coefficient of $t^{\ell(n!)+a_\ell}$ in $\sigma^i\gamma_i$ as a power series in $D[[x]]$ is either zero or has order greater than $m_\ell + a_\ell$.

PROOF. As in the proof of Claim 28.29, all of the terms in $\sigma^i\gamma_i$ of the form $bt^{\ell(n!)+a_\ell}$, for some $b \in D[[x]]$, appear in the product

$$(\delta_i + t^{i(n!)})(\sum_{j=0}^{\ell(n!)+a_\ell} f_{ij}(x)(xt)^j) = \sum_{j=0}^{\ell(n!)+a_\ell} f_{ij}(x)(xt)^j(\delta_i + t^{i(n!)}).$$

Since $\deg_t(\delta_i + t^{i(n!)}) = i(n!)$, each term in $f_{ij}(x)(xt)^j(\delta_i + t^{i(n!)})$ has degree in t at most $j + i(n!)$. Thus each term in this product of the form $bt^{\ell(n!)+a_\ell}$, for some nonzero $b \in D[[x]]$, has

$$j \ge \ell(n!) + a_\ell - i(n!) \ge n! + a_\ell > m_\ell + a_\ell.$$

Thus $\operatorname{ord}_x b \ge j > m_\ell + a_\ell$. This completes the proof of Claim 28.30. Hence $\gamma \notin D[[x, xt]]$. This completes the proof of Proposition 28.28. □

QUESTION AND REMARKS 28.31. (1) As we show in Proposition 28.12, the embeddings from Equation 1 involving two mixed polynomial-power series ringss of dimension two over a field k with inverted elements are TGF. In the article [**87**] we ask whether this is true in the three-dimensional case. For example, is the embedding θ below TGF?

$$k[x,y]\,[[z]] \overset{\theta}{\hookrightarrow} k[x,y,1/x]\,[[z]]$$

Yasuda shows the answer for this example is "No" in [**188**]. Dumitrescu establishes the answer is "No" in more generality; see Theorem 28.32.

(2) For the four-dimensional case, as observed in the discussion of Question 28.1, it follows from a result of Heinzer and Rotthaus [**70**, Theorem 1.12, p. 364] that the extension $k[x,y,u]\,[[z]] \hookrightarrow k[x,y,u,1/x]\,[[z]]$ is not TGF. Theorem 28.32 yields a direct proof of this fact.

We close this chapter with a result of Dumitrescu that shows many extensions involving only one power series variable are not TGF.

THEOREM 28.32. [**41**, Cor. 4 and Prop. 3] *Let D be an integral domain and let x, y, z be indeterminates over D. For every subring B of $D[[x,y]]$ that contains $D[x,y]$, the extension $B[[z]] \hookrightarrow B[1/x][[z]]$ is not TGF.*

PROOF. Let K be the field of fractions of D and let $\theta(z) \in D[[z]]$ be algebraically independent over $K(z)$.

CLAIM 28.33. The elements xz and $x\theta(z) \in K[[x,z]]$ are analytically independent over $K[[x]]$, and $x\theta(z)$ is analytically independent over $K[[x, xz]]$.

Proof of Claim. Let v and w be indeterminates over $K[[x]]$ and consider the $K[[x]]$-algebra homomorphism $\varphi : K[[x,v,w]] \to K[[x,z]]$ where $\varphi(v) = xz$ and $\varphi(w) = x\theta$. Let $g \in K[[x,v,w]]$ and write $g = \sum_{n \geq 0} g_n(x,v,w)$, where g_n is a form of degree n in x, v, w with coefficients in K. Thus $\varphi(g) = \sum_{n \geq 0} \varphi(g_n)$ and

$$g_n = \sum_{i+j+k=n} c_{ijk} x^i v^j w^k \implies \varphi(g_n) = \sum_{i+j+k=n} c_{ijk} x^n z^j \theta^k.$$

If $\varphi(g) = 0$, then $\varphi(g_n) = 0$ for each n. Since z and θ are algebraically independent over K, each $c_{ijk} = 0$. Thus φ is injective. By Definition 28.26, Claim 28.33 holds.

It follows from Claim 28.33 that $\lambda := x\theta(z/x) \in zD[x, 1/x][[z]]$ is analytically independent over $D[[x,z]]$.

Consider the $D[[x]][1/x][[z]]$-algebra homomorphism

$$\pi : D[[x]]\,[\frac{1}{x}]\,[[z,y]] \to D[[x]]\,[\frac{1}{x}]\,[[z]], \quad \text{where} \quad \pi(y) = \lambda.$$

Let $\mathfrak{p} = \ker \pi$. Then $y - \lambda$ is a nonzero element of $\mathfrak{p} \cap D[x,y][1/x][[z]]$, and so $0 \neq \mathfrak{p} \cap B[1/x][[z]]$. We show $\mathfrak{p} \cap D[[x,y,z]] = (0)$, and so also $\mathfrak{p} \cap B\,[[z]] = (0)$.

The restriction of π to $D[[x,z,y]]$ is injective because λ is analytically independent over $D[[x,z]]$. Therefore $\mathfrak{p} \cap D[[x,z,y]] = (0)$. This completes the proof of Theorem 28.32. □

Exercises

(1) Let k be a field and let $\aleph_0 = |\mathbb{N}|$. Prove that $\alpha = |k| \cdot \aleph_0$ and $\beta = |k|^{\aleph_0}$ in Theorem 28.14.

Suggestion. Notice that every polynomial of the form $x - a$, for $a \in k$, generates a maximal ideal of $k[x]$ and also that $|k[x]| = |k| \cdot \aleph_0$, since $k[x]$ can be considered as an infinite union of polynomials of each finite degree.

(2) Let y denote an indeterminate over the ring of integers $\mathbb{Z}$, and let $A = \mathbb{Z}[[y]]$.
  (a) Prove that every maximal ideal of A has height two.
  (b) Describe and make a diagram of the partially ordered set $\operatorname{Spec} A$.
  (c) Let $B = A[\frac{1}{y+2}]$. Describe the partially ordered set $\operatorname{Spec} B$. Prove that B has maximal ideals of height one, and deduce that $\operatorname{Spec} B$ is not order-isomorphic to $\operatorname{Spec} A$.
  (d) Let $C = A[\frac{y}{2}]$. Describe the partially ordered set $\operatorname{Spec} C$. Prove that C has precisely two non-maximal prime ideal of height one that are an intersection of maximal ideals, while each of A and B has precisely one non-maximal prime ideal of height one that is an intersection of maximal ideals. Deduce that $\operatorname{Spec} C$ is not order-isomorphic to either $\operatorname{Spec} A$ or $\operatorname{Spec} B$.

CHAPTER 29

Extensions of local domains with trivial generic fiber

We consider injective local maps from a Noetherian local domain R to a Noetherian local domain S such that the generic fiber of the inclusion map $R \hookrightarrow S$ is trivial, that is $P \cap R \neq (0)$ for every nonzero prime ideal P of S. [1] Recall that S is said to be a *trivial generic fiber extension of* R, or more briefly, a *TGF extension*, if each nonzero ideal of S has a nonzero intersection with R, or equivalently, if each nonzero element of S has a nonzero multiple in R. We present in this chapter examples of injective local maps involving power series that are TGF, and other examples that fail to be TGF extensions.

Let $R \hookrightarrow S$ be an injective map of integral domains. The ideals of S maximal with respect to having empty intersection with $R \setminus \{0\}$ are prime ideals of S. Thus S is a TGF extension of R if and only if $P \cap R \neq (0)$ for each nonzero prime ideal P of S; Equivalently, S is a TGF extension of $R \iff U^{-1}S$ is a field, where $U = R \setminus (0)$.

Our work in this chapter is motivated by Question 26.4 asked by Melvin Hochster and Yongwei Yao. In this connection, we often use the following setting.

SETTING 29.1. Let $(R, \mathfrak{m}) \hookrightarrow (S, \mathfrak{n})$ be an injective local homomorphism of complete Noetherian local domains such that S is a TGF extension of R.

By Remark 26.5, in the equicharacteristic zero case of Setting 29.1 such extensions arise as a composition

$$(29.1.0) \qquad R = K[[x_1, \dots, x_n]] \hookrightarrow T = L[[x_1, \dots, x_n, y_1, \dots, y_m]] \to T/P = S,$$

where K is a subfield of L, the x_i, y_j are indeterminates, and P is a prime ideal of T maximal with respect to being disjoint from the image of $R \setminus \{0\}$.

We discuss several topics and questions related to Question 26.4. Previous workconcerning homomorphisms of formal power series rings appears in articles of Matsumura, Rotthaus, Abhyankar, Moh, van der Put [**158**],[**3**] [**11**], [**12**] among others. In particular, in [**11**, Section 3], Abhyankar and Moh consider an extension $R = k[[x, xy]] \subsetneq k[[x, y]] = S$ where x and y are indeterminates over an algebrically closed field k.

[1]The material in this chapter is adapted from our paper [**88**] dedicated to Phil Griffith in honor of his contributions to commutative algebra.

29.1. General remarks about TGF extensions

PROPOSITION 29.2. *Let* $(R, \mathfrak{m}) \hookrightarrow (S, \mathfrak{n})$ *be an injective local homomorphism of complete Noetherian local domains with* S *a TGF-extension of* R.

(1) *Assume that* $\dim R = 1$. *Then:*
   (a) $\dim S = 1$ *and* $\mathfrak{m}S$ *is* $\mathfrak{n}$*-primary.*
   (b) *If* $[S/\mathfrak{n} : R/\mathfrak{m}] < \infty$, *then* S *is a finite integral extension of* R.
(2) *If* $R \hookrightarrow S$ *has finite residue extension and* $\dim S \geq 2$, *then* $\dim R \geq 2$.

PROOF. By Krull's Altitude Theorem 2.23, $\mathfrak{n}$ is the union of the height-one primes of S. If $\dim S > 1$, then S has infinitely many height-one primes. Since S is Noetherian, each nonzero element of $\mathfrak{n}$ is contained in only finitely many of these height-one primes. If $\dim S > 1$, then the intersection of the height-one primes of S is zero. For item 1, since $\dim R = 1$ and $R \hookrightarrow S$ is TGF, every nonzero prime of S contains $\mathfrak{m}$. Thus $\dim S = 1$ and $\mathfrak{m}S$ is $\mathfrak{n}$-primary. Moreover, if $[S/\mathfrak{n} : R/\mathfrak{m}] < \infty$, then S is finite over R by Theorem 3.16. Item 2 follows from item 1. □

REMARKS 29.3. (1) There exist extensions S of R as in Setting 29.1 that have an arbitrarily large extension of residue field. For example, if k is a subfield of a field F and x is an indeterminate over F, then $R := k[[x]] \subseteq S := F[[x]]$ is an injective local homomorphism of complete Noetherian local domains and S is a TGF-extension of R.

(2) Let $(R, \mathfrak{m}) \hookrightarrow (T, \mathfrak{q})$ be an injective local homomorphism of complete Noetherian local domains, and let $P \in \operatorname{Spec} T$. Then $S := T/P$ is a TGF-extension of R as in Setting 29.1 if and only if P is an ideal of T maximal with respect to the property that $P \cap R = (0)$.

REMARKS 29.4. Let $X = \{x_1 \dots, x_n\}$, $Y = \{y_1 \dots, y_m\}$ and $Z = \{z_1 \dots, z_r\}$ be finite sets of indeterminates over a field k, where $n \geq 2$, $m, r \geq 1$. Set $R := k[[X]]$ and let P be a prime ideal of $k[[X, Y, Z]]$ that is maximal with respect to $P \cap R = (0)$. Then we have the inclusions:

$$R := k[[X]] \overset{\sigma}{\hookrightarrow} S := k[[X, Y]]/(P \cap k[[X, Y]]) \overset{\tau}{\hookrightarrow} T := k[[X, Y, Z]]/P.$$

By Remark 29.3.2, $\tau \cdot \sigma$ is a TGF extension. By Proposition 28.8.3, $S \hookrightarrow T$ is TGF.

(1) If the map $\operatorname{Spec} T \to \operatorname{Spec} S$ is surjective, then $\sigma : R \hookrightarrow S$ is TGF by Proposition 28.8.2.
(2) If $R \hookrightarrow T$ is finite, then $R \hookrightarrow S$ is also finite, and so $\sigma : R \hookrightarrow S$ is TGF.
(3) If $R \hookrightarrow T$ is not finite, then $\dim T = 2$ by Theorem 27.16.
(4) If $P \cap k[[X, Y]] = 0$, then $S = R[[Y]]$ and $R \hookrightarrow S$ is not TGF. (Example 29.16 shows that this can occur.)

REMARKS AND QUESTION 29.5. (1) With notation as in Remarks 29.4 and with $Y = \{y\}$, a singleton set, it is always true that $\operatorname{ht}(P \cap R[[y]]) \leq n - 1$. (See Theorem 27.15.) Moreover, if $\operatorname{ht}(P \cap R[[y]]) = n - 1$, then $R \hookrightarrow S$ is TGF. Thus if $n = 2$ and $P \cap R[[y]] \neq 0$, then $R \hookrightarrow S$ is TGF.

(2) With notation as in (1) and $n = 3$, it can happen that $P \cap k[[X, y]] \neq (0)$ and $R \hookrightarrow R[[y]]/(P \cap R[[y]])$ is not a TGF extension. To construct an example of such a prime ideal P, we proceed as follows: Since $\dim(k[[X, y]]) = 4$, there exists a prime ideal Q of $k[[X, y]]$ with $\operatorname{ht} Q = 2$ and $Q \cap k[[X]] = (0)$, see Theorem 27.15. Let $\mathfrak{p} \subset Q$ be a prime ideal with $\operatorname{ht} \mathfrak{p} = 1$. Since $\mathfrak{p} \subsetneq Q$ and $Q \cap k[[X]] = (0)$,

the extension $k[[X]] \hookrightarrow k[[X, y]]/\mathfrak{p}$ is not a TGF extension. In particular, it is not finite. Let $P \in \operatorname{Spec}(k[[X, y, Z]])$ be maximal with respect to $P \cap k[[X, y]] = \mathfrak{p}$. By Corollary 29.10 below, $\dim(k[[X, y, Z]]/P) = 2$. Hence P is maximal in the generic fiber over $k[[X]]$.

(3) If $(R, \mathfrak{m}) \hookrightarrow (S, \mathfrak{n})$ is a TGF local extension of complete Noetherian local domains and $S/\mathfrak{n}$ is finite algebraic over $R/\mathfrak{m}$, can the transcendence degree of S over R be finite but nonzero?

(4) If $(R, \mathfrak{m}) \hookrightarrow (S, \mathfrak{n})$ is a TGF extension as in (3) with R equicharacteristic and $\dim R \geq 2$, then by Corollary 29.9 below it follows that either S is a finite integral extension of R or $\dim S = 2$.

PROPOSITION 29.6. *Let $A \hookrightarrow B$ be a TGF extension, where B is a Noetherian integral domain. For each $Q \in \operatorname{Spec} B$, we have $\operatorname{ht} Q \leq \operatorname{ht}(Q \cap A)$. In particular, $\dim A \geq \dim B$.*

PROOF. If $\operatorname{ht} Q = 1$, it is clear that $\operatorname{ht} Q \leq \operatorname{ht}(Q \cap A)$ since $Q \cap A \neq (0)$. Let $\operatorname{ht} Q = n \geq 2$, and assume by induction that $\operatorname{ht} Q' \leq \operatorname{ht}(Q' \cap A)$ for each $Q' \in \operatorname{Spec} B$ with $\operatorname{ht} Q' \leq n - 1$. Since B is Noetherian,

$$(0) = \bigcap \{Q' \mid Q' \subset Q \quad \text{and} \quad \operatorname{ht} Q' = n - 1\}.$$

Hence there exists $Q' \subset Q$ with $\operatorname{ht} Q' = n - 1$ and $Q' \cap A \subsetneq Q \cap A$. We have $n - 1 \leq \operatorname{ht}(Q' \cap A) < \operatorname{ht}(Q \cap A)$, and so $\operatorname{ht}(Q \cap A) \geq n$. $\square$

29.2. TGF extensions with finite residue field extension

SETTING 29.7. Let $n \geq 2$ be an integer, let $X = \{x_1, \ldots, x_n\}$ be a set of independent variables over the field k and let $R = k[[X]]$ be the formal power series ring in n variables over the field k.

THEOREM 29.8. *Let $R = k[[X]]$ be as in Setting 29.7. Assume that $R \hookrightarrow S$ is a TGF local extension, where $(S, \mathfrak{n})$ is a complete Noetherian local domain and $S/\mathfrak{n}$ is finite algebraic over k. Then either $\dim S = n$ and S is a finite integral extension of R or $\dim S = 2$.*

PROOF. It is clear that if S is a finite integral extension of R, then $\dim S = n$. Assume S is not a finite integral extension of R. Let $b_1, \ldots, b_m \in \mathfrak{n}$ be such that $\mathfrak{n} = (b_1, \ldots, b_m)S$, and let $Y = \{y_1, \ldots, y_m\}$ be a set of independent variables over R. Since S is complete the R-algebra homomorphism $\varphi : T := R[[Y]] \to S$ such that $\varphi(y_i) = b_i$ for each i with $1 \leq i \leq m$ is well defined. Let $Q = \ker \varphi$. We have

$$R \hookrightarrow T/Q \hookrightarrow S.$$

By Theorem 3.16 S is a finite module over T/Q. Hence $\dim S = \dim(T/Q)$ and the map $\operatorname{Spec} S \to \operatorname{Spec}(T/Q)$ is surjective, and so by Proposition 28.8(3) $R \hookrightarrow T/Q$ is TGF. By Corollary 27.18, $\dim(T/Q) = 2$, and so $\dim S = 2$. $\square$

COROLLARY 29.9. *Let $(A, \mathfrak{m})$ and $(S, \mathfrak{n})$ be complete equicharacteristic Noetherian local domains with $\dim A = n \geq 2$. Assume that $A \hookrightarrow S$ is a local injective homomorphism and that the residue field $S/\mathfrak{n}$ is finite algebraic over the residue field $A/\mathfrak{m} := k$. If $A \hookrightarrow S$ is a TGF extension, then either $\dim S = n$ and S is a finite integral extension of A or $\dim S = 2$.*

PROOF. By [**122**, Theorem 29.4(3)], A is a finite integral extension of $R = k[[X]]$, where X is as in Setting 29.7. We have $R \hookrightarrow A \hookrightarrow S$. By Proposition 28.8(1), $R \hookrightarrow S$ is TGF. By Theorem 29.8, either $\dim S = n$ and S is a finite integral extension of A or $\dim S = 2$. $\square$

For example, if $R = k[[x_1, \dots, x_4]]$ and $S = k[[y_1, y_2, y_3]]$, then every k-algebra embedding $R \hookrightarrow S$ fails to be TGF.

COROLLARY 29.10. *Let $R = k[[X]]$ be as in Setting 29.7. Let $Y = \{y_1, \dots, y_m\}$ be a set of m independent variables over R and let $S = R[[Y]]$. If $P \in \operatorname{Spec} R$ is such that* $\dim(R/P) \geq 2$ *and $Q \in \operatorname{Spec} S$ is maximal with respect to $Q \cap R = P$, then either*

(i) $\dim(S/Q) = 2$, *or*

(ii) *$R/P \hookrightarrow S/Q$ is a finite integral extension and* $\dim(R/P) = \dim(S/Q)$.

PROOF. Let $A := R/P \hookrightarrow S/Q =: B$, and apply Corollary 29.9. $\square$

GENERAL EXAMPLE 29.11. It is known that, for each positive integer n, the power series ring $R = k[[x_1, \dots, x_n]]$ in n variables over a field k can be embedded into a power series ring in two variables over k. The construction is based on the fact that the power series ring $k[[z]]$ in the single variable z contains an infinite set of algebraically independent elements over k. Let $\{f_i\}_{i=1}^{\infty} \subset k[[z]]$ with $f_1 \neq 0$ and $\{f_i\}_{i=2}^{\infty}$ algebraically independent over $k(f_1)$. Let $(S := k[[z, w]], \mathfrak{n} := (z, w))$ be the formal power series ring in the two variables z, w. Fix a positive integer n and consider the subring $R_n := k[[f_1 w, \dots, f_n w]]$ of S with maximal ideal $\mathfrak{m}_n = (f_1 w, \dots, f_n w)$. Let $x_1, \dots, x_n$ be new indeterminates over k and define a k-algebra homomorphism $\varphi : k[[x_1, \dots, x_n]] \to R_n$ by setting $\varphi(x_i) = f_i w$ for $i = 1, \dots, n$.

CLAIM 29.12. (See [**193**, pp. 219-220]). φ is an isomorphism.

PROOF. For $g \in k[[x_1, \dots, x_n]]$, write $g = \sum_{m=0}^{\infty} g_m$, where g_m is a form of degree m in $k[x_1, \dots, x_n]$. Then

$$\varphi(g) = \sum_{m=0}^{\infty} \varphi(g_m) \quad \text{and} \quad \varphi(g_m) = g_m(f_1 w, \dots, f_n w) = w^m g_m(f_1, \dots, f_n),$$

where $g_m(f_1, \dots, f_n) \in k[[z]]$. Iff $\varphi(g) = 0$, then $g_m(f_1, \dots, f_n) = 0$ for each m. Thus

$$0 = g_m(f_1, \dots, f_n) = \sum_{i_1 + \cdots + i_n = m} a_{i_1, \dots, i_n} f_1^{i_1} \cdots f_n^{i_n},$$

where the $a_{i_1, \dots, i_n} \in k$ and the i_j are nonnegative integers. Our hypothesis on the f_j implies that each of the $a_{i_1, \dots, i_n} = 0$, and so $g_m = 0$ for each m. $\square$

PROPOSITION 29.13. *With notation as in Example 29.11, for each $n \geq 2$, the extension $(R_n, \mathfrak{m}_n) \hookrightarrow (S, \mathfrak{n})$ is a non-finite TGF extension with trivial residue field extension. Moreover* $\operatorname{ht}(P \cap R_n) \geq n - 1$, *for each nonzero prime $P \in \operatorname{Spec} S$.*

PROOF. We have $k = R_n/\mathfrak{m}_n = S/\mathfrak{n}$, so the residue field of S is a trivial extension of that of R_n. Since $\mathfrak{m}_n S$ is not $\mathfrak{n}$-primary, S is not finite over R_n. If $P \cap R_n = \mathfrak{m}_n$, then $\operatorname{ht}(P \cap R_n) = n \geq n-1$. Since $\dim S = 2$, if $\mathfrak{m}_n$ is not contained in P, then $\operatorname{ht} P = 1$, S/P is a one-dimensional local domain, and $\mathfrak{m}_n(S/P)$ is primary for the maximal ideal $\mathfrak{n}/P$ of S/P. It follows that $R_n/(P \cap R_n) \hookrightarrow S/P$ is a finite integral extension by Theorem 3.16. Therefore $\dim(R_n/(P \cap R_n)) = 1$. Since R_n is catenary and $\dim R_n = n$, $\operatorname{ht}(P \cap R_n) = n - 1$. $\square$

COROLLARY 29.14. *Let X and $R = k[[X]]$ be as in Setting 29.7. Then there exists an infinite properly ascending chain of two-dimensional TGF local extensions $R =: S_0 \hookrightarrow S_1 \hookrightarrow S_2 \hookrightarrow \cdots$ such that each S_i is a complete Noetherian local domain that has the same residue field as R, and S_{i+1} is a non-finite TGF local extension of S_i for each i.*

PROOF. Example 29.11 and Proposition 29.13 imply that R can be identified with a proper subring of the power series ring in two variables so that $k[[y_1, y_2]]$ is a TGF local extension of R and the extension is not finite. Now Example 29.11 and Proposition 29.13 can be applied again, to $k[[y_1, y_2]]$, and so on. □

EXAMPLE 29.15. A particular case of Example 29.11.

For $R := k[[x, y]]$, the extension ring $S := k[[x, y/x]]$ has infinite transcendence over R by Sheldon's work; see [**173**]. The method used in [**173**] to prove that S has infinite transcendence degree over R is by constructing power series in y/x with 'special large gaps'. Since $k[[x]]$ is contained in R, it follows that S is a TGF extension of R. To show this, it suffices to show $P \cap R \neq (0)$ for each $P \in \operatorname{Spec} S$ with $\operatorname{ht} P = 1$. This is clear if $x \in P$, while if $x \notin P$, then $k[[x]] \cap P = (0)$, and so $k[[x]] \hookrightarrow R/(P \cap R) \hookrightarrow S/P$ and S/P is finite over $k[[x]]$. Therefore $\dim(R/(P \cap R)) = 1$, and so $P \cap R \neq (0)$.

Notice that the extension $k[[x, y]] \hookrightarrow k[[x, y/x]]$ is, up to isomorphism, the same as the extension $k[[x, xy]] \hookrightarrow k[[x, y]]$.

Example 29.16 shows that the situation of Remark 29.4.4 does occur.

EXAMPLE 29.16. Let k, $X = \{x_1, x_2\}$, $Y = \{y\}$, $Z = \{z\}$ and $R = k[[x_1, x_2]]$ be as in Remarks 29.4. Let $f_1, f_2 \in k[[z]]$ be algebraically independent over k. Let P denote the ideal of $k[[x_1, x_2, y, z]]$ generated by $(x_1 - f_1 y, x_2 - f_2 y)$. Then P is the kernel of the k-algebra homomorphism $\theta : k[[x_1, x_2, y, z]] \to k[[y, z]]$ obtained by defining $\theta(x_1) = f_1 y$, $\theta(x_2) = f_2 y$, $\theta(y) = y$ and $\theta(z) = z$. In the notation of Remark 29.4,

$$T = k[[x_1, x_2, y, z]]/P \cong k[[y, z]].$$

Let $\varphi := \theta|_R$ and $\tau := \theta|_{R[[y]]}$. The proof of Claim 29.12 shows that φ and τ are embeddings. Hence $P \cap R[[y]] = (0)$. Thus

$$R \overset{\sigma}{\hookrightarrow} S = \frac{R[[y]]}{P \cap R[[y]]} = R[[y]] \overset{\tau}{\hookrightarrow} T = \frac{R[[y, z]]}{P} \cong k[[y, z]],$$

where $\sigma : R \hookrightarrow S$ is the inclusion map. By Proposition 29.13, φ and τ are TGF. Since $yS \cap R = (0)$, the map $\sigma : R \hookrightarrow S$ is not TGF, as in Remark 29.4.4.

QUESTIONS AND REMARKS 29.17. Let R and S be complete Noetherian local domains and let $\varphi : R \hookrightarrow S$ be a TGF local non-finite extension with finite residue field extension. Let y be an indeterminate over S.

(1) Is it always true that φ can be extended to a TGF local non-finite extension $R[[y]] \hookrightarrow S$?
(2) It is natural to ask: Does $R[[y]] \hookrightarrow S[[y]]$ have the TGF property? The answer is "No" in general. To see this, use Example 29.11, Claim 29.12, and Proposition 29.13 in the case $n = 3$ to define a TGF local non-finite extension using indeterminates x_1, x_2, x_3, z, w over a field k:

$$\varphi : R := k[[x_1, x_2, x_3]] \overset{\varphi}{\hookrightarrow} S := k[[z, w]].$$

Then $\dim(R[[y]]) = 4$ and $\dim(S[[y]]) = 3$. Theorem 29.8 implies that $R[[y]] \hookrightarrow S[[y]]$ is not TGF.

(3) A related question is whether the given $R \overset{\varphi}{\hookrightarrow} S$ is *extendable* to an injective local homomorphism $\psi : R[[y]] \hookrightarrow S$. That is, φ is the composite map:

$$R \overset{\subseteq}{\hookrightarrow} R[[y]] \overset{\psi}{\hookrightarrow} S.$$

(4) For a non-complete example, the extension

$$\theta : k[[x_1]][y]_{(x_1,y)} \hookrightarrow k[y][[x_1]]_{(x_1,y)}$$

is TGF, if k is a field. Can θ be extended to a local injective map

$$k[[x_1]][y][[x_2]]_{(x_1,x_2,y)} \hookrightarrow k[y][[x_1]]_{(x_1,y)}?$$

Proposition 29.18 shows that these questions are connected: the answer to (29.17.2) is 'no' if the answer to (29.17.3) is "yes", that is, If the given $R \hookrightarrow S$ is extendable to an injective local homomorphism $R[[y]] \hookrightarrow S$, then $R[[y]] \hookrightarrow S[[y]]$ is not TGF. In Example 29.19, we present an example where this occurs.

PROPOSITION 29.18. *Let $\varphi : R \hookrightarrow S$ be a TGF local extension of complete Noetherian local domains and let y be an indeterminate over S. If φ is extendable to an injective local homomorphism $\psi : R[[y]] \hookrightarrow S$, then $R[[y]] \hookrightarrow S[[y]]$ is not TGF.*

PROOF. Let $a := \psi(y)$ and consider the ideal $Q = (y - a)S[[y]]$. Therefore $Q \cap R[[y]] = (0)$ and $R[[y]] \hookrightarrow S[[y]]$ is not TGF. □

EXAMPLE 29.19. Let $R := R_n = k[[f_1 w, \dots, f_n w]] \hookrightarrow S := k[[z, w]]$ be as in Example 29.11 with $n \geq 2$. Define the extension $\varphi : R[[y]] \hookrightarrow S$ by setting $\varphi(y) = f_{n+1} w \in S$. By Proposition 29.13, $\varphi : R[[y]] \hookrightarrow S$ is a TGF local extension . Thus by Proposition 29.18, $R[[y]] \hookrightarrow S[[y]]$ is not TGF.

REMARK AND QUESTIONS 29.20. Let $(R, \mathfrak{m}) \hookrightarrow (S, \mathfrak{n})$ be a TGF local extension of complete Noetherian local domains. Assume that $[S/\mathfrak{n} : R/\mathfrak{m}] < \infty$ and that S is not finite over R. By Theorem 3.16, $\mathfrak{m}S$ is not $\mathfrak{n}$-primary. Thus $\dim S > \operatorname{ht}(\mathfrak{m}S)$. Therefore $\dim S > 1$, and so by Proposition 29.2, $\dim R > 1$.

(1) If $(R, \mathfrak{m})$ is equicharacteristic, then by Corollary 29.9, $\dim S = 2$. Is it true in generals that $\dim S = 2$?
(2) Is it possible to have $\dim S - \operatorname{ht}(\mathfrak{m}S) > 1$?

EXAMPLES 29.21. (1) Let $R := k[[x, xy, z]] \hookrightarrow S := k[[x, y, z]]$. We show this is not a TGF extension. By Example 29.15, $\varphi : k[[x, xy]] \hookrightarrow k[[x, y]]$ is TGF. By Proposition 29.18, it suffices to extend φ to an injective local homomorphism of $k[[x, xy, z]]$ to $k[[x, y]]$. Let $f \in k[[x]]$ be such that x and f are algebraically independent over k, and so $(1, x, f)$ is not a solution to any nonzero homogeneous form over k. As in Proposition 29.8 and Example 29.11, the extension of φ obtained by mapping $z \to fy$ is an injective local homomorphism.

(2) The extension $R = k[[x, xy, xz]] \hookrightarrow S = k[[x, y, z]]$ is not a TGF extension, since $R = k[[x, xy, xz]] \hookrightarrow k[[x, xy, z]] \hookrightarrow S = k[[x, y, z]]$ is a composition of two extensions that are not TGF by part (1). Now apply Proposition 28.8.

29.3. The case of transcendental residue extension

In this section we address but do not fully resolve the following question.

QUESTION 29.22. Let $(R, \mathfrak{m}) \hookrightarrow (S, \mathfrak{n})$ be a TGF local extension of complete Noetherian local domains. If $S/\mathfrak{n}$ is transcendental over $R/\mathfrak{m}$ does it follow that $\dim S \le 1$?

Proposition 29.23 shows that every Noetherian local domain of positive dimension is dominated by a one-dimensional complete Noetherian local domain that is a TGF local extension.

PROPOSITION 29.23. *Let $(R, \mathfrak{m})$ be a Noetherian local domain of positive dimension. Then there exists a one-dimensional complete Noetherian local domain $(S, \mathfrak{n})$ that is a TGF local extension of R.*

PROOF. By Chevalley's Theorem 2.35, there exists a discrete rank-one valuation domain $(S, \mathfrak{n})$ that dominates R. The $\mathfrak{n}$-adic completion $\widehat{S}$ of S is a discrete rank-one valuation domain that dominates R. Since $(S, \mathfrak{n})$ is a one-dimensional Noetherian local domain and dominates the Noetherian local domain $(R, \mathfrak{m})$ of positive dimension, it follows that $R \hookrightarrow S$ is TGF local. Also $S \hookrightarrow \widehat{S}$ is TGF local. By Proposition 28.8, $R \hookrightarrow \widehat{S}$ is a TGF local extension, as desired. □

SETTING 29.24. Let $n \ge 2$ be an integer, let $X = \{x_1, \ldots, x_n\}$ be a set of independent variables over the field k and let $R = k[[X]]$ be the formal power series ring in n variables over the field k. Let z, w, t, v be independent variables over R.

PROPOSITION 29.25. *Let notation be as in Setting* 29.24.

1. *There exists a TGF embedding $\theta : k[[z, w]] \to k(t)[[v]]$ defined by $\theta(z) = tv$ and $\theta(w) = v$.*
2. *Moreover, the composition $\psi = \theta \circ \varphi$ of θ with $\varphi : R \to k[[z, w]]$ given in General Example* 29.11 *is also TGF.*

PROOF. Suppose $f \in \ker \theta$. Write $f = \sum_{n=0}^{\infty} f_n(z, w)$, where f_n is a homogeneous form of degree n with coefficients in k. We have

$$0 = \theta(f) = \sum_{n=0}^{\infty} f_n(tv, v) = \sum_{n=0}^{\infty} v^n f_n(t, 1)\,.$$

This implies $f_n(t, 1) = 0$ for each n. Since t is algebraically independent over k, we have $f_n(z, w) = 0$ for each n. Thus $f = 0$ and θ is an embedding. Since θ is a local homomorphism and $\dim(k(t)[[v]]) = 1$, it is clear that θ is TGF. The second statement is clear since a local embedding or a local domain into a one-dimensional local domain is TGF. □

As a consequence of Proposition 29.25, we prove:

COROLLARY 29.26. *Let $R = k[[X]]$ be as above and let $A = k(t)[[X]]$. There exists a prime ideal $P \in \operatorname{Spec} A$ in the generic fiber over R with $\operatorname{ht} P = n - 1$. In particular, the inclusion map $R = k[[X]] \hookrightarrow A = k(t)[[X]]$ is not TGF.*

PROOF. Define $\varphi : R \to k[[z, w]] := S$, by

$$\varphi(x_1) = z, \quad \varphi(x_2) = h_2(w)z, \quad \ldots, \quad \varphi(x_n) = h_n(w)z,$$

where $h_2(w), \dots, h_n(w) \in k[[w]]$ are algebraically independent over k. Also define $\theta : S \to k(t)[[v]] := B$ by $\theta(z) = tv$ and $\theta(w) = v$. Consider the following diagram

$$\begin{array}{ccc} R = k[[X]] & \xrightarrow{\subset} & A = k(t)[[X]] \\ \varphi \downarrow & & \Psi \downarrow \\ S = k[[z,w]] & \xrightarrow{\theta} & B = k(t)[[v]], \end{array}$$

where $\Psi : A \to B$ is the identity map on $k(t)$ and is defined by

$$\Psi(x_1) = tv, \quad \Psi(x_2) = h_2(v)tv, \quad \dots, \quad \Psi(x_n) = h_n(v)tv.$$

Notice that $\Psi|_R = \psi = \theta \circ \varphi$. Therefore the diagram is commutative. Let $P = \ker \Psi$. Since Ψ is surjective, $\operatorname{ht} P = n - 1$. Commutativity of the diagram implies that $P \cap R = (0)$. □

DISCUSSION 29.27. Let us describe generators for the prime ideal $P = \ker \Psi$ given in Corollary 29.26. Under the map Ψ, $x_1 \mapsto tv$, and so $\frac{x_1}{t} \mapsto v$. Since also $x_2 \mapsto h_2(v)tv, \dots, x_n \mapsto h_n(v)tv$, we see that

$$(x_2 - h_2(\frac{x_1}{t})x_1,\ x_3 - h_3(\frac{x_1}{t})x_1, \dots, x_n - h_n(\frac{x_1}{t})x_1)A \subseteq P$$

(that is, $\Psi(x_2 - h_2(x_1/t)x_1) = h_2(v)tv - h_2(v)tv = 0$ etc.) Since the ideal on the left-hand-side is a prime ideal of height $n - 1$, the inclusion is an equality. Thus we have generators for the prime ideal $P = \ker \Psi$ resulting from the definitions of φ and θ given in the corollary.

On the other hand, in Corollary 29.26 if we change the definition of θ and we define $\theta' : k[[z,w]] \to k(t)[[v]]$ by $\theta'(z) = v$ and $\theta'(w) = tv$ (but we keep φ as above), then Ψ' defined by $\Psi'|_R = \theta' \cdot \varphi$ maps $x_1 \to v$, $x_2 \to h_2(tv)v$, $\dots$, $x_n \to h_n(tv)v$. In this case,

$$(x_2 - h_2(tx_1)x_1,\ x_3 - h_3(tx_1)x_1, \dots, x_n - h_n(tx_1)x_1)A \subseteq \ker \Psi' = P'.$$

Again the ideal on the left-hand-side is a prime ideal of height $n - 1$, and so we have equality. This yields a different prime ideal P'.

In this case one can also see directly for

$$P' = (x_2 - h_2(tx_1)x_1,\ x_3 - h_3(tx_1)x_1, \dots, x_n - h_n(tx_1)x_1)A$$

that $P' \cap R = (0)$. The map $\Psi' : A \to A/P' = k(t)[[v]]$. Suppose $f \in R \cap P'$. Write $f = \sum_{\ell=0}^{\infty} f_\ell(x_1, \dots, x_n)$, where $f_\ell \in k[x_1, \dots, x_n]$ is a homogeneous form of degree ℓ. Then

$$0 = \Psi'(f) = \sum_{\ell=0}^{\infty} f_\ell(v, h_2(tv)v, \dots, h_n(tv)v) = \sum_{\ell=0}^{\infty} v^\ell f_\ell(1, h_2(tv), \dots, h_n(tv)).$$

This implies $f_\ell(1, h_2(tv), \dots, h_n(tv)) = 0$, for each ℓ. Since $h_2, \dots, h_n$ are algebraically independent over k, each of the homogeneous forms $f_\ell(x_1, \dots, x_n) = 0$. Hence $f = 0$.

QUESTION 29.28. With notation as in Corollary 29.26, does every prime ideal of the ring A maximal in the generic fiber over R have height $n - 1$?

THEOREM 29.29. *Let $(A, \mathfrak{m}) \hookrightarrow (B, \mathfrak{n})$ be an extension of two-dimensional regular local domains. Assume that B dominates A and that $B/\mathfrak{n}$ as a field extension of $A/\mathfrak{m}$ is not algebraic. Then $A \hookrightarrow B$ is not TGF.*

PROOF. Since $\dim A = \dim B$, the assumption that $B/\mathfrak{n}$ is transcendental over $A/\mathfrak{m}$ implies that B is not algebraic over A by Cohen's Theorem 2.26 on extensions. If $\mathfrak{m}B$ is $\mathfrak{n}$-primary, then B is faithfully flat over A [**122**, Theorem 23.1], and a result of Heinzer and Rotthaus, [**70**, Theorem 1.12], implies that $A \hookrightarrow B$ is not TGF in this case.

If $\mathfrak{m}B$ is principal, then $\mathfrak{m}B = xB$ for some $x \in \mathfrak{m}$ since B is local. It follows that $\mathfrak{m}/x \subset B$. Localizing $A[\mathfrak{m}/x]$ at the prime ideal $\mathfrak{n} \cap A[\mathfrak{m}/x]$ gives a local quadratic transform $(A_1, \mathfrak{m}_1)$ of A, see Definition 13.1. If $\dim A_1 = 1$, then $A_1 \hookrightarrow B$ is not TGF because only finitely many prime ideals of B can contract to the maximal ideal of A_1. Hence $A \hookrightarrow B$ is not TGF if $\dim A_1 = 1$. If $\dim A_1 = 2$, then $(A_1, \mathfrak{m}_1)$ is a 2-dimensional regular local domain dominated by $(B, \mathfrak{n})$ and the field $A_1/\mathfrak{m}_1$ is finite algebraic over $A/\mathfrak{m}$, and so $B/\mathfrak{n}$ is transcendental over $A_1/\mathfrak{m}_1$. Thus we can repeat the above analysis: If $\mathfrak{m}_1 B$ is $\mathfrak{n}$-primary, then as above $A \hookrightarrow B$ is not TGF. If $\mathfrak{m}_1 B$ is principal, we obtain a local quadratic transform $(A_2, \mathfrak{m}_2)$ of A_1. If this process does not end after finitely many steps, we have a union $V = \cup_{n=1}^{\infty} A_n$ of an infinite sequence A_n of quadratic transforms of a 2-dimensional regular local domains. By [**2**], the integral domain V is a valuation domain of rank at most 2 contained in B, and so at most finitely many of the height-one primes of B have a nonzero intersection with V. Therefore $V \hookrightarrow B$ is not TGF and hence also $A \hookrightarrow B$ is not TGF.

Thus by possibly replacing A by an iterated local quadratic transform A_n of A, we may assume that $\mathfrak{m}B$ is neither $\mathfrak{n}$-primary nor principal. Let $\mathfrak{m} = (x, y)A$. There exist $f, g, h \in B$ such that $x = gf, y = hf$ and g, h is a regular sequence in B. Hence $(g, h)B$ is $\mathfrak{n}$-primary. Let $f = f_1^{e_1} \cdots f_r^{e_r}$, where $f_1 B, \ldots f_r B$ are distinct height-one prime ideals and the e_i are positive integers. Then $f_1 B, \ldots, f_r B$ are precisely the height-one primes of B that contain $\mathfrak{m}$.

Let $t \in B$ be such that the image to t in $B/\mathfrak{n}$ is transcendental over $A/\mathfrak{m}$. Modifying t if necessary by an element of $\mathfrak{n}$ we may assume that t is transcendental over A. We have $\mathfrak{n} \cap A[t] = \mathfrak{m}[t]$. Let $A(t) = A[t]_{\mathfrak{m}[t]}$. Notice that $A(t)$ is a 2-dimensional regular local domain with maximal ideal $\mathfrak{m}A(t)$ that is dominated by $(B, \mathfrak{n})$. We have

$$A \hookrightarrow A[t] \hookrightarrow A(t) \hookrightarrow B.$$

For each positive integer i, let $P_i = (xt^i - y)A(t)$. Since t is transcendental over A, we have $P_i \cap A = (0)$ for each $i \in \mathbb{N}$. Notice that $P_i B = (gft^i - hf)B = f(gt^i - h)B$. If $i \neq j$, the element $t^i - t^j$ is a unit of B. Hence $(gt^i - h, gt^j - h)B = (g, h)B$ is $\mathfrak{n}$-primary if $i \neq j$. Therefore a height one prime Q of B contains $gt^i - h$ for at most one integer i. Hence there exists a positive integer n such that if Q is a minimal prime of $(gt^n - h)B$, then $Q \notin \{f_1 B, \ldots, f_r B\}$. It follows that $Q \cap A(t)$ has height one. Since $P_n \subseteq (gt^n - h)B \subseteq Q$, we have $Q \cap A(t) = P_n$. Thus $Q \cap A = (0)$. This completes the proof. □

Corollary 29.30 is an immediate consequence of Theorem 29.29.

COROLLARY 29.30. *Let x, y, z, w, t be indeterminates over the field k and let*

$$\varphi : R = k[[x, y]] \hookrightarrow S := k(t)[[z, w]]$$

be an injective local k-algebra homomorphism. Then $\varphi(R) \hookrightarrow S$ is not TGF.

In relation to Question 29.22, Example 29.31 is a TGF extension $A \hookrightarrow B$ that is not complete for which the residue field of B is transcendental over that of A and $\dim B = 2$.

EXAMPLE 29.31. Let $A = k[x, y, z, w]_{(x,y,z,w)}$, where k is a field and $xw = yz$. Thus A is a 3-dimensional normal Noetherian local domain with maximal ideal $\mathfrak{m} := (x, y, z, w)A$ and residue field $A/\mathfrak{m} \cong k$. Since $y/x = w/z$, we have

$$C := A[\frac{y}{x}] = k[\frac{y}{x},\ x,\ z]$$

is a polynomial ring in 3 variables over k. Thus $B := C_{(x,z)}$ is a 2-dimensional regular local domain with maximal ideal $\mathfrak{n} = (x, z)B$. Notice that $(B, \mathfrak{n})$ birationally dominates $(A, \mathfrak{m})$. Hence $(A, \mathfrak{m}) \hookrightarrow (B, \mathfrak{n})$ is a TGF extension. Also $B = k(y/x)[x, z]_{(x,z)}$, and so $k(y/x)$ is a coefficient field for B. The image t of y/x in $B/\mathfrak{n}$ is transcendental over k and $B/\mathfrak{n} = k(t)$. The completion of A is the normal local domain $\widehat{A} = k[[x, y, z, w]]$, where $xw = yz$. A form of Zariski's Subspace Theorem [**3**, (10.6)] implies $\widehat{A}$ is dominated by $\widehat{B}$. Also $\widehat{B}$ is isomorphic to $k(t)[[x, z]]$, where t is transcendental over k, and $\varphi : \widehat{A} \hookrightarrow \widehat{B}$, where $\varphi(x) = x, \varphi(z) = z, \varphi(y) = tx, \varphi(w) = tz$, and $\varphi(xw) = xtz = \varphi(yz)$.

Exercise

(1) With notation as in Example 29.31, prove that $\widehat{A} \hookrightarrow \widehat{B}$ is not a TGF extension. Equivalently, prove that the inclusion map

$$R := k[[x, z, tx, tz]] \hookrightarrow k(t)[[x, z]] := S$$

is not a TGF extension.

CHAPTER 30

Constructions and examples discussed in this book

In Section 30.1, we list the construction techniques used in the book . We give a partial list of examples with a brief description of each in Section 30.2.

30.1. Construction techniques

• **Intersection Construction 1.3** is the most general form of the constructions given in the book. In every construction the constructed ring is an intersection of a field with a homomorphic image of a completion. All the constructions use this basic format.

INTERSECTION CONSTRUCTION 30.1 (See (1.3)). Construct the *Intersection Domain*

$$A \;:=\; L \cap (R^*/I),$$

where R is an integral domain, R^* is a completion of R, such as an ideal-adic or multi-adic completion, I is an ideal of R^* and L is a subfield of the total ring of fractions of R^*/I such that $R \subseteq L$. [1] If R^* is a principal ideal-adic completion of R with certain additional hypotheses, then Construction 1.5 expands Construction 1.3 to include an approximation domain.

• **Inclusion Construction 5.3** is introduced in Chapter 4. The formal definition is in Chapter 5. For this construction R is an integral domain with field of fractions K, $x \in R$ is a nonzero nonunit, and

- R is separated in the x-adic topology, that is, $\bigcap_{n\in\mathbb{N}} x^n R = (0)$,
- the x-adic completion R^* of R is a Noetherian ring,
- $\tau_1, \ldots, \tau_s \in xR^*$ are algebraically independent elements over R, and
- $K(\tau_1, \ldots, \tau_s) \subseteq \mathcal{Q}(R^*)$, the total ring of fractions of R^*.

INCLUSION CONSTRUCTION 30.2. (See (5.3).) Define A to be the *Intersection Domain*

$$A := K(\tau_1, \ldots, \tau_s) \cap R^*.$$

An Approximation Domain for Inclusion Construction 5.3 is defined in Section 5.2; see Definition 5.7. Noetherian Flatness Theorem 6.3 gives conditions that imply A is Noetherian.

• **Insider Construction 10.7** is introduced in Chapter 6 and presented in general in Chapter 10. It involves an iteration of Inclusion Construction 5.3.

[1] This requires restrictions on I.

First apply Inclusion Construction 5.3 to the following situation:

- R is a Noetherian integral domain with field of fractions K,
- x is a nonzero nonunit of R and R^* is the x-adic completion of R,
- $\tau_1, \dots, \tau_s \in xR^*$ are algebraically independent over K, and $\underline{\tau}$ abbreviates the list $\tau_1, \dots, \tau_s$,
- $\mathcal{Q}(R)(\underline{\tau}) \subseteq \mathcal{Q}(R^*)$,
- $T := R[\underline{\tau}] = R[\tau_1, \dots, \tau_s] \overset{\psi}{\hookrightarrow} R^*[1/x]$ is flat, and
- $D := K(\underline{\tau}) \cap R^*$, as in Inclusion Construction 5.3.

Since the map ψ is flat, Noetherian Flatness Theorem 6.3 implies the integral domain D is a Prototype, as in Definition 10.3.

INSIDER CONSTRUCTION 30.3. (See (10.7).) To construct "insider" examples inside D:

- Choose polynomials $f_1, \dots, f_m$ in $T = R[\underline{\tau}]$ that are algebraically independent over K, so $m \leq n$.
- As in Inclusion Construction 5.3, there is no loss of generality in assuming that each $f_i \in (\underline{\tau})T \subseteq xR^*$, and abbreviate $f_1, \dots, f_m$ by $\underline{f}$.
- Define the Insider Intersection Domain

$$A := K(\underline{f}) \cap R^*.$$

There is an Approximation Domain B corresponding to the domain A using $\underline{f}$ instead of $\underline{\tau}$, as in Section 5.2. Then $B \subseteq A \subseteq D$. The ring B is Noetherian if the embedding $\varphi : S := R[\underline{f}] \overset{\varphi}{\hookrightarrow} T := R[\underline{\tau}]$ is flat. See Corollary 10.11.

• **Homomorphic Image Construction 17.2** is defined in Chapter 17. For this construction:

- R is an integral domain with field of fractions $K := \mathcal{Q}(R)$.
- $x \in R$ is a nonzero nonunit such that $\bigcap_{n \geq 1} x^n R = (0)$.
- the x-adic completion R^* is Noetherian, and x is a regular element of R^*.
- I is an ideal of R^* such that $P \cap R = (0)$ for each $P \in \operatorname{Spec} R^*$ that is associated to I.

HOMOMORPHIC IMAGE CONSTRUCTION 30.4. (See (17.2).) Define the *Intersection Domain*

$$A = A_{\text{hom}} := K \cap (R^*/I).$$

There is an Approximation Domain B corresponding to A; see Section 17.2. By Noetherian Flatness Theorem 17.13, A is Noetherian and $A = B$ if and only if the inclusion map $R \hookrightarrow (R^*/I)[1//x]$ is flat.

• **Multi-adic Inclusion Construction 21.5** is an adaptation of Inclusion Construction 5.3 where a multi-adic completion is used in place of an ideal-adic completion; see Chapter 21.

30.2. Examples considered in the book

(1) Let $(A, \mathfrak{n})$ be a Noetherian local domain having a coefficient field k, and having the property that the field of fractions L of A is finitely generated over k. There exists a Noetherian local subring R of A such that
  (a) A birationally dominates R.

(b) R is essentially finitely generated over k.

(c) There exists an ideal I in the completion $\widehat{R}$ of R such that L is a subring of the total quotient ring of $\widehat{R}/I$ and $A = L \cap (\widehat{R}/I)$.

Thus every Noetherian local domain A with these properties is realizable as an intersection, where R is a Noetherian local domain essentially finitely generated over a field and I is an ideal in $\widehat{R}$; see Corollary 4.3.

(2) The "simplest" example of a Noetherian local domain A on an algebraic function field L/k of two or more variables that is not essentially finitely generated over its ground field k, that is, A is not the localization of a finitely generated k-algebra; see Example 4.7.

(3) A two-dimensional regular local domain A that is a nested union of three-dimensional regular local domains that A birationally dominates; see Example 4.10.

(4) A two-dimensional regular local domain A that is a nested union of four-dimensional regular local domains that A birationally dominates; see Example 4.11.

(5) A one-dimensional Noetherian local domain A that is the local coordinate ring of a nodal plane curve singularity; see Example 4.13. The integral closure of A has precisely two maximal ideals and is a homomorphic image of a regular Noetherian domain of dimension two; see also Example 22.38.

(6) A two-dimensional normal Noetherian local domain D that is analytically reducible; see Example 4.15 and Remarks 4.16.1, [**137**, Example 7, pp. 209-211]. There exists a two-dimensional regular local domain S that birationally dominates D such that S is not essentially finitely generated over D [**66**, page 670], and the inclusion map $D \hookrightarrow S$ extends to a surjective map $\widehat{D} \to \widehat{S}$. This demonstrates the necessity of the analytically irreducible assumption in [**137**, Theorem 37.4].

(7) A two-dimensional regular local domain A that is not Nagata and thus not excellent. The ring A contains a prime element f that factors as a square in the completion $\widehat{A}$ of A, that is, $f = g^2$ for some element $g \in \widehat{A}$; see Example 4.15, Remarks 4.16.2, Proposition 6.19 and Remark 6.20, [**137**, Example 7, pp. 209-211].

(8) A three-dimensional regular local domain A that is Nagata but not excellent. The formal fibers of A are reduced but not regular; see Examples 4.17 and 6.23 and Remark 4.18, [**155**].

(9) An example of Inclusion Construction 5.3 where the approximation domain B is equal to the intersection domain A; see Remark 4.20, Local Prototype Example 4.26 and Example 12.21.

(10) A non-excellent DVR obtained by Prototype Theorem 10.2; see Proposition 10.4.

(11) A two-dimensional non-excellent regular local domain obtained by Prototype Theorem 10.2; see Remark 10.5.

(12) An example of a regular local domain A of dimension three having a prime ideal P with $\operatorname{ht} P = 2$ such that the extension of P to the completion of A is not integrally closed; see Theorem 11.11. It follows that A is not a Nagata ring and A is not excellent.

More generally, for each integer $n \geq 2$, and every integer r with $2 \leq r \leq n$, there exists a regular local domain A with $\dim A = n + 1$

having a prime ideal P with $\operatorname{ht} P = r$ such that the extension of P to the completion of A is not integrally closed; see Example 11.13.

(13) A non-Noetherian three-dimensional local Krull domain $(B, \mathfrak{n})$ such that $\mathfrak{n}$ is two-generated, the $\mathfrak{n}$-adic completion of B is a two-dimensional regular local domain, and B birationally dominates a four-dimensional regular local domain; see Theorem 12.3 and Example 12.7.

(14) For every integer $m > 0$, an example of Insider Construction 10.7 where $B \subsetneq A$ and B is a non-Noetherian 3-dimensional local UFD such that B is not catenary, B has precisely m prime ideals of height 2, $B_{\mathfrak{p}}$ is Noetherian for every non-maximal prime ideal $\mathfrak{p}$ of B, the prime ideals of B that are not finitely generated are precisely the prime ideals of height 2, the maximal ideal of B is 2-generated and the completion $\widehat{B}$ of B is a power series ring in two variables over a field; see Examples 14.1, Theorem 14.3 and Proposition 14.5.

An analysis of the spectrum of B and the map $\operatorname{Spec} \widehat{B} \to \operatorname{Spec} B$ for the cases $m = 1$ and $m = 2$ is given in Section 14.2.

(15) For every $m, n \in \mathbb{N}$ with $n \geq 4$, an example of Insider Construction 10.7 where the approximation domain B is equal to the intersection domain A, $\dim B = n$, and B has exactly m prime ideals of height $n-1$. The domain B is a non-catenary non-Noetherian UFD, and every prime ideal of B of height $n-1$ is not finitely generated; see Theorem 16.2 and Example 10.15.

(16) An example of Insider Construction 10.7 where the approximation domain B is equal to the intersection domain A. The domain B is a non-catenary non-Noetherian four-dimensional local UFD that is very close to being Noetherian. The ring B has exactly one prime ideal Q of height 3; the ideal Q is not finitely generated; see Examples 16.4, 10.15 and 6.24.

Another example of a non-catenary non-Noetherian four dimensional local UFD using Insider Construction 10.7 is given in Theorem 15.11. The intersection domain A in this example is a Local Prototype as in Definition 4.28 and is a 3-dimensional RLR.

(17) A strictly descending chain of one-dimensional analytically ramified Noetherian local domains that birationally dominate a polynomial ring in two variables over a field; see Example 17.18.

(18) For each pair of positive integers r, n, a Noetherian local domain A with $\dim A = r$ and a principal ideal-adic completion A^* of A such that A^* has nilradical with nilpotency index n; see Example 17.27.

(19) A two-dimensional Noetherian local domain B with geometrically regular formal fibers such that B is not universally catenary, and B birationally dominates a three-dimensional regular local domain. The completion of B has two minimal primes, one of dimension one and one of dimension two. The ring B is not a homomorphic image of a regular local ring; see Example 18.15.

More generally Example 18.20 shows: given a positive integer t and nonnegative integers n_r, for each r with $1 \leq r \leq t$ and $n_1 \geq 1$, there exists a t-dimensional Noetherian local domain A with geometrically regular formal fibers such that A birationally dominates a $t+1$-dimensional RLR and the completion $\widehat{A}$ of A has exactly n_r minimal primes of dimension $t+1-r$, for each r.

(20) An Ogoma-like example having the properties of Ogoma's famous example, that is, a normal Nagata local domain C such that the completion $\widehat{C}$ of C has a minimal prime ideal of dimension 3 and a minimal prime ideal of dimension 2. Hence C is not formally equidimensional, and so C is not universally catenary by Ratliff's Equidimension Theorem 3.26. Thus C is a counterexample to Chain Conjecture 18.2; see Examples 19.4 and 19.13 and Theorem 19.15.

(21) An example of Insider Construction 10.7, where the approximation domain B is properly contained in the intersection domain A, and neither A nor B is Noetherian. The local domain B is a UFD that fails to have Cohen-Macaulay formal fibers; see Examples 19.6 and 6.26 and Theorems 19.8 and 19.9.

(22) Let $(R, \mathfrak{m})$ be a countable excellent normal local domain with $\dim R \geq 2$, and let $\widehat{R}$ denote the completion of R There exists a subfield L of the field of fractions of $\widehat{R}$ such that the intersection domain $S = \widehat{R} \cap L$ is an infinite-dimensional non-Noetherian local Krull domain.

In particular, if k is a countable field and $R = k[x, y]_{(x,y)}$ is a localized polynomial ring, then there exists a subfield L of the field of fractions of the power series ring $k[[x, y]]$ such that $S = k[[x, y]] \cap L$ is an infinite-dimensional non-Noetherian local Krull domain; see Corollary 22.23.

(23) Let k be the algebraic closure of the field $\mathbb{Q}$ and let $R = k[x, y, z]_{(x,y,z)}$, where $z^2 = x^3 + y^z$. Then R is a normal Noetherian local UFD with $\dim R = 2$, and $\mathfrak{p} = (z - xy)R$ is a principal prime ideal of R. In the completion $\widehat{R}$ of R there exist two height one primes $\widehat{p}$ and $\widehat{\mathfrak{q}}$ that lie over $\mathfrak{p}$ in R, and $\widehat{\mathfrak{p}}$ is not the radical of a principal ideal. Therefore the divisor class group of the 2-dimensional normal Noetherian local domain $\widehat{R}$ is not a torsion group; see Example 22.36.

(24) Let $\mathbb{Q}[[x, y, z]]$ denote the formal power series ring in the variables x, y, z. There is an excellent regular local domain $(R, \mathfrak{m})$ with $\mathfrak{m} = (x, y, z)R$ and $\widehat{R} = \mathbb{Q}[[x, y, z]]$ for which the following hold: with R^* the y-adic completion of R, there exists an element $\tau \in yR^*$ that is algebraically independent over R and
   (a) τ limit-intersecting and residually limit-intersecting in y over R.
   (b) τ is not primarily limit-intersecting in y over R;

In this example, $B = A$ and B is non-Noetherian; see Theorem 24.16.

(25) An example of Basic Construction 1.3 where the result of the construction is the base ring. The notation is from Example 24.25 and Theorem 12.18. Let $S = \mathbb{Q}[x, y, z]_{(x,y,z)}$ and $P = (z - \tau)\widehat{S}$, a prime ideal of $\widehat{S}$. Then $\mathcal{Q}(S) \cap (\widehat{S}/P) = S$, and P is in the generic formal fiber of S, but P is not a maximal element of the generic formal fiber of S.

(26) For integer $n \geq 3$, an example of a non-finite TGF local embedding of a power series ring in n variables over a field k into a power series ring in two variables over k; see Example 29.11 and Section 29.2. A particular case is given in Example 29.15.

(27) An example where R and T are power series rings in 2 variables and S is a power series ring in 3 variables all over the same field field k, and

$\sigma : R \hookrightarrow S$ is an inclusion map, $\tau : S \hookrightarrow T$ is a TGF-embedding, and $\tau \cdot \sigma = \varphi : R \hookrightarrow T$ is TGF, but $\sigma : R \hookrightarrow S$ is not TGF; see Examples 29.16.

(28) An example where $(A, \mathfrak{m})$ is a 3-dimensional normal Noetherian local domain and $(B, \mathfrak{n})$ is a 2-dimensional regular local domain that dominates A, the inclusion map $(A, \mathfrak{m}) \hookrightarrow (B, \mathfrak{n})$ is a TGF extension and the canonical map on completions $\widehat{A} \to \widehat{B}$ is injective, but $\widehat{A} \hookrightarrow \widehat{B}$ is not TGF; see Example 29.31 and Exercise 1 of Chapter 29. In this example the residue field of B is transcendental over that of A.

Bibliography

[1] Shreeram Abhyankar, *Two notes on formal power series*, Proc. Amer. Math. Soc. **7** (1956), 903–905, DOI 10.2307/2033558. MR80647

[2] Shreeram Abhyankar, *On the valuations centered in a local domain*, Amer. J. Math. **78** (1956), 321–348, DOI 10.2307/2372519. MR82477

[3] Shreeram Shankar Abhyankar, *Resolution of singularities of embedded algebraic surfaces*, Pure and Applied Mathematics, Vol. 24, Academic Press, New York-London, 1966. MR0217069

[4] S. S. Abhyankar, *Lectures on algebra. Vol. I*, World Scientific Publishing Co. Pte. Ltd., Hackensack, NJ, 2006, DOI 10.1142/9789812773449. MR2253320

[5] Shreeram S. Abhyankar, *Inversion and invariance of characteristic terms: Part I*, The legacy of Alladi Ramakrishnan in the mathematical sciences, Springer, New York, 2010, pp. 93–168, DOI 10.1007/978-1-4419-6263-8_8. MR2744259

[6] Shreeram S. Abhyankar, William Heinzer, and Paul Eakin, *On the uniqueness of the coefficient ring in a polynomial ring*, J. Algebra **23** (1972), 310–342, DOI 10.1016/0021-8693(72)90134-2. MR306173

[7] Shreeram S. Abhyankar and William J. Heinzer, *Ramification in infinite integral extensions*, J. Algebra **170** (1994), no. 3, 861–879, DOI 10.1006/jabr.1994.1367. MR1305267

[8] Shreeram S. Abhyankar, William J. Heinzer, and Avinash Sathaye, *Translates of polynomials*, A tribute to C. S. Seshadri (Chennai, 2002), Trends Math., Birkhäuser, Basel, 2003, pp. 51–124. MR2017580

[9] Shreeram S. Abhyankar, William Heinzer, and Sylvia Wiegand, *On the compositum of two power series rings*, Proc. Amer. Math. Soc. **112** (1991), no. 3, 629–636, DOI 10.2307/2048682. MR1057740

[10] Shreeram S. Abhyankar and Ben Kravitz, *Two counterexamples in normalization*, Proc. Amer. Math. Soc. **135** (2007), no. 11, 3521–3523, DOI 10.1090/S0002-9939-07-09010-7. MR2336566

[11] S. S. Abhyankar and T. T. Moh, *On analytic independence*, Trans. Amer. Math. Soc. **219** (1976), 77–87, DOI 10.2307/1997583. MR414546

[12] S. S. Abhyankar and M. van der Put, *Homomorphisms of analytic local rings*, J. Reine Angew. Math. **242** (1970), 26–60. MR260729

[13] Y. Akizuki, *Einige Bemerkungen über primäre Integritatsbereiche mit Teilerkettensatz*, Proc. Phys-Math. Soc. Japan 17 (1935), 327-336.

[14] Leovigildo Alonso Tarrío, Ana Jeremías López, and Joseph Lipman, *Correction to the paper: "Duality and flat base change on formal schemes" [in Studies in duality on Noetherian formal schemes and non-Noetherian ordinary schemes, 3–90, Amer. Math. Soc., Providence, RI, 1999; MR1716706 (2000h:14017)]*, Proc. Amer. Math. Soc. **131** (2003), no. 2, 351–357, DOI 10.1090/S0002-9939-02-06558-9. MR1933323

[15] M. F. Atiyah and I. G. Macdonald, *Introduction to commutative algebra*, Addison-Wesley Publishing Co., Reading, Mass.-London-Don Mills, Ont., 1969. MR0242802

[16] D. D. Anderson, J. Matijevic, and W. Nichols, *The Krull intersection theorem. II*, Pacific J. Math. **66** (1976), no. 1, 15–22. MR435062

[17] Maurice Auslander and D. A. Buchsbaum, *Unique factorization in regular local rings*, Proc. Nat. Acad. Sci. U.S.A. **45** (1959), 733–734, DOI 10.1073/pnas.45.5.733. MR103906

[18] James Ax, *On Schanuel's conjectures*, Ann. of Math. (2) **93** (1971), 252–268, DOI 10.2307/1970774. MR277482

[19] Gorô Azumaya, *On maximally central algebras*, Nagoya Math. J. **2** (1951), 119–150. MR40287

[20] David J. Benson, Srikanth B. Iyengar, and Henning Krause, *Colocalizing subcategories and cosupport*, J. Reine Angew. Math. **673** (2012), 161–207, DOI 10.1515/crelle.2011.180. MR2999131

[21] R. Berger, R. Kiehl, E. Kunz, and Hans-Joachim Nastold, *Differentialrechnung in der analytischen Geometrie* (German), Lecture Notes in Mathematics, No. 38, Springer-Verlag, Berlin-New York, 1967. MR0224870

[22] N. Bourbaki, *Algèbre Commutative*, Hermann, Paris, 1965.

[23] Nicolas Bourbaki, *Commutative algebra. Chapters 1–7*, Elements of Mathematics (Berlin), Springer-Verlag, Berlin, 1989. Translated from the French; Reprint of the 1972 edition. MR979760

[24] Jim Brewer, *The ideal transform and overrings of an integral domain*, Math. Z. **107** (1968), 301–306, DOI 10.1007/BF01110018. MR236158

[25] J. W. Brewer and E. A. Rutter, *$D + M$ constructions with general overrings*, Michigan Math. J. **23** (1976), no. 1, 33–42. MR401744

[26] M. Brodmann and C. Rotthaus, *Local domains with bad sets of formal prime divisors*, J. Algebra **75** (1982), no. 2, 386–394, DOI 10.1016/0021-8693(82)90046-1. MR653898

[27] Markus Brodmann and Christel Rotthaus, *A peculiar unmixed domain*, Proc. Amer. Math. Soc. **87** (1983), no. 4, 596–600, DOI 10.2307/2043342. MR687624

[28] M. P. Brodmann and R. Y. Sharp, *Local cohomology: an algebraic introduction with geometric applications*, Cambridge Studies in Advanced Mathematics, vol. 60, Cambridge University Press, Cambridge, 1998, DOI 10.1017/CBO9780511629204. MR1613627

[29] Winfried Bruns and Jürgen Herzog, *Cohen-Macaulay rings*, Cambridge Studies in Advanced Mathematics, vol. 39, Cambridge University Press, Cambridge, 1993. MR1251956

[30] Lindsay Burch, *Codimension and analytic spread*, Proc. Cambridge Philos. Soc. **72** (1972), 369–373, DOI 10.1017/s0305004100047198. MR304377

[31] P. Charters and S. Loepp, *Semilocal generic formal fibers*, J. Algebra **278** (2004), no. 1, 370–382, DOI 10.1016/j.jalgebra.2004.01.011. MR2068083

[32] Stephen U. Chase, *Direct products of modules*, Trans. Amer. Math. Soc. **97** (1960), 457–473, DOI 10.2307/1993382. MR120260

[33] Claude Chevalley, *On the theory of local rings*, Ann. of Math. (2) **44** (1943), 690–708, DOI 10.2307/1969105. MR9603

[34] C. Chevalley, *La notion d'anneau de décomposition* (French), Nagoya Math. J. **7** (1954), 21–33. MR67866

[35] I. S. Cohen, *On the structure and ideal theory of complete local rings*, Trans. Amer. Math. Soc. **59** (1946), 54–106, DOI 10.2307/1990313. MR16094

[36] I. S. Cohen, *Commutative rings with restricted minimum condition*, Duke Math. J. **17** (1950), 27–42. MR33276

[37] I. S. Cohen, *Lengths of prime ideal chains*, Amer. J. Math. **76** (1954), 654–668, DOI 10.2307/2372708. MR62116

[38] Steven Dale Cutkosky, *Local factorization of birational maps*, Adv. Math. **132** (1997), no. 2, 167–315, DOI 10.1006/aima.1997.1675. MR1491444

[39] Steven Dale Cutkosky, *Local monomialization and factorization of morphisms* (English, with English and French summaries), Astérisque **260** (1999), vi+143. MR1734239

[40] John David, *A characteristic zero non-Noetherian factorial ring of dimension three*, Trans. Amer. Math. Soc. **180** (1973), 315–325, DOI 10.2307/1996667. MR325604

[41] Tiberiu Dumitrescu, *Generic fiber of power series ring extensions*, Comm. Algebra **37** (2009), no. 3, 1098–1103, DOI 10.1080/00927870802271755. MR2503196

[42] David S. Dummit and Richard M. Foote, *Abstract algebra*, 3rd ed., John Wiley & Sons, Inc., Hoboken, NJ, 2004. MR2286236

[43] Paul M. Eakin Jr. and William J. Heinzer, *Some open questions on minimal primes of a Krull domain*, Canadian J. Math. **20** (1968), 1261–1264, DOI 10.4153/CJM-1968-122-6. MR241398

[44] Paul Eakin and William Heinzer, *Non finiteness in finite dimensional Krull domains*, J. Algebra **14** (1970), 333–340, DOI 10.1016/0021-8693(70)90109-2. MR254023

[45] David Eisenbud, *Commutative algebra*, Graduate Texts in Mathematics, vol. 150, Springer-Verlag, New York, 1995. With a view toward algebraic geometry, DOI 10.1007/978-1-4612-5350-1. MR1322960

[46] Renée Elkik, *Solutions d'équations à coefficients dans un anneau hensélien* (French), Ann. Sci. École Norm. Sup. (4) **6** (1973), 553–603 (1974). MR345966

[47] E. Graham Evans Jr., *A generalization of Zariski's main theorem*, Proc. Amer. Math. Soc. **26** (1970), 45–48, DOI 10.2307/2036800. MR260716

[48] Robert M. Fossum, *The divisor class group of a Krull domain*, Springer-Verlag, New York-Heidelberg, 1973. Ergebnisse der Mathematik und ihrer Grenzgebiete, Band 74. MR0382254

[49] Daniel Ferrand and Michel Raynaud, *Fibres formelles d'un anneau local noethérien* (French), Ann. Sci. École Norm. Sup. (4) **3** (1970), 295–311. MR272779

[50] László Fuchs and Luigi Salce, *Modules over non-Noetherian domains*, Mathematical Surveys and Monographs, vol. 84, American Mathematical Society, Providence, RI, 2001, DOI 10.1090/surv/084. MR1794715

[51] Stefania Gabelli and Evan Houston, *Coherentlike conditions in pullbacks*, Michigan Math. J. **44** (1997), no. 1, 99–123, DOI 10.1307/mmj/1029005623. MR1439671

[52] Robert Gilmer, *Prüfer-like conditions on the set of overrings of an integral domain*, Conference on Commutative Algebra (Univ. Kansas, Lawrence, Kan., 1972), Springer, Berlin, 1973, pp. 90–102. Lecture Notes in Math., Vol. 311. MR0340245

[53] Robert Gilmer, *Multiplicative ideal theory*, Marcel Dekker, Inc., New York, 1972. Pure and Applied Mathematics, No. 12. MR0427289

[54] Robert Gilmer, *Multiplicative ideal theory*, Queen's Papers in Pure and Applied Mathematics, vol. 90, Queen's University, Kingston, ON, 1992. Corrected reprint of the 1972 edition. MR1204267

[55] Robert Gilmer and William Heinzer, *The Noetherian property for quotient rings of infinite polynomial rings*, Proc. Amer. Math. Soc. **76** (1979), no. 1, 1–7, DOI 10.2307/2042904. MR534377

[56] Robert Gilmer and William Heinzer, *Jónsson ω_0-generated algebraic field extensions*, Pacific J. Math. **128** (1987), no. 1, 81–116. MR883379

[57] Sarah Glaz, *Commutative coherent rings*, Lecture Notes in Mathematics, vol. 1371, Springer-Verlag, Berlin, 1989, DOI 10.1007/BFb0084570. MR999133

[58] Sarah Glaz, *Finite conductor rings*, Proc. Amer. Math. Soc. **129** (2001), no. 10, 2833–2843, DOI 10.1090/S0002-9939-00-05882-2. MR1840085

[59] Shiro Goto, *Integral closedness of complete-intersection ideals*, J. Algebra **108** (1987), no. 1, 151–160, DOI 10.1016/0021-8693(87)90128-1. MR887198

[60] A. Granja, *Valuations determined by quadratic transforms of a regular ring*, J. Algebra **280** (2004), no. 2, 699–718, DOI 10.1016/j.jalgebra.2004.07.010. MR2090059

[61] Silvio Greco, *A note on universally catenary rings*, Nagoya Math. J. **87** (1982), 95–100. MR676588

[62] A. Grothendieck, *Tores maximaux, groupe de Weyl, sous-groupes de Cartan, centre réductif des schémas en groupes lissés et affines* (French), Schémas en Groupes (Sém. Géométrie Algébrique, Inst. Hautes Études Sci., 1963/64), Inst. Hautes Études Sci., Paris, 1964, pp. 69. MR0232779

[63] A. Grothendieck and J. A. Dieudonné, *Eléments de géométrie algébrique. I* (French), Grundlehren der Mathematischen Wissenschaften [Fundamental Principles of Mathematical Sciences], vol. 166, Springer-Verlag, Berlin, 1971. MR3075000

[64] Robin Hartshorne, *Algebraic geometry*, Springer-Verlag, New York-Heidelberg, 1977. Graduate Texts in Mathematics, No. 52. MR0463157

[65] William Heinzer, *On Krull overrings of a Noetherian domain*, Proc. Amer. Math. Soc. **22** (1969), 217–222, DOI 10.2307/2036956. MR254022

[66] William Heinzer, Craig Huneke, and Judith D. Sally, *A criterion for spots*, J. Math. Kyoto Univ. **26** (1986), no. 4, 667–671, DOI 10.1215/kjm/1250520834. MR864469

[67] William Heinzer and David Lantz, *Exceptional prime divisors of two-dimensional local domains*, Commutative algebra (Berkeley, CA, 1987), Math. Sci. Res. Inst. Publ., vol. 15, Springer, New York, 1989, pp. 279–304, DOI 10.1007/978-1-4612-3660-3_14. MR1015523

[68] William Heinzer and Moshe Roitman, *Generalized local rings and finite generation of powers of ideals*, Non-Noetherian commutative ring theory, Math. Appl., vol. 520, Kluwer Acad. Publ., Dordrecht, 2000, pp. 287–312. MR1858167

[69] William Heinzer and Moshe Roitman, *The homogeneous spectrum of a graded commutative ring*, Proc. Amer. Math. Soc. **130** (2002), no. 6, 1573–1580, DOI 10.1090/S0002-9939-01-06231-1. MR1887039

[70] William Heinzer and Christel Rotthaus, *Formal fibers and complete homomorphic images*, Proc. Amer. Math. Soc. **120** (1994), no. 2, 359–369, DOI 10.2307/2159870. MR1189544

[71] William Heinzer, Christel Rotthaus, and Judith D. Sally, *Formal fibers and birational extensions*, Nagoya Math. J. **131** (1993), 1–38, DOI 10.1017/S0027763000004529. MR1238631

[72] William Heinzer, Christel Rotthaus, and Sylvia Wiegand, *Idealwise algebraic independence for elements of the completion of a local domain*, Illinois J. Math. **41** (1997), no. 2, 272–308. MR1441678

[73] William Heinzer, Christel Rotthaus, and Sylvia Wiegand, *Noetherian rings between a semilocal domain and its completion*, J. Algebra **198** (1997), no. 2, 627–655, DOI 10.1006/jabr.1997.7169. MR1489916

[74] William Heinzer, Christel Rotthaus, and Sylvia Wiegand, *Building Noetherian domains inside an ideal-adic completion*, Abelian groups, module theory, and topology (Padua, 1997), Lecture Notes in Pure and Appl. Math., vol. 201, Dekker, New York, 1998, pp. 279–287. MR1651173

[75] William Heinzer, Christel Rotthaus, and Sylvia Wiegand, *Building Noetherian domains inside an ideal-adic completion. II*, Advances in commutative ring theory (Fez, 1997), Lecture Notes in Pure and Appl. Math., vol. 205, Dekker, New York, 1999, pp. 411–419. MR1767426

[76] William Heinzer, Christel Rotthaus, and Sylvia Wiegand, *Intermediate rings between a local domain and its completion*, Illinois J. Math. **43** (1999), no. 1, 19–46. MR1665712

[77] William Heinzer, Christel Rotthaus, and Sylvia Wiegand, *Noetherian domains inside a homomorphic image of a completion*, J. Algebra **215** (1999), no. 2, 666–681, DOI 10.1006/jabr.1998.7763. MR1686210

[78] William Heinzer, Christel Rotthaus, and Sylvia Wiegand, *Approximating discrete valuation rings by regular local rings*, Proc. Amer. Math. Soc. **129** (2001), no. 1, 37–43, DOI 10.1090/S0002-9939-00-05492-7. MR1694346

[79] William Heinzer, Christel Rotthaus, and Sylvia Wiegand, *Intermediate rings between a local domain and its completion. II*, Illinois J. Math. **45** (2001), no. 3, 965–979. MR1879247

[80] William Heinzer, Christel Rotthaus, and Sylvia Weigand, *Corrigendum to our paper: "Intermediate rings between a local domain and its completion" [Illinois J. Math.* ***43*** *(1999), no. 1, 19–46; MR1665712 (99m:13047)]*, Illinois J. Math. **44** (2000), no. 4, 927–928. MR1804309

[81] William Heinzer, Christel Rotthaus, and Sylvia Wiegand, *Building Noetherian and non-Noetherian integral domains using power series*, Ideal theoretic methods in commutative algebra (Columbia, MO, 1999), Lecture Notes in Pure and Appl. Math., vol. 220, Dekker, New York, 2001, pp. 251–264. MR1836605

[82] William Heinzer, Christel Rotthaus, and Sylvia Wiegand, *Catenary local rings with geometrically normal formal fibers*, Algebra, arithmetic and geometry with applications (West Lafayette, IN, 2000), Springer, Berlin, 2004, pp. 497–510. MR2037106

[83] William Heinzer, Christel Rotthaus, and Sylvia Wiegand, *Examples of integral domains inside power series rings*, Commutative ring theory and applications (Fez, 2001), Lecture Notes in Pure and Appl. Math., vol. 231, Dekker, New York, 2003, pp. 233–254. MR2029829

[84] William Heinzer, Christel Rotthaus, and Sylvia Wiegand, *Non-finitely generated prime ideals in subrings of power series rings*, Rings, modules, algebras, and abelian groups, Lecture Notes in Pure and Appl. Math., vol. 236, Dekker, New York, 2004, pp. 333–341. MR2050721

[85] William Heinzer, Christel Rotthaus, and Sylvia Wiegand, *Integral closures of ideals in completions of regular local domains*, Commutative algebra, Lect. Notes Pure Appl. Math., vol. 244, Chapman & Hall/CRC, Boca Raton, FL, 2006, pp. 141–150, DOI 10.1201/9781420028324.ch10. MR2184794

[86] William Heinzer, Christel Rotthaus, and Sylvia Wiegand, *Generic fiber rings of mixed power series/polynomial rings*, J. Algebra **298** (2006), no. 1, 248–272, DOI 10.1016/j.jalgebra.2005.07.015. MR2215127

[87] William Heinzer, Christel Rotthaus, and Sylvia Wiegand, *Mixed polynomial/power series rings and relations among their spectra*, Multiplicative ideal theory in commutative algebra, Springer, New York, 2006, pp. 227–242, DOI 10.1007/978-0-387-36717-0_14. MR2265811

[88] William Heinzer, Christel Rotthaus, and Sylvia Wiegand, *Extensions of local domains with trivial generic fiber*, Illinois J. Math. **51** (2007), no. 1, 123–136. MR2346190

[89] William Heinzer, Christel Rotthaus, and Sylvia Wiegand, *Multi-ideal-adic completions of Noetherian rings*, Michigan Math. J. **57** (2008), 427–438, DOI 10.1307/mmj/1220879416. Special volume in honor of Melvin Hochster. MR2492460

[90] William Heinzer, Christel Rotthaus, and Sylvia Wiegand, *Examples of non-Noetherian domains inside power series rings*, J. Commut. Algebra **6** (2014), no. 1, 53–93, DOI 10.1216/JCA-2014-6-1-53. MR3215561

[91] William Heinzer, Christel Rotthaus, and Sylvia Wiegand, *Excellent normal local domains and extensions of Krull domains*, J. Pure Appl. Algebra **219** (2015), no. 3, 510–529, DOI 10.1016/j.jpaa.2014.05.010. MR3279371

[92] William Heinzer, Christel Rotthaus, and Sylvia Wiegand, *Formal fibers of prime ideals in polynomial rings*, Commutative algebra and noncommutative algebraic geometry. Vol. II, Math. Sci. Res. Inst. Publ., vol. 68, Cambridge Univ. Press, New York, 2015, pp. 213–222. MR3496866

[93] William Heinzer and Judith D. Sally, *Extensions of valuations to the completion of a local domain*, J. Pure Appl. Algebra **71** (1991), no. 2-3, 175–185, DOI 10.1016/0022-4049(91)90146-S. MR1117633

[94] William Heinzer and Sylvia Wiegand, *Prime ideals in two-dimensional polynomial rings*, Proc. Amer. Math. Soc. **107** (1989), no. 3, 577–586, DOI 10.2307/2048152. MR982402

[95] Raymond C. Heitmann, *Examples of noncatenary rings*, Trans. Amer. Math. Soc. **247** (1979), 125–136, DOI 10.2307/1998777. MR517688

[96] Raymond C. Heitmann, *Characterization of completions of unique factorization domains*, Trans. Amer. Math. Soc. **337** (1993), no. 1, 379–387, DOI 10.2307/2154327. MR1102888

[97] Raymond C. Heitmann, *Completions of local rings with an isolated singularity*, J. Algebra **163** (1994), no. 2, 538–567, DOI 10.1006/jabr.1994.1031. MR1262718

[98] Raymond C. Heitmann, *A noncatenary, normal, local domain*, Rocky Mountain J. Math. **12** (1982), no. 1, 145–148, DOI 10.1216/RMJ-1982-12-1-145. MR649747

[99] M. Hochster, *Prime ideal structure in commutative rings*, Trans. Amer. Math. Soc. **142** (1969), 43–60, DOI 10.2307/1995344. MR251026

[100] Jun-ichi Igusa, *On a theorem of Lueroth*, Mem. Coll. Sci. Univ. Kyoto Ser. A. Math. **26** (1951), 251–253, DOI 10.1215/kjm/1250777938. MR48860

[101] Cristodor Ionescu, *More properties of almost Cohen-Macaulay rings*, J. Commut. Algebra **7** (2015), no. 3, 363–372, DOI 10.1216/JCA-2015-7-3-363. MR3433987

[102] Ming-Chang Kang, *Almost Cohen-Macaulay modules*, Comm. Algebra **29** (2001), no. 2, 781–787, DOI 10.1081/AGB-100001541. MR1841999

[103] Irving Kaplansky, *Commutative rings*, Allyn and Bacon, Inc., Boston, Mass., 1970. MR0254021

[104] Daniel Katz, *On the number of minimal prime ideals in the completion of a local domain*, Rocky Mountain J. Math. **16** (1986), no. 3, 575–578, DOI 10.1216/RMJ-1986-16-3-575. MR862283

[105] Keith A. Kearnes and Greg Oman, *Cardinalities of residue fields of Noetherian integral domains*, Comm. Algebra **38** (2010), no. 10, 3580–3588, DOI 10.1080/00927870903200893. MR2760675

[106] Wolfgang Krull, *Beiträge zur Arithmetik kommutativer Integritätsbereiche* (German), Math. Z. **41** (1936), no. 1, 545–577, DOI 10.1007/BF01180441. MR1545640

[107] Wolfgang Krull, *Beiträge zur Arithmetik kommutativer Integritätsbereiche* (German), Math. Z. **42** (1937), no. 1, 745–766, DOI 10.1007/BF01160110. MR1545707

[108] Ernst Kunz, *Kähler differentials*, Advanced Lectures in Mathematics, Friedr. Vieweg & Sohn, Braunschweig, 1986, DOI 10.1007/978-3-663-14074-0. MR864975

[109] Yves Lequain, *Differential simplicity and complete integral closure*, Pacific J. Math. **36** (1971), 741–751. MR284422

[110] Graham J. Leuschke and Roger Wiegand, *Cohen-Macaulay representations*, Mathematical Surveys and Monographs, vol. 181, American Mathematical Society, Providence, RI, 2012, DOI 10.1090/surv/181. MR2919145

[111] Joseph Lipman, *Relative Lipschitz-saturation*, Amer. J. Math. **97** (1975), no. 3, 791–813, DOI 10.2307/2373777. MR417169

[112] S. Loepp, *Constructing local generic formal fibers*, J. Algebra **187** (1997), no. 1, 16–38, DOI 10.1006/jabr.1997.6768. MR1425557

[113] S. Loepp, *Excellent rings with local generic formal fibers*, J. Algebra **201** (1998), no. 2, 573–583, DOI 10.1006/jabr.1997.7306. MR1612339

[114] S. Loepp and C. Rotthaus, *Some results on tight closure and completion*, J. Algebra **246** (2001), no. 2, 859–880, DOI 10.1006/jabr.2001.9006. MR1872128

[115] S. Loepp and C. Rotthaus, *Generic formal fibers of polynomial ring extensions*, Rocky Mountain J. Math. **34** (2004), no. 1, 253–262, DOI 10.1216/rmjm/1181069903. MR2061129

[116] Saunders Mac Lane and O. F. G. Schilling, *Zero-dimensional branches of rank one on algebraic varieties*, Ann. of Math. (2) **40** (1939), 507–520, DOI 10.2307/1968935. MR158

[117] Jean Marot, *Sur les anneaux universellement japonais*, Bull. Soc. Math. France **103** (1975), no. 1, 103–111. MR406995

[118] Eben Matlis, *Cotorsion modules*, Mem. Amer. Math. Soc. **49** (1964), 66. MR178025

[119] Eben Matlis, *Torsion-free modules*, The University of Chicago Press, Chicago-London, 1972. Chicago Lectures in Mathematics. MR0344237

[120] Hideyuki Matsumura, *Commutative algebra*, 2nd ed., Mathematics Lecture Note Series, vol. 56, Benjamin/Cummings Publishing Co., Inc., Reading, Mass., 1980. MR575344

[121] Hideyuki Matsumura, *On the dimension of formal fibres of a local ring*, Algebraic geometry and commutative algebra, Vol. I, Kinokuniya, Tokyo, 1988, pp. 261–266. MR977763

[122] Hideyuki Matsumura, *Commutative ring theory*, 2nd ed., Cambridge Studies in Advanced Mathematics, vol. 8, Cambridge University Press, Cambridge, 1989. Translated from the Japanese by M. Reid. MR1011461

[123] Stephen McAdam, *Two conductor theorems*, J. Algebra **23** (1972), 239–240, DOI 10.1016/0021-8693(72)90128-7. MR304371

[124] Stephen McAdam, *Intersections of height 2 primes*, J. Algebra **49** (1977), no. 2, 315–321, DOI 10.1016/0021-8693(77)90242-3. MR480481

[125] D. Mumford, *Introduction to Algebraic Geometry*, Preliminary version of first chapters, 1968.

[126] D. Mumford and J. Tate, "Can one explain schemes to biologists", an obituary of Grothendieck and description of his work, with commentary, at the website: `http://www.dam.brown.edu/people/mumford/blog/2014/Grothendieck.html`.

[127] K. R. Nagarajan, *Groups acting on Noetherian rings*, Nieuw Arch. Wisk. (3) **16** (1968), 25–29. MR229628

[128] Masayoshi Nagata, *On the structure of complete local rings*, Nagoya Math. J. **1** (1950), 63–70. MR41829

[129] Masayoshi Nagata, *On the theory of Henselian rings*, Nagoya Math. J. **5** (1953), 45–57. MR51821

[130] Masayoshi Nagata, *Some remarks on local rings*, Nagoya Math. J. **6** (1953), 53–58. MR58577

[131] Masayoshi Nagata, *On the theory of Henselian rings. II*, Nagoya Math. J. **7** (1954), 1–19. MR67865

[132] Masayoshi Nagata, *The theory of multiplicity in general local rings*, Proceedings of the international symposium on algebraic number theory, Tokyo & Nikko, 1955, Science Council of Japan, Tokyo, 1956, pp. 191–226. MR0082966

[133] Masayoshi Nagata, *An example of normal local ring which is analytically ramified*, Nagoya Math. J. **9** (1955), 111–113. MR73572

[134] Masayoshi Nagata, *On the chain problem of prime ideals*, Nagoya Math. J. **10** (1956), 51–64. MR78974

[135] Masayoshi Nagata, *An example of a normal local ring which is analytically reducible*, Mem. Coll. Sci. Univ. Kyoto Ser. A. Math. **31** (1958), 83–85, DOI 10.1215/kjm/1250776950. MR97395

[136] Masayoshi Nagata, *On the theory of Henselian rings. III*, Mem. Coll. Sci. Univ. Kyoto Ser. A. Math. **32** (1959), 93–101, DOI 10.1215/kjm/1250776700. MR109835

[137] Masayoshi Nagata, *Local rings*, Interscience Tracts in Pure and Applied Mathematics, No. 13, Interscience Publishers a division of John Wiley & Sons New York-London, 1962. MR0155856

[138] M. Nagata, *Lectures on the fourteenth problem of Hilbert*, Tata Institute of Fundamental Research, Bombay, 1965. MR0215828

[139] Amnon Neeman, *Colocalizing subcategories of* $\mathbf{D}(R)$, J. Reine Angew. Math. **653** (2011), 221–243, DOI 10.1515/CRELLE.2011.028. MR2794632

[140] Jun-ichi Nishimura, *Note on Krull domains*, J. Math. Kyoto Univ. **15** (1975), no. 2, 397–400, DOI 10.1215/kjm/1250523071. MR384783

[141] Jun-ichi Nishimura, *On ideal-adic completion of Noetherian rings*, J. Math. Kyoto Univ. **21** (1981), no. 1, 153–169, DOI 10.1215/kjm/1250522110. MR606317

[142] Jun-ichi Nishimura, *A few examples of local rings, I*, Kyoto J. Math. **52** (2012), no. 1, 51–87, DOI 10.1215/21562261-1503754. MR2892767

[143] Jun-ichi Nishimura, *A few examples of local rings, II*, Kyoto J. Math. **56** (2016), no. 1, 49–96, DOI 10.1215/21562261-3445156. MR3479318

[144] D. G. Northcott, *Lessons on rings, modules and multiplicities*, Cambridge University Press, London, 1968. MR0231816

[145] Bruce Olberding and Serpil Saydam, *Ultraproducts of commutative rings*, Commutative ring theory and applications (Fez, 2001), Lecture Notes in Pure and Appl. Math., vol. 231, Dekker, New York, 2003, pp. 369–386. MR2029839

[146] Tetsushi Ogoma, *Noncatenary pseudogeometric normal rings*, Japan. J. Math. (N.S.) **6** (1980), no. 1, 147–163, DOI 10.4099/math1924.6.147. MR615018

[147] Tetsushi Ogoma, *Cohen Macaulay factorial domain is not necessarily Gorenstein*, Mem. Fac. Sci. Kôchi Univ. Ser. A Math. **3** (1982), 65–74. MR643928

[148] Christian Peskine, *Une généralisation du "main theorem" de Zariski* (French), Bull. Sci. Math. (2) **90** (1966), 119–127. MR206037

[149] Gabriel Picavet, *Endomorphismes polynomiaux plats* (French, with English summary), J. Algebra **138** (1991), no. 1, 195–225, DOI 10.1016/0021-8693(91)90197-G. MR1102574

[150] Heinz Prüfer, *Untersuchungen über Teilbarkeitseigenschaften in Körpern* (German), J. Reine Angew. Math. **168** (1932), 1–36, DOI 10.1515/crll.1932.168.1. MR1581355

[151] L. J. Ratliff Jr., *On quasi-unmixed local domains, the altitude formula, and the chain condition for prime ideals. II*, Amer. J. Math. **92** (1970), 99–144, DOI 10.2307/2373501. MR265339

[152] Louis J. Ratliff Jr., *Chain conjectures in ring theory*, Lecture Notes in Mathematics, vol. 647, Springer, Berlin, 1978. An exposition of conjectures on catenary chains. MR496884

[153] Michel Raynaud, *Anneaux locaux henséliens* (French), Lecture Notes in Mathematics, Vol. 169, Springer-Verlag, Berlin-New York, 1970. MR0277519

[154] D. Rees, *A note on analytically unramified local rings*, J. London Math. Soc. **36** (1961), 24–28, DOI 10.1112/jlms/s1-36.1.24. MR126465

[155] Christel Rotthaus, *Nicht ausgezeichnete, universell japanische Ringe*, Math. Z. **152** (1977), no. 2, 107–125, DOI 10.1007/BF01214184. MR427319

[156] Christel Rotthaus, *Universell japanische Ringe mit nicht offenem regulärem Ort* (German), Nagoya Math. J. **74** (1979), 123–135. MR535964

[157] Christel Rotthaus, *Komplettierung semilokaler quasiausgezeichneter Ringe* (German), Nagoya Math. J. **76** (1979), 173–180. MR550860

[158] Christel Rotthaus, *On rings with low-dimensional formal fibres*, J. Pure Appl. Algebra **71** (1991), no. 2-3, 287–296, DOI 10.1016/0022-4049(91)90152-R. MR1117639

[159] Christel Rotthaus, *Homomorphic images of regular local rings*, Comm. Algebra **24** (1996), no. 2, 445–476, DOI 10.1080/00927879608825580. MR1373487

[160] Christel Rotthaus, *Excellent rings, Henselian rings, and the approximation property*, Rocky Mountain J. Math. **27** (1997), no. 1, 317–334, DOI 10.1216/rmjm/1181071964. MR1453106

[161] Christel Rotthaus and Liana M. Şega, *On a class of coherent regular rings*, Proc. Amer. Math. Soc. **135** (2007), no. 6, 1631–1640, DOI 10.1090/S0002-9939-06-08679-5. MR2286070

[162] P. Samuel, *Unique Factorization Domains*, Tata Inst. of Fundamental Research, Bombay (1964).

[163] Sean Sather-Wagstaff and Richard Wicklein, *Support and adic finiteness for complexes*, Comm. Algebra **45** (2017), no. 6, 2569–2592, DOI 10.1080/00927872.2015.1087008. MR3594539

[164] Friedrich Karl Schmidt, *Mehrfach perfekte Körper* (German), Math. Ann. **108** (1933), no. 1, 1–25, DOI 10.1007/BF01452819. MR1512831

[165] Friedrich Karl Schmidt, *Über die Erhaltung der Kettensätze der Idealtheorie bei beliebigen endlichen Körpererweiterungen* (German), Math. Z. **41** (1936), no. 1, 443–450, DOI 10.1007/BF01180433. MR1545632

[166] A. Seidenberg, *On the dimension theory of rings. II*, Pacific J. Math. **4** (1954), 603–614. MR65540

[167] Jean-Pierre Serre, *Géométrie algébrique et géométrie analytique* (French), Ann. Inst. Fourier (Grenoble) **6** (1955/56), 1–42. MR82175

[168] I. R. Shafarevich, *Basic algebraic geometry*, Springer Study Edition, Springer-Verlag, Berlin-New York, 1977. Translated from the Russian by K. A. Hirsch; Revised printing of Grundlehren der mathematischen Wissenschaften, Vol. 213, 1974. MR0447223

[169] Chandni Shah, *Affine and projective lines over one-dimensional semilocal domains*, Proc. Amer. Math. Soc. **124** (1996), no. 3, 697–705, DOI 10.1090/S0002-9939-96-03159-0. MR1301048

[170] David L. Shannon, *Monoidal transforms of regular local rings*, Amer. J. Math. **95** (1973), 294–320, DOI 10.2307/2373787. MR330154

[171] R. Y. Sharp and P. Vámos, *Baire's category theorem and prime avoidance in complete local rings*, Arch. Math. (Basel) **44** (1985), no. 3, 243–248, DOI 10.1007/BF01237858. MR784093

[172] Peter Donald Shelburne, *Low-dimensional formal fibers in characteristic p greater than 0*, ProQuest LLC, Ann Arbor, MI, 1994. Thesis (Ph.D.)–Michigan State University. MR2691861

[173] Philip B. Sheldon, *How changing* $D[[x]]$ *changes its quotient field*, Trans. Amer. Math. Soc. **159** (1971), 223–244, DOI 10.2307/1996008. MR279092

[174] Richard G. Swan, *Néron-Popescu desingularization*, Algebra and geometry (Taipei, 1995), Lect. Algebra Geom., vol. 2, Int. Press, Cambridge, MA, 1998, pp. 135–192. MR1697953

[175] Craig Huneke and Irena Swanson, *Integral closure of ideals, rings, and modules*, London Mathematical Society Lecture Note Series, vol. 336, Cambridge University Press, Cambridge, 2006. MR2266432

[176] Hiroshi Tanimoto, *Some characterizations of smoothness*, J. Math. Kyoto Univ. **23** (1983), no. 4, 695–706, DOI 10.1215/kjm/1250521430. MR728156

[177] Hiroshi Tanimoto, *Smoothness of Noetherian rings*, Nagoya Math. J. **95** (1984), 163–179, DOI 10.1017/S002776300002105X. MR759472

[178] Diana Kahn Taylor, *IDEALS GENERATED BY MONOMIALS IN AN R-SEQUENCE*, ProQuest LLC, Ann Arbor, MI, 1966. Thesis (Ph.D.)–The University of Chicago. MR2611561

[179] Simplice Tchamna Kouna, *The ideal completion of a Noetherian local domain*, Comm. Algebra **44** (2016), no. 6, 2513–2530, DOI 10.1080/00927872.2015.1053901. MR3492171

[180] P. Thompson, *Cosupport computations for finitely generated modules*, `arXiv:1702.03270v1`.

[181] Paolo Valabrega, *On two-dimensional regular local rings and a lifting problem*, Ann. Scuola Norm. Sup. Pisa Cl. Sci. (3) **27** (1973), 787–807 (1974). MR364228

[182] Stuart Sui Sheng Wang, *A Jacobian criterion for separability*, J. Algebra **65** (1980), no. 2, 453–494, DOI 10.1016/0021-8693(80)90233-1. MR585736

[183] Dana Weston, *On descent in dimension two and nonsplit Gorenstein modules*, J. Algebra **118** (1988), no. 2, 263–275, DOI 10.1016/0021-8693(88)90021-X. MR969672

[184] Roger Wiegand, *Homeomorphisms of affine surfaces over a finite field*, J. London Math. Soc. (2) **18** (1978), no. 1, 28–32, DOI 10.1112/jlms/s2-18.1.28. MR491732

[185] Roger Wiegand, *The prime spectrum of a two-dimensional affine domain*, J. Pure Appl. Algebra **40** (1986), no. 2, 209–214, DOI 10.1016/0022-4049(86)90041-1. MR830322

[186] Sylvia Wiegand and Roger Wiegand, *The maximal ideal space of a Noetherian ring*, J. Pure Appl. Algebra **8** (1976), no. 2, 129–141, DOI 10.1016/0022-4049(76)90011-6. MR409446

[187] Roger Wiegand and Sylvia Wiegand, *Prime ideals in Noetherian rings: a survey*, Ring and module theory, Trends Math., Birkhäuser/Springer Basel AG, Basel, 2010, pp. 175–193, DOI 10.1007/978-3-0346-0007-1_13. MR2744272

[188] Hiroshi Yamada, *On formal schemes*, Number theory, algebraic geometry and commutative algebra, in honor of Yasuo Akizuki, Kinokuniya, Tokyo, 1973, pp. 243–295. MR0379491

[189] Oscar Zariski, *Local uniformization on algebraic varieties*, Ann. of Math. (2) **41** (1940), 852–896, DOI 10.2307/1968864. MR2864

[190] Oscar Zariski, *The concept of a simple point of an abstract algebraic variety*, Trans. Amer. Math. Soc. **62** (1947), 1–52, DOI 10.2307/1990628. MR21694

[191] Oscar Zariski, *Applicazioni geometriche della teoria delle valutazioni* (Italian), Univ. Roma Ist. Naz. Alta Mat. Rend. Mat. e Appl. (5) **13** (1954), 51–88. MR63704

[192] Oscar Zariski and Pierre Samuel, *Commutative algebra, Volume I*, The University Series in Higher Mathematics, D. Van Nostrand Company, Inc., Princeton, New Jersey, 1958. With the cooperation of I. S. Cohen. MR0090581

[193] Oscar Zariski and Pierre Samuel, *Commutative algebra, Volume I*, The University Series in Higher Mathematics, D. Van Nostrand Company, Inc., Princeton, New Jersey, 1958. With the cooperation of I. S. Cohen. MR0090581

[194] Daniel Zelinsky, *Complete fields from local rings*, Proc. Nat. Acad. Sci. U.S.A. **37** (1951), 379–381, DOI 10.1073/pnas.37.6.379. MR41830

Index

Abhyankar, 1, 38, 109, 113, 163, 166, 231, 386, 391
Affine algebraic variety, 104
Affine ring, 104, 109
Akizuki, 1–3, 17, 49, 141
Algebra
étale, *111*, 142
essentlally finite, *11*
essentlally of finite type, *11*
finite, *11*
finite type, *11*
quasi-normal, *141*
Rees, *140*
Algebraic independence, *12*, 87
Algebraic retract, *43*
Almost Cohen-Macaulay module, *256*
Alonzo-Tarrio, 375, 376
Alonzo-Tarrio, Jeremias-Lopez and Lipman Question: formal schemes, 375, 376
Analytically independent, *386*
Analytically irreducible Noetherian local ring, *33*, 49
Analytically normal Noetherian local ring, *33*, 49, 113, 154
Analytically ramified Noetherian local ring, *34*
Analytically reducible Noetherian local ring, *34*, 48, 50
Analytically unramified Noetherian local ring, *33*, 34, 109, 141
Anderson, Dan, 180
Annihilator ideal, of an elementl, *11*
Approximation domain, 3–6, 59, 60, 62, 63, 70, 71, 73, 81, 83, 84, 118, 127, 131, 136, 137, 143, 150, 152, 154, 174, 183, 186, 189, 190, 198, 200, 201, 206, 209, *211*, *213*, 215, 237, 239, 307, 402
Homomorphic Image Construction, *213*
Inclusion Construction, *61*
Approximation methods
Homomorphic Image Construction, 211
Approximation theorem for Krull domains, 15
Associated prime ideal, *11*, 206, 244
embedded, *11*
Auslander, 14
Avramov, 244, 246
Ax, 106, 154, 158, 162
Azumaya, 36

Benson, 381
Berger, 113, 374
Birational
domination, *10*, 46, 150, 180, 181, 210, 239, 242
extension, 1, 7, *10*, 209, 211, 218, 229, 235
Bourbaki, 15, 117
Brewer, 125, 204, 206
Brodmann, 2, 4, 210, 231, 258, 261, 262
Buchsbaum, 14
Burch, 283, 330

Catenary ring, 5, 7, 25, *34*, 41, 69, 70, 85, 173, 175, 176, 186, 224, *231*, 232, 234–236, 240, 242, 244
Cauchy sequence, 28
Chain condition for prime ideals, 232
Chain Conjecture, 232, 246, 252
Change of coefficient ring, 20
Change of variables, *345*, *346*, 359, 365
Charters, 341
Chase, 206
Chevalley, 18, 32, 109, 397
Theorem: Every Noetherian local domain is birationally domiinated by a DVR, 18
Christel's example, 51, 73, 137
Classical examples, 2, 5
Closed ideal
in the I-adic topology, *25*, 28, 32, 42
Closed singular locus, 109
Coefficient field for a local ring, *33*
Cohen, 17, 23, 32, 33, 43, 109, 152, 177, 179, 180, 231, 342, 399
Structure theorems for complete local rings, 33, 342

Theorem: A ring is Noetherian if each prime ideal is finitely generated, 17, 152
Theorem: Every complete Noetherian local domain is a finite integral extension of a complete regular local domain, 33
Theorem: Every complete Noetherian local ring is a homomorphic image of a complete regular local ring, 33
Theorem: Every equicharacteristic complete Noetherian local ring has a coefficient field k and is a homomorphic image of a formal power series ring over k, 33
Cohen's Theorem 8, *32*
Cohen-Macaulay
fibers, 6, *88*, 89
formal fibers, 245, 249, 250
ring, 35, *87*, 87, 88, 101, 110, 140, 144
Coherence, 204
Coherent regular ring, 5, 206
Coherent ring, *206*
Coherent sequences, *27*, 30, 257–261
Commutative diagram, 259, 260, 263
Complete intersection, *10*, 144, 246
Completely integrally closed, *15*
Completion, 2, 5–7
I-adic, *25*
associated to a filtration, 25
coherent sequences, 259
ideal-adic, 2, 3, *25*, *26*, 150, 257
multi-adic, *257*, 257, 258, 262, 265, 269, 271
multi-ideal-adic, 6, 7, *257*
uncountable, 30
Compositum of two power series rings, 154
Conrad, 375, 377
Question: formal schemes, 375, 376
Constructed ring, 1, *2*
Construction
Homomorphic Image, 209, 211, 213, 216, 237, 239, 240, 245
limit-intersecting, 213
Inclusion, 59, 60, 65, 73, 78, 81, 127, 209, 210, 220, 229, 245, 401
limit-intersecting, *63*
Insider, 5, 81, 127, *130*, *131*, *401*, *402*
Intersection, 2–4, 59, 401
limit-intersecting, 115, 309
primarily limit-intersecting, *310*
residually limit-intersecting, *309*
Remark: Every Inclusion Construction is up to isomorphism a Homomorphic Image Construction, 220
Construction Properties Theorem, 3, 59, *64*, 66, 68, 69, 128, 131, 137, 138, 145, 153, 175, 189, 191, 198, *213*, 216, 217, 221, 237, 241, *272*, 309, 311
Contraction of an ideal, *35*
Cosupport, *381*
of a ring, *381*
of a finitely generated module, *381*, 381
Cutkosky, 166

D+M construction, 184, 205, 207
David, 173
Delta ideal, *92*
Depth
of a module, *38*, 244
of a ring, *38*, 232, 244, 256
Derivation, 92, 106
Derived categories of chain complexes, 381
Derived normal ring, *12*, 232, 242, 244
Differential morphism, 92
Dilatation of R by I along V, *165*
Dimension
of a prime ideal, *236*
of a ring, *10*
Dimension formula, 167, 225
Direct limit of subrings, 3, 4, *47*, 47
Direct limits, 25
Directed union of subrings, 3, *47*
Discrete rank-1 valuation domain, 7, *11*, 165
as a directed union of RLRs, 167
Distinguished monic polynomial, *343*, 344, 356
Divisor Class Group, *16*
Divisorial ideal, *15*
Dominates, for local rings, *10*, 42
Dumitrescu, 41, 376, 386, 389
DVR, *11*, 13–15, 17, 21, 41, 47, 49, 55, 127–129, 156, *165*, 167, 168, 170, 172, 175, 177, 180, 181, 184, 218, 223, 240, 348, 349, 358, 360–362, 364, 367, 370

Eakin, 18, 163
Eakin-Nagata, 18
Theorem: If B is a Noetherian ring and A is a subring such that B is a finitely generated A-module, then A is a Noetherian ring, 18
Eisenbud, 105
Element of a ring
prime, *10*
regular element, *9*
zerodivisor, *9*
Element of extension ring
idealwise independent, *275*, 276, 277, 280, 285, 293, *294*
limit-intersecting, 3, 309
primarily independent, *277*, 277, 280, 283, 285, 287, *294*, 294
existence of, 283, 293

residually algebraically independent, 154, *277*, 277, 280, 285–287, 289, 293, *294*
Elkik, 92
Elkik ideal, *92*, 93, 94, 105, 253
Embedded associated prime ideal, *11*
Embedded local uniformization of R along V, *166*, 167
Endpiece notation, *60*, 67, 128, 150, 174, 188, 190, 200, 212, 220, 221, 223
Endpiece Recursion Relation, *61*
Equicharacteristic local ring, *33*, 342
Equidimensional ring, 6, *34*, 85, 228, 234, 235, 240
formally, *233*
Essential valuation rings for a Krull domain, *15*, 122, 125
Essentially finite extension of commutative rings, *11*
Essentially finitely generated over a field, 355, 371, 372
Essentially finitely generated over a commutative ring, *11*, 370
Essentially finitely presented over a commutative ring, *11*, *91*
Essentially of finite type extension, *11*
Essentially smooth morphism, *91*, 93, 95, 96
Etale, 111
algebra, *111*, 142
extension, *111*, 169, 170, 302
homomorphism, *111*
neighborhood, *111*, 112
Evans, 182
Example
a non-catenary four-dimensional local UFD with maximal ideal generated by three elements that has precisely one prime ideal of height three and this prime ideal is not finitely generated, 200
an injective local map of normal Noetherian local domains that satisfies LF_{d-1} but not LF_d, 295
a 2-dimensional Noetherian local domain whose generic formal fiber is not Cohen-Macaulay, 250
a 3-dimensional RLR $(A, \mathfrak{n})$ having a height-two prime ideal $I = (f, g)A$ for which the extension $I\widehat{A}$ to the $\mathfrak{n}$-adic completion is not integrally closed, 5, 142
a non-universally catenary 2-dimensional Noetherian local domain with geometrically regular formal fibers that birationally dominates a 3-dimensional regular local domain, 240
a non-universally catenary 2-dimensional Noetherian local domain with geometrically regular formal fibers that birationally dominates a 3-dimensional regular local domain., 404
a regular local ring over a non-perfect field k that is not smooth over k, 106
a residually algebraically independent element need not be primarily independent, 287
an excellent ring that fails to satisfy Jacobian criteria, 106
an idealwise independent element may fail to be residually algebraically independent, 290
where the constructed domains A and B are equal and are not Noetherian, 5, 135
where the constructed domains A and B are not equal but A is Noetherian, 3, 5
where the constructed domains A and B are not Noetherian and are not equal, 248
Examples
classical, 2, 5
Discrete valuation rings that are not Nagata rings, 129
iterative, 149
Non-catenary 3-dimensional local UFDs with maximal ideal generated by two elements that have m prime ideals of height two and each prime ideal of height two not finitely generated, 5, 173
Non-universally catenary n-dimensional Noetherian local domains with geometrically regular formal fibers, 5, 231
related to integral closure, 2
Excellence, 5–7, *40*, 41, 104, 106, 107, 109, 110, 127, 130, 136, 263
preservation of under completion, 265
Excellent ring, 2, 3, 5–7, 25, *40*, 50, 51, 84, 103, 104, 106, 109, 110, 113, 127, 129, 130, 136, 137, 145, 153, 156, 222, 224, 229, 240, 245, 252, 255, 257, 265, 293, 302, 327, 330–332, 340, 341
Exponential functions, 154
Extension
trivial generic fiber/TGF, 369, 373
Extension of commutative rings
height-one preserving, *115*
weakly flat, *115*
essentially finite, *11*
essentially finitely generated, *11*
essentially finitely presented, *11*
essentially of finite type, *11*
finite, *11*
finite type, *11*
height-one preserving, 123

integral closure in, *12*
integral extension, *12*
LF_d, *115*, 294
LF_{d-1}, 123, 134, 135
Extension of integral domains
birational, 1, *10*, 60, 209–211
trivial generic fiber/TGF, *6*, *342*, 377–379, 385, 386, 389
Extension of Krull domains
height-one preserving, 276, 308
locally flat in height r, or LF_r, 308
weakly flat, 276, 308, 309
PDE, *15*, *276*, 286, 307
Extension of local rings
étale, *111*, 169, 170, 302
birationally dominates, *10*
domination, *10*
Extensions of Krull domains
a weakly flat extension is height-one preserving, 119
height-one preserving and PDE imply weakly flat, 121
height-one preserving does not imply weakly flat, 120
PDE does not imply height-one preserving, 121
PDE does not imply weak flatness, 121
PDE is equivalent to LF_1, 121
weakly flat does not imply PDE, 121, 122

Factorial ring, *10*
Faithful flatness, *18*, 19, 37, 132, 137, 262, 264, 265, 350, 399
Faithfully flat module, *18*, 26
Ferrand, 2
Fiber ring, *36*
Fibers, 1, 88, 89, 234
Cohen-Macaulay, 6, *88*, 245, 328
formal, *339*, 340, 341, 371
generic, 339–341, 355, 371
geometrically normal, 5, 7, *39*, 231, 244
geometrically reduced, *39*
geometrically regular, 5, 7, *39*, 42, 231, 232, 239, 240, 243, 340
normal, *39*
of a ring homomorphism, *36*
reduced, *39*
regular, *39*, 88, 89, 340
trivial generic fiber, 6
Filtration of ideals, *257*
I-adic, *25*
ideal-adic, *25*
multi-adic, *257*
multiplicative, *25*, *257*
not multiplicative, 257
Finite
algebra extension, *11*
module, *11*, 370, 374
ring extension, *11*
Finite conductor domain, *206*
Finite presentation, for an algebra extension, *11*, 91
Finite type, for an algebra extension, *11*
Flat extension, 4–6, *18*, 375, 379
conditions for, 83, 87
Flat homomorphism, *19*, 20, 105, 189, 375, 379
Flat locus
of a polynomial ring extension, 87, 97
Flat module, *18*, 37
Flatness, 4–6, *18*, 19, 22, 30, 40, 42, 73, 74, 81–83, 87, 89, 100, 101, 129, 130, 154, 211, 215, 217, 233, 237, 238, 293, 350, 351
elementwise criterion, 19
localization, 19
relations to the Jacobian ideal, 95
Formal fibers, 5, *339*, 340, 341, 371
Cohen-Macaulay, 246, 249, 250
for a Noetherian local ring, *40*
geometrically normal, 5, 112, 113, *231*, 233, 234, 244
geometrically reduced, 109, 112
geometrically regular, 110, 112, 113, *231*, 232, 239, 240, 243, 340, 404
Formal power series ring, *13*, 22
leading form, *13*
Formal scheme, 375
Formal spectrum, *376*
Formally equidimensional Noetherian local ring, *233*
Fossum, 15
Fractional ideal, *15*
Frontpiece notation, *211*, 212, 213, 220, 222, 223, 229
Fuchs, 262

G-ring, *40*, 40, 136, 265
Gabelli, 184, 205
Gamma function, 157
$\gcd(a, b)$, *21*
GCD-domain, *21*
Generalization, stable under, *232*, *235*
Generalized local ring, *180*, 180
Generalized power series, *170*, 171
Generic fiber, 6, 398
trivial generic fiber, 6, *342*, 378
Generic fiber ring, *340*
mixed polynomial-power series rings, 339, 355
Generic formal fiber, 6, *322*, 355, 371
Theorem: height of maximal ideals, 341
Generic formal fiber ring, *339*, *340*, 340, 361, 373
Geometrically normal fibers, *39*, 40

Geometrically normal formal fibers, 5, 7, 233, 234
Geometrically reduced fibers, *39*, 40, 109
Geometrically regular fibers, 5, 7, *39*, 39, 41, 42, 105, 108, 109
Geometrically regular formal fibers, 41, 232, 239, 240, 265, 340, 404
Gff, 355, 370, 371, 373
Gff (R), *339*, *340*, 340, 373
Gilmer, 113, 204, 375
Glaz, 5, 206
Going-down property, 19, 95, 120, 232, 236, 249, 331, 337, 352
Going-up Theorem, 183, 235, 358, 360
Goto, 144
Grade
  of a module, *38*
  of a ring, *38*, 256
Graded ring, with respect to an ideal, 217
Granja, 166
Greatest common divisor, *21*
Greco, 240
Griffith, 391
Grothendieck, 103, 106, 109, 110, 234, 240

Hartshorne, 104
Hausdorff, 150, 151
Height of a prime ideal, *10*
Height-one preserving, *115*, 119–121, 123, 124, *276*
Height-one prime ideals are radicals of principal ideals, 117, 122–124
Heinzer, 46, 113, 163, 181, 340, 376, 380, 389, 399
Heitmann, 2, 4, 38, 77, 210, 232, 246, 258, 261, 262, 340
  How to adjoin a transcendental element preserving an ideal-adic completion, 4
  Theorem: Every complete Noetherian local ring $(T, \mathfrak{n})$ of depth at least two such that no nonzero element in the prime subring of T is a zerodivisor on T is the completion of a Noetherian local UFD, 232
Hensel's Lemma, 36, 111
Henselian affine partially ordered set, 380
Henselian ring, 6, 25, 36, 37, 103, *111*, 111, 169, 233, 257, 265, 267, 278, 302, 374
Henselization, 5, *37*, 103, *111*, 111–113, 142, 169, 170, 231, 233–235, 242, 243, 267, 291, 293, 302
Hilbert Nullstellensatz, 97
Hironaka, 109
Hochster, 6, 36, 257, 342, 377, 391
Hochster and Yao
  Question: trivial generic fibers, 342, 377
Homomorphic image
  of a regular local ring, 35, 240
Homomorphic Image Construction, *209*, 211, 213, 216, 220, 237, 240, 402
  Approximation domain, *213*
  approximation methods, 211
Homomorphism
  étale, *111*
  essentially smooth, *91*
  flat, *19*, 19, 20
  local, *10*
  normal, *40*
  regular, *40*, 51, 91, 101, 137, 234, 240
  smooth, *91*
Houston, 184, 204, 205
Huckaba, 140
Huneke, 46, 139, 140
Hypertranscendental element, *157*

Ideal
  annihilator of an element, *11*
  closed in the I-adic topology, *25*, 28, 32, 42
  completion, 262
  contraction of, *35*
  element integral over, *139*, *140*
  extension in a ring homomorphism, *36*
  filtration, *25*, 26, 257–259, 261, 263
  integral closure, *139*, 140
  integrally closed, *139*, *140*, 140–143, 145–147
  Jacobian, 146
  normal, *139*, *140*, 140
  order function, *13*
  radical, 146, 147
  reduction, *139*, *140*, 140, 142
  topology, *262*, 268
Ideal transforms, 125
Ideal-adic
  completion, 2, 3, *25*, *26*, 30, 37, 45, 48, 72, 150, 209, 220, 257, 265–267, 375, 376
  filtration, *25*
  ideal separated for, 87
  topology, 1, *25*, 32, 37, 87, 150, 371, 376, 378
Idealwise independent, *6*, *275*, 276–280, 285, 287, 293, 295, 298, 300, 303, 315
Igusa, 163
Immersion of schemes, 375, *376*
  closed, 375, *376*
  open, 375, *376*
Inclusion Construction, *59*, *60*, 60, 65, 70, 73, 78, 81, 127, 153, 155, 186, *209*, 209, *210*, 211, 220, 229, 307, *401*
  Approximation Domain, *61*
Independence
  idealwise independence, 6, *275*, *276*, 278, 280, 285, 287, 293, 295, 298, 300, 303

primary independence, 6, *277*, 280, 283, 285, 287, 293, 302, 305
residual algebraic independence, 154, *277*, 278, 280, 285–287, 289, 293
algebraic, *12*, 87
analytic, *386, 387*
Insider Construction, 401
Insider Construction, 5, 73, 82, 87, 127, 130, 131, 133, 136, 139, 173, 186, 197, 402, 404
Insider Inclusion Construction, 73
Integral closure, 2, *12*, 233
examples related to, 2
of an ideal, *139, 140*
Integral domain, *10*
integral closure, *12*
integrally closed, 5, *12*
Krull domain, *14*
regular local ring, 2, 3, 5, *14*
unique factorization domain, *10*
valuation domain, *11*
Integral element, *12*
Integral extension of commutative rings, *12*
Integral over an ideal, *139*, 140
Integrally closed domain, 5, *12*, 140
Integrally closed ideal, *139, 140*, 141–143, 145–147
Intersection Construction, 2–4, 6, 59, 209, 401
approximation domain, 3
equation, 4
intersection domain, 3
universality, 2, *46*
Intersection domain, 2, 3, 6, 46, 59, *60, 61*, 62, 63, 81, 84, 118, 130, 131, 136, 154, 174, 189, 198, 200, 209, 210, 219, 238, 240, 307, 402
Noetherian limit, *76*, *127*, 225
Inverse limits, 25
Ionescu, 256
Iterative examples, 149, 153, 154
Iyengar, 381

J-2 property of a Noetherian ring, *266*
Jacobian criterion, 106
for smoothness, 103, 105, 106
Jacobian ideal
of a polynomial map, 87, *94*, 95, 98, 136–138, 145, 146
Jacobian matrix, 92, 106, 253
Jacobson radical of a ring, *11*, 16, 18, 26, 29, 61, 257–261, 264
Jacobson ring, 340
Jeremias-Lopez, 375, 376

Kang, 256
Katz, 34, 49
Kearnes, 380
Kiehl, 103, 109, 113, 374
Kim, Youngsu, 371
Krause, 381
Kravitz, 38
Krull, 16, 17, 162, 204, 231, 392
Krull Altitude Theorem, *17*, 22, 35, 392
Krull dimension, *10*
Krull domain, *14*, 14, 15, 17, 21, 23, 50, 117, 118, 149–151, 154, 199, 293, 307, 308, 310, 312, 325, 382
essential valuation rings, *15*, 122
Krull Intersection Theorem, 16, 162
Krull-Akizuki theorem, *17*, 46
Kunz, 113, 154, 163, 374
Theorem: The ring compositum of $k[[x]]$ and $k[[y]]$ in $k[[x, y]]$ is not Noetherian, 154, 163

$\mathrm{lcm}(a, b)$, *21*
Least common multiple, *21*
Lequain, 2
Leuschke, 184, 205, 283, 330
LF_d, 115, 123, 134, 135, 294, 308, 310–315, 324
Limit-intersecting, *3*, 63, 115, *309*, 309, 310, 315
primarily, *310*
residually, *309*
for *Homomorphic Image Construction*, *213*
for *Inclusion Construction*, *63*
Limits
Direct, 25
inverse, 25
Linear topology, *25*
Linearly disjoint field extension, 149
Lipman, 30, 141, 142, 375–377
Local Flatness Theorem, 73, 192
Local Homomorphic Image Prototype, 240
Local homomorphism, *10*
Local Prototype, 54, *55*, 73, 77, 82, 83, 137, 138, 153, 174, 239, 316
Local Prototype Theorem, 130, 138, 142, 145, 153, 174, 175, 190, 200, 223, 240, 316
Local quadratic transform, 49, *165*, 166–169, 171, 184, 203, 399
iterated, *165*
Local ring, *10*
étale neighborhood, *111*
coefficient field, *33*
coherent regular, 5, 206
equicharacteristic, *33*
Henselian, 6, *36, 37*, 233, 267
Henselization, 5, *37*
regular local ring, *14*
Local uniformization of R along V, *166*
Localization, *9*
Localized polynomial ring, *10*, 275

Loepp, 147, 340, 341
Lost prime ideal, *178*, 179–181, 183, 184
Luroth, 163
Luroth theorem, 163

MacLane, 171
Marley, 114
Marot, 373
Matijevic, 180
Matlis, 262
Matsumura, 9, 20, 22, 91, 162, 266, 340, 341, 351, 356, 371, 391
  Question: dimension of generic formal fiber in excellent local rings, 340
  Theorem: dimension of generic formal fiber rings, 340, 355, 356
McAdam, 2, 206, 373
Minimal prime divisor, 17
Mixed polynomial-power series rings, 1, 6, *339*, 342, 355, 371, 373, 375, 385, 389
  relations among spectra, 379
  spectra, 375, 379
Module
  I-adically ideal-separated, 87
  elementwise criterion for flatness, 19
  faithfully flat, *18*
  finite, *11*
  flat, *18*, 19, 20
  regular sequence on, *38*
  separated for the I-adic topology, *87*, 87
  torsionfree, *20*
Moh, 386, 391
Monoidal transform, *166*
Multi-adic completion, *257*, 257, 258, 262, 265, 269, 271
  Noetherian flatness, 269
Multi-adic Inclusion Construction, 272
Multi-ideal-adic completion, 6, 7, *257*, 262
Mumford, 18, 104, 111

Nagata, 1, 2, 4, 16–18, 50, 73, 81–83, 103, 106, 108–111, 113, 137, 141, 151, 210, 231, 233
  Example: a 2-dimensional analytically reducible normal Noetherian local domain, 50, 82
  Example: a 2-dimensional regular local ring that is not a Nagata ring, 50
  Example: a 2-dimensional regular local ring that is not a Nagata ring, 2, 4, 82, 137
  Example: a 2-dimensional regular local ring with a prime element that factors as a square in the completion, 51, 82
Nagata domain, 2, 51, 108, 129, 262
Nagata Polynomial Theorem, 16, 109
Nagata ring, *16*, 51, 84, 103, 109, 110, 113
Nakayama's lemma, 32
Nastold, 103, 113, 374

Neeman, 381
Nested union of subrings, 3, *47*
Nichols, 180
Nilradical of a ring, *10*
Nishimura, 2, 18, 23, 110, 258, 261, 262
  Theorem: If R is a Krull domain and R/P is Noetherian for each height-one prime P, then R is Noetherian, 18
Noether normalization, 372
Noetherian flatness
  Multi-adic completion, 269
Noetherian Flatness Theorem, 4, 73, 74, 77, 83, 85, 127, 129, 131, 132, 134, 153, 154, 175, 191, 221, 237, 240, 309, 312, 401, 402
  for *Homomorphic Image Construction*, 215
  for *Inclusion Construction*, 74
Noetherian limit intersection domain, *76*, *127*, 186, 225, 226
Noetherian local ring
  analytically irreducible, *33*
  analytically normal, *33*, 113, 154
  analytically ramified, *34*
  analytically reducible, *34*
  analytically unramified, *33*, 34, 109
  Cohen-Macaulay, *87*, 144
  depth, *38*, 38
  formal fibers, *40*
  formally equidimensional, *233*, 246
  geometrically normal formal fibers, *231*, 244
  geometrically regular formal fibers, *231*, 239, 240, 243
  quasi-unmixed, 46, *233*
Noetherian ring, 9
  excellent, *40*
  G-ring, *40*
  J-2, 266
Noetherian spectrum, *9*, 70, 177, 185, 201
Non-catenary ring, 173, 175, 176, 186, 197, 198, 209, 211, 221, 234, 244, 404
Non-flat locus, *20*, 79, 80, 122, 131–134, 175, 176, 186, 189, 190, 198, 201–203, 227
  of *Insider Construction*, 131
  defines, *20*, *95*
  determines, *20*
  of a homomorphism, *20*
  of a polynomial map, 87, *94*, 98, 190, 198
  of *Insider Construction*, 80, 132
Non-Noetherian ring, 1, 5, 7, 173, 175, 190, 197, 198, 404
Non-smooth locus, 92, 105
  of a polynomial map, 87, 95, 100
Non-uniqueness of representation
  of power series expressions, 62
Nonconstant coefficients, *83*

ideal generated by, *83*
Nor R, 110
Normal domain, 140
Normal fibers, *39*
Normal ideal, *139*, *140*, 140
Normal locus, 110
Normal morphism, *40*, 141
Normal ring, *12*, 38, 39, 110, 141
Northcott, 25

Ogoma, 2, 4, 6, 210, 232, 245, 246, 252, 258, 261, 262
Example: a 3-dimensional Nagata local domain whose generic formal fiber is not equidimensional, 6, 246
Example: a normal non-catenary Noetherian local domain, 211, 232, 245, 246
Ogoma-like example, 251, 255, 405
Olberding, 7, 180, 183, 206
Oman, 380
Order function of an ideal, *13*, 22, 343

P-morphism, *234*
Parameter, regular, 147
Partial derivative, 92, 100
PDE extension, *15*, 121, 275, *276*, 277, 279, 280, 285, 286, 288, 289, 292
Peskine, 182
Picavet, 87
PID, *11*, 20, 72
Popescu, 91
Power series
generalized, *170*, 171
Power series pitfall, 31
Preservation of excellence
under multi-adic completion, 263, 265
with Insider Construction, 127, 136
Preservation of Noetherian
under multi-adic completion, 262
Primarily independent, 6, *277*, *280*, 283, 285, 287, 293, 295, 298–300, 302, 305, 315
Primarily limit-intersecting, 4, *310*, 310, 312, 315, 318, 324, 325, *327*, 327, 330, 332–334, 336, 337
Prime spectrum, 192
Prime divisor, *11*
Prime element, *10*
Prime ideal
associated to a module, *11*, 207, 244
dimension of, *10*
height, *10*
lost, *178*, 179–181, 183
structure, 6, 7
symbolic power, *11*
Prime spectrum, *9*, 36, 173, 190, 200, 202, 375–377, 379, 384, 385
diagram, 181, 183, 193, 202, 381, 382

Principal fractional ideal, *15*
Principal ideal domain, 11
Properties $\mathcal{A}$.1-4
relevant for excellence, *104*
Prototype, 52, 54, *55*, 131, 209, 217, 221, 223, 225, 228, 240, 402
Prototype Example
Intersection form, *55*
Prototype Theorem, 5, 127, 145, 312
Inclusion Version, *129*
Local Version, 240
Local Version, *130*, 143, 200
Pruefer, 205
Pseudo-geometric ring, *16*
Pullback, 184, 205

Quadratic transform, local, 49, *165*, 166–169, 171, 184, 203, 399
iterated, *165*
Quasi-normal algebra, *141*
Quasi-normal morphism, *141*
Quasi-regular sequence, *217*
Quasi-smooth
algebra over a ring, *91*, 93
Quasi-unmixed Noetherian local ring, *34*, 46, *233*

R-completion, 262
R-topology, *262*
Radical ideal, *10*, 146, 147
Ratliff, 2, 35, 85, 232, 233, 240, 282
Ratliff's Equidimension Theorem, *35*, 85, 228, 233, 240, 246
a Noetherian local ring R is universally catenary if and only if the completion of R/P is equidimensional for every minimal prime P of R, 35
Raynaud, 2, 111, 112
Reduced fibers, *39*
Reduced ideal, *10*
Reduced ring, *10*, 33, 34, 38, 39, 110, 233
Reduction of an ideal, *139*, *140*, 140, 142
Rees, 34, 109
Rees algebra, *140*
Rees Finite Integral Closure Theorem, *34*, 34, 38, 109
necessary and sufficient conditions for a Noetherian local ring to be analytically unramified, 34
Reg, of a ring, *40*
Regular element, *9*, 42, 73, 131, 218, 402
Regular fibers, *39*, 39, 88, 89, 340
Regular for a power series
in the sense of Zariski-Samuel, *160*
Regular integral domain, *222*
Regular local domain, 14
Regular local ring, 2, 5, 7, *14*, 33, 38, 40, 47, 88, 110, 130, 165, 167, 168, 170,

172, 196, 222, 223, 239, 243, 262, 340, 343, 362, 404
homomorphic image of, 35, 240
Regular morphism, *40*, 51, 91, 101, 136, 137, 141, 145, 146, 234, 240, 265
Regular parameter, 147
Regular sequence, *9*, 87, 89, 141, 144, 145, 147, 217, 222, 345, 350, 359
on a module, *38*
Regular system of parameters, *14*, 196
Residually algebraically independent, 154, *277*, 277, 278, 280, 285–287, 289, 293, 295, 298, 300, 303, 304, 315
Residually limit-intersecting, *309*, 312–315, 318, 320, 322, 324
Residue field of a ring at a prime ideal, *36*
Riemann Zeta function, 157
Ring
catenary, *34*, *231*, 232, 234–236, 240, 242, 244
coherent regular, 5, 206
discrete rank-1 valuation ring, *11*
equidimensional, *34*
Henselian, 6, *36*, 233, 257, 265
Jacobson radical, *11*, 61
Krull dimension, *10*
Krull domain, *14*
local ring, *10*
Nagata, *16*
normal ring, *12*
not universally catenary, 240, 242–244
of fractions, *10*
pseudo-geometric, *16*
regular local, 2, 5, 7, *14*
unique factorization domain, *10*
universally catenary, *34*, *231*, 232, 234–236, 240, 242–244, 266
valuation domain, *11*
Ringed space, 376
RLR, *14*, 14, 33, 35, 39, 138–140, 142–144, 147, 152, 153, 167, 168, 170, 173, 175, 176, 184, 240, 265, 341, 362, 364, 367, 398, 399
Roitman, 43
Rond, 76
Rotthaus, 2–5, 51, 73, 81, 84, 108, 137, 147, 162, 206, 210, 239, 258, 261, 262, 323, 340, 341, 359, 376, 389, 391, 399
Theorem: dimension of generic formal fiber in excellent rings, 340
Example: a 3-dimensional regular local domain that is a Nagata domain and is not excellent, 4, 51, 84, 137
Example: a 3-dimensional regular local domain that is a Nagata domain and is not excellent., 3
Example: A Nagata domain for which the singular locus is not closed, 108, 262
Theorem: If R is a Noetherian semilocal ring with geometrically regular formal fibers and I_0 is an ideal of R contained in the Jacobson radical of R, then the I_0-adic completion of R also has geometrically regular formal fibers, 239
Rutter, 204, 206

Salce, 262
Sally, 1, 46, 181, 323, 340
Question: What rings lie between a Noetherian integral domain and its field of fractions?, 1, 46
Samuel, 15, 39, 160, 386
Sathaye, 163
Sather-Wagstaff, 381
Saydam, 206
Schanuel, 154
conjecture, 158
Scheme, 104, 375, 376
affine, *376*
formal, 376
immersion of, 375, *376*, 376, 377
Schilling, 171
Schmidt, 1, 2, 49, 113, 141, 374
Sega, 5, 206
Seidenberg, 205
Semilocal ring, *10*
Separated for the I-adic topology, *87*, 88
Serre, 14, 18
Serre's conditions, *14*
Seydi, 267
Shah, 380
Shannon, 50, 166
Sharp, 231, 283, 330, 370
Sheaf, *376*
Sheldon, 395
Simple PS-extension, 77
Singular locus of a Nagata domain not closed, 108, 262
Singular locus of a Noetherian ring, 103, *104*, 266
closed, *104*, 105, 107, 109, 110
Singularities of algebraic curves, 48
Smooth
0-smooth, *91*
algebra over a ring, *91*, 253
morphism, *91*
quasi-smooth, *91*
Smoothness
relations to the Jacobian ideal, 95
Spec, *9*
Spec map, 36
Spectral map, *36*, 182, 375, 377–379
Stable under generalization, *232*, 235

Subspace, *154*
topology, *42*, *154*, 163, 371
Swan, 91, 93
Swanson, 139, 140
Symbolic power of a prime ideal, *11*

Tanimoto, 91
Tate, 104, 109
Taylor, 145
Tchamna, 262
Tensor product of modules, 19
TGF extension, *6*, *342*, 369, 373, 377–379, 385, 387, 389, 391–400, 405, 406
Thompson, 381
Tight closure, 147
Topology
ideal-adic, *25*, 32, 37, 42
linear, *25*
subspace, *42*, *154*, 163, 371
Zariski, 9
Torsionfree module, *20*, 155
Total ring of fractions, *10*, 401
Transcendence degree
uncountable, 30
Trivial generic fiber extension, *6*, 7, *342*, 378, 391

UFD, *10*, 14, 17, 21, 59, 71, 117, 151, 173, 175, 180, 182, 190, 191, 198, 248–250
Theorem: For R a UFD, x a prime element of R and R^* the x-adic completion of R, the Approximation Domain B is a UFD, 71
Unique factorization domain, 5, *10*, 71
Universality of the construction, 2, 45, *46*, 402
Universally catenary ring, 5, 7, *34*, 40, 41, 85, 104, 110, 176, 186, 211, 217, 224, *231*, 232–236, 240, 242–244, 266, 267

Valabrega, 47, 48, 84, 151, 174, 276, 343, 362, 367
consequence of his theorem, 84, 151
Theorem: C a DVR and L a field with $C[y] \subset L \subseteq \mathcal{Q}(C[[y]])$ implies $L \cap C[[y]]$ is an RLR, 47, 151, 317, 325, 343, 362, 367
Valuation, 12
Valuation domain, *11*, 21, 181, 184, 203
discrete rank-1, 7, *11*, 165, 167
Value group, 12
Vamos, 283, 330
van der Put, 391

Wang, 87
Weak Flatness Theorem
homomorphic image version, 218
inclusion version, 118
Weakly flat, *115*, 118–121, 123, 124, 275, *276*, 277, 279, 280, 309
Weierstrass, 343
Weierstrass Preparation Theorem, 339, *343*, 347
Weston, 2, 210, 258, 261, 262
Wicklein, 381
Wiegand, R., 31, 73, 184, 205, 283, 330, 370, 380, 383
a crucial flatness lemma, 73
Wiegand, S., 283, 330, 380, 383

Yao, 6, 342, 377, 391
Yasuda, 378, 389
non-TGF extension, 378, 389

Zariski, 39, 49, 106, 109, 112, 160, 166, 167, 182, 386, 400
Zariski Subspace Theorem, 400
Zariski topology, *9*, 104, 376
closed sets, *9*
open sets, *9*
Zariski's Jacobian criterion
for regularity in polynomial rings, 106
Zariski's Main Theorem, 50, 112, 182
Zariski-Samuel Commutative Algebra II
erroneous theorem, 39
Zelinsky, 262
Zero-smooth, *91*
Zerodivisor, *9*
Zorn's Lemma, 22

Selected Published Titles in This Series

259 **William Heinzer, Christel Rotthaus, and Sylvia Wiegand,** Integral Domains Inside Noetherian Power Series Rings, 2021

256 **Michio Jimbo, Tetsuji Miwa, and Fedor Smirnov,** Local Operators in Integrable Models I, 2021

255 **Alexandre Boritchev and Sergei Kuksin,** One-Dimensional Turbulence and the Stochastic Burgers Equation, 2021

254 **Karim Belabas and Henri Cohen,** Numerical Algorithms for Number Theory, 2021

252 **Julie Déserti,** The Cremona Group and Its Subgroups, 2021

251 **David Hoff,** Linear and Quasilinear Parabolic Systems, 2020

250 **Bachir Bekka and Pierre de la Harpe,** Unitary Representations of Groups, Duals, and Characters, 2020

249 **Nikolai M. Adrianov, Fedor Pakovich, and Alexander K. Zvonkin,** Davenport–Zannier Polynomials and Dessins d'Enfants, 2020

248 **Paul B. Larson and Jindrich Zapletal,** Geometric Set Theory, 2020

247 **István Heckenberger and Hans-Jürgen Schneider,** Hopf Algebras and Root Systems, 2020

246 **Matheus C. Bortolan, Alexandre N. Carvalho, and José A. Langa,** Attractors Under Autonomous and Non-autonomous Perturbations, 2020

245 **Aiping Wang and Anton Zettl,** Ordinary Differential Operators, 2019

244 **Nabile Boussaïd and Andrew Comech,** Nonlinear Dirac Equation, 2019

243 **José M. Isidro,** Jordan Triple Systems in Complex and Functional Analysis, 2019

242 **Bhargav Bhatt, Ana Caraiani, Kiran S. Kedlaya, Peter Scholze, and Jared Weinstein,** Perfectoid Spaces, 2019

241 **Dana P. Williams,** A Tool Kit for Groupoid C^*-Algebras, 2019

240 **Antonio Fernández López,** Jordan Structures in Lie Algebras, 2019

239 **Nicola Arcozzi, Richard Rochberg, Eric T. Sawyer, and Brett D. Wick,** The Dirichlet Space and Related Function Spaces, 2019

238 **Michael Tsfasman, Serge Vlăduţ, and Dmitry Nogin,** Algebraic Geometry Codes: Advanced Chapters, 2019

237 **Dusa McDuff, Mohammad Tehrani, Kenji Fukaya, and Dominic Joyce,** Virtual Fundamental Cycles in Symplectic Topology, 2019

236 **Bernard Host and Bryna Kra,** Nilpotent Structures in Ergodic Theory, 2018

235 **Habib Ammari, Brian Fitzpatrick, Hyeonbae Kang, Matias Ruiz, Sanghyeon Yu, and Hai Zhang,** Mathematical and Computational Methods in Photonics and Phononics, 2018

234 **Vladimir I. Bogachev,** Weak Convergence of Measures, 2018

233 **N. V. Krylov,** Sobolev and Viscosity Solutions for Fully Nonlinear Elliptic and Parabolic Equations, 2018

232 **Dmitry Khavinson and Erik Lundberg,** Linear Holomorphic Partial Differential Equations and Classical Potential Theory, 2018

231 **Eberhard Kaniuth and Anthony To-Ming Lau,** Fourier and Fourier-Stieltjes Algebras on Locally Compact Groups, 2018

230 **Stephen D. Smith,** Applying the Classification of Finite Simple Groups, 2018

229 **Alexander Molev,** Sugawara Operators for Classical Lie Algebras, 2018

228 **Zhenbo Qin,** Hilbert Schemes of Points and Infinite Dimensional Lie Algebras, 2018

227 **Roberto Frigerio,** Bounded Cohomology of Discrete Groups, 2017

226 **Marcelo Aguiar and Swapneel Mahajan,** Topics in Hyperplane Arrangements, 2017

For a complete list of titles in this series, visit the
AMS Bookstore at **www.ams.org/bookstore/survseries/**.